Advanced Sciences and Technologies for Security Applications

Indexed by SCOPUS

The series Advanced Sciences and Technologies for Security Applications comprises interdisciplinary research covering the theory, foundations and domain-specific topics pertaining to security. Publications within the series are peer-reviewed monographs and edited works in the areas of:

- biological and chemical threat recognition and detection (e.g., biosensors, aerosols, forensics)
- crisis and disaster management
- terrorism
- cyber security and secure information systems (e.g., encryption, optical and photonic systems)
- traditional and non-traditional security
- energy, food and resource security
- economic security and securitization (including associated infrastructures)
- transnational crime
- human security and health security
- social, political and psychological aspects of security
- recognition and identification (e.g., optical imaging, biometrics, authentication and verification)
- smart surveillance systems
- applications of theoretical frameworks and methodologies (e.g., grounded theory, complexity, network sciences, modelling and simulation)

Together, the high-quality contributions to this series provide a cross-disciplinary overview of forefront research endeavours aiming to make the world a safer place.

The editors encourage prospective authors to correspond with them in advance of submitting a manuscript. Submission of manuscripts should be made to the Editor-in-Chief or one of the Editors.

Tünde Anna Kovács · Róbert Gábor Stadler ·
Norbert Daruka
Editors

The Impact of the Energy Dependency on Critical Infrastructure Protection

Proceedings of the 5th International Conference on Central European Critical Infrastructure Protection (ICCECIP 2023), Budapest, Hungary

Springer

Editors
Tünde Anna Kovács
Banki Donat Faculty of Mechanical and Safety Engineering
Óbuda University
Budapest, Hungary

Róbert Gábor Stadler
Banki Donat Faculty of Mechanical and Safety Engineering
Óbuda University
Budapest, Hungary

Norbert Daruka
Banki Donat Faculty of Mechanical and Safety Engineering
Óbuda University
Budapest, Hungary

ISSN 1613-5113 ISSN 2363-9466 (electronic)
Advanced Sciences and Technologies for Security Applications
ISBN 978-3-031-78543-6 ISBN 978-3-031-78544-3 (eBook)
https://doi.org/10.1007/978-3-031-78544-3

This Springer imprint is published by the registered company Springer Nature Switzerland AG
The registered company address is: Gewerbestrasse 11, 6330 Cham, Switzerland

Foreword

We are honoured to host the "Central European Critical Infrastructure Protection" international conference at the Banki Donat Faculty of Mechanical and Safety Engineering at Obuda University.

In our daily lives, energy manifests itself in many ways. As long as it is available, we will not notice its absence. In recent years, however, we have realized that energy availability is not a given but a dependency. For critical infrastructures, this means even more exposure and dependence. This book also focuses on "The Impact of the Energy Dependency on Critical Infrastructure Protection".

Developing technologies, cyber technologies, and digitalization in our daily lives have brought about new security challenges in critical infrastructure protection and security science. Our fundamental goal is to demonstrate the unity of our region (Central Europe) and provide a platform for experienced and young researchers and Ph.D. students from around the world to share their latest results with the global public, thereby highlighting the global impact of our collective efforts.

The volume contains lectures at the ICCECIP 2023 conference, which cover the broad research area of security science, including cyber security, individual security, physical protection and political security. The book's information and latest research results, with their practical applicability, are helpful and empowering for Ph.D. students, researchers, professionals, experts, and teachers in all science areas.

Nowadays, when protecting critical infrastructures is a top priority worldwide, we are delighted to be able to host this conference in Central Europe. This region is at the forefront of security research and innovation.

Today's conference is also an excellent opportunity for students at the Doctoral School on Safety and Security Sciences to discuss their research and publish their results.

I want to thank the organizers for their unwavering dedication in bringing this conference to life, where researchers from several countries presented their results. Your hard work and commitment have not gone unnoticed and are deeply appreciated.

The presentations also cover cyber security, individual security, physical protection and political security.

See you again at the ICCECIP conference in Budapest!

Budapest, Hungary Zoltan Rajnai

Organization

Scientific Guarantors

Prof. Zoltán Rajnai, Ph.D., Obuda University, Budapest, Hungary
Ing. Tünde Anna Kovács, Ph.D., Obuda University, Budapest, Hungary
Ing. László Tóth, Ph.D., Obuda University, Budapest, Hungary

The International Scientific Committee

Prof. Miklós Maróth, D.Sc., Eötvös Loránd Research Network, Hungary
Prof. Mihály Réger, D.Sc., Obuda University, Hungary
Prof. emer. Tamás Réti, D.Sc., Obuda University, Hungary
Prof. Lívia Cvetityánin, D.Sc., Obuda University, Hungary
Prof. László Kovács, D.Sc., National University of Public Service, Hungary
Prof. emer. Lajos Berek, C.Sc., Obuda University, Hungary
Prof. Kornélia Lazányi, Ph.D., Obuda University, Hungary
Prof. Pál Michelberger, Ph.D., Obuda University, Hungary
Prof. János Besenyő, Ph.D., Obuda University, Hungary
Prof. Róbert Szabolcsi, Ph.D., Obuda University, Hungary
Prof. Pavel Manas, Ph.D., University of Defence in Brno, Czech Republic
Prof. David Rehak, Ph.D., Technical University of Ostrava, Czech Republic
Prof. Petrica Vizureanu, Ph.D., Gheorghe Asachi Technical University of Iasi, Romania
Prof. Miroslaw Banasik, Ph.D., Jan Kochanowski University in Kielce, Poland
Prof. Michel Arrigoni, Ph.D., ENSTA Bretagne, France
Prof. Alexis Rusinek, Ph.D., French Embassy in Budapest, France
Prof. Valeriu Gabriel Ghica, Ph.D., University Politehnica of Bucharest, Romania
Prof. Mircea Voda, Ph.D., "Politehnica" University of Timisoara, Romania

Prof. Ján Dusza, DrSc, PROMATECH Institute of Materials Research of SAS, Slovakia
Prof. Zdenek Dvorak, Ph.D., University of Žilina, Slovakia
Prof. Josip Stapanic, Ph.D., University of Zagreb, Croatia
Prof. Zlatko Čović, Ph.D., Subotica Tech—College of Applied Sciences, Serbia
Prof. Sanja Maravić Čisar, Ph.D., Subotica Tech—College of Applied Sciences, Serbia
Prof. Petar Čisar, Ph.D., University of Criminal Investigation and Police Studies, Serbia
Prof. Darko Trifunovic, Ph.D., Institute for National and International Securit, Serbia
Prof. Ilias Majdouline, Ph.D., Universiapolis, International University of Agadir, Morocco
Prof. Enikő Bitay, Ph.D., Sapientia Hungarian University Of Transylvania, Romania
Robert C. Castel, Ph.D., Israel National Parks Authority, Israel
Ing. Igor Fürstner, Ph.D., Subotica Tech—College of Applied Sciences, Serbia
Ing. Róbert Pintér, Ph.D., Subotica Tech—College of Applied Sciences, Serbia
Ing. János Simon, Ph.D., Subotica Tech—College of Applied Sciences, Serbia
Ing. László Gogolák, Ph.D., Subotica Tech—College of Applied Sciences, Serbia
Livia Szedmina, Ph.D., Subotica Tech—College of Applied Sciences, Serbia
Ing. Dalibor Dobrilovic, Ph.D., University of Novi Sad Technical faculty "Mihajlo Pupin" Zrenjanin, Serbia
Ing. Maciej Klosak, Ph.D., Universiapolis, International University of Agadir, Morocco
Ing. Amine Bendarma, Ph.D., Universiapolis, International University of Agadir, Morocco
Ing. Lucia Figuli, Ph.D., University of Žilina, Slovakia
Ing. Yuriy Klyuchka, Dr.Sc., National University of Civil Defence of Ukraine, Ukraine
Ing. Kostiantyn Afanasenko, Ph.D., National University of Civil Defence of Ukraine, Ukraine
Ing. Jovan Trajkovski, Ph.D., University of Ljubljana, Slovenia
Ing. Chiara Bedon, Ph.D., University of Triest, Italy
Agnieszka Rogozinska, Ph.D., Jan Kochanowski University in Kielce, Poland
Ing. Marcin Adamiak, Ph.D., Silesian University of Technology, Gliwice, Poland
Ing. Umesh Mhatre, Ph.D., Surface Modification Technologies Pvt. Ltd. India
Ing. Martin Hromoda, Ph.D., Tomas Bata University in Zlin, Czech Republic
Ing. Levente Dimen, Ph.D., University of Alba Iulia, Romania
Ing. Christophe Czarnota, Ph.D., Lorraine University, France
Ing. Slim Bahi, Ph.D., Lorraine University, France
Ing. Hajej Zied, Ph.D., Lorraine University, France
Ing. Nidhal Rezg, Ph.D., Lorraine University, France
Ing. Péter Pinke, Ph.D., Obuda University, Hungary
Ing. Endre Szűcs, Ph.D., Obuda University, Hungary
Ing. Norbert Daruka, Ph.D., Obuda University, Hungary
Irén Zabóné Varga, Ph.D., Obuda University, Hungary

The International Organization Board

Contents

Data-Driven Strategies for Enhancing Supply Chain and Transportation Infrastructure Resilience

Judit Nagy and Pavel Foltin

Abstract The presentation "Data-Driven Strategies for Enhancing Supply Chain and Transportation Infrastructure Resilience" underscores the crucial role of data-driven approaches and Key Performance Indicators (KPIs) in fortifying supply chains and transportation infrastructure. It emphasizes the importance of monitoring socio-economic factors and implementing resilience measures to address the complex impacts of disruptions. KPIs such as on-time delivery, inventory turnover, and capacity utilization play a pivotal role in assessing and improving resilience, ensuring business continuity, economic stability, and community well-being. The presentation is structured around a comprehensive framework, including an in-depth analysis of supply chain and transportation infrastructure resilience, highlighting the significance of integrating socio-economic indicators and supply chain KPIs. Through case studies, the methodology section demonstrates the practical application of these data-driven strategies, offering valuable insights and recommendations for enhancing resilience. The conclusion succinctly summarizes key findings their implications for business and policy, and outlines future research directions.

Keywords Transportation infrastructure · Resilience · Big data · Economic indicators · GDP per capita

1 Introduction

Supply chains create a bridge between available resources and a place of consumption, overcoming the distance between the place of production and the place of demand. Managing such complex, sometimes globally extensive chains has many

J. Nagy (✉)
Corvinus University of Budapest, 8 Fővám tér, Budapest 1093, Hungary
e-mail: judit.nagy@uni-corvinus.hu

P. Foltin
University of Defence, Kounicova 65, 662 10 Brno, Czech Republic
e-mail: pavel.foltin@unob.cz

T. A. Kovács et al. (eds.), *The Impact of the Energy Dependency on Critical Infrastructure Protection*, Advanced Sciences and Technologies for Security Applications, https://doi.org/10.1007/978-3-031-78544-3_1

challenges, especially when the business environment is turbulent, and threats of different kinds of disruptions appear. In order to ensure the constant supply of society with goods and services, companies of the supply chains lay more and more emphasis on tools and methods to help achieve business continuity and resilience. To achieve this, they need different resources and capabilities.

This paper will focus on transportation as a critical process of global supply chains. The costs and the efficiency of transportation substantially influence the efficiency of the supply chains, as well as the condition and reliability of the transportation infrastructure, which is crucial both from civilian and military perspectives. In this paper, we are dealing with the critical transport infrastructure (CTI) and admit its role in the sustainable development of society and the economy, and we also call the attention that from the military perspective, CTI is crucial for maintaining effective, long-term military logistics capabilities.

As the civilian and military uses of critical infrastructure are inseparable, it is necessary to take an interdisciplinary approach to the subject. In our study, we therefore apply both supply chain and supply chain resilience approaches, civilian and military interpretations of critical transport infrastructure and its logistical aspects, and the science of data analysis.

The remainder of the paper is structured as follows: first, we present the theoretical background, global trends, and the basic terms and concepts used in the study, and then we present the data analytics framework that we apply to the paper. We present the source of the data and the results of the analysis carried out and conclude the study.

2 Theoretical Background

In recent years, we have witnessed the emergence of a number of crises threatening business and society, such as the epidemic, the closure of the Suez Canal, war in Europe and the energy crisis. So, we are seeing a number of threats to the security of supply chains. Supply chain disruption can happen inside or outside of the supply chain [1], intentionally or un-intentionally [2], in supply chain flows like materials flow, information flow, knowledge flow, control and coordination flow [1, 3–9]. The consequences of catastrophic events are also investigated by supply chain research teams [10], similarly consequences of socio-economic, political, man-made and natural disasters [11].

In handling disruptions and being prepared for the challenges of the (business) environment, resilience has a big role. In general, resilience is the speed at which a system resists and returns to its original equilibrium state [12]. Supply chain resilience is the adaptive capability of the supply chain to prepare for unexpected events, respond to disruptions, and recover from them by maintaining continuity of operations at the desired level of connectedness and control over structure and function [13]. But how can we ensure that supply chain operations are resilient and able to withstand the challenges that constantly arise? On the one hand, the processes need to be rethought, as, for example, the Just in Time solution across the supply chain makes supply chains extremely vulnerable due to low stock levels and cannot ensure continuity of supply in a crisis, even with intensive communication [14]. There is a need to identify, monitor and prepare for potential threats by putting in place business continuity plans to ensure that the level of performance of the operation can be maintained and restored in case of a disruptive event [15].

To prepare for the global trends of the economic environment organizations of the supply chain can use several approaches. Supply chain resilience analysis can be an important step to understanding the complexity [16]. Its significant part, where all flows of goods are managed, is the transportation network and for this reason the analytical framework should be extended to analyze specifically the transportation infrastructure resilience [17]. Preliminary assessment of the possible impacts of various disruption scenarios and applying resilience measures in supply chain performance assessment are also help preparations [18, 19]. Technological innovations and digitalization can also be exploited while preparing the supply chain for future challenges [20, 21].

Society is also affected by threats to supply chains and critical transport infrastructure, their basic needs must be guaranteed, and transport infrastructure plays a critical role in this. CTI Supports the economic and social well-being of a society, facilitates the movement of goods and people provide access to critical services like healthcare or education and supports economic growth. CTI is also crucial to maintain supply chain security and to support the military mobility concept [22–24].

In recent years, we have seen several variants of threats to transport infrastructure during the epidemic, such as closed factories, closed ports and borders which were previously unimaginable, and what civilians have experienced is a shortage of goods in shops [24]. However, transport infrastructure is used not only by civil society but also by military logistics, so preserving its condition and maintaining its operational reliability is very important from both perspectives.

Transport infrastructure resilience is dependent on the sources of disruptions. Disrupted supply chains usually mean shortage of essential goods and services, which might also cause economic losses. Also, emergency response is limited, both in mobility and services availability meaning [1].

The resilience of CTI is strongly influenced by the specific investment policies that states follow when thinking about critical infrastructure. Sharing data on risks and hazards between the countries concerned will help to make transport infrastructure resilient. Critical transport infrastructure databases should be kept up-to-date at national and local level and on a continuous basis [25]. Research on and development of CTI are of paramount importance not only at local, national but also at regional level [26, 27].

Numerous applications of digital technologies within supply chain facilitate enhanced communication, collaboration, and information sharing, thereby bolstering efficiency [28, 29]. The widespread adoption of platforms contributes to heightened agility, reliability, and efficiency in risk management [30]. The generation of large volumes of big data across various organizational processes, characterized by high velocity and variety, surpasses the capabilities of conventional data processing systems [31]. Consequently, Big Data Analysis encompasses the general capabilities of company ERP systems, personnel skills, and analytical methodologies aimed at structuring and processing data [32] to supply decision-makers with the necessary information [33].

In the assessment of the likelihood and ramifications of disruptive events, identifying risks that jeopardize the integrity of the CTI assumes paramount importance [34]. Management must undertake a comprehensive evaluation encompassing the magnitude, nature, complexity, and interconnectivity of potential consequences, in anticipation of possible occurrences [35]. Factors such as volatility, temporal dynamics of threats, and the efficacy of extant control mechanisms necessitate careful consideration [36].

This paper emphasizes the role of Big Data Analytics in facilitating these processes, delineating the requisite types of data and their respective sources to be collected and processed as Early Warning Signals. Such insights are instrumental in preempting risks and fortifying CTI against disruptions, thereby fostering resilience. Data-driven approaches and an analytical framework might help building a resilient transportation infrastructure and to better understand the sources of potential disruption.

Most of the literature studied deals with critical infrastructure in general or critical infrastructure of the energy supply [13]. The papers focusing on CTI either focus their research solely on a single transportation mode or a specific threat concerning the CTI. For this reason, concentration on CTI as a whole is important from both a civil and military perspective. We also did not find any paper that discusses the importance of Big Data analysis in predicting disruptive events to increase CTI resilience. Nor has there been a study that seeks to use data from publicly available databases to predict risk factors affecting CTI, using them as early warning signals to watch for disruptive events. The absence in application of Big Data in the area of increasing the CTI resilience was identified as a main research gap.

Our study is therefore important from both an academic and a practical point of view. From an academic perspective, we explore and prove a new relationship between CTI and big data analysis, with introducing a possible practical application in conceptional level. From a practical perspective, we provide a tool for civil and military organizations to analyze threats to their supply chains and processes through publicly available, free data (Fig. 1).

3 Goal and Methodology

Drawing upon the observations of the previous chapter, that most of the research papers concern mostly the individual transportation modes or specific threats within Critical Transportation Infrastructure, our objective is to uncover the interrelations between disruptions in transportation infrastructure and their impacts on diverse

Fig. 1 Main sources of disruption [37, 38]

socio-economic indicators. Formulated as a guiding inquiry, our research question seeks to elucidate the connections between transportation disruptions and their ensuing economic ramifications.

3.1 Defining the Research Goals

The presented research is focused on the identification of the ties between transportation infrastructure disruptions and their effects on various socio-economic indicators. These lead to formulate following research question: What are the ties between disruptions in the critical transportation infrastructure and the various economic consequences? The initial basic research concept leads to the identification of the potential use of Big Data Analytics, more precisely, early warning signals to predict supply chain disruptions.

Numerous case studies, as demonstrated by Li et al. [7], Beulens et al. [28], Knemeyer et al. [18], and Xu et al. [39], highlight the efficiency of using Early Warning Systems (EWS) in predicting both quantity and quality issues and in gauging the magnitude of associated risks. Within the fashion industry, Sumei [40] underscores the versatility of EWS in forecasting an array of risks encompassing informational, demand-related, supply chain, and environmental factors, while also serving to mitigate customer attrition, as indicated by Lo and Hong [27]. Furthermore, EWS finds application in diverse sectors such as banking and insurance for detecting customer churn [22], assessing risks to production control [21], and anticipating unforeseen events within supply management [41]. However, despite these specific applications, there remains a notable gap in the literature concerning the absence of supply chain-level prediction and systematic analysis of disruptive events.

What data sources can be used as EWS? Almost anything. Economic data obviously, e.g. GDP, inflation, employment [42]. In addition, user-generated content (UGC), e.g. Facebook, Instagram, Twitter posts can be very useful. There can also be

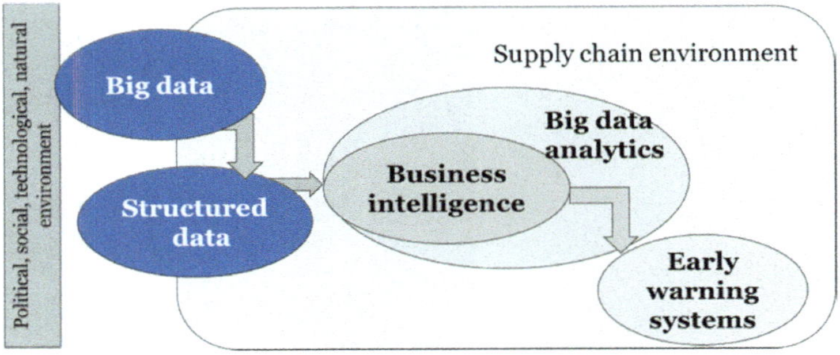

Fig. 2 Structure of an early warning system. *Source* Own edition

interactive maps, frequently updated databases, indices to characterize a sector, the environment or society, which can be accessed and downloaded for free and analyzed to produce forecasts. In one of our publications, we introduced the model of an EW System, which can be seen in Fig. 2.

Our goal is to uncover the interrelations between disruptions in critical transportation infrastructure and their impacts on diverse socio-economic indicators. By finding the most important indicators, we identify data sources which can help monitoring the factors affecting the CTI and create an EWS to predict disruptions.

3.2 Methodology

The research started with the Data Collection and preparation phase to find appropriate data for building up the Early Warning System. We gather historical data on the chosen indicators (Total Inland freight transport, GDP per capita, Business Confidence Index, Consumer Confidence Index) from OECD or other reliable sources. We had to be sure that the data we collected cover the period what we selected to examine the significant socio-economic disruptive events, such as financial crises, pandemic and international conflicts.

In second step, we visualized the data. We created plot time series graphs for each indicator to observe trends and patterns over the targeted period of time. We overlaid the major socio-economic events on the graphs to identify any correlations or patterns.

In the third step, we calculated the correlation coefficients between the transportation indicators and the economic indicators to quantify their relationships. We paid special attention to the direction and strength of correlations, and we considered both short-term and long-term effects.

In the next stage we started the event analysis. We identified major socio-economic disruptive events such as financial crises in 2001 and 2009, or the COVID-19 pandemic. We analyzed how these events have impacted the critical transportation infrastructure and the economic indicators. We were looking for any observable sign of resilience or vulnerability in transportation infrastructure during these events.

In the fifth phase of the predictive model creation, we utilized statistical or machine learning models to predict the resilience of the transportation infrastructure based on economic indicators and past disruptive events. We incorporated features such as GDP per capita, confidence indices, and historical event data into the predictive model.

We have though some considerations regarding our model:

- Causality versus correlation: Remember, that correlation does not imply causation. While we may find strong correlations between indicators, establishing causality requires deeper analysis.
- Data Quality and Completeness: we have to ensure that our data is reliable, consistent, and covers relevant time periods.
- Contextual Understanding: we should consider a broader socio-economic and geopolitical context when interpreting the results of the analysis.
- External Factors: we have to acknowledge the influence of external factors beyond economic and transportation indicators, such as government policies, technological advancements, and environmental factors.

3.3 Data and Analysis

In Fig. 3 we show an example for how to choose indicators to build up an EWS in case of a special SC disruption. First, the threat must be identified what we want to analyze, then possible indicators have to be collected, including macro and micro-level socio-economic or industrial indices. To ensure about their reliability and ability to be used as EWS, we tested them on past disruptions concerning CTI. We observed whether they really indicate the approach of a threat or not, and selected the ones, giving the most appropriate prediction.

In data collection we strived to find constantly refreshed, up-to-date data sources, indices, user-generated contents, which can be downloaded free of charge or if higher precision is needed, from paid sources. The exact data and the structure depend on the disruption itself, its sources and consequences. As main data sources, which could be used also as an EWS, were identified the following databases:

- **Gross Domestic Product per capita** (**GDP p.c.**) [in USD, in 2019 prices and PPPs]: Is used as a base indicator of economic development of the OECD countries, is a measure of the average economic output per person within a country. It is calculated by dividing the total GDP of a country by its population. GDP per capita provides insight into the economic prosperity and standard of living within a given country, as it reflects the average income and production level per individual. This metric is often used to compare the economic performance of different countries and to assess changes in living standards over time [43].

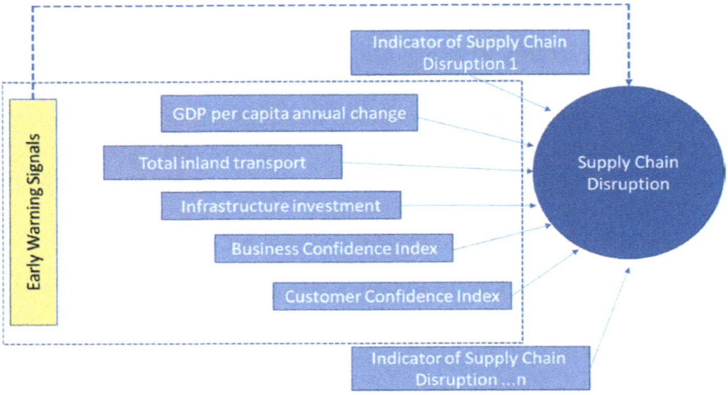

Fig. 3 Construction of an EWS. *Source* Own edition

- **Transportation—Total Inland freight transport** [tons-kilometers, million]: This indicator of OECD refers to all road, rail and inland waterway freight transport and any movement of goods using these vehicles or a given road, rail and inland water network. From OECD countries were counted European continental counties, without the Albania, Belgium, Bosnia-Hercegovina, Denmark, Lichtenstein, Montenegro [44].
- **Infrastructure Investment**: Investment in infrastructure encompasses expenditures directed towards the construction of new transportation facilities and enhancements to existing networks, serving as a crucial factor influencing performance within the transport sector. Inland infrastructure, covering roadways, railways, inland waterways, maritime ports, and airports, considers various financing sources. Efficient transport infrastructure yields economic and social advantages for both developed and developing economies, fostering enhanced market accessibility, productivity, regional economic equilibrium, employment opportunities, labor mobility, and community connectivity. This indicator is gauged as a portion of GDP for overall inland investment and in euros for specific components, including road, rail, air, inland waterways, and maritime segments. For data analysis there were selected investments into inland total. The time frame is yearly, from 2000 till the latest available data 2021, and counted in EUR [45].
- **Business Confidence Index (BCI)**: This indicator of business confidence offers insights into prospective trends, derived from surveys gauging perceptions of production, orders, and finished goods stocks within the industry sector. It serves as a tool for tracking output expansion and predicting shifts in economic activity. Values exceeding 100 indicate heightened optimism regarding forthcoming business performance, while values below 100 signal pessimism regarding future prospects [46].
- **Consumer Confidence Index (CCI)**: The consumer confidence index offers insights into the anticipated trajectories of household consumption and saving

habits, drawing from responses concerning expected financial well-being, perceptions of the broader economic climate, unemployment outlook, and saving capacity. A reading surpassing 100 indicates enhanced confidence among consumers regarding future economic conditions, potentially leading to reduced savings and increased spending on significant purchases over the subsequent 12 months. Conversely, values below 100 denote a pessimistic outlook toward future economic trends, potentially prompting a proclivity for heightened saving and diminished consumption [47].

4 Results and Discussion

During our analysis we selected a period between 2001 and 2021. During this period, we highlighted four disruptive events: the DotCom crisis in 2001–02, the financial crisis in 2009, the bankruptcy of Hanjin Shg in 2016 and the Covid-19 pandemic in 2020.

We analyzed the data of the 3.3 chapter to find connection between the selected indicators. In order to better identify the trend changes and to identify appropriate datasets that could be used to predict potential future constraints or disruptions in transport capacity, it is advisable to examine the relationships between individual indicators separately.

The correlation between total transport performance and changes in GDP p.c. can be seen in Fig. 4. By looking at the year-by-year change in GDP p.c. and Total Inland Transport, it is evident how the transport outputs are driven by the economy, i.e. it is very sensitive to changes in trends. At the same time, they show a slight inertia and react with a slight lag, e.g. in 2008–2009 when the global economic crisis slowed down the economic growth. At the same time, however, in the recovery phase, year-by-year transport performance seemed to pull the economy towards growth. This phenomenon is confirmed by the fact that goods had to be transported from producers to consumers, who then made payments, i.e. the transaction only became part of the economic indicators after the transport performance had been realized. However, the situation was different in the COVID-19 period when transport performance was reduced (for the sample of countries examined).

The next step was to examine the relationship between year-by-year changes in economic output and infrastructure investments. Figure 5 shows how infrastructure investments have been used as one of the accelerators of GDP growth for the sample of countries. The graph shows limited interconnection between infrastructure investment and the GDP p.c. change of the economies in the selected group of countries in the highlighted periods, e.g. 2000 and 2001, 2005–2007. Conversely, in case of recessions, a lag of about one year in infrastructure investment is visible in 2007 and 2008, and at the beginning of the COVID-19 period being it has been a significant driver contributing to the recovery of the selected economies. For this reason, we determine that the volume changes of investments in infrastructure could not be

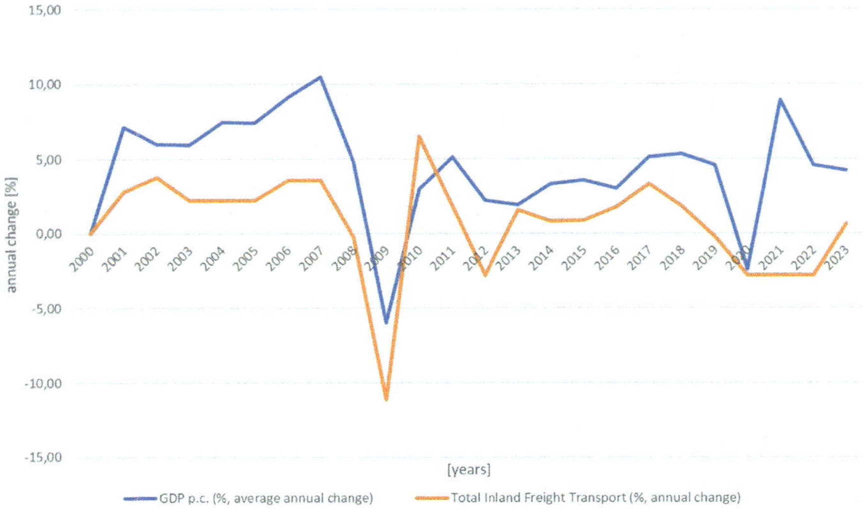

Fig. 4 Ties of annual changes of GDP p.c. and total inland transport [%]. *Source* Own edition

considered a useful measure to forecast supply chain disruptions. It either reacts too slowly to economic downturn or is used to generate growth.

A similar trend can be seen in the relationship between the annual level of infrastructure investment and total transport performance is shown in Fig. 6, which shows similar trends of change to the relationship between the rate of change in GDP p.c.

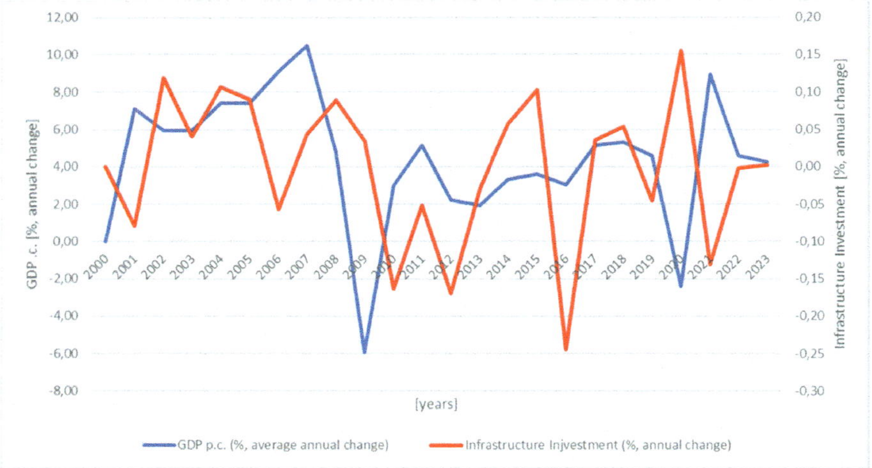

Fig. 5 Ties of GDP p.c. and infrastructure investments [%, annual change]. *Source* Own edition

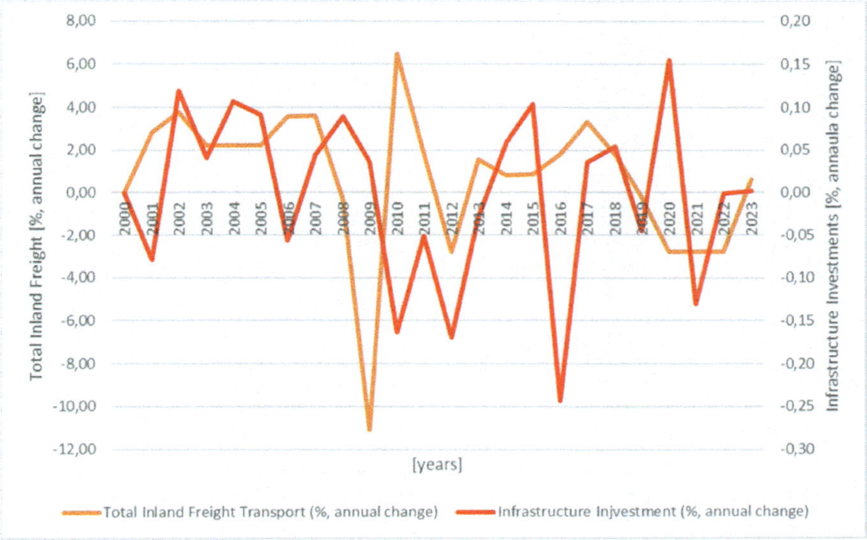

Fig. 6 Ties of total inland freight transport and investments info infrastructure [%, annual change]. *Source* Own edition

and the rate of change in infrastructure investment. We can see similarly no ties between transportation capabilities and investments in infrastructure.

However, the inherent relationship between GDP p.c., transport capacity expressed as total inland transport (%, annual change) and its link to the availability and throughput of distribution chains in the case of examining past and potential future disruptions is limited. For this reason, it is appropriate to include rational expectations, both from the supply side, i.e. producers, economic agents, and from the demand side, i.e. consumers. BCI and CCI have been chosen as these factors. The relationship between the annual change in GDP p.c., BCI and CCI is shown in Fig. 7.

From Fig. 7 shows a significant interlinkage of the annual change trends of the GDP p.c., BCI and CCI, when, however, GDP p.c. shows more significant increases in these year-on-year changes.

However, if replaced by GDP p.c. annual %-change for Total Inland Freight Transport annual %-change, it is possible to identify a more significant connection between transport performance, producer and consumer confidence. This interdependence is shown in Fig. 8.

Overall interconnections, of annual changes in economic, transportation capabilities and consumers/producent perceptions, presented through indicators of the GDP p.c., infrastructure investments, total inland freight transport, BCI and CCI, with World Bank inactive containership capacity [in mil. TEU], highlighting on background the selected disruptive events in monitored period, which were DotCom crisis

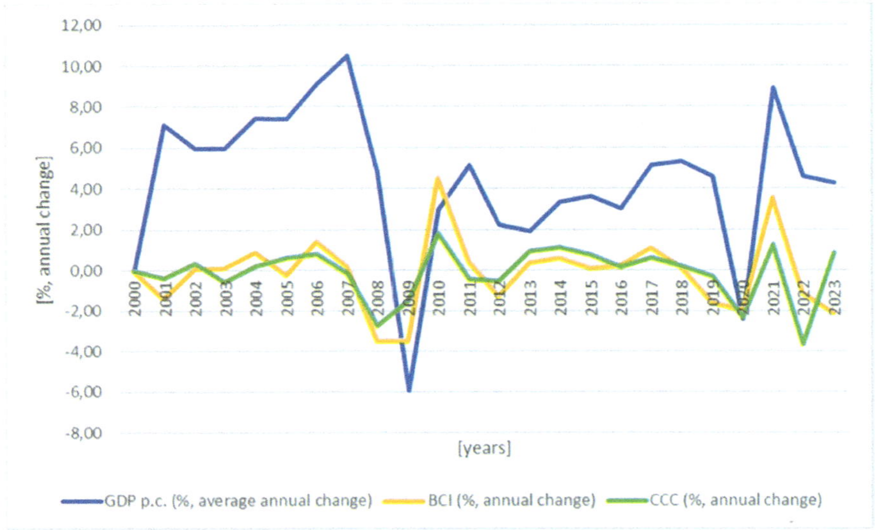

Fig. 7 Ties of annual changes in GDP p.c., BCI and CCI [%, annual change]. *Source* Own edition

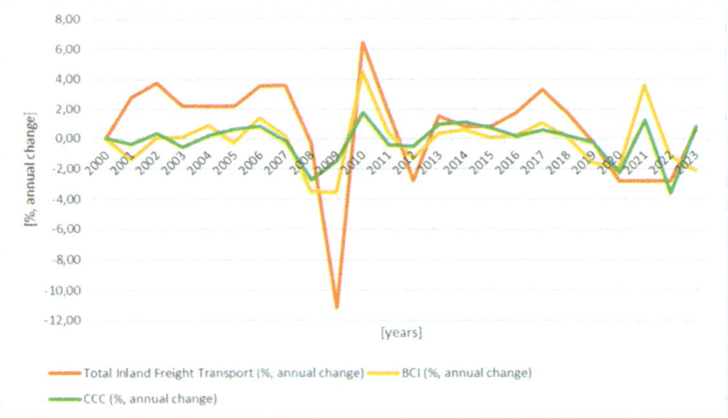

Fig. 8 Ties of annual changes in total inland freight transport, BCI and CCC [%]. *Source* Own edition

in 2001/2002, global financial crisis in 2009, Hanjin Shg bankruptcy and 2019–2021 COVID-19 pandemic (Fig. 9).

Our data collection can have some limitations. Due to lack of data and the similarities of the methodology how they were acquired, we had to limit OECD countries for European continental countries. In some cases there were not data available for all of the selected indicators for all of the countries. For this reason, from the whole group of OECD countries we could analyze only those where whole datasets were

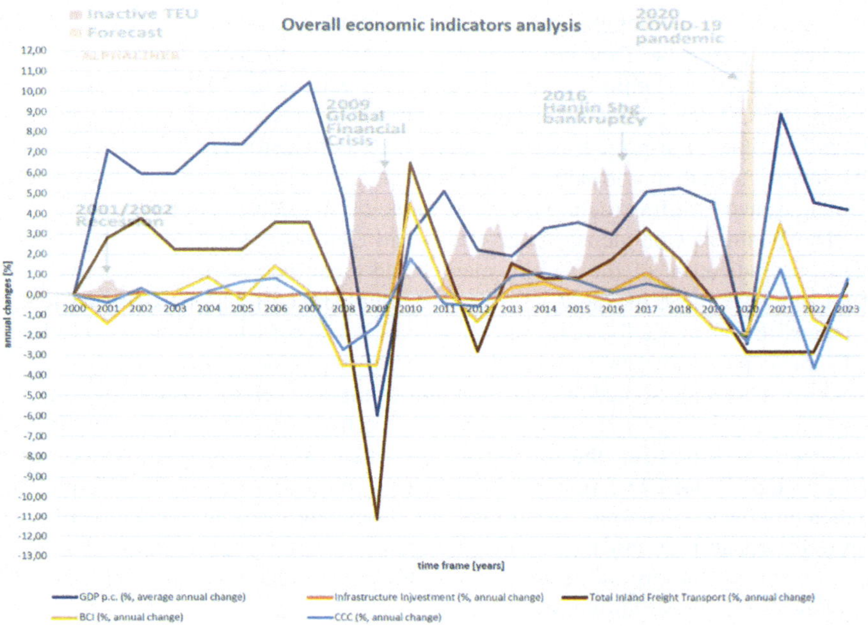

Fig. 9 Overall economic indicators analysis. *Source* Own edition

available. In case of the Consumer Confidence Index (CCI) for Latvia, Lithuania and Poland for the period between January 2000 and April 2001 only approximated CCI was available on the initial level of 100. While the CCI is refreshed on a monthly basis, we calculated a yearly average for the rest of the countries.

As a conclusion of the analysis, we can determine that BCI and CCI metrics, together with Total Inland Freight Transport, can be considered appropriate metrics for identifying changes in transportation chains. These indicators offer a higher level of visibility of potential disruptions within supply chains.

5 Conclusion

In our study, we investigated how big data analysis can be used to increase the resilience of critical transport infrastructure. From the category of big data analysis, we highlighted the importance of early warning signals and investigated which publicly available databases can be used to predict threats that threaten CTI. We have selected a CTI risk which causes that international container freight traffic will drop dramatically. By examining selected data from the past 20+ years (2001–2021), we looked for correlations between how well the chosen indicators could have predicted the crises that occurred during this time period.

There have been four examples of a decline in international container freight over the last 21 years, the DotCom crisis of 2001–02, the financial recession of 2009, the bankruptcy of Hanjin Shg in 2016 and the pandemic of 2020. Our chosen indicators (OECD data) were the change in GDP p.c., Infrastructure Investments, Total Inland Freight Transport, Business Confidence and Customer Confidence Indices. We investigated their correlations and their potential role in predicting disruption to critical transport infrastructure on the example of 16 selected countries.

The Business Confidence Index (BCI) and Consumer Confidence Index (CCI) serve as indicators of the supply market and customer market, respectively. As depicted in Fig. 8, the analysis reveals significant correlations between annual changes in transportation and both BCI and CCI, surpassing those observed with GDP per capita annual changes or other selected criteria such as investments in infrastructure. This underscores the crucial relationship between the availability of transportation capabilities and economic confidence from both the supply and demand perspectives. Moreover, the findings suggest that over a long-term period across 16 selected EU/OECD countries, these indices could be deemed effective tools for enhancing overall economic performance. Consequently, the augmentation of transportation capabilities emerges as a viable strategy for fostering economic dynamism and improving the overall performance of the supply chain, thereby enhancing the resilience of transportation infrastructure.

Our research has limitations. First, we now have a sample of 16 countries for which all relevant data were available during the period under study. The lack of data has significantly reduced the number of countries that can be included in the analysis. Further analysis is also needed on what additional indicators should be included to provide a more accurate forecast and whether they should be weighted differently when building an EWS system. Although our approach is theoretical and experimental, it would be very interesting to test the possibility of forecasting a disruption relevant to a specific company on the basis of real company and market data.

References

1. Mukherjee S, Padhi SS (2022) Sourcing decision under interconnected risks: an application of mean–variance preferences approach. Ann Oper Res 313:1243–1268
2. Foltin P (2011) Security of logistics chains against terrorist threats. In: 17th International conference the knowledge-based organization, conference proceedings 1: management and military sciences. Nicolae Balcescu-Land Forces Academy, Sibiu, pp 100–105
3. Neiger D, Rotaru K, Churilov L (2009) Supply chain risk identification with value-focused process engineering. J Oper Manag 27:154–168
4. Demeter K, Losonci D, Szalavetz A, Baksa M (2023) Strategic drivers behind the digital transformation of subsidiaries: a longitudinal approach. Post Communist Econ 35:744–769
5. European Union (2008) Council directive on the identification and designation of European critical infrastructures and the assessment of the need to improve their protection. Council Directive, pp 1–8

6. Golgeci I, Ponomarov SY (2013) Does firm innovativeness enable effective responses to supply chain disruptions? An empirical study. Supply Chain Manag Int J 18:604–617
7. Li Y, Kramer MR, Beulens AJM, van der Vorst JGAJ (2010) A framework for early warning and proactive control systems in food supply chain networks. Comput Ind 61:852–862
8. Narasimhan R, Talluri S (2009) Perspectives on risk management in supply chains. J Oper Manag 27:114–118
9. Pospisil K, Fric J (2016) Implementation of strategic management tools to operation of highway agencies. In: 3rd International multidisciplinary scientific conference on social sciences and arts, SGEM 2016. STEF92 Technology Ltd, Bulgaria. WOS:000395727100044
10. Kozine I, Petrenj B, Trucco P (2018) Resilience capacities assessment for critical infrastructures disruption: the READ framework (part 1). IJCIS 14:199
11. Singh NP, Singh S (2019) Building supply chain risk resilience: role of big data analytics in supply chain disruption mitigation. BIJ 26:2318–2342
12. Stone J, Rahimifard S (2018) Resilience in agri-food supply chains: a critical analysis of the literature and synthesis of a novel framework. SCM 23:207–238
13. Ponomarov SY, Holcomb MC (2009) Understanding the concept of supply chain resilience. Int J Logistics Manag 20:124–143
14. Ye Y, Suleiman MA, Huo B (2022) Impact of just-in-time (JIT) on supply chain disruption risk: the moderating role of supply chain centralization. IMDS 122:1665–1685
15. Vilko JPP, Hallikas JM (2012) Risk assessment in multimodal supply chains. Int J Prod Econ 140:586–595
16. Katsaliaki K, Galetsi P, Kumar S (2022) Supply chain disruptions and resilience: a major review and future research agenda. Ann Oper Res 319:965–1002
17. Rehak D, Splichalova A, Hromada M, Walker N, Janeckova H, Ristvej J (2023) Critical entities resilience failure indication. Saf Sci 170. https://doi.org/10.1016/j.ssci.2023.106371
18. Knemeyer AM, Zinn W, Eroglu C (2009) Proactive planning for catastrophic events in supply chains. J Oper Manag 27:141–153
19. Olivares-Aguila J, Vital-Soto A (2021) Supply chain resilience roadmaps for major disruptions. Logistics 5:78
20. Ambrogio G, Filice L, Longo F, Padovano A (2022) Workforce and supply chain disruption as a digital and technological innovation opportunity for resilient manufacturing systems in the COVID-19 pandemic. Comput Ind Eng 169:108158
21. Genc E, Duffie N, Reinhart G (2014) Event-based supply chain early warning system for an adaptive production control. Procedia CIRP 19:39–44
22. Er Kara M, Oktay Fırat SÜ, Ghadge A (2020) A data mining-based framework for supply chain risk management. Comput Ind Eng 139:105570
23. ITF (2021) Data-driven transport infrastructure maintenance, pp 1–32
24. Kurth M et al (2020) Lack of resilience in transportation networks: economic implications. Transp Res Part D Transp Environ 86
25. Park Y, Hong P, Roh JJ (2012) Supply chain lessons from the catastrophic natural disaster in Japan. Bus Horiz 56:75–85
26. Kudlak A, Urban R, Hoskova-Mayerova S (2020) Determination of the financial minimum in a municipal budget to deal with crisis situations. Soft Comput 24(12):8607–8616
27. Lo WS, Hong TP (2012) A three-level multiple-agent early warning mechanism for preventing loss of customers in fashion supply chains. In: Choi T-M (ed) Fashion supply chain management: industry and business analysis. IGI Global, USA. https://doi.org/10.4018/978-1-60960-756-2
28. Beulens AJM, Li Y, Kramer MR, van der Vorst JGAJ. Possibilities for applying data mining for early warning in food supply networks. In: Workshop on methodologies and tools for complex system modelling and integrated policy assessment
29. Ruzicka J, Tichy J, Hajciarova E (2022) Big data application for urban transport solutions. In: Smart city symposium Prague (SCSP). IEEE, New York, USA. https://doi.org/10.1109/SCSP54748.2022.9792538

30. Tseng M-L, Bui T-D, Lim MK, Lewi S (2021) A cause and effect model for digital sustainable supply chain competitiveness under uncertainties: enhancing digital platform. Sustainability 13:10150
31. Wang G, Gunasekaran A, Ngai EWT, Papadopoulos T (2016) Big data analytics in logistics and supply chain management: certain investigations for research and applications. Int J Prod Econ 176:98–110
32. Wamba SF et al (2017) Big data analytics and firm performance: effects of dynamic capabilities. J Bus Res 70:356–365
33. Modgil S, Singh RK, Hannibal C (2021) Artificial intelligence for supply chain resilience: learning from Covid-19. IJLM (ahead-of-print)
34. Ivanov D, Dolgui A, Sokolov B (2019) The impact of digital technology and Industry 4.0 on the ripple effect and supply chain risk analytics. Int J Prod Res 57:829–846
35. Dvorak Z, Rehak D, David A, Cekerevac Z. Qualitative approach to environmental risk assessment in transport. Int J Environ Res Public Health 17(15)
36. Birkel HS, Hartmann E (2020) Internet of things—the future of managing supply chain risks. Supply Chain Manag 25:535–548
37. Haimes YY (2002) Risk of terrorism to cyber-physical and organizational-societal infrastructures. Public Works Manag Policy 6:231–240
38. Tonn G, Kesan J, Zhang L, Czajkowski J (2019) Cyber risk and insurance for transportation infrastructure. Transp Policy 79:103–114
39. Xu L, Dong Q, Xiao K (2010) Research on early-warning model for food supply chain risk based on logistic regression. In: 2010 International conference on logistics engineering and intelligent transportation systems. IEEE, Wuhan, China, pp 1–4. https://doi.org/10.1109/LEITS.2010.5665012
40. Sumei, L (2010) Research on risk early-warning mechanism of apparel supply chain. In: 2010 3rd International conference on information management, innovation management and industrial engineering. IEEE, Kunming, China, pp 574–577. https://doi.org/10.1109/ICIII.2010.143
41. Jahns IC, Hartmann E, Moder M (2006) Managing supply risks: a system theory approach to supply early warning systems. In: Managing risks in supply chains: how to build reliable collaboration in logistics. Erich Schmidt Verlag, Berlin, pp 195–212
42. Aczel P, Szanto Z, Csak J (2019) Social futuring: an innovation in measuring countries' fitness for the future. In: 19th SGEM International multidisciplinary scientific geoconference EXPO proceedings
43. OECD (2024) Infrastructure investment (indicator). https://doi.org/10.1787/b06ce3ad-en. Accessed on 02 Feb 2024. https://stats.oecd.org/Index.aspx?datasetcode=SNA_TABLE1_ARCHIVE
44. OECD (2024) Transport measurement—total inland freight transport in ton-km per one thousand units of current USD GDP. https://stats.oecd.org/Index.aspx?DataSetCode=MEI_BTS_COS
45. OECD (2024) Total inland transport infrastructure investment. https://stats.oecd.org/Index.aspx?DataSetCode=MEI_BTS_COS
46. OECD (2024) Business confidence index (BCI) (indicator). https://doi.org/10.1787/3092dc4f-en. Accessed on 02 Feb 2024. https://data.oecd.org/leadind/business-confidence-index-bci.htm
47. OECD (2024) Consumer confidence index (CCI) (indicator). https://doi.org/10.1787/46434d78-en. Accessed on 10 Feb 2024. https://data.oecd.org/leadind/consumer-confidence-index-cci.htm

Are There Still Holes in the Swiss Cheese?—Exploring the Inherent Limits of Safety and the Human Factor

Kornélia Lazányi

Abstract The article delves into the enduring conundrum of safety in complex systems and asks the pivotal question: "Are there still holes in the Swiss cheese?" Drawing inspiration from James Reason's Swiss cheese model, the article examines the inherent limitations of achieving complete safety within any system, highlighting the perpetual challenges posed by the human element. In analysing the layers of defence and the vulnerabilities that persist despite relentless efforts, the paper underscores the crucial reality that absolute safety remains elusive due to the fundamental imperfections in human behaviour, cognition, and decision-making. By acknowledging these persistent "holes" in the safety paradigm, the article emphasises the need for ongoing vigilance, adaptation, and a more nuanced understanding of safety that integrates the inherent fallibility of human actors within complex systems. This exploration ultimately underscores the importance of continuous improvement and a proactive approach to risk mitigation, as well as the vital recognition that while perfection in safety may be unattainable, diligent efforts can minimize the alignment of those critical vulnerabilities and enhance overall security and resilience.

Keywords Safety · Human factor · Consciousness · Swiss cheese model

1 Introduction

The roots of safety as a science can be traced back to the tumultuous landscape of the nineteenth century, characterized by rapid industrialization, burgeoning factories, and a significant surge in workforce engagement. This period marked a pivotal juncture where the adverse consequences of unchecked industrial growth began to manifest in the form of workplace accidents and hazards. Taylor [1], a key figure in the early development of organisational management principles, inadvertently

K. Lazányi (✉)
John von Neumann Faculty of Informatics, Obuda University, Budapest, Hungary
e-mail: lazanyi.kornelia@nik.uni-obuda.hu

T. A. Kovács et al. (eds.), *The Impact of the Energy Dependency on Critical Infrastructure Protection*, Advanced Sciences and Technologies for Security Applications, https://doi.org/10.1007/978-3-031-78544-3_2

laid the groundwork for safety considerations. Taylor, having come from an engineering background approached organisations with a rigorous mind advocating for the systematic study of work processes, emphasising the need for standardized procedures to enhance efficiency [2]. Taylor's principles of scientific management—while primarily focused on improving industrial efficiency and productivity—indirectly contributed to the emergence of safety as a scientific discipline.

Of course, it was not only Taylor, who has realised the need for more safe and secure work environments. In the early twentieth century, diverse disciplines contributed to the evolving landscape of safety, each offering unique perspectives on the definition and implementation of safety practices. Hamilton [3], a pioneering occupational physician and toxicologist, approached safety from a health and medical standpoint. Through her investigations into workplace conditions, particularly in factories and mines, Hamilton played a crucial role in advocating for improved safety regulations and occupational health practices [4]. Additionally, the contributions of Gilbreth and Gilbreth [5], known for their work in time and motion studies, influenced safety through the lens of efficiency and ergonomics [6]. Having worked together with Taylor, their primary focus was on motion efficiency. Their principles indirectly contributed to the reduction of physical strain in various industrial processes, aligning with broader safety considerations [7]. Meanwhile, Fayol [8], a French mining engineer and management theorist, emphasised the organisational aspects of safety. His administrative management principles, which included coordination, discipline, and order, contributed to the development of a framework for enhancing workplace safety [9]. Collectively, these figures from different disciplines enriched the understanding of safety, reflecting the multifaceted nature of safety definitions that emerged in the early twentieth century.

However, the growing recognition of workplace hazards and the escalating frequency of accidents necessitated a more concerted effort to address safety concerns. As the world progressed into the twentieth century, the concept of safety evolved from mere accident prevention to a more sophisticated understanding of risk and hazard management. The turn of the century witnessed the emergence of safety as a distinct field of study. Pioneering figures such as Herbert William Heinrich made significant contributions with his groundbreaking work, "Industrial Accident Prevention: A Scientific Approach" [10]. Heinrich's research laid the foundation for understanding the correlation between workplace accidents, injuries, and the broader concept of safety. Heinrich's work marked a shift from a reactive approach—responding to accidents after they occurred—to a more proactive stance, advocating for the identification and elimination of root causes to prevent accidents [11]. Heinrich's "Accident Pyramid," suggesting that major accidents are preceded by a larger number of minor accidents and near misses. The adoption of such a scientific approach represented a transformative moment in the evolution of safety, aligning it more closely with systematic inquiry and evidence-based practices [12].

In line with this, the mid-twentieth century saw a paradigm shift with the introduction of systems exploring different aspects of safety. Deming's [13] Total Quality Management principles emphasized a holistic approach to quality and safety, considering the entire organisation as an interconnected system [14], while Perrow [15], in his seminal work "Normal Accidents," posited that complex systems inherently harbour the potential for catastrophic failure, challenging the then-prevailing notions of fail-safe systems [16].

The next big cornerstone in safety theory was Reason's Swiss Cheese Model [17]. This model underscored the concept of latent conditions and active failures within complex systems. Reason's work highlighted that accidents result from a unique alignment of systemic vulnerabilities, revolutionizing the approach towards organisational safety by focusing on multiple layers of defence.

2 The Swiss Cheese Model

Bridging the historical evolution of organisational safety with the Swiss Cheese Model provides a comprehensive understanding of how our approach to risk management has matured over time. The model encapsulates the culmination of lessons learned from the industrial age to the information era, synthesising past safety philosophies, from the early focus on individual errors to the complex interplay of systemic vulnerabilities. This progression not only reflects a deepening understanding of what constitutes safety but also sets the stage for examining the critical role of the human factor in this sophisticated safety paradigm.

The Swiss Cheese Model [18] compares the layers of defences to slices of Swiss cheese, each slice representing a layer of defence with potential flaws or 'holes'. The central message of this model is that accidents occur when the holes in these layers momentarily align, allowing a trajectory of hazard to materialize into a catastrophic event [19].

The slices of the Swiss cheese can be broadly categorized into latent conditions and active failures. While latent conditions are the inherent weaknesses in the system, often going unnoticed until they contribute to an accident like design flaws, inadequate training, organisational culture issues, and maintenance deficiencies, active failures are direct and observable actions or omissions that lead to an unsafe situation. They represent the last line of defence where human operators can either prevent an accident or be the direct cause of it [20].

While in theory latent conditions and active failures can be easily distinguished, it is important to understand their interconnected nature and its implications [21]. An example of this delicate interplay is when a suboptimal decision is made—owing to lack of information and communication—because of an unsupportive culture of unconditional obedience—where subordinates are afraid of, or do not feel encouraged to share their insight with the higher ups who are making decisions [22]. All in all, the interaction between different layers—how organisational culture impacts human performance, or how system design can mitigate or exacerbate human error—forms the crux of this model.

Although the model's preliminary purpose is to call attention to the existence of the multiple slices and to highlight their interconnectedness, it may also serve as a tool for both analysis and prevention. It aids in dissecting past incidents to understand how—because of what layers—did the defences fail and also encourages and empowers decision makers to create comprehensive risk management strategies [23].

The Swiss Cheese's slices connected to the Deepwater Horizon Oil Spill—an ecological catastrophe of the Gulf of Mexico, stemming from the blowout of the Macondo well may be identified as follows [24, 25]:

- Cost-cutting measures
- Technical failures (blowout preventer)
- Lack of emphasis on safety culture
- Lack of awareness
- Lack of preparedness
- Compromised safety protocols
- Lack of risk assessment
- Lack of enforcement
- Inadequate decision (to proceed).

From the above list it is easy to understand, how the different slices highlight different aspects of an organisation's functioning, but how these aspects are still interconnected, and very often form a vicious cycle, such as in case of safety culture—awareness—preparedness layers [26].

3 The Role of the Human Factor

The Swiss Cheese Model, offering profound insights into the intricate layers that constitute safety mechanisms within complex systems accentuates the critical role of human factors, illustrating how they interact with and influence each layer of defence. Central to the model's human aspect are cognitive and psychological factors, which significantly influence human performance. As cognitive psychologist Kahneman [27] have highlighted, human cognition is susceptible to biases and heuristics that can precipitate errors. The interplay of attention, perception, memory, and decision-making processes, particularly under stress and fatigue, often exacerbates the likelihood of errors, aligning with the model's conceptualization of safety as a series of potentially flawed defences [28]. Kahneman's work on fast and slow thinking elucidates how these cognitive processes can lead to systematic errors in high-pressure environments, a critical consideration in the Swiss Cheese Model.

Moving beyond cognitive elements, the role of skill-based factors in determining the efficacy of safety measures is undeniable. The Dreyfus [29] model of skill acquisition sheds light on how expertise, developed through knowledge and experience, is instrumental in mitigating risks. However, the variance in skill levels, particularly in high-risk environments, stands as a testament to the model's assertion that human error is an inevitable, though manageable, component of any complex system [30]. Soft skills such as adaptability, problem-solving, and emotional intelligence are indispensable in regards to organisational safety. These skills enable individuals to effectively respond to unforeseen situations, a concept that is central to the theory of High Reliability Organisations [31]. Operating in complex, high-risk environments, HROs rely heavily on these soft skills to maintain safety, resonating with the Swiss Cheese Model's perspective on the unpredictability and dynamic nature of risks [32].

Communication and teamwork also emerge as vital cogs in the safety machinery. The breakdown of these elements often leads to breaches in safety protocols. The principles of Crew Resource Management (CRM) in aviation [33], for instance, underscore the necessity of effective communication, situational awareness, and collaborative decision-making. These principles are not just confined to aviation but resonate across various industries, aligning with the Swiss Cheese Model's emphasis on the systemic nature of safety [34].

As already indicated in the previous chapter, organisational culture, especially the notion of safety culture, plays a paramount role in shaping the effectiveness of safety systems. Schein's [35] model of organisational culture highlights the significance of underlying assumptions and beliefs in influencing behaviour. A safety culture that promotes error reporting, continuous learning, and open communication can fortify the layers of defence against potential mishaps, echoing the Swiss Cheese Model's stance on the systemic nature of failures [36].

Despite the acknowledgment of these human factors, integrating them effectively into safety models is fraught with challenges. The variability and unpredictability of human behaviour make consistent safety performance a daunting task. Furthermore, the resistance to cultural change, especially in established industries, is a significant hurdle [37]. The need for continuous training and skill development, coupled with overcoming communication barriers, presents additional challenges in realizing the model's full potential in enhancing safety [38].

Of course, the human aspects, and before others the organisational culture as a key variable cannot be separated from the structural and other hard elements of the organisational reality. They are intertwined and constantly interact with each other. These aspects of organisational behaviour underline the Swiss Cheese Model's assertion that safety is not merely the absence of active failures but also the presence of robust and proactive defences.

4 Developing the Human Factor to Improve Organisational Safety

The discourse on enhancing resilience in organisational safety often orbits around the human factor, given its pivotal role in the dynamics of safety systems. In light of the Swiss Cheese Model, this chapter critically examines the strategies to improve resilience through the lens of the human element, weaving through the arguments and counterarguments presented in international literature regarding the importance of training, communication and teamwork, and providing refutations where necessary.

Training and Development

Training plays a pivotal role in enhancing safety-conscious behaviour within organisations, serving as a cornerstone for promoting a culture of safety and mitigating risks in the workplace. However, training and organisational development can be approached from multiple angles.

Social Learning Theory [39] posits that individuals learn by observing others and modelling their behaviour. Training programs provide opportunities for individuals to observe safe practices, learn from experienced colleagues, and receive feedback on their performance. By reinforcing safety messages and behaviours, training can effectively promote safety-conscious behaviour [40]. Along these lines Behaviour-Based Safety Theory (BBS) emphasizes the analysis and modification of observable behaviours to improve safety outcomes [41]. Training programs aligned with BBS principles emphasize the identification of critical safety behaviours, reinforcement of desired behaviours through positive feedback, and involvement of employees in setting safety goals. By focusing on observable behaviours, training can directly impact safety-conscious behaviour in the workplace [42].

On the other hand, training interventions focused on understanding how human capabilities and limitations interact with the design of systems and environments aim to enhance individuals' understanding of their cognitive and physical limitations, as well as how these factors interact with the work environment [18]. By providing knowledge and skills tailored to human capabilities, training can mitigate errors and promote safer behaviour.

Cognitive-Behavioural Theory [43] posits that behaviour is influenced by individual beliefs, attitudes, and perceptions. Training interventions based on Cognitive-Behavioural principles aim to modify individuals' beliefs and attitudes towards safety, challenge unsafe norms, and promote a culture of safety. By targeting cognitive processes such as risk perception and decision-making, training can effectively influence safety-conscious behaviour. Hence, one primary approach to bolstering safety conscious behaviour involves addressing cognitive and psychological factors [44]. Kahneman [27] suggests that training aimed at recognizing and mitigating these biases can enhance decision-making under pressure. However, a counterargument emerges from the works of researchers like Dekker [45], who argue that such training can oversimplify the complexity of human cognition, potentially leading to a false sense of security. To refute this, it's crucial to combine cognitive training with real-world scenario simulations, ensuring a more holistic enhancement of cognitive resilience [37].

The emphasis on skill-based training and continuous learning is widely advocated [29]. This perspective suggests that regular, updated training programs can significantly elevate skill levels and adaptability. Nonetheless, critics like Weick and Sutcliffe [31] argue that over-reliance on structured training can lead to rigidity, limiting individuals' ability to adapt to novel situations. This criticism may be addressed by integrating principles of High Reliability Organisations (HROs) [46, 47] into training, which focus on flexibility and the capacity to respond to the unexpected, thus maintaining a balance between structured skills and adaptive capabilities.

In conclusion, training serves as a vital mechanism for bolstering safety-conscious behaviour within organisations. By integrating principles from various theoretical frameworks, training programs can effectively shape individual behaviours, attitudes, and organisational culture to enhance safety outcomes in the workplace. However, while training is essential for imparting knowledge and skills, it primarily focuses on individual-level factors such as understanding safety procedures and using safety equipment, safety-conscious behaviour is influenced by a wide range of factors, including organisational culture, leadership, and the work environment, which training alone may not address adequately [48]. In line with this emphasizing training as the primary approach to bolstering safety-conscious behaviour overlooks the systemic factors within organisations that shape employee attitudes and behaviours. Factors such as management practices, resource allocation, and reward systems can influence employees' perceptions of safety and their willingness to prioritize it in their daily [49]. On top of this, training often occurs in controlled environments, which may not fully replicate the complexities and pressures of real-world work settings. Even well-trained individuals may struggle to apply their knowledge and skills effectively in high-stress or ambiguous situations, where decision-making plays a significant role in safety-conscious behaviour [50].

Over-reliance on training can reinforce a "blame-the-worker" mentality, where incidents or accidents are attributed solely to individual error or lack of competence [51, 45]. This approach fails to address underlying systemic issues within organisations that may contribute to safety lapses and limits opportunities for organisational learning and improvement. It also emphasises the validity of the Swiss Cheese approach that while training initiatives may yield positive results in the short term, such as increased safety knowledge or awareness, without ongoing reinforcement, support, and integration into daily work practices, the impact of training on behaviour may diminish over time, leading to a regression to previous habits or practices [52]. Accordingly, training and development initiatives cannot be considered the golden way to organisational safety and shall only be regarded as an important slice of the Swiss Cheese.

Effective Communication

The role of communication in high-risk industries such as aviation, healthcare, and maritime operations, as a central element of Crew Resource Management [33], cannot be overstated in enhancing safety resilience. Effective communication is seen as a panacea for many safety issues for multiple reasons. Clear, concise, and assertive communication is crucial for preventing errors. Misunderstandings or misinterpretations can lead to mistakes that compromise safety. Open and respectful communication allows team members to address and resolve conflicts before they impact safety.

Interactive communication ensures that everyone on the team has the correct information to perform their tasks accurately. On top of this, in critical situations, the quality of decision-making often depends on the accuracy and completeness of the information available [53]. Well-functioning communication channels ensure that decision-makers have access to timely and relevant information, allowing them to assess risks accurately and choose the best course of action. Feedback from all levels of the organisation is essential for identifying safety gaps and implementing corrective actions. Communication enables feedback loops where safety practices are continuously monitored, evaluated, and improved [54].

Efficient communication is key to developing a positive safety culture within an organisation. It encourages the reporting of safety concerns and near-misses, facilitating a learning environment where mistakes are analysed, and lessons are shared to prevent future incidents [36]. It ensures that safety procedures, regulations, and any changes are clearly understood and implemented by all team members. This compliance is critical for maintaining safety standards and minimizing risks. In addition, strategic communication helps organisations adapt to these changes by disseminating new safety policies, procedures, and training efficiently across the workforce [55].

In essence, communication is the lifeblood of CRM and organisational safety, enabling the flow of critical information, ensuring team cohesion, facilitating decision-making, and fostering a culture of safety. Without effective communication, the risk of errors increases, potentially leading to accidents and incidents that could have been prevented. Organisations that prioritize and continuously improve their communication practices are better equipped to manage safety and achieve operational excellence. However, as scholars like Roberts and Smith [32] have noted, communication strategies that work well in one cultural or organisational context may not translate effectively into another. This necessitates a tailored approach to communication, considering the unique cultural and organisational contexts, thereby ensuring its effectiveness across diverse settings.

It also has to be noted that while communication is undoubtedly important for CRM and organisational safety, it is not the sole factor. Theoretical frameworks like systems theory, human factors and ergonomics, HRO theory [31], resilience engineering [56, 57], and sociotechnical systems theory [58] highlight that structural, technological, and adaptive capabilities are also critical. These frameworks suggest that focusing exclusively on communication might overlook other essential strategies for enhancing safety and effectiveness in complex organisational settings [59]. The optimal way of approaching communications is to regard it as an important slice of the Swiss Cheese, one that is interconnected with multiple others, but which is far from being the most important, or one that can stop hazards alone from causing organisational issues.

Teamwork

There cannot be an effective CRM without well working teams, and teamwork. On the one hand, it is easy to understand how effective communication among team members and the sense of belonging to a team encouraging the sharing of information with peers are two sides of the same coin [54]. In a team setting, individual members contribute to a collective understanding of the environment, operations, and potential hazards. This shared situational awareness is critical for early detection of safety issues and timely interventions, reducing the likelihood of accidents or incidents [60]. On top of this, a well-functioning team encourages mutual support, continuous learning, and collective responsibility for safety outcomes. Teams that regularly communicate, debrief, and learn from each other's experiences are better positioned to identify latent safety threats and implement improvements. Working in teams can also result in a more even distribution of tasks among team members, reducing the cognitive and physical workload on individuals, which does not only minimize the risk of errors due to fatigue or overload but also helps in maintaining a calm and focused approach in emergency situations [61]. What is more, teams that work well together can adapt more quickly to changing situations, an essential capability in dynamic and complex environments. Teamwork enables flexible responses to unforeseen challenges, with members supporting each other to modify plans, roles, or actions as required to maintain safety.

In summary, teamwork encourages the pooling of knowledge, skills, and experiences, leading to more informed and robust decision-making processes. By working collaboratively, teams can evaluate risks more accurately, consider a wider range of solutions, and choose the most effective course of action to maintain or enhance safety [62]. It underpins the ability of groups to operate more effectively, safely, and cohesively, especially in contexts where the margin for error is small and the consequences of failure are significant [40]. By prioritizing and investing in teamwork, organisations can significantly enhance their overall safety performance and resilience.

However, while teamwork holds promise for enhancing safety performance and resilience, its efficacy is contingent upon mitigating the inherent limitations caused by the negative side effects of groups. Janis's [63] concept of groupthink highlights the potential for cohesive groups to prioritize harmony and consensus over critical evaluation of alternative viewpoints. In safety–critical contexts, such as aviation or healthcare, this can lead to conformity bias, where dissenting voices are suppressed, and risks are underestimated [64, 65]. Despite the pooling of knowledge and experiences, groupthink inhibits teams from considering unconventional perspectives or questioning established norms, thereby compromising the robustness of decision-making processes [66]. Of course, it is not indisputable that group decisions will be drawn to conservative solutions. Research on group polarization suggests that group discussions tend to amplify pre-existing attitudes, leading to more extreme decisions than those initially favoured by individual members [67]. In safety contexts, this can manifest as a tendency towards either overly cautious or excessively risky decision-making, depending on the dominant risk orientation within the team [68]. Group polarization may diminish the team's ability to strike a balanced approach to risk management, resulting in suboptimal safety outcomes.

Another potential threat to efficient group work is the phenomenon of social loafing. Social loafing theory suggests that individuals may exert less effort in a group setting, relying on others to compensate for their contributions [69]. In safety–critical environments, this phenomenon can lead to a diffusion of responsibility, where team members assume that others will address safety concerns. Consequently, the collective vigilance and proactive risk management expected from teamwork may be undermined by individual complacency or disengagement [70].

Organisational interventions that promote psychological safety, encourage diverse perspectives, and provide structured team training can help mitigate the adverse effects of groupthink, social loafing, and group polarization. Furthermore, recognizing the complexities of team coordination and investing in supportive technologies and procedures can enhance the reliability of teamwork in safety–critical contexts. In summary, while teamwork offers numerous benefits for organisational safety, including enhanced decision-making, communication, and resilience, it also presents challenges such as groupthink, social loafing, and coordination difficulties. Organisations must strive to mitigate these drawbacks through targeted interventions, such as fostering a culture of psychological safety, providing structured team training, and promoting diversity of thought within teams.

5 Conclusion

The evolution of organisational safety from the nineteenth century to the present day reflects a profound shift in understanding and approach. From the initial focus on individual safety measures to the complex interplay of systems thinking, and finally to the contemporary emphasis on resilience, this journey mirrors the broader evolution of industrial and organisational practices. As we continue to explore and understand the intricate tapestry of factors influencing safety, the commitment to creating more resilient systems becomes increasingly paramount.

The Swiss Cheese Model, with its layers of defence and emphasis on the human factor, encapsulates the essence of organisational safety. The evolution of safety from its historical roots to the present day highlights the deep-seated complexities within systems. The human factor, often the linchpin of safety, requires a dynamic and nuanced approach, acknowledging its multifaceted nature. As organisations grapple with the challenges and intricacies, the quest for safety remains an eternal journey—a journey that demands perpetual vigilance, adaptation, and a commitment to fostering a culture where safety is not merely a goal but an inherent part of organisational DNA.

In contemplating the unending quest for organisational safety, it is imperative to recognize that the complexities and challenges encapsulated within the Swiss Cheese Model render safety a continuous and evolving narrative. The historical evolution, theoretical frameworks, human factors, and strategies for resilience collectively paint a picture of perpetual improvement. The quest for safety is not a static goal achieved at a particular point in time but an ongoing process of learning, adaptation, and refinement.

As we navigate the human factor within the Swiss Cheese Model, it becomes evident that the quest for organisational safety pivots significantly on understanding and addressing organisational safety as a system, which is more than the mere sum of its components, which is an intricate, interconnected and interdependent system of the various layers. The cognitive and psychological aspects of the human factor underscore the inherent vulnerabilities within these layers. The integration of skill-based factors, communication dynamics, and teamwork, all influenced by organisational culture, forms the tapestry of the model. The model serves not merely as a static representation but as a dynamic framework, acknowledging the constant flux and interdependence of these factors. The approaches proposed in current paper to enhance resilience delve into the heart of these human factors. However, the journey is not without its challenges. The variability in human behaviour, resistance to cultural change, and the intricate dynamics of communication present hurdles that require nuanced solutions. By addressing these challenges, organisations can forge a more robust and resilient safety system, aligning with the principles encapsulated by the Swiss Cheese Model.

While the Swiss Cheese Model has significantly advanced our understanding of system safety and the multifaceted nature of failures, it also emphasizes the inherent limits of safety. The quest for a completely risk-free system is Sisyphean. The model reminds us that safety is a dynamic and ongoing process, not a static goal. It requires continuous vigilance, learning, and adaptation. What is more, despite its widespread acceptance and utility, the Swiss Cheese Model is not without criticism. A pivotal limitation is its implication that with enough layers, safety can be absolute. However, in reality, no system can be completely foolproof. The dynamic nature of the 'holes'—constantly opening, closing, and varying in size—underlines the unpredictable and evolving nature of risk in complex systems. Moreover, the model's focus on failures might overshadow the importance of understanding why systems work well. Nonetheless, in acknowledging the imperfections and intricacies within each layer of defence, the model not only offers a framework for analysing accidents but also a philosophical perspective on the nature of safety in complex systems. Thus, while we strive for improvement, the Swiss Cheese Model soberly reminds us that perfection, much like the elusive completely solid slice of Swiss cheese, remains an ideal to aspire to, rather than a practical reality.

References

1. Taylor FW (1911) The principles of scientific management. Harper Brothers, USA
2. Roberts D (2018) Revisiting Taylor's principles of scientific management: implications for workplace safety. Int J Occup Saf Ergon 24(3):321–335
3. Hamilton A (1911) Industrial poisons in the United States. Atlantic Monthly 108(5):643–656
4. Jones A (2019) Occupational health and safety in historical context: insights from the work of Alice Hamilton. J Saf Res 35(2):78–91
5. Gilbreth FB, Gilbreth LM (1917) Motion study for the crippled soldier. In: Applied motion study: a collection of papers on the efficient method to industrial preparedness, New York, pp 138–3
6. Smith J (2020) Ergonomic principles in organizational safety: a review of historical foundations. J Occup Health Saf 25(3):112–125
7. Brown C (2020) Taylorism and its impact on organizational safety: a contemporary analysis. Saf Manag J 42(4):215–228
8. Fayol H (1916) General and industrial management. Pitman Publishing, London

9. Smith L (2021) Henri Fayol's administrative theory and its relevance to organizational safety. Saf Sci J 38(3):123–136
10. Heinrich HW (1931) Industrial accident prevention: a scientific approach. McGraw-Hill, USA
11. Garcia M, Patel R (2020) Modernizing Heinrich's pyramid: a data-driven approach to industrial accident prevention. Saf Sci 45(4):321–335
12. Smith L (2019) Revisiting Heinrich's industrial accident prevention model: lessons for contemporary safety management. J Occup Health Saf 30(2):87–101
13. Deming WE (1986) Out of the crisis. MIT Press, USA
14. Smith JA, Johnson MB (2021) The impact of teamwork on safety performance in high-risk industries: a systematic review. Saf Sci 45(3):201–215
15. Perrow C (1999) Normal accidents: living with high-risk technologies. Princeton University Press, USA
16. Garcia LM, Chen Y (2020) Revisiting normal accidents theory in the age of complex systems: implications for risk management. J Risk Anal 25(4):301–315
17. Reason J (1990) Managing the risks of organizational accidents. Ashgate Publishing Ltd, UK
18. Reason J (2000) Human error: models and management. BMJ Publishing Group, UK
19. Smith EL, Johnson MA (2021) Applying the Swiss cheese model to healthcare: a contemporary analysis of patient safety incidents. J Patient Saf 15(3):201–215
20. Johnson A (2019) Applying the Swiss cheese model to risk management: lessons from aviation safety. Risk Anal Rev 22(2):115–130
21. Dababneh AJ (2024) A cognitive model for teaching and training in occupational safety and health. J Eng Project Prod Manag 14(2)
22. Shabani T, Jerie S, Shabani T (2023) A comprehensive review of the Swiss cheese model in risk management. Saf Extreme Environ 1–15
23. Puthillath B, Marath B, Ayappan BC (2023) Analyzing the cause of human electrical accidents using Swiss cheese model. Vilakshan-XIMB J Manag 20(1):193–208
24. Hayes J, Hopkins A (2012) Deepwater horizon—lessons for the pipeline industry. J Pipeline Eng 11(3):145
25. Kim SD (2017) Characterization of unknown unknowns using separation principles in case study on deepwater horizon oil spill. J Risk Res 20(1):151–168
26. Williams DA, Glasmeier AK (2023) Evaluation of error across natural gas pipeline incidents. Risk Anal 43(5):1079–1091
27. Kahneman D (2011) Thinking, fast and slow. Farrar, Straus and Giroux, USA
28. Garcia LM, Patel R (2021) Cognitive biases in safety decision-making: implications for organizational safety culture. J Occup Saf Health 25(3):201–215
29. Dreyfus HL, Dreyfus SE (1980) A five-stage model of the mental activities involved in directed skill acquisition. Operations Research Center, University of California, Berkeley
30. Dehn P, Munch Simonsen S, Olesen ML (2023) Multidimensional factors determine skill acquisition development in guided self-determination: a qualitative study. Scand J Caring Sci 37(2):549–560
31. Weick KE, Sutcliffe, KM (2011) Managing the unexpected: assuring high performance in an age of complexity. Jossey-Bass, USA
32. Roberts DA, Smith JL (2020) Navigating complexity: lessons from high-reliability organizations in the digital age. J Appl Psychol 45(3):201–215
33. Helmreich RL, Merritt AC (2001) Safety and error management: the role of crew resource management. Int J Aviat Psychol 11(4):341–352
34. Herin B (2023) Retaining nurses with crew resources management. Sinclair School of Nurses. Downloaded from: https://mospace.umsystem.edu/xmlui/bitstream/handle/10355/95181/Herin_RetainingNurseswithCrewResourceManagement.pdf?sequence=1
35. Schein EH (1985) Organizational culture and leadership. Jossey-Bass, USA
36. Tear MJ, Reader TW (2023) Understanding safety culture and safety citizenship through the lens of social identity theory. Saf Sci 158:105993
37. Rass L, Treur J, Kucharska W, Wiewiora A (2023) Adaptive dynamical systems modelling of transformational organizational change with focus on organizational culture and organizational learning. Cogn Syst Res 79:85–108

38. Tejay GP, Mohammed ZA (2023) Cultivating security culture for information security success: a mixed-methods study based on anthropological perspective. Inf Manag 60(3):103751
39. Bandura A (1977) Social learning theory. J Psychol 1(1):25–35
40. Smith JA, Johnson MB (2020) Applying social learning theory to leadership development: insights from contemporary research. Leadersh Quart 30(2):201–215
41. Geller ES (2001) Working safe: how to help people actively care for health and safety. Van Nostrand Reinhold, New York
42. Toma A, Negura FG, Moraru RI, Mureşan F, Racautanu M (2024) Behavior-based safety, the right long-term approach to address workplace safety. In: MATEC web of conferences, vol 389. EDP Sciences, France, p 00003
43. Beck AT (1976) Cognitive therapy and the emotional disorders. Penguin, UK
44. Odu JO, Hamedon TR, Mahmud A, Baharudin MR (2023) Systematic review of intervention programs to improve the level of knowledge, attitude, and practices towards work safety culture among office workers. Malays J Med Health Sci 19
45. Dekker S (2011) Just culture: balancing safety and accountability. Ashgate Publishing Ltd, UK
46. Hassandoust F, Johnston AC (2023) Peering through the lens of high-reliability theory: a competencies driven security culture model of high-reliability organisations. Inf Syst J 33:1212–1238
47. Dwyer J, Karanikas N, Sav A (2023) Scoping review of peer-reviewed empirical studies on implementing high reliability organisation theory. Saf Sci 164:106178
48. Conchie SM, Donald IJ, Robertson IT (2017) The associations between safety climate and safety behaviours in the construction industry: a systematic review. J Saf Res 61:105–117
49. Rosen MA, DiazGranados D, Dietz AS, Benishek LE, Thompson D, Pronovost PJ, Weaver SJ (2018) Teamwork in healthcare: key discoveries enabling safer, high-quality care. Am Psychol 73(4):433–450
50. Salas E, Tannenbaum SI, Kraiger K, Smith-Jentsch KA (2012) The science of training and development in organizations: what matters in practice. Psychol Sci Public Interest 13(2):74–101
51. Hopkins A (2009) The blame game—a foundational myth of risk and safety management. Australian OHS Education Accreditation Board
52. Joo BK, Yoon SK, Galbraith D (2023) The effects of organizational trust and empowering leadership on group conflict: psychological safety as a mediator. Organ Manag J 20(1):4–16
53. Stroeve S, Kirwan B, Turan O, Kurt RE, van Doorn B, Save L, Jonk P, de Maya BN, Kilner A, Verhoeven R, Farag YBA, Demiral A, Bettignies-Thiebaux B, de Wolff L, de Vries V, Ahn SI, Pozzi, S (2023) SHIELD human factors taxonomy and database for learning from aviation and maritime safety occurrences. Safety 9(1):14
54. Curcuruto M, Griffin MA (2023) Upward safety communication in the workplace: how team leaders stimulate employees' voice through empowering and monitoring supervision. Saf Sci 157:105947
55. Hopcraft R, Tam K, Misas JDP, Moara-Nkwe K, Jones K (2023) Developing a maritime cyber safety culture: improving safety of operations. Marit Technol Res 5(1)
56. Karltun A, Karltun J, Coelho DA, Havemose K, Kjellström S (2023) The work of first line managers—a key to resilience in manufacturing. Appl Ergon 110:103993
57. Hollnagel E, Woods DD, Leveson N (eds) (2006) Resilience engineering: concepts and precepts. Ashgate Publishing Ltd, UK
58. von Bertalanffy L (1968) General system theory: foundations, development, applications. George Braziller, New York
59. Karanikas N, Weber D, Bruschi K, Brown S (2022) Identification of systems thinking aspects in ISO 45001: 2018 on occupational health safety management. Saf Sci 148:105671
60. Laurila-Pant M, Pihlajamäki M, Lanki A, Lehikoinen A (2023) A protocol for analysing the role of shared situational awareness and decision-making in cooperative disaster simulations. Int J Disaster Risk Reduction 86:103544
61. Rosekind MR, Flynn-Evans EE, Gregory KB (2023) The future of fatigue management: strategies, policies, and societal expectations. In: The handbook of fatigue management in transportation. CRC Press, USA, pp 553–568

62. Schulman PR (2020) Organizational structure and safety culture: conceptual and practical challenges. Saf Sci 126:104669
63. Janis IL (1972) Victims of groupthink: a psychological study of foreign-policy decisions and fiascoes. Houghton Mifflin, USA
64. Gaba DM (2000) Anaesthesiology as a model for patient safety in health care. BMJ 320(7237):785–788
65. Nguyen TM, Malik A, Budhwar P (2022) Knowledge hiding in organizational crisis: the moderating role of leadership. J Bus Res 139:161–172
66. Samuels T (2022) How to construct a safety management system (SMS) that promotes safety culture in your organization. Int J Aviation Aeronaut Aerosp 9(2):3
67. Myers DG, Lamm H (1976) The group polarization phenomenon. Psychol Bull 83(4):602–627
68. Fiset J (2023) Multilingual organisations: employee motives and human resource management adaptive strategies. In: Understanding the dynamics of language and multilingualism in professional contexts: advances in language-sensitive management research, vol 102
69. Karau SJ, Williams KD (1993) Social loafing: a meta-analytic review and theoretical integration. J Pers Soc Psychol 65(4):681–706
70. Hadlington L, Binder J, Stanulewicz N (2021) Exploring role of moral disengagement and counterproductive work behaviours in information security awareness. Comput Hum Behav 114:106557

Using of Advance Technologies for the Protection of Critical Infrastructures

Michal Miške and Zdeněk Dvořák

Abstract In an era characterized by rapid technological progress, the assessment of critical infrastructure resilience has entered a new age as well. The inclusion of advanced technologies into the evaluation process has revolutionized how we measure the capacity of critical structures to resist and recover from various threats, including natural disasters, cyberattacks, and other unforeseen disruptions. This transformation is of paramount importance, given the crucial role that critical infrastructure presents in supporting modern societies. These technologies enable a multifaceted approach to resilience assessment, providing a more comprehensive and accurate understanding of vulnerabilities and potential points of failure. One of the primary advantages of these technologies is their ability to process huge amounts of data in real time. Big Data analytics can aggregate and analyse data from various sources, including sensors, remote monitoring systems, and historical records. This data-driven approach offers a real-time view of the critical infrastructure's performance, identifying deviations from normal operation and potential weaknesses. Furthermore, simulations and modelling, powered by these technologies, offer a virtual testing ground for assessing resilience. Engineers and planners can simulate various crisis scenarios, evaluating the infrastructure's ability to withstand and recover from extreme conditions. This virtual experimentation enables the development of more effective contingency plans and investment strategies to enhance resilience. In conclusion, the integration of advanced technologies in the assessment of critical structure resilience has ushered in a new era of precision and proactive planning. These tools empower decision-makers to make informed choices, reduce vulnerabilities, and improve the resilience of critical infrastructure. In a world characterized by evolving threats, leveraging the advantages of these technologies is paramount in safeguarding the stability and functionality of the infrastructure that underpins our societies.

Keywords Critical infrastructure · Resilience · Advanced technologies · Assessment

---

M. Miške · Z. Dvořák (✉)
University of Žilina, Univerzitná 8215, 010 26 Žilina, Slovakia
e-mail: zdenek.dvorak@uniza.sk

T. A. Kovács et al. (eds.), *The Impact of the Energy Dependency on Critical Infrastructure Protection*, Advanced Sciences and Technologies for Security Applications, https://doi.org/10.1007/978-3-031-78544-3_3

1 Introduction

The issue of protecting significant infrastructure objects accompanies all military conflicts. Therefore, in the past, Law 319/2002 Coll. on the defence of the Slovak Republic was defined for the defence needs of the Slovak Republic. This law defines objects of special importance and other important objects (Law 319/2002 Coll.). In the first decade of the twenty-first century, Europe, the United States, and Australia began to create an agenda for the defence and protection of critical infrastructure. The whole process culminated in the issuance of Directive 2008/114/EC on the identification and designation of European critical infrastructure and the assessment of the need to improve their protection [1]. This directive served as the foundational document for the creation of Law 45/2011 Coll. on Critical Infrastructure [2]. The second decade of the twenty-first century confirmed the importance of protecting critical infrastructure. Around the year 2016, the concept of resilience of critical infrastructure emerged in scientific and professional circles. As part of a security research project in the Czech Republic, a certified methodology focused on the resilience (sturdiness, recoverability, and adaptability) of critical infrastructure was developed. This methodology underwent multiple practical tests and verifications [3]. In an era characterized by rapid technological advancement, the assessment of critical infrastructure resilience has entered a new age as well. After 2020, a professional discussion was initiated, leading to a change in approach, and today, the goal is to strengthen the resilience of critical entities, as outlined in Directive 2557/2022 on the resilience of critical entities [4]. The integration of cutting-edge technologies into the assessment process fundamentally changes the way we evaluate the ability of critical elements and entities to withstand and recover from various threats, including natural disasters, cyberattacks, and other unexpected occurrences. This transformation is of paramount importance considering the pivotal role that critical infrastructure plays in supporting modern societies.

The main goal of the papers is to write about new project APVV Remake 3D in Slovakia. Identify what new directives of EU means for our project, what kind of risks we must deal with and introduce our research project in general.

2 Terminology and Related Projects

Critical subject A subject can be either from the public or private sector, and it is identified by a member state in accordance with Article 6 of EU Directive 2022/2557. It must be categorized within one of the critical infrastructure sectors [4].

Critical infrastructure is primarily identified as objects of special importance, as well as other significant entities, specifically selected information and communication systems, water production and distribution facilities, electricity, oil, and natural gas infrastructure. This also includes other assets owned by the state, corporations, and individuals that have been defined by the government of the Slovak Republic or other relevant state authorities. These assets are critical for managing crisis situations, protecting the population and property, ensuring the continuity of economic and state operations, as well as guaranteeing national and internal security. Their proper functioning is essential for safeguarding national security interests [5].

Resilience of critical infrastructure Řehák et al. V Metodike honoceni resilience prvku kriticke infrastruktury (CIERA), is defined as the internal preparedness of critical infrastructure subsystems for undesirable events. In other words, it refers to the ability of these subsystems to ensure the sustainability of their functions in the presence of adverse external or internal factors [3].

With the advancement of digital technologies, it is essential to consider **cyber security**, which is defined in Slovakia by Law 69/2018 Coll. as the state in which networks and information systems are capable of resisting, to a certain degree of reliability, any action that, in any way, threatens the data stored, transmitted, or processed, as well as related services provided or accessible through networks and information systems. The National Cybersecurity Strategy states that cyber security is a state in which information systems and services are resilient to current threats and vulnerabilities. It also emphasizes that entities must be prepared to detect and address cyber security incidents, recover data and processes, and minimize the consequences. However, cyber security is not solely the concern of specific organizations and entities that protect their assets through appropriate measures. It also affects every individual who, for example, uses social networks, computers, or other digital technologies.

According to EU Directive 2022/2555 [4], a **network and information system** is an electronic communication network, any device or group of interconnected or related devices, of which at least one automatically processes digital data based on a program or digital data that is stored, processed, acquired, or transmitted through the aforementioned two elements.

Network and information system security, as defined by EU Directive 2022/2555, is the ability of networks and information systems to resist, to a certain degree of reliability, any event that may threaten the availability, authenticity, integrity, or confidentiality of the data stored, transmitted, or processed, or the services provided or accessible through these networks and information systems [4].

The National Cybersecurity Strategy, defined in EU Directive 2022/2555, is a coherent framework of a member state in which strategic objectives and priorities in the field of cybersecurity are determined, along with a management system for their achievement in that member state [4].

Advanced technologies are technologies that are either not yet mature and exist as prototypes but hold the promise of significant improvements or technologies that are technically advanced but still have relatively few users. Currently, you can include artificial intelligence, agents, speech and handwriting recognition, virtual reality, 3D visualization, smart cards, real-time collaboration, knowledge and information management, and more.

Projects related to the topic

After many recent natural disasters in countries like Italy, Greece, Libya, as well as anthropogenic activities that led to the destruction of critical infrastructure such as the Kachov Dam, the concept of critical infrastructure is gaining increasing attention. However, the world has been addressing this issue for quite some time, and various projects are already in that field:

- DESDEMONA (No 800687) [6],
- IRIS G5924 (NATO SPS) [7],
- APRIORI G6140 (NATO SPS) [8],
- REMAKE 3D (APVV-22-0562) [9].

These projects are using different advanced technologies, including robots and drones for scanning critical infrastructure objects to create what is known as digital twins nowadays. These digital twins are then used to examine the impact of external influences on the subject. The output from these technologies can be utilized for various purposes, such as numerical simulations, visualization, or simple data analysis. The primary tools used in these projects include drones, cable robots, robots capable of operating underwater for scanning subjects that partially stand in water, like bridges, and, of course, 3D scanners are deployed on every of these as it is visualized on Fig. 1 [6–9].

The Faculty of Security Engineering at UNIZA actively participates in these projects, collaborating with various entities not only in Slovakia but also internationally. The IRIS project, financed by NATO's SPS program, focuses on creating digital twins and its main goals include [7]:

Fig. 1 Types of robots used to scan features CI

- Development of new technologies for fully automating the use of robotized systems and sensor networks for data acquisition in survey, inspection, and monitoring,
- The interaction between data acquisition and storage is managed through the construction of advanced models representing real-time Digital Twins of infrastructure,
- To ensure continuous Non-destructive Evaluation (NDE) of infrastructure, various information is integrated through the data fusion process to develop powerful digital models that offer a comprehensive and realistic description of the examined facility throughout its service life, even post-disaster,
- Data and models form the basis for identifying and describing defects and degradation, especially concerning possible performance reduction in existing structures,
- All acquired knowledge, adequately managed, serves as input for automated or partially automated decision-making processes useful in managing facilities and infrastructure.

The participation in international research projects motivated the preparation of a national research project proposal for applied research in APVV. This project was approved and financially supported. The main goal of the REMAKE 3D project is to examine and analyze critical infrastructure in the vicinity of Púchov. Various top technical devices will be used within this project. The acquired images will subsequently be used to create digital twins of these infrastructure subjects. Figure 2 displays specific parts of the critical infrastructure that will be addressed and scanned in the context of this project, including a railway bridge and the Nosická Dam.

The aim of this project is to develop and validate an integrated system for assessing the resilience of critical infrastructure subjects, focusing on evaluating their resilience

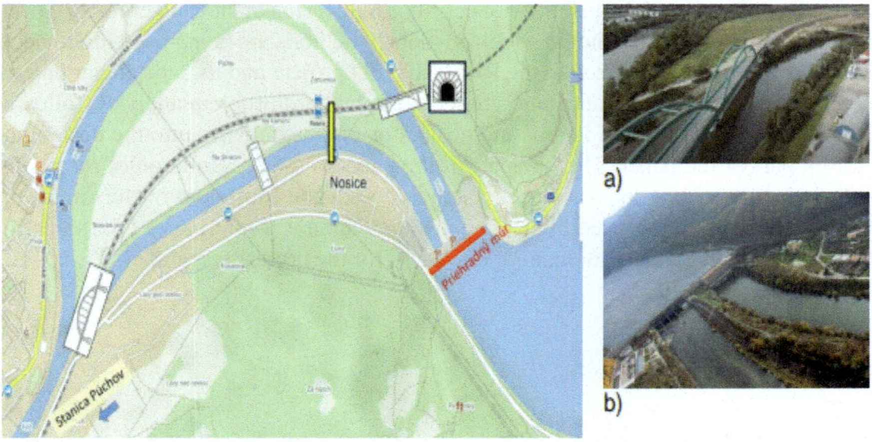

Fig. 2 Remake 3D environment view

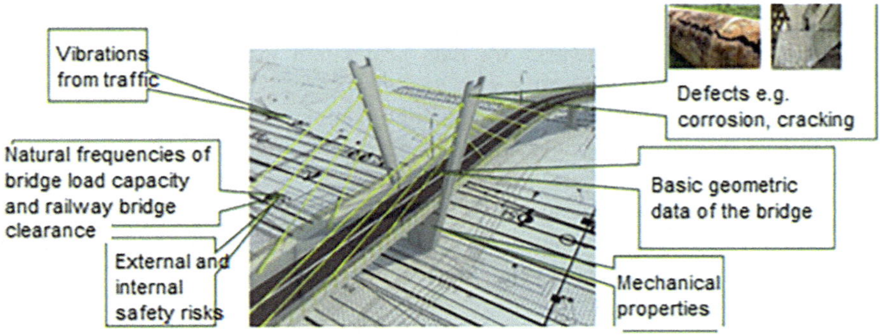

Fig. 3 Expected output from AResIM

against threats and security risks. The expected outcome of the solution will be a pilot version of the Area Resilience Information Modeling (AResIM) information system and a methodology for assessing the resilience level of selected types of objects in the area of interest, which is the surroundings of the Púchov town. Figure 3 illustrates one of the outputs that AResIM should provide.

3 Methodology

The introduction of this article provides definitions closely related to the discussed topic. These definitions are publicly available, with the most up-to-date versions found in the European Union Directives 2022/2555 and 2022/2557 [4]. The article then mentioned current projects focused on critical infrastructure issues, with the active participation of the Faculty of Security Engineering at the University of Žilina. In the results section, it is noted which advanced methods are being used in Slovakia, and how they will be utilized in the REMAKE 3D project and its subsequent phases.

The methodology of the paper is based on a standard problem analysis, followed by an in-depth analysis of lateral and project resources, moving into the design and testing part of the paper. The key part of the paper is the presentation of the results and their subsequent discussion. In a standard way, the fulfilment of the aim of the paper is evaluated, followed by the conclusion, which is a summary of the scientific and technical part and a statement of new tasks for the project researchers [3, 10–12].

4 Results

The REMAKE-3D project was approved in 2023, and the need for such a project is highly relevant when considering that in the third decade of the twenty-first century, outdated technologies for designing and managing critical infrastructure assets are still in use in practice. The current global trend of urbanization poses numerous challenges. Additionally, the aging infrastructure complicates matters. That is why, a modern society requires equally modern technologies, which have already been invented but need to be brought to Slovakia. For instance, digital 3D models are already used in water management by companies like Bentley abroad. The ongoing research project brings many benefits, including:

- Greater precision and more detailed data, reducing the number of errors,
- Facilitating collaboration by allowing multiple people to work on the same element in real time in the online space,
- Improved cost predictability, resulting in lower costs for enhancing resilience or recovery,
- Easier detection of errors thanks to the 3D model, as assessors can immediately see the real condition and deficiencies,
- Process automation, saving time,
- Better understanding and visualization using virtual or augmented reality, ultimately saving time,
- Safer transportation, both in terms of technical security and usability, which increases road safety.

The significance of this initiative, in which the Faculty of Security Engineering collaborates with the Faculties of Management and Informatics, the Faculty of Civil Engineering, and a team of experts from the field of road transport, is that AResIM will be a final product for various state public administration authorities with competence in the field of critical infrastructure. This product can also be utilized by agencies responsible for managing roadways, such as regional self-governing bodies, the National Highway Company, and the Slovak Road Administration.

The APVV REMAKE-3D research project is in its early stages and is preparing an innovative approach to assessing the resilience of critical infrastructure in the transportation sector. Its main goal is to create an information system that incorporates outputs from 3D modelling, scanning, and sensing of critical infrastructure objects in outside are of the town of Púchov. The application of new technologies will be tested on individual elements using Building Information Modelling (BIM) modelling software. The outputs will then be used to develop the AResIM information system and methodologies for assessing the resilience of critical infrastructure elements. This new approach in Slovakia will primarily benefit public administration and entities responsible for managing roads in Slovakia. It introduces a novel approach and technology aimed at simplifying the decision-making process based on the outputs from 3D models and digital twins of critical infrastructure subjects. These will provide assessors, operators, or responsible individuals with a real-time status of specific critical infrastructure subjects in the transportation sector. The purpose of this project is to enhance the resilience of critical infrastructure elements and, consequently, positively influence safety in Slovakia [9].

4.1 Identification of Threats

The biggest risk of modern technologies and information systems is cyber-attacks, which can significantly disrupt their operation and activities. In the cyber environment, an infinite number of threats emerge daily, often transitioning into actual attacks. Although these threats and vulnerabilities have diverse characteristics, and their precise quantification is nearly impossible, several specific types of threats and vulnerabilities can be identified. If these are not addressed in the mid-term horizon, there is a risk of a severe disruption of the principles of cybersecurity. From a strategic perspective, these are challenges that require appropriate responses [13–16].

The use of modern technologies to protect critical infrastructure brings numerous benefits but is also associated with certain risks. Some of them is:

- Cyber Threats—With the increasing digitization and connected devices, critical infrastructure is exposed to a heightened risk of cyber-attacks. Hackers and cybercriminals may specifically target information systems and attempt to disrupt critical operations. So, in this case is necessary to have IT specialists or at least antivirus software which can identify danger and cyber-attacks,
- Sudden Outages and Failures this kind of dependence on modern technologies poses the risk of sudden outages or failures. For instance, the failure of critical infrastructure supported by artificial intelligence (AI), which is using in digital twins 3.0, 4.0 and 5.0 or automated systems can have severe consequences. In some cases, we can lose the data, which could be crucial for critical infrastructure subject. Sometimes it can be caused by massive storms or landslide or fallen trees on high voltage cables and so on, but there is always a possibility that failures are anthropogenic not only natural character.
- Data Management and Protection—With the growing volume of data being collected and processed, it is crucial to pay attention to secure data storage and protection against unauthorized access or misuse. One way of this is cloud projecting and using this kind of storage, which is online, but there is still several threats which should be solved.
- Social Engineering—Attackers may leverage the human factor through social engineering to gain access to crucial information or induce erroneous decisions from cybersecurity personnel. This risk could be least common, but sometimes with biggest outcome, because of losing know-how and such a crucial information without any feeling of danger. If it is happened in real life there is no such a software or antivirus which can stop or identify potential risk.

For the effective protection of critical infrastructure (CI), it is essential to systematically identify these risks and implement robust security measures and protocols for increasing resilience of the CI [3, 17–19].

4.2 Assessment of Risks

Risk assessment is a key activity that must be carried out as part of enhancing the protection of critical infrastructure. Identified risk sources tell the story of the current state of threats, potential consequences and risks. Life and all processes are time-varying and often stochastic. For this reason, the application of the approach defined by ISO 31000 is also crucial for the critical infrastructure assets under assessment. The current change in the philosophy of CI protection towards enhancing the resilience of critical entities brings another very important factor into the process, namely human behaviour. A comprehensive or integrated approach to the solution is now a prerequisite for the correct setup of processes in resilience building. According to the CIERA methodology, resilience has two main parts, namely technical and organisational resilience. The technical resilience examines the agility and recoverability of the object, and the organisational resilience examines its adaptability.

The REMAKE-3D project is expected to make a significant advance in resilience-building using advanced technology. If we are able to measure stochastic effects on an object in real time and then test and verify these in a 3D model, we will be able to provide the key information needed to keep regionally important objects in a serviceable state in advance [11, 20].

5 Discussion

The advancement of human knowledge has always been linked to the promotion of science and research, or to a direct threat. We have been motivated by a long history of active work to improve the theoretical foundations for critical infrastructure protection. Gradually, we have been working on national and international research projects at our department. The first was the APVV project Critical Infrastructure Protection in the transport sector. This project was created at the time of the approval of the law 45/2011 on critical infrastructure and its fulfilled ambition was to create a theoretical basis in the field of security and protection of critical infrastructure.

Then there were the international projects COBACORE and RAIN, solved in cooperation with top Hungarian universities. Both projects were aimed at the use of modern information and communication technologies in the process of risk assessment with a focus on energy and transport infrastructure. Another attribute was to improve public information on potential impacts on the local community [21, 22].

Later we had a share in the Czech security research project RESILIENCE—Dynamic Resilience Assessment of Critical Infrastructure Interconnected Subsystems. This project gave us the opportunity to form a theoretical basis for resilience. The main result was a certified methodology aimed at describing robustness, adaptability and recoverability [12, 23].

Our main goal was about the new project APVV Remake 3D in Slovakia. Identify what new directives of EU means for our project, what kind of risks we have to deal with and introduce our research project. This goal was accomplished by mentioned project APVV Remake 3D, which is still in early state of research. We mentioned risks which can collide with our project from perspective of new EU directives, but we realize that is not detailed and specify enough, because of early state of project as was mentioned before. We also showed some figures, where you can see located area of project realisation.

Few challenges which stand ahead of us are detailed risk management and risk assessment, creation digital twins of targeted subject of the critical infrastructure, collecting data, analysing the data, creation of data driven model, choosing right technologies and their implementation and implement data to cloud storage. Some others supportive activities are preparing as well in second year of the research project APVV Remake 3D and for any news we already have website https://remake3d.uni za.sk/.

6 Conclusion

A globalised society that depends on huge transport flows (oil, natural gas and LNG, iron ore, coal and construction materials as bulk substrates, and the transport of goods (mostly in containers and on pallets)) has enabled the transfer of labour-dependent production to countries with low labour costs. Thus, automation-based manufacturing—Industry 4.0—has taken a back seat in countries with high labour costs. The global production of goods and services and the increase in the number of people on Earth (currently 8.1 billion) requires an increase in energy production, in the developed countries of Western Europe mainly from renewable sources, in other parts of the world still dependent on the burning of fossil fuels.

Changes in technology are also affecting the displacement of rural areas, and there is currently an increase in the number of people in cities and urban agglomerations. The shift to cities, increasing their size, is associated with the implementation of information and communication technologies in the urban environment. The Internet of Things has become the present day, with an estimated 25 billion people, devices and processes currently connected to the Internet, according to CISCO. This number is growing by one billion new devices and processes every year. For this reason, it is necessary to create the prerequisites for the efficient use of existing and newly installed sensors and sensors all around us.

Connecting research teams in Europe is a great opportunity to exchange data, information and, most importantly, knowledge. In the COBACORE, RAIN, DESDEMONA, IRIS and APRIORI projects there is an efficient exchange of know-how in different areas of research and development. Thanks to the financial support of the European Union and NATO, we have a great opportunity to share knowledge between countries, generations, academics, and practice, with appropriate information for the whole of society. The new REMAKE-3D project just presented focuses on strengthening the resilience of transport infrastructure at regional level. It aims at preparing database structures for efficient storage of measured data, their automated evaluation transfer to the newly created AResIM information system. In the first year of the solution, it was important to link the research team and practitioners so that the planned results and outputs would be accepted and realistically used in practice. I believe that cooperation at the regional level with the support of know-how gained from international projects will move our companies and institutions towards a modern and resilient society.

Acknowledgements The article was supported by The Ministry of Education, Science, Research and Sport of the Slovak Republic and Slovak research and development agency grant number APVV-22-0562 Strengthening the Resilience Management of Key infrastructure Elements using advances in 3D modelling.

References

1. Directive 2008/114/EC on the identification and designation of European critical infrastructures and the assessment of the need to improve their protection
2. Act 45/2011 Coll. Critical infrastructure (https://www.zakonypreludi.sk/zz/2011-45)
3. Řehák D, Flynnova L, Hromada M, Fuggini C (2023) The importance of resistance in the context of critical infrastructure resilience: an extension of the CIERA method. Systems 11(10):506
4. Directive 2557/2022 of the European Parliament and of the Council on the resilience of critical actors
5. Bezpečnostná rada Slovenskej republiky (2017) Terminologický slovník krízového riadenia. MV SR, Bratislava
6. Research project DESDEMONA
7. IRIS research project
8. APRIORI research project
9. Research project APVV-22-0562 REMAKE-3D
10. Dvořák Z, Chovančíková N, Leitner B (2021) Practical use of drones for energy infrastructure monitoring in Slovakia. In: Modern technologies enabling safe and secure UAV operation in urban airspace. NATO science for peace and security series 2021. IOS Press, Amsterdam, pp 32–46
11. Figuli L, Dvořák Z (2023) Threats on the critical infrastructure elements. In: Monitoring and protection of critical infrastructure by unmanned systems, p 238. ISBN 978-1-64368-377-5
12. Hromada M, Rehak D, Skobiej B, Bajer M (2023) Converged security and information management system as a tool for smart city infrastructure resilience assessment. Smart Cities 6(5):2221–2244
13. National Cyber Security Strategy of the Slovak Republic 2023–2030
14. Řehák D, Patrman D, Brabcová V, Dvořák Z (2020) Identifying critical elements of road infrastructure using cascading impact assessment. Transport 35(3):300–314
15. Dvořák Z, Řehák D, Čekerevac Z (2020) Qualitative approach to environmental risk assessment in transport. Int J Environ Res Public Health 17(15):1–18, 5494
16. Figuli L, Papán D, Vavák B, Ottaviano E, Daponte P (2023) Long life structural health monitoring of selected bridges. Transp Res Procedia 74:1524–1531
17. Daponte P (2023) Monitoring and protection of critical infrastructure by unmanned systems
18. Dvořák Z, Chovančíková N, Hoterová K, Szatmári M (2021) Assessment of static resilience of objects in the rail transport. Commun Sci Lett Univ Žilina 23(3):F96–F108

19. Řehák D, Patrman D, Foltin P, Dvořák Z, Skrickij V (2022) Negative impacts from disruption of road infrastructure element performance on dependent subsystems: methodological framework. Transport 36(6):510–524
20. Dvořák Z, Hoterová K (2022) Resilience as a feature of intelligent railway infrastructure. In: Lecture notes in intelligent transportation and infrastructure, Part F1395, pp 741–750
21. COBACORE research project
22. Research project RAIN
23. Řehák D, Šenovský P, Hromada M, Pidhaňuk L, Dvořák Z, Loveček T, Ristvej J, Leitner B, Sventekova E, Mariš L (2018) Methodology for resilience assessment of critical infrastructure elements (CIERA 2018)

The Other Sources

24. Research project Critical Infrastructure Protection in the Transport Sector
25. Act 319/2002 Coll. on the Defence of the Slovak Republic

Challenges in Using Photovoltaic Solar Energy System in Serbia as an Individual Producer–Consumer

Igor Fürstner

Abstract As the global shift toward cleaner and more sustainable energy sources continues, the adoption of solar energy is becoming increasingly prominent. However, the Republic of Serbia, like many other nations, confronts unique obstacles when striving to unlock the full potential of solar energy, particularly within the domain of individual producers–consumers. This research explores the intricate challenges that an individual producer–consumer encounter in Serbia as they endeavor to embrace solar power as a renewable energy source. An insight into the distinctive landscape for individual producer–consumer will be presented, encompassing the current regulatory framework. Through a case study that includes a producer–consumer household, experiences of using photovoltaic solar energy system in Serbia will be discussed, including the economic viability by evaluating the financial aspects of solar power investments, including initial costs and the return on investment. Ultimately, the main objective of this research is to provide valuable insights into the challenges confronted by individual producers–consumers in Serbia regarding the adoption of solar energy. It aims to inspire constructive discussions and facilitate the development of solutions for a sustainable energy future in the country.

Keywords Photovoltaic solar energy system · Individual producer–consumer

1 Introduction

Solar energy has gained significant attention as a renewable and sustainable source of electric energy, not only in industry, i.e., in the sector of electric energy production, but also in individual households. With advancements in solar technologies

I. Fürstner (✉)
Banki Donat Faculty of Mechanical and Safety Engineering, Obuda University, Budapest, Hungary
e-mail: furstner.igor@bgk.uni-obuda.hu; ifurst@vts.su.ac.rs

Subotica Tech—College of Applied Sciences, Subotica, Serbia

T. A. Kovács et al. (eds.), *The Impact of the Energy Dependency on Critical Infrastructure Protection*, Advanced Sciences and Technologies for Security Applications, https://doi.org/10.1007/978-3-031-78544-3_4

and increasing environmental concerns, many studies have explored the feasibility, benefits, and challenges associated with using solar energy in individual households.

Researchers have indicated that solar energy systems can offer long-term cost savings compared to traditional electricity sources, reducing electricity bills, and providing a return on investment over time [1–3]. Moreover, solar energy adoption contributes to the reduction of greenhouse gas emissions and dependence on non-renewable energy sources, promoting sustainability [4, 5].

Also, recent advancements in solar energy system technologies have significantly improved the efficiency and effectiveness of residential solar systems. The development of high-performance solar panels, advanced tracking systems, and energy storage solutions, enhance the overall performance of individual household solar systems [6, 7]. These advancements have increased the feasibility and attractiveness of solar energy adoption in individual households [8, 9].

The economic feasibility of solar energy adoption in individual households has been extensively studied too. Research suggests that the cost-effectiveness of solar systems is influenced by factors such as government incentives, the availability of net metering programs, and the cost of alternative electricity sources [10–16]. Studies have also emphasized the importance of accurately assessing the financial benefits and payback periods associated with solar energy systems [17–19].

The environmental benefits of solar energy adoption in individual households are well-documented as well. Solar energy systems generate clean and renewable energy, reducing carbon emissions and mitigating climate change [20, 21]. Additionally, the use of solar energy in individual households can contribute to the overall sustainability of the electricity grid by reducing peak demand and promoting decentralized power generation [22, 23].

Despite the numerous advantages of using solar energy in individual households, several challenges and barriers must also be addressed. Researchers have highlighted issues such as system intermittency, limited rooftop space, high upfront costs, and complex installation processes as potential obstacles to widespread adoption [24–28]. Additionally, policy and regulatory frameworks, grid integration challenges, and public perception can affect the rate of solar energy adoption [29–31].

This points towards the presented research, which aims at providing valuable insights into the challenges confronted by individual producers–consumers in Serbia regarding the adoption of solar energy for electric energy production. In the following paragraphs, the legal background will be presented, followed by the introduction of the cost calculation algorithm. The use of a photovoltaic solar energy system will be presented on a case study showing the energy production, the costs, the possible savings, and the return of investment in one year, namely during 2023. Following this, a short discussion and conclusions will be presented.

2 Legal Background

To be able to understand the context of being a producer–consumer of electric energy from photovoltaic panels as a household in Republic of Serbia, two legal documents, the Law on the Use of Renewable Energy Sources, and the Regulation on Criteria, Conditions, and Method of Calculation of Claims and Obligations between Producer–Consumer and Supplier should be considered [32, 33].

The law which regulates the use of energy from renewable sources in Republic of Serbia is rather new, and dates to 2021. It states that the use of energy from renewable sources is in the public interest of the Republic of Serbia and is of special significance.

The law regulates the use of energy from renewable sources and includes measures and activities taken to achieve long-term goals, such as:

- Reduction of the use of fossil fuels and increased use of renewable energy sources to protect the environment.
- Long-term reduction of dependence on energy imports.
- Integration of electric energy from renewable sources into the electricity market.
- Ensuring the stability of the electricity market, considering the costs of integrating renewable energy sources into the system and network stability.
- Sustainable and independent development through the maximum utilization of national scientific-research, technological-development, and human capacities in the planning process of increasing the use of renewable energy sources, etc.

The law defines a specific producer category of electricity production from renewable sources, which deals with the production for own consumption. In this case the producer–consumer category is defined. The producer–consumer as an end customer has the right to:

- Produce electrical energy for own consumption.
- Store electrical energy for own needs.
- Deliver excess produced electrical energy to the transmission system, distribution system, or closed distribution system.

The producer–consumer cannot use incentive measures in the form of market premiums and feed-in tariffs. The installed capacity of the production facility of the producer–consumer cannot exceed the approved power of the end customer's connection. Additionally, the installed capacity of the production facility of the producer–consumer cannot exceed 10.8 kW if the producer–consumer is a household.

Some expressions used in the legislative are explained below, to help understanding the specific aspects of being a household producer–consumer:

- Guaranteed supplier is a supplier that provides public service of guaranteed supply in accordance with the law regulating energy.
- Household is a community of individuals, or a single person, whose members live together, eat together, and spend earned incomes together, in accordance with the law.

- Delivered electrical energy is active electrical energy delivered to the power grid by the facility for the production of electrical energy from renewable sources.
- End customer is a legal or natural person or entrepreneur who buys electrical energy for their own needs or for the needs of a household.
- Net electrical energy, or in the reminder of the text the used electrical energy, represents the difference between the total taken and total delivered electrical energy of the producer–consumer in the system for one month, determined in kWh based on meter readings of electrical energy meters.
- Net metering is the method of calculating net electrical energy, where the excess delivered electrical energy for one month is reduced by the quantity of net electrical energy during the next billing period, in accordance with the law.
- Billing period is usually one calendar month.
- Full supply is the sale of electrical energy in which the quantity of electrical energy for the billing period is not determined by the supply contract, but the end customer has the right to determine the quantity based on actual consumption at the delivery point in accordance with the law regulating energy.
- Taken electrical energy is the electrical energy taken by the buyer–producer from the power system.
- Production facility is a facility for the production of electrical energy from renewable sources connected to the internal installation of the producer–consumer.
- Storage of electrical energy is the storage of produced electrical energy in a storage facility until the moment it will be used, in accordance with the law.
- Supplier is a participant in the market of electrical energy who performs the energy supply activity with electrical energy.

The producer–consumer has no right to claims for the quantity of delivered electrical energy that exceeds the quantity of consumed electrical energy in the alignment period.

The operator of the transmission, distribution, or closed distribution system is obliged to prioritize the acquisition of electrical energy from the producer–consumer, except when the safety of the system's operation is jeopardized.

The operator of the transmission, distribution, or closed distribution system is obliged to install a meter at the point of transfer of electrical energy from the producer–consumer that allows separate measurement of the quantity of delivered and acquired electrical energy.

If the producer–consumer is a household, the supplier is obliged to settle claims through net metering and arrange obligations with the producer–consumer by a full supply agreement with net metering.

To be able to obtain the status of producer–consumer, several general conditions should be fulfilled by the household:

- The installed capacity of the production facility must not be greater than the approved capacity.
- The production facility and the measuring point must meet all technical and safety requirements in accordance with the rules of the competent system operator.

- The producer–consumer can install and use electrical energy storage for their own needs in combination with the production facility, with the condition that the storage installation can take energy only from the production facility.
- The producer–consumer is obliged to manage the technical equipment and devices used for the production and delivery of electrical energy produced from renewable sources (photovoltaic panels, cables, electrical and measuring equipment, etc.).
- The installed capacity is the installed capacity of the inverter, in kW, at $\cos \varphi = 1$, when the nominal capacity of the inverter is less than or equal to the sum of the installed capacities of the photovoltaic panels, or the sum of the installed capacities of the photovoltaic panels when the nominal capacity of the inverter is greater than the sum of the installed capacities of the photovoltaic panels.
- The end customer is obliged to submit a request to the supplier for the full supply agreement with net metering. The end customer is obliged, before submitting a request for the conclusion of a full supply agreement with net metering, to, in accordance with regulations and standards, build a production facility and adapt the measuring point.
- In the process of connecting the production facility, the end customer is obliged to provide the system operator with the original statement from the contractor, with confirmation from the responsible contractor, stating that the devices, installations, and the measuring point are correct and constructed in accordance with regulations and standards.

The competent system operator reads the metering devices in accordance with the law regulating the energy sector and provides the data to the supplier, who has concluded a full supply agreement with net metering with the producer–consumer, within the prescribed deadlines. The provided data includes:

- The electrical energy taken by the producer–consumer.
- The electrical energy delivered by producer–consumer.

The period for the alignment of claims and obligations between the producer–consumer and the supplier is one year, lasting from April 1 of the current year to March 31 of the next year.

For each settlement period, the supplier is obliged to:

- calculate the consumed electrical energy for the producer–consumer as a positive value of the difference in net electrical energy determined based on the times of applying the tariff for active energy and the surplus of electrical energy from the previous period determined based on the times of applying the tariff for active energy.
- Determine the surplus of electrical energy based on the times of applying the tariff for active energy, which is carried over to the next period.
- Calculate taxes, excises, fees, and other applicable obligations determined by special regulations based on the consumed electrical energy.

The surplus of electrical energy can be transferred to subsequent settlement periods within the period for the alignment of claims and obligations and cannot affect previous settlement periods. The costs of system access are calculated for the producer–consumer based on the methodology for determining the access price.

3 Consumption Cost Calculations

3.1 Inputs

In the case of a household, where the billing period is one calendar month, the cost calculation is rather complex, and consists of several fixed and variable costs. Also, a dual tariff system must be considered. The first tariff component is based on energy consumption levels. There are three levels, namely the green tariff limit for up to 350 kWh/month (for a 30-day month), the blue tariff limit for up to 1600 kWh (for a 30-day month), and the red tariff limit for more than 1600 kWh (for a 30-day month). The second tariff component is based on the time of the day, the high tariff from 07 a.m. to 11 p.m., and the low tariff from 11 p.m. to 07 a.m. The distribution of electric energy into the green, blue, and red zones according to the higher or lower tariff is proportionate to the total consumed electric energy according to the higher or lower tariff. The inputs for cost calculation regarding one month are presented in Table 1.

Based on the inputs for cost calculation, the inputs for the total cost (TC) calculation can be defined, which are presented in Table 2.

Having in mind the inputs from Tables 1 and 2, the total cost can be calculated by using Eq. (1).

$$\begin{aligned} TC = {} & CCP + GSF + CUE + CIPEP + IEEF \\ & + FADS + CET + CVAT + FPMS \end{aligned} \tag{1}$$

3.2 Case Study Regarding One Household

The case study presented is a household that produces electric energy by using photovoltaic elements (PV) with a total string capacity of 12 kWp (32 modules—Monocrystalline—CHSM60M-HC (BF) 166), with a 10 KW inverter (SUN2000-10KTL-M1) and with no storage.

The PV production is shown in Fig. 1, while the total consumption, the energy taken from supplier that was produced by the PV, the direct consumption from the supplier, i.e., the used electric energy, as well as the direct consumption from the PV is presented in Fig. 2.

Figure 3 shows the total costs of electric energy consumed by using PV, and the total costs that would have occurred if there was no PV, as well as the difference

Table 1 Inputs for cost calculation

Input	Abbreviation	Amount	Unit	Amount validity (date)
Cost per unit of contracted power	CPUCP	54.258	[RSD/kW]	–
Contracted power (for the household presented in the case study)	CP	11.04	[kW]	–
Guaranteed supplier fee	GSF	145.22	[RSD]	2023/01–2023/11
		160.67		2023/12
Fee for public media service	FPMS	299	[RSD]	–
Compensation for the incentive of privileged electricity producers—per unit	CIPEPU	0.801	[RSD/kWh]	–
Improvement of energy efficiency—per unit	IEEFU	0.015	[RSD/kWh]	–
Fee for the access to the distributions system for the difference between taken and used electricity—per unit	FADSU	3.88	[RSD/kWh]	
Excise tax	ET	7.5	[%]	–
Value added tax	VAT	20	[%]	–
High tariff—green	T_1	7.584	[RSD/kWh]	2023/01–2023/03
		8.336		2023/04–2023/10
		9.1092		2023/11–2023/12
Low tariff—green	T_2	1.896	[RSD/kWh]	2023/01–2023/03
		2.084		2023/04–2023/10
		2.2773		2023/11–2023/12
High tariff—blue	T_3	11.376	[RSD/kWh]	2023/01–2023/03
		12.504		2023/04–2023/10
		13.6638		2023/11–2023/12
Low tariff—blue	T_4	2.844	[RSD/kWh]	2023/01–2023/03
		3.126		2023/04–2023/10
		3.416		2023/11–2023/12
High tariff—red	T_5	22.752	[RSD/kWh]	2023/01–2023/03
		25.008		2023/04–2023/10
		27.3276		2023/11–2023/12
Low tariff—red	T_6	5.688	[RSD/kWh]	2023/01–2023/03
		6.252		2023/04–2023/10
		6.8319		2023/11–2023/12

(continued)

Table 1 (continued)

Input	Abbreviation	Amount	Unit	Amount validity (date)
Used electricity high tariff—green	UE_1	Variable	[kWh]	–
Used electricity low tariff—green	UE_2	Variable	[kWh]	–
Used electricity high tariff—blue	UE_3	Variable	[kWh]	–
Used electricity low tariff—blue	UE_4	Variable	[kWh]	–
Used electricity high tariff—red	UE_5	Variable	[kWh]	–
Used electricity low tariff—red	UE_6	Variable	[kWh]	–
Taken electricity high tariff—green	TE_1	Variable	[kWh]	–
Taken electricity low tariff—green	TE_2	Variable	[kWh]	–
Taken electricity high tariff—blue	TE_3	Variable	[kWh]	–
Taken electricity low tariff—blue	TE_4	Variable	[kWh]	–
Taken electricity high tariff—red	TE_5	Variable	[kWh]	–
Taken electricity low tariff—red	TE_6	Variable	[kWh]	–

between the total costs, while in Fig. 4 the average costs for 1 kWh of electric energy by using PV, and the costs that would have occurred if there was no PV, as well as the difference between the average costs.

Figure 5 shows the ratio of the return of investment, having in mind that the initial investment into the solar system was 1.700.000 [RSD].

3.3 *Discussion*

The presented case study shows that the production of electric energy from the PV system was 13.442 kWh (Fig. 1). The total consumption was 14.395 kWh (Fig. 2). This consumption could be divided into direct consumption from PV (4.041 kWh), taken energy from the supplier produced by PV and previously delivered to the supplier (5.890 kWh), and used energy directly from the supplier (4.464 kWh). This has shown that the PV production was higher than the amount consumed directly and indirectly by 3.511 kWh. The main reason for the occurrence of excess energy is

Table 2 Inputs for cost calculation [RSD]

Input	Abbreviation	Equation
Cost of contracted power	CCP	$CPUP * CP$
Cost of used electricity	CUE	$\sum_{i=1}^{6} (T_i * UE_i)$
Cost of taken electricity	CTE	$\sum_{i=1}^{6} (T_i * TE_i)$
Compensation for the incentive of privileged electricity producers	CIPEP	$CIPEPU * \sum_{i=1}^{6} TE_i$
Improvement of energy efficiency fee	IEEF	$IEEFU * \sum_{i=1}^{6} TE_i$
Fee for the access to the distribution system	FADS	$FADSU * \sum_{i=1}^{6} (TE_i * UE_i)$
Cost of excise tax (until Aug 2023)	CET	$(CCP + GSF + CTE + CIPEP + IEEF) * ET$
Cost of excise tax (from Aug 2023)	CET	$(CCP + GSF + FADS + CIPEP + IEEF + CUE) * ET$
Cost of value added tax	CVAT	$(CCP + GSF + CUE + CIPEP + IEEF + FADS + CET) * VAT$

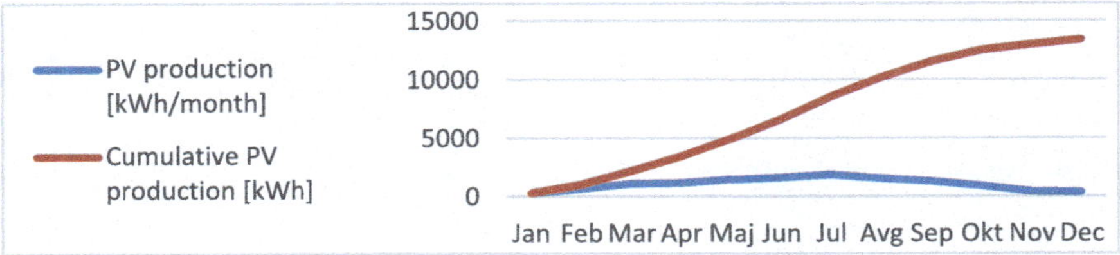

Fig. 1 PV production

because the PV produced energy during the high tariff, and according to the regulation the delivered energy could be than taken back only during that tariff.

The total cost of electric energy by using the PV system was 72.588 [RSD], while the total cost of electric energy if there was no PV would have been 192.725 [RSD].

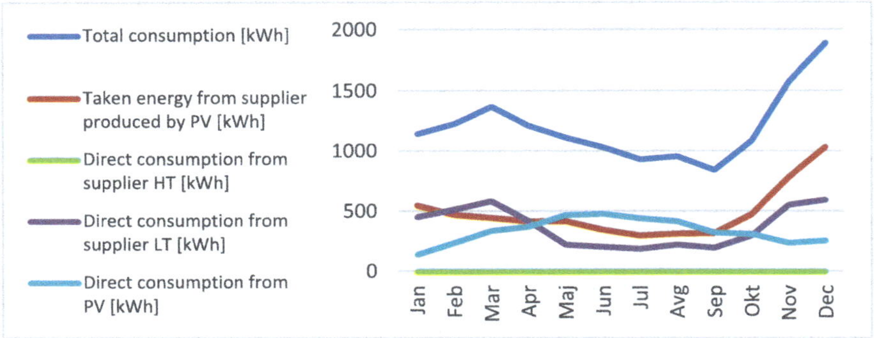

Fig. 2 Electric energy consumption

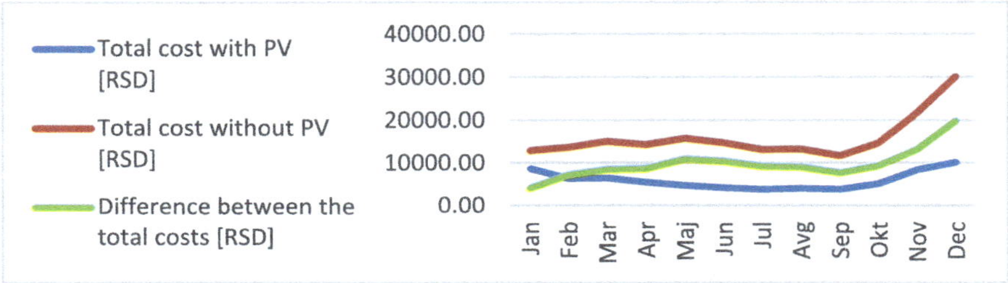

Fig. 3 Total costs of electric energy with and without PV

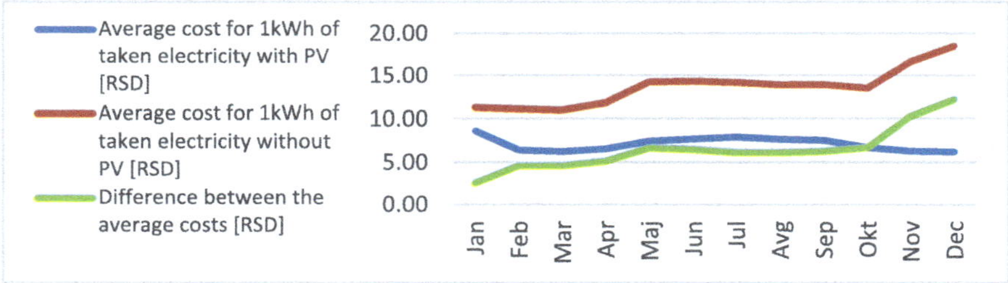

Fig. 4 Average costs of electric energy with and without PV

This leads to the saving of 120.137 [RSD]. Having in mind the initial investment into the system of 1.700.000 [RSD], the yearly return of investment was 7.07%, which leads to the conclusion that the period of total return is more than 15 years and is too long. It is shown in Fig. 5, that the return of investment has risen during the end of the year. This shows that higher prices and higher consumption result in a larger difference in the total cost of energy without the use of a PV system compared to the case when the PV system is used (Figs. 2, 3 and 4).

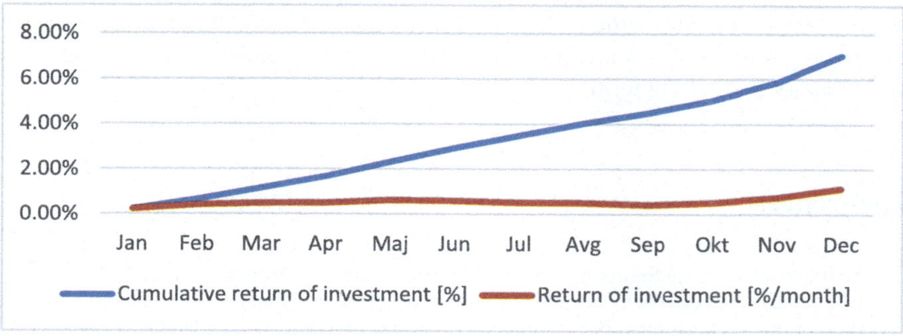

Fig. 5 Return of investment

4 Conclusions

This research on using solar energy to produce electric energy in individual households demonstrates the potential for significant benefits in terms of cost savings, environmental impact, and sustainability. However, challenges such as economic viability, technological limitations, and policy support need to be addressed to facilitate widespread adoption. Continued research and development efforts, along with supportive policies and financial incentives, are crucial for the successful integration of solar energy systems into individual households.

The legal framework for households in the Republic of Serbia that aim to become producers–consumers of electric energy from photovoltaic panels include the Law on the Use of Renewable Energy Sources and the Regulation on Criteria, Conditions, and Method of Calculation of Claims and Obligations between Producer–Consumer and Supplier. The law, dating back to 2021, emphasizes the public interest and environmental significance of using renewable energy sources. It outlines long-term goals such as reducing fossil fuel use, integrating renewable energy into the electricity market, and ensuring market stability. The law defines the producer–consumer category, allowing end customers to produce, store, and deliver excess electrical energy. However, they are not eligible for certain incentives like market premiums and feed-in tariffs.

The presented case study provides valuable insights into the performance of the photovoltaic system for electric energy production in the case of a household. The financial evaluation demonstrates a significant cost-saving aspect, however, the calculated return on investment is suggesting a lengthy payback period. It is noteworthy that the rate of the return of investment increased towards the end of the year, indicating that higher electricity prices and consumption contribute to a more favorable economic outcome.

In summary, while the PV system proves to be economically beneficial in terms of cost savings and performance, the extended payback period raises considerations for the feasibility of the investment. Further exploration of potential efficiency improvements or government incentives could enhance the economic viability of adopting solar energy in the long term.

References

1. Santiago I, Lopez-Rodriguez MA, Trillo-Montero D, Torriti J, Moreno-Munoz A (2014) Activities related with electricity consumption in the Spanish residential sector: variations between days of the week, autonomous communities and size of towns. Energy Build 79:84–97
2. Yunusov T, Torriti J (2021) Distributional effects of time of use tariffs based on electricity demand and time use. Energy Policy 156:112412
3. Qin X, Xu B, Lestas I, Guo Y, Sun H (2023) The role of electricity market design for energy storage in cost-efficient decarbonization. Joule 7(6):1227–1240
4. Chang Y, Wei Y, Zhang J, Xu X, Zhang L, Zhao Y (2021) Mitigating the greenhouse gas emissions from urban roadway lighting in China via energy-efficient luminaire adoption and renewable energy utilization. Resour Conserv Recycl 164:105197
5. Shahsavari A, Akbari M (2018) Potential of solar energy in developing countries for reducing energy-related emissions. Renew Sustain Energy Rev 90:275–291
6. Mârza C, Moldovan R, Corsiuc G, Chisăliță G (2023) Improving the energy performance of a household using solar energy: a case study. Energies 16(18):6423
7. Sekitou M, Tanaka K, Managi S (2018) Household electricity demand after the introduction of solar photovoltaic systems. Econ Anal Policy 57:102–110
8. Schulte E, Scheller F, Sloot D, Bruckner T (2022) A meta-analysis of residential PV adoption: the important role of perceived benefits, intentions and antecedents in solar energy acceptance. Energy Res Soc Sci 84:102339
9. Bach L, Hopkins D, Stephenson J (2020) Solar electricity cultures: household adoption dynamics and energy policy in Switzerland. Energy Res Soc Sci 63:101395
10. O'Shaughnessy E (2022) Rooftop solar incentives remain effective for low- and moderate-income adoption. Energy Policy 163:112881
11. Eslami M, Nahani P (2021) How policies affect the cost-effectiveness of residential renewable energy in Iran: a techno-economic analysis for optimization. Utilities Policy 72:101254
12. Poponi D, Basosi R, Kurdgelashvili L (2021) Subsidisation cost analysis of renewable energy deployment: a case study on the Italian feed-in tariff programme for photovoltaics. Energy Policy 154:112297
13. Xin-gang Z, Yi Z, Hui W, Zhen W (2022) How can the cost and effectiveness of renewable portfolio standards be coordinated? Incentive mechanism design from the coevolution perspective. Renew Sustain Energy Rev 158:112096
14. Jia X, Du H, Zou H, He G (2020) Assessing the effectiveness of China's net-metering subsidies for household distributed photovoltaic systems. J Clean Prod 262:121161
15. Saquib FR, Ankon C, Atik YR, Fariha HS, Shiblee MSAAF, Mahfuzar RR (2021) A case study on the cost-effectiveness of net energy metering of residential grid-connected photovoltaic in the context of Bangladesh. In: 5th International conference on electrical engineering and information communication technology. IEEE, pp 1–6
16. Perdigones A, García I, Porras-Prieto CJ, Gómez-Villarino MT, Baptista F, García JL (2021) Evaluation of the cost-effectiveness of grid-connected photovoltaic solar energy in wineries. Ciênc Téc Vitivinícola 36(1):45–54
17. Delapedra-Silva V, Ferreira P, Cunha J, Kimura H (2022) Methods for financial assessment of renewable energy projects: a review. Processes 10(2):184
18. Cui Y, Zhu J, Meng F, Zoras S, McKechnie J, Chu J (2020) Energy assessment and economic sensitivity analysis of a grid-connected photovoltaic system. Renew Energy 150:101–115

19. Schultz HS, Carvalho M (2022) Design, greenhouse emissions, and environmental payback of a photovoltaic solar energy system. Energies 15(16):6098
20. Rabaia MKH, Abdelkareem MA, Sayed ET, Elsaid K, Chae KJ, Wilberforce T, Olabi AG (2021) Environmental impacts of solar energy systems: a review. Sci Total Environ 754:141989
21. Chen XH, Tee K, Elnahass M, Ahmed R (2023) Assessing the environmental impacts of renewable energy sources: a case study on air pollution and carbon emissions in China. J Environ Manage 345:118525
22. Strielkowski W, Civín L, Tarkhanova E, Tvaronavičienė M, Petrenko Y (2021) Renewable energy in the sustainable development of electrical power sector: a review. Energies 14(24):8240
23. Kabeyi MJB, Olanrewaju OA (2022) Sustainable energy transition for renewable and low carbon grid electricity generation and supply. Front Energy Res 9:1032
24. Asiaban S, Kayedpour N, Samani AE, Bozalakov D, De Kooning JD, Crevecoeur G, Vandevelde L (2021) Wind and solar intermittency and the associated integration challenges: a comprehensive review including the status in the Belgian power system. Energies 14(9):2630
25. Bakht MP, Salam Z, Gul M, Anjum W, Kamaruddin MA, Khan N, Bukar AL (2022) The potential role of hybrid renewable energy system for grid intermittency problem: a techno-economic optimisation and comparative analysis. Sustainability 14(21):14045
26. Sinsel SR, Riemke RL, Hoffmann VH (2020) Challenges and solution technologies for the integration of variable renewable energy sources—a review. Renew Energy 145:2271–2285
27. Zander KK (2020) Unrealised opportunities for residential solar panels in Australia. Energy Policy 142:111508
28. Adenle AA (2020) Assessment of solar energy technologies in Africa—opportunities and challenges in meeting the 2030 agenda and sustainable development goals. Energy Policy 137:111180
29. Inês C, Guilherme PL, Esther MG, Swantje G, Stephen H, Lars H (2020) Regulatory challenges and opportunities for collective renewable energy prosumers in the EU. Energy Policy 138:111212
30. Lazdins R, Mutule A, Zalostiba D (2021) PV energy communities—challenges and barriers from a consumer perspective: a literature review. Energies 14(16):4873
31. Abdullahi D, Renukappa S, Suresh S, Oloke D (2022) Barriers for implementing solar energy initiatives in Nigeria: an empirical study. Smart Sustain Built Environ 11(3):647–660
32. Zakon o korišćenju obnovljivih izvora energije (2023). https://www.paragraf.rs/propisi/zakon-o-koriscenju-obnovljivih-izvora-energije.html. Last accessed 28 Jan 2024
33. Уредба о критеријумима, условима и начину обрачуна потраживања и обавеза између купца—произвођача и снабдевача (2022). https://www.eps.rs/cir/snabdevanje/Documents/kupac/uredba_o_kriterijumima_uslovima_kupac_proizvodac_Sl_glasnik_74_2022.pdf. Last accessed 28 Jan 2024

The Increase in the Production of the Defence Industry in Central Europe and the Risks Arising from It

Ladislav Pásztor and Vladimír Kavický

Abstract The deterioration of the security situation in Ukraine, with an impact on Europe, forced European countries to reconsider their relationships with defence industry. Political declarations and the allocation of funds to increase the production capacities of weapons and ammunition in Europe bring, on the one hand, positive investments in new technologies; on the other hand, they increase the risk of industrial accidents associated with incorrect handling during production, transportation, or storage. Other threats include sabotage, theft and illegal sale of weapons and ammunition to criminal and terrorist groups and their subsequent misuse.

Keywords Defence industry · Russo-Ukrainian war · Act in support of ammunition production

1 Introduction, and the Antecedents of the Russo-Ukrainian War

The Central European Defence Industry continuously undergoes significant changes influenced by the shift in the political system in the Central and Southeast European countries. The wave of revolutions at the end of the eighties of the twentieth century, the German reunification, and the collapse of the Soviet Union resulted in changes not only in the political system but also in economy. The European states, which had belonged in the Soviet sphere of influence, gradually transitioned from a Soviet-type of planned economy to a market economy. These states gradually became members of the European Union (EU) and the North Atlantic Treaty Organisation (NATO). In addition to indisputable economic benefits, increasing well-being,

L. Pásztor (✉)
Doctoral School of Military Sciences, Ludovika University of Public Service, Budapest, Hungary
e-mail: Pasztor.Ladislav@uni-nke.hu

V. Kavický
University of Žilina, Žilina, Slovak Republic

T. A. Kovács et al. (eds.), *The Impact of the Energy Dependency on Critical Infrastructure Protection*, Advanced Sciences and Technologies for Security Applications, https://doi.org/10.1007/978-3-031-78544-3_5

freedom of travel, and democratic political plurality, the process also had adverse effects, primarily manifested in the downsizing of military capabilities and the arms (defence) industry. The post-bipolar period has begun. Military equipment, arms, ammunition, and explosives (AA&E) that had become redundant were further sold, and domestic arms production was entirely or partially privatized.

The Central and Southeast European states' politicians put defence issues on the back burner. They focused on "building democracy", trying to catch up with Western economies, and on full integration into the collective defence umbrella of NATO. While democratic changes took place in Central and Southeast European countries, Ukraine proclaimed its independence from the Soviet Union a bit later, on 24th August 1991. This proclamation was approved by 90.32% of the Ukrainian electorate in a referendum on 1st December [1]. The Ukrainian independence led to the dissolution of the Soviet Union on 26th December 1991. The independent Ukraine inherited the borders of the former Ukrainian Soviet Socialist Republic. Similarly to other former socialist states, Ukraine had many Soviet-type AA&E, moreover, it possessed an extensive arsenal of nuclear weapons. Given the political instability, high-level corruption, and economic difficulties of the newly established state, there was a real danger that Ukraine would not be able to maintain the nuclear weapons that remained on its territory after the collapse of the Soviet Union. The second security problem was solved by the "Memorandum on security assurances concerning Ukraine's accession to the Treaty on the Non-Proliferation of Nuclear Weapons" (NPT), signed in Budapest on 5th December 1994 [2]. Under the NPT, Russia, the United States, and the United Kingdom confirmed their recognition of Ukraine becoming parties to the NPT, and effectively removing all Soviet nuclear weapons from Ukrainian soil. The three nuclear-weapon states (NWS) signing the NPT document "reaffirm their obligation to refrain from the threat or use of force against the territorial integrity or political independence of Ukraine, and that none of their weapons will ever be used against Ukraine except in self-defence or otherwise in accordance with the Charter of the United Nations" [2, article 2]. Between 1994 and 1996, there were Soviet nuclear weapons stationed in Ukraine but their disarmament took place under Russian operational control. It seemed that Kant's democratic peace theory [3] could become applicable in practice.

Dialogue and cooperation between the newly independent Ukraine and NATO started when Ukraine joined the NAC (1991), the Partnership for Peace (PfP) program (1994), and in 1997 Ukrainian and Allied Heads of State and Government signed the "NATO-Ukraine Charter on a Distinctive Partnership" [4]. The association agreement with the EU was initiated in 2012. However, then President of Ukraine pro-Russian Viktor Yanukovych did not sign the Association Agreement with the European Union on 21st November 2013, moreover, he sought closer economic bonds with Russia, Belarus, and Kazakhstan. This decision led to widespread protests known as the Euromaidan movement in Ukraine. Demonstrators advocated for closer relations with the EU and opposed Yanukovych's turn towards Russia [5]. The rejection of the Association Agreement ultimately contributed to the escalation of tensions in Ukraine, leading to significant political changes, including Yanukovych fleeing the country and the annexation of Crimea by Russia in March 2014, as well as the subsequent war in eastern Ukraine. The Euromaidan protests marked a pivotal moment in Ukrainian history and had far-reaching consequences for the region. Two self-proclaimed internationally unrecognized republics emerged in the eastern part of Ukraine: the Donetsk and Luhansk People's Republics, while the Crimean Peninsula has been under Russian occupation since 2014. The Russo-Ukrainian hybrid war began when the Minsk Agreements (Trilateral Contact Group on Ukraine, consisting of Ukraine, Russia, and the Organization for Security and Cooperation in Europe (OSCE), with mediation by the leaders of France and Germany) unsuccessfully attempted to settle the conflict. Then, on 24th February 2022, Russia launched its full-scale invasion of Ukraine.

However, Russia's hybrid war against Ukraine's ongoing internal political and economic problems did not allow Ukraine to achieve the necessary economic growth and democratization changes. Based on an independent Ukrainian study from 2020, Ukraine, given the financial and economic crisis, external military aggression, weakness and instability of democratic institutions, and the prevalence of a culture of poverty, is a developing country. In 2020, it took the 74th place on the Human Development Index, and together with Moldova, it is the poorest country in Europe and suffers from very high degree of poverty and corruption [6]. After the appointment of a new, pro-Western government on 27th February 2014, led by Arseniy Yatsenyuk, negotiations with the EU continued. On 16th September 2014, the Association Agreement was ratified by the Ukrainian Parliament and consent was given by the European Parliament. On 8th November 2023, the European Commission issued a recommendation to open accession negotiations with Ukraine. On 14th and 15th December 2023, EU leaders (except for the PM of Hungary) made a decision on opening accession negotiations with Ukraine [7]. After the full-scale Russian invasion of Ukraine, to ensure its defence, Ukraine continued to approach NATO. The NATO-Ukraine Council was launched at the 2023 NATO Summit in Vilnius as part of a three-part package of support bringing Ukraine closer to NATO. The package also includes a multi-year programme of assistance to help the Ukrainian armed forces become fully interoperable with Allied forces and a reaffirmation that Ukraine will become a member of NATO when the Allies agree, and conditions are met.

After the collapse of the Soviet Union, fifteen independent states emerged. Three Baltic states (Estonia, Latvia, and Lithuania) fully integrated into the Euro-Atlantic structures. Georgia, the Republic of Moldova, and Ukraine also achieved a new level of EU integration. Armenia and Azerbaijan still want to become EU member states. The effort to integrate Belarus into the EU has come under a comprehensive, in-depth review due to the 9th August 2020 Presidential election, which was neither free nor fair. Moreover, in 2022, Belarus supported the Russian invasion of Ukraine [8]. Former Soviet republics Armenia, Azerbaijan, Georgia, Kazakhstan, the Kyrgyz Republic, the Republic of Moldova, Tajikistan, Turkmenistan, Ukraine, and Uzbekistan cooperate with NATO in the framework PfP. Apart from these states, in the framework of the PfP program NATO cooperates also with other countries belonging in Russia's sphere of influence, such as Bosnia and Herzegovina, and Serbia. The partnership with Belarus and Russia is suspended following North Atlantic Council (NAC) decisions related to the security environment connected not only with the Russo-Ukrainian war [9].

Russia, as a successor state of the Soviet Union, despite its internal political (1991s Soviet coup attempt and its consequences and 1993s constitutional crisis) as well as economic (nationwide privatization and emergence of Russian oligarchy, 1998 financial crisis) problems, has not come to terms with losing its superpower status. To prevent further disintegration of the Commonwealth of the Independent States (CIS) and the Russian sphere of influence, Russia became an actor in the war in Chechnya (1994–1996, 1999–2009), Nagorno-Karabakh (1988–1994, 2020), Georgia (2008) and sent peacekeeping forces to Moldova. Russia's military and political interventions had clear objectives—to sustain its slowly disappearing superpower positions and to make efforts to restore bipolarity or the old balance of (military) power in Europe.

After the USSR's internal economic, political, and geographic disintegration, the USA considered itself the only global superpower. It left behind Kissinger's policy of *balance of power* among the superpowers and failed to support the stabilization of Russia as a geopolitical partner. Besides Russia, Ukraine is the second largest country in Europe in terms of area, with critical access to the Sea of Azov and the strategic port of Sevastopol on the Black Sea. In 2013, Russia accused the West, and specifically the USA, that through the guise of humanitarian aid (directly in Crimean Sevastopol and Simferopol mainly through philanthropic activities, such as building and rebuilding of schools and hospitals) they were creating conditions for future NATO naval bases, which would be directed against Russia. If Russia wants to remain in the position of a global power player, from the military-strategic point of view it cannot lose these two critical geopolitical areas. If that were to happen, Russia would lose access to the Black Sea, the Mediterranean Sea, and consequently to the Indian and Atlantic Oceans, which would significantly weaken its position in international affairs. Russia perceives the Western support for the Ukrainian reforms as expanding the West's, namely of the USA, zone of influence via the enlargement of NATO, as going way beyond what was expected in the 1990s, for example, to the territories of Moldova, Ukraine and Georgia. The West is accusing Russia of aggression against Ukraine and the illegal annexation of Crimea, while Moscow argues that it was just a response to the crisis triggered by the EU and the USA and Russia is simply protecting its national interests and zone of influence against the spread of Western and NATO influence [10].

2 Military Support to Ukraine

The Russo-Ukrainian war escalated on 24th February 2022, when Russia launched a full-scale invasion of Ukraine (see Fig. 1) after all its previous (political, economic and hybrid) efforts to maintain a sphere of influence had failed. The war of principles and a war fought by proxies between the West and Russia became conventional.

Ukraine and its people are defending themselves against large-scale Russian aggression. In this struggle for survival, given the facts mentioned above, without the help of Western allies, the country will not be able to withstand the Russian military power.

During the Russo-Ukrainian War, which began in 2014, several countries and organizations provided military assistance to Ukraine to support its defence against Russian-backed separatists in Eastern Ukraine. One of the earliest and most significant military equipment donors for Ukraine was the United States. The United States plays a crucial role in providing military aid to Ukraine, including non-lethal and lethal assistance. Non-lethal assistance included communication equipment, medical supplies, vehicles, and philanthropic activities, mentioned above. Lethal assistance was weapons like small arms and anti-tank missiles.

Military aid to Ukraine was more visible after the initiation phase of the escalated Russo-Ukrainian War. More than 50 countries of the world, referred to as the "global West", started providing humanitarian, financial and military aid to defend Ukraine immediately after the open Russian military campaign in 2022. Countries providing military aid have created the Ukraine Defence Contact Group, also called the Ramstein Format. The first meeting took place on 26th April 2022 at the US Air Force Base in Ramstein, that is why the contact group is labelled Ramstein Format [12].

The help from the West, already provided after Russia's start of hybrid operations in 2014 and continuing to supply weapons, military equipment, ammunition, and humanitarian aid after 24th February 2022 (see Fig. 2), helped the Ukrainians achieve the initial victories. Nevertheless, despite massive military support, the war in Ukraine essentially froze and turned into a war of attrition (see Fig. 3).

Fig. 1 The Russo-Ukrainian war escalation (initial invasion, as of 25th February 2022) [11]

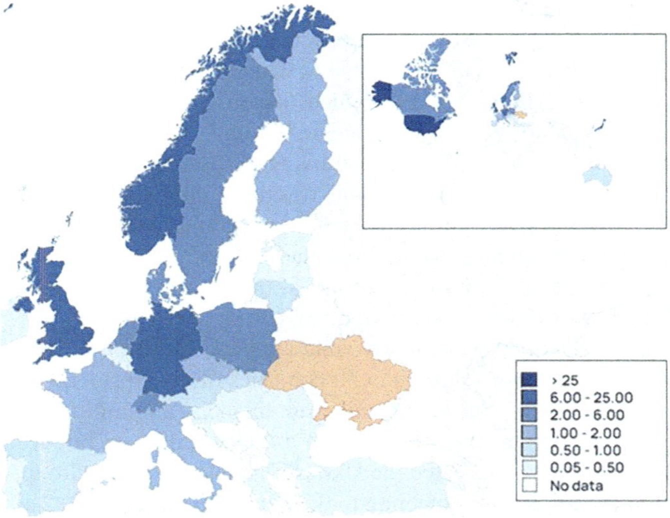

Fig. 2 Total bilateral aid: government commitments in billion Euros [13] according "Ukraine Support Tracker Data" (24th February 2022–31st October 2023) [14]

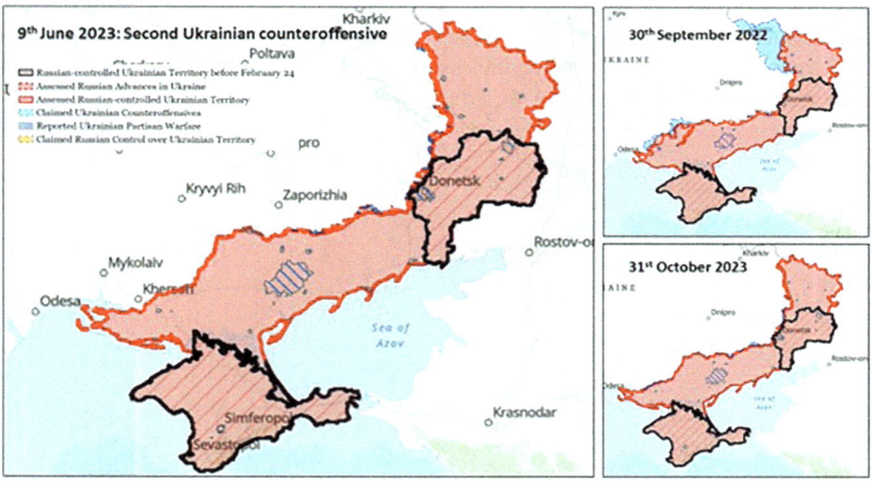

Fig. 3 The situation in the battlefield on Eastern and South Ukraine from first counteroffensive in 2022 to end of October 2023 [11]

As we have already pointed out in the study's introduction, Russia is trying to promote its power interests even concerning Ukraine. After the failure of the hybrid war launched against Ukraine in 2014 (political efforts to bring Ukraine into the Russian sphere of interest failed), on 24th February 2022, Russia launched an armed attack—according to the Russian wording "a special military operation"—against neighbouring Ukraine. Using Clausewitz's basic idea of war, this means nothing more than the continuation of Russia's new imperial policy, which was manifested against Ukraine in the spring of 2014, only by other violent means [15].

Analysing the Russo-Ukrainian War from a strategic point of view, the war could be divided into four phases, such as (1) the *initial invasion* (24th February–March 2022); (2) the *southeastern front* (April–August 2022); (3) the *first counteroffensive* (August–November 2022); and the protracted *war of attrition* (since November 2023).

The goal of the first phase was to implement a political regime change by the use of military force. It began with the seizure of the Antonov (Hostomel) airport on the outskirts of Kyiv after the electronic jamming of the Ukrainian air defence and delivering attacks with means of precision strike. Starting from there, the airborne and special forces were supposed to occupy the Kyiv government quarter and the military command centres, as well as eliminate the political leadership of the country. In other parts of Ukraine, the task of Russian forces advancing in a total of four or five main strategic directions was to capture facilities of critical infrastructure, occupy local administration buildings, and isolate and eradicate pockets of resistance. However, the execution of these plans failed, and the Russian forces, suffering significant losses, got stuck in the suburbs of Kiev. The Russian military leadership was forced to withdraw its forces and redeploy them to the Donbass, thereby opening up the south-eastern front [16].

The objective of the second phase of the war was to systematically grind down the Ukrainian forces, through Russian long-range strikes inter alia. In the beginning of March 2022, the attacks on the communication and information infrastructure began, and from June, the destruction of the fuel storage sites, refineries and railway infrastructure started. As the war dragged on, the West gradually crossed the red lines drawn by Moscow through transferring or donating more and more heavy artillery weapons to Ukraine. In other words, while the strategic strikes managed to destroy a significant part of the military industrial capacity of Ukraine, the military industrial base of the Ukrainian war efforts moved outside Ukraine, into the Western countries. It is important to point out that Russia is unable, or able only to a limited extent, to attack Western arms shipments in the territory of Western Ukraine [17].

After the Russian offensive in Donbas petered out in the summer of 2022, the Ukrainian forces managed to seize the operational initiative in late August. A counteroffensive was launched, in the preparation of which the HIMARS and M270 multiple launch rocket systems played a crucial role. These artillery systems enabled strikes at Russian ammunition depots and command posts deep in the rear, as well as the destruction of the bridges critical for the supply of the occupying Russian forces in the Kherson bridgehead. The Russian forces were forced to withdraw from the battlefield (in a disorganized and panic-stricken manner) and began partial mobilization [17].

The beginning of November 2022 marked the beginning of the fourth stage of the Russo-Ukrainian war. Having withdrawn their forces in the framework of an organized retreat the Russian leadership surrendered Herson's territories west of the Dnieper River. This shortened the front section, and with the redeployment, and the arrival of some of the newly mobilized personnel, the density of the Russian defence increased significantly. The Russians also started building defensive lines on several sections of the front. Simultaneously, from the beginning of October 2022, an extensive, systematic attack on the Ukrainian electrical infrastructure and other civil targets began with unmanned aerial vehicles and Shahed kamikaze drones purchased from Iran, in order to break the Ukrainian societal will.

The fourth phase of the war, despite the fact that in June 2023 Ukraine launched a not very successful second counteroffensive, was still in progress as late as the end of 2023. In other words, there has been no major success on either side for more than a year, except for minor tactical successes and failures (see Fig. 3). It can be stated that from the fall of 2022, the Russo-Ukrainian conflict developed into a protracted war of attrition.

From the point of view of the defence industry and the protection of critical infrastructure, the main lesson of the Russo-Ukrainian war is that military strategy has returned to the twentieth century with the massive deployment of artillery. The warring parties used millions of rounds of 120–155 mm calibre artillery and mortar ammunition. Because of its economic, political, and social situation, Ukraine did not have (and does not have) sufficient capability to replenish its ammunition stocks and increase production capacities. That is, the country inexorably needs the support of the democratic West.

As mentioned above, there was minimal willingness on the part of NATO countries to deliver heavy and high-performance systems before February 2022, since they were anxious of both Russian countermeasures and international reactions. However, the Russian military invasion of Ukraine and the way of the fighting, as well as the lessons learned from them, changed the attitudes of some NATO member states and of the countries of the "global West". The most significant change in the pacifist attitude occurred in the case of Germany, which (right after the United States) became the second-largest donor and supplier of military equipment and materiel to Ukraine. (see Fig. 4) During the first phase of the arms deliveries, approximately between March and June 2022, the eastern wing of NATO countries began the transfer of their existing Soviet-made weapons, including rocket and missile systems. This trend was encouraged by the compensation (Ringtausch) system developed by Germany, in which Germany promised to deliver modern military equipment. Since the German economy is profit-based, only German-made military equipment (e.g. Leopard2 A4s) has been supplied to countries in the former Soviet sphere of influence in exchange for Soviet-type military equipment donated to Ukraine. From the western wing and the central part of NATO and the other countries of the "global West" Western-made

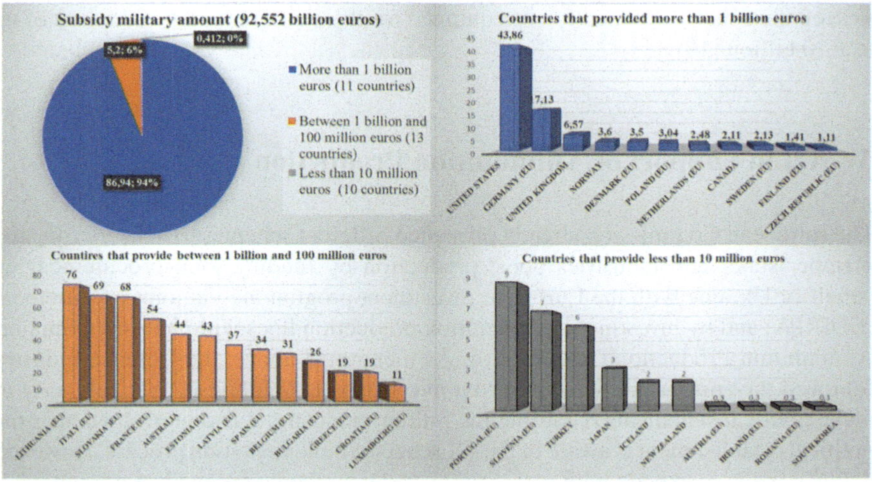

Fig. 4 Military supplies to Ukraine between 24th January 2022 and 31st October 2023 [14]

military vehicles, mainly light combat, and transport, as well as MRAP category vehicles were delivered, which have been proven during the wars in Iraq and Afghanistan. Heavy combat equipment and Western-type artillery systems were supplied during the second phase of military equipment deliveries, roughly from mid-2022, to the beginning of 2023 [17]. These arms deliveries could not have been sustainable in the long term due to the rapid depletion of supplies in the eastern wing of NATO and the difficulties in supplying ammunition and spare parts for Western combat equipment. The possibilities of the arms industry of the countries that provide military support for Ukraine were exhausted in the spring of 2023. The increase of production may take several years, not because of reluctance to support Ukraine but because of limited production capacities. Another reason is that defence companies are bound by the fulfilment of deliveries resulting from previously concluded supply contracts.

Even though the political representatives of Ukraine are critical of the dynamics of supplies and the number of delivered weapon systems, the countries supporting Ukraine in its fight against the Russian aggressor are going to the limits of their economic, political, and social possibilities. In terms of the EU, this is the ongoing support of the representatives of the union, who transformed The European Peace Facility (EPF) into a fund that primarily supports Ukraine in the form of partial reimbursement of weapons and military equipment supplies to Ukraine. By the end of 2023, the EPF had 12 billion euros, of which 5.6 billion were earmarked for the support to Ukraine. In addition, another 255 million euros were allocated for the operations of the EU Military Assistance Mission in support of Ukraine (EUMAM Ukraine). As of March 2023, within the allocated financial resources, one million euros were allocated exclusively for procuring artillery ammunition [18]. From June 2023, there were ongoing discussions to set up a dedicated "Ukraine Defence Fund" to keep the country's armed forces equipped for the next four years at the cost of up to €20 billion [19].

3 Act in Support of Ammunition Production

The European Commission already presented different schemes to help coordinated member states and industries boost production by funding joint procurement on behalf of Ukraine with the European investment programme via joint procurement (EDIRPA) and by investing in ammunition production lines with the Act in Support Ammunition Production (ASAP) [20]. As mentioned above, the Defence Industry followed the current political and economic situation. While it is relatively easy to make political statements regarding the military industry, it is doubtful whether the political establishment is aware of the subsequent implementation process. However, defence industry experts with real experience know them very well. Accordingly, it can be stated that political statements are not always in line with reality, therefore it is not guaranteed that they match the possibilities of the economy (and within that, those of the defence industry).

ASAP is a direct response to the Council's call to urgently deliver ammunition and, if requested, missiles to Ukraine and to help the member states refill their stocks by introducing targeted measures. The Act ensures that the EU can ramp-up its production capacity. The main goal of ASAP is that the EU states are to have supplied a million rounds of artillery ammunition to Ukraine, which is facing Russian aggression, by March 2024. EU politicians made a promise in the spring of 2023 [21] while ASAP highlighted the shortcomings and the actual state of the European defence industry and aligned expectations to reality.

In the autumn of 2023, the EU member states delivered only 300 thousand rounds of artillery ammunition. European defence companies (see Fig. 5) are less interested in contracts for Ukraine than the EU politicians expected because these manufacturers must keep their earlier clients supplied. Thierry Breton, Commissioner for the Internal Market of the EU, stated that the European defence industry had increased its capacity by 30% compared to 2022. However, ammunition is still exported to other countries to the detriment of Ukraine, as companies want to retain their vital non-European customers. Jiří Šedivý, director of the European Defence Agency (EDA), stated that in the past, European defence companies had to re-orient themselves to other than domestic markets because of the long-term weak demand. In 2022, only 35% of EU-produced munitions remained in Europe. Another problem for expanding ammunition production (and also for the entire defence industry) is represented by banks and other financial institutions unwilling to provide loans to increase production capacities. From the banks' perspective, the defence sector is not considered sustainable in the long term from the point of view of the so-called Environmental, Social, and Corporate Governance (ESG). EDA, which coordinated the supply of ammunition to Ukraine until the end of November 2023, signed only nine contracts with EU defence companies on ammunition supply. These contracts represent only a quarter of a billion euros. The Bloomberg agency wrote that supplies from the EU are "necessary" for Ukraine to keep up with Russia regarding ammunition. According to Bloomberg, the Russians will be able to produce two million rounds in 2024 and continuously receive additional ammunition from North Korea [22].

Based on the above information, the capacities of the European arms industry, compared to the end of the 1990s, reached roughly half of its previous capacity. This is also because the European defence industry focused on demilitarizing and liquidating excess NATO and Warsaw Pact stockpiles over the past 30 years. Production capacities either disappeared or were reduced to a minimum. This downsizing trend was also impacted by the ageing and loss of professional staff, the loss of technological capabilities, and R&D reduction. The capacity of the European defence industry to produce artillery ammunition is roughly 250,000 rounds per year, including the supply of its armed forces.

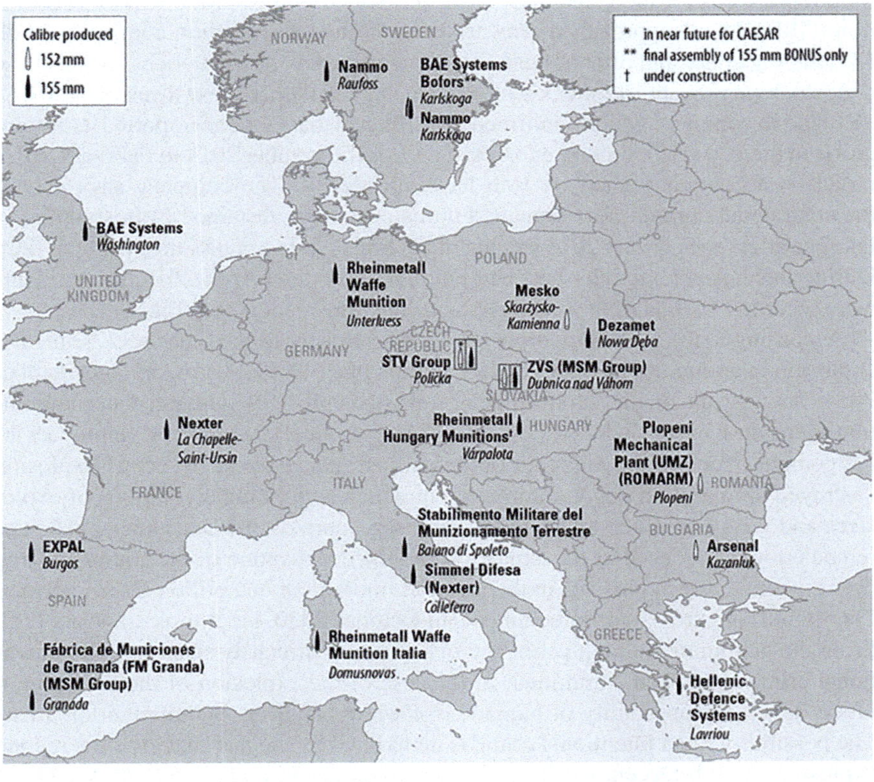

Fig. 5 Selected European producers and production sites for 155 and 152 mm artillery ammunition [23]

On the other hand, Russia has switched to a war economy regime and achieves a substantial part of its production through the military-industrial complex, while war-torn Ukraine is more and more dependent on the aid of the "global West" [24].

4 Case Studies

As early as 2014, Russia started an operation of diversionary activities, namely the explosions of ammunition depots in the Czech Republic and Bulgaria. At this time, only these two countries and Slovakia were supposed to supply Ukraine with ammunition of Warsaw Pact calibres partially. In addition to these three mentioned countries Serbia, Bosnia and Herzegovina had (and still have) production capacities. Due to their above-standard relations with Russia, the last two countries are unlikely to supply munitions to Ukraine.

According to Netherlands-based investigative journalism group *Bellingcat*, the Main Directorate of the General Staff of the Armed Forces of the Russian Federation (GRU) agents approached arms traders who had ammunition compatible with Ukraine's artillery and offered them "significantly above market prices" to purchase their stocks to prevent Ukraine's access to the market. One of these firms was EMCO, a Bulgarian company, which confirmed to Bellingcat that Ukraine imported its ammunition in the 120–152 mm range for Ukraine from December 2014 to February 2015, based on a contract signed on 10th November 2014. The company says part of the contractual commitment remained unrealized, as it abstained from shipping to Ukraine after the February 2015 signing of the Minsk-2 de-escalation agreement [26]. During this business activity between February 2014 and April 2015, six incidents were recorded at Bulgarian weapons plants and arms depots (see Fig. 6).

According to the official report of the Czech Ministry of the Interior [27], the first explosion in ammunition depot number 16 took place at 10:43 a.m. on 16th October 2014. As a result of the explosion, 57 tons of explosives integrated in ammunition were destroyed, and two people (UXO personnel), who were employees of the company renting the ammunition depot, lost their lives. The second explosion destroyed ammunition depot number 12, in which approximately 98 tons of explosives and weapons were stored, in the morning hours of 3rd December 2014. The remediation activities after the explosion and the deactivation of the equipment and devices containing explosives in the area cost more than one billion Czech korunas (39.40 million euros) and lasted until 13th October 2020, i.e. it took six years [28]. From the beginning, in both cases, the investigators primarily suspected that intentional crimes had been committed. In the case of the explosion of the first ammunition depot, the possibility of human error while handling the ammunition arose. The possibility of an intentional crime is underlined by the fact that after the second explosion, on 9th December 2014, the investigation was taken over by the Organized Crime Investigation Unit of the Czech national police. On 14th October 2015, the then President of the Republic—Miloš Zeman—stated that the explosions at the ammunition depots had certainly been not accidental cases but had probably been

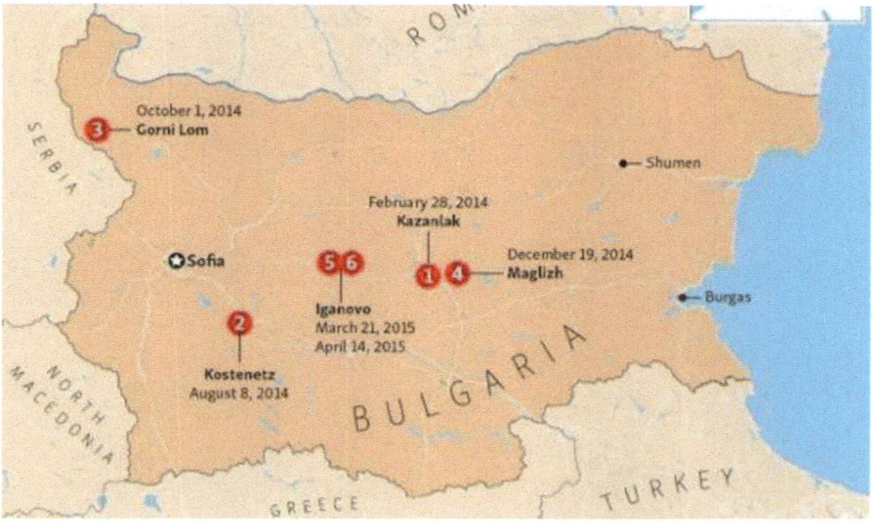

Fig. 6 Bulgarian weapons plants and arms depots sabotages between 2014 and 2015 [25]

deliberate criminal acts, and at the same time he requested the police authorities to act more rapidly [29]. The police continued to investigate the explosions of ammunition depots as intentional crimes.

The two warehouses that exploded were rented by the Czech limited liability company IMEX Group from the Ministry of Defence's Institute of Military Technology. In addition, the company rented another four buildings from 2011 [30] (Fig. 7).

As early as after the explosion of the first ammunition depot, the then Minister of Finance and later Czech Prime Minister Andrej Babiš stated in the media that the owner of the IMEX Group (Petr Bernatík Jr.) had inspected the ammunition storage site one day before the explosion. This fact was confirmed by the lawyer of the IMEX Group, adding that it was a standard procedure before dispatching stores from the storage site. During the inspection, the company owner found deficiencies, and the following day (16th October 2014) he delegated two employees of the company, who were killed in the explosion later, to eliminate them [30]. According to an article published in May 2021 by the Czech investigative weekly Respekt [32], the owner of the IMEX Group, one day before the explosion of the first weapons storage site, i.e. on the morning of 15th October 2014, was presumably in the Hotel Corrado in Ostrava for 40 min, or within its 100 m radius. There would not have been anything special about the business owner's stay in or in the vicinity of the hotel if he had gone there directly from his place of residence. However, it happened in a very different way. Bernatík left his place of residence in his car that morning and after approximately 30 min, he suddenly turned on the road and parked his car in front of or near the Hotel Corrado. At that time, there were another two men in the hotel

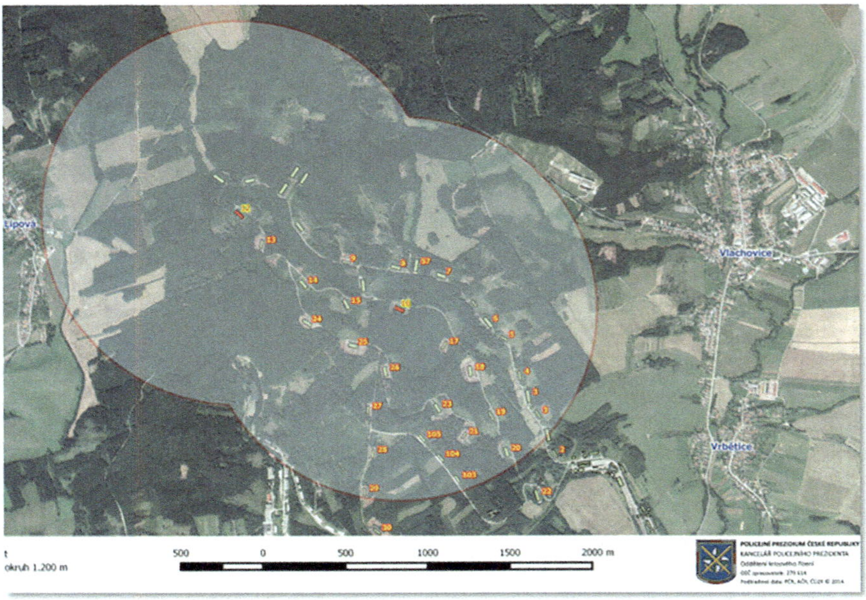

Fig. 7 The location of the ammunition depots in Vrbětice and the area of danger of explosion created by the blast [31]

who had passports issued to Alexander Petrov (likely real name: Alexandr Mishkin) and to Ruslan Boshirov (likely real name: Anatoly Chepiga). The second suspicious circumstance is that Bernatík drove from the hotel straight to the ammunition depots in Vrbětice, where he arrived at 11:45. He entered the munitions depots alone for an inspection, which is the third suspicious circumstance. According to the guards, never before had the owner entered the ammunition warehouse alone for inspection. In the ammunition warehouses the manager stayed for approximately three hours, which would not be unusual if there had been other people in the ammunition depots with him, the investigative weekly Respekt reports [33].

The fourth suspicious circumstance was that according to the statement of the Imex Group, which stored the ammunition, Mishkin and Chepiga made an appointment to visit ammunition depots in Vrbětice in the autumn of 2014. As usual in the case of arms dealers, they first sent copies of their passports—but as detectives nowadays know, these were fake documents issued by Moldova and Tajikistan. However, they did not arrive in Vrbětice on the agreed date. According to materials supplied by the Security Information Service (the Czech counter-intelligence service), these Russian agents were moving around the area of the Vrbětice ammunition storage site not long before the explosion. In 2020, the Czech police linked information from their English colleagues to the explosion in the ammunition depots in Vrbětice, thus obtaining convincing evidence of the participation of Russian GRU agents in these explosions [34].

In February 2021, the police arrested Bernatík for 48 h and searched his residence as well as the headquarters of the IMEX Group. This suggests that Bernatík was not only regarded as a witness by the police, but also as a suspect who might conceal something in connection with the explosions in the ammunition depots [33].

In 2014 and 2015, the GRU must have blown up the ammunition depots in order to prevent the weapons from reaching the Ukrainian forces, whose troops were fighting with our Russian separatists in the Donetsk and Luhansk regions. Allegedly it would have been the Ukrainian Armed Forces to which Gebrev Emilian, a Bulgarian businessman, was to sell the shipment. The operation carried out abroad, in the Czech Republic and Bulgaria had one single purpose—to gain an advantage for the Russian Armed Forces in the ongoing military activities against Ukraine, which clearly indicates hybrid warfare.

The subversive attack clearly indicates a Russian covert operation, which was carried out on the territory of a NATO country with the objective to gain a military advantage in an armed conflict with another country—most probably Ukraine. According to the theory of Mark M. Lowenthal—who held a leading position in several important institutions of the United States intelligence community and currently works as an intelligence analyst and trainer—this covert operation act matches the definition of "sabotage", which is between economic activity and a coup d'état on the pyramid of covert operations. According to the theory, propaganda is at the bottom of the pyramid of covert operations, followed by political and economic activities, sabotage, and coup d'état, while paramilitary operations are at the top [35].

Investigative journalists of Bellingcat, based on open-source information, connected the explosions of ammunition depots in Bulgaria and the Czech Republic with the GRU diversion groups [36]. The findings of the investigators are in the Fig. 8.

Fig. 8 GRU diversion unit members who participated in the missions in Bulgaria and Czechia [36]

Table 1 Explosions in the Czech and Slovak Republics from 2007 to 2022

Date	Place	Country	Reasons
March 2007	Nováky	Czech Republic	Industrial accident
March 2007	Polička	Czech Republic	Industrial accident
April 2011	Pardubice	Czech Republic	Industrial accident
October 2013	Nováky	Slovak Republic	Military accident
October 2014	Vrbětice	Czech Republic	Sabotage
December 2014	Vrbětice	Czech Republic	Sabotage
February 2017	Polička	Czech Republic	Industrial accident
November 2019	Pardubice	Czech Republic	Industrial accident

Own editing

For the sake of completeness, we present the statistics of ammunition depot explosions on the territories of the Czech and Slovak Republics from 2007 to 2022. The overview also includes industrial accidents, the most serious of which occurred in March 2007 in Slovakia (Table 1).

The explosion of the ammunition elaboration hall in the Slovak town of Nováky occurred on 2nd March 2007, around 4:30 p.m. To date, it represents one of the most severe industrial accidents in Slovakia's history. Eight people were killed as a consequence of the blast (four bodies were found on the spot, and four remained missing). Seventeen people were hospitalized, and medics treated another 30 people with minor injuries on the spot [37].

5 Conclusion

The Russo-Ukrainian war proves the claim of *democratic peace* that democratic states show more willingness to preserve peace and international cooperation than states with authoritarian political systems. On the other hand, considering the theory criticisms, *democratic peace* is not based on the values of democracy but on shared economic, financial, and political interests. As mentioned above several times, in addition to politics, the arms industry is also influenced by the economic situation of a particular country or an alliance, while each factor is interconnected.

In connection with the country's economic situation, in this case, Ukraine, and the facts mentioned at the beginning of this study, it must also be taken into consideration that Ukraine is at war. Donor states keep records of donations, but there is no central registry. Despite the massive contributions, no authority has accurate information on the movement of weapons and ammunition [38].

In our conclusion, we focus on the risks arising from the current need to increase the defence industry's production in Central Europe considering the Russo-Ukrainian war while also considering our practical experience. From the defence industry management perspective, the increase in the production of the Central European defence industry involves the following risks and difficulties, and the expansion of production is associated with several limiting factors.

Firstly, there are many issues about maintaining the safety distance of production enterprises from civilian objects and critical infrastructure. Safety distances are much tighter than in the past (e.g. at the end of the twentieth century), and the increased amounts of explosive materials increase the risk and danger of explosion. A lot of acids, cellulose, glycerine, and other components are required for production, generally classified as hazardous materials. The production of these components is energy-intensive and also burdens the environment.

It must be emphasized that the supply of mechanical parts depends on the technical mastery of producing bullets and cartridges. As a result of disarmament, many countries have lost this capability. This capacity was lost not only in the field of technology but also at personnel level, having ultimately led to the loss of technological skills. Technologies for filling shells with explosives are demanding to produce, and delivery times are at the level of 14–18 months (one production line is usually capable of filling up 20–50,000 rounds per year). Due to the specific features of working with explosives, the range of manufacturers of such devices is limited. For this reason, non-European capacities are currently being purchased. The production of ammunition is also connected with the production of detonators and fuses, whose production is also limited.

It is also essential to ensure a smooth and safe supply of mechanical parts and modern and safe technologies for filling shells and for secure storage. In connection with the storage of ammunition, we must face the fact that central Europe has insufficiently secured storage capacities. The ammunition storages are sold, and existing ones are in a desolate state or no longer meet the current requirements for ammunition storage. Finally, only reliable suppliers and supply chains must be involved in the supply of explosive components.

The construction of new factories to produce explosives is almost unfeasible in Europe due to ecological and safety standards. In this context, the activities of civic and NGOs, which have a relatively strong position in functioning democracies, should not be forgotten. In the spirit of environmental protection, these organisations can successfully block the development of road infrastructure and car or battery factories, and they also will probably rightfully stay with the population that opposes the construction or modernisation of ammunition production.

In conclusion, Europe can reach the required number of one million supplied artillery rounds per year within 4–5 years at the earliest, assuming significant investments in technological development, at costs estimated around 500 million euros. The low interest in the production of artillery ammunition for Ukraine (only nine concluded contracts in more than half a year) indicates that the increase in production capacity is dependent only on the management of defence industry companies.

The conclusion drawn from the case studies is that ammunition storage sites have insufficient physical security, control of entry and movement of persons and neglected intelligence protection. The cause of the industrial disasters in Czech Republic and Slovakia could be a violation of safety regulations or a human factor, e.g. improper handling of a fuse, its fall, or the initiation of a detonator.

6 Recommendations

Based on conclusion and professional experience, we recommend, that each country should have:

(1) Consequence Management documentation for objects Attractive to Criminals and Terrorist Organization (ACTO);
(2) National Personal Protection Guidelines against Explosive Attacks;
(3) National Facility Protection Guidelines against Explosive Attacks;
(4) National Guidelines for EOD consequence management.

Moreover, each country and/or headquarters, including the armed forces, must have a Primary Warning/Awareness System against threats. Safety units must have a functional Communication System (secured line) among rescue units. This communication system should not be dependent on the mobile operator but, if possible, completely independent, including backup energy sources. Finally, after-blast forensic work must be the responsibility of a single authority, e.g., the Ministry of Interior/Police/Forensic Authority.

We also recommend using the UN SaferGuard Programme for safe and secure ammunition management. The program was established because over the past five decades, in more than 100 countries poorly-managed ammunition stockpiles have resulted in explosions, often bringing about humanitarian disasters. In addition to the humanitarian and socio-economic consequences induced by unsecured or poorly managed ammunition stockpiles or fuel insecurity. Massive diversion of ammunition to illicit markets has been a catalyst for armed conflicts, increase in organized crime and terrorism. Moreover, diverted ammunition is increasingly used to assemble improvised explosive devices. In response to these concerns, the General Assembly requested the United Nations to develop guidelines for adequate ammunition management to ensure that the United Nations consistently delivers high-quality counselling and support. As a result, the International Ammunition Technical Guidelines (IATG) were developed in 2011, and the UN SaferGuard Programme was established as the corresponding knowledge management platform [39].

References

1. Центральний державний архів вищих органів влади та управління України. https://tsdavo.gov.ua/gmedia/5-1-28-144-6-jpg/. Last accessed 19 Jan 2024
2. United Nations Treaty Collection. https://treaties.un.org/doc/Publication/UNTS/Volume%203007/Part/volume-3007-I-52241.pdf. Last accessed 19 Jan 2024

3. Russett B, Antholis W, Ember CR, Ember M, Maoz Z (1993) Grasping the democratic peace: principles for a post-cold war world. Princeton University Press, Princeton, pp 3–5
4. NATO Relation with Ukraine. https://www.nato.int/cps/en/natolive/topics_37750.htm. Last accessed 19 Jan 2024
5. BBC News Europe. http://www.bbc.com/news/world-middle-east-26248275. Last accessed 19 Jan 2024
6. Vox Ukraine. https://voxukraine.org/en/why-is-ukraine-poor-look-to-the-culture-of-poverty. Last accessed 19 Jan 2024
7. EU relation with Ukraine. https://www.consilium.europa.eu/en/policies/eastern-partnership/ukraine/#candidate. Last accessed 19 Jan 2024
8. European Union External Action. https://www.eeas.europa.eu/eeas/eastern-partnership_en. Last accessed 19 Jan 2024
9. NATO Partners. https://www.nato.int/cps/en/natohq/51288.htm#fn10. Last accessed 10 Jan 2024
10. Kavický V, Vršanská I (2016) The conflict in Eastern Ukraine—the new geopolitical source of terrorism. Int J Humanit Soc Sci 6(3):95–103
11. Institute for the Study of War and Critical Threats. https://experience.arcgis.com/experience/c5eeec0f7ce5469ab6fc70176327fb0e/. Last accessed 19 Jan 2024
12. Bundesministerium der Verteidigung. https://www.bmvg.de/de/das-ramstein-format. Last accessed 19 Jan 2024
13. Ukraine Support Tracker Map (EN). https://app.23degrees.io/view/x67vE7NsM3NeQu7z-atlas-slideshow_v4-atlantic/MN8juinnuWJ7fecU-choro-ukraine-aid-tracker-final-data. Last accessed 19 Jan 2024
14. Trebesch C, Antezza A, Bushnell K, Bomprezzi P, Dyussimbinov Y, Frank A, Frank P, Franz L, Kharitonov I, Kumar B, Rebinskaya E, Schade C, Schramm S, Weiser L. The Ukraine support tracker: which countries help Ukraine and how? Kiel working paper, no. 2218, pp 1–75
15. Clausewitz C (2013) A háborúról. Zrínyi, Budapest, p 11
16. Rusi. https://rusi.org/explore-our-research/publications/special-resources/preliminary-lessons-conventional-warfighting-russias-invasion-ukraine-february-july-2022. Last accessed 19 Jan 2024
17. Varga TC, Deák A, Egeresi Z, Háda B, Jójárt K, Kemény J, Tálas P (2023) Az orosz-ukrán háború első évének tanulságai. NKE Stratégiai Védelmi Kutatóintézet Eelemzések 3
18. European Council, Council of the European Union. https://www.consilium.europa.eu/en/policies/european-peace-facility/. Last accessed 19 Jan 2024
19. Euractiv. https://www.euractiv.com/section/defence-and-security/news/eu-member-states-to-explore-e20-billion-war-fund-option-for-ukraine/. Last accessed 19 Jan 2024
20. Euractiv. https://www.euractiv.com/section/defence-and-security/news/breton-pitches-e100-billion-fund-for-defence-industry-cooperation/. Last accessed 19 Jan 2024
21. European Commission Defence Industry and Space. https://defence-industry-space.ec.europa.eu/eu-defence-industry/act-support-ammunition-production-asap_en. Last accessed 19 Jan 2024
22. SME. https://svet.sme.sk/c/23247040/ukrajina-vojna-naboje-dodavky.html. Last accessed 19 Jan 2024
23. International Institute for Strategic Studies. https://www.iiss.org/online-analysis/survival-online/2023/06/the-guns-of-europe-defence-industrial-challenges-in-a-time-of-war/. Last accessed 19 Jan 2024
24. Euractiv. https://www.euractiv.com/section/europe-s-east/news/whats-next-for-ukraine-over-to-eu-and-washington/. Last accessed 19 Jan 2024
25. Global Security. https://www.globalsecurity.org/intell/world/russia/gru-ops.htm. Last accessed 19 Jan 2024
26. Bellingcat. https://www.bellingcat.com/news/uk-and-europe/2021/04/26/how-gru-sabotage-and-assassination-operations-in-czechia-and-bulgaria-sought-to-undermine-ukraine/. Last accessed 19 Jan 2024

27. Ministerstvo vnitra České republiky. https://www.mvcr.cz/sluzba/docDetail.aspx?docid=21877825&docType=ART. Last accessed 19 Jan 2024
28. Česká televize. https://ct24.ceskatelevize.cz/clanek/regiony/rok-od-vybuchu-ve-vrbeticich-bez-vinika-a-odskodneni-125938. Last accessed 19 Jan 2024
29. Lidovky. https://www.lidovky.cz/domov/zeman-vybuchy-ve-vrbeticich-jsou-umyslnym-trestnym-cinem.A151014_130625_ln_domov_ele. Last accessed 19 Jan 2024
30. iDnes. https://www.idnes.cz/zpravy/domaci/babis-munice-sklad-vrbetice-imex.A141207_125558_domaci_hv. Last accessed 19 Jan 2024
31. Lidovky. https://www.lidovky.cz/domov/odvoz-munice-z-arealu-u-vrbetic-zacne-nejdrive-v-pulce-ledna.A141223_085601_ln_domov_mmu. Last accessed 19 Jan 2024
32. Respekt. https://www.respekt.cz/agenda/vybuch-ve-vrbeticich-majitel-imexu-byl-40-minut-u-hotelu-v-nemz-byli-ubytovani-agenti-gru. Last accessed 19 Jan 2024
33. Deník. https://www.denik.cz/z_domova/vybuch-vrbetice-marie-benesova-vysetrovani-20210517.html. Last accessed 19 Jan 2024
34. Seznam Zprávy. https://www.seznamzpravy.cz/clanek/agenti-podle-bis-pronikli-primo-do-vrbetic-zeman-tvrdil-ze-to-neni-jiste-152733. Last accessed 19 Jan 2024
35. Lowenthal M. Hírszerzés—a titkoktól a politikai döntésekig. Antall József Tudásközpont, Budapest, pp 289–295
36. Euractive. https://www.euractiv.com/section/global-europe/news/bellingcat-connects-the-dots-between-czech-explosion-and-bulgaria-poisoning/. Last accessed 19 Jan 2024
37. SME. https://domov.sme.sk/c/3176247/v-novakoch-stale-piati-nezvestni.html. Last accessed 19 Jan 2024
38. Euractive. https://euractiv.sk/section/obrana-a-zahranicie/video/diskusia-kto-pocita-zbrane-ktore-idu-na-ukrajinu/?utm_source=traqli&utm_medium=email&utm_campaign=6888&pnespid=DKZ471BH_zweywmZ6YreBBkWqw8o2rJ6sAISHb9SL5HK1Q41TuY3HBL0fje7NQAsyLPU4Fyp. Last accessed 19 Jan 2024
39. International Ammunition Technical Guidelines. https://unsaferguard.org/. Last accessed 19 Jan 2024

Enhancing Safety and Efficiency: Human–Cobot Interaction in Critical Infrastructure

Nada El Yasmine Aichaoui and Tünde Anna Kovács

Abstract Collaborative robots (cobots) are transforming critical infrastructure sectors, from electricity production to transportation, emphasizing the crucial need for human–robot interaction with user-friendly interfaces, safety measures, and training programs. With the assistance of artificial intelligence, cobots adapt to human preferences and enhance their cooperation capabilities, showcasing their transformative impact on infrastructure security and efficiency. Meanwhile, achieving autonomy in robots remains a major challenge, categorized into human–robot collaborations based on various criteria aimed at improving necessary domains. The robot undergoes four standard stages: perception, decision-making, motion execution, and evaluation. This paper focuses on optimizing collaboration by enhancing the robot's decision-making process, assessed through various performance metrics. The developed framework enhances human–robot collaboration by considering diverse metrics, independently of human behavior.

Keywords Cobots · Critical infrastructure · Safety measures · Artificial intelligence · Autonomy · Decision-making

1 Introduction

In an age of technological innovation and automation, the integration of collaborative robots, often referred to as "cobots," into critical infrastructure is fundamentally changing the way we manage and protect our critical systems. From power generation and distribution to transportation networks, these dynamic machines have emerged

N. El Yasmine Aichaoui (✉)
Doctoral School on Safety and Security Sciences, Obuda University, Budapest, Hungary
e-mail: nada.aichaoui@phd.uni-obuda.hu

T. A. Kovács
Banki Donat Faculty of Mechanical and Safety Engineering, Obuda University, Budapest, Hungary
e-mail: kovacs.tunde@bgk.uni-obuda.hu

T. A. Kovács et al. (eds.), *The Impact of the Energy Dependency on Critical Infrastructure Protection*, Advanced Sciences and Technologies for Security Applications, https://doi.org/10.1007/978-3-031-78544-3_6

as indispensable assets, augmented the capabilities of human workers, and enhanced the overall safety and efficiency of critical infrastructure operations. The introduction of robots into industry is not new. Indeed, since the mid-1970s, the use of robots has gradually developed for carrying out basic, repetitive operations (palletizing, pick and place, etc.), requiring significant forces (handling and setting up heavy parts, forced insertion, etc.) or even dangerous ones (welding, adhesive operations, handling of hot parts or processes generating smoke or dust, etc.). To ensure the safety of the operators involved on or near these often fast and powerful robots, the protection strategy adopted was to separate the robot's space from that of the operators using physical protection (protectors, grilles, etc.). Consequently, in the production phase, interactions between operators and robots are almost non-existent [1].

Critical infrastructure encompasses a broad array of sectors crucial for the operation of today's society, such as telecommunications, energy generation, transportation, healthcare, and water distribution, among others. Ensuring the dependable and secure functioning of this infrastructure is of paramount importance, given that any interruptions or malfunctions can deeply impact public safety, economic health, and national security. Within this framework, the most significant shift lies in the adoption of collaborative robots, which present fresh avenues and prospects for managing and maintaining infrastructure effectively [2].

At the heart of this transformation, the evolution of technologies in different fields such as mechanics, electronics, automation, and IT, has made it possible to gradually broaden the spectrum of applications of industrial robots by entrusting them with increasingly complex operations [1]. These developments highlight the evolving partnership between operators and robots in crucial infrastructure scenarios, particularly for tasks that demand specific expertise or human-guided decision-making [3].

The term "collaborative" in the realm of robotics is divided into three stages, depending on the desired level of engagement between the operator and the robot. The goal is to amplify human capabilities, not replace them. Yet, such progress raises pertinent concerns about employee health and safety. On the one hand, they accentuate certain risks due to the proximity sought between the robot and the operators: physical risks, in the event of contact or psycho-social risks, in the event of lack of control or understanding of the movements of the robot by the operators. On the other hand, they call into question the protection strategy [4]. In response to all or part of these questions, the standard NF EN ISO 10218-1 [5] relating to the security of industrial robots, proposes to implement one or more of the following four principles: nominal controlled safety stop: the robot must stop when a human being is in the cooperative workspace; manual guidance: the operator must use a validation device to authorize the guided movement of the robot; control of the speed and separation distance: when the operator is in the robot's working zone, the latter must maintain a determined speed and respect a separation distance from the operator, to avoid any collision; limitation of power and force: force reduction mechanisms must be integrated into the robot, to reduce the intensity of potential contacts between the operator and the robot.

Manufacturers today offer collaborative robotic solutions to industrialists that respect these principles. Faced with the diversity of the offer and the security issues, it is important to ensure that these solutions meet the needs of industrial users of robots. This represents that Human–cobot interaction is central to the successful integration of these machines into the critical infrastructure landscape, as it influences safety, efficiency, and overall productivity. The importance of human–cobot interaction becomes evident when one considers the multifaceted roles these machines play [6]. Collaborative robots are capable of accessing hard-to-reach areas, performing repetitive tasks with exceptional precision, and minimizing human exposure to dangerous conditions. In nuclear power plants, for instance, cobots assist in maintaining and inspecting equipment within high-radiation environments, reducing the risk to human personnel. Similarly, in the transportation sector, cobots are deployed for the inspection and repair of bridges, tunnels, and railway tracks, tasks that often involve tight spaces and heights [7].

Effectively harnessing the potential of cobots in critical infrastructure necessitates a holistic approach. Key components include designing user-friendly interfaces, implementing rigorous safety measures, and establishing training programs for human operators [8]. These elements are interconnected, as intuitive interfaces enable human operators to communicate with cobots seamlessly, safety measures protect both humans and machines in dynamic work environments, and training programs equip operators with the skills and knowledge needed to work alongside cobots safely and effectively [9, 10]. Furthermore, incorporating artificial intelligence (AI) into cobot frameworks amplifies their functionalities. Through machine learning algorithms, cobots can adjust based on human preferences, assimilate insights from human operators, and enhance their overall efficiency. AI-enabled cobots not only perform designated tasks but also engage actively with human counterparts, rendering them increasingly versatile and responsive to changing operational demands [11].

In this article, we delve into the world of human–robot interaction in critical infrastructure. We explore the design principles behind user-friendly interfaces, examine the critical safety measures that underpin successful cobot deployment, and discuss the importance of tailored training programs for human operators. As the relentless march of technological progress continues, human–cobot interaction stands at the forefront of reshaping critical infrastructure operations, enhancing safety, and driving efficiency.

2 Design Intuitive Interface and Safety Protocols

2.1 Optimizing Human–Robot Interaction in Critical Infrastructure

Within the context of human–robot interaction in critical infrastructure, the creation of intuitive interfaces stands out as a crucial element influencing the efficiency and safety of cobot integration. This interface serves as the conduit enabling human operators to interact with and direct cobots effortlessly, playing an essential role in fostering seamless collaboration between humans and robots within intricate operational landscapes [12]. Furthermore, such interfaces are vital for ensuring that human intentions are accurately conveyed to cobots. A user-friendly interface enables operators to issue precise commands, oversee cobot activities, and swiftly adapt to evolving circumstances [13]. Consequently, it optimizes operational processes and reduces the potential for errors stemming from communication gaps or misconceptions.

Although designing user-friendly interfaces for human–robot interaction involves several principles that revolve around simplicity, clarity, and adaptability, it needs to be straightforward, with clearly defined controls and feedback mechanisms. Visual and auditory cues should be designed to convey information effectively, allowing operators to comprehend the cobot's status and actions intuitively [12, 14].

In this context, cobots are used in diverse environments with unique demands, and designing interfaces that are adaptable to these various conditions can be complex [14]. However, ongoing innovations in interface design, such as the integration of augmented reality (AR) and virtual reality (VR) technologies, are advancing the state of the art. AR and VR interfaces provide operators with immersive experiences, enhancing their ability to control cobots and make informed decisions in real time [15].

2.2 Challenges and Safety Considerations in Welding Environments

Moving to the Welding environment, it is crucial to consider the challenges factors, and hazards that can be caused to the health of humans which introduces process-related hazards associated with heat, sparks, chemicals, and electrical currents. Such hazards can cause respiratory issues, burns, and irritations to the eyes, skin, and throat [16]. Notably, ultraviolet (UV) radiation exposure poses a risk of skin ailments, including cancer. The welding environment further raises concerns of asphyxiation from inhaling gases and smoke, while the released particulates can lead to lung diseases. Electrical shocks, burns, and fall-related injuries are also prevalent, with human errors, like haste or improper techniques, contributing significantly to welding accidents [17, 18].

Therefore, Welding robots promise consistent not only weld quality and repetitive task achievement but also maintain safety concerns [19]. However, it cannot turn a blind eye to the possibility of unintended contact with certain surfaces that might trigger electrical or thermal issues. Additionally, sparks from the robot's nozzle can cause severe injuries. Misusing robots might even induce electrical fires [20]. Thus, while manufacturing guidelines offer some safety measures, it's essential to continually assess and manage risks associated with these robots. UV radiation control and safety protocol adherence are crucial for risk mitigation [21]. Moreover, welding robots might not always match the precision of traditional welding, where seasoned welders oversee the entire process [22]. With robots, operators mainly adjust speed, power, and equipment settings. Although robots provide adaptability, they necessitate rigorous training. Proximity to sparks during robot operation mandates safety gear use. Yet, ensuring weld quality might still require manual interventions, particularly if operators lack adequate training [23].

3 Integrated Control Paradigm by Enhancing Welding Robotics with Operator Interaction and Safety Measures

A primary objective revolves around comprehending the operational behaviour of the welding robot, particularly in scenarios where an object, often a human, may inadvertently enter the robot's hazardous zone during its ongoing operation Fig. 1. Understanding the intricacies of the robot's working mechanism is crucial for ensuring both safety and efficiency in industrial settings. In typical situations, potential risks arise when objects, including human operators, approach the designated danger zones while the robot is in operation. This necessitates a comprehensive understanding of the robot's response mechanisms and the implementation of effective safety measures to prevent accidents or injuries. By delving into the nuances of the welding robot's operational dynamics, engineers and operators can establish protocols and safeguards that mitigate risks associated with proximity to the robot during its active phases. This overarching goal aligns with the broader objective of creating a secure working environment, where the welding robot seamlessly integrates with human activities without compromising safety standards. The continuous pursuit of this goal not only enhances the overall safety culture within industrial operations but also contributes to the optimization of the robot's performance in dynamic and potentially unpredictable workspaces.

The implementation of the safety and quality requirement should pass as the flowchart in Fig. 2. Where this system starts when a human comes closer to the danger zone, the monitoring system shall detect this action and provide an immediate response relative to the welding robot control system, this manner could be included in the following setups:

1. The controller can press the "Stop" button.
2. The system checks if the robot is in a critical area.
3. If the robot is in a critical area, it continues welding until it exits the critical area, at which point it stops.
4. If the robot is not in a critical area, it stops immediately.
5. If the operator doesn't press the "Stop" button, the robot continues welding.
6. An external emergency stop can be pressed at any time to trigger an immediate stop of the robot.

The presented flowchart outlines a comprehensive control system for a welding robot in an industrial environment, emphasizing safety, efficiency, and human–robot interaction. The clarity of the flowchart allows for a clear understanding of

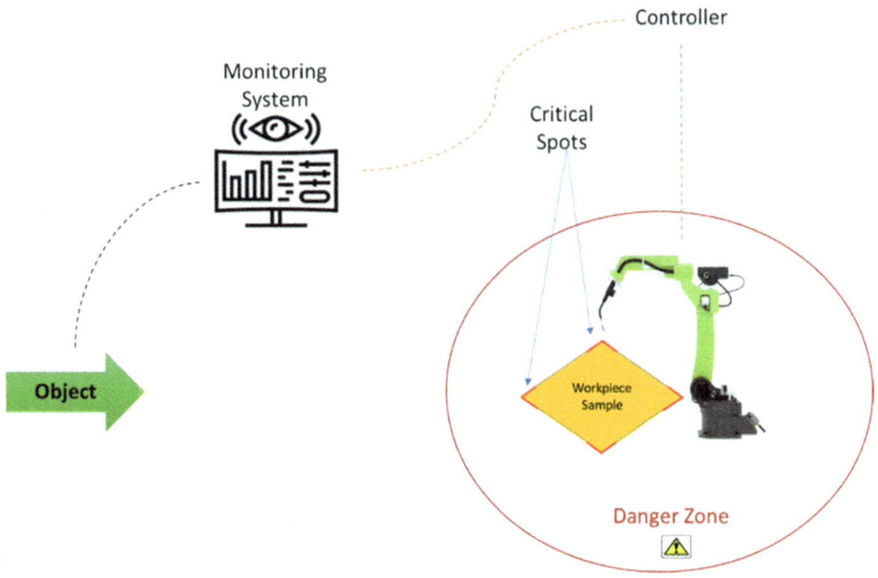

Fig. 1 The general implementation of an object approaching to welding robot danger zone

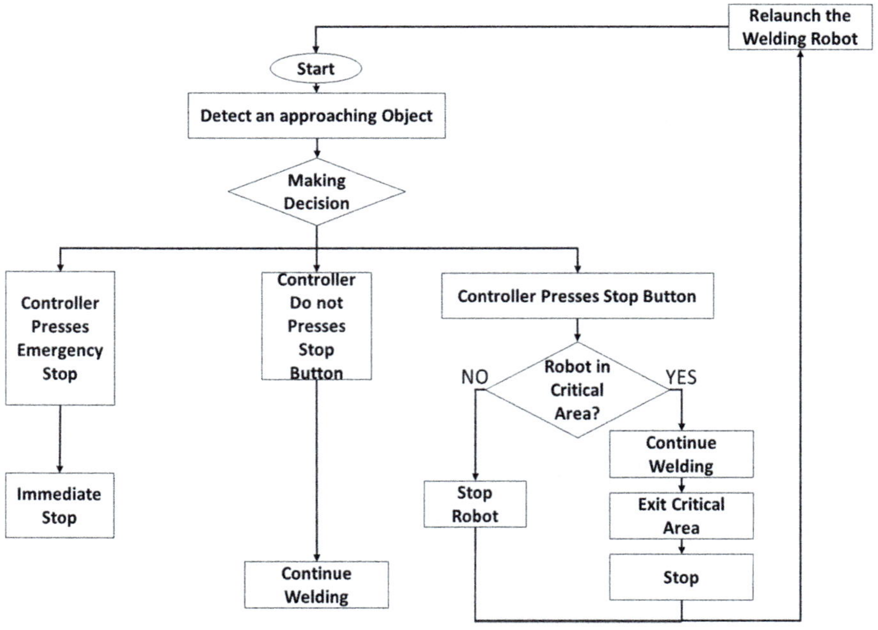

Fig. 2 Flowchart of the decision-making process by the controller

the decision-making process, showcasing the robot's immediate response to the operator's "Stop" button, and ensuring prompt safety measures. Additionally, the flowchart illustrates the robot's intelligent behaviour in continuing welding when situated in a critical area, thus preventing production interruptions and optimizing the welding process. The incorporation of an emergency stop mechanism adds an extra layer of safety, ensuring the robot can be halted instantly in unforeseen circumstances.

Considering the potential integration of deep learning techniques into the fast and real-time control of the welding robot, several benefits and challenges should be considered. Deep learning algorithms can enhance the robot's decision-making capabilities by enabling it to learn and adapt to various scenarios, potentially improving its responsiveness to dynamic and complex environments. For instance, a deep learning model could be trained to recognize specific patterns or conditions in the welding environment and adjust the robot's behaviour accordingly.

However, the integration of deep learning into real-time control systems poses challenges. The process of training deep learning models typically demands substantial datasets and computational resources, and ensuring the safety and reliability of a deep learning-controlled system within an industrial context is paramount. The model must exhibit robustness to handle diverse operating conditions and possess the ability to generalize its learning to novel situations. The subsequent phases of incorporating deep learning techniques involve several key steps. First, comprehensive data collection and preprocessing are essential, encompassing various scenarios the welding robot may encounter. Subsequently, a deep learning model is developed and trained using the amassed dataset, implementing algorithms for real-time data processing and decision-making. The model's performance is rigorously evaluated through simulation and real-world testing, ensuring its accuracy and adaptability to dynamic conditions. Following successful validation, the model is seamlessly integrated into the existing control system. Robust safety measures, including fail-safe mechanisms and continuous system monitoring, are implemented to mitigate potential risks. The final steps encompass continuous optimization and fine-tuning of the deep learning model based on real-world performance, ensuring its adaptability to evolving conditions while maintaining a high level of accuracy.

By integrating deep learning techniques into the control system, the welding robot can potentially become more adaptive and intelligent, enhancing its overall efficiency and safety in dynamic industrial environments. However, careful consideration must be given to the challenges associated with training and deploying deep learning models in real-time control applications. Regular updates and improvements to the model, along with ongoing safety assessments, will be essential for successful implementation.

4 Conclusion

In the dynamic field of critical infrastructure, ensuring the reliable operation of essential systems is of the utmost importance, the interaction between humans and cobots emerges as a transformative force, reshaping the management and protection of these crucial facilities [24].

This paper revealed the reason behind the interest of manufacturers and the industrial sectors in human–robot coactivity in terms of gaining productivity and optimizing the efficiency process, flexibility in terms of mobility and reuse of the robot, increased responsiveness to change, and easier adaptation of the line, and the improvement in working conditions. However, it is crucial to consider the effect of negative consequences of this collaboration which are sometimes considered difficult to reduce or even unacceptable and may prove incompatible with the process (speed, radiation, etc.) and sometimes even counterproductive (risk of untimely stopping of the robot due to the regular presence of operators in the its immediate environment). This shows that manufacturers are generally aware that the use of collaborative robotics cannot be considered without, on the one hand, a "functional" analysis to ensure the relevance of the solution, concerning the constraints of implementation and production, on the other hand, a risk analysis specific to the cell concerned [10]. Although the needs expressed are generally covered by the methods of collaboration proposed by European standardization, the technological offer for their implementation is still reduced and numerous developments and research work are carried out, to develop systems for protection allowing easier implementation of this type of installation and thus ensuring the health and safety of employees who have to work near the robots.

In this regard, the emphasis on user-friendly interfaces is critical to ensuring effective communication and control, with a focus on intuition, clarity, and adaptability. Fundamental safety measures, such as collision detection and emergency stop mechanisms, along with robust risk assessments, are essential to protect both humans and cobots in dynamic operational environments [14].

Real-world applications in power generation, transportation, and healthcare infrastructure demonstrate tangible benefits in enhancing safety, reliability, and efficiency. Challenges in customization, interoperability, and compatibility with legacy systems are being addressed through innovations in interface design, including augmented reality and virtual reality. The necessity for flexibility in various crucial infrastructure environments is emphasized by real-world applications, creating opportunities to improve system resilience through increased automation and remote operations [12]. Furthermore, as technology progresses, the interaction between humans and collaborative robots is expected to play a more significant role in critical infrastructure. Thus, ongoing advancements in artificial intelligence, safety protocols, and interface design will continue to shape this field, where the incorporation of AI enhances collaboration by allowing cobots to adjust to human preferences and make immediate decisions, thereby enhancing efficiency and responsiveness. Ethical considerations, cybersecurity, and regulatory frameworks will evolve to address challenges associated with deploying cobots in critical infrastructure [11, 25].

In this ever-evolving landscape, human–cobot interaction has the potential to revolutionize critical infrastructure operations, making them safer, more efficient, and more reliable. The collaborative partnership between humans and cobots promises a future where critical infrastructure is not only resilient but also adaptable to the changing needs and challenges of our increasingly complex world.

References

1. Grau A, Indri M, Bello LL, Sauter T (2020) Robots in industry: the past, present, and future of a growing collaboration with humans. IEEE Ind Electron Mag 15(1):50–61
2. Berkeley AR, Wallace M, Coo C (2010) A framework for establishing critical infrastructure resilience goals. Final report and recommendations by the council. National Infrastructure Advisory Council, pp 18–21
3. Lattanzi D, Miller G (2017) Review of robotic infrastructure inspection systems. J Infrastruct Syst 23(3):04017004
4. Rani P, Sarkar N (2005, Apr) Operator engagement detection and robot behavior adaptation in human-robot interaction. In: Proceedings of the 2005 IEEE international conference on robotics and automation. IEEE, pp 2051–2056
5. Sghaier A, Charpentier P (2012) La problématique de l'utilisation des robots industriels en matière de sécurité. Réalités Industrielles 1:24–31
6. Tihay D, Perrin N (2018) Robotique collaborative: perception et attentes des industriels. Hyg Sécurité Travail INRS 250:50–57
7. Romanov D, Korostynska O, Lekang OI, Mason A (2022) Towards human-robot collaboration in meat processing: challenges and possibilities. J Food Eng 331:111117
8. Hollerer S, Fischer C, Brenner B, Papa M, Schlund S, Kastner W, Fabini J, Zseby T (2021) Cobot attack: a security assessment exemplified by a specific collaborative robot. Procedia Manuf 54:191–196
9. Khalid A, Khan ZH, Idrees M, Kirisci P, Ghrairi Z, Thoben KD, Pannek J (2022) Understanding vulnerabilities in cyber-physical production systems. Int J Comput Integr Manuf 35(6):569–582
10. Adriaensen A, Costantino F, Di Gravio G, Patriarca R (2022) Teaming with industrial cobots: a socio-technical perspective on safety analysis. Hum Factors Ergon Manufact Serv Ind 32(2):173–198
11. Abishek BA, Kavyashree T, Jayalakshmi R, Tharunkumar S, Raffik R (2023, June) Collaborative robots and cyber security in industry 5.0. In: 2023 2nd International conference on advancements in electrical, electronics, communication, computing and automation (ICAECA). IEEE, pp 1–6
12. Yip HM, Wang Z, Navarro-Alarcon D, Li P, Cheung TH, Greiffenhagen C, Liu YH (2020) A collaborative robotic uterine positioning system for laparoscopic hysterectomy: design and experiments. Int J Med Robot Comput Assist Surg 16(4):e2103
13. Segura P, Lobato-Calleros O, Ramírez-Serrano A, Hernández-Martínez EG (2022) Safety assurance in human-robot collaborative systems: a survey in the manufacturing industry. Procedia CIRP 107:740–745
14. Frijns HA, Schmidbauer C (2021) Design guidelines for collaborative industrial robot user interfaces. In: Human-computer interaction—INTERACT 2021: 18th IFIP TC 13 international conference, Bari, Italy, 30 Aug–3 Sept 2021, Proceedings, Part III 18. Springer International Publishing, Berlin, pp 407–427
15. Gallala A, Hichri B, Plapper P (2021, Dec) Human-robot interaction using mixed reality. In: 2021 International conference on electrical, computer and energy technologies (ICECET). IEEE, pp 1–6

16. Mehrifar Y, Zeverdegani SK, Faraji M, Rismanchian M (2019) Risk assessment of welders exposure to the released contaminated gases in different types of welding processes in a steel industry. Health Scope 8(1)
17. Bakri SFZ, Hariri A, Ismail M (2020) Occupational health risk assessment of inhalation exposure to welding fumes. Int J Emerg Trends Eng Res 8(1.2):90–97
18. Takahashi J, Nakashima H, Fujii N, Okuno T (2020) Comprehensive analysis of hazard of ultraviolet radiation emitted during arc welding of cast iron. J Occup Health 62(1):e12091
19. Kah P, Shrestha M, Hiltunen E, Martikainen J (2015) Robotic arc welding sensors and programming in industrial applications. Int J Mech Mater Eng 10(1):1–16
20. Antonelli D, Astanin S (2016) Qualification of a collaborative human-robot welding cell. Procedia CIRP 41:352–357
21. Schramkó M, Kafi A, Gyura L, Kovács TA (2022) An experimental study of the gas metal arc welding ultraviolet effect as a function of distance. In: Vehicle and automotive engineering. Springer International Publishing, Cham, pp 917–924
22. Wang B, Hu SJ, Sun L, Freiheit T (2020) Intelligent welding system technologies: state-of-the-art review and perspectives. J Manuf Syst 56:373–391
23. Liu Y (2016) Toward intelligent welding robots: virtualized welding-based learning of human welder behaviors. Weld World 60:719–729
24. Biringer B, Vugrin E, Warren D (2013) Critical infrastructure system security and resiliency. CRC Press, USA
25. Michalos G, Karagiannis P, Dimitropoulos N, Andronas D, Makris S (2022) Human-robot collaboration in industrial environments. In: The 21st-century industrial robot: when tools become collaborators, pp 17–39

5G Infrastructure Standardization, Integration, Industry 4.0 Applications in EU Precisely Germany, and the Future of Industry 5.0 and 6G: A Comprehensive Overview

Haya Altaleb and Zoltan Rajnai

Abstract The process of 5G standardization, essential for the global implementation of fifth-generation (5G) wireless communication networks, involves multiple international organizations such as the ITU, 3GPP, and IEEE. This paper explores the intricate landscape of 5G standardization and its impact on the technology's development and deployment. Beginning with an overview of the organizations driving 5G standardization, it delves into the research and development efforts in China, Korea, Japan, and other regions. The paper also emphasizes the role of publications, patents, and standards in shaping 5G technology. It highlights the top 5G publishing and funding organizations, revealing a strong Asian presence, particularly in China. Furthermore, it underscores the strategic importance of standards in 5G technology development and the need for coordination between patents and standards. German contributions to 5G standardization are discussed, focusing on organizations like Deutsche Telekom, VDE/DKE, and initiatives like the 5G-ACIA. The EU's 5G spectrum strategy and Germany's allocation plans across different frequency bands are examined in detail. The paper explores 5G's significance in connected and automated driving, as exemplified by the 5GCroCo project, which aims to ensure seamless cross-border communication for cooperative, connected, and automated mobility. Finally, the integration of 5G in Industry 4.0 is explored, emphasizing the role of 5G in enabling smart factories, IoT, industrial automation control, logistics tracking, and remote diagnosis, operation, and maintenance services. It underscores the transformative potential of 5G technology in enhancing efficiency and connectivity within the manufacturing sector, and we discussed the Future of Industry 5.0 and 6G briefly.

H. Altaleb (✉) · Z. Rajnai
Banki Donat Faculty of Mechanical and Safety Engineering, Obuda University, Budapest, Hungary
e-mail: haya.altaleb@uni-obuda.hu

Z. Rajnai
e-mail: rajnai.zoltan@bgk.uni-obuda.hu

T. A. Kovács et al. (eds.), *The Impact of the Energy Dependency on Critical Infrastructure Protection*, Advanced Sciences and Technologies for Security Applications, https://doi.org/10.1007/978-3-031-78544-3_7

Keywords 5G standardization · Industry 4.0 · Critical infrastructure · 5G integration · 6G · Industry 5.0

1 Introduction

The term "5G standardization" refers to the process of creating and applying a set of guidelines for the implementation of fifth-generation (5G) wireless communication networks. This process involves establishing technical specifications and performance criteria for 5G networks, such as data transfer rates, latency, frequency bands, and security features. International organizations, including the International Telecommunication Union (ITU), the 3rd Generation Partnership Project (3GPP), and the Institute of Electrical and Electronics Engineers (IEEE), are involved in the 5G standardization process. The ITU defines the overall structure of 5G networks, including frequency bands and spectrum allocation. The 3GPP develops the technical specifications for 5G, such as the network architecture [1, 2], radio access technology, and protocols. IEEE sets the standards for wireless technologies used in 5G networks, like Wi-Fi and Bluetooth. In 2013, China and Korea formed the IMT-2020 Promo Group (PG) and the 5G Forum to collaborate on investigating 5G technology [3]. In addition to this, China also has the Future Technology of Universal Radio Environment FORUM (POTENTIAL FORUM) for 5G research, At the same time, Japan's Association of Radio Industries and Services (ARIB) started the 2020 and Beyond Ad Hoc Group (20B AH) to explore the potential of 5G and its technologies, publishing their findings in a white paper in October 2014. The Telecommunication Technology Board (TTC) has also released a white paper on the future direction of mobile networks and the technical challenges that must be addressed for their realization. In September 2014, the Fifth Age Mobile Communications Promo Forum (5GMF) was established to research networks and applications beyond 5G wireless access technology [4].

The 5G standardization process started in 2015 and concluded in 2019, with the finalization of the 3GPP Release 15 and Release 16 specifications. These specifications specify the technical requirements for 5G networks, including new frequency bands, the use of small cells, and advanced radio technologies such as massive MIMO and beamforming. Standardization is crucial in 5G technology to ensure that different 5G networks globally can communicate seamlessly, allowing for the widespread adoption of 5G and its potential applications, such as smart cities, remote surgery, and autonomous vehicles [5]. When considering the economic, geopolitical, strategic, and security aspects, it becomes evident that the importance attributed to 5G networks is significant. However, the literature highlights certain weaknesses in the EU's positioning. These weaknesses encompass substantial disparities in public support and the extent of initiatives, along with a noticeable shortage of proficient experts in the design and development of 5G network solutions. Some studies indicate a relatively small market share in communication electronics DEC-EC [6], and a "significant" reliance on external sources for cutting-edge chips [7, 8]. Nevertheless, Buggenhagen and Blind identify Sweden and Finland as top leaders in terms of patents, and various European countries excel in scientific publications related to 5G [9].

The authors of "5G Development in China: From Policy Strategy to User-Oriented Architecture" examine the development of 5G technology in China, highlighting its shift from a strategy policy to a user-oriented approach that considers the interplay between technology and society. The paper recommends proposals for future technical development and industrial regulation, focusing on improving throughput and spectrum efficiency through channel assignment and load balancing. To address the issue of spectrum waste and self-interference, two solutions were proposed: a hybrid routing protocol called TOHRP (tree-based and on-demand hybrid routing protocol) and a distributed channel assignment algorithm known as LBCA (Load Based Channel Assignment) in multichannel environments [10].

Patenting and standardization are crucial aspects of technology development, as they involve documenting and sharing knowledge with external entities. They are strategic tools for organizations to enhance their technology portfolios, making them valuable indicators for analyzing 5G reports. Nevertheless, an all-encompassing technology collection extends beyond mere patents and typical contributions. Strategic publishing, alternatively referred to as defensive publishing, is widely acknowledged as a crucial strategy for maximizing the benefits of intellectual property [11, 12]. Regrettably, current research neglects the importance of publications, their contributions to advancing 5G technology, and their strategic use, value, and impact on organizations. Considering the significant roles that publications, patents, and standard contributions play in developing 5G, this discourse will explore the diverse motivations for employing these tools and the resulting benefits for organizations.

2 5G Standardization and Publishing Worldwide

Standardization is becoming increasingly important in the public discourse. A standard document serves as a reference point, providing a common set of guidelines, measurements, conventions, or designs to ensure consistency in practices or behaviours. Standardization, on the other hand, refers to the process of adapting, creating, and adhering to these standards [13]. The collaborative process entails stakeholders sharing information about current measurements or technologies, with the objective of achieving consensus in the form of a standardized document. This document compiles the knowledge and information provided by participating stakeholders, ensuring accessibility to external users with an interest in the field. Beyond relying on publications or patents alone, organizations can disseminate their technological knowledge by contributing to or adhering to standards.

Standards are established technological norms that encompass scientific knowledge, although they may not be at the forefront of the technology. Unlike patents, contributors to standards don't receive exclusive usage rights, aiming to accommodate a broad user base. Companies involved in standardization must coordinate with their patent departments if seeking patent protection for research outcomes. Like scientific publications, patent applications should be filed before knowledge is disclosed in standards. However, publishing standards or contributions can also help secure freedom to operate (FTO) rights.

2.1 Top 5G Publishing Organizations

The prevalence of Asian organizations in the realm of publications underscores their remarkable productivity, with eight of them securing positions in the top 20 list. Beijing University of Posts Telecommunications stands out as the foremost publishing organization in the field of 5G, followed by Nokia, Huawei, and Ericsson, as outlined in Table 1. Following these, prominent Chinese research institutions such as the Chinese Academy of Sciences, Southeast University China, University of Electronic Science Technology of China, and Tsinghua University, held notable positions before the first European research institution, the University of London, appeared on the list. Overall, the top 20 list predominantly comprises diverse research institutions, with only a quarter representing companies, including Nokia, Huawei, Ericsson, Samsung, and Intel (refer to Table 1).

Table 1 Leading organizations in publishing related to 5G technology [14]

Organization (country)	Number of publications	Share of total 5G publications (%)
Beijing University of Posts Telecommunications (China)	866	3.09
Nokia (Finland)	852	3.04
Huawei (China)	755	2.69
Ericsson (Sweden)	529	1.89
Chinese Academy of Sciences (China)	483	1.72
Southeast University China (China)	458	1.63
University of Electronic Science Technology of China (China)	410	1.46
Tsinghua University (China)	403	1.44

2.2 Top 5G Funding Organizations

The extent of dedication and support for the advancement of 5G research is evident through the substantial number of financially endorsed scientific publications. The leading contributor in funding is the National Natural Science Foundation of China, which has sponsored 4226 publications. The European Commission closely follows, supporting 2343 publications, and the National Science Foundation ranks third with funding for 942 publications. Among the top 15 funding organizations, nine are in Asia, three in Europe, and two in North America.

Figure 1 depicts the geographical distribution of noteworthy 5G funding organizations across various regions. The figure showcases the sources of funding organizations that have sponsored at least five publications. Identification of these organizations was accomplished through the Web of Science database, and their respective locations were determined through manual investigations. The results were subsequently compared and presented in Fig. 1. The illustration highlights that a significant proportion of 5G funding organizations are primarily located in Asia (51%), followed by Europe (30%), and North America (14%). The remaining (5%) accounts for funding organizations from other parts of the world.

As illustrated in Fig. 2, China leads the way in releasing 5G articles, with the USA, United Kingdom, Germany, and South Korea closely following. China alone contributes 25% of all 5G publications, the USA contributes 15.5%, the United Kingdom contributes 8.1%, and Germany with South Korea contribute 6.3% each. Collectively, the top 10 countries make up 69% of all 5G publications. The substantial publication outputs from China and South Korea emphasize Asia's significant research contribution advantage, particularly in terms of publication volume.

The breakdown of 5G publications by country can be summarized as follows:

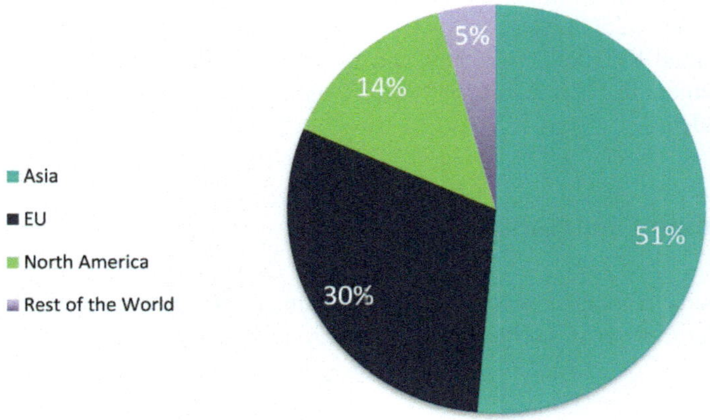

Fig. 1 The distribution of significant 5G funding organizations, each having funded at least five publications, can be examined from a geographical standpoint. This analysis provides insights into the regional concentration of funding efforts in the 5G domain [14]

Fig. 2 5G publication percentage by county [14]

- China takes the lead in 5G publications, with a significant number of research articles in this field.
- Following China, the United States, United Kingdom, Germany, and South Korea also make substantial contributions to 5G publications.
- China accounts for approximately a quarter of all 5G publications, indicating a strong research output in this country.
- The United States holds the second position, contributing significantly to the global body of 5G research.
- The United Kingdom, Germany, and South Korea also demonstrate noteworthy involvement in 5G research, with each country making valuable contributions to the field.

3 German Contribution

This section covers the key local German organizations involved in 5G deployment and standardization in Germany and the EU, as well as Germany's 5G spectrum strategy.

3.1 Local German Organizations' Contribution

The earliest relevant organizations and regulatory bodies involved in 5G deployment and standardization in Germany are listed as the following:

1. Deutsche Telekom: Deutsche Telekom, one of Germany's major telecommunications providers, has been actively engaged in 5G standardization. The company has contributed to defining technical requirements and use cases for 5G networks, as well as participating in trials and testing of 5G technologies [15].
2. VDE/DKE: The German Association for Electrical, Electronic & Information Technologies (VDE) and its division, the German Commission for Electrical, Electronic & Information Technologies (DKE), play a significant role in standardization activities related to 5G in Germany. They collaborate with international standardization bodies and contribute to developing standards for 5G technologies.
3. Industrial Collaborations: Germany has emphasized the importance of 5G in enabling Industry 4.0 and has fostered collaborations between industry and academia to drive 5G standardization efforts. For example, initiatives like the 5G Alliance for Connected Industries and Automation (5G-ACIA) bring together various stakeholders to define the requirements and standards for 5G in industrial applications [16].
4. Spectrum Allocation: The German regulatory authority, the Federal Network Agency (Bundesnetzagentur), is responsible for spectrum allocation and management [17]. It has been involved in ensuring the availability of suitable frequency bands for 5G deployment in Germany, including both sub-6 GHz and millimeter-wave (mm Wave) frequencies.

3.2 EU and German 5G Spectrum Strategy

The European Union (EU) has unveiled its strategy for the 5G spectrum and is actively pursuing opportunities for 5G deployment. On November 10, 2016, the Radio Spectrum Policy Group (RSPG) of the European Commission announced the European 5G spectrum strategy, delineating the planned allocation of frequency bands for 5G deployment across Europe. The strategy prioritizes the "3400–3800 MHz" band as the key frequency range for 5G deployment before 2020, aiming for extensive coverage using the "700–1000 MHz" band. For higher-frequency bands, the "26 GHz (24.25–27.5 GHz)" band has been designated for the initial deployment of 5G in Europe to meet the market demand for 5G by 2020. Additionally, the European Union will conduct further investigations into the utilization of other high-frequency bands, such as the "32 GHz (31.8–33.4 GHz)" and "40 GHz (40.5–43.5 GHz)" bands, to support the future development of 5G technology [18].

The German 5G spectrum strategy encompasses four frequency bands, covering high, middle, and low frequencies. On July 13, 2017, Germany revealed its national

5G strategy, which includes detailed plans for spectrum allocation across these bands. The "2 GHz" band, spanning frequencies "1920–1980 MHz" and "2110–2170 MHz," is currently allocated for 3G services in Germany. However, with the expiration of the existing license in 2025, Germany aims to repurpose these frequencies as the operational band for 5G. The "3.4–3.8 GHz" frequency band, predominantly used for mobile communication services, underwent an auction in June 2015. The next step involves allocating the "738–753 MHz" range within this band for 5G usage as Supplemental Downlink (SDL). In the case of the "26–28 GHz" bands, Germany has made a distinct decision in contrast to the European Union. It has chosen to utilize the 28 GHz band for 5G, specifically employing the frequencies "27.8285–28.4445 GHz" and "28.9485–29.4525 GHz." However, Germany has not entirely excluded the 26 GHz band and will continue to use it for research purposes alongside the selected 28 GHz band [18].

3.3 5G Connected and Automated Driving

To realize the vision of Cooperative, Connected, and Automated Mobility (CCAM) across Europe, it is crucial to implement coordinated solutions that streamline cross-border traffic. There are substantial opportunities for innovative businesses to provide CCAM services as vehicles move across national borders. Nonetheless, addressing technical challenges to ensure smooth connectivity and continuous delivery of real-time services across borders can be achieved through the adoption of 5G technologies.

The complexity of the situation is amplified by the participation of numerous countries, operators, telecommunications vendors, car manufacturers, and diverse generations in any cross-border setup. Recognizing this intricate landscape, the 5GCroCo project has been initiated with a total budget of 17 million euros, partially funded by the European Commission. Its primary objective is to evaluate the effectiveness of 5G technologies in the cross-border corridor connecting Metz, Merzig, and Luxembourg, which spans the borders of France, Germany, and Luxembourg [19].

The validation efforts of the 5GCroCo project will primarily focus on three use cases:

1. Tele-operated driving: This use case involves the remote operation of vehicles, utilizing 5G connectivity to ensure safe and efficient mobility across borders. High-definition map generation and distribution for automated vehicles: 5G technologies will be employed to create and disseminate detailed maps that cater specifically to the needs of automated vehicles, ensuring precise navigation across national boundaries.
2. Anticipated Cooperative Collision Avoidance: Through the utilization of 5G capabilities, this use case aims to enhance safety by enabling cooperative collision avoidance systems that proactively anticipate and prevent potential accidents.

The outcomes of the 5GCroCo project's validation activities will play a crucial role in minimizing uncertainties associated with cross-border eV2X (everything-to-ever thing) communications in Europe. These efforts serve as a preparatory step toward the commercial deployment of 5G technology, ensuring its successful integration into cross-border CCAM operations [19].

4 Industry 4.0 and 5G Integration

Industry 4.0 revolutionizes manufacturing by connecting various devices through data interaction and Cyber-Physical Systems (CPS), creating a unified entity that encompasses both the internal and external aspects of a factory. This integration aims to achieve decentralized manufacturing, shifting from centralized control to decentralized enhanced control. Decentralized production, facilitated by Industry 4.0, offers greater flexibility than traditional assembly line automation methods [20].

Smart factories represent more than mere automation. They are dynamic systems capable of optimizing the performance of the entire network, swiftly adapting to and learning from new environmental conditions in real-time or near real-time. These factories can automatically manage the complete production process and are not limited to the factory floor. They continuously evolve towards a seamless integration with the Internet, incorporating objects, data, and services. Media reports indicate that Huawei has engaged in partnerships with manufacturing companies in Germany within the realm of intelligent manufacturing [21]. One such collaboration involved Festo, focusing on developing a cloud robot project based on the 5G slicing network. The project aimed to assess the closed-loop control system's real-time performance and high reliability by leveraging the capabilities of the 5G ultra-reliable low-latency communication (uRLLC) slicing network. View Fig. 3.

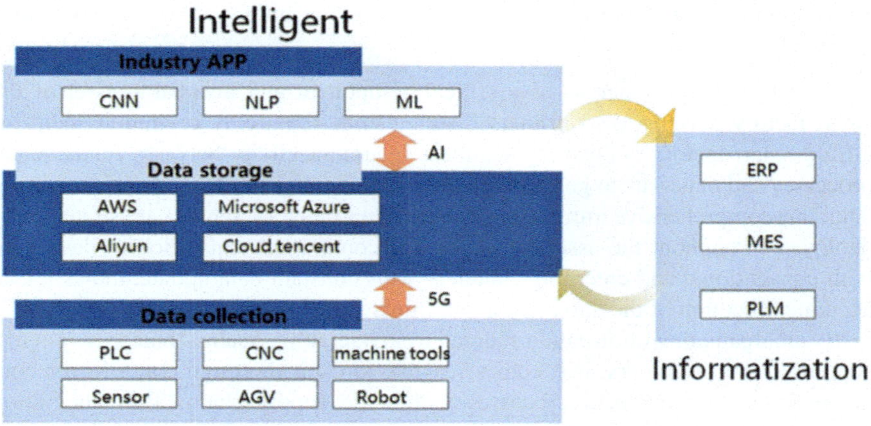

Fig. 3 Smart factory-based uRLLC [18]

With the aid of 5G technology, industrial robots have transcended the confines of glass cabinets and now traverse the workshop autonomously day and night. They undertake various tasks, including equipment inspection and repair, feeding, quality inspection, and complex production operations. These robots have assumed the roles of mid-level and grassroots managers, coordinating production and making informed decisions through information processing and precise judgment. They have become invaluable assistants to humans, taking on challenging tasks that are typically beyond human capabilities. In this way, humans and robots can coexist harmoniously within factory environments like:

a. Internet of Things (IoT): The intelligent transformation of factories has propelled the significance of IoT as a critical technology for interconnecting individuals, machinery, and equipment. Enterprises are placing considerable emphasis on IoT, thereby fostering the advancement of 5G technology and facilitating the widespread adoption of IoT applications.
b. Industrial automation control: Within manufacturing facilities, industrial automation control represents a fundamental application wherein a closed-loop control system is pivotal. Leveraging 5G technology, it becomes feasible to establish wireless network connectivity for closed-loop control applications, thanks to its capabilities to provide extremely low latency, high reliability, and extensive network connectivity.
c. Logistics tracking: In the domains of warehouse management and logistics distribution, there is a need for technology that provides extensive coverage, minimal power consumption, widespread connectivity, and affordability. Moreover, achieving seamless integration of the virtual factory throughout the product life cycle requires linking widely dispersed sold goods through a network characterized by low power consumption, economical operation, and broad coverage [22]. Additionally, horizontal integration within and between enterprises calls for an omnipresent network, which can be effectively addressed by the capabilities of a 5G network.

The 5G network emerges as a critical component and invaluable asset of the smart factory, serving as its primary distinguishing feature. A key requirement for a truly smart factory is ensuring seamless interconnectivity between fundamental processes and materials to generate diverse and essential real-time decision-making data. In a comprehensive implementation of a smart factory, sensors are strategically deployed throughout the assets, enabling the continuous acquisition of data from both conventional and emerging channels. This constant data update allows for an accurate reflection of the current state of affairs [23].

By amalgamating information sourced from operating systems, business systems, suppliers, and customers, it becomes feasible to exert thorough control over both upstream and downstream processes within the supply chain. This integration empowers the intelligent factory to augment the overall efficiency of the supply network, thereby enhancing the overall performance of the entire supply chain, as depicted in Fig. 4.

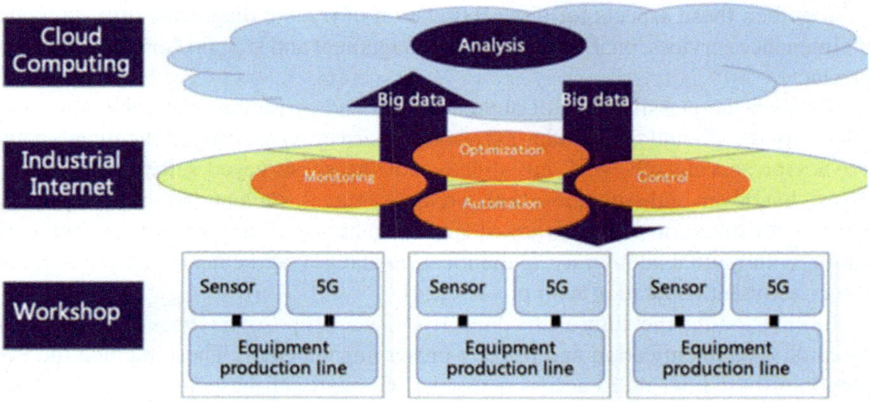

Fig. 4 Massive machine type communication (mMTC) amplifies the remote control capability for operating and maintaining equipment [18]

4.1 Remote Diagnosis, Operation, and Maintenance Services Key Aspects

The remote service for diagnosing, operating, and maintaining involves four fundamental components. These include [18]:

1. Creation of a standardized information acquisition and control system: This entails establishing a system for gathering and managing standardized information. It encompasses elements like automated diagnostic systems, expert system-driven fault prediction models, and a knowledge base containing fault indices.
2. Achievement of remote unmanned control for equipment (products): This facet facilitates the remote management of equipment and products without requiring physical presence. It encompasses functionalities such as early detection of the working environment, monitoring operational status, fault diagnosis, and self-repair capabilities.
3. Creation of product life cycle analysis platform: This platform facilitates the analysis of a product's entire life cycle. It includes sub-platforms for analyzing core accessories and models for capturing user usage habits. These components contribute to a comprehensive understanding of a product's life cycle and its optimization.
4. Provision of intelligent equipment (products) services: This aspect focuses on providing various services to enhance the functionality of intelligent equipment. It includes health monitoring, formulation and implementation of virtual equipment maintenance schemes, push for optimal usage schemes, and the facilitation of innovative applications.

Together, these aspects form the foundation of remote diagnosis, operation, and maintenance services, enabling efficient management and support for equipment and products.

Through the classification of authentication and privacy models, the authors of "Security for 4G and 5G cellular networks: A survey of existing authentication and privacy-preserving schemes" investigate categorized surveyed schemes into seven types: handover authentication with privacy, mutual authentication with privacy, RFID authentication with privacy, deniable authentication with privacy, authentication with mutual anonymity, authentication and key agreement with privacy, and three-factor authentication with privacy [24].

In envisioning the future of connectivity, we propose six open directions for research on authentication and privacy-preserving schemes. These include the Fog paradigm-based 5G radio access network, 5G small cell-based smart grids, SDN/NFV-based architecture in 5G scenarios, datasets for intrusion detection in 5G scenarios, UAV systems in 5G environments, and 5G small cell-based vehicular crowdsensing [24].

4.2 Enabling Functionalities Through Remote Diagnosis, Operation, and Maintenance Services

In 5G scenarios, the essential integration of AI, big data, and automation technologies is imperative to enable remote diagnosis, operation, and maintenance services. These technologies facilitate the following functions [18]:

1. Possessing open data interfaces: The system must incorporate capabilities for data collection, communication, and remote control. It leverages the industrial Internet, supporting both IPv4 and IPv6 technologies, to gather and upload equipment status, operational data, environmental conditions, and other relevant information. It should also be adaptable to remote instructions, allowing for flexible adjustment of equipment operation parameters.
2. Effective data management and service provision: The system efficiently screens, sorts, stores, and manages the uploaded equipment and product data. Through data mining and analysis, it offers users services such as daily operation and maintenance guidance, online detection, predictive maintenance, fault warning, diagnosis and repair, operation optimization, remote upgrades, and more.
3. Sharing information with management systems: The system accomplishes smooth information sharing with the product life cycle management system, customer relationship management system, and product development management system. This integration improves coordination and streamlines processes across diverse domains.
4. Expert libraries and consulting systems: The system establishes relevant expert libraries and expert consulting systems. These resources provide intelligent decision support for remote diagnosis of intelligent equipment and products. They

offer operation and maintenance solutions to users, leveraging the expertise of the system and its knowledge base.
5. An intelligent service system that is both efficient and secure is established through ongoing enhancements. The services offered facilitate real-time and effective interaction with products, thereby significantly elevating the integrated application level of embedded systems, mobile Internet, big data analysis, and intelligent decision support systems. Refer to Fig. 4 for details.

By integrating these capabilities, 5G-enabled remote diagnosis, operation, and maintenance services can effectively harness the power of AI, big data, and automation technologies to provide comprehensive and intelligent assistance to users.

5 Future of Industry 5.0 and 6G

An examination of previous research on technologies facilitating the implementation of 6G, spanning from 2020 to 2023, primarily concentrated on reviews. This survey specifically considered prominent academic databases such as IEEE Explore, Google Scholar, MDPI, Science Direct, Hindawi, and Springer. These databases are recognized for housing reputable and high-quality research reports, including those in journals, conference proceedings, and review articles. The search terms "6G" and "Industry 5.0" were employed to search these databases, resulting in the retrieval of a total of 35,400 English-language articles from the specified sources, 7800 of them being Reviewed articles.

In the realm of scientific discourse, the forthcoming 6G technology is poised to explore novel communication mechanisms, unencumbered by existing network paradigms or technological limitations. It incorporates entirely compatible new concepts, architectures, protocols, and solutions to underpin both current and prospective scenarios. Specifically, the 6G system encompasses intelligent connectivity, deep connectivity, holographic connectivity, and ubiquitous connectivity.

Intelligent connectivity permeates every facet of the communication system, encompassing the intelligence embedded within network elements, network architecture, connected objects (such as terminal devices), and the information transmitted to support intelligent services. Deep connectivity delves into profound sensing, learning, and cognitive capabilities. Holographic connectivity is characterized by pervasive communication at any location and time, high-fidelity interactions, and seamless coverage of augmented reality (AR), virtual reality (VR), and extended reality (XR). Ubiquitous connectivity establishes an all-encompassing, multi-dimensional coverage connection that transcends terrains and spaces [25].

The adoption of "Industry 5.0" technologies is becoming more widespread, and rather than diminishing the value of human contributions, it is poised to foster a synergistic integration of human and machine intelligence within a collaborative framework. "Industry 5.0" is anticipated to eliminate the boundaries between the physical and virtual realms, although the precise implications for businesses remain uncertain.

The collaboration between individuals and intelligent systems, such as robots, is expected to intensify in the context of "Industry 5.0," particularly within the industrial sector. At this level, mundane and repetitive tasks will be delegated to machines, allowing humans to channel their creative abilities and assume greater responsibilities in supervising systems to enhance overall production quality. This shift is facilitated by the growing demand for more intricate human–machine interfaces, amplified robot automation, and the intellectual prowess of human minds.

Industry 5.0, the latest industrial revolution, leverages human–machine interfaces for streamlined tasks and emphasizes personalization through virtual environments, cutting-edge technology, and IT. This integration of AI, Big Data, Cloud Computing, IoT, COBOTS, innovation, and creativity is expected to create high-value jobs, boost innovation, and enhance labour productivity. However, the challenge lies in workforce skill development due to highly automated systems. While Industry 5.0 raises cybersecurity concerns with its increased connectivity, it is projected to revolutionize manufacturing by fostering human–robot collaboration. In India, initiatives like Make in India, Skill India, and Start-up India aim to position the country as a leader in intelligent and cooperative production systems [26].

6 Conclusion

In summary, exploring 5G evolution and standardization has underscored its pivotal role in the global wireless communication landscape. The intricate standardization process, involving international organizations such as ITU, 3GPP, and IEEE, along with contributions from various regions, notably Asia, has shaped the trajectory of 5G. The key drivers of 5G development, including publications, patents, and standards, reveal Asia's dominance, particularly China, and highlight specific strategies of German entities aligning with Industry 4.0. Examining the EU's 5G spectrum strategy and Germany's allocation plans showcases concerted efforts to position Europe at the forefront of 5G deployment. The role of 5G in connected and automated driving, exemplified by the 5GCroCo project, emphasizes the importance of cross-border communication in achieving cooperative mobility. Furthermore, the integration of 5G in Industry 4.0 unveils transformative potential in various sectors, with practical applications in intelligent manufacturing showcased through partnerships between Huawei and German companies. In conclusion, the journey from 5G standardization to its integration in Industry 4.0 signifies a paradigm shift in connectivity and efficiency. Realizing the full potential of 5G necessitates collaborative efforts among nations, organizations, and regulatory bodies, with Germany's strategic contributions, particularly in Industry 4.0, positioning the nation at the forefront of this technological revolution. The analysis of 6G technologies from 2020 to 2023 indicates a transformative impact on communication. Industry 5.0, integrating human–machine interfaces and innovation, erases physical-virtual boundaries, especially in industry. Challenges in workforce skill development and cybersecurity persist, but initiatives like Make in India position the country as a leader in intelligent production systems.

References

1. Imam-Fulani YO et al (2023) 5G Frequency standardization, technologies, channel models, and network deployment: advances, challenges, and future directions. Sustainability 15(6). https://doi.org/10.3390/su15065173
2. Altaleb H, Zoltán R (2023) Addressing cybersecurity challenges in 5G-enabled IoT and critical infrastructures: a comprehensive overview. In: 27th IEEE international conference on intelligent engineering systems (INES), pp 000131–000136. https://doi.org/10.1109/ines59282.2023.10297774
3. Roh W et al (2014) Millimeter-wave beamforming as an enabling technology for 5G cellular communications: theoretical feasibility and prototype results. IEEE Commun Mag 52(2):106–113. https://doi.org/10.1109/MCOM.2014.6736750
4. Nguyen VG, Brunstrom A, Grinnemo KJ, Taheri J (2018) 5G Mobile networks: requirements, enabling technologies, and research activities. In: Comprehensive guide to 5G security, pp 31–57. https://doi.org/10.1002/9781119293071.ch2
5. Dahlman E, Parkvall S, Sköld J (2018) 5G NR: the next generation wireless access technology. [Online]. Available: https://wom1n.blogspot.com/2019/02/5g-nr-next-generation-wireless-access.html. Accessed: 13 May 2023
6. De Xie J et al (2020) Highly concentrated carbonate electrolyte for Li-ion batteries with lithium metal and graphite anodes. J Power Sources 450. https://doi.org/10.1016/j.jpowsour.2019.227657
7. Nordin J, Stünkel L (2022) EU-Taiwan semiconductor cooperation: lopsided priorities?
8. Rühlig T, Seaman J, Voelsen D (2019) 5G and the US-China tech rivalry—a test for Europe's future in the digital age. How can Europe shift from back foot to front foot? SWP Comment 29:1–9
9. Buggenhagen M, Blind K (2022) Development of 5G—identifying organizations active in publishing, patenting, and standardization. Telecommun Policy 46(4):102326. https://doi.org/10.1016/j.telpol.2022.102326
10. Shafi M et al (2017) 5G: a tutorial overview of standards, trials, challenges, deployment, and practice. IEEE J Sel Areas Commun 35(6):1201–1221. https://doi.org/10.1109/JSAC.2017.2692307
11. Johnson JP (2014) Defensive publishing by a leading firm. Inf Econ Policy 28(1):15–27. https://doi.org/10.1016/j.infoecopol.2014.05.001
12. Rauber J (2014) Connecting standardization, patenting and strategic publishing in the framework of firms' innovation and intellectual property management
13. Hawkins R, Blind K (2017) Introduction: unravelling the relationship between standards and innovation. In: Handbook of innovation and standards, pp 1–18. https://doi.org/10.4337/9781783470082.00007
14. Buggenhagen M, Blind K (2022) Development of 5G—identifying organizations active in publishing, patenting, and standardization. Telecommun Policy 46(4). https://doi.org/10.1016/j.telpol.2022.102326.
15. Lehr W, Queder F, Haucap J (2021) 5G: a new future for mobile network operators, or not? Telecommun Policy 45(3). https://doi.org/10.1016/j.telpol.2020.102086

16. Lei W et al (2021) 5G industrial IoT. In: Wireless networks (United Kingdom), pp 515–532. https://doi.org/10.1007/978-3-030-73703-0_9
17. Kokkinen T, Kokkinen H, Yrjola S, Kliks A (2020) A regulators perspective on digital twinning for mobile communications
18. Wang X, Gao L (2020) When 5G meets industry 4.0
19. Hetzer D, Muehleisen M, Kousaridas A, Alonso-Zarate J (2019) 5G Connected and automated driving: use cases and technologies in cross-border environments. In: 2019 European conference on networks communications (EuCNC), pp 78–82. https://doi.org/10.1109/EuCNC.2019.8801993
20. Morgan J, Halton M, Qiao Y, Breslin JG (2021) Industry 4.0 smart reconfigurable manufacturing machines. J Manuf Syst 59:481–506. https://doi.org/10.1016/j.jmsy.2021.03.001
21. Ryalat M, El Moaqet H, Al Faouri M (2023) Design of a smart factory based on cyber-physical systems and internet of things towards industry 4.0. Appl Sci 13(4). https://doi.org/10.3390/app13042156
22. Tan WC, Sidhu MS (2022) Review of RFID and IoT integration in supply chain management. Oper Res Perspect 9. https://doi.org/10.1016/j.orp.2022.100229
23. Saafi S, Fodor G, Hosek J, Andreev S (2021) Cellular connectivity and wearable technology enablers for industrial mid-end applications. IEEE Commun Mag 59(7):61–67. https://doi.org/10.1109/MCOM.001.2000988
24. Ferrag MA, Maglaras L, Argyriou A, Kosmanos D, Janicke H (2018) Security for 4G and 5G cellular networks: a survey of existing authentication and privacy-preserving schemes. J Netw Comput Appl 101:55–82. https://doi.org/10.1016/j.jnca.2017.10.017
25. Lu Y, Zheng X (2020) 6G: a survey on technologies, scenarios, challenges, and the related issues. J Ind Inf Integr 19. https://doi.org/10.1016/j.jii.2020.100158
26. Dharmadhikari S, Joglekar S, Kadam S (2023) Industry 5.0: analysis, applications and prognosis. Online J Distance Educ e-Learn 11(1)

As Many Organizations, as Many Compliance Solutions—Enforcement of Special Protection in Practice

Tímea Antal and Róza Számadó

Abstract I consider the financial and capital market sectors, the pharmaceutical industry and telecommunications compliance-sensitive sectors. For sectors with high risk, the operation is determined by numerous sectoral legislations, international standards, professional guidelines, and recommendations. In my presentation, I will analyze the publicly available compliance documents of OTP Bank Plc., Gedeon Richter Plc. and Magyar Telekom Plc. Long-term corporate and organizational success and compliance-based operations are closely linked. Legislation, professional and quality standards, and ethical rules can all contribute to ensuring that compliance works not only in principle but also in practice. Nowadays, resilience and the ability to adapt play a central role. The development of a compliance culture is of strategic importance and is intertwined with risk management and effective organizational management, which can lead to the practical application of compliance as a value.

Keywords Compliance · High risk · Best practices · Pharmaceutical industry · Telecommunication

1 Compliance is Everywhere

I consider the financial and capital market sector, the pharmaceutical industry and telecommunications to be compliance-sensitive sectors. For sectors with high risk, the operation is determined by numerous sectoral legislation, international standards, professional guidelines, and recommendations.

T. Antal (✉)
Doctoral School on Safety and Security Sciences, Obuda University, Budapest, Hungary
e-mail: timea.antal0@gmail.com

R. Számadó
Banki Donat Faculty of Mechanical and Safety Engineering, Obuda University, Budapest, Hungary
e-mail: szamado.roza@bgk.uni-obuda.hu

T. A. Kovács et al. (eds.), *The Impact of the Energy Dependency on Critical Infrastructure Protection*, Advanced Sciences and Technologies for Security Applications, https://doi.org/10.1007/978-3-031-78544-3_8

As uncertainty and instability continue to roil global markets, corporate risk and compliance teams face a wide range of challenges, from cost pressures and security concerns to ever-evolving regulatory changes, fraud prevention strategies, and more.

In my article, I will analyze the publicly available compliance documents of OTP Bank Plc., Gedeon Richter Plc. and Magyar Telekom Plc. Long-term corporate and organizational success and compliance-based operations are closely linked. Legislation, professional and quality standards, and ethical rules can all contribute to ensuring that compliance works not only in principle but also in practice. Nowadays, resilience and the ability to adapt play a central role. The development of a compliance culture is of strategic importance and is intertwined with risk management and effective organizational management, which can lead to the practical application of compliance as a value. Compliance activities could help as a cost-cutting measure for a long-term period: how best to manage emerging compliance risks and strategic priorities such as time management and specialization, and how best to pivot the team's role from check-the-box risk managers to one of strategic advisors, etc.

Risk management is the basis of an effective and efficient compliance system, i.e. the risky operational and activity areas of the organization must be mapped. In line with Jacsó's categorization, compliance risks include legal, regulatory or other official sanctions, financial loss, reputational damage, and loss of trust [1]. The compliance function aims to identify, mitigate and manage compliance risks.

Compliance is not an effective control tool on its own but is part of a more complex control framework comprising three lines of defence. Based on the Institute of Internal Auditors' concept of three lines of defence [2], the first line of defence is the departments responsible for ensuring business operations and carrying out operational tasks. The second line of defence includes compliance units with compliance and risk expertise proactively monitoring risk factors and supporting the first line of defence. A key compliance area is the achievement of risk management objectives such as compliance with laws, regulations, and the definition of acceptable ethical conduct, as well as supporting internal governance, information and technology security, sustainability, and quality assurance.

The first two lines of defence can only perform their functions effectively if they are both based on a risk-based approach at the organizational level. The third line is independent of the first and second lines, with a reactive role for internal control linked to risk. The governing body of the organization oversees compliance with legal, regulatory and ethical requirements, taking into account compliance aspects, while the management ensures that these are put into practice. The work of the entire organization is supported by external assurance providers such as auditors.

The compliance function, therefore, performs both control and support functions:

- identifying and managing risks and conflicts of interest;
- support the proper functioning of the organization;
- taking measures and making proposals to address shortcomings;
- spread awareness of compliance at the internal and organizational levels;
- reporting to senior management;
- preventing corruption;

- managing conflicts of interest;
- respecting restrictions on the flow of information;
- preventing market manipulation (insider trading, unfair exchange rate manipulation);
- combating money laundering and financing of terrorism;
- preventing external and internal fraud;
- confidentiality and data protection (commercial, banking, securities and insurance confidentiality, protection of personal data),
- ensure transparent operation;
- ensuring fair treatment of customers, complaint complaint-handling activities;
- maintaining reputation and goodwill;
- carry out checks;
- liaising with the authorities.

According to the survey of Occupational Fraud 2022: A Report to the Nations [3]—the study targeted organizations in 23 distinct industry categories in 133 countries, based on 2110 cases of occupational fraud—banking and financial services, health care, and information sectors are in the top ten list of victim organization's industry. The industry affected by the greatest number of fraud cases was banking and financial services. Fraud doesn't happen in a vacuum; organizations often enact specific internal controls designed to prevent and detect any attempted wrongdoing.

2 Financial and Capital Sector

Financial compliance ensures the economic system's stability by compelling institutions to operate ethically and responsibly. This safeguards the industry's integrity and protects investors and consumers from fraud and unfair practices.

The financial sector is a high-risk sector, the operation of which is governed by a number of sectoral laws, international standards, professional guidelines and recommendations. The need for compliance is strong in this sector. In addition to anti-money laundering, combating terrorist financing, anti-corruption, ethical issues, conflict of interest regulation, market manipulation investigations, and investor and consumer protection are essential compliance areas. The financial organization is expected to develop and operate internal lines of defence which facilitate the financial organization.

(a) prudent, reliable and efficient, in accordance with the laws and internal regulations, proper functioning,
(b) assets, owners and customers related to the financial organization, the protection of its interests and social goals,
(c) smooth and effective operation, trust in the financial organization maintenance.

The most important task of compliance is to be preventive and proactively contribute to the fulfilment of these goals by, during operation, potential problems and deficiencies in the earliest possible phase, as early as they are identified and treated when they occur, but if possible even before that, thereby ensuring the speed and efficiency of the solution.

Typical tasks addressed in this sector:

- combating money laundering and terrorist financing;
- anti-corruption;
- managing conflicts of interest;
- the separation of financial and investment service activities;
- respecting the restrictions on the flow of information;
- preventing market manipulation (insider trading, exchange rate manipulation);
- preventing external and internal fraud;
- confidentiality and data protection (commercial, banking, securities and insurance confidentiality, protection of personal data);
- the financial organization, group members and employees' own-account transactions;
- ensuring fair treatment of customers, including complaint handling;
- maintaining a good reputation and goodwill.

Moreover, compliance plays a pivotal role in upholding the reputation of financial institutions. A compliance breach can lead to negative consequences, including substantial fines, legal repercussions, and severe damage to an institution's reputation.

Recommendation 12/2022 (VIII.11.) of the National Bank of Hungary on the establishment and operation of internal lines of defence and on the management and control functions of financial institutions was a significant progress, which sets out obligations for financial institutions in five areas. From 1 January 2023, it is mandatory to establish a fraud prevention and management culture and internal audit trail, and to review the risk management and compliance function procedures and related policies, taking into account conflicts of interest regarding close relatives. From 1 January 2014, the designation of a governing body/senior management member responsible for compliance with the requirements related to the prevention of money laundering and terrorist financing [4] is mandatory.

Best Practice—Compliance Policy of OTP Bank Plc [5]

The definition of compliance risk is the potential legal risk, the risk of supervisory or other official sanctions, of significant financial losses, or of reputational damage due to a failure to adhere to legislation or other non-legislative standards and internal rules applicable to the financial organization and pertaining to its service activities.

Compliance policy is the framework of 7 programmes, including a Data Protection Policy, Anti-Corruption Policy, Social Media Policy, Consumer Protection Compliance Program, Internal lines of defence—ESG, Sanctions Policy, and Policy on Financing of the Defence Industry.

The Compliance function covers the Bank as a whole, including all of its organizational units and activities, and everybody must meet compliance requirements and standards as well. The Bank's internal lines of defence comprise the responsible internal governance and internal control functions. The internal control functions include three functions: risk control, compliance, and internal audit function. The policy determines the principles of operating the compliance function to create a lawful and ethical corporate culture that ensures the prudential and ethical operation of the Bank in the long term:

- independence
- integrity
- operation without interference objectivity
- preventive and proactive approach
- risk-based approach
- proportionality
- high level of professional care and competence
- full coverage
- efficiency, and rationalization of compliance costs.

Specific provisions specify the critical areas of compliance and the general principles and requirements (Table 1).

3 Health Care

In the healthcare sector, the focus is on protecting patients, and clear and unambiguous information is essential, particularly in view of the information and knowledge gaps that patients have. An important principle is to respect doctors' therapeutic freedom, avoid undue influence, ensure that the promotion of medicines complies with the law, and ensure fair market conduct. It is a highly regulated industry with a high level of regulatory oversight and significant industry self-regulation. Sector-specific compliance risks may include legal transparency in interactions with healthcare professionals, communication, information and advertising about medicines, and generally, health data impose great risks in this sector.

Typical tasks addressed in this sector:

- anti-corruption;
- managing conflicts of interest;
- respecting the restrictions on the flow of information;
- preventing external and internal fraud;
- confidentiality and data protection;
- ensuring fair treatment of customers, including complaint handling;
- maintaining a good reputation and goodwill.

Best Practice—Compliance Policy of Gedeon Richter [6]

Long-term thinking and continuous innovation are stable building blocks of Gedeon Richter.

Ethics and compliance are part of sustainable business growth and value creation. Conducting business in a responsible manner is an essential part of Richter's strategy.

Table 1 Key areas of compliance—messages (own editing)

Key areas of compliance	Message: principles and procedural mechanism
1. Processing and protection of personal data—GDPR	System for regulation, implementation and auditing that provides for the adequate protection of personal data by meeting the criteria specified in applicable legislation and safeguarding its core business interests
2. Conflict of interest, ethics, prohibition of corruption and bribery	Identifying clear principles and policies to prevent and manage disputes, whistleblowing system in practise
3. Compliance with restrictions on information flows between financial and investment service activities	Internal organizational, operational and procedural mechanisms to ensure that the data and information flow
4. Prevention of market abuse (insider dealing, unfair price manipulation)	A system for carrying out analyses and examinations of such incidents
5. Fair treatment of customers, consumer protection	Enforcement of consumers' interests
6. Ensuring compliance with the requirements pertaining to investment service activities as set out in legal regulations and regulatory—particularly supervisory—standards	Providing investment services and personal transactions
7. Corporate governance (ESG)	Evaluating its activity from the aspect of the activity's environmental impact (E), social fairness (S), and the related corporate governance issues (G), as well as its compliance with the relevant legislative requirements
8. Compliance with international tax agreements	Full compliance with the customer identification and reporting requirements set out in international and local tax arrangements
9. Compliance with the requirements of international sanctions and mitigation of risks associated with sensitive transactions	Group-level sanctions policy on its general principles of the application of international sanctions, and a group-level policy on financing of the defence industry. System for market position and good reputation, supporting international collaboration and complying with legal regulations

It generates value that can, on the one hand, be reinvested in the business and, on the other hand, provide wider societal benefits since healthy people and communities are essential to building healthy societies. Gedeon Richter's Compliance Program is a global program that covers Europe and Latin America. Sustainability is the key message. The code of ethics is the fundamental document of compliance policy. At the same time, a complex regulation system is developed:

- Anti-corruption manual

- Manual on interactions with healthcare professionals
- Pharma law manual
- Transparency regulation
- Competition law compliance manual
- Corporate Communications Manual
- Website content manual
- Conflict of interest policy
- Regulation on the protection of trade secrets.

The Code of Ethics determined a very practical manual for the employees and partners of both Gedeon Richter Plc. and its affiliates:

- Is it legal?
- Is it ethically sound?
- Is it in line with Richter's values?
- Would your line manager or the Company's CEO agree that this is the right decision?
- Would you and your coworkers be at ease if it was widely publicized?

These questions could guide everybody to exercise their discretionary powers responsibly.

The special key areas of compliance at Richter are research and development and the protection of IP rights, quality requirements, general rules on providing product information, and compliance with trade control rules. The Code of Ethics emphasizes protecting Richter's reputation, Richter and the community. It is obvious that sector-specific rules and global aspects are the main framework of ethics and compliance. I think compliance culture is based on the regulation–ethics–compliance triangle at Richter.

4 Telecommunication Sector

The focus of compliance is on anti-corruption, ethical issues, and internal communication. Typically, an integrated management system is applied instead of sectoral compliance standards. Compliance tends not to be an independent organizational unit. It usually functions in conjunction with other areas (e.g. legal and risk management).

Typical tasks addressed in this sector:

- anti-corruption;
- managing conflicts of interest;
- respecting the restrictions on the flow of information;
- confidentiality and data protection;
- ensuring fair treatment of customers, including complaint handling;
- maintaining a good reputation and goodwill.

Best Practice—Compliance Policy of Telekom [7]

Magyar Telekom Corporate Compliance Program has been designed to ensure that the Group conducts its business to the highest standards of awareness, training, commitment, and adherence to applicable laws and regulations. The implementation of compliance policy is based on mechanisms for reporting, investigating, monitoring, and correcting cases of suspected or actual non-compliance.

Parts of compliance: Anti-corruption measures, training system, effective system of internal controls, monitoring of the Compliance Program, external verification and certification, internal communication with employees—Ask me portal. I would highlight e-learning programs on business ethics and anti-corruption compliance, case management and the internal communication system. These tools could help to raise awareness, realize prevention, detect and respond to risk.

Compliance Messages Based on the 3 Best Practices

The following organizational governance steps are recommended to follow:

1. Specific regulations and tasks must be considered.
2. Review of the operating model based on the strategy, processes, tasks, and controls in the light of risks, i.e. applying a continuous plan-do-check-act logic.
3. Establishment and operation of a complex monitoring system.
4. Develop internal rules: codes of conduct (ethics), procedures and data protection.
5. Introduction of new ICT solutions such as automated business processes, digital risk assessment, and testing procedures.
6. Providing training to staff members, and acknowledging talents to promote proactivity.
7. Appropriate levels of authorization and decision-making powers need to be defined.
8. Ensuring a permanent supportive presence.
9. Building a network of helpers: compliance ambassadors, professional communities.
10. Use of ad hoc and systematic tests.
11. Identify and present best practices to the organization.

Based on the above, it can be stated that without a compliance function within the organization, business success cannot be achieved in the long term, and its role is becoming increasingly important in almost all sectors. The development of a compliance culture and the application of a risk management system can represent additional operational support tools that can significantly contribute to the development and maintenance of a resilient operation.

5 Summary

In this paper, I summarised the concept of compliance and outlined its place in practice. I described some compliance-sensitive sectors, including the financial and capital market sector, the pharmaceutical industry and telecommunications. Long-term corporate and organizational success and compliance-based operations are closely connected. Legislation, professional, quality standards, and ethical rules can all contribute to compliance not only in principle but also in practice. All three compliance programmes are based on six topics: GDPR, ethical rules, restrictions on the flow of information, market conduct, fair treatment of customers, anti-corruption, and sector-specific rules are typical. The OTP's compliance framework is part of the internal control functions, Richter's compliance policy is part of a global, ethical-based program, and Telekom's compliance system is focused on multi-level knowledge sharing.

The aim is to prevent compliance risks in the organization in the framework of a complex model. There are "as many compliance solutions as there are organizations", i.e. the specific industry, typical processes, risks, organizational structure, etc., are pillars on which to build a compliance structure commensurate with the risks involved. A compliance culture does not bring immediate benefits in the short term; however, establishing a resilient organization provides a competitive advantage and a good reputation. In the twenty-first century, flexibility and adaptability play a central role. The development of a compliance culture is of strategic importance and is intertwined with risk management and effective organizational governance, which can lead to the practical application of compliance as a value.

References

1. Jancso J (2019) A pénzmosás compliance a hazai és európai dimenzióban a társadalmi innováció tükrében (Money laundering compliance in the domestic and European dimension in the light of social innovation). Miskolc Law Rev 397:397
2. The Institute of Internal Auditors (2023) The IIA's three lines model, an update of the three lines of defence. https://www.theiia.org/globalassets/site/about-us/advocacy/three-lines-model-updated.pdf. Last accessed 25 Apr 2023
3. Association of Certified Fraud Examiners (2023) Occupational fraud 2022: a report to the nations, p 32. https://legacy.acfe.com/report-to-the-nations/2022/. Last accessed 31 Oct 2023
4. Recommendation No. 12/2022 (VIII.11.) of the Hungarian National Bank on the establishment and operation of internal lines of defence, the management and control functions of financial institutions. https://www.mnb.hu/letoltes/12-2022-belso-vedelmi-vonalak-ajanlas.pdf. Last accessed 25 Apr 2023
5. OTP Bank Plc (2023) EXTRACT from the compliance policy of OTP Bank Plc. https://www.otpbank.hu/static/portal/sw/file/OTP_Extract_from_Compliance_Policy_2022.pdf. Last accessed 31 Oct 2023
6. Gedeon Richter. https://www.gedeonrichter.com/en/sustainability/ethics-and-compliance. Last accessed 31 Oct 2023
7. Magyar Telekom's compliance principles. https://www.telekom.hu/about_us/about_magyar_telekom/principles/compliance. Last accessed 31 Oct 2023

Methodology for Risk Analysis in the Design Phase of High-Rise Building

Ahmmad Bakhtyar Saleh

Abstract The design phase of high-rise buildings is a critical stage of construction and affects the entire life cycle of the project. As the demand for high-rise structures increases, the complexity of design and construction tasks increases. This article reviews a comprehensive risk analysis methodology for the design phase of high-rise buildings, emphasizing the importance of early risk identification, stakeholder collaboration, and effective risk mitigation strategies. The study integrates both qualitative and quantitative approaches to risk assessment, drawing on the extensive literature on construction risk management. He recognizes the dynamic nature of the construction industry and advocates adaptable methods tailored to high-rise projects. The article contributes to the evolving discourse on risk management and focuses on the unique challenges associated with the design of high-rise buildings.

Keywords High-rise buildings · Risk analysis · Safety

1 Introduction

The design phase of high-rise buildings is a critical juncture in the construction process, where decisions made can reverberate throughout the entire project lifecycle. As the demand for towering structures continues to rise, so does the complexity of challenges associated with their design and construction. Effective risk management during this phase is paramount to ensure the structural integrity, safety, and financial viability of these ambitious architectural endeavours [1].

Risk analysis in the design phase involves the systematic identification, assessment, and mitigation of potential risks that may impact the project. This proactive approach allows project stakeholders to anticipate challenges, make informed decisions, and implement strategies that enhance the resilience of high-rise buildings. In recent years, the field of construction risk management has seen a growing emphasis

A. B. Saleh (✉)
Doctoral School on Safety and Security Sciences, Obuda University, Budapest, Hungary
e-mail: bakhtyar.ahmman@phd.uni-obuda.hu

T. A. Kovács et al. (eds.), *The Impact of the Energy Dependency on Critical Infrastructure Protection*, Advanced Sciences and Technologies for Security Applications, https://doi.org/10.1007/978-3-031-78544-3_9

on integrating robust methodologies specifically tailored to the unique characteristics of high-rise structures [2].

This article seeks to explore and elucidate a comprehensive methodology for risk analysis during the design phase of high-rise buildings. Drawing upon insights from a rich body of literature on construction risk management, this research aims to provide a structured framework that integrates both quantitative and qualitative approaches to risk assessment. The significance of early risk identification, stakeholder collaboration [3], and the development of effective risk mitigation strategies will be highlighted throughout the discussion.

As we embark on this exploration, it is crucial to recognize the dynamic nature of the construction industry and the need for adaptable methodologies that address the unique challenges posed by high-rise building projects. Through a comprehensive examination of existing literature and practical insights, this article aims to contribute to the evolving discourse on risk management in the realm of high-rise building design [4].

2 Overview

In the realm of high-rise building design, the importance of thorough risk analysis cannot be overstated [5]. The towering nature of these structures introduces complexities that necessitate a proactive and comprehensive approach to risk management, encompassing structural integrity, safety, financial viability, and regulatory compliance [1]. Delays, unforeseen challenges, and design modifications can have profound financial implications, impacting the economic feasibility of projects [2]. Additionally, stringent building codes and regulatory standards require meticulous attention to compliance to avoid legal consequences and project disruptions [4]. Beyond financial considerations, the ambitious timelines associated with high-rise construction demand effective risk analysis to identify and mitigate potential delays, ensuring the timely completion of projects [3]. Moreover, the prominence of high-rise buildings in urban landscapes underscores the importance of public perception and reputation management, emphasizing the need for risk analysis to maintain a positive image [5]. The integration of advanced technologies, such as Building Information Modeling (BIM), further contributes to risk analysis by providing a holistic understanding of the building's behaviour, facilitating proactive mitigation strategies [6]. Overall, risk analysis in high-rise building design emerges as a multidimensional process crucial for project success and the sustained growth of the construction industry.

3 Qualitative and Quantitative Methods of Risk Analysis

The progression to manage risks in projects involves the crucial step of risk analysis. The fundamental objective is to scrutinize identified risks in a manner that leverages insights into existing critical events, their associated risks, and their impacts on subsequent project stages. This analytical process aims to inform the selection of promising concepts. The assessment of risks can manifest either qualitatively or quantitatively. It is pivotal to consider both the nature and the numeric aspects of risks in order to develop a comprehensive understanding that guides decision-making in project development.

3.1 The Quantitative Risk Assessment

In quantitative risk assessment, the focus lies in defining the likelihood of critical events and their subsequent impacts within the project, serving as a crucial factor for informed decision-making. Throughout the risk assessment process, it is valuable to consider the varied consequences within the project, such as potential time extensions, cost escalations, or compromises in quality and performance [7]. Identifying situations where critical events may exert influence on other components is essential to preemptively avert a chain reaction that could impede project progress.

Numerous methods are available for risk analysis, contingent on the risk level and project stages. While different methodologies may yield diverse formats of results, the underlying risk level remains constant, irrespective of the chosen method. The decision of which method to employ is not a straightforward one. Commencing with simpler risk analysis methods and progressively incorporating more advanced approaches when necessary is considered an optimal approach [7]. This approach allows for a gradual and strategic development of the risk analysis process tailored to the specific needs and complexities of the project.

3.2 Probability Analysis

Probability analysis involves determining the likelihood of critical events, requiring specialized calculations tailored to the project's complexity, the nature of the risk, and the conditions under which critical events may occur. In instances such as the conceptual design phase, quantifying risks proves challenging as numerical calculations are often insufficient. Instead, judgments about probability lean more on intuition and past experiences than on strict mathematical formulas. However, for the analysis of common technical risks in structures, such as the risk of explosions or crack formation in concrete structures, specialized computer programs come into play. These programs contribute precision to the assessment process, particularly where numerical analysis is more applicable.

3.3 Monte Carlo Technique

The Monte Carlo technique earns its name from its emulation of the randomness found in a roulette wheel, and it stands out as one of the most widely used methods in probability analysis. This approach involves a random data development process, selecting various values from a range predefined by a probability distribution. For each uncertain variable (input), a range is determined, considering the statistical properties of that variable. A selected value is then subjected to a model through an iterative process, with the results presented in histograms.

Selecting a representative probability distribution is crucial, taking into consideration both the range and the expected variations in variable values. The choice of the probability distribution shape often relies on historical data. However, when historical data is unavailable, the decision regarding the shape of the probability distribution depends on the judgment of risk analysts and other stakeholders involved in the risk analysis and the project.

3.4 Qualitative Risk Assessment

Qualitative risk assessment relies on various factors that delineate the essence of risks. It is essential, at this juncture, to contemplate the project stages during the manifestation of a critical event. Elements of the project susceptible to impact, factors influencing the risk, relationships with other risks, the probability of occurrence, and the resultant consequences on the project all warrant consideration. The comprehension of the timing of critical events and the appropriate allocation within the project is facilitated by the aforementioned considerations, leading to more informed decision-making.

3.5 The Delphi Method

The Delphi method can be regarded as either a qualitative or quantitative approach to risk assessment. The general procedure involves assembling a team of experts representing various facets of the project. This team formulates a definition of the risk under consideration. Each expert is then tasked with individually assessing the risks, estimating their probability of occurrence, gauging their impact on different phases of the project, and optionally, incorporating cost estimations into the analysis. Subsequently, the experts are apprised of all estimates and requested to provide revised assessments. This iterative process continues until a consensus estimate is reached. It is imperative for the team of experts to be interdisciplinary to ensure objectivity and valuable results.

Given that the Delphi method is inherently subjective, relying on the opinions of presumed experts, caution is warranted in interpreting the solutions. The procedure is time-consuming, requiring each expert to provide multiple individual analyses. Although consolidating meetings may expedite the process, scheduling difficulties arising from varying expert availabilities could pose a challenge. Additionally, the presence of project team members during estimation is advisable for hearing expert opinions and providing necessary comments. This method is particularly useful when a company has limited experience in implementing similar projects.

4 Relevance of Risk Analysis in the Design Phase

The relevance of risk analysis in the design phase of any construction project, particularly in the context of high-rise buildings, cannot be overstated. This phase serves as the bedrock upon which the entire project is built, and decisions made during design have far-reaching implications for both the construction process and the eventual functionality of the structure. Risk analysis at this early stage is crucial for several reasons:

Proactive Problem Identification: Risk analysis allows project stakeholders to proactively identify potential issues and challenges that may arise during the construction and operation of the high-rise building. By anticipating these risks, designers can develop strategies to mitigate or eliminate them before they escalate, preventing costly issues later in the project lifecycle [8].

4.1 Cost Containment

The design phase is where the majority of project costs are established. Effective risk analysis helps identify potential cost overruns, allowing for the development of realistic budgets and financial plans. This proactive approach aids in cost containment, preventing unexpected financial burdens during construction [9].

4.2 Timeline Management

Delays in construction timelines often have cascading effects on project schedules and budgets. By conducting risk analysis during the design phase, project managers can identify potential sources of delay and implement strategies to keep the project on track. This is especially critical for high-rise buildings, where timely completion is often a key project objective [10].

4.3 Safety and Structural Integrity

High-rise buildings have unique structural challenges due to their height and complexity. Risk analysis during the design phase is essential for identifying potential structural vulnerabilities and safety concerns. Addressing these issues early ensures that the final design meets stringent safety standards and building codes [11].

4.4 Regulatory Compliance

Construction projects, especially high-rise buildings, must comply with a myriad of regulations and building codes. Risk analysis helps in identifying regulatory risks, ensuring that the design aligns with all legal requirements. This not only prevents legal complications but also contributes to the overall sustainability and longevity of the structure [12].

The relevance of risk analysis in the design phase lies in its ability to foresee and manage potential challenges before they materialize. This proactive approach enhances the overall success of the project, from its financial aspects to safety and compliance, ensuring that high-rise buildings are not only iconic but also structurally sound and economically viable.

5 Risk Analysis in the Design Structural Reinforcement Concrete of High-Rise Buildings

Constructing tall buildings involves intricate engineering processes that require careful attention, particularly in reinforcing structures with concrete. Ensuring the safety and stability of these towering structures is crucial. Risk analysis in designing structural reinforcement for tall buildings involves examining key factors, methodologies, and its critical role in reducing potential hazards. Reinforcing structures in tall buildings is a crucial construction element as it directly impacts the building's ability to withstand forces like wind, seismic activity, and vertical loads. Reinforced concrete, a widely used material due to its strength and durability, combines steel and concrete. Examining risk factors in the design of tall buildings includes assessing various elements that can pose potential hazards. This involves a comprehensive analysis of key factors and methodologies to ensure the safety and stability of these structures, significant basic of risks are:

5.1 Structural Loading and Gravity

In the field of high-rise building design, the profound influence of structural loading, in particular self-loading, which includes the self-weight of materials, and the ubiquitous force of gravity, is of paramount importance. As buildings rise to great heights, vertical gravitational pull becomes an increasingly dominant force. The dead load, consisting of the self-weight of building components such as concrete and steel, requires careful consideration when designing foundations and structural systems. The challenge is to effectively manage the vertical distribution of loads to ensure stability while combating the inexorable influence of gravity. In addition, the self-weight of materials requires a judicious selection of building materials with high strength-to-weight ratios to optimize structural integrity while minimizing overall static loads. This complex interaction between structural load and gravity is the cornerstone of high-rise building design, where engineers must find a delicate balance to withstand the unrelenting force of gravity and ensure the stability and safety of these tall structures [13].

5.2 Earthquake Considerations

Seismic considerations are an important aspect of the design of high-rise buildings. Structural engineers must assess the seismic risk in the region and implement strengthening strategies to ensure that the building can withstand ground movement without compromising its integrity. Seismic resistance requirements for high-rise buildings include strength and deformation requirements. The design of high-rise buildings in seismically active regions varies greatly from region to region, and best practices for the construction of high-rise buildings are not represented by traditional design codes such as the Uniform Building Code or its successor, the International Building Code.

Seismic hazard assessment is an important step in the design process. It includes the determination of response spectra, site response, and selection and modification of earthquake history for response analysis. Foundation effects such as geotechnical parameters, rocking and movement, soil-structure interaction, and foundation burial must also be considered.

Structural analysis and modelling procedures are also critical in the design of high-rise buildings. Elastic analysis and nonlinear response history analysis are two common methods used for structural analysis. Energy dissipation components such as dampers are also used to reduce seismic forces on a building [14].

5.3 Material and Construction Risks

The design of high-rise buildings requires careful consideration of material and construction risks to ensure structural integrity and safety. The selection and implementation of appropriate materials are critical given the unique challenges associated with the height and environmental impacts of high-rise buildings. Engineers must consider elements such as wind loads, seismic.

Activity and ground conditions to select materials that can withstand these forces. High-strength concrete and modern steel alloys have become popular in high-rise construction, providing improved structural performance. However, the integration of such innovative materials carries inherent risks, such as potential compatibility issues between different materials, which may affect the overall structural integrity of the building [15].

Structural risks when building high-rise buildings complicate the design process. The vertical nature of these structures presents logistical challenges for material transportation and on-site assembly and requires careful planning and coordination. Safety aspects in construction are of utmost importance and require comprehensive measures to protect workers from accidents and deaths. Additionally, longer construction schedules expose projects to external factors such as adverse weather conditions and unforeseen site-specific issues [15].

The use of advanced technologies is essential to address these challenges. Building Information Modeling (BIM) acts as a strategic tool to reduce material and construction risks. BIM provides a comprehensive digital representation of a building and allows engineers to simulate different scenarios and identify potential problems before construction begins. Real-time monitoring systems during construction also help reduce risk by monitoring material performance and identifying deviations from design specifications [6].

5.4 Foundation Stability and Soil-Structure Interaction

The foundation is a critical component of the stability of high-rise buildings. In civil engineering, a thorough understanding of foundation stability, soil-structure interaction, and soil-bearing capacity is critical to the safety and durability of buildings. The construction process begins with a thorough risk analysis focusing on soil conditions, which have a significant impact on foundation construction. The key to foundation design is to predict the likely settlement over time to ensure that it remains within acceptable limits. Foundation stability depends on the dynamic interaction between the foundation and the surrounding soil, which is influenced by factors such as soil type and moisture content. In addition, engineers evaluate the bearing capacity of the soil, that is, the maximum load that the soil can withstand without destruction. This understanding is critical to the design of foundations that can support the complex loads of high-rise building.

5.5 *Wind Load Analysis*

The vulnerability of high-rise buildings to wind forces poses a major challenge to maintaining their structural stability. The dynamic interaction between high-rise buildings and wind can cause sway and potentially compromise the integrity of the building. To solve this problem, engineers conduct careful analysis of wind loads, an important step in designing structures that can withstand external forces. Wind load analysis evaluates the pressure distribution on a building's surface, taking into account factors such as wind speed, building shape, and terrain. This analysis helps to understand how wind forces affect the structure and helps to take necessary strengthening measures. Engineers can propose damping or bracing systems to mitigate the effects of wind forces, thereby improving the overall resistance of high-rise buildings to dynamic loads. The knowledge gained from wind load analysis makes a significant contribution to the planning and construction of high-rise structures and ensures their durability and structural integrity. Notable studies such as those of Simiu and Scanlan [16] provide valuable information and methods for comprehensive wind load analysis in the field of civil engineering.

6 Conclusion

The complexity of designing high-rise buildings requires a thorough and proactive approach to risk analysis during the design phase. This article explores an integrated methodology that integrates both qualitative and quantitative methods, taking into account the dynamic landscape of the construction industry and the unique challenges associated with high-rise structures.

Proactive problem identification, cost containment, planning, safety considerations and regulatory compliance have highlighted the critical role of risk analysis in the development of successful high-rise building projects. The discussion emphasized the importance of early risk identification, stakeholder engagement, and developing effective risk mitigation strategies.

Research into risk factors in the design of structural strengthening for high-rise buildings has also identified multidimensional issues related to materials, construction processes, foundation stability, soil-structure interactions, and wind loads. The introduction of advanced technologies such as Building Information Modeling has proven to be a strategic tool to address these challenges and reduce risks during the construction phase.

References

1. Smith AC (2003) Construction risk management: concepts and practice. J Constr Eng Manag

2. Chapman RM (2019) Effective project risk management: beyond risk modeling. Wiley, Hoboken
3. Morris PWG (2013) The management of projects. Thomas Telford
4. Hillson D (2012) Practical project risk management: the ATOM methodology. Management Concepts
5. Yap NS, Rowlinson B (2014) Factors influencing corporate reputation in the construction industry. Construction Management and Economics
6. Eastman C et al (2018) BIM handbook: a guide to building information modeling for owners, managers, designers, engineers, and contractors. Wiley, USA
7. Goral J (2007) Risk management in the conceptual design phase of building projects
8. Schexnayder CE, Mayo RD (2011) Risk and decision analysis in projects. Wiley, USA
9. Brandon PS, Hallowell TS (2004) Cost overruns in large construction projects: the impact of project phase and risk shocks. Int J Proj Manag
10. Kwak YW et al (2009) Delays in large public construction projects: the case of Korea. J Constr Eng Manag
11. Al-Mutairi MS et al (2011) Risk factors in construction projects: the case of Kuwait. Int J Proj Manag
12. Garrett JH, Risher JM (2014) Building code compliance and enforcement: the value of third-party reviews. J Perform Constr Facil
13. Skempton AW (2001) The structural engineer's contribution to twentieth-century architecture. Struct Eng Int 11(1):22–25
14. Madhumathi J, Lavanya G (2019) Seismic analysis and design of high rise building by using ETABS in different seismic zones. Int Res J Eng Technol (IRJET) 6(5):1–8
15. Smith S et al (2018) Material selection criteria for high-rise buildings. J Struct Eng 35(2):123–145
16. Simiu E, Scanlan RH (1996) Wind effects on structures: fundamentals and applications to design. Wiley, USA

Temperature-Dependencies and Dynamic Mechanical Behavior of Polymer and Metal Materials Under High Impact Velocity: A Review of Our Original Experimental Devices Development

A. Bendarma, T. Jankowiak, A. Rusinek, T. Lodygowski, R. Bernier, J. Bin, S. Bahi, Hamza Bouchta, and M. Klosak

Abstract Ballistic studies have been the subject of numerous analyses and research activities at the theoretical, experimental, or numerical levels. Many studies related to the perforation problem and impact loading of thin metallic specimens have been presented in the literature. Analytical models like (Recht–Ipson) have previously been developed to determine the relationship between the residual velocity of a projectile and its initial velocity in the case of metal plates. The perforation process of a thin plate by projectiles of different shapes (e.g., flat, hemispherical, or conical) has been analyzed in numerous works with a wide range of analyzed materials: metals and polymers. A list of selected articles in the literature that served as scientific inspiration and were cited in the presented works is provided. It is worth mentioning the names of the authors of fundamental works on the theoretical, experimental, and numerical

A. Bendarma (✉)
Arts et Métiers Campus de Rabat Technopolis, Rocade de Rabat-Salé, 11100 Sala Al Jadida, Morocco
e-mail: amine.bendarma@artsetmetiers.ma

A. Bendarma · H. Bouchta
Laboratory for Sustainable Innovation and Applied Research, Universiapolis, Technical University of Agadir, Bab Al Madina, Qr Tilila, BP 8143, 80000 Agadir, Morocco

A. Bendarma · T. Jankowiak · T. Lodygowski · M. Klosak
Institute of Structural Analysis, Poznan University of Technology, Piotrowo 5, 60-965 Poznan, Poland

A. Rusinek · R. Bernier · S. Bahi
Laboratory of Microstructure Studies and Mechanics of Materials LEM3, University of Lorraine, 7 Rue Félix Savart, 57070 Metz, France

J. Bin
Beijing Research Institute of Automation for Machinery Industry, Beijing 100120, China

H. Bouchta
Thermodynamics and Energetics Laboratory (LTE), Department of Physics, Faculty of Science, Ibn Zohr University, BP 8106, 80006 Agadir, Morocco

T. A. Kovács et al. (eds.), *The Impact of the Energy Dependency on Critical Infrastructure Protection*, Advanced Sciences and Technologies for Security Applications, https://doi.org/10.1007/978-3-031-78544-3_10

approach to ballistic tests: Recht, Ipson, Atkins, Borvik, Holmen, Gupta, Clausen, Rusinek, and Jankowiak. This paper focuses on the experimental methodologies used in ballistic testing for dynamic perforation using a pneumatic gas gun. The development of a thermal chamber is presented, designed to improve perforation analysis on different materials and over a wide range of experimental temperatures, from T0 = 293 K to 533 K. Low-temperature perforation tests performed at 113 K (−160 °C) using an original set-up are also included. In addition, constitutive relations and failure criteria are proposed to contribute to the understanding of perforation phenomena.

Keywords Impact response · Thermal chamber · Failure mechanisms · Failure · Perforation · Hopkinson bar

1 Introduction

Construction materials subjected to loading rates exhibit phenomena of hardening or softening, typically leading to failures and fracture events. This process can occur rapidly for both quasi-static and dynamic loading. To define the behavior of materials intended for industrial applications, various tests are necessary to understand the diverse effects of loading based on strain rates and temperatures of the material. As depicted in Table 1, different techniques may be used to study a wide range of strain rates. Integrating data from various tests helps to characterize the material's sensitivity to strain rates and temperature. The Split Hopkinson Pressure Bar (SHPB) or Taylor test is crucial for achieving high deformation rates. For lower deformation rates, a high-rate servo-hydraulic machine may be used, and a screw machine for quasi-static loadings.

Experimental techniques for the mechanical characterization of materials are diverse. The selection of these techniques is not arbitrary but rather aimed at specific goals, depending on industrial needs in civil or military domains, for example. For static problems, the system remains elliptic, while for dynamic problems, it becomes hyperbolic. The discussion on the type of loading, quasi-static or dynamic, should be balanced based on your own experience compared to specific strain rates for typical processes, as shown in Table 1 [1].

Materials testing and structures are an important part of the overall design process for infrastructure elements and buildings. Since threats are dynamic in nature, studies should focus on experiments dealing with high strain rates and high temperatures [2, 3]. The dynamic behavior characterizing materials at different temperatures is crucial for a diverse range of industrial applications, particularly in the design of strong infrastructure and structures. This introduction aims to synthesize the results of several previous works on the dynamic behavior of materials, focusing on the effects of temperature on perforation and mechanical properties of materials. Numerous experimental, analytical, and numerical studies are now available in the international literature. Work on the penetration and perforation of structures by projectiles has

Table 1 Deformation rates for specific physical phenomena [1]

Type of phenomenon	Velocity of deformation [s^{-1}]
Creep	From 10^{-10} to 10^{-5}
Creep beyond the yield criteria	From 10^{-5} to 10^{-1}
Hot drawing	From 10^{-1} to 10^{1}
High speed drawing	From 10^{1} to 10^{3}
Machining	From 10^{3} to 10^{5}
Drawing with use of explosion	$> 10^{5}$
Earthquake	From 10^{-3} to 10^{-1}
Car crash	From 10^{-2} to 10^{0}
Plane crash	From 5×10^{-2} to 2×10^{0}
Hard hit	From 10^{-0} to 5×10^{1}
Projectile hitting	From 10^{2} to 10^{6}
Explosive loading	From 10^{6} to …
Geological movements	$\approx 10^{-10}$
Creeping	$\approx 10^{-6}$
One axial tension test	$\approx 10^{-4}$
Drilling, rolling, drawing	$\approx 10^{0}$
Test with Hopkinson's bar	$\approx 10^{3}$
High-velocity impacts	$\approx 10^{6}$

been carried out by researchers such as Goldsmith [4], Corbett et al. [5], Backman et al. [6], Børvik et al. [7] and the impact velocity [8–10]. The influence of the thickness of the metal plate and the geometry of the projectile on the process has been examined. The current challenge is to correlate experimental, analytical and numerical aspects to provide a detailed analysis of the behaviour of structures subjected to rapid dynamic loading [11]. As example, Durrenberger et al. [12] developed a phenomenological constitutive model to describe the visco-plastic behaviour of high-performance steels subjected to quasi-static and dynamic loading. The flow stress is considered to be the sum of an internal stress and an effective stress. The evolution of internal stress with plastic strain is regulated by a single internal parameter, δ, which controls strain-hardening effects. The material parameters in the law of evolution of δ are temperature-dependent. Fras et al. [13] investigated the ballistic performance of the AA7020-T651 aluminium alloy against simulated fragment projectiles, with very high impact velocities ranging from 900 to 1500 m/s. The thick target plates show an evolution of failure modes from plugs to discs as the impact velocity increases. The stress pattern leading to disc formation is detailed and numerically simulated. In addition, the mechanical, thermal and micro-structural properties of the material are examined to understand the mechanisms under highly dynamic loading. These findings enrich our understanding of the response of materials to ballistic impacts, particularly under conditions of thick targets and high-impact velocities.

In the first work, an experimental setup was used for dynamic and perforation tests on PMMA at high strain rates and temperatures up to 130 °C. The transition between brittleness and ductility was defined, as was an analytical model based on the thermal activation of molecules. In the second work, the effect of the initial temperature on the perforation of 304 stainless steel was to consider the phase transformation from austenite to martensite, showing an increase in the number of petals and the ballistic velocity limit at lower temperatures. Finally, in the third work, dynamic compression and tensile tests were used to study the impact of temperature on the ballistic properties of brass alloy, showing significant variation with initial temperature [14–16]. This synthesis offers an integrated perspective on the impact of temperature on the dynamic behavior of materials, highlighting its importance for structural design and safety.

2 Experimental Set-Up

2.1 Dynamic Compression Tests Using the Split Hopkinson Pressure Bar (SHPB)

At high deformation rates, the most widely used technique to study the dynamic properties of materials relies on the Split Hopkinson Pressure Bar (SHPB). This method was initially developed by Kolsky [17] and has undergone several improvements. Its principle is based on the theory of elastic wave propagation in bars, allowing for deformation speeds between 10^2 and 10^4 s^{-1}. The principle of Hopkinson bars is well-known and has been the subject of numerous publications [3, 10, 11, 18, 19].

The assembly used for these studies is illustrated in Fig. 1. Specimens are placed between two identical bars of high tensile strength relative to the material under test. The first and second bars are referred to as the "Input bar" and "Output bar" respectively. The input bar is the one of ends that receives who is impacted by the blunt cylindrical impactor.

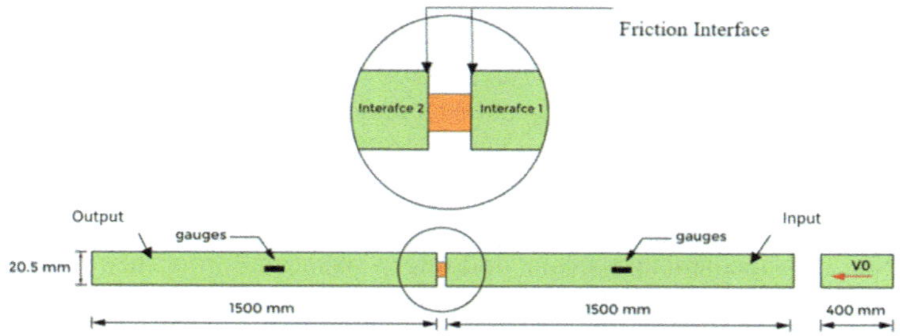

Fig. 1 Schematic description of the Hopkinson bar device [18, 19]

Sometimes, friction at the bar/sample interfaces can lead to a change in the uniaxial stress state and may yield incorrect results [20]. Various authors, including [2, 21, 22], who have worked on this subject, demonstrated that the sample geometry, specifically the ratio L_0/d_0 (where d_0 is the initial diameter of the sample), determines the significance of the friction effect on test results and the puncture effect. To mitigate this factor's impact, it is recommended to lubricate the bar/sample interfaces and maintain the L_0/d_0 ratio between 0.5 and 1 [20, 23]. To consider temperature effects, experimental tests were conducted using a specific furnace fixed around the specimen to determine material behavior at high strain rates (up to $\dot{\varepsilon} \geq 1000\ \text{s}^{-1}$) and within a temperature range (20 °C $\leq T_0 \leq$ 170 °C). The device ensures a uniform temperature across the entire sample volume, avoiding temperature gradients along the sample, as discussed by Lennon and Ramesh [20, 24]. The setup developed and used for dynamic compression tests is presented in Fig. 2. Tests are performed at different strain rates and temperatures, ranging from room temperature to 170 °C [2].

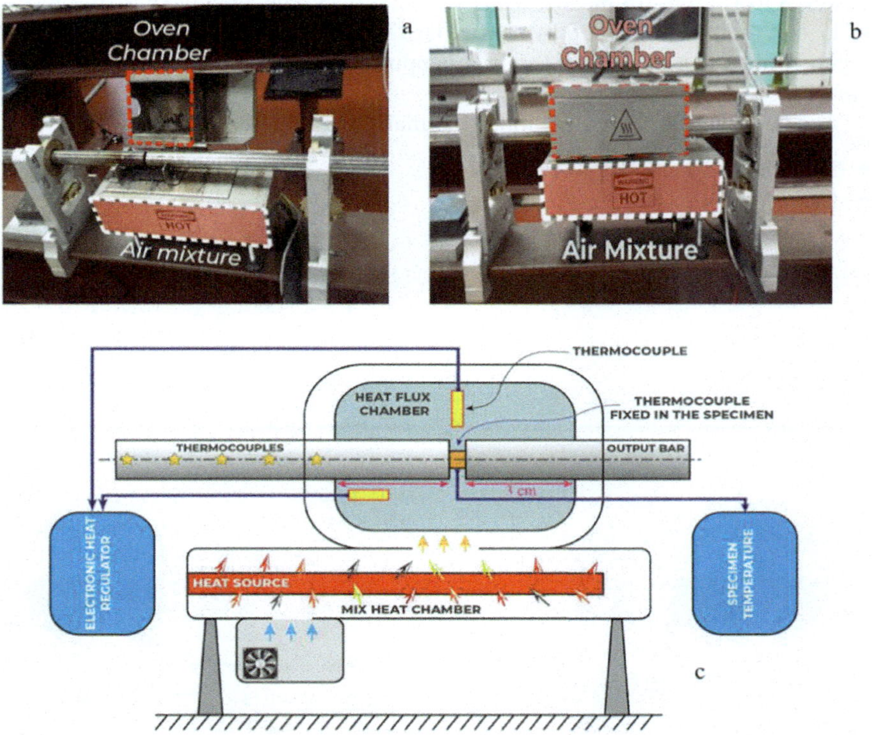

Fig. 2 **a** Split Hopkinson pressure bars used in dynamic tests with a furnace; **b** description of the internal part; **c** description of the airflow [2, 23, 25]

In this work, the sample used has a thickness of 4 mm and a diameter of 8 mm. A specific furnace has been designed to achieve and maintain a uniformly distributed temperature using an electronic controller, Fig. 2. Thus, a constant volume is heated to an initial temperature T_0. This volume is larger than the chamber volume where the sample is sandwiched between the two Hopkinson bars, Fig. 2c. To maintain a constant temperature in the sample chamber, a fan is used to evacuate the hot air flow as needed.

2.2 Perforation Test from Low to High Temperatures

A schematic view of the high-pressure pneumatic gas gun is presented in Fig. 3. For this research review, generally, conical projectiles at with 72° angle with a diameter of 11.5 mm are used, although results obtained with blunt projectiles (flat end) and hemispherical projectiles are also presented. The tested plates had a thickness varying from 0.8 to 1.2 mm for metals, while for other materials, the thickness reached up to 11 mm (composites and polymers). The 130 mm × 130 mm sample plates were securely fixed in a special holder, and clamping on the periphery of the specimen to embed it.

The gas gun (Fig. 3) has the following main characteristics:

- Launch tube diameter: 13 mm.
- Launch tube length: 5 m.

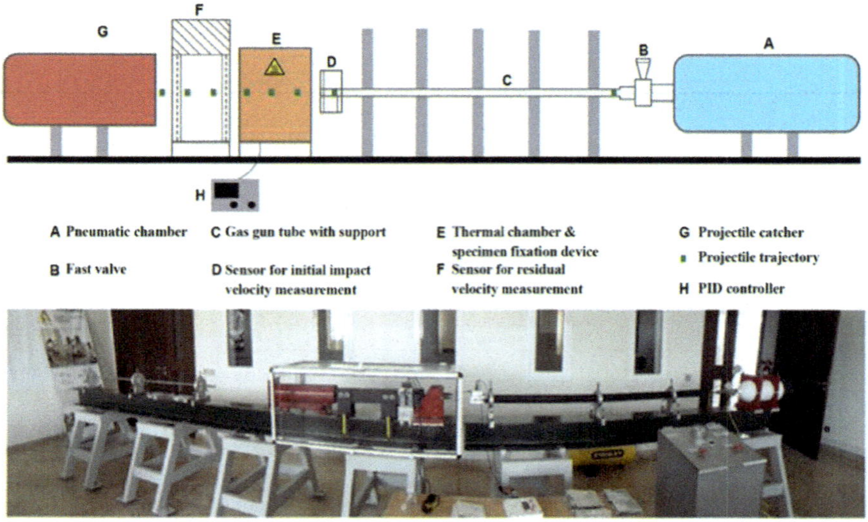

Fig. 3 Gas gun configuration used for high-velocity impact and high-temperature perforation tests [2]

- Impact velocity range: 30–180 m/s for a projectile mass of 28 g.

The projectile was launched using a pneumatic gas gun and accelerated in the tube to an initial impact velocity V_0. The projectile then impacted the plate, producing partial or complete perforation depending on the amount of kinetic energy transferred to the test object. The residual velocity V_R of the projectile was measured after perforation. A pair of laser sensors was used to measure the initial impact velocity V_0 (position "D" in Fig. 4a), and a laser barrier attached behind the sample measured the residual velocity V_R (position "F" in Fig. 4a). The projectile mass was 28 g. The material used for the projectile was specially heat-treated maraging steel to achieve a high tensile strength close to 2 GPa—no "mushrooming effect" was observed for such a robust projectile. The projectile showed no visible permanent deformation during the perforation process and could be described as a rigid body [2, 3]. The apparatus was equipped with a thermal chamber "E" (Fig. 3) in which the sample "I" was fixed. The temperature can vary from room temperature (20 °C ~ 293 K) to the maximum temperature (approximately 300 °C ~ 573 K). External air was introduced into the chamber through a fan that also contained heating coils. The sample "I" was positioned inside a thermal chamber ensuring a uniform temperature distribution, so that both sides of the sample were heated with the same intensity. Due to the thermal conductivity of the sample, it could reach the defined initial temperature as a whole—heating was regulated by the PID controller "H". The electronic heat regulator (PID type) allowed the setting of the required temperature and control of temperature increase. Temperature values in the thermal chamber and the sample were continuously measured by two thermocouples. An integrated thermocouple continuously measured the temperature inside the chamber.

To conduct low-temperature perforation tests, a cooling device has been designed and developed at the LEM3 laboratory in Metz—France. As shown in Fig. 4a, the cooling device is attached to the ballistic impact device. On the one hand, liquid nitrogen is circulated from a liquid nitrogen tank through an aluminum pipe to the cooling device; on the other hand, the temperature inside the cooling device is monitored by a thermocouple connected to a temperature controller. Once the temperature inside the box reaches the defined value, the temperature regulator cuts off the power to the pump to stop the flow of nitrogen gas. Using this method, low temperatures between − 90 and − 20 °C can be achieved. Additionally, by directly filling the cooling box with liquid nitrogen, − 163 °C can be reached, Figs. 5b and 6 [15].

3 Analysis of the Mechanical Properties of PMMA

This study aims to characterize the mechanical behavior of polymethyl-methacrylate (PMMA) using all previous original techniques under high-velocity impact conditions across a broad range of test temperatures. To achieve this, the mechanical

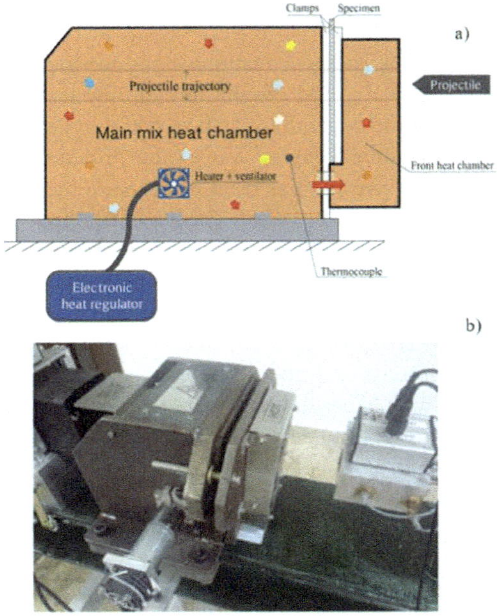

Fig. 4 Thermal chamber for heating target plate samples, **a** schematic **b** overall view of the apparatus (Patent Number MA41357A) [2, 3]

response to uniaxial compression is investigated under quasi-static and dynamic conditions (Fig. 7), spanning test temperatures below, at, and above the glass transition. A pseudo-brittle to ductile transition in PMMA failure is observed at a temperature-dependent threshold and deformation rate. This analysis provides insights into interpreting impact perforation tests and elucidating the primary deformation and failure mechanisms.

By analyzing its response to uniaxial compression under quasi-static and dynamic conditions. Tests are conducted over a range of temperatures, including those below, at and above the material's glass transition. A major observation of this study is the transition of PMMA's behavior from a pseudo-fragile to a ductile state at fracture, with thresholds depending on both the test temperature and the material's strain rate.

The results in terms of ballistic limit $V_B(T_0)$ as a function of the initial temperature is presented in Fig. 8. Different slopes can be observed by analyzing the results using the ratio V_B/T_0. In this regard, a slight continuous increase is observed with the test temperature up to around 80 °C. Subsequently, a marked increase in the ballistic limit occurs for temperatures ranging from 80 to 100 °C. This change in trend can be explained by a transition from quasi-brittle to ductile failure. Then, above the glass transition, the material undergoes a significant decrease in stiffness, leading to a complete change in V_B/T_0, with a sharp decrease in the ballistic limit with the test

Fig. 5 Cooling device setup for tests: **a** between − 90 and − 20 °C and **b** at − 163 °C: (1) ballistic impact device, (2) thermocouple, (3) temperature regulator, (4) pump, (5) pipe, (6) liquid nitrogen reservoir [15]

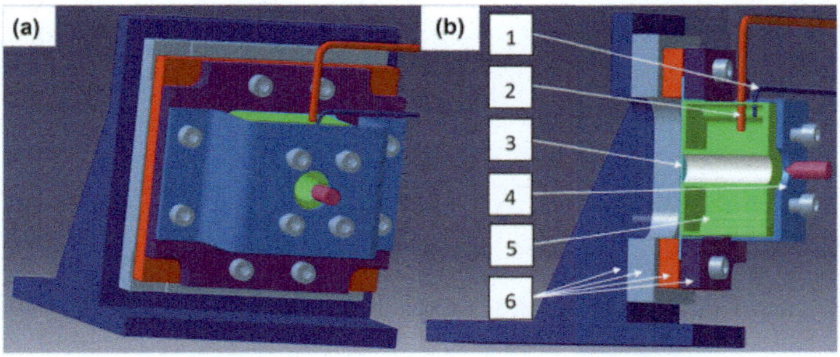

Fig. 6 The arrangement of the cooling device on the ballistic impact device: **a** isometric view and **b** sectional view: (1) thermocouple, (2) pipe, (3) target, (4) cooling box holder, (5) cooling box, (6) target holder [15]

temperature. For test temperatures close to the glass transition, there is strong competition in terms of energy absorption capacity between thermal softening (reducing energy absorption) and an increase in ductility (improving energy absorption), as shown in Fig. 8.

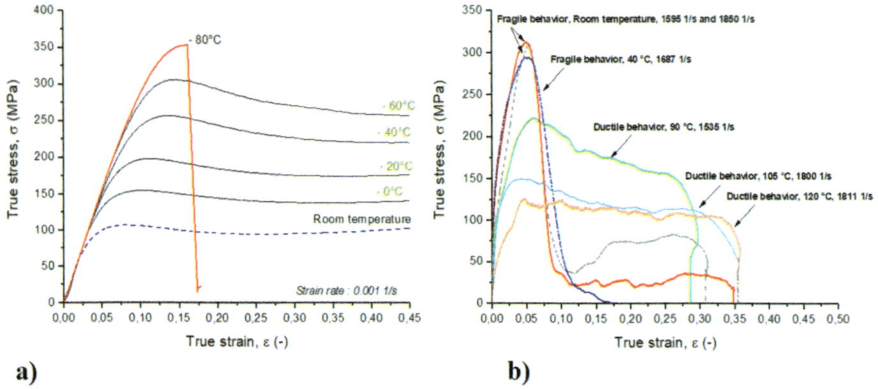

Fig. 7 Material behavior description: **a** temperature effect under quasi-static loading; **b** temperature effect under dynamic loading [14]

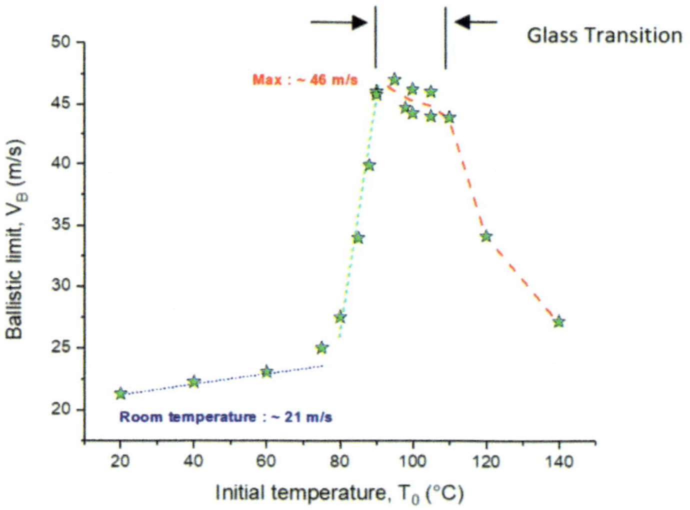

Fig. 8 Description of the ballistic limit as a function of the initial temperature, transitioning from brittleness to ductility [14]

The prediction of absorbed energy compared to impact velocity for different test temperatures is compared to experimental results in Fig. 9. This analytical approximation captures trends from room temperature up to $T_0 = 90$ °C. However, at the glass transition temperature and above, the model does not accurately predict experimental values at high impact velocities ($V_0 > 120$ m/s). Under these specific conditions, the plate cannot absorb a significant amount of energy.

Concerning this latter case, Fig. 10d–e, for a test temperature above the glass transition, the kinematics of the perforation process can be described in 4 states

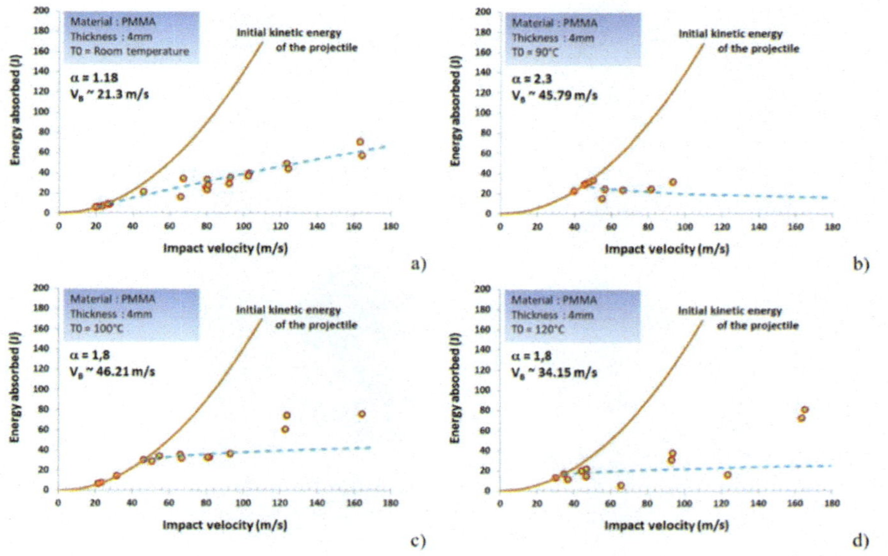

Fig. 9 Energy absorption for different initial temperatures, comparison between experiments and the analytical model: **a** room temperature; **b** 90 °C; **c** 100 °C; **d** 120 °C [14]

as proposed in Fig. 11. The first corresponds to the projectile impact, generating a small hole due to the projectile's conical shape and a few petals. Then, the projectile penetrates the plate, inducing a radial expansion of the initial hole until reaching the diameter of the projectile, $\phi_p = 12.8$ mm. As the projectile passes through the entire thickness of the plate, the hole begins to close due to elastic recovery resulting from the rubbery state of the amorphous chains within the PMMA. Finally, the hole closes completely with a few reduced radial cracks.

4 Mechanical Properties of Stainless Steel 304 from Low to High Conditions

The effect of temperature on the perforation behavior of austenitic stainless steel 304 plates was experimentally studied. Perforation tests were conducted at speeds ranging from 80 to 180 m/s and temperatures between − 163 and 200 °C. Based on the experimental results, the primary failure mode, initial-residual velocity curves, ballistic limit velocities, and energy absorption capacity at different temperatures were analyzed. The main observed failure mode during the perforation process was petaling. The average number of petals increased continuously from three at 20 or 200 °C to five at − 163 °C. The ballistic limit (V_B) was also influenced by initial temperature. It increased slightly from 93 m/s at 200 °C to 103 m/s at − 20 °C, then

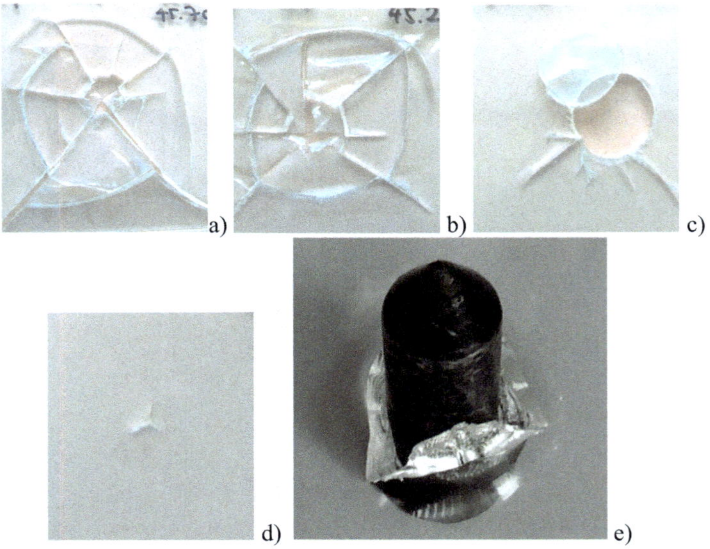

Fig. 10 Failure mode of PMMA at the ballistic limit at different temperatures: **a** room temperature; **b** 60 °C; **c** 90 °C; **d** 105 °C; **e** 105 °C and an initial velocity close to V_b [14]

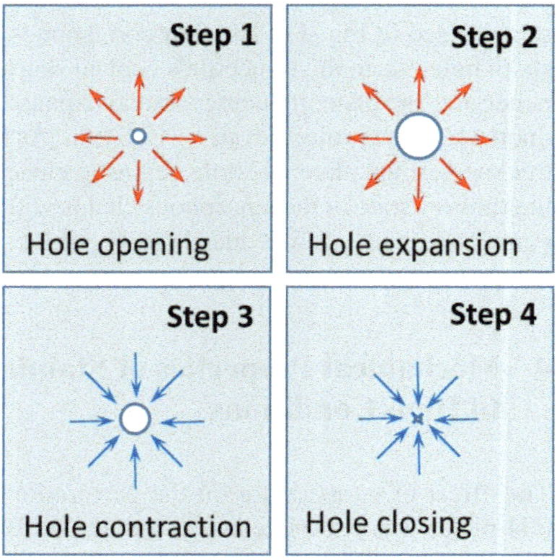

Fig. 11 Kinematics of the failure mode, perforation and structural behaviour of PMMA at high temperature [14]

held constant at lower temperatures. The material showed an increased capacity for energy absorption at low temperatures.

As shown in Fig. 12, the equation of Recht–Ipson (Eq. 1) [26] correctly defines the shape of the ballistic curve.

Fig. 12 Ballistic curves for austenitic 304 stainless steel plates with a thickness of 1.5 mm impacted at different temperatures [15]

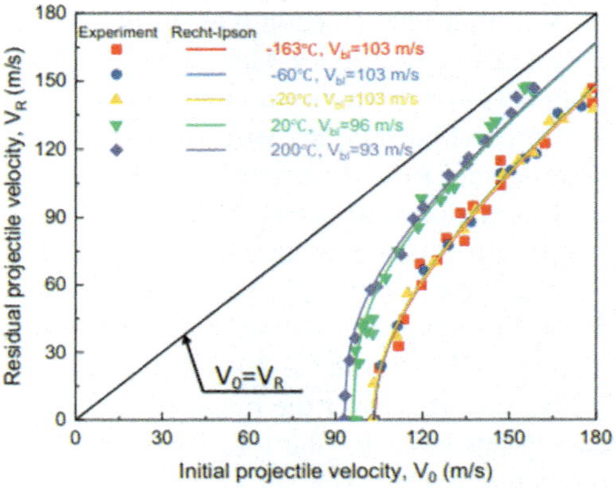

$$V_R = \left(V_0^k - V_B^k\right)^{1/k} \tag{1}$$

For tests at − 20, − 60, and − 163 °C, the ballistic limit velocity remains the same at 130 m/s. For tests at higher temperatures, V_B decreases with increasing temperature. At a test temperature of 200 °C, the ballistic limit velocity is 93 m/s. Although the ballistic limit velocity at 200 °C is lower than that at 20 °C, the difference between the two curves decreases with increasing impact velocity. At ballistic impact velocities exceeding 150 m/s, the two curves coincide. The adjustment parameter α is also influenced by the test temperature, with the value at room temperature and high temperatures being significantly lower than at lower temperatures.

The results of energy absorption as a function of V_0 for tests at different temperatures are shown in Fig. 13. Firstly, energy absorption for low-temperature tests (− 163, − 60 and − 20 °C) is significantly higher than at 20 °C, while energy absorption at 200 °C is lowest. This is consistent with the observed fracture pattern: more petals and cracks at low temperatures.

Fig. 13 Energy absorbed by the target as a function of the projectile's initial velocity for different test temperatures [15]

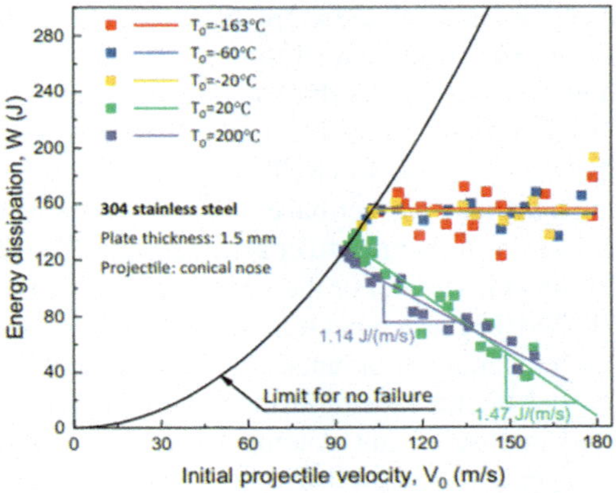

A similar phenomenon was observed in low-velocity perforation tests of TRIP 1000 steel by Rodriguez-Martinez et al. [27]: the ballistic limit velocity V_B changed from 2.6 to 3.1 m/s as the test temperature decreased from 15 to − 60 °C. According to Rodriguez-Martinez analysis, the improvement in energy absorption at low temperatures comes from the temperature sensitivity of TRIP 1000 steel, and no martensitic transformation was observed during the perforation process. Using different initial temperatures failure modes were changed as follows.

Another interesting phenomenon concerning the energy absorption capacity of SS 304 is the evolution of absorbed energy with impact speed. For the low-temperature tests (− 163, − 60 and − 20 °C), the initial projectile velocity V_0 does not affect the energy absorbed over the range of impact speeds tested (see Fig. 14). The average energy absorbed for the three preceding temperatures is 156 J, 154 J and 155 J, respectively. However, for tests at 20 and 200 °C, the energy absorbed decreases linearly with increasing impact velocity. Although the absorbed energy at 20 °C is slightly higher than at 200 °C, it decreases more rapidly and becomes lower at impact velocities above 137 m/s.

The VB ballistic limit is also strongly influenced by the initial test temperature. It increases slightly from 93 m/s at 200 °C and 103 m/s at − 20 °C, then remains constant at even lower temperatures. Ballistic curves V0–VR are also influenced by initial test temperature. At 20 and 200 °C, the ballistic curves appear to converge when the initial projectile velocity is much higher than the ballistic limit. The ballistic curves at − 20, − 60 and − 163 °C are almost the same and marginally lower than those at 20 and 200 °C.

5 Mechanical Properties of Brass Alloys

To comprehensively characterize the mechanical properties of a common brass alloy under dynamic loading conditions, this study extends the initial perforation experiments to include dynamic compression and tensile tests, aiming to observe material behavior and define failure modes [28, 29]. The material tested was a common brass alloy. This study aimed to observe the behavior of the sample material and define failure modes under dynamic impact loading conditions. For tensile tests, the strain rates available ranged from 0.001 to 0.01 1/s, and the initial temperature T_0 varied between 293 and 453 K. The results of the stress–strain curves are shown in Fig. 15a. A slight negative thermal effect is observed, resulting in a downward shift of the curve, while no strain rate effect is perceptible. The effect of temperature can be seen in the strain at fracture. For the strain rate of 0.001 1/s, true strain at break decreases from 0.37 to 0.32 as temperature increases from 293 to 453 K. For the 0.01 strain rate, this effect is more pronounced at nominal values, varying from 0.445 to 0.375.

For the perforation using a thermal chamber to heat brass thin metal specimens and impose an initial temperature higher than room temperature. The maximum temperature achieved with this original device is 260 °C. This level is effective for exploring the behavior of structures where the thermal softening effect is observed.

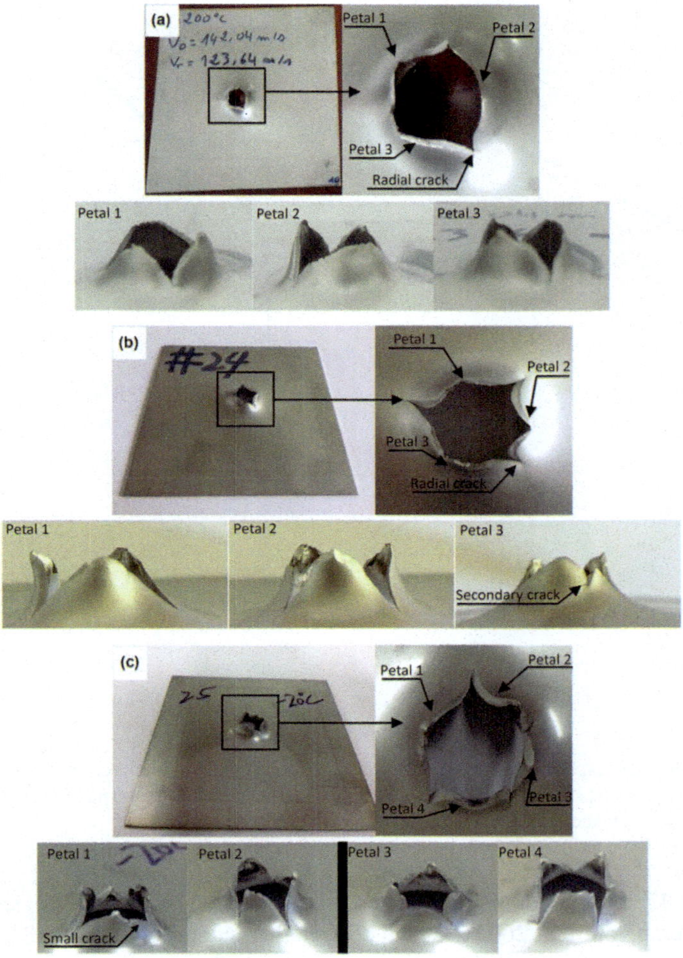

Fig. 14 Failure pattern for different test temperatures: **a** 200 °C, $V_0 = 124$ m/s, **b** 20 °C, $V_0 = 127$ m/s, **c** − 20 °C, $V_0 = 134$ m/s, **d** − 60 °C, $V_0 = 152$ m/s, and **e** − 163 °C, $V_0 = 167$ m/s [15]

A brass alloy was selected for this work, as the sensitivity to the deformation rate can be considered negligible. In this case, it is possible to consider only one dependence of the material behavior, namely temperature sensitivity. The experimental results obtained were compared to data available in the literature [7, 9]. At high speeds, strong plastic deformation localization occurs in a very small shear zone, with plastic deformation limited to the immediate vicinity of the impact point (Fig. 16b). Similar conclusions were drawn in [25]. However, for impact speeds close to the ballistic limit (no complete perforation), plastic deformation of the entire active zone is observed (Fig. 16a). For comparison purposes, this phenomenon is shown for the conical projectile (Fig. 16c) where more localized plastic deformation is observed.

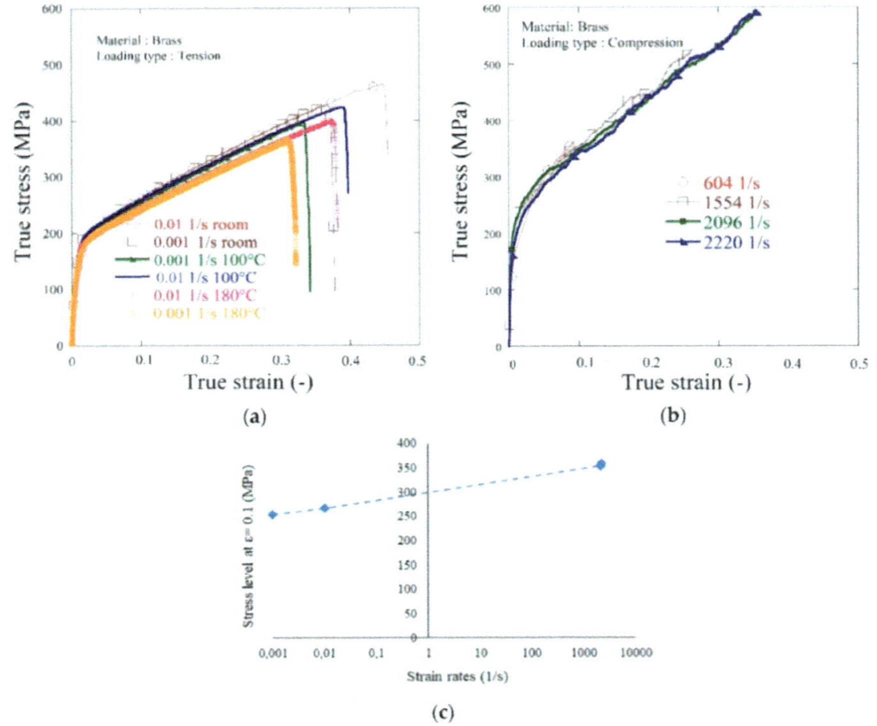

Fig. 15 Dynamic experimental results for selected strain rates and temperatures: **a** tensile tests: true stress versus true strain curves; **b** compression tests: true stress versus true strain curves; **c** effect of strain rate on plastic flow [16]

Figure 17 shows a comparison between experimental results at room temperature and at a temperature of 260 °C. It can be seen that increasing the initial temperature of the sample shifts the ballistic limit (non-perforation state) to lower values: the ballistic limit obtained is around 43 m/s (20 °C) and 40 m/s (260 °C) for the conical projectile. The Recht–Ipson [26] estimate would suggest a slightly lower ballistic limit at 260 °C, of around 37 m/s. For the round-ended projectile, the ballistic limit is 63 m/s (20 °C) and 60 m/s (260 °C), respectively. The other measured points are also displaced when higher V_R residual velocities are measured at elevated temperatures.

The typical failure mode involves the formation of 3 or 4 petals. Up to 6 petals have been observed at lower impact velocities (close to the ballistic limit) and higher temperatures. The failure modes observed during the experiments are presented in Fig. 18. According to the theoretical considerations of the authors in [30–33], the number of petals strongly depends on the failure strain parameter. This parameter changes with temperature, meaning that material elongation is more significant at higher temperatures, directly increasing the number of petals.

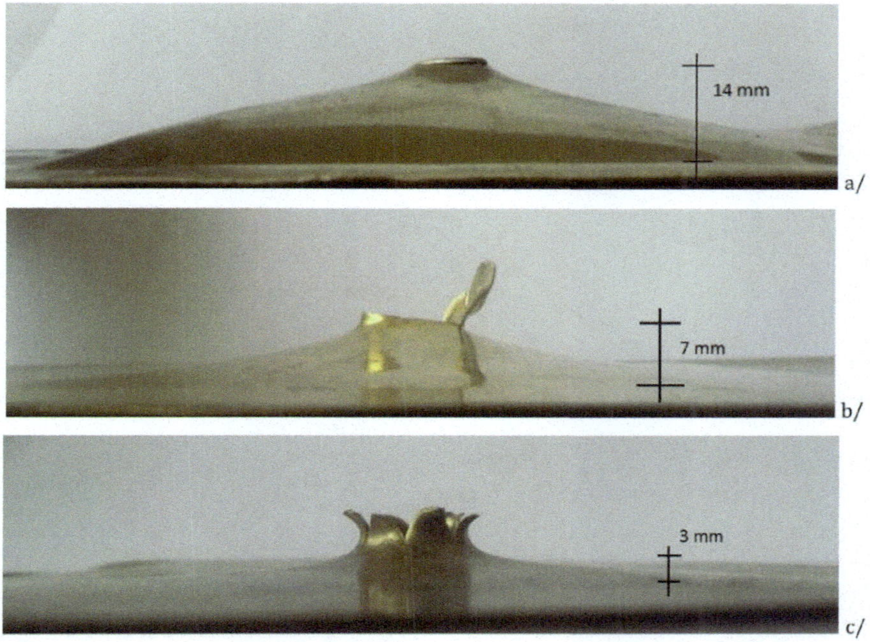

Fig. 16 Overall deformation of the target plates' active zone: **a** no perforation with a round-nose projectile at T = 20 °C and $V_0 = 60.39$ m/s, **b** perforation with a round-nose projectile at T = 20 °C and $V_0 = 74.18$ m/s, **c** petal-shaped perforation obtained with a conical projectile s at T = 200 °C and $V_0 = 54.94$ m/s [16]

6 Conclusion and Discussion

This study aimed to characterize the mechanical behavior of polymethyl methacrylate (PMMA) under high-velocity impact over a wide range of test temperatures. The results highlight a transition in the behavior of PMMA from a pseudo-fragile state to a ductile state at fracture, dependent on both test temperature and material strain rate. This transition is crucial for the accurate prediction of the material's performance under impact and for the optimization of its design in various applications.

Analysis of the results in terms of the ballistic limit V_B (T_0) shows distinct trends as a function of test temperature. An increase in the ballistic limit is observed with temperature up to a critical threshold, followed by a marked decrease above the glass transition, underlining the importance of temperature on the material's ballistic properties.

The cooling device enables perforation tests of austenitic stainless steel 304 plates at low temperatures ranging from – 163 to – 20 °C. The uniformity of plate surface temperature is verified both experimentally and by numerical simulation. After a waiting time of 1200 s, the maximum temperature fluctuations at the plate surface are 2 °C, 3 °C, 4 °C and 8 °C for test temperatures of – 20 °C, – 40 °C, – 60 °C and –

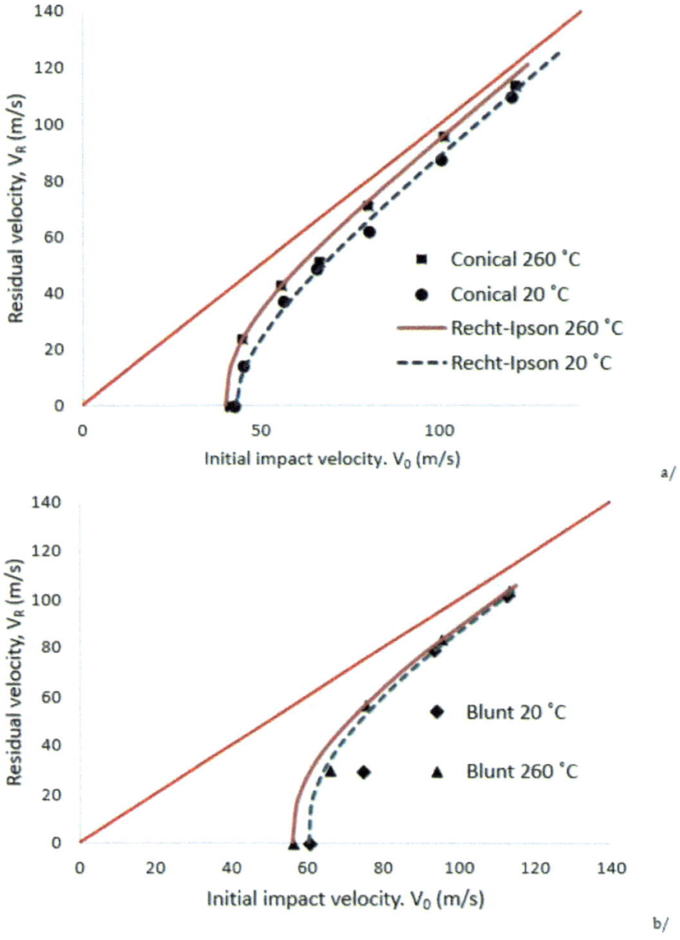

Fig. 17 Initial impact velocity V_0 versus residual velocity V_R—experimental results for T = 20 °C and T = 260 °C; **a** conical projectile, **b** round-nose projectile [16]

163 °C, respectively. Considering only the center of the plate (with a radius of 20 mm) where plate deformation mainly occurs, the maximum temperature fluctuations are 0.3 °C, 0.4 °C, 0.6 °C and 1.4 °C respectively.

At the test temperatures considered, petaling is the failure mode during the perforation process. The number of petals increases with decreasing test temperature. The average number of petals is three at 20 or 200 °C, rising continuously to five at –163 °C. The shape of the petals is also influenced by the test temperature, resembling a regular triangle at 20 and 200 °C, while becoming rough and discontinuous at lower temperatures.

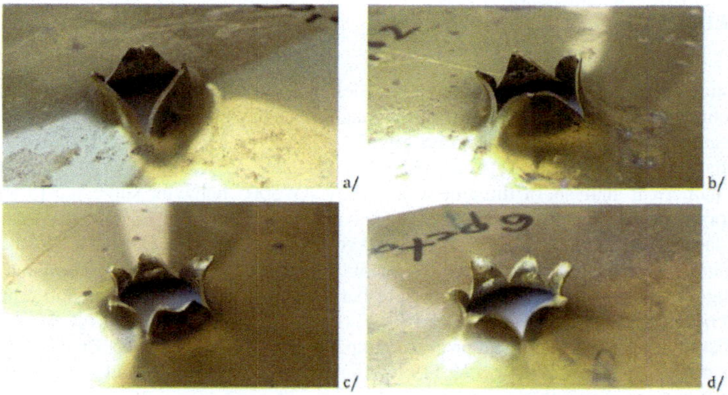

Fig. 18 Effect of initial temperature and impact velocity during perforation. Experimental observations of petal-shaped failure mode, **a** 3 petals at T = 20 °C and V_0 = 12,018 m/s, **b** 4 petals at T = 100 °C and V_0 = 6527 m/s, **c** 5 petals at T = 200 °C and V_0 = 5494 m/s, **d** 6 petals at T = 260 °C and V_0 = 4488 m/s [16]

For the analysis of the dynamic behavior of brass, the main objective of the study was to confirm the effectiveness of the new experimental technique enabling perforation tests to be carried out over a wide temperature range, from room temperature to 260 °C. The main aspects on which attention was focused included:

- The reproducibility of experimental results with heated specimens (temperature stabilization in the thermal chamber),
- Observation of failure patterns during perforation related to thermal softening of brass.

Significant results were obtained for high-temperature brass. The failure mode for the petal-shaped conical projectile proved the analytical considerations that the number of petals varied from 3 to 6. Initial impact velocity and residual impact velocity studies reproduced typical material behavior. The energy absorbed during perforation remained almost constant over the speed range studied (up to 120 m/s). For the conical projectile, the measured value was 43 m/sat ambient temperature and decreased to 40 m/sat 260 °C. For the pointed projectile, the ballistic limit evolved from 63 to 60 m/s under the same temperature conditions.

In conclusion, this review provides valuable information on the dynamic behavior of PMMA, austenitic 304 stainless steel, and brass at different temperatures, offering promising prospects for more efficient and safer use of these materials in a variety of industrial applications.

A specialized oven has been developed with a maximum temperature of 300 °C to assess and analyze the ballistic response behavior of both polymers and metals at elevated temperatures (Patent Number MA41357A). This thermal chamber was designed and developed in collaboration with the LEM3 laboratory in Metz, France, and the LIDRA Laboratory at Universiapolis Agadir, Morocco.

References

1. Voyiadjis GZ, Łodygowski T, Jankowiak T, Sielicki PW (2011) Safety of concrete and masonry structures under unusual loadings. In: Damage mechanics and micromechanics of localized fracture phenomena in inelastic solids, pp 379–411
2. Bendarma A, Jankowiak T, Łodygowski T, Rusinek A, Klósak M (2017) Experimental and numerical analysis of the aluminium alloy AW5005 behavior subjected to tension and perforation under dynamic loading. J Theor Appl Mech 55(4):1219–1233
3. Bendarma A, Jankowiak T, Rusinek A, Lodygowski T, Jia B, Miguélez MH, Klosak M (2020) Dynamic behavior of aluminium alloy Aw 5005 undergoing interfacial friction and specimen configuration in split Hopkinson pressure bar system at high strain rates and temperatures. Materials 13(20):4614
4. Goldsmith W (1999) Non-ideal projectile impact on targets. Int J Impact Eng 22(2–3):95–395
5. Corbett GG, Reid SR, Johnson W (1996) Impact loading of plates and shells by free-flying projectiles: a review. Int J Impact Eng 18(2):141–230
6. Backman ME, Goldsmith W (1978) The mechanics of penetration of projectiles into targets. Int J Eng Sci 16(1):1–99
7. Børvik T, Hopperstad OS, Langseth M, Malo KA (2003) Effect of target thickness in blunt projectile penetration of Weldox 460 E steel plates. Int J Impact Eng 28(4):413–464
8. Arias A, Rodríguez-Martínez JA, Rusinek A (2008) Numerical simulations of impact behaviour of thin steel plates subjected to cylindrical, conical and hemispherical non-deformable projectiles. Eng Fract Mech 75(6):1635–1656
9. Børvik T, Langseth M, Hopperstad OS, Malo KA (2002) Perforation of 12 mm thick steel plates by 20 mm diameter projectiles with flat, hemispherical and conical noses: Part I: Experimental study. Int J Impact Eng 27(1):19–35
10. Rusinek A, Rodríguez-Martínez JA, Arias A, Klepaczko JR, López-Puente J (2008) Influence of conical projectile diameter on perpendicular impact of thin steel plate. Eng Fract Mech 75(10):2946–2967
11. Bendarma A, Jankowiak T, Rusinek A, Lodygowski T, Czarnota C, Bernier R, Khalfaoui M (2023) Experimental and numerical analysis of aluminum-polyethylene composite structure subjected to tension and perforation under dynamic loading for a wide range of temperatures. J Dyn Behav Mater 1–24
12. Durrenberger L, Molinari A, Rusinek A (2008) Internal variable modeling of the high strain-rate behavior of metals with applications to multiphase steels. Mater Sci Eng A 478(1–2):297–304
13. Fras T, Colard L, Lach E, Rusinek A, Reck B (2015) Thick AA7020-T651 plates under ballistic impact of fragment-simulating projectiles. Int J Impact Eng 86:336–353
14. Garcia-Gonzalez D, Rusinek A, Bendarma A, Bernier R, Klosak M, Bahi S (2020) Material and structural behaviour of PMMA from low temperatures to over the glass transition: quasi-static and dynamic loading. Polym Testing 81:106263
15. Jia B, Rusinek A, Bahi S, Bernier R, Pesci R, Bendarma A (2019) Perforation behavior of 304 stainless steel plates at various temperatures. J Dyn Behav Mater 5:416–431
16. Klosak M, Jankowiak T, Rusinek A, Bendarma A, Sielicki PW, Lodygowski T (2020) Mechanical properties of brass under impact and perforation tests for a wide range of temperatures: experimental and numerical approach. Materials 13(24):5821
17. Kolsky H (1949) An investigation of the mechanical properties of materials at very high rates of loading. Proc Phys Soc Section B 62(11):676
18. Lindholm US (1964) Some experiments with the split Hopkinson pressure bar∗. J Mech Phys Solids 12(5):317–335
19. Chen WW, Song B (2010) Split Hopkinson (Kolsky) bar: design, testing and applications. Springer Science & Business Media, Berlin
20. Ramesh KT (2008) High rates and impact experiments. In: Handbook of experimental solid mechanics, p 933
21. Zhao H (2003) Material behaviour characterisation using SHPB techniques, tests and simulations. Comput Struct 81(12):1301–1310

22. Klepaczko J, Malinowski Z (1979) Dynamic frictional effects as measured from the split Hopkinson pressure bar. In: High velocity deformation of solids: symposium Tokyo/Japan, 24–27 Aug 1977, pp 403–416. Springer, Berlin
23. Jankowiak T, Rusinek A, Lodygowski T (2011) Validation of the Klepaczko-Malinowski model for friction correction and recommendations on split Hopkinson pressure bar. Finite Elem Anal Des 47(10):1191–1208
24. Lennon AM, Ramesh KT (1998) A technique for measuring the dynamic behavior of materials at high temperatures. Int J Plast 14(12):1279–1292
25. Rusinek A, Bernier R, Boumbimba RM, Klosak M, Jankowiak T, Voyiadjis GZ (2018) New devices to capture the temperature effect under dynamic compression and impact perforation of polymers, application to PMMA. Polym Testing 65:1–9
26. Recht R, Ipson TW (1963) Ballistic perforation dynamics. J Appl Mech 30(3):384–390
27. Rodríguez-Martínez JA, Pesci R, Rusinek A, Arias A, Zaera R, Pedroche DA (2010) Thermo-mechanical behaviour of TRIP 1000 steel sheets subjected to low velocity perforation by conical projectiles at different temperatures. Int J Solids Struct 47(9):1268–1284
28. Bendarma A, Gourgue H, Jankowiak T, Rusinek A, Kardellass S, Klosak M (2020) Perforation tests of composite structure specimens at wide range of temperatures and strain rates-experimental analysis. Mater Today Proc 24:7–10
29. Bendarma A, Jankowiak T, Rusinek A, Lodygowski T, Klosak M (2019) Perforation tests of aluminum alloy specimens for a wide range of temperatures using high-performance thermal chamber-experimental and numerical analysis. IOP Conf Ser Mater Sci Eng 491(1):012027
30. Landkof B, Goldsmith W (1985) Petalling of thin, metallic plates during penetration by cylindro-conical projectiles. Int J Solids Struct 21(3):245–266
31. Jankowiak T, Rusinek A, Voyiadjis GZ (2020) Modeling and design of SHPB to characterize brittle materials under compression for high strain rates. Materials 13(9):2191
32. Jankowiak T, Rusinek A, Bendarma A (2018) Protocol to define material behaviour and failure strain level at low and high strain rates based on compression test. J Theor Appl Mech 56(2):471–481
33. Bendarma A, Rusinek A, Jankowiak T, Lodygowski T, Jia B (2021) Experimental analysis of the aluminum alloy sheet subjected to impact and perforation process. Mater Today Proc 36:88–93

Water Safety Plan as a Tool for Water Supply Safety

Tamás Berek

Abstract Drinking water supply is a complex process. Experiences and regulations highlight the need for coordinated action in the event of water quality deterioration. This is also a requirement for the implementation of sustainable water consumption. According to the recommendations of the World Health Organization (WHO), the best method to ensure the safety of drinking water distribution systems is to develop and maintain water safety plans. Meeting these objectives may be ensured by a drinking water safety plan based on the relevant hazard analysis and risk assessment.

Keywords Sustainability · Water safety · Drinking water safety plan · Risk assessment

1 Introduction

Currently, 54% of the available freshwater is utilized by mankind and this ratio is expected to reach 70% by 2025 [1]. Water issues are typically found in developing countries. Nearly 18% of the population, i.e. 1.1 billion people, does not have access to healthy drinking water.

In 2002, the United Nations organized a World Summit on Sustainable Development in Johannesburg, South Africa where the WEHAB (water and sanitation, energy, health, agriculture and biodiversity) frameworks were initiated. It identifies the impacted areas and the concluding report of the summit highlights that, among others, measures are needed to be taken to efficiently address water scarcity and achieve sustainable water utilization.

While meeting communal needs, sustainable water consumption may also ensure groundwater sustainability which is essential for both water-related ecosystems (aquatic habitat) and economic activities (agriculture). The quantitative and qualitative aspects of water protection define several goals, including efficient water

T. Berek (✉)
Ludovika University of Public Service, Budapest, Hungary
e-mail: berek.tamas@uni-nke.hu

T. A. Kovács et al. (eds.), *The Impact of the Energy Dependency on Critical Infrastructure Protection*, Advanced Sciences and Technologies for Security Applications, https://doi.org/10.1007/978-3-031-78544-3_11

conservation and prevention of water pollution. For this reason, it is important to treat and remediate wastewater, collect, clean and return to the water cycle as much wastewater of freshwater origin as possible by following the strict regulations [2].

In terms of water supply per capita, Hungary is—compared to other EU member states—considered a water-conscious country with its 60 m^3 consumption per person [3]. After the regime change, due to the closure of significant industrial plants and the introduction of water tariffs, the registered water consumption shows a considerable decline, but at the same time, the rate of illegal water connections has increased.

Even though, according to official data, access to drinking water is fully available in Hungary, 40% of the population still lives in settlements where the water is of questionable quality. Due to the contamination of sediment, soil, and groundwater, the causes and effects as well as the results of interventions become visible only with a significant delay. In certain places, several problems may occur simultaneously and of different natures that need to be addressed [4].

The consequences of climate change adversely affect both the quantity and quality of our groundwater. This raises the risk that, in the event of an incident in the water supply process leading to pollution of the water source, it will become increasingly challenging to replace it with an alternative water source. In light of this, the role of all security measures supporting the security of the drinking water supply will become more critical.

The Water Safety Plan (WSP) is a tool used to detect and reduce the risks to water supply and thereby improve public health and safety, which includes responses to the above risks at water extraction sites, water treatment plants, distribution networks, and consumer points, from the transition from the use of reserve water outlets to the redundant subsystems of water transport, while taking into account a number of intervention points. However, damage to the water supply system is possible because it forces the water supplier to switch to another, temporary, water supply.

2 Risks and Forecasts

Hungary is located in one of the most enclosed basins on the globe where waters enter the country from three different directions and leave to the south. By analyzing the water balance of the country, interesting conclusions may be drawn [5].

A significant amount of rainwater can be expected even if its distribution is not consistent across the country. Due to this phenomenon, the importance of water management significantly increases, especially in terms of agricultural water demand. Additionally, the evaporation rate should also be noted. The quantity of usable water resources heavily depends on the runoff viability.

In Hungary, most precipitation falls in May and June, while the least precipitation may be observed in January and February. The precipitation quantity fluctuates from year to year, meaning that in the wettest year, three times more rain could fall than in the driest one [6].

On the long term, drought and other extreme weather conditions, such as torrential rains may cause serious ecological and economic problems. Due to climate change, these events now happen more frequently both globally and nationally as well. The effects of drought and the lack of irrigation on agricultural productivity are visible.

On one hand side, drought increases water demand, while on the other hand side, water scarcity limits water consumption and irrigation. Damages may be severe in agriculture, energy production, tourism, and other sectors [7].

In the global context, factors such as population growth, urbanization, economic conditions, and social factors are likely to play a greater role in influencing water availability than climate change. However, in economically developed regions, where the population and water consumption per capita are relatively constant, climate change will have a greater impact on water-related problems. In addition to the regional and temporal changes, the qualitative and quantitative changes in water resources may also influence water availability.

The impacts of climate change will be reflected in the future in both the development of water demand and the characteristics of water resources.

Despite volatility, the general temperature rise in the past 40 years is documented. From the beginning of the twentieth century until 2010, the global average mean temperature had increased by 0.74%. This rise is more intense on the continent than it is in the ocean and this is also true of the northern hemisphere, which is more densely covered with land. Between 2001 and 2010 the average mean temperature has passed the degree measured between 1960 and 1990 by 0.46%. The past decade was the warmest ever recorded. Based on current trends, the annual mean temperature in Budapest has increased by 1%, which is mainly due to the significant urbanization effects. In Budapest [8].

Regarding the impacts of climate change in the Carpathian Basin, it is important to point out that the Mediterranean characteristics of the area are increasing and that will be reflected in gradual warming and less average precipitation. Due to the continuously changing environment, extreme weather conditions appear more often. For instance, in the case of a flood, this will mean early floods, larger peaks, and faster flow. Nevertheless, extreme conditions will have a negative impact as well: extremely low water levels and drying will occur more often. As a result, the amount of usable water resources will decrease which would also affect the irrigation, and water supply of fish ponds and would eventually lead to further restrictions. As a result of a smaller runoff in the riverbed, water quality may deteriorate and reservoirs would be filled much slower. Additionally, the water supply necessary for irrigation and for fish ponds may also decrease. Furthermore, the longer the water is stored in a reservoir, the more the quality may deteriorate [9].

Our waters are burdened more and more, and at the same, there is a continuous increase in all areas of usage for good quality potable water. The impact of climate change has a negative influence on both the quality and the quantity of underground and groundwater. It indicates a future danger that in case of an incident in the process of the water supply that can lead to water pollution the solution of which with an alternative water source would be more and more difficult. As a result, the role of all security measures connected with the security of potable water supply will be more and more valuable.

Besides creating a comprehensive program for the handling of situations endangering security it is necessary, for the security of water supply for the population, for the protection of water bases and water management facilities to plan and create a systematic approach.

From the point of view of the establishment of security, it is necessary to establish the types of possible risks. It is necessary to account for the following:

- natural risks that are independent of human activities and which occur as natural disasters
- risks of civilization, and technology that is connected to human activities, due to improper human intervention, carelessness, or because of technical or planning errors
- due to the danger from deliberate or harmful activities.

Water supply facilities are critical infrastructure, the operator must work out an operational security plan that is established by the branch authorities including both content and formal requirements, and where he/she has to determine the organization and means that provide the security aspects [10].

The operational regulations of the water supply system have to include technical, technological, and health directives connected with the proper and secure operation of the establishment, including security technology and personnel provisions.

The integrated physical protection of the water provision facilities needs to ensure that the protection of the production processes through the means and materials used would be achieved.

When establishing the system protecting water production facilities physically it is necessary to identify the inner and outer factors that endanger security, to map their characteristics, and after their assessment to plan the establishment of the defence.

An exceptionally important element of the limitation of the risks of security should be the physical protection of the facilities of operational protection.

Activities performed during water production, from removal of the water to filling it into the network, the amount of the used materials, their value; and dangerousness can be relatively well identified. During the establishment of the water supply facilities, it is necessary to consider some specifications.

In December 2012, Parliament adopted Act CLXVI of 2012 on the Identification, Designation, and Protection of Vital Systems and Facilities, Annex 1 which also refers to the water subsector.

According to the law, the Government must submit an annual report to the European Commission, which must include, on the one hand, the number of vital constituents per sector designated as European vital constituents and, on the other hand, the number of Member States of the European Union that depend on European vital constituents. A report should also be drawn up on the types of vulnerabilities, threats, and risks in sectors where a European vital system component has been designated.

Approaching the issue from the point of view of objects providing drinking water supply, the goal is to maintain not only the hazard-free state of the given object but also of the activity carried out there. This hazard-free state, assuming the trouble-free operation of the safety system—which can only be assumed—may seem constant in time, but it is only apparent. In almost all cases, the activity performed, and among others, the means of activity, the quantity, value, and hazard of the materials used, can be determined relatively well. The change in the harm-free state can be predicted in certain respects, based on, among other things, the function of the object, the activity carried out inside, and the technologies and materials used [11].

During the planning of the complex defence system, during the exploration and analysis of the danger arising from the environment of water production activities, the purpose and object of protection must be determined, the sources of danger must be determined, and the protection system must be designed and built in light of these, specifying the values and activities to be protected in detail.

The internal protection zone shall be fenced and, if necessary, kept safeguarded. The Regulation also declares the definition of those eligible for entry and the need to establish such admission.

When designing and designing protection, one of the primary aspects, in order to ensure high-quality operating conditions of mechanical, electronic, and manned subsystems in the area of the object to be protected, is the design of the planned safety subsystems in accordance with the purpose of the given water production facility.

In view of the fact that the permanent manned protection of wells operating on the territory of the water abstraction facility would be very expensive, the construction, in their area, of the intrusion detection and access control subsystems of the electronic component, can ensure their direct physical protection. To minimize the possibility of unauthorized access, the maintenance and inspection of wells and other activities related to their operation can be carried out by recording access by proxy card and thus by recording the person entering and the time of entry. Unauthorized or unauthorized intrusion can be prevented by the armed security guard if appropriate authority is in place, provided that appropriate authority is in place [12].

However, after the design and development of the system, meteorological effects caused by climate change must be taken into account, the extreme manifestations of which, mainly in the natural environment, can cause energy supply problems, which will take more or less time to correct. Weather conditions such as strong winds and pouring rain often cause overhead power lines to fail.

When constructing a system for the physical protection of water production facilities, as in all other cases, external and internal factors compromising the state of safety must be identified, their nature mapped, and after their evaluation, responses and protection architecture must be determined. An emphatically important element of measures aimed at reducing security threats should be the physical protection of facilities and assets that underpin operational security.

Careful coordination of the requirements of the water safety plan, including the incident management procedure, and the design of the property protection system is necessary to maintain the safety of water quality and the safety of the water production facility and the activities carried out there.

3 Role of Water Safety Plan in Sustainable Water Management

To achieve social, agricultural, and water sustainability, it is required to meet the growing demand for water, including a water supply of adequate quality and quantity.

According to the recommendations of the World Health Organization (WHO), the best method to ensure the safety of drinking water distribution systems is to develop and maintain water safety plans that are based on risk assessment and hazard analysis.

The Government Decree No. 201/2001 (X. 25) on the quality requirements for drinking—modified by the Government Decree No. 430/2013 (XI. 15.)—states that a drinking water safety plan shall be prepared for all water supply systems with a capacity of more than 1000 m^3/day or with a capacity of more than 50 persons. The plan must include a water safety and management system of the water supply system. The drinking water safety plans are to be based on a comprehensive risk analysis and assessment, which should apply to each element of the water supply chain, from water procurement to water consumption. Furthermore, it shall contain hazard analysis, risk assessment, control measures, and a monitoring system as well as a list of corrective and preventive actions and—if needed—an emergency plan, documented throughout the whole water supply system.

A well-structured safety plan will, directly and indirectly, contribute effectively to the goals of sustainable water management through the implementation of a system that is grouped around the elements listed [13] (Fig. 1).

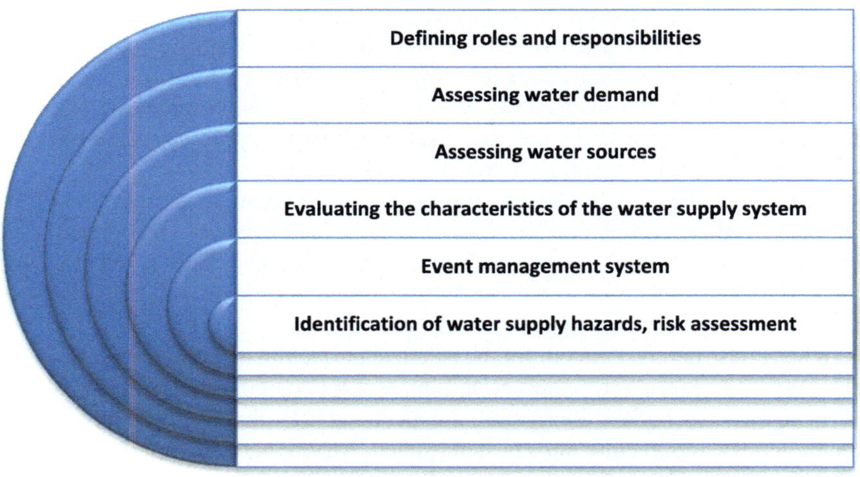

Fig. 1 Role of water safety plan in sustainable water management. *Edit* Berek Tamás

Defining Roles and Responsibilities

In order to ensure a secure drinking water supply, in addition to the relevant control measures built into the water supply process, special attention must be paid to guarantee the safe execution of those additional tasks that are required to maintain good water quality. These are mainly monitoring control measures. Similarly to other security systems, the efficiency of a water safety system is defined by the efficiency of its weakest element. Without the necessary precautions, human factors are often considered as the weakest element of these systems. In order to maintain the security of facilities and operational safety, and in addition to the definition of roles and responsibilities, it is essential to ensure the right control, establish a regulatory system, including regulations of omissions or sabotage, and define liability and accountability.

Measuring Water Demand

According to the Council of the European Union Council Directive 98/83/EC, "water intended for human consumption shall mean all water either in its original state or after treatment, intended for drinking, cooking, food preparation or other domestic purposes, regardless of its origin and whether it is supplied from a distribution network, from a tanker, or in bottles or containers".

Water production capacity shall be in line with the actual demand, however, it must also take into account the long-term needs. It is also necessary to define the purpose of water utilisation in order to ensure that the water safety plan is following the quality requirements.

Regarding the supplied settlements—number of households and inhabitants -, both the average and peak water consumption and water capacity should be measured and documented.

Measuring Water Resources

Concerning the installation of an adequate water abstraction system, it is important to measure the typical level of water resource, its highest and lowest rates as well as the extent of contamination.

Water supply companies seek to meet the drinking water quality criteria by introducing the necessary water protection measures to maintain the purity of surface and underground water. These measures are outlined in the water resource protection plan.

For long-term and sustainable water production and economical operation, the mode of operation, depth, and capacity of the well as well as the water production needed to be analysed well by well.

Classifying water quality is essential to determine the adequate water treatment technology. The overall water quality conditions and the microbiological characterisation of produced water serve as a basis for the development of disinfection technology. As long as it is not necessary, expensive investments should be avoided. The question of disinfection by-products (DBPs) is another important factor.

In Hungary, the most widespread materials used for disinfection are sodium hypochlorite solution, chlorine, and chlorine dioxide. During the disinfection process, as a result of chemical reactions, trihalomethanes (in the case of hypochlorite or chlorine) and chlorite (in the case of chlorine dioxide) are formed. These by-products are harmful to human health [14].

According to the aforementioned EU directive, "member states shall take all measures necessary to ensure that, where disinfection forms part of the preparation or distribution of water intended for human consumption, the efficiency of the disinfection treatment applied is verified, and that any contamination from disinfection by-products is kept as low as possible without compromising the disinfection" [15].

Assessing the Main Characteristics of the Water Supply System

The pressure, the type of water tank and pipeline, the material of the network and the diameters provide useful information that helps to predict possible network fractures and issues. Identifying and assessing the mechanical vulnerability of the pipeline network is of great importance. In the case of several decades-old waterworks systems, the majority of the operational issues are due to pipe bursts.

The risk of contaminating drinking water during the occurrence of burst pipes and the subsequent repair work is very high. The assessment-based reconstruction work is cost efficient, and predictable and reduces the number of unexpected events in a given section. The same applies to soil testing as well. The soil aggressiveness assessment allows the application of protection procedures that increase the lifespan of a given network segment under adverse soil conditions. Together with the risk assessment of various impacting factors, such as biofilm formation, and pollution), the quality characteristics of water distribution networks are also important elements.

Based on the EU directive on the quality of water intended for human consumption, "member states shall take all measures necessary to ensure that regular monitoring of the quality of water intended for human consumption is carried out".

The characteristics of the distribution system (lifts, basins, high tanks, applied materials, and pipe length) must also be defined. Generally, the Operating Rules and Regulations document contains a detailed description of these items.

Identifying Threats to Water Supply, Risk Assessment

Both the WHO recommendations and the cited EU document state that the determination of specific values of water intended for human consumption should be based on public health considerations and risk assessment methods. Then, the defined parameters should be kept at the point where water intended for human consumption is made available to the appropriate user.

The risk assessment process aims to determine the likelihood of adverse effects on drinking water safety about sources of exposure. The sources of exposure may be of biological, physical, or chemical nature as well as of industrial or other origin. Following a detailed description of the system, a hazard analysis may be performed by identifying all potential hazards.

During the execution of hazard analysis, and in addition to the actual sources of danger threatening the security of the drinking water supply, it is important to explore the possible events and activities that may lead to the contamination of drinking water. Threats to water security may be categorized and described based on their probability of occurrence. Regarding a secure drinking water supply, it is important to note that not only the incidental hazardous impacts are taken into account, but the evaluation of processes with long-term impact as well.

To maintain a hygienic environment and the basic conditions for drinking water treatment, safeguard technologies shall be incorporated.

The water safety plan identifies the critical points in the supply chain where technological failure, negligence, sabotage, etc. may occur. Additionally, in these cases, regulation, periodic inspection, or monitoring is necessary the ensure a safe, clean drinking water resource and to prevent contamination. These critical control points are determined with great attention in both the technological process of drinking water production and the storage and transportation phase, as unjustified determination of critical control points would increase operating costs while under-planning would weaken the quality management system of the production.

Concerning the water resource, further examination must be carried out in terms of raw water and protection area.

Event Management System

Besides the protection of catchment areas, water resources, and water management facilities, to ensure the safety of drinking water supply systems, a program must be implemented to provide a protocol for comprehensive management of security incidents and events. With this program, the aim is to prevent or reduce the occurrence of events that may accidentally contaminate water.

Information and regulations relevant to preventive actions, controlling processes, and corrective measures are determined in the water safety plan and are assigned to a specific hazard.

Taking into account the safety risks that are threatening the operation of the drinking water supply system and the quality of drinking water as well as the responses given to this threat are of great importance for maintaining a secure water supply. A well-designed and constructed event management system—which is a prerequisite for the water safety plan—may ensure the protection of the water supply. The ranking of such events facilitates the management of incidents, which are threatening drinking water security and differ from the normal state. The classification of these events helps to determine whether the water service provider is capable of solving the issue with its resources or if it would require the involvement of external assistance as the problem goes beyond its capabilities. From detection through classification to the determination of response and recovery tasks induced by events or emergency situations, the event management processes must be integrated into a complex system so that it defines both the activities to perform and the decision-making framework. Certainly, the operation of this system must be maintained and frequently tested.

4 Conclusion

Water scarcity and related public health issues continue to create severe situations in several regions across the globe.

The Earth has lost—mainly because of climate change—half of its wetlands in the past few decades, while the water demand kept rising and it may even increase by 50% for future generations. This unfavourable ratio between the two phenomena may intensify further. The population is therefore likely to move towards more preferable areas, and this migration will cause a significant additional burden to the coastal zones.

Approaching this field from a global perspective, in order to ensure water supply as a priority service, it is necessary to provide water protection through an appropriate water safety philosophy, which also enables the sustainable development of water management, as a whole.

The major rivers in Hungary are sourced outside the borders, hence water management greatly depends on the neighbouring countries as well. The cooperation with neighbouring countries is good, but neither the problem of the Bős powerplant nor the cyanide pollution have been satisfactorily resolved. Both cases indicate that Hungary does not fully control the freshwater flow and, in times of crisis, it heavily depends on foreign partners.

Planning a sustainable water management approach is greatly contributing to the fight against climate change. According to forecasts, water management will also face extreme rainfall conditions in the future, which could lead to both flood risks and droughts at the same time. Flood control in smaller river basins and protection against flash floods in low-lying areas and settlements are currently challenging, although in view of recent cases, one must prepare for hydrometeorological events involving flood-like precipitation. The development of efficient irrigation systems in areas hit by drought may balance the lack of precipitation. It may also increase the water retention capacity in Hungary and ensure a more focused protection of water resources.

Sustainability is an aspect of a rational policy that defines the long-term management of environmental resources, taking into account economic and social interactions. It also determines that the purpose of development is to ensure social welfare with available tools such as the economy, while it is limited by the utilisation of natural resources.

Drinking water supply is a complex process and experiences show that, as highlighted also by the regulations, fighting against water quality deterioration requires coordinated measures to comply with the health and other parameters. Water protection measures should be designed by the principle of complexity to be consistent with the measures implemented to regulate activities that have an indirect impact on water supply. These measures include for instance the application of Council Directive 91/414/EEC concerning the placing of plant protection products on the market and of Directive 98/8/EC of the European Parliament and the Council concerning the placing of biocidal products on the market. The Directive 2000/60/EC of the European Parliament and of the Council declares that "further integration of protection and sustainable management of water into other Community policy areas such as energy, transport, agriculture, fisheries, regional policy, and tourism is necessary".

The climate sensitivity and climate vulnerability of society is much higher today than it was 100–150 years ago. This fact also justifies the fact that it was caused by the weather paying close attention to security risks.

The water safety plan lists the efficient actions that may guarantee the security of drinking water supply from the extraction of raw water to the supply of households. Because of this, it may become an efficient tool for water quality management. The main purposes of the water safety plan are, among others, the exploration of the entire water supply process, the determination of critical control points based on threat analysis and risk assessment, the effective event management. It also aims to ensure that the service providers are dedicated in terms of drinking water supply and that they seek to maintain the safe condition of drinking water. In addition to the recommendations and guidelines, the structure and the content of the water safety plan must be determined as a system. To achieve this, it is necessary to eliminate the identified issues of water safety planning in the future and to ensure universal conditions of development.

References

1. Bula M, Gyulai I, Ónodi G, Pajer J, Pestiné Rácz Éva V, Radnainé Gyöngyös Z, Rédey Á, Zseni A (2008) Environmental health assessment, environmental status of Hungary, monitoring (2008)
2. Sustainable Development Indicators in Hungary (2014) Hungarian Central Statistical Office, 2015. ISSN: 2064-0307
3. Bulla M, Guzli P (2015) Sustainable development indicators in Sustainable development in Hungary strategic research—Hungary. Új Mandátum Publishing. ISBN 963 9609 38 2
4. Somlyódy L (2011) Quo vadis domestic water management? Strategic summary in public body strategic programmes. In: Somlyódy L (eds) Hungary water management: state of play and strategic tasks. Hungarian Academy of Sciences, Budapest (2011)
5. Sustainable Development Indicators in Hungary (2015) Hungarian Central Statistical Office, 2015. ISSN: 2064-0307
6. Somlyódy L (2008) Musings about water, Magyar Tudomány 2008/04. http://www.matud.iif.hu/08apr/09.html
7. Glatz, F.: Water management in the Carpathian Basin, Executive summary, Centre for Social Sciences of the Hungarian Academy of Sciences. http://www.mta.hu/fileadmin/2009/11/07_Vizgazdalkodas.pdf
8. Act CLXVI of 2012 on the identification, designation and protection of critical systems and facilities
9. Berek T, Bodrácska G (2011) Use of physical protection tools to support manned protection of construction works Bolyai Szemle 20(2). ISSN: 1416-1443 (2011)
10. Berek T, Rácz L (2013) Protection of water resources as national vital constituents (Vízbázis mint nemzeti létfontosságú rendszerelem védelme). Hadmérnök **8**(2), 120–133 (2013)
11. Zsuzsanna D (2011) Difficulties in drawing up drinking water safety plans, National Public Health Institute, Water Safety Department, abstract, 8th Forum of Young Hygienists conference, Gödöllő

12. Kis A, Borsányi M, Brodmann T, Dávidovits Z, Dömény J, Hüse P, Soósné Harsányi S, Szebényiné Vince B, Professional recommendation No. 6/2014 of the Hungarian Water Utility Association for the preparation of drinking water safety plans
13. Council Directive 98/83/EC of 3 November 1998 on the quality of water intended for human consumption
14. Berek T, Dávidovits Z (2012) The role of a water safety plan in the drinking water supply security system. Hadmérnök 7(3):14–25
15. Padányi J (2022) Challenges, risks, responses. In: Climate change challenges and their impact on military power. Ludovika Egyetemi Kiadó. ISBN 9789635318131

Africa and the Russian-Ukrainian War and Its Impact on Africa

János Besenyő

Abstract Although the Russian-Ukrainian conflict has been going on with varying degrees of intensity since 2013, it has not had a significant impact on the international community. However, this changed in the summer of 2022 when open war broke out between the opposing sides. The conflict, which started as a regional one, quickly became an international one, and the members of the international community were also forced to take a stand. Although African and Middle Eastern countries are not directly involved in the conflict, it affects them in several ways. In my research, I will briefly present the ongoing conflict, its causes and the parties interested in it. After that, I will describe how the opposing parties were/are in relation to the African countries and how these countries reacted to the outbreak of the war. After that, I introduce the impact of the "Europeans war" on the African countries, how it affects them, and what challenges and opportunities it presents. I deal separately with the political, diplomatic, economic (especially with food, fertilizer and fuel insecurity) and security effects of the conflict, and the African reactions to it.

Keywords Russian-Ukrainian war · 3F (food fertilizer and fuel) insecurity · Wagner group

1 Introduction

In an article in 1963, Edward Lorenz pointed out how sensitive some so-called non-linear systems are to small—but ever-increasing—deviations of the initial conditions due to the dynamics inherent in the system ("sensitive dependence on initial conditions"), i.e.,—for example, in the case of the behaviour of the weather and the atmosphere—how disordered, chaotic, in their case predictions for the future are limited, and the reason for this is not the rudimentary nature of the available tools,

J. Besenyő (✉)
Banki Donat Faculty of Mechanical and Safety Engineering, Obuda University, Budapest, Hungary
e-mail: besenyo.janos@bgk.uni-obuda.hu

T. A. Kovács et al. (eds.), *The Impact of the Energy Dependency on Critical Infrastructure Protection*, Advanced Sciences and Technologies for Security Applications, https://doi.org/10.1007/978-3-031-78544-3_12

but the basic properties of the system itself [1]. His work is the basic premise of the so-called chaos theory based on the above and the related butterfly effect, according to the latter of which a local event can have a lasting effect on the outcome of events at a greater distance in time [2]. The theory basically developed in the interpretation of the natural sciences. The question was to what extent they can be used in a social-theoretical approach, i.e., among other things, to describe and understand economic and social processes. In this field, the interdisciplinary usability of the theory and its continuing popularity and relevance to this day is proved by the fact that there are those who wish to explain the changes and dynamics of the international system using the theory [3]. In addition, given that in European history following the Second World War, through the Russo-Ukrainian war that broke out in 2022, the European nations and the international system were confronted with violence not seen for a long time, it is understandable that there are those who want to use the theory to explain the war also to the resulting global effects [4].

Despite the fact that both Russia and Ukraine account for only two per cent of the world's economic output [2], the global, systemic consequences of Moscow's aggression and the resulting war between the two countries are now evident, whether these are political or social effects, the slowdown of economic growth, inflation, the increase in food and energy prices, disruptions in supply chains, or the intensification of migration processes [5]. But here, we can also mention the direct and indirect climate and environmental security issues from local to global scale that came to the fore in connection with the war. What is more, it also comprises systemic changes that include the transformation of the decades-old practice of outsourcing domestic production to fundamentally developing states with lower labour costs, i.e., the abolition of offshoring and its relocation to states that provide a more predictable political and economic environment (back-shoring or friend-shoring), as the United States and Italy did [3].

Although the war is primarily taking place on the European battlefield, its global implications are felt in the international system as a whole, especially in the developing states. As one of the winners of previous globalization processes, Africa is one of the biggest victims of the current war since the conflict, in addition to generating new waves of migration [6], which fundamentally complicates the food supply and slows down economic growth [7]. The consequences of this also appear in the fields of health and education [8]. It is no coincidence that the African Union Development Agency initiated a comprehensive dialogue involving many groups of society to create a more resilient social and economic unit [9], as well as the fact that the UN gave a comprehensive assessment of the situation regarding the context of the conflict in Africa [10]. The long-term consequences of the Russo-Ukrainian war, such as the poverty rate and social inequalities as well as the increase in the debt stock of the countries, can also be forecasted, so it is, therefore, necessary to evaluate the effects of the events at the systemic level in addition to the local European effects [11].

Some clearly highlight the implications for food and energy security and trends in armed conflict as a direct consequence of the war [12]. Others emphasize the impacts felt in the financial sector in addition to energy prices and their supply and trade as three main areas [13]. In addition to the above, there are also those who mention the rise of inflation as a key challenge with regard to Africa [11]. Faced with a kind of livelihood crisis, in the light of rising energy prices and disruptions in supply chains—primarily in relation to petroleum, for example—, the question is how Africa is able to organize itself and how it is able to face the consequences of the war and make the most of it, equally in terms of security policy, political, and economic development [14]. In my study, I try to get an answer to how the Russo-Ukrainian conflict affects the African continent in the short and long term. Also, what consequences the resulting changes in Africa may have for the rest of the world.

2 The Russo-Ukrainian Conflict

Ukraine gained its independence with the dissolution of the Soviet Union in 1991, then renounced its nuclear weapons in 1994, in return for which the USA, Great Britain, and Russia guaranteed its territorial integrity and independence in the Budapest Memorandum [15]. In the following years, the country's leadership was sometimes influenced by Russia and sometimes by the West, and its leaders created policy accordingly. This brought more and more tension not only between the political forces within Ukraine, the oligarchs and other groups, but also between Ukraine and Russia and between Russia and Western (European) countries, as the Russians feared that the US, the EU, and NATO would draw Ukraine into their sphere of interest [16]. In the heated situation, the opposing parties increasingly resorted to underhanded means, such as election fraud, discrediting or possibly poisoning political opponents, and using physical violence against each other. Until 2014, governments that were pro-Russian and others that were closer to the West alternated. Then, the pro-Russian Viktor Yanukovych was swept away by the "Maidan Uprising" that began in November 2013, and a pro-Western government came to power [17]. In February 2014, the Russian leadership, which was dissatisfied with the changes in Ukraine, provided financial and military support to separatists of Russian nationality who wanted to secede from Ukraine in the eastern and southern regions. At the same time, they attacked and occupied the Crimean Peninsula, which was soon annexed [18]. Then, with Russian support, the separatists created the Donetsk People's Republic and the Luhansk People's Republic, which, however, were not recognized by anyone. The opposing parties first concluded a ceasefire in September 2014 (the first Minsk Agreement) [19], then in February 2015, the second Minsk Agreement was concluded, which, however, did not fulfil the hopes attached to it, and the hostilities continued as a kind of proxy war, albeit at a low intensity [20].

The next phase of the conflict began at the beginning of 2021 when the Russian leadership carried out troop consolidations on the Russo-Ukrainian border and then protested with increasingly bellicose rhetoric against the eastward expansion of NATO and the possible accession of Ukraine to NATO. In fact, they demanded that Ukraine could never enter the defence organization and asked for guarantees

[21]. After they did not receive this, they recognized the two breakaway "states" as independent states, and then, on February 24, 2022, they launched a "special operation" against Ukraine. Although the Russian leadership expected quick successes and trusted in the support of the Russian-speaking population living in Ukraine, the "lightning war" stalled due to the desperate resistance of the Ukrainians, whom they underestimated. The Ukrainians were also supported by a large part of the international community, which provided them with weapons and financial support [22], and economic sanctions were introduced against Russia [23]. Since then, a kind of stalemate has developed. Although the Russians were not able to capture Kyiv and replace the pro-Western government, they are occupying—and have even annexed—parts of the eastern and southern territories of Ukraine. On the other hand, although the Ukrainians fought heroically and made good use of the military equipment provided by the USA, the EU, and NATO as support, they were not able to recapture the territories previously occupied by the Russians and push the Russian troops out of the country. The open war that has been going on for more than a year has had an impact not only on Russia and Ukraine or on the Central and Eastern European region but also on the international community. These effects and the consequences of the conflict have been studied by several people, including the impact it has had on the African continent from political, social, economic, security, and other points of view.

3 Russia's Relationship with Countries of the African Conflict at the Outbreak

The presence of Russians in Africa has become increasingly visible in recent decades. Although after the end of the Cold War, it seemed that they gave up some of their positions—which other emerging countries like India, China, Turkey, Japan, Brazil, Iran, etc. tried to acquire –they never completely left the continent. Moreover, Russia's relationship capital—the basis of its later influence—was preserved, as several of those with Russian education became presidents or influential political, economic, and military leaders, many of whom are still in positions today [24]. After Putin came to power, he dealt with the oligarchies, strengthened and centralized the government, started the development of the army, and launched a much stronger Africa policy than his predecessors. In the 2010s, it regained a significant part of its previous influence [25]. Although it was less visible in the economic field than the USA or China, the Russia-Africa Business Council was nevertheless established in 2002, and several economic programmes were subsequently organized (e.g. Russia-South Africa Business Forum, etc.). In these years, the Russians regained their old positions extremely quickly and built up political, diplomatic, economic, and security policy cooperations, especially in those countries with which they maintained good relations even during the bipolar world order, but, of course, they also tried to build good relations with other African countries [26]. The Russian leadership launched generous aid, financial, and educational projects and established military/police cooperation with several African countries [27]. The change in African politics was also observable at the level of official communication. Russian foreign policy in the material titled "A Survey of Russian Federation Foreign Policy" and published in 2007, called Africa a continent plagued by problems, but important from a world political and economic point of view, where Russian influence must be strengthened. The leaders of the country also dealt with Africa in the National Security Strategy adopted in 2009, and subsequently, based on the Foreign Policy Concept of the Russian Federation approved by President Putin in 2013, they further strengthened bi- and multilateral relations with African states as well as with the BRICS states [28] and established the Russian-African Forum (RAF), where they try to promote cooperation between Russian and African economic, educational, and scientific organizations. Russia participates in the work of the Pan African Forum and the New Partnership for African Development (NEPAD), sent an ambassador to the African Union, and cooperates with such African regional organizations as ECOWAS, SADC, EAC, IGAD, and other intergovernmental organizations [29]. Russia's interest in Africa is multifaceted. In contrast to other countries, it is not only economic—because it itself has quite a large stock of raw materials—[30], but rather political, to which it often subordinates other aspects as well, as the continent of 54 countries provides an opportunity for the country to break out of its isolation by Western countries and even take effective countermeasures [31]. The Russians do not deny that—similarly to the Chinese—cooperation with Africans is desirable not from a principled, but a pragmatic point of view (intelligent pragmatism/razumniy pragmatism) [32]. They consider it important to be as vulnerable as possible to the Western liberal states and their economic sanctions, so they also try to diversify and strengthen their economic relations with African countries [33]. They condemn the practice imposed by the Western world, which wants to impose liberal-democratic principles on African nations during their activities in Africa [34]. The situation of the Russians was strengthened by the fact that the USA was not only less active on the continent under the Trump administration but even reduced its presence [35]. That is why more and more African states have looked to Russia as a possible cooperation partner and have established good relations with it. The most important partner countries of Russian diplomacy at the outbreak of the conflict were Algeria, Egypt, Morocco, South Africa, Angola, Guinea, Namibia, the Democratic Republic of the Congo, Gabon, Zimbabwe, Ethiopia, Cameroon, Ivory Coast, Kenya, Mozambique, and Tanzania [36].

4 Political Reactions of African States and Attempts to Win Them Over

The Russo-Ukrainian war was a test of African unity. The different reactions to the conflict clearly proved that African institutions and regional organizations (African Union, ECOWAS, SACD, etc.) were unable to form a unified African position regarding the war. After the attack by the Russian troops, Kenya, Gabon, and Ghana immediately condemned the Russian invasion, while South Africa only made this decision after a long hesitation. On the other hand, although Nigeria expressed surprise at the conflict, it did not condemn Russia. During the voting of UN resolutions related to the conflict, it was also seen that the African countries were not able to act in unison, but each one voted in accordance with its own interests or even abstained from voting [37]. On March 2 2022, when the UN voted on the Russian invasion, 28 of the 54 African countries (51%) voted for the resolution condemning Russia, 17 (31%) African countries were present but did not vote, 8 (15%) African country did not even go to the vote, while Eritrea (2%) was the only African country to vote against the resolution condemning Russia [38].

On April 7 2022, when they wanted to suspend Russia's membership in the UN Human Rights Council, only 10 (18%) of the African countries supported it, 9 (16%) opposed it, and 24 (44%) did not vote while 11 (20%) did not attend the vote [39].

On October 12 2022, the World Organization held a vote again, where a large majority of the participating states condemned Russia's "illegal annexation" of Ukrainian territories. The African states could not form a unified position here either. The resolution condemning Russia was voted for by 29 (54%) African countries, 19 (31%) countries abstained, and 5 (9%) countries did not show up for the vote [40]. According to some experts, states with democratic traditions supported the resolutions condemning the Russians, while anti-democratic/authoritarian countries either did not support the resolutions or abstained from voting.

In these votes and from the results of the United Nations General Assembly votes held in November 2022, February 2023, and March 2023, it was clearly proven that the "Africa policy" pursued by the Russian leadership has borne fruit, and a stable group of African countries support Russia's presence in Africa, as it is seen as an alternative to the United States of America, China, and the former colonial countries (France, Great Britain, Germany, Spain, Portugal, and Italy) [41]. In addition, in several African countries, Russia is seen as a supporter of previous liberation movements and thus the independence of African countries, which also does not interfere in their internal affairs and does not force them/hold them to account for Western democratic values [42]. The term "multipolar world with different poles of influence" voiced by the Russians is also positively received, especially in light of the fact that some Africans see Western countries as colonizers who exploit the countries of the continent with other, less violent methods compared to the former colonizers. The leak of a classified EU document also clearly indicates that European decision-makers also see the division of African countries, so they are trying to influence them in several ways (aid, cooperation, etc.) to put on the side of Ukraine those countries that did not condemn Russia in the previous votes [43]. To facilitate this, Germany, France, and the USA sent high-level delegations to several African countries to further strengthen their existing relations with them [44]. Between 13 and 15 December 2022, the American president organized another US-Africa Leaders Summit after an 8-year absence, in which 49 African leaders participated [45]. However, the African countries participating in the meeting still did not line up behind the US and its Western allies, and several of them still maintain good relations with Russia [46]. It is probably also due to this that, compared to before, the US wants to play a much more active role on the African continent through various programmes in order to somehow reduce the rise of China and Russia [47]. However, this will not be so simple, as shown by the fact that 49 African delegations participated in the second Russia-Africa Summit, which began on July 27 2023, only 17 of which were led by the country's highest position leader. This also shows that several African countries still want to cooperate closely with Russia despite the fact that the US and its Western allies are doing everything to support Ukraine against Russia [48].

Ukraine, which so far has not paid much attention to the African continent, is also trying to win over African countries. As a first step, on June 20 2022, Volodymyr Zelenskiy organized an online meeting with the leaders of African countries, where he would have discussed the Russo-Ukrainian conflict. However, only four African leaders took part in the meeting, despite the fact that the Ukrainian president has stated in several forums that he views African countries as strategic partners and wants to strengthen Ukraine's presence in Africa [49]. To prove this, he appointed Maksym Subkh as a special envoy to African and Middle Eastern countries, whose activities are supported by American and Polish diplomacy [50]. For the second time, in October 2022, he sent Foreign Minister Dmytro Kuleba to several African countries to win their support. At that time, the "Grain from Ukraine Humanitarian Programme" was established with international support, within the framework of which Ukraine provided free grain to several African countries [51]. This attempt proved more successful [52], so much so that on June 16 2023, six African states (Comoros, Egypt, Senegal, South Africa, Uganda, and Zambia) sent a delegation to Kyiv who wanted to promote the Russo-Ukrainian settlement, but more serious results were not achieved [53]. Although the Ukrainians pursued a more active foreign policy on the African continent than in the previous period, some African countries still did not terminate their partnership with Russia. Of course, Russia is also trying to strengthen its influence in African countries, and not even without success [54]. This is clearly visible in the Central African Republic, Burkina Faso, Guinea, and Mali, where demonstrations against the local elite and the Western powers took place in which Russian flags were waved and Russia was cheered by the demonstrators. The leaders of the mentioned countries, referring to the fact that the UN, the AU, and the international community were unable to act effectively against the various terrorist groups, therefore asked for help to consolidate their security from the Russians, namely the Wagner group—being closely connected to the government—instead of the peacekeeping/training forces of the former, mainly Western countries [55]. The consequence of this is that Mali was abandoned by the French, and the Wagner Group's militants arrived in their place [56]. In the past period, the leaders of Mali called on the UN operation in the country (MINUSMA) to leave the country [57]. In the meantime, a military coup was also carried out in Niger against the democratically elected government, and the new military leadership has already contacted the Wagner group, with whom they are preparing for closer cooperation [58]. It seems that the Wagner group has already established a foothold in the Democratic Republic of the Congo, which could also pose security risks in the region [59]. Although the Wagner Group's position has weakened after the failed coup against the Russian military leadership, this is less evident in their African activities, and they even seem to continue to represent Russia's interests on the continent [60]. Russia's position is expected to be strengthened by the change in the refugee policy of Western states since the outbreak of the Russo-Ukrainian war. The previous permissive migration policy was criticized by many people, mainly from a security point of view, but since the skilled and unskilled labour force was needed, no major strictures were made. However, after the outbreak of the Russian-Ukrainian war, more than eight million Ukrainians and hundreds of thousands of Russian refugees arrived in Western countries [61], who, in many cases, could integrate more easily than African refugees from certain countries. Ukrainian and Russian refugees generally come from the same cultural (European) background and have the appropriate educational and professional background to settle down and take up work. Many people consider the admission of Ukrainian refugees to be a good opportunity, as they can reduce the number of unfilled jobs, and they can integrate more easily and become "net taxpayers" relatively quickly. Of course, difficulties may arise in accepting so many refugees, which few people have talked about so far [62]. This may mean that some of the African migrants are pushed out of the labour market of Western countries and may have to return to their countries of origin. At the same time, European countries will tighten immigration, employment, and other opportunities for them. This means that more people stay in their countries, where they cannot support themselves and their families without adequate work and earnings. This, however, will result in new economic and security challenges for African countries [63].

5 Food, Fertilizer, and Fuel Insecurity

The majority of African states are unable to provide for their populations on their own, so they need to import mainly food products, fertilizers, fuel (the "three F's"), and several other products. This can be clearly seen when we look at the wheat imports of African states between 2007 and 2019, which increased by more than 68%, reaching 47 million tons. However, this is not enough. According to some estimates, the wheat demand of African states will reach 76.5 million tons by 2025, of which 63.4% will come from outside the continent [64]. Based on the 2020 data, the two countries involved in the Russo-Ukrainian conflict account for 53% of the sunflower oil/seed, 27% of wheat, 23% of barley, 16% of rapeseed, and 14% of corn available in international trade. However, since the outbreak of the conflict, these products have not reached the international market, and thus the African market, due to acts of war, economic sanctions, and other reasons. This is interesting in light of the fact that 42% of the wheat purchased by the countries of the African continent is provided by Russia and Ukraine, so the war that broke out between them had an immediate negative impact on food prices [65]. Due to the sanctions introduced by the Western states and the Russian actions in response (blocking Ukrainian grain exports in the Black Sea), much less wheat, corn, and sunflower oil reached the countries of the continent, which led to shortages and price increases. The rise in prices was facilitated by the fact that many parts of the continent (e.g. Horn of Africa, West Africa, etc.) have been affected by years of drought, which means that they are able to produce less grain than usual, so they have a much greater need for food from abroad, the price of which is rising rapidly. This led to more and more people not being able to access basic types of food, so famine developed in several places [66].

The largest receiving markets for Russian and Ukrainian agricultural products are the North African countries (Algeria, Egypt, Libya, Morocco, and Tunisia) [67], where the states try to keep the prices of cereals and cooking oil low with state subsidies, as essential foodstuffs are vital for the extremely fast-growing population. In addition, Ethiopia, Sudan, Kenya, Nigeria, and South Africa also purchase most of their cereals and cooking oil from here. According to the UNCTAD report, 16 African states obtain more than 50% of their wheat needs from Russia and Ukraine. Among them, Somalia and Benin get 100%, Egypt gets 82%, and Sudan gets 75% of the wheat it needs from the mentioned countries [68]. Russian and Ukrainian grain imports are vital for African countries, as they are unable to produce enough grain for their own population [69]. That is why, under the pressure of the international community, Russia and Ukraine agreed on July 22 2022, in Istanbul (Black Sea Grain Initiative/BSGI) to ensure, regardless of the war, that the agricultural products of both countries (mainly cereals and fertilizers) pass through the Black Sea and reach the international and African markets. With this, it was possible to stabilize the international grain markets, and although the increase in food and energy prices did not stop, the supposed "hunger riots" did not occur [70]. Although the Black Sea Grain Initiative was extended several times, it faced continuous challenges and problems until Russia withdrew from the agreement in July 2023, so the delivery of Ukrainian agricultural products to the market will again face obstacles, and another rise in grain prices is expected [71].

The agriculture of several African countries (Ghana, Ivory Coast, Mauritania, Cameroon, Senegal, Kenya, etc.) was also negatively affected by the fact that they could not buy fertilizers (fertilizer/nitrogen fertilizer, potassium and phosphorous fertilizers) and other plant protection agents from Russia, which is one of the largest exporters in the world thereof, which caused crop yields to drop and thus meant they could produce less food to feed the population. According to one analysis, there could be as much as a 20% drop in food production in Africa due to the lack of fertilization [72]. Since the natural soil fertility of African countries is low, fertilizing the lands is essential for an adequate crop yield [73]. The world market price of urea and phosphate, which are the main components of fertilizers, has already increased significantly in 2021 [74]. This trend continued even after the outbreak of the conflict, and even Russia and China, the largest exporters of fertilizers, introduced export bans to meet the needs of their own agriculture [75]. Later, Ukraine also introduced similar measures, which led to further price increases on the international and African markets [76]. Among the African states, Morocco, Nigeria, and Algeria export fertilizer and are able to increase production and thus the market share, which can mean extra income for the mentioned countries. This was done by Nigeria as well, where the continent's largest fertilizer factory was opened shortly after the outbreak of the Russo-Ukrainian war. With this, Nigeria is not only able to supply its own economy with fertilizer, but also receives significant export income [77]. Similarly, Morocco also significantly increased the production and sales of phosphate [78]. However, other African countries need imports, and thus, their agriculture is exposed to shortages and increasingly higher purchase prices. For this reason, several African countries (Algeria, Tunisia, and Burkina Faso) introduced an export ban on some food products and fertilizers [76]. Interestingly enough, the countries of the African continent would be able to supply themselves with fertilizer since they have the production capacity, but for various reasons, they have not been able to organize production and trade until now [79].

The Russians are the third largest exporter of crude oil after the USA and Saudi Arabia, and they are the second largest exporter of natural gas. Its customers include a significant part of African countries, so the sanctions imposed on natural gas and oil industry products also had a negative effect on their economy. The amount of available fuel decreased, and its price increased, which made food production and transportation more expensive, so the price of food increased significantly further. The lack of Russian crude oil has caused problems in the energy sector of Cameroon, Nigeria, and Equatorial Guinea, and Lukoil, which is very active in the region, has frozen or minimized its African activities, which also has had a negative impact on the local economy.

The weakening of food security is aided by the extremely dry weather of recent periods, as a result of which millions of Africans in Somalia, Ethiopia, Kenya [80], Mali, Burkina Faso, Mauritania, Niger, and even Nigeria are starving and in need of food aid, which the INGO/NGO organizations are unable to provide for them, or only in a limited capacity, due to the rising price. In addition, some of these organizations as well as the donors who support them, shifted the centre of gravity of their activities to the war-torn Ukraine, so African countries receive less aid [81]. In addition, due to the worsening situation in Ukraine and the winter, even more support is needed for the Ukrainian population [82], which may further reduce the resources allocated to the treatment of famine in Africa. If African governments and the international community are not able to handle the famine in a satisfactory manner, it may also contribute to the intensification of conflicts in Africa [83]. This is proven by the fact that in the weeks after the outbreak of the Russo-Ukrainian war in Kenya, due to the increased food prices, more and more groups of society expressed their dissatisfaction on social media platforms [84]. In the summer of 2022, several demonstrations against the government took place in South Africa due to skyrocketing food and fuel prices, which have not been completely eradicated since [85]. A similar situation occurred in Nigeria, where the economic crisis since 2016 was further deepened by the war, as a result of which more people protested in several cases due to the steeply rising food prices and the increasingly difficult living conditions, to the point where protesters started robbing granaries and food stores [86]. Prices also rose significantly in Senegal in August 2023, which is why experts expect further conflicts and violent demonstrations [87]. As a kind of push factor, the fall in food prices can increase internal migration in Africa and emigration to Europe, for which neither the African nor the European states were really prepared [88].

Many people believe that the Russo-Ukrainian conflict is an excellent opportunity to develop the agriculture of African countries and thereby reduce the continent's food dependence since 60% of the world's arable land is located on the African continent [89]. An analysis cites Zimbabwe as a positive example, as the country increased its wheat production by 250,000 tons to meet its domestic needs, saving more than $100 million [90]. However, the analysis did not mention that previously, Zimbabwean agriculture used to satisfy its own internal needs with agricultural products and exports to several other African countries. However, the Mugabe regime destroyed and then chased away the white farmers, so the previously world-famous Zimbabwean agriculture collapsed, and a significant part of the population was in need of food aid [91]. Although the development of agriculture has started in recent years with government and international support, the unfavourable structure of the agricultural sector (the predominance of small producers), years of drought, the financial situation of producers, limited loans available to them, transportation difficulties, etc. still pose a risk, and the country is still far from reaching the level of agriculture from before Mugabe [92]. The study citing Zimbabwe as a positive example also mentions Ethiopia as a country that has significant potential for the development of its agriculture; however, the country is struggling with similar challenges as Zimbabwe, and the civil war situation that developed in 2022 is not conducive to the initiatives proposed by the author [93]. The two countries' examples also clearly show that in the current economic and security situation, the development of agriculture in most African countries is not cost-effective, as this would require several years. Production methods should be transformed, infrastructural improvements should be implemented, agricultural training should be improved, and logistical/financial support should be provided to farmers and local merchants so that local products reach local markets at affordable prices, as agricultural products imported from Europe are often cheaper than those produced locally [94]. In addition, it is important to create political stability and legal certainty as well as to strengthen regional cooperation since African countries can achieve these independently much more slowly. Although the African Union is currently planning to develop and strengthen the "African Food System" and the oil and gas sectors so that they are not at the mercy of external actors (Russia, Ukraine, etc.), no concrete steps have been taken in this regard so far [64]. Additionally, we cannot forget that the COVID epidemic caused serious economic damage to the continent's countries, which they have not yet fully recovered from [95]. For this reason, African countries primarily strive to maintain their living standards rather than develop their agriculture.

Several African countries see an opportunity to supply crude oil and natural gas to European countries themselves instead of Russian companies hit by sanctions [96].

Although several African countries, such as Nigeria, Mozambique, Tanzania, and Senegal, have significant recoverable natural gas, the necessary infrastructures for extraction and transportation have not yet been developed [97]. On the other hand, Algeria and Niger have been supplying natural gas to several European countries for years, while Nigeria provides 6% of the crude oil imported by them [98]. These countries have undertaken to increase the amount of natural gas and crude oil supplied to EU countries, but this requires further expansions/developments of their existing pipeline systems, which involve significant costs. Of course, not only the previously mentioned possibilities arose. South Africa, the world's second-largest producer of palladium [99], which is the metal required for producing automobiles and electronic devices, also produces a significant amount of gold, the price of which has continuously increased since the outbreak of the conflict. The sanctions also provide an opportunity—because the Russians cannot market the gold they have mined due to the sanctions nickel, palladium, titanium, aluminium, copper, and uranium inventory—to increase production and sales [100].

As I indicated earlier, some African countries do not support the sanctions imposed on Russia and want to continue cooperating and trading with it. This can be partly achieved by evading the sanctions or by trying to become independent from the SWIFT payment system and actively participate in the operation of the Pan-African Payment and Settlement System (PAPSS) created by the African Import–Export Bank (Afrexim). This way, they can avoid the possible sanctions that could be applied against them due to their cooperation with the Russians. In addition, there are other possibilities. An excellent example of this is the case of the African regional power Algeria, which made a deal with its foreign partners in such a way that they could choose every six months in which currency they wished to pay the price of Algerian natural gas [101]. Nevertheless, this step can significantly strengthen African countries' economic cooperation and integration [97].

6 Russian Military Presence, Weapons [102]

In recent years, Russia has established military/security cooperation with more than 20 African countries (such as Egypt, Libya, Zimbabwe, Sudan, South Africa, Algeria, etc.). A significant part of them also buy/have bought a large number of Russian weapons, in the operation and maintenance of which they count on the experience of Russian experts [103]. Today, Russia is considered a serious player in the security field on the continent, which not only provides weapons and military equipment to its African partners but also trains its armed and law enforcement forces, provides them with intelligence and other information, and is even slowly regaining the positions it owned during the Cold War in terms of military bases, airports, and seaports in several African countries [104]. The strengthening of the Russian presence is helped by the fact that those international players who previously supported African peace and security activities with substantial amounts have transferred a significant part of them to Ukraine so that it can defend itself against the Russians. For example, the

EU has reduced the support provided through the European Peace Facility (EPF), which also recently contributed to the cooling of relations between the EU and the AU. In addition, there are ongoing professional discussions about whether it is worth supporting African countries in this area and, if so, how and with what amount [105]. Understandably, some African states are disappointed with their former partners, so in their eyes, the Russians can fill the void in the security field caused by the reduction of Western resources. Thus, we return to the techniques used during the Cold War, when African and other countries also successfully "swung" between the Soviet Union and the Western powers, depending on who they received more support from. Therefore, I believe the presence of Russian military-industrial products/military programmes in Africa will strengthen in the coming period.

In addition to the official cooperation, as I mentioned earlier, the Wagner group and other Russian PMC organizations are present and have gained significant influence in the following countries: Mali, Central African Republic, Burkina Faso, Sudan, Guinea, and Libya. This is mainly due to the fact that the activities of terrorist groups have strengthened in several African countries (Mali, Burkina Faso, Niger, Mozambique, Somalia), against whom neither the governments of the respective countries nor the peace operations arranged by international organizations (EU, AU) were quite effective. The Wagner group, representing Russia's African interests, offered a kind of "alternative" to them, which so far has not proven its reason to exist. In Mozambique, the organization was defeated by a local terrorist group linked to the Islamic State, and in other countries, they have not been able to push back the terrorist groups to this day. On the other hand, violations of human rights (murders, executions of civilians, rape, etc.) were committed in several countries, which provoked protests both in the respective countries and internationally. Since the aforementioned African countries continue to arm themselves due to various security challenges, and the costs of Wagner or other Russian PMCs must be paid, the amounts that can be spent on food security and the independence of agriculture have decreased in their budgets. Because of this, the amount that can be spent on feeding the population in the given countries will further decrease, so the given countries may face additional economic and security challenges [106].

7 Conclusions

Based on my own research, it can be seen that the Russo-Ukrainian War had a significant impact on the countries of the African continent in several ways. On the one hand, the conflict can have negative effects (economic difficulties, decreasing financial support and aid, famine, political conflicts, etc.), while on the other hand, positive opportunities for some countries as well (entry into the international market, strengthening of positions, etc.). Which of these will have the strongest effects depends on many things and only partly on the African countries. Although many people envision some reorganization or the ousting of Russia from the continent, in my opinion, this is not to happen. Whether the war ends soon or becomes a prolonged battle,

Russia will continue cultivating economic, political and security relationships with African countries. Without significant resources to address the challenges facing many African countries, such as governance issues, poverty and terrorism, Russia will continue to focus on using low-cost tools at its disposal to exert influence.

References

1. Lorenz E (1963) Deterministic nonperiodic flow. J Atmos Sci 20:130–141
2. Khanna V (2022) The' butterfly effect' of the Russia-Ukraine war on the global economy. The Straits Times, 6 June 2022. https://www.straitstimes.com/world/europe/the-butterfly-effect-of-the-russia-ukraine-war-on-the-global-economy. Last accessed 18 Jan 2023
3. Mannu G (2023) The new international scenarios and the butterfly effect. R&P, 26 Oct 2022. Available at: https://www.roncucciandpartners.com/en/2022/10/26/the-new-international-scenarios-and-the-butterfly-effect/. Last accessed 18 Jan 2023
4. Yudina N (2022) Economic "butterfly" and futurology of the war in Ukraine. Econ Bull NTUU Kyiv Polytech Inst 23:77–82
5. Dibb P (2022) The geopolitical implication of Russia's invasion of Ukraine. Australian Strategic Policy Institute (2022). https://www.aspi.org.au/report/geopolitical-implications-russias-invasion-ukraine. Last accessed 18 Jan 2023. Ruta M (ed) (2022) The impact of the war in Ukraine on global trade and investment. Washington, DC. © World Bank. https://openknowledge.worldbank.org/handle/10986/37359. Last accessed 18 Jan 2023. United Nations Development Programme (2022) The impact of the war in Ukraine on sustainable development in Africa. Rapid Assessment by the Regional Bureau for Africa. https://www.undp.org/africa/publications/impact-war-ukraine-sustainable-development-africa. Last accessed 18 Jan 2023
6. Delgado C, Tschunkert K, Smith D (2023) Food security in Africa: drivers and solutions. SIPRI Research Policy Paper, Jan 2023. https://www.sipri.org/sites/default/files/2023-01/2301_sipri_rpp_food_insecurity_in_africa_0.pdf. Last accessed 20 Jan 2023
7. Hassen BT, El Bilali H (2022) Impacts of the Russia-Ukraine war on global food security: towards more sustainable and resilient food systems? Foods 11(15).
8. Lusigi A (2022) Africa and the Russia-Ukraine conflict: seizing the opportunity in the crisis. United Nations Africa Renewal, 30 June 2022. https://www.un.org/africarenewal/magazine/africa-and-russia-ukraine-conflict-seizing-opportunity-crisis. Last accessed 18 Jan 2023
9. More on this see homepage of AUDA-NEPAD. https://www.nepad.org/microsite/impact-of-russia-ukraine-conflict-african-food-systems. Last accessed 18 Jan 2023
10. United Nations Development Programme (2022) The impact of the war in Ukraine on sustainable development in Africa. Rapid Assessment by the Regional Bureau for Africa. https://www.undp.org/africa/publications/impact-war-ukraine-sustainable-development-africa, last accessed 18 Jan 2023
11. Artuc E, Falcone G, Port G, Rijkers B, War-induced food price inflation imperils the poor. In: Garicano L, Rohner D, di Mauro BW (eds) Global economic consequences of the war in Ukraine Sanctions, supply chains and sustainability. CEPR Press, pp 155–161
12. Soko V, Sakor BZ (2022) The Ukraine crisis and its impact on Africa's geopolitics: what do we know so far? Peace Research Institute Oslo (PRIO) (Blog). https://blogs.prio.org/2022/09/the-ukraine-crisis-and-its-impact-on-africas-geopolitics-what-do-we-know-so-far/, last accessed 18 Jan 2023
13. Garicano L, Rohner D, di Mauro BW (eds) Global economic consequences of the war in Ukraine sanctions, supply chains and sustainability. CEPR Press, London
14. Soko V, Sakor BZ (2022) The Ukraine crisis and its impact on Africa's geopolitics: what do we know so far? Peace Research Institute Oslo (PRIO) (Blog). https://blogs.prio.org/2022/09/the-ukraine-crisis-and-its-impact-on-africas-geopolitics-what-do-we-know-so-far/, last accessed

18 Jan 2023. National Resource Governance Institute. Implications of the Russia-Ukraine conflict on Africa's extractive sectors, energy transition and economic Outlook. NRGI (Web Event), 13 July 2022. https://resourcegovernance.org/events/web-event/implications-russia-invasion-ukraine-africa-extractives-energy-transition-economic-outlook, last accessed 18 Jan 2023
15. D'Anieri P (2019) Ukraine and Russia: from civilized divorce to uncivil war. Cambridge University Press, pp 87–91. Yekelchyk S (2015) The conflict in Ukraine: what everyone needs to know. Oxford University Press, pp 67–71
16. Arel D, Driscoll J (2023) Ukraine's unnamed war: before the Russian invasion of 2022. Cambridge University Press, p 13
17. D'Anieri P (2019) Ukraine and Russia: from civilized divorce to uncivil war. Cambridge University Press, pp 207–224. Ishchenko V (2020) Insufficiently diverse: the problem of nonviolent leverage and radicalization of Ukraine's Maidan uprising, 2013–2014. J, Eurasian Stud. 11(2):201–215
18. Yekelchyk S (2015) The conflict in Ukraine: what everyone needs to know. Oxford University Press, pp 128–132
19. Hahn GM (2018) Ukraine over the edge: Russia, the west and the "new cold war". McFarland, 273 (2018)
20. Arel D, Driscoll J (2023) Ukraine's unnamed war: before the Russian invasion of 2022. Cambridge University Press, pp 172–188 (2023). Kuzio T (2022) Why Russia invaded Ukraine, Horizons. J. Int. Relat. Sustain. Dev. 21:40–51 (Summer 2022). https://www.cirsd.org/en/horizons/horizons-summer-2022-issue-no.21/why-russia-invaded-ukraine, last accessed 18 Jan 2023
21. Minic D (2022) Russia's invasion of Ukraine. A political-strategic break? Russie.Nei.Visions 126, Ifri, May 2022. https://www.ifri.org/sites/default/files/atoms/files/d.minic_invasion_russe_ukraine_2022_us.pdf. Last accessed 18 Jan 2023. Mohsen, K.J., Al Khudari, M.N., Al-Taher MA (2022) The future of the world order after the Russian-Ukrainian war. J Southwest Jiaotong Univ 57(5)
22. Brown D, Horton J, Ahmedzade T (2023) Ukraine weapons: what military equipment is the world giving? BBC News, 16 Jan 2023. https://www.bbc.com/news/world-europe-62002218, last accessed 18 Jan 2023. Lillis KB, Liebermann O (2023) How Ukraine became a testbed for Western weapons and battlefield innovation. CNN, 16 Jan 2023. https://edition.cnn.com/2023/01/15/politics/ukraine-russia-war-weapons-lab/index.html, last accessed 18 Jan 2023. Rathbone JP, Pfeifer S, Chávez S (2022) Military briefing: Ukraine war exposes 'hard reality' of West's weapons capacity. Financial Times, 2 Dec 2022. https://www.ft.com/content/a781fb71-49bb-4052-ab05-a87386bf3d5e, last accessed 18 Jan 2023
23. Demertzis M, Hilgenstock B, McWilliams B, Ribakova E, Tagliapietra S (2022) How have sanctions impacted Russia? Policy contribution. Bruegel, p 18. https://www.bruegel.org/policy-brief/how-have-sanctions-impacted-russia, last accessed 18 Jan 2023. Reynolds P (2022) Are sanctions on Russia working? The lessons from a new era of economic warfare. Economist, 25 Aug 2022. https://www.economist.com/leaders/2022/08/25/are-sanctions-working, last accessed 18 Jan 2023
24. Akrhangelskaya A, Shubin V (2013) Russia's Africa policy. SAIIA occasional paper. SAIIA, Sept 2013, p 157. http://www.saiia.org.za/occasional-papers/429-russia-s-africa-policy/file, last accessed 18 Jan 2023. Arkhangelskaya A, Shubin V (2023) Is Russia back? Realities of Russian engagement in Africa. http://www.lse.ac.uk/IDEAS/publications/reports/pdf/SR016/SR-016-ArkanghelShubin.pdf, last accessed 18 Jan 2023
25. Paczynska CA (2020) Russia in Africa: is great power competition returning to the continent?. In: Deutsches Institut für Entwicklungspolitik, Briefing Paper p 15. Besenyő J (2019) The Africa policy of Russia. Terror Polit Violence 31(1):132–153. Taylor I, Africa rising? BRICS—diversifying dependency, vol 64
26. The Russian interest is shown by the fact that in 2011, Russia maintained 43 diplomatic missions in African countries!. Shinn DH (2011) Emerging powers in Africa. International

Policy Digest, 12 July 2011. https://intpolicydigest.org/2011/07/12/emerging-powers-in-africa/, last accessed 18 Jan 2023
27. Donelli F (2019) Russia's engagement with Africa: Understanding mutual Russia-Africa interests in an era of global competition, TRENDS Research & Advisory, 16. January 2023, https://trendsresearch.org/research.php?id=86&title=Russia%E2%80%99s_engagement_with_Africa:_Understanding_mutual_Russia-Africa_interests_in_an_era_of_global_competition, last accessed 18 Jan 2023. Besenyő J (2019) The Africa policy of Russia. Terror Polit Violence 31(1):132–153
28. Wilson JD (2015) Resource powers? Minerals, energy and the rise of the BRICS. Third World Quart 36(2):223–239. https://www.academia.edu/11247946/Resource_powers_Minerals_energy_and_the_rise_of_the_BRICS, last accessed 18 Jan 2023
29. Olivier G, Suchkov D, Russia is back in Africa, p 161
30. Unlike China and India (2013) Russia's economic development is not primarily dependent on African raw material resources, so it is much less vulnerable to political bargaining and can negotiate more favourable terms. However, like many African states, a significant part of its revenues comes from its raw material resources and it is therefore more exposed to the risks that this entails. Akrhangelskaya A, Shubin V (2013) Russia's Africa policy, SAIIA occasional paper 157, SAIIA, Sept 2013. http://www.saiia.org.za/occasional-papers/429-russia-s-africa-policy/file, last accessed 18 Jan 2023
31. Peunova M (2015) Russia currently has 49 missions in 54 African countries. Russia's push to Africa: cooperation or competition? Glob Risk Inside, 20 Dec 2015. http://globalriskinsights.com/2015/12/russias-push-to-africa-cooperation-or-competition/, last accessed 18 Jan 2023
32. Fidan H, Aras B (2023) The return of Russia-Africa relations. Bilig 52, 47–68 (Winter 2010). http://sam.gov.tr/wp-content/pdfs/2.pdf, last accessed 18 Jan 2023
33. Arkhangelskaya A, Dodd N (2022) Guns and poseurs: Russia returns to Africa, p 167. Singh P (2022) Russia–Africa relations in an age of renewed great power competition, Africa report 42. Institute For Security Studies, Nov 2022. https://issafrica.s3.amazonaws.com/site/uploads/ar-42-v3.pdf, last accessed 18 Jan 2023
34. Korendyasov E, Vasilyev A (2013) Russian-African relations in ten years: a new start, Russian International Affairs Council, 21 June 2013. http://russiancouncil.ru/en/inner/?id_4=2002#top-content, last accessed 18 Jan 2023
35. Kalika A (2019) Russia's "great return" to Africa? Ifri, Apr 2019. https://www.ifri.org/en/publications/notes-de-lifri/russieneivisions/russias-great-return-africa, last accessed 18 Jan 2023
36. Olivier G, Suchkov D, Russia is back in Africa, pp 150–151
37. Raafat O (2023) Walking the tightrope: the Ukrainian crisis through the eyes of Africa. Euromesco Policy Brief 128, Mar 2023. https://www.euromesco.net/publication/walking-the-tightrope-the-ukrainian-crisis-through-the-eyes-of-africa/, last accessed 18 Jan 2023. Szalai M (2023) Between Hedging and Bandwagoning—interpreting the reactions of Middle Eastern and North African States to the Russian-Ukrainian War. Euromesco Policy Brief 123, Feb 2023. https://www.euromesco.net/publication/between-hedging-and-bandwagoning-interpreting-the-reactions-of-middle-eastern-and-north-african-states-to-the-russian-ukranian-war/, last accessed 18 Jan 2023
38. Hood L (2022) Russia-Ukraine war: decoding how African countries voted at the UN, The Conversation, 8 Mar 2022. https://theconversation.com/russia-ukraine-war-decoding-how-african-countries-voted-at-the-un-178663, last accessed 18 Jan 2023. Ryder H, Kebret E (2022) Why African countries had different views on the UNGA Ukraine resolution, and why this matters. Center for Strategic and International Studies, 15 Mar 2022. https://www.csis.org/analysis/why-african-countries-had-different-views-unga-ukraine-resolution-and-why-matters, last accessed 18 Jan 2023
39. https://www.dw.com/en/russias-reengagement-with-africa-pays-off/a-61064011, last accessed 18 Jan 2023
40. How have African countries voted on Russia and Ukraine at the United Nations? Development reimagined. https://developmentreimagined.com/2022/04/08/africanvotesonukraine/, last accessed 18 Jan 2023

41. Besenyő J (2023) The Africa policy of Russia. Terror Polit Violence 31(1):132–153. https://doi.org/10.1080/09546553.2018.1555976. Gopaldas R (2023) Will the invasion of Ukraine change Russia-Africa relations? Carnegie Endowment for International Peace, 26 Apr 2023. https://carnegieendowment.org/2023/04/26/will-invasion-of-ukraine-change-russia-africa-relations-pub-89596, last accessed 18 Jan 2023
42. Matusevich M (2019) Russia in Africa: a search for continuity in a post-cold war era. In: Insight Turkey, 13 Mar 2019, pp 25–40. Nzuki C (2023) Africa's peace delegation: a new chapter for Africa and the Ukraine War, Center for Strategic & International Studies, 16 June 2023. https://www.csis.org/analysis/africas-peace-delegation-new-chapter-africa-and-ukraine-war, last accessed 18 Jan 2023
43. Chadwick V (2022) Exclusive: internal report shows EU fears losing Africa over Ukraine, devex.com, 22 July 2022. https://www.devex.com/news/exclusive-internal-report-shows-eu-fears-losing-africa-over-ukraine-103694, last accessed 18 Jan 2023
44. https://www.ispionline.it/en/pubblicazione/war-ukraine-calls-increased-african-unity-37118, last accessed 18 Jan 2023. Matibe P (2022) Evolving US-Africa relations since the Russia-Ukraine War, Africa up close/Wilson Center, 22 Sept 2022. https://africaupclose.wilsoncenter.org/us-africa-relations-russia-ukraine-war/, last accessed 18 Jan 2023
45. Hruby A (2023) The US has gotten the day to day right in Africa policy. Time to think bigger. Atlantic Council, 13 Mar 2023, https://www.atlanticcouncil.org/blogs/africasource/the-us-has-gotten-the-day-to-day-right-in-africa-policy-time-to-think-bigger/, last accessed 18 Jan 2023
46. Gruzd S (2023) Why African leaders have a blind spot for Russia, Council of Councils, 24 Feb 2023. https://saiia.org.za/research/the-invasion-that-shook-the-world/, last accessed 18 Jan 2023. Gopaldas R (2023) Will the invasion of Ukraine change Russia-Africa relations? Carnegie Endowment For International Peace, 26 Apr 2023. https://carnegieendowment.org/2023/04/26/will-invasion-of-ukraine-change-russia-africa-relations-pub-89596, last accessed 18 Jan 2023. Akeredolu F (2023) African perception of the United States in an evolving geopolitical landscape. Wilson Center, 27 July 2023. https://www.wilsoncenter.org/blog-post/african-perception-united-states-evolving-geopolitical-landscape, last accessed 18 Jan 2023
47. Wekesa B (2023) U.S.-Africa University partnerships: possibilities and realities, vol 8. The Africa-US Forum, Aug 2023. https://www.africa-usforum.africa/u-s-africa-university-partnerships-possibilities-and-realities/, last accessed 18 Jan 2023
48. Gbane NC (2023) African leaders should compel Russia to get the grain deal online-US rep. to NATO, Africanews, 15 Aug 2023. https://www.africanews.com/2023/07/26/african-leaders-should-compel-russia-to-get-the-grain-deal-online-us-amb-to-un/, last accessed 18 Jan 2023. Ashby H, Rupert J, Sharad K (2023) Russia's Africa summit—and a future for Wagner. The United States Institute of Peace, 3 Aug 2023. https://www.usip.org/publications/2023/08/russias-africa-summit-and-future-wagner, last accessed 18 Jan 2023
49. Klomegah KK (2022) Ukraine and Africa in emerging multipolar world, modern diplomacy, 26 Dec 2022. https://moderndiplomacy.eu/2022/12/26/ukraine-and-africa-in-emerging-multipolar-world/, last accessed 18 Jan 2023
50. Klyszcz IUK (2023) Ukraine is not powerless in the Middle East, The International Centre for Defence and Security (ICDS), Estonia, 4 Apr 2023. https://icds.ee/en/ukraine-is-not-powerless-in-the-middle-east/, last accessed 18 Jan 2023
51. Owino V (2022) Ukraine to donate grain to hunger-stricken African countries, The East African, 29 Nov 2022. https://www.theeastafrican.co.ke/tea/news/world/ukraine-to-donate-grain-to-african-countries-4036596, last accessed 18 Jan 2023
52. Ashby H (2023) Ukraine's Africa visit shows its fight against Russia goes beyond the battlefield, United States Institute of Peace, 3 Nov 2022. https://www.usip.org/publications/2022/11/ukraines-africa-visit-shows-its-fight-against-russia-goes-beyond-battlefield, last accessed 18 Jan 2023
53. Nzuki C (2023) Africa's peace delegation: a new chapter for Africa and the Ukraine war. Center for Strategic and International Studies, 16 June 2023. https://www.csis.org/analysis/africas-peace-delegation-new-chapter-africa-and-ukraine-war, last accessed 18 Jan 2023

54. Fabricius P (2023) Sergey Lavrov back to Africa with a vengeance. Institute For Security Studies, 20 Jan 2023. https://issafrica.org/iss-today/sergey-lavrov-back-to-africa-with-a-vengeance, last accessed 18 Jan 2023
55. Fasanotti FS (2023) Russia's Wagner group in Africa: influence, commercial concessions, rights violations, and counterinsurgency failure. Brookings Institution, 8 Feb 2022. https://www.brookings.edu/blog/order-from-chaos/2022/02/08/russias-wagner-group-in-africa-influence-commercial-concessions-rights-violations-and-counterinsurgency-failure/, last accessed 18 Jan 2023. Wouters P (2023) Strategic priorities for the Russian PMC WAGNER: geopolitics, propaganda and mercenary business. Egmont Policy Brief 303, Apr 2023. https://www.egmontinstitute.be/app/uploads/2023/04/Patrick-Wouters_Policy_Brief_303.pdf?type=pdf, last accessed 18 Jan 2023
56. Although they tried to keep it a secret for a long time, Wagner's gunmen were already present in Mali from the summer of 2021, although they were officially in the country only from December 2021. The European peacekeepers in the country repeatedly reported the growing Russian activity through various channels, but apart from threats by European countries to withdraw from peace operations, nothing else happened. Interviews with two French and Hungarian peacekeepers from EUTM Mali, who asked to keep their names and ranks confidential.
57. Mcallister E, Lewis D (2023) Analysis: Mali faces spectre of anarchy after demanding UN's departure. Reuters, 19 June 2023. https://www.reuters.com/world/africa/mali-faces-spectre-anarchy-after-demanding-uns-departure-2023-06-18/, last accessed 18 Jan 2023
58. Romaniuk SN, Besenyő J (2023) Wagner mercenaries: a potential lifeline for the Niger Junta. Geopolitical Monitor, 14 Aug 2023. https://www.geopoliticalmonitor.com/wagner-mercenaries-a-potential-lifeline-for-the-niger-junta/, last accessed 18 Jan 2023
59. Schlindwein S (2023) Are white mercenaries fighting in the DRC conflict? Deutsche Welle, 17 Jan 2023. https://www.dw.com/en/are-white-mercenaries-fighting-in-the-drc-conflict/a-64407711, last accessed 18 Jan 2023
60. Margolin J (2023) The new Russian mercenary marketplace. Riddle, 21 Aug 2023. https://ridl.io/the-new-russian-mercenary-marketplace/, last accessed 18 Jan 2023. After the failed coup in Russia: what's new for the Wagner group in Africa? Istituto per gli Studi di Politica Internazionale (ISPI), 1 Aug 2023. https://www.ispionline.it/en/publication/after-the-failed-coup-in-russia-whats-new-for-the-wagner-group-in-africa-137936, last accessed 18 Jan 2023
61. Ukraine Refugee Situation, UNHCR Operational Data Portal, 27 Mar 2023, https://data.unhcr.org/en/situations/ukraine, last accessed 18 Jan 2023. Update: situation at the EU's land border with Russia, Frontex, 27 Sept 2022. https://frontex.europa.eu/media-centre/news/news-release/update-situation-at-the-eu-s-land-border-with-russia-X3iWuK, last accessed 18 Jan 2023
62. The potential contribution of Ukrainian refugees to the labour force in European host countries. OECD, 27 July 2022. https://www.oecd.org/ukraine-hub/policy-responses/the-potential-contribution-of-ukrainian-refugees-to-the-labour-force-in-european-host-countries-e88a6a55/, last accessed 18 Jan 2023. What we know about the skills and early labour market outcomes of refugees from Ukraine. OECD, 6 Jan 2023. https://www.oecd.org/ukraine-hub/policy-responses/what-we-know-about-the-skills-and-early-labour-market-outcomes-of-refugees-from-ukraine-c7e694aa/#boxsection-d1e1631, last accessed 18 Jan 2023. Karasapan O (2022) Ukrainian refugees? Challenges in a welcoming Europe. Brookings Institution, 14 Oct 2022. https://www.brookings.edu/articles/ukrainian-refugees-challenges-in-a-welcoming-europe/, last accessed 18 Jan 2023
63. Kumar C, Knoll A, Veron P (2022) How the war in Ukraine is impacting Africa-Europe relations and the migration dialogue. European Think Thanks Group (ETTG), 6 Dec 2022. https://ettg.eu/blog-posts/how-the-war-in-ukraine-is-impacting-africa-europe-relations-and-the-migration-dialogue/, last accessed 18 Jan 2023
64. https://www.un.org/africarenewal/magazine/may-2022/how-russia-ukraine-conflict%C2%A0impacts-africa, last accessed 18 Jan 2023

65. The impact on trade and development of the war in Ukraine. UNCTAD, 16 Mar 2022. https://unctad.org/system/files/official-document/osginf2022d1_en.pdf, last accessed 18 Jan 2023. Rose A, Chen Z, Wei D (2022) The economic impacts of Russia-Ukraine war export disruptions of grains and metals, 26 Nov 2022. https://doi.org/10.2139/ssrn.4286716, last accessed 18 Jan 2023
66. Nzuki C (2023) Africa's peace delegation: a new chapter for Africa and the Ukraine war. Center for Strategic and International Studies, 16 June 2023. https://www.csis.org/analysis/africas-peace-delegation-new-chapter-africa-and-ukraine-war, last accessed 18 Jan 2023. EAST AFRICA—maize regional supply and market outlook, Famine Early Warning Systems Network, 30 Nov 2022. https://fews.net/sites/default/files/documents/reports/EA_RSMO_2022.pdf, last accessed 18 Jan 2023
67. https://www.global-counsel.com/insights/blog/russia-ukraine-war-threatens-menas-food-security, last accessed 18 Jan 2023
68. These countries are, in descending order, Somalia, Benin, Egypt, Sudan, Democratic Republic of Congo, Senegal, United Republic of Tanzania, Rwanda, Madagascar, Congo, Burkina Faso, Gambia, Libya, Tunisia, Uganda, Burundi. Impact of the war in Ukraine on trade and development, UNCTAD, 16 Mar 2022, https://unctad.org/system/files/official-document/osginf2022d1_en.pdf, last accessed 18 Jan 2023
69. Toulmin C, Soulé F (2023) The impact of the war in Ukraine on West Africa requires a disaggregated analysis. Institute for New Economic Thinking, 12 Dec 2022. https://www.ineteconomics.org/perspectives/blog/turning-africas-food-crisis-into-a-development-opportunity, last accessed 18 Jan 2023
70. Stewart TR (2023) Russian blackmail and the Black sea grain initiative, FIIA briefing paper, Apr 2023. https://www.fiia.fi/wp-content/uploads/2023/04/bp360_russian-blackmail-and-the-black-sea-grain-initiative.pdf, last accessed 18 Jan 2023. Kumaria R (2023) Improving global food security: the impact of the Black sea grain initiative. Manohar Parrikar Institute for Defence Studies and Analyses, New Delhi, 27 Mar 2023. https://www.idsa.in/issuebrief/improving-global-food-security-rkumaria-270323, last accessed 18 Jan 2023. Welsh C, Kirschenbaum A (2022) Beyond Black sea grains: the global malnutrition crisis caused by Russia's war in Ukraine. Center for Strategic and International Studies, 16 Dec 2022. https://www.csis.org/analysis/beyond-black-sea-grains-global-malnutrition-crisis-caused-russias-war-ukraine, last accessed 18 Jan 2023
71. Glauber J, McNamara B, Olivetti E (2023) Russia terminates the Black Sea grain Initiative: what's next for Ukraine and the world? International Food Policy Research Institute 20 July 2023. https://www.ifpri.org/blog/russia-terminates-black-sea-grain-initiative-whats-next-ukraine-and-world, last accessed 18 Jan 2023
72. Shah S (2023) Africa needs more, not less, fertilizer. For Policy 8 Oct 2022. https://foreignpolicy.com/2022/10/08/fertilizer-war-climate-shortage-food-agriculture-africa-europe/, last accessed 18 Jan 2023
73. FAO Regional Office for Africa, boosting Africa's soils, food and agriculture organization of the United Nations (2016). https://www.fao.org/3/i5532e/i5532e.pdf, last accessed 18 Jan 2023
74. Fertilizer prices spike and will continue rallying next year, IndexBox, 17 Dec 2021. https://www.indexbox.io/blog/global-fertilizer-market-2021-key-insights/, last accessed 18 Jan 2023
75. Baffes J, Koh WC (2023) Soaring fertilizer prices add to inflationary pressures and food security concerns, worldbank.org. https://blogs.worldbank.org/opendata/soaring-fertilizer-prices-add-inflationary-pressures-and-food-security-concerns, last accessed 18 Jan 2023. Steuer C, Rieker P (2023) Food security in the context of the war in Ukraine. Institute of International Relations Prague, Policy Paper, 24 May 2023. https://www.iir.cz/food-security-in-the-context-of-the-war-in-ukraine, last accessed 18 Jan 2023
76. Laborde D (2023) Food & fertilizer export restrictions tracker, public.tableau.com, 26 Nov 2022. https://public.tableau.com/app/profile/laborde6680/viz/ExportRestrictionsTracker/FoodExportRestrictionsTracker, last accessed 18 Jan 2023

77. Busari S (2022) People are begging us to sell. Africa's richest man opens huge fertilizer plant as food crisis looms. CNN Business, 24 Mar 2022. https://edition.cnn.com/2022/03/23/business/dangote-fertilizer-plant-food-crisis-lgs-intl/index.html, last accessed 18 Jan 2023
78. Hebebarndt C, Glauber J (2023) The Russia-Ukraine war after a year: impacts on fertilizer production, prices, and trade flows, 9 Mar 2023. https://www.ifpri.org/blog/russia-ukraine-war-after-year-impacts-fertilizer-production-prices-and-trade-flows, last accessed 18 Jan 2023
79. Yohannes-Kassahun B (2023) One year later: the impact of the Russian conflict with Ukraine on Africa. United Nations Office of the Special Adviser on Africa, 3 Apr 2023. https://www.un.org/osaa/news/one-year-later-impact-russian-conflict-ukraine-africa, last accessed 18 Jan 2023
80. Halakhe A (2022) We were warned: unlearned lessons of famine in the Horn of Africa, Refugees International, 9 Dec 2022. https://www.refugeesinternational.org/reports/2022/12/7/we-were-warned-unlearned-lessons-of-famine-in-the-horn-of-africa, last accessed 18 Jan 2023. Delgado C, Tschunkert K, Smith D (2023) Food Insecurity in Africa: Drivers and Solutions, SIPRI Research Policy Paper, Stockholm International Peace Research Institute, Jan 2023. https://doi.org/10.55163/GISR2785, last accessed 18 Jan 2023
81. https://www.pbs.org/newshour/world/africa-faced-extreme-weather-and-an-international-community-distracted-by-ukraine-in-2022, last accessed 18 Jan 2023
82. https://www.crisisgroup.org/europe-central-asia/eastern-europe/ukraine-russiaus/ukraine-needs-wests-help-winter-black-white, last accessed 18 Jan 2023
83. Delgado C, Tschunkert K, Smith D (2023) Food insecurity in Africa: drivers and solutions, SIPRI research policy paper, Stockholm International Peace Research Institute, Jan 2023. https://doi.org/10.55163/GISR2785, last accessed 18 Jan 2023
84. Bilala DD, Nyachieo VN (2022) Four ways to address the rising food prices in Kenya, KIPRA, 29 Aug 2022. https://kippra.or.ke/four-ways-to-address-the-rising-food-prices-in-kenya/, last accessed 18 Jan 2023
85. Worsening hunger crisis could threaten Africa's stability, PSC report. Institute for Security Studies, 5 Dec 2022. https://issafrica.org/pscreport/psc-insights/worsening-hunger-crisis-could-threaten-africas-stability, last accessed 18 Jan 2023
86. Nigeria: curfew in Adamawa state after massive looting. Africanews, 31 July 2023. https://www.africanews.com/2023/07/31/nigeria-curfew-in-adamawa-state-after-massive-looting/, last accessed 18 Jan 2023
87. With the increase in food prices, Senegal could experience new turbulence, 14 Aug 2023. https://lejournaldelafrique.com/en/with-the-increase-in-the-price-of-foodstuffs%2C-senegal-could-experience-new-turbulence/, last accessed 18 Jan 2023
88. Braw E (2022) Russia is taking advantage of the invasion-stirred migration crisis. As food problems worsen, new refugees head for Europe. For Policy, 18 July 2022. https://foreignpolicy.com/2022/07/18/russia-ukraine-war-migration-food-crisis-putin/, last accessed 18 Jan 2023
89. Cornock O (2023) Agriculture in Africa 2019: special report. Oxford Business Group, https://oxfordbusinessgroup.com/sites/default/files/blog/specialreports/949525/africa_2019_special_report.pdf, last accessed 18 Jan 2023. Akiwumi P (2022) Revitalizing African agriculture: time for bold action. In: United Nations conference on trade and development, 30 Sept 2022. https://unctad.org/news/blog-revitalizing-african-agriculture-time-bold-action, last accessed 18 Jan 2023
90. Oduro MEO (2022) The Russia-Ukraine war: a silver lining for Africa? The Ghana Center for Democratic Development (CDD-Ghana), 1 Dec 2022. https://cddgh.org/the-russia-ukraine-war-a-silver-lining-for-africa/, last accessed 18 Jan 2023
91. Takirambudde P, Levine I (2003) ZIMBABWE: not eligible: the politicization of food in Zimbabwe. Human Rights Watch, Oct 2003 15(17(A)). https://www.hrw.org/reports/2003/zimbabwe1003/zimbabwe1003.pdf, last accessed 18 Jan 2023. Kapuya T, Meyer FH, Kirsten JF (2013) Modelling the impact of the 'fast track' land reform policy on Zimbabwe's Maize sector. Dev South Afr 30(3):417–436

92. Zimbabwe (2019) Agriculture sector disaster risk assessment. World Bank Group, Mar 2019. https://www.gfdrr.org/sites/default/files/publication/Zimbabwe%20Agriculture%20Sector%20Disaster%20Risk%20Assessment%20Report.pdf, last accessed 18 Jan 2023
93. Wendimu GY, Moral MT (eds) The challenges and prospects of Ethiopian agriculture. Cogent Food Agric 7(1)
94. Livingstone E (2018) How EU milk is sinking Africa's farmers, Politico, 8 Apr 2018. https://www.politico.eu/article/eus-milk-scramble-for-africa/, last accessed 18 Jan 2023. Rwabwogo O (2022) Lopsided EU trade agreements are harming Africa, African Business, 3 Feb 2022. https://african.business/2022/02/agribusiness-manufacturing/lopsided-eu-trade-agreements-are-harming-africa/, last accessed 18 Jan 2023
95. Sen AK (2022) Russia's war in Ukraine is taking a toll on Africa, United States Institute of Peace, 15 June 2022. https://www.usip.org/publications/2022/06/russias-war-ukraine-taking-toll-africa, last accessed 18 Jan 2023. Lakemann T, Lay J, Tafese T (2020) Africa after the Covid-19 lockdowns: economic impacts and prospects. German Institute for global and area studies, vol 6. https://www.giga-hamburg.de/en/publications/giga-focus/africa-after-the-covid-19-lockdowns-economic-impacts-and-prospects, last accessed 18 Jan 2023
96. https://www.brookings.edu/blog/africa-in-focus/2022/02/25/what-does-the-war-in-ukraine-mean-for-africa/, last accessed 18 Jan 2023
97. Adibe J (2022) The Russo-Ukrainian war's impact: Africa and a 'neo-cold war', E-International Relations, 20 Dec 2022. https://www.e-ir.info/2022/12/20/the-russo-ukrainian-wars-impact-africa-and-a-neo-cold-war/, last accessed 18 Jan 2023
98. https://ec.europa.eu/eurostat/cache/infographs/energy/bloc-2c.html, last accessed 18 Jan 2023
99. https://newagemetals.com/will-palladium-continue-its-golden-performance/, last accessed 18 Jan 2023
100. Rose A, Chen Z, Wei D (2022) The economic impacts of Russia-Ukraine war export disruptions of grains and metals, 26 Nov 2022
101. Ghilès F (2023) Global South does not buy western stance on Ukraine, CIDOB, May 2023. https://www.cidob.org/es/publicaciones/serie_de_publicacion/opinion_cidob/2023/global_south_does_not_buy_western_stance_on_ukraine, last accessed 18 Jan 2023
102. https://guardian.ng/news/how-russia-ukraine-conflict-violence-in-sahel-compound-terrorism-by-buhari/, last accessed 18 Jan 2023
103. Raafat O (2023) Walking the tightrope: the Ukrainian crisis through the eyes of Africa, Euromesco policy brief, No. 128, Mar 2023. https://www.euromesco.net/publication/walking-the-tightrope-the-ukrainian-crisis-through-the-eyes-of-africa/, last accessed 18 Jan 2023. Hedenskog J (2018) Russia is stepping up its military cooperation in Africa, FOI Memo 6604, Swedish Defence Research Agency, Dec 2018. https://www.foi.se/rest-api/report/FOI%20MEMO%206604, last accessed 18 Jan 2023
104. Besenyő J (2019) The Africa policy of Russia. Terror Polit Violence 31(1):132–153
105. Sophie Desmidt S (2023) One year on: the ripple effects of Russia's war in Ukraine on the EU-Africa peace and security partnership, 27 Feb 2023. https://ecdpm.org/work/one-year-ripple-effects-russias-war-ukraine-eu-africa-peace-and-security-partnership, last accessed 18 Jan 2023
106. https://www.linkedin.com/feed/update/urn:li:activity:7011733652371841024/, last accessed 18 Jan 2023

Optimized Multi-cloud Service Orchestration in Cloud Computing

Abel Kamagara, Susan Babirye, and Döníz Borsos

Abstract This paper presents an optimization model for service orchestration in multi-cloud environments. The objective is to minimize the total cost of deploying a set of services over multiple clouds while ensuring that the Quality of Service (QoS) requirements are met. The proposed model considers the heterogeneity of cloud resources and the interdependence among the services. A case study is presented to demonstrate the effectiveness of the proposed approach. The results show that the proposed model can achieve significant cost savings while satisfying QoS requirements. The execution time of the algorithm is also analyzed, and it is found that it increases with the number of network elements. This study provides a framework for efficient service orchestration in multi-cloud environments, which can be extended to include additional constraints and objectives. The findings of this study are promising for practical applications for cloud service providers and users, who can benefit from the proposed optimization algorithm to achieve cost-effective service orchestration while meeting QoS requirements.

Keywords Integer linear programming · Multi-cloud computing · Service orchestration

1 Introduction

In recent years, the rapid proliferation of cloud computing technologies has revolutionized the way organizations deploy and manage their IT infrastructure [1]. Cloud environments offer numerous benefits, including scalability, flexibility, and

A. Kamagara · S. Babirye
Department of Electrical and Electronics Engineering, Faculty of Engineering, Kyambogo University, Kampala, Uganda
e-mail: akamagara@kyu.ac.ug

D. Borsos (✉)
Kando Kalman Faculty of Electrical Engineering, Obuda University, Budapest, Hungary
e-mail: borsos.doniz@uni-obuda.hu

T. A. Kovács et al. (eds.), *The Impact of the Energy Dependency on Critical Infrastructure Protection*, Advanced Sciences and Technologies for Security Applications, https://doi.org/10.1007/978-3-031-78544-3_13

cost-efficiency. However, with the increasing complexity and heterogeneity of cloud resources, orchestrating services across multiple clouds has become a daunting task. The need to optimize service orchestration while meeting Quality of Service (QoS) requirements has emerged as a critical challenge [2].

The optimization of service orchestration has garnered significant attention from researchers and industry practitioners. The goal is to minimize costs while ensuring that QoS requirements, such as response time, reliability, and availability, are met. Efficient service orchestration not only reduces operational expenses but also improves the overall performance and user experience.

In this paper, we present a case study to demonstrate the effectiveness of the proposed optimization model. The results highlight the significant cost savings achieved while ensuring QoS requirements are met. Furthermore, the execution time of the algorithm is analyzed, showcasing its scalability with the number of network elements. With this in mind, this paper sets out to achieve the following objectives:

- Examine an optimization model for service orchestration in multi-cloud environments that considers the interdependencies among services, the heterogeneity of cloud resources, and the dynamic nature of workload patterns.
- Adopting the Mixed-Integer Linear Programming (MILP) algorithm approach of formulating the orchestration problem and solving it using the YALMIP solver in MATLAB.
- Minimize costs associated with resource allocation and service placement while meeting QoS requirements, such as response time, throughput, availability, and reliability.

The remainder of this paper is organized as follows: Sect. 2 provides a literature review on related works in service orchestration and optimization. Section 3 presents the proposed optimization model in detail. Section 4 describes the case study setup and presents the experimental results. Section 5 further concludes this paper, discusses the conclusions, and future directions of this research.

2 Literature Review

Service orchestration refers to the process of integrating and coordinating multiple services to deliver a desired outcome [3]. It involves managing the deployment, configuration, and execution of services across a distributed infrastructure. In multi-cloud environments, where services are sourced from multiple cloud providers, orchestrating these services becomes even more complex. Each cloud provider may offer different specifications, pricing models, and Quality of Service guarantees, making the decision-making process challenging [4, 5].

2.1 Multi-cloud Environment and Challenges

Multi-cloud environments involve the utilization of multiple cloud service providers to meet an organization's computing needs. These environments offer a range of advantages, including improved reliability, flexibility, and cost optimization. Each cloud provider may have different strengths, such as specialized services, geographic presence, or pricing models [6]. Organizations leverage these diverse cloud offerings to create a hybrid infrastructure that best suits their requirements.

In a multi-cloud environment, organizations can distribute workloads and resources across multiple clouds, ensuring redundancy and mitigating the risks of service disruptions. This distributed architecture enables workload scalability and load balancing, leading to improved performance and availability [7]. Additionally, multi-cloud environments provide opportunities for workload optimization by selecting the most suitable cloud provider for specific tasks based on factors like cost, performance, and compliance [8].

Service orchestration in multi-cloud environments presents several challenges that need to be addressed to ensure seamless coordination and efficient resource utilization. The following challenges can be highlighted [9–13]:

- Resource Heterogeneity: Each cloud provider may offer a different set of services, APIs, and deployment models. Managing and orchestrating resources across these diverse platforms requires dealing with variations in terminology, interfaces, and compatibility.
- Service Interdependencies: In multi-cloud environments, applications often rely on multiple services provided by different cloud providers. Coordinating the interdependencies among these services, such as data sharing, security, and synchronization, becomes complex.
- Dynamic Workload Patterns: Workload patterns in multi-cloud environments can be highly dynamic, with varying demands and resource requirements. Managing workload fluctuations and ensuring optimal resource allocation in response to changing needs is a challenge.
- Data Management and Security: Multi-cloud environments involve the distribution of data across different cloud providers. Managing data consistency, integrity, and security in a distributed and heterogeneous environment poses challenges.

2.2 Existing Approaches and Techniques

Several research efforts have focused on addressing the challenges of service orchestration in multi-cloud environments. These approaches aim to optimize resource allocation, service placement, and workload management. By leveraging various techniques, researchers have proposed innovative solutions to enhance the efficiency and effectiveness of service orchestration.

One prominent approach is the use of machine learning and artificial intelligence techniques for resource allocation and workload prediction [14, 15]. By analyzing historical workload patterns and resource utilization data, these approaches can make intelligent decisions regarding resource allocation, scaling, and optimization [16]. Examples include reinforcement learning-based algorithms, genetic algorithms, and predictive modeling techniques.

Another approach involves the use of optimization models, such as linear programming and mixed-integer programming, to solve resource allocation and service placement problems [17–19]. These models take into account various constraints, such as resource availability, cost, and performance requirements, to determine the optimal deployment configuration. These optimization techniques provide a systematic and mathematical approach to orchestrate services across multiple clouds [20].

Existing approaches for service orchestration in multi-cloud environments exhibit varying strengths and limitations. Machine learning-based techniques provide adaptive and self-learning capabilities but may require extensive training data and computational resources. Optimization models offer a systematic and optimal solution but may have limitations in scalability and real-time decision-making. Heuristic-based approaches provide simplicity and efficiency but may lack global optimization guarantees.

The choice of approach depends on the specific requirements, constraints, and trade-offs of the organization. A comprehensive analysis of these approaches is necessary to identify the most suitable technique based on factors such as scalability, performance, cost, adaptability, and ease of implementation.

2.3 *Effectiveness and Applicability in Real-World Scenarios*

The effectiveness and applicability of optimization models for service orchestration in multi-cloud environments depend on several factors, including the complexity of the environment, scalability, and real-time decision-making requirements. While these models offer a systematic and rigorous approach, their practical implementation may face challenges related to computational complexity and scalability.

Real-world scenarios require considering additional factors, such as the dynamic nature of workload patterns, resource heterogeneity, and service interdependencies. The effectiveness of optimization models depends on their ability to handle these complexities efficiently [21]. Further research is needed to enhance the scalability, adaptability, and real-time decision-making capabilities of these models to ensure their applicability in large-scale multi-cloud environments.

3 Methodology

The proposed model aims to address the challenge of dynamically allocating and managing resources to meet service demands while ensuring cost efficiency and high performance. The theory is based on the concept of mixed integer linear programming to formulate the optimization problem and solve it using MATLAB's optimization toolbox. The MILP model seeks to minimize the total cost of deploying services subject to resource constraints and service level agreements (SLAs). The optimization problem involves selecting appropriate virtual machines (VMs) for hosting services and routing service requests to the selected VMs while maintaining SLA requirements.

3.1 Problem Formulation

In order to optimize the service orchestration process in multi-cloud environments, it is essential to formulate the problem in a mathematical framework. The problem formulation aims to capture the key elements of resource allocation, service interdependencies, and quality of service (QoS) optimization. By representing the problem mathematically, we can leverage optimization techniques to find the optimal solution.

3.2 Modelling the Problem

Let $A = \{S1, S2, \ldots\}$ denote an IoT application as a set of component service functions $S_j = 1, 2, \ldots$ A service function can be selected at many network elements and, therefore represented as a tuple $S_i = \langle s_{i1}, s_{i2}, \ldots \rangle$, where s_{ij} is the jth service at network element i. We define the cost of using and deploying as c_{ij} and d_j respectively. We define the binary variable x_{ij} whose value is 1 if service j is selected or used at network element i, and 0 otherwise. Another variable y_j, is used which is 1 if service j is deployed, and 0 otherwise [2].

Furthermore, we realize that a service can only be used if it's deployed. The number of services that can be deployed is constrained by a constant, k. Also, the bandwidth used by each service b_{wj}, should not exceed the capacity of the network element. Lastly, we need to ensure that each network element uses only one service.

The objective is to minimize the overall cost and maximize QoS satisfaction by finding an optimal allocation of resources that meets the QoS targets and satisfies the service interdependencies. The formulated objective function can be represented as follows:

$$\textbf{\textit{minimize}}\,C(A) = \sum (c_{ij} * x_{ij}) + \sum (d_j * y_j) \tag{1}$$

where $C(A)$ is the total cost of the services used, c_{ij} is the cost of using service j at network element i, x_{ij} is a binary variable that is 1 if service j is used at network element i, and 0 otherwise, d_j is the cost of deploying service j, y_j is a binary variable that is 1 if service j is deployed, and 0 otherwise.

Subject to:

Service Constraints

$$\sum (x_{ij}) = 1 \quad \text{for each } i, \tag{2}$$

where

$$i \in \{1, \ldots, n\} \quad \text{and} \quad j \in \{1, \ldots, m\}$$

This constraint ensures that each network element uses only one service.

Capacity Constraints

$$\sum (x_{ij} * b_{wj}) \leq 1 \quad \text{for each } i, \tag{3}$$

where

$$i \in \{1, \ldots, n\} \quad \text{and} \quad j \in \{1, \ldots, m\}$$

This constraint ensures that the bandwidth used by each service does not exceed the capacity of the network element.

Deployment Constraints

$$x_{ij} \leq y_j \quad \text{for each } j, \quad j \in \{1, \ldots, m\}, \tag{4}$$

This constraint ensures that a service can only be used if it is deployed.

Resource Constraints

$$\sum (y_j) \leq k \tag{5}$$

This constraint limits the number of services that can be deployed to k.

3.3 The Objective Function

The objective function in our proposed model represents the goal we aim to optimize. In the context of service orchestration in multi-cloud environments, our objective is to minimize the total cost of resource allocation while meeting the QoS requirements for the deployed services.

Let's denote the objective function as J, which can be formulated as follows:

$$J = \sum_{i=0}^{N} \sum_{j=1}^{M} (C_{ij} * x_{ij}) + (d_j * y_j) \tag{6}$$

In this equation, N represents the total number of services to be deployed, and M represents the available cloud providers. The decision variable x_{ij} is a binary variable that indicates whether service i is allocated to cloud provider j. C_{ij} denotes the cost associated with deploying service i on cloud provider j.

The objective function aims to minimize the overall cost by summing up the costs of all the services deployed on their respective cloud providers. By finding the optimal allocation of services to cloud providers, we can reduce the total cost of resource utilization and achieve cost-efficient service orchestration.

3.4 Virtual Experiment

The optimization algorithm plays a crucial role in solving the proposed model by formulating it as a MILP problem and finding an optimal solution. In this study, we will utilize the YALMIP solver in MATLAB to solve the MILP problem. The general steps of the optimization algorithm are as follows (Fig. 1):

- ***Step 1:*** Initialization:
  - Initialize the decision variables, such as the allocation matrix A, with initial values or feasible solutions.
- ***Step 2:*** Formulate the MILP Problem:
  - Convert the resource allocation and service orchestration problem into a MILP problem using mathematical modeling.
  - Define the decision variables, constraints, and objective function based on the defined objectives and constraints discussed earlier.
- ***Step 3:*** Solve the MILP Problem using YALMIP solver:
  - Utilize the YALMIP modelling language in MATLAB to define the MILP problem.
  - Set up the objective function and constraints using YALMIP syntax.
  - Invoke the YALMIP solver to solve the MILP problem and obtain the optimal or best feasible solution.
- ***Step 4:*** Obtain the Allocation Results:

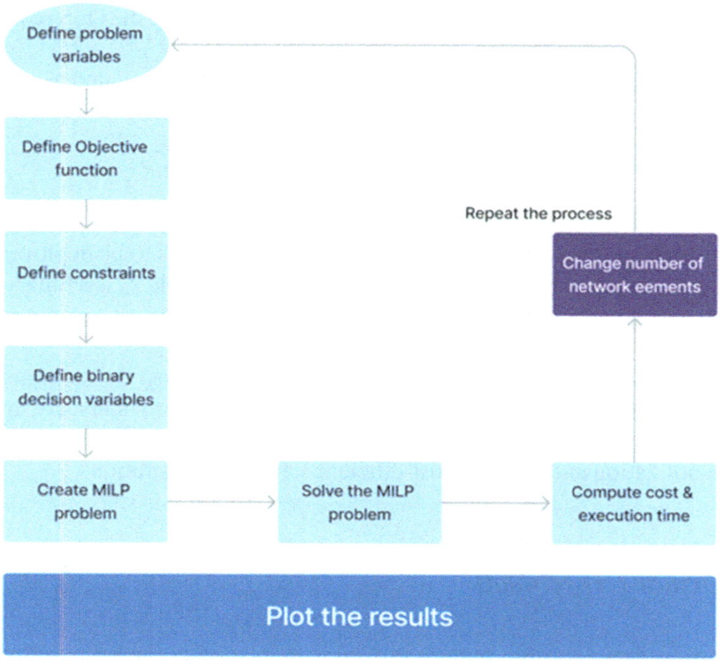

Fig. 1 Flowchart showing steps taken to run simulations

- Retrieve the optimal allocation results from the solver, including the allocation matrix A.
- Analyze the results to determine the allocation of each service to a specific cloud provider.

- ***Step 5:*** Evaluate the Results:
  - Assess the achieved objectives, such as QoS satisfaction, resource utilization, and other relevant performance measures.
  - Analyze the impact of the allocation on the overall system efficiency and cost-effectiveness.

By utilizing the YALMIP solver in MATLAB and following these steps, we can effectively solve the MILP problem formulated in the proposed model, leading to an optimal allocation of services to cloud providers while considering various objectives and constraints.

4 Results and Discussion

4.1 Total Cost Versus Number of Network Elements

The cost plot (Fig. 2) shows a 'zigzag' trend. The cost rises steadily between 50 and 55 network elements after which there is a steep increase of cost for the next additional 2–3 network elements. The cost stabilizes for another 3 network elements then shoots up again. This trend continues throughout and produces what we see as a rising zigzag line.

Initially, as the number of network elements increases from 50 to 57, the cost appears to rise steadily. This can be attributed to the inherent complexity and resource requirements associated with setting up and maintaining a larger network. As the network grows, more resources, such as hardware, software, and maintenance, are needed, leading to increased costs.

However, the slight drop in cost after reaching a certain threshold (around 58 network elements) suggests that there might be an optimal point where the network achieves better resource utilization and cost efficiency. At this point, the added network elements contribute to more efficient service deployment options, leading to improved resource allocation and cost reduction.

The observed zigzag trend in the cost plot indicates the need for careful analysis and optimization when scaling network infrastructure. It highlights the importance of finding the right balance between network size, resource utilization, and cost efficiency. By understanding these cost patterns, network administrators and

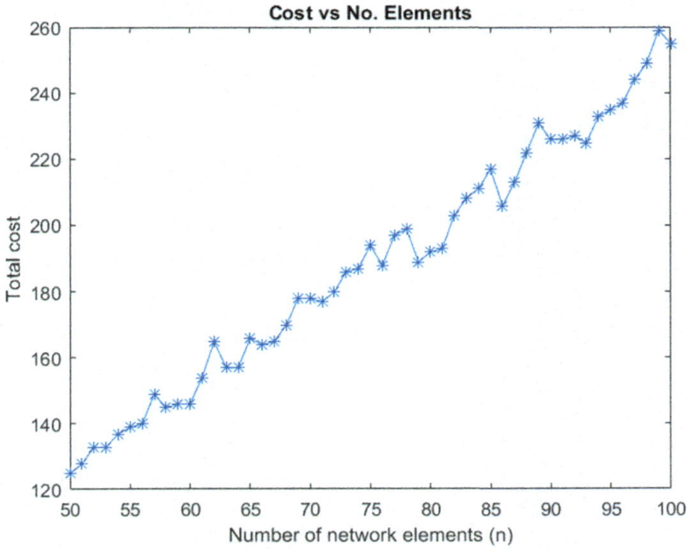

Fig. 2 Plot of total cost against number of network elements

planners can make informed decisions to optimize resource allocation, enhance cost-effectiveness, and ensure optimal service orchestration.

Further research and analysis can delve deeper into the factors contributing to these cost patterns, explore alternative network configurations, and develop strategies to minimize costs while maintaining optimal network performance.

4.2 Execution Time versus Number of Network Elements

The execution time plot (Fig. 3) shows a slightly different trend. The execution time stabilizes and becomes more or less the same for the next 15 additional network elements before it starts to gradually increase again as more additional elements are added.

The observed trend in the execution time plot, particularly the steep fall followed by stabilization and gradual increase, provides insights into the relationship between the number of network elements and the time required for executing network operations. Let's analyze this trend.

The significant and rapid decrease in execution time between 50 and 55 network elements suggests a notable improvement in the efficiency of network operations. This can be attributed to several factors, such as optimization techniques, enhanced algorithms, or improved hardware/software configurations. The network might have reached a critical point where the existing resources and algorithms can handle the workload more efficiently, resulting in a substantial reduction in execution time.

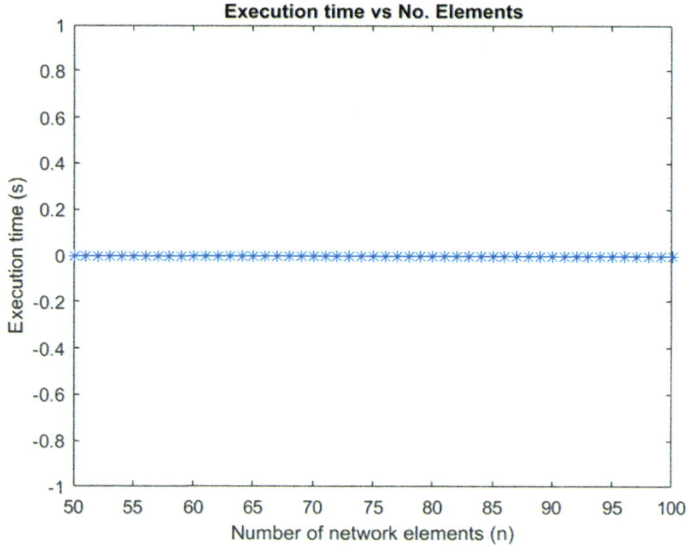

Fig. 3 Plot of execution time against number of network elements

Following this steep fall, the execution time stabilizes for the next 15 additional network elements. This stabilization indicates that the network infrastructure is able to maintain consistent performance and handle the increased workload without significant variations in execution time. It suggests that the existing resources and algorithms are sufficient to handle the added network elements without compromising efficiency.

However, as the number of additional network elements continues to increase beyond this stabilization point, the execution time gradually starts to increase again. This can be attributed to the growing complexity and increased computational requirements associated with larger network sizes. The additional network elements introduce more interactions, dependencies, and calculations, leading to longer execution times.

It is worth noting that the gradual increase in execution time might not be as steep as the initial fall. This can be attributed to the network infrastructure's ability to adapt and optimize its operations to accommodate the growing workload. Nevertheless, the increasing execution time indicates that additional resources or optimization strategies may be necessary to maintain acceptable performance levels as the network continues to scale.

To further optimize the execution time, future research could focus on developing advanced algorithms, exploring parallel computing techniques, or leveraging specialized hardware accelerators. These approaches can help mitigate the increasing execution time and ensure efficient network operations even with larger network sizes.

5 Conclusion

In conclusion, the results obtained from the cost and execution time plots provide valuable insights into the behavior of the proposed network optimization model. The cost plot shows a zigzag pattern, with a rise in cost between certain network elements and subsequent reductions. On the other hand, the execution time plot exhibits a steep fall followed by stabilization and a gradual increase.

The analysis of these results reveals that the proposed model exhibits scalability and efficiency characteristics. The cost reduction between specific network element ranges indicates that the model is effective in optimizing resource utilization and minimizing overall costs. The execution time's steep fall and subsequent stabilization suggest that the model efficiently handles network operations for a significant number of network elements, resulting in improved execution efficiency.

However, the gradual increase in execution time beyond a certain point indicates that as the network size further expands, additional computational resources or optimization strategies may be required to maintain optimal performance. This finding highlights the need for ongoing research and development to enhance the model's scalability and address the potential performance limitations associated with larger network sizes.

Overall, the results validate the effectiveness of the proposed model in optimizing network resources and achieving cost reductions. The findings also underline the importance of continuous improvement efforts to address the challenges posed by increasing network complexities. With further enhancements and optimizations, the proposed model holds significant potential for real-world implementation, leading to more efficient and cost-effective network management and service orchestration.

6 Applications

Optimized service orchestration can be applied in many other areas where efficient allocation and management of resources are essential for achieving optimal outcomes. In context-aware power systems, optimization techniques can be used to determine the most efficient allocation of energy resources to meet user demand while minimizing waste. For instance, the use of advanced sensors to monitor energy usage can be combined with optimization algorithms to enable real-time demand-response mechanisms, where energy consumption can be adjusted in response to varying energy prices and user preferences. This can help reduce energy consumption during peak hours, leading to cost savings and a reduced carbon footprint.

Acknowledgements We would like to express our sincere gratitude to Kyambogo University Research Grants Committee for the funding towards this project.

References

1. Miorandi D, Sicari S, De Pellegrini F, Chlamtac I (2012) Internet of things: vision, applications and research challenges. Ad Hoc Netw 10(7):1497–1516
2. Nguyen D, Nguyen K, Cheriet M (2017) Optimized IoT service orchestration. In: IEEE 28th annual international symposium on personal, indoor, and mobile radio communications. IEEE, Montreal, pp 1–6
3. Erl T (2005) Service-oriented architecture: concepts, technology, and design. Pearson Education, India
4. Basu S, Soumyakanti C, Megha S (2015) Pricing cloud services—the impact of broadband quality. Omega 50:96–114
5. Laatikainen G, Ojala A, Mazhelis O (2013) Cloud services pricing models. In: Herzwurm G, Margaria T (eds) Software business. From physical products to software services and solutions. ICSOB. Lecture notes in business information processing, vol 150, pp 117–129. Springer, Berlin
6. Toosi AN, Calheiros RN, Buyya R (2014) Interconnected cloud computing environments: challenges, taxonomy, and survey. ACM Comput Surv 47(1):7
7. Ardagna D (2015) Cloud and multi-cloud computing: current challenges and future applications. In: 2015 IEEE/ACM 7th international workshop on principles of engineering service-oriented and cloud systems (PESOS), Florence, Italy
8. Dickinson M et al (2021) Multi-cloud performance and security driven federated workflow management. IEEE Trans. Cloud Comput 9(1):240–257

9. Kertesz A (2014) Characterizing cloud federation approaches. In: Mahmood Z (eds) Cloud computing. computer communications and networks. Springer, Cham, pp 277–296
10. Alshammari M, Alwan A, Nordin A, Al-Shaikhli I (2017) Disaster recovery in single-cloud and multi-cloud environments: Issues and challenges. In: 2017 4th IEEE international conference on engineering technologies and applied sciences (ICETAS). IEEE, Salmabad, pp 1–7
11. Imran HA et al (2020) Multi-Cloud: a comprehensive review. In: 2020 IEEE 23rd international multitopic conference (INMIC). IEEE, Bahawalpur, pp 1–5
12. Ramamurthy A, Saurabh S, Gharote M, Lodha S (2020) Selection of cloud service providers for hosting web applications in a multi-cloud environment. In: 2020 IEEE international conference on services computing (SCC). IEEE, Beijing, pp 202–209
13. Raj P, Raman A (2018) Multi-cloud management: technologies, tools, and techniques. in: software-defined cloud centers. In: Computer communications and networks. Springer, Cham, pp 219–240
14. Salman T, Bh D, Erbad A, Jain R, Samaka M (2017) Machine learning for anomaly detection and categorization in multi-cloud environments. In: 2017 IEEE 4th international conference on cyber security and cloud computing (CSCloud). IEEE, New York, pp 97–103
15. Abusitta A, Bellaiche M, Dagenais M, Halabi T (2019) A deep learning approach for proactive multi-cloud cooperative intrusion detection system. Futur Gener Comput Syst 98:308–318
16. Tsakalidou VN, Mitsou P, Papakostas GA (2023) Machine learning for cloud resources management—an overview. In: Smys S, Lafata P, Palanisamy R, Kamel KA (eds) Computer networks and inventive communication technologies, vol 141. Lecture notes on data engineering and communications technologies. Springer, Singapore, pp 903–915
17. Ciavotta M, Ardagna D, Gibilisco GP (2017) A mixed integer linear programming optimization approach for multi-cloud capacity allocation. J Syst Softw 123:64–78
18. Mohammadi S, Pourkarimi L, Pedram H (2019) Integer linear programming-based multi-objective scheduling for scientific workflows in multi-cloud environments. J Supercomput 75:6683–6709
19. Mohammadi S, Pedram H, Pourkarimi L (2018) Integer linear programming-based cost optimization for scheduling scientific workflows in multi-cloud environments. J Supercomput 74:4717–4745
20. Masdari M, Nabavi S, Ahmadi V (2016) An overview of virtual machine placement schemes in cloud computing. J Netw Comput Appl 66:106–127
21. Amani P, Bastani S, Landfeldt B (2015) Towards optimal content replication and request routing in content delivery networks. In: 2015 IEEE international conference on communications (ICC). IEEE, London, pp 5733–5739

Thermodynamic Assessments of the Ce–Te and Ce–Te–Bi Systems

H. Bouchta, N. Selhaoui, M. A. Boukideur, A. Bendarma, A. Khadija, A. Marjaoui, and M. Zanouni

Abstract In the present article, a re-evaluation of the thermodynamic parameters and phase diagram of the Ce–Te binary system and an initial assessment of the Ce–Te–Bi ternary system have been carried out. The new modelling of the Ce–Te system contains two modifications of the liquid phase and the extended phase of the solid compound ($CeTe_2$), and the optimized diagram shows satisfactory agreement with experimental data in the literature. For an initial assessment of the Ce–Te–Bi system, the current description is made by analogy with the ternary Ce–Fe–Bi system to confirm the part of the Ce–Bi binary system missing in literature from the ternary Ce–Te–Bi system and using thermodynamic descriptions of limiting binary systems, Ce–Bi and Bi–Te reported in the literature. The results we found are in very good agreement with existing literature data.

Keywords Binary system · Ternary system · Associate model equation · Thermodynamic assessment · CALPHAD approach

H. Bouchta · N. Selhaoui · M. A. Boukideur · A. Khadija
Thermodynamics and Energetics Laboratory (LTE), Department of Physics, Faculty of Science, Ibn Zohr University, BP8106, 80006 Agadir, Morocco
e-mail: hamza.bouchta@e-polytechnique.ma

H. Bouchta · A. Bendarma (✉)
InterDisciplinary Applied Research Laboratory—LIDRA, International University of Agadir—Universiapolis, Agadir, Morocco
e-mail: amine.bendarma@artsetmetiers.ma

A. Marjaoui · M. Zanouni
Materials, Systems and Energy Engineering Laboratory (MaSEEL), ERCMN (UAE/U01FST), Abdelmalek Essaâdi University, Tetouan, Morocco

A. Bendarma
Arts et Métiers Campus de Rabat Technopolis, Rocade de Rabat-Salé, 11100 Sala Al Jadida, Morocco

T. A. Kovács et al. (eds.), *The Impact of the Energy Dependency on Critical Infrastructure Protection*, Advanced Sciences and Technologies for Security Applications, https://doi.org/10.1007/978-3-031-78544-3_14

1 Introduction

The fascinating materials in the realm of advanced materials research are chalcogenides, particularly diamonds composed of elements from the chalcogenide group, with tellurium being a prominent example. These materials hold significant importance due to their extraordinary properties, especially in the context of phase-change materials and thermoelectric devices [1–3]. In the field of photovoltaic cells and infrared detection systems, CdTe and Na_2Te serve as fundamental materials [4, 5]. Various dopants [4] have been investigated to enhance performance, including Ce [6] and Te [7], Concurrently, studies such as perforation tests of composite structure specimens at different temperatures and strain rates, as well as the mechanical properties of brass under impact and perforation tests at different temperatures, enrich our understanding of materials in varied environments, contributing to a better grasp of their behavior and potential applications. For example, the works of Bendarma et al. [8, 9] further elucidate the complexities of material responses under diverse conditions, providing valuable insights for advancing various technological applications.

The creation of Te-based Ce chalcogenides requires precise data on phase equilibria and their thermodynamic properties [10]. Conversely, rare-earth chalcogenides have been acknowledged as highly efficient solutions for high-temperature thermoelectric materials [11]. Ce_3Te_4 has been identified as a potential high-temperature thermoelectric material with exceptional performance [12].

In the ongoing pursuit of progress, it is crucial to explore new avenues in chalcogenide research. For instance, investigating alternative dopants and their impact on the properties of these materials could open new possibilities for applications in diverse technological fields. Additionally, delving deeper into the synthesis techniques and structural aspects of chalcogenides may unveil unprecedented properties, paving the way for innovative applications in emerging technologies.

Thermodynamic modeling of the Ce–Te–Bi system requires knowledge of the thermodynamic parameters of the binary systems Ce–Te, Ce–Bi and Te–Bi. The aim of the present work is to establish a thermodynamic model of the Ce–Te binary system and the Ce–Te–Bi ternary system using the thermo-Calc code and the CALPHAD method developed by Sundman et al. [13] and Jansson [14].

2 Literature Review

2.1 Ce–Te System

Chukalin et al. [15] conducted a comprehensive study of the Ce–Te system utilizing X-ray analysis, metallography, and thermal analysis with 99.5% pure Ce and 99.99% pure Te. However, their reported eutectic temperature in the Te-rich region (455 °C) exceeded the accepted melting point of Te (449.6 °C), suggesting potential impurities in the Te stock [16], particularly impacting the Te-rich section of the diagram.

Pardo and Flahaut [17], examining the system between 65 and 100 at.% Te using X-ray and thermal analysis, proposed a more accurate Te-rich eutectic temperature of 435 °C and 99 at.% Te. The rest of the diagram relied on Chukalin et al.'s [15] findings. Okamoto [18], in agreement with Gschneidner et al. [16], assessed the Te–Ce system, identifying seven intermediate phases (CeTe, Ce_3Te_4, Ce_2Te_3, Ce_4Te_7, Ce_2Te_5, $CeTe_2$ and $CeTe_3$). Congruently formed phases include CeTe and Ce_3Te_4 Chukalin et al. [15] reported the existence of solid solutions ranging from Ce_3Te_4 to Ce_4Te_7 (57.1–60.0 at.% Te) and $CeTe_{\sim 1.9}$ to $CeTe_2$ (66.0–66.67 at.% Te).

Philipp et al. [19] evaluated the Te–Ce system using the ChemSage program [20], based on Okamoto's [18] results. However, they questioned the accuracy of adjusting eutectic reactions as presented by Okamoto [18], particularly the one involving CeTe and Ce_3Te_4. Instead, they assumed that Ce_3Te_4 formed through a peritectic reaction, similar to other tellurium-rich compounds. Additionally, for simplification, Philipp et al. [19] disregarded the homogeneity of $Ce_{3-x}Te_4$ and $CeTe_2$. Enthalpies of formation for all phases were computed by fixing the CeTe phase as a reference value during modeling in ChemSage [20].

Thermodynamic data for the Ce–Te system are scarce due to the high volatility of Te [21], posing challenges for experimental accuracy and feasibility. Mills et al. [22] experimentally determined thermodynamic data for CeTe through vapor pressure measurements using direct calorimetry. They also provided estimated data for Ce_2Te_3 ($Ce_{3-x}Te_4$). Furthermore, Ferro et al. [23] measured the standard enthalpy of the CeTe and $Ce_{3-x}Te_4$ phases using the direct calorimetry method.

Dong et al. [24] assessed the Te–Ce system by using Thermo-Calc program [25] on the basis of the result from Okamoto [18]. The homogeneity of $CeTe_2$ is ignored in his work for simplification considering its narrow and insignificant range.

The phase diagram of the binary Ce–Te system was thermodynamically reassessed in our previous work, the obtained phase diagram is given in Fig. 1, Furthermore, the crystallographic data of all intermediate compounds formed in the cerium–Tellurium system are illustrated in Table 1.

2.2 *Ce–Bi System*

The Ce–Bi phase diagram, compiled by Gschneidner et al. [32], relies on boundary data provided by Vogel et al. [33], Pleasance et al. [34], and Yoshihara et al. [35]. Initially, Vogel et al. [33] conducted measurements on the phase equilibrium of the Ce–Bi binary system, identifying stable compounds such as Ce_3Bi, Ce_4Bi_3 CeBi and $CeBi_2$ with Ce_4Bi_3 being the only compound melting congruently. However, concerns arose about the purity of Vogel et al.'s raw materials, casting doubt on various aspects of the reported phase diagram. Pleasance et al. [34] investigated the solubility of Ce in liquid Bi, while Yoshihara et al. [35] determined the complete range of Ce–Bi alloys using X-ray diffraction and metallographic techniques. They confirmed that Ce_2Bi, not Ce_3Bi as reported by Vogel et al. [33], is the first compound on the Ce-rich side and identified the crystal structure of Ce_5Bi_3 (Mn_5Si_3-type). The

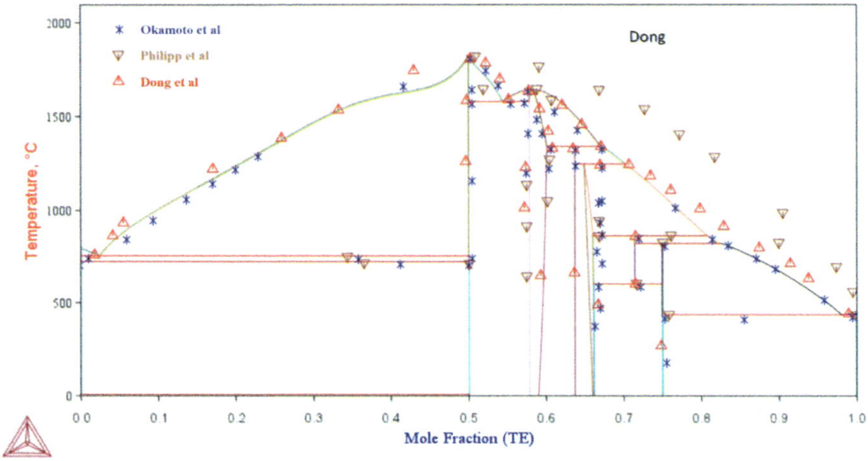

Fig. 1 The phase diagram of the Ce–Rh binary system optimized in the present work compared with experimental data of the literature [18, 19, 24]

Ce–Bi binary system includes five compounds: Ce_2Bi, Ce_5Bi_3, Ce_4Bi_3, CeBi and $CeBi_2$, all formed through peritectic reactions except Ce_4Bi_3. These compounds exhibit homogeneity ranges of less than 70.5 at.% and are treated as stoichiometric compounds. Additionally, the system contains other phases—liquid, γCe(Fcc_A_1), δCe(Bcc_A_2), and Bi($Rhomb\_A7$).

The phase diagram of the binary Ce–Bi system was thermodynamically assessed by [36] the obtained phase diagram is given in Fig. 2.

2.3 Bi–Te System

The binary Bi–Te system, as per Mao et al. [37], comprises seven phases: liquid, Rhombohedral_A7 (Bi), Hexagonal_A8 (Te), Bi_2Te_3, *Beta*, Bi_2Te and Bi_4Te_3. However, earlier sources exhibited inconsistencies in describing the phase diagram for concentrations ranging from approximately 25 to about 60 atomic percent (at.%) of Te. Hansen et al. [38] suggested an intermediate phase separated from terminal phases by eutectic reactions, while Brown and Lewis [39] proposed a broad *Beta* phase with a homogeneity range varying from 20 to 60 at.% Te. Glazov et al. [40] proposed four intermediate phases: Bi_7Te_3, Bi_2Te, BiTe, and Bi_2Te_3. Following Glazov et al. [40], Okamoto and Tanner [41] identified seven specific intermediate phases and ten uncertain linear intermetallic compounds (IMCs).

The Bi–Te binary system was modeled by Gierlotka [42]. The obtained phase diagram is represented in Fig. 3, the values of the calculated parameters are integrated in the ternary database of the Bi–Ce–Te system.

Table 1 Crystallographic data for the intermediate phases of the Ce–Te system

Compound	Structure type	Lattice constants (A)			Reference
		a	b	c	
(γCe)	$CF4$ Cu ($Fm\overline{3}m$)	5.1612	5.1612	5.1612	[26]
(δCe)	$cI2$ W ($Im\overline{3}m$)	4.12	4.12	4.12	[27]
CeTe	$cF8$ NaCl ($Fm\overline{3}m$)	6.36	6.36	6.36	[28]
$Ce_{3-x}Te_4$	$cI28$ Th_3Te_4 ($I\overline{4}3d$)	9.57	9.57	9.57	[29]
Ce_4Te_7	CrB ($oC8-Cmcm$)	–	–	–	[15]
$CeTe_2$	$tP22$ ($P4mbm$)	4.653	4.653	9.305	[30]
Ce_2Te_5	$oS28$ Nd_2Te_5 ($Cmcm$)	4.444	4.454	4.444	[29]
$CeTe_3$	$oS16$ $CeTe_3$ ($Ama2$)	4.398	4.398	25.99	[29]
(Te)	$hP3$ γSe ($P3_121$)	4.44	4.44	5.92	[31]

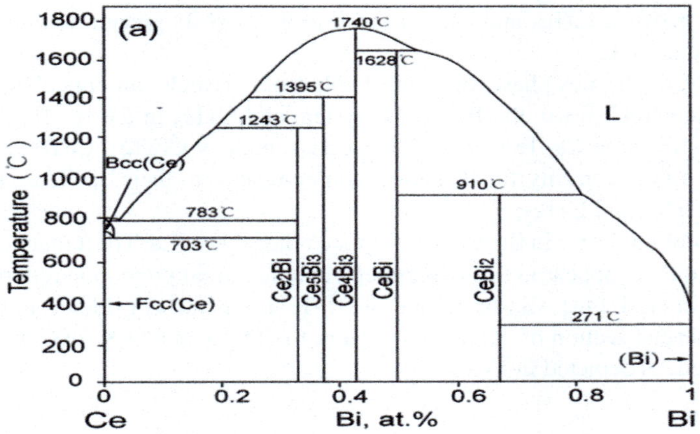

Fig. 2 Thermodynamic calculated Ce–Bi phase diagram [36]

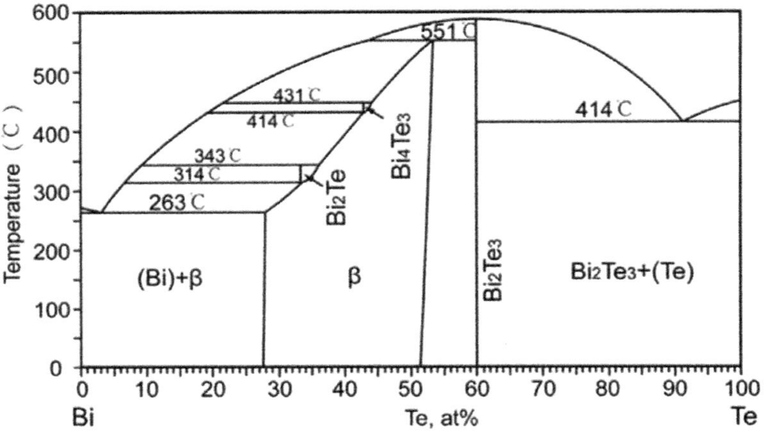

Fig. 3 Calculated Bi–Te phase diagram [42]

2.4 Ce–Te–Bi Ternary System

The Bi–Ce–Te ternary system was studied at 673 K by Tan et al. [43] using electron probe micro-analysis (EPMA) and powder X-ray diffractometry (XRD).

Due to the similarity in the physical and chemical properties of La and Ce, as well as the reported similar phase relationships in the sub-binary and ternary systems Bi–Te–La and Bi–Te–Ce, it is logical to investigate these two systems collectively. Microstructures and XRD diagrams of the samples allow us to confirm the presence of four three-phase regions and one two-phase region in the isothermal section of Bi–Te–Ce at 673 K. These phase equilibria closely resemble those of the Bi–Te–La system. The morphological difference between $CeTe_3$ and Te is more pronounced than that between $LaTe_3$ and Te, attributed to the slightly greater atomic weight of Ce compared to La.

Clearly, the porous phase, distinguishable from Te under an optical microscope, corresponds to $CeTe_3$ in the Bi–Te–Ce system and $LaTe_3$ in the Bi–Te–La system. Additionally, three Ce–Te intermetallic phases, namely Ce_2Te_3, $CeTe_2$ and $CeTe_3$, exhibit a higher capacity for Bi dissolution compared to their La–Te counterparts: La_2Te_3, $LaTe_2$ and $LaTe_3$.

The solubility of Ce in Bi_2Te_3 or β is greater than that of La. The ternary compound $CeBiTe_3$ did not appear in this experiment, and the liquid projection reported by the Mamedova et al. [44], similar to that of Bi–Te–La, is not considered in this study. The isothermal section of the ternary system Bi–Te–Ce at 673 K (400 °C) has been established, as depicted in Fig. 4.

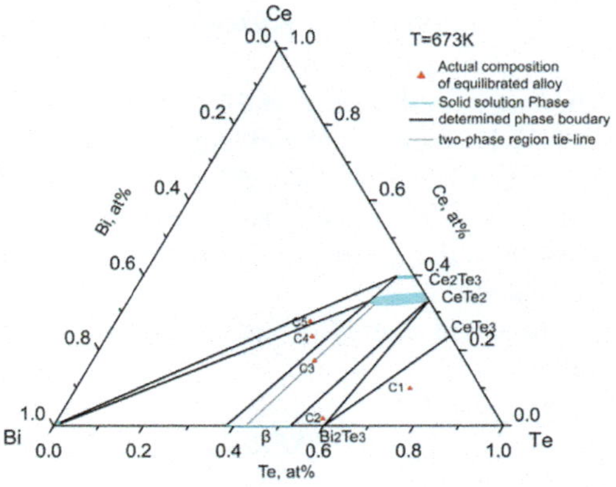

Fig. 4 Experimentally determined Bi–Te–Ce isothermal section at 673 K [43]

3 Thermodynamic Models

No intermediate phases were detected in the ternary Ce–Ti–Bi system. The Gibbs free energies of the liquid, primary and intermetallic phases of the Ce–Te binary system are described by the Redlich–Kister polynomial, as follows:

3.1 Pure Elements (Reference States)

As reference states for the optimization, the pure elements Ce and Te were chosen in their SER states at 298.15 K (Stable Element Reference) as recommended by the SGTE [43]. The Gibbs free energies of the pure elements Ce and Te as a function of temperature were given by Dinsdale in the following equation:

$$G_i^{0,\varphi}(T) = a + b.T + c.T.\ln(T) + d.T^2 + \frac{e}{T} + f.T^3 + i.T^7 + j.T^{-9} \quad (1)$$

where T is the absolute temperature in K, and a, b, c, d, e, f, g and h are coefficients; in different temperature ranges, different sets of coefficients can be used.

3.2 Liquid and Solution Phases

The solution phases concerning liquid, solid solution phases, (Te), (αCe), (βCe), and (δCe) were described by the substitutional solution model, The Gibbs energy function $G_i^{\varphi}(T) = {}^0G_i^{\varphi} - H_i^{SER}$ (298.15 K) for the element i ($i =$ Ce, Te) in the phase (*Fcc_A*1, *Bcc_A*2, and liquid phase) is given by the following formula:

$$G_i^{\varphi}(T) = a + bT + cT + dT^2 + eT^3 + fT^7 + gT^{-1} + hT^{-9} \quad (2)$$

where a, b, c, d, e, f, g, and h are the coefficient to be optimized. Moreover H_i^{SER} (298.15 K) is the molar enthalpy of the element i at 298.15 K in its standard element reference (SER) state, fcc_A1 for Ce and Te and T is the absolute temperature.

In this work, the Gibbs energy functions of the pure elements Ce and Te are taken from the compilation of the SGTE (Scientific Group Thermo data Europe) of Dinsdale [45].

The Gibbs energy of the element i, in its SER state, is denoted by $GHSER_i$ as follows:

$$GHSER_{\mathrm{Ce}} = G_{\mathrm{Ce}}^{FCC\_A1} - H_{\mathrm{Ce}}^{FCC\_A1}(298.15\ \mathrm{K}) \quad (3)$$

$$GHSER_{\mathrm{Te}} = G_{\mathrm{Te}}^{FCC\_A1} - H_{\mathrm{Te}}^{FCC\_A1}(298.15\ \mathrm{K}) \quad (4)$$

3.3 Associate Solution Model

To treat the liquid phase of the Ce–Te binary system, we used the associates model [46]. The phase diagram of this system shows a strong attractive interaction. This model was developed in 2000 by Sommer et al. [47] and was used for this system with the constituent species Ce, Te and CeTe as associated compounds in the liquid. The Gibbs free energies of the binary liquid phase for a 1 mol is given:

$$G^{Liq} = \sum_{i=\mathrm{Ce,Te,CeTe}} x_i G_i^{Liq} + RT \sum_{i=\mathrm{Ce,Te,CeTe}} x_i \ln(x_i) + G_{exc}^{Liq} \quad (5)$$

where x_i is the mole fraction of each species i (i = Ce, Te and CeTe) in the liquid phase. G_{exc}^{Liq} is the excess Gibbs free energy term of the liquid phase, which will be described as follows:

$$G_{exc}^{Liq} = \sum_i \sum_{i \neq j} x_i x_j \sum_{v=0}^{n} {}^{v}L_{i,j}^{Liq} (x_i - x_j)^v \quad (6)$$

With ${}^{v}L_{i,j}^{Liq}$ is the binary interaction parameter, which is described by the Redlich–Kister model as follows:

$${}^{v}L_{i,j}^{Liq} = a + b * T \quad (7)$$

3.4 *Intermetallic Compounds*

Stoichiometric compounds

The intermetallic phases CeTe, Ce_4Te_7, Ce_2Te_5 and $CeTe_3$ were optimized as stoichiometric phases. The Gibbs energy per atomic mole of the compound Ce_xTe_y can be expressed as follows:

$$^0G_{Ce_xTe_y} = \frac{x}{x+y}\,^0G_{Ce} + \frac{y}{x+y}\,^0G_{Te} + a + b * T \tag{8}$$

where $^0G_{Ce}$ and $^0G_{Te}$ are the Gibbs energy in the reference state of the pure elements Ce and Te, respectively, a and b are two variables to be optimized for each phase of the Ce–Te system.

Intermediate phase with homogeneity range Ce_3Te_4 and $CeTe_2$ phases

Compounds Ce_3Te_4 and $CeTe_2$ are non-stoichiometric phases whose homogeneity rank lies on the Te-rich side of the strict stoichiometry Ce_3Te_4 and $CeTe_2$, respectively. These phases have been treated as $(Ce, Te)_{0.428}$: $(Te)_{0.572}$ and $(Ce)_{0.333}$: $(Te)_{0.667}$ by the heard phase model (two-subarray model [48]). The Gibbs energy expression of these phases is given by the following equation:

$$\begin{aligned} ^0G_m^{Ce_xTe_y} &= x_{Ce}\,^0G_{Ce:Te}^{Ce_xTe_y} + x_{Te}\,^0G_{Ce:Te}^{Ce_xTe_y} + RT(x_{Ce}\ln x_{Ce} + x_{Te}\ln x_{Te}) \\ &+ x_{Ce}x_{Te}\sum {}^{v}L_{Ce:Ce,Te}(x_{Ce} - x_{Te})^{v} \end{aligned} \tag{9}$$

With x_{Ce} and x_{Te} are the part of the site of Ce and Te constituents located on two sublattices, and the term $^0G_m^{Ce_xTe_y}$ gives the Gibbs energy of the constituent when both sublattices are occupied by Ce or Te elements. $^0G_{Ce:Te}^{Ce_xTe_y}$ is the interaction parameter between Ce and Te elements in the first or second sublattice, when the other sublattice is occupied by Ce or Te.

4 Optimization Strategy

4.1 *The Ce–Te System*

The thermodynamic optimization of the excess Gibbs energy part ^{exc}G using the CALPHAD method [14] was carried out by the PARROT module of the Thermo-Calc code [13].

The Parrot module minimizes the sum of squares of the difference between the input and output parameters. Optimization begins with the introduction of experimental data from the literature as starting parameters for the enthalpy of formation for each solution of the binary system. Next, the enthalpy and entropy of all phases were optimized based on the data available in the literature.

A group of optimized thermodynamic parameters are obtained, and each experimental input value is carefully weighted according to its merits and weights and modified by error and trial-and-error during modeling, to achieve reasonable agreement with the experimental data within the expected uncertainty.

4.2 The Ce–Te–Bi System

The optimization of the Ce–Te–Bi ternary system was conducted as follows: initially, thorough verification of the thermodynamic parameters for the liquid phase, primary phases, and intermetallic compounds of the Ce–Bi [32] and Bi–Te [37] subsystems was performed, drawing directly from existing literature. Subsequently, the Ce–Te–Bi ternary system was extrapolated from the three binary systems, taking into consideration all of their phases (liquid, solid, and intermediate). The thermodynamic optimization of the Ce–Te–Bi ternary diagram was carried out by incorporating all optimized thermodynamic interaction parameters from the three binary subsystems (Ce–Te–Bi). The biphasic domains and junction lines between calculated binary phases were compared with experimental diagrams to validate the accuracy of the thermodynamic optimization.

5 Results and Discussion

5.1 The Ce–Te System

The phase diagram of the Ce–Te system optimized in this work is shown in Fig. 1 compared with the experimental phase diagrams in the literature by Chukalin [15], Pardo and Flahant [17] and Okamoto [18] and evaluated by Dong [24] and Philipp [19]. The optimized liquid phase is very similar to that plotted experimentally for all invariant reactions to that previously optimized by Dong [24] and Philipp [19]. As can be seen, two small differences are noted in the Ce-rich part (composition range 10–50 at.% Te) and the Te-rich part (composition range 95–100 at.%Te). For the former, the liquidus is a little more basic than the experimental data of Okamoto [18], while for the latter the eutectic reaction in this part is a little degenerate. In addition, the two non-stoichiometric phases are well treated by the sub-lattice model and are in reasonable agreement with the experimental diagram of Okamoto. Table 2 illustrates the comparison of the optimized invariant reaction with that of the literature. This comparison shows very good agreement.

The thermodynamic parameters calculated in the present optimization of the Ce–Te system are summarized in Table 3. Table 4 and Fig. 5 summarize the enthalpies

Table 2 Comparison between the calculated invariant reactions and the experimental data from the literature

Invariant reactions	Reaction type	This work		Literature [18]	
		x(Te)/at	T(K)	x(Te)/at	T(K)
Liq ⇔ CeTe + (Ce)	Eutectic	2.4	1031	2.2	1033
Liq ⇔ CeTe	Congruent	50	2091	50	2093
Liq ⇔ CeTe + Ce_3Te_4	Eutectic	54.36	1856	55	1858
Liq ⇔ Ce_3Te_4	Congruent	57.1	1917	57.1	1918
Liq + Ce_3Te_4 ⇔ Ce_4Te_7	Peritectic	67.01	1617	67.0	1618
Liq + Ce_4Te_7 ⇔ $CeTe_2$	Peritectic	70.20	1522	69.2	1523
Liq + $CeTe_2$ ⇔ Ce_2Te_5	Peritectic	80.77	1140	80.40	1143
Ce_2Te_5 ⇔ $CeTe_2$ + $CeTe_3$	Peritecticoid	–	880	–	880
Liq + Ce_2Te_5 ⇔ $CeTe_3$	Peritectic	82.52	1099	83.1	1102
Liq + $CeTe_3$ ⇔ (Te)	Eutectic	98.09	713.6	98.9	708

of formation of the intermetallic compounds of the X-T binary system calculated in this work by the Calphad method, the theoretically calculated enthalpies presented in online databases such as (Materials project [49], OQMD [50] and Jarvis [51]), those determined experimentally by [23]. In general, the enthalpies of formation calculated and optimized in this work are in good agreement with the literature [23], as shown in Fig. 5.

To test the stability of the liquid phase, we have plotted in Fig. 6 the curves of the enthalpy of mixing of the liquid phase calculated at different temperatures (2000 K–4000 K–6000 K) using the associates model. These curves show that the enthalpy of mixing of the liquid phase is inversely proportional to increasing temperature. The liquid phase is very stable at high temperatures, so no reverse miscibility gap was detected at this temperature as indicated by the authors [18, 19, 24].

5.2 *The Ce–Te–Bi System*

After optimizing the Ce–Te system and collecting optimizations for the Ce–Bi and Bi–Te binary systems. The results collected are important for optimizing the ternary Ce–Te–Bi system. In fact, all the databases created during the optimization of the three binary diagrams are directly included in another database to describe the different phases (unary, binary) of the Ce–Bi–Te system. While no intermetallic phases were detected in the ternary Ce–Bi–Te system, no thermodynamic parameters were calculated for the ternary phases in this system.

The isothermal cross-section of the Bi–Te–Ce system calculated at 673 K is shown in Fig. 4, compared with experimental results from the literature of Tan et al. [43] and Mamedova et al. [44]. This section shows a small difference when compared

Table 3 Thermodynamic parameters calculated in this work

Phases	Thermodynamic models	Parameters
Liquid	$(Ce, CeTe, Te)_1$	${}^{0}L^{Liq}_{CeTe} = -324{,}569 - 5 * T$
		${}^{0}L^{Liq}_{Ce,CeTe} = +21{,}631 - 1.10 * T$
		${}^{1}L^{Liq}_{Ce,CeTe} = +14{,}995 - 13.87 * T$
		${}^{0}L^{Liq}_{CeTe,Ce} = +29{,}780.1 - 118.52 * T$
		${}^{1}L^{Liq}_{CeTe,Ce} = -6024.9 + 45.46 * T$
		${}^{2}L^{Liq}_{CeTe,Ce} = -47{,}615 + 28.24 * T$
Fcc_A1	(Ce)	No excess term optimized
Fcc_A1	(Ce, Te)	No excess term optimized
CeTe	$(Ce)_{0.5} : (Te)_{0.5}$	$G^{CeTe}_{Ce,Rh} = 0.5 * {}^{0}G^{Fcc}_{Ce} + 0.5 * {}^{0}G^{Fcc}_{Te} - 168{,}025 - 14.41 * T$
$Ce_{3-x}Te_4$	$(Ce)_{0.42} : (Te)_{0.58}$	$G^{Ce_{3-x}Te_4}_{Ce,Rh} = 0.42 * {}^{0}G^{Fcc}_{Ce} + 0.58 * {}^{0}G^{Fcc}_{Te} - 151{,}120 - 21.57 * T$
Ce_4Te_7	$(Ce)_{0.36} : (Te)_{0.64}$	$G^{Ce_4Te_7}_{Ce,Rh} = 0.36 * {}^{0}G^{Fcc}_{Ce} + 0.64 * {}^{0}G^{Fcc}_{Te} - 135{,}001 - 23.99 * T$
$CeTe_2$	$(Ce)_{0.333} : (Te)_{0.667}$	$G^{CeTe_2}_{Ce,Rh} = 0.333 * {}^{0}G^{Fcc}_{Ce} + 0.667 * {}^{0}G^{Fcc}_{Te} - 127{,}544 - 21.80 * T$
Ce_2Te_5	$(Ce)_{0.28} : (Te)_{0.72}$	$G^{Ce_2Te_5}_{Ce,Rh} = 0.28 * {}^{0}G^{Fcc}_{Ce} + 0.72 * {}^{0}G^{Fcc}_{Te} - 115{,}217 - 17.34 * T$
$CeTe_3$	$(Ce)_{0.25} : (Te)_{0.75}$	$G^{CeTe_3}_{Ce,Rh} = 0.25 * {}^{0}G^{Fcc}_{Ce} + 0.75 * {}^{0}G^{Fcc}_{Te} - 108{,}067 - 11.01 * T$

with the Bi–Te–Ce system diagram concerning the lines of conjunction between the binary compounds on the Ce–Bi rich side and the binary domains on the Bi–Te rich side. These differences are due to the presence of the extended binary phases of the Ce–Te and Bi–Te dykes and also to the reactions between the primary and binary phases.

The isothermal section given by Tan et al. [43] at 673 shows the existence of three binary phases on the Ce–Te side instead of six compounds (Fig. 1), two phases in the Bi–Te-rich side instead of three compounds (Fig. 2), and no compounds in the Bi–Ce binary system part. However, the results found in the present work for the Bi–Te-rich side are very reliable when compared with the results of Tan et al. [43] and Mamedova et al. [44].

In Fig. 7, an isothermal section of the ternary diagram of the Bi–Te–Ce system at 1500 K is extrapolated, allowing the existence levels of the binary phases to be verified. This section shows that most of the binary phases disappear at this temperature and that the most stable phases (those with the highest melting point) appear (See Fig. 8).

Table 4 Comparison between the optimized thermodynamic parameters of the intermetallic compounds derived in this work and those drawn from the literature

Phases	Enthalpy of formation (KJ/mol. at)	Technique used	Reference
CeTe	– 168.025	Optimisation	This work
	– 168	Calorimetry	[23]
	– 151.666	Ab-initio	[49]
	– 140.474	Ab-initio	[50]
	– 134.749	Ab-initio	[51]
Ce_3Te_4	– 151.120	Optimisation	This work
	– 163	Calorimetry	[23]
	– 158.613	Ab-initio	[49]
	–	Ab-initio	[50]
	– 143.669	Ab-initio	[51]
Ce_4Te_7	– 135.001	Optimisation	This work
	–	Ab-initio	[49]
	–	Ab-initio	[50]
	–	Ab-initio	[51]
$CeTe_2$	– 127.544	Optimisation	This work
	– 144.527	Ab-initio	[49]
	–	Ab-initio	[50]
	– 121.857	Ab-initio	[51]
Ce_2Te_5	– 115.217	Optimisation	This work
	– 168.743	Ab-initio	[49]
	–	Ab-initio	[50]
	–	Ab-initio	[51]
$CeTe_3$	– 108.067	Optimisation	This work
	– 144.527	Ab-initio	[49]
	– 97.251	Ab-initio	[50]
	– 91.409	Ab-initio	[51]

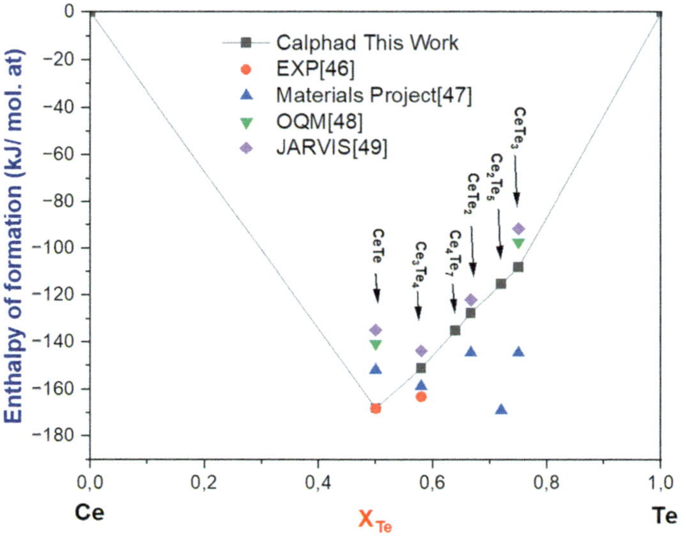

Fig. 5 The enthalpies of formation calculated in this work using the Calphad approach compared with the results measured in the present work by calorimetry [23] and with the values predicted by [49–51]

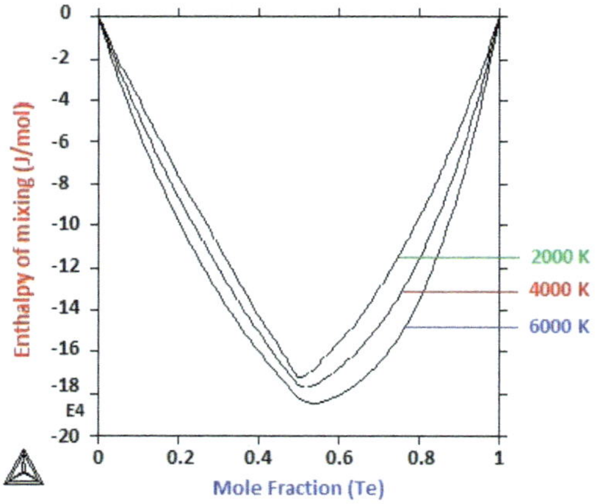

Fig. 6 The enthalpy of mixing of the liquid phase calculated at different temperatures (2000–4000–6000 K) using the association solution model [41, 42]

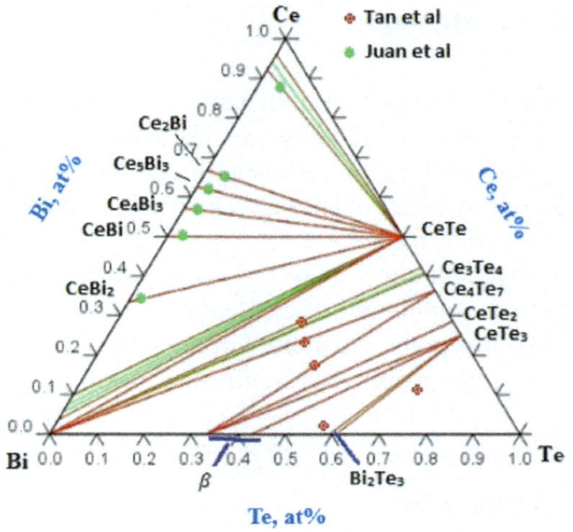

Fig. 7 Calculated Bi–Te–Ce isothermal section at 673 K compared with data from the literature [43] and [44]

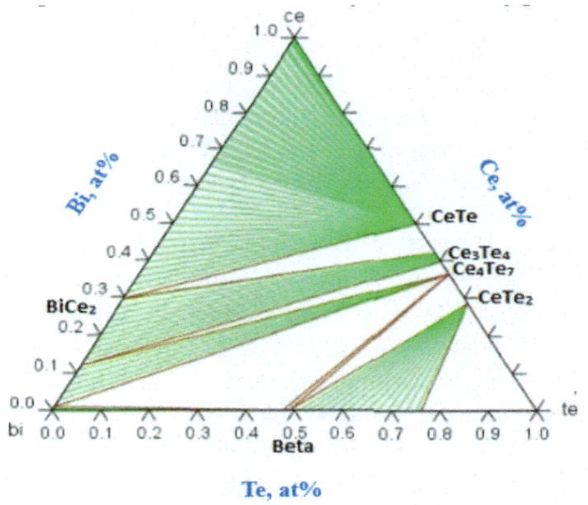

Fig. 8 Calculated Bi–Te–Ce isothermal section at 1500 K

6 Conclusion

A self-cohering thermodynamic database was issued for the Ce–Te binary system using thermo-calc software. The thermodynamic parameters of the Ce–Te binary system were optimized using literature experimental results on phase diagram and thermodynamic quantities. The overall results obtained are in good agreement with the literature data.

By modeling the Ce–Te system, we were able to extrapolate the Bi–Ce–Te ternary system completely. The calculated cross-sections agree well with the experimental diagrams in the literature. For the missing part of the diagram, further experimental studies are required to confirm our result.

References

1. Sharma YC, Purohit A (2016) Tellurium based thermoelectric materials: new directions and prospects. J Integr Sci Technol 4:29–32
2. Lyeo H-K, Cahill DG, Lee B-S, Abelson JR, Kwon M-H, Kim K-B, Bishop SG, Cheong B-K (2006) Thermal conductivity of phase-change material $Ge_2Sb_2Te_5$. Appl Phys Lett 89:151904. https://doi.org/10.1063/1.2359354
3. Wuttig M (2005) Towards a universal memory? Nat Mater 4:265–266. https://doi.org/10.1038/nmat1359
4. Wong HSP, Salahuddin S (2015) Memory leads the way to better computing. Nat Nanotechnol 10:191–194. https://doi.org/10.1038/nnano.2015.29
5. Rao F, Ding K, Zhou Y, Zheng Y, Xia M, Lv S, Song Z, Feng S, Ronneberger I, Mazzarello R, Zhang W, Ma E (2017) Reducing the stochasticity of crystal nucleation to enable subnanosecond memory writing. Science 358:1423. https://doi.org/10.1126/science.aao3212
6. Levin E, Cook B, Harringa J, Bud'ko S, Venkatasubramanian R, Schmidt-Rohr K (2011) Analysis of Ce-and Yb-doped TAGS-85 materials with enhanced thermoelectric figure of merit. Adv Funct Mater 21:441–447.https://doi.org/10.1002/adfm.201001307
7. Zhu M, Wu L, Rao F, Song Z, Ren K, Ji X, Song S, Yao D, Feng S (2014) Uniform tidoped Sb_2Te_3 materials for high-speed phase change memory applications. Appl Phys Lett 104:053119. https://doi.org/10.1063/1.4863430
8. Bendarma A, Gourgue H, Jankowiak T, Rusinek A, Kardellass S, Klosak M (2020) Perforation tests of composite structure specimens at wide range of temperatures and strain rates-experimental analysis. Mater Today Proc 24:7–10
9. Klosak M, Jankowiak T, Rusinek A, Bendarma A, Sielicki PW, Lodygowski T (2020) Mechanical properties of brass under impact and perforation tests for a wide range of temperatures: experimental and numerical approach. Materials 13(24):5821
10. Liu Z-K (2020) Computational thermodynamics and its applications. Acta Mater 200:745–792. https://doi.org/10.1016/j.actamat.2020.08.008
11. Wood C (1988) Materials for thermoelectric energy conversion. Rep Prog Phys 51:459–539
12. Wang X, Yang R, Zhang Y, Zhang P, Xue Y (2011) Rare earth chalcogenide Ce_3Te_4 as high efficiency high temperature thermoelectric material. Appl Phys Lett 98:222110. https://doi.org/10.1063/1.3597409
13. Sundman B, Jansson B, Andersson J-O (1985) The thermo-calc data bank system. J Calphad 9:153–190
14. Jansson B (1984) Division of physical metallurgy. PhD thesis, Royal Institute of Technology, Stockholm, Sweden

15. Chukalin VI, Yaremhash EL, Villenski AL (1967) Phase diagram of the Ce–Te system. Inorg Mater 3:1341–1347
16. Gschneidner Jr KA, Verkade ME (1974) Selected cerium phase diagrams. Iowa State University of Science and Technology, Rare-Earth Information Center, Ames, USA
17. Pardo MP, Flahaut J (1967) Les tellurures superieurs des éléments des terres rares, de formules L_2Te_5 et LTe_3. Bull Soc Chim Fr 3658–3664
18. Okamoto H (1990) In: Massalski TB (ed) Binary alloy phase diagrams, 2nd edn, vol 2. ASM International, Materials Park, pp 1118–1120 (editor-in-chief)
19. Philipp F, Pinkert K, Schmidt P (2009) Phosphidtelluriden auf der spur: zum system Ce/Te/P. Z Anorg Allg Chem 635:1420–1429. https://doi.org/10.1002/zaac.200801396
20. Eriksson G, Hack K, Philipps M, Fullerton-Batten C (1998) ChemSage version 4.1. GTT Technologies mbH, Herzogenrath, Germany
21. Chiba R, Funakoshi N (1988) Uniform composition Te–Se film preparation from alloy sources. Thin Solid Films 157:307–313. https://doi.org/10.1016/0040-6090(88)90011-9
22. Mills KC (1974) Thermodynamic data for inorganic sulphides, selenides and tellurides. Butterworths, London, pp 327–332
23. Ferro R, Borzone G, Parodi N (2001) Comments on the formation thermodynamics of selected groups of rare earth compounds. J Alloys Compd 321:248–260
24. Dong C, Wang J, Hu B, Zhu L, Wu Q, Du Y (2021) Thermodynamic modeling of the Te–X (X = Zr, Ce, Eu) systems. Calphad. https://doi.org/10.1016/j.calphad.2021.102281
25. Perdew JP, Chevary JA, Vosko SH, Jackson KA, Pederson MR, Singh DJ, Fiolhais C (1992) Phys Rev B 46:6671
26. Speight JD, Harris IR, Raynor GV (1968) Alloys of cerium with neodymium, samarium and terbium, and of praseodymium with terbium. J Less Common Met 15:317–330. https://doi.org/10.1016/0022-5088(68)90191-4
27. Spedding FH, Hanak JJ, Daane AH (1961) High temperature allotropy and thermal expansion of the rare-earth metals. J Less Common Met 3:110–124. https://doi.org/10.1016/0022-5088(61)90003-0
28. Smolensky G, Adamjan V, Loginov G (1968) Antiferromagnetic properties of light rare earth monochalcogenides. J Appl Phys 39:786–790. https://doi.org/10.1063/1.2163619
29. Lefevre C (1969) Etude mineralogique et chimique des laves et pegmatitoides du volcan des rives (Herault). Bull Soc Geol Fr S7-XI:419–425. https://doi.org/10.2113/gssgfbull.S7-XI.3.419
30. Kwon YS, Park TS, Lee KR, Kim JM, Haga Y, Suzuki T (1995) Transport and optical properties of $CeTe_2$. J Magn Magn Mater 1173–1174. https://doi.org/10.1016/0304-8853(94)01297-0
31. Ohmasa Y, Yamamoto I, Yao M, Endo H (1995) Structure and electronic properties of Te–Se mixtures under high pressure. J Phys Soc Jpn 64:4766–4789. https://doi.org/10.1143/jpsj.64.4766
32. Gschneidner KA, Calderwood FW (1989) Bull Alloy Phase Diagram 10:427–430
33. Vogel R, Anorg Z (1913) Allg Chem 84:323–327
34. Pleasance RJ (1959) J Inst Met 88:45–50
35. Yoshihara K, Taylor JB, Calvert LD, Despault JG (1975) J Less-Common Met 41:239–245
36. Li J, Cai Y, Chen Q, Zhang H, Liu HS, Jin ZP (2012) Thermodynamic modeling of Fe–Ce–Bi system and explanation of interfacial reaction in Fe-45 at%–Ce/Bi couple. Calphad 38:133–139
37. Mao C, Tan M, Zhang L, Wu D, Bai W, Liu L (2018) Experimental reinvestigation and thermodynamic description of Bi–Te binary system. Calphad 60:81–89
38. Hansen M, Onderko K, Salzberg HW (1958) Constitution of binary alloys. J Electrochem Soc 105:260–261
39. Brown A, Lewis B (1962) The system bismuth-tellurium and antimony-tellurium and the synthesis of the minerals hedleyite and wehrlite. J Phys Chem Solids 23:1597–1604
40. Glazov VM, Pavlova LM, Yatmanov YV (1984) A thermodynamic assessment and experimental verification of the possibility of diffusionless crystallization of alloys in the Bi–Te system. Russ J Phys Chem 58:176–180

41. Okamoto H, Tanner LE (1990) Bi–Te (bismuth-tellurium) binary alloy phase diagrams. Materials Park, OH, USA (T. B. Massalski, H. Okamoto)
42. Gierlotka W (2018) A new thermodynamic description of the binary Bi–Te system using the associate solution and the Wagner-Schottky models. Calphad
43. Tan M, Mao C, Zhang L, Bai W, Liu L (2018) Experimental investigation of phase relations in Bi–Te–RE (Yb, La, Ce) ternary systems. Calphad
44. Mamedova SG, Sadygov FM, Il'yasly TM, Ismailov ZI (2009) T–x–y diagram of the Ce–Bi–Te system. Russ J Inorg Chem
45. Dinsdale AT (1991) SGTE data for pure elements. Calphad 15:317–425
46. Sommer F, Metallkd Z (1982) Association model for the description of thermodynamic functions of liquid alloys. J Mater Res 73:72–86
47. Krull HG, Singh RN, Sommer F (2000) Generalized association model. Z Metallkd/Mater Res Adv Tech 91(5):356–365
48. Sundman B et al (1981) J Phys Chem Solid 42:297–301
49. The materials project. https://materialsproject.org. Accessed 13 July 2022
50. The open quantum materials database. http://oqmd.org. Accessed 13 July 2022
51. Choudhary K, Garrity KF, Reid AC et al (2020) The joint automated repository for various integrated simulations (JARVIS) for data-driven materials design. npj Comput Mater 6:1–13

Application Possibilities of PCB Rework at Large Manufacturers

Béla Bődi and Viktor Gonda

Abstract In automotive electronics industry, the rework of solder joints of Printed Circuit Assemblies (PCA) is generally not permitted, except in a few individual cases. The present work is focused on trying to list the reasons why the users forbid the PCA manufacturers to rework the non-complying products' solder joints. Case studies are compared in order to classify the physical phenomena, as well as classifying by the cost of rework process for the volume of the products, and reasons of the customer-manufacturer relations are assessed. Conclusions are drawn based on the review of the relevant literature as well as on discussions with quality assurance and technical professionals working in the automotive electronics industry.

Keywords Rework · Solder joints · Fatigue · Lead-free solder

1 Introduction

A commonly applied practice in the automotive electronics industry to avoid rework of electronic products that failed any tests during production, due to the fact that customers do not accept reworked products, while zero-failure production is non-achievable. In the majority of cases, the reason for failing the tests is only a missing or displaced electronic component, a solder bridge or an open solder joint at the terminations or leads of the components. If these errors had occurred in commercial electronic products (e.g., entertainment- and consumer-electronics) and not in automotive electronic products, rework would be applied without further problems

---

B. Bődi
Doctoral School on Materials Sciences and Technologies, Obuda University, Bécsi út 96/b, Budapest 1034, Hungary
e-mail: bela.bodi@elas.hu

V. Gonda (✉)
Banki Donat Faculty Mechanical and of Safety Engineering, Obuda University, Népszínház u. 8, Budapest 1081, Hungary
e-mail: gonda.viktor@bgk.uni-obuda.hu

T. A. Kovács et al. (eds.), *The Impact of the Energy Dependency on Critical Infrastructure Protection*, Advanced Sciences and Technologies for Security Applications, https://doi.org/10.1007/978-3-031-78544-3_15

during production. Where the cost of rework is less than the value of the electronic assembly, the rework is widely used in commercial electronic production. However, automotive customers prohibit reworking and these products are usually scrapped. In an auto electronics factory, the prime cost of scrapped reworkable products can range from a few thousand euros to as much as a hundred thousand euros per month. Also, car electronics manufacturers are under increasing pressure to reduce the amount of e-waste that comes from production.

Rework is defined in IPC-T-50 [1] as the act of reprocessing a non-complying product, through the use of original or alternate equivalent processing, in a manner that assures compliance of the product with applicable drawings or specifications. The rework is fundamentally different from repair that is such kind of action on a nonconforming product to make it conform to form, fit and functional attributes, but will not meet all drawing requirements.

Based on the rework definition, either if a faulty component is desoldered, removed, and a new one is installed; or replacing a missing one on the electronics circuit assembly, these activities fulfill definition of rework if:

- the used processes (soldering, component placement) are equivalent with the originals,
- the result of the processes assures compliance of the product with the original drawings and specifications.

If a damaged copper trace is restored with a jumper wire or a torn (damaged) pad is replaced with a copper foil (e.g., repair copper foil covered by thermoplastic adhesive film), they restore the functionality of the electronics, but these solutions already deviated from the original drawings and documentation, so these methods fall into the category of repair.

As for the maximum rework cycle, the IPC-7711/21 (Rework, modification and repair of electronic assemblies) standard—widely used in the electronic industry—does not limit the maximum number of rework actions to a PCA.

The user's (entity that receives a service or product from the supplier) requirements are reflected in the procurement documentations (e.g., contract) and master drawings. According to standard industry practice, the rework feasibility is a key point of the negotiation between User and Supplier. In other words, the usage of rework is an important question of the discussion between User and Supplier because it has a financial impact of unit price as well. On the one hand, the manufacturer wants to negotiate a higher unit price if the electronic assembly cannot be reworked. On the other hand, the buyer is interested in the lowest possible unit price.

In this work, based on the reference studies, the lifetime results of the solder joints of several component packages (chip, TSOP, BGA) are overviewed. Subsequently, the technical reasons for banning reworking are discussed. Reasoning whether it is possible/reasonable to rework automotive products without risking the reliability of the product is summarized.

2 Characteristics of Manual Rework and Problems with the Process

In this section, the main steps of the manual rework process are summarized and two fundamental process errors are emphasized that can lead to failure due to the worker's careless work. The rework is summarized below, as it is a mostly a manual, non-automated process.

2.1 Main Process Steps of Manual Rework

In the following, the main steps and possible procedures of manual rework of a SMD (Surface Mount Device) are described, covering the possible causes of errors.

A general rework process has three main steps:

1. component desoldering,
2. solder lands (pads) preparation,
3. component installation.

For component desoldering, either a conductive heat transfer method can be used (e.g. removal tips with tweezer handpiece); or convective heat transfer method (e.g. hot air pencil); or focused infrared (IR) heating method.

After component removal, solder braid and the continuous vacuum methods are used for removing old solder from the pads. If necessary, blade tip method is employed for pad releveling. After the old, residual solder has been completely removed from the lands, the next step is to clean the old flux residues from the lands and their surroundings.

The component installation can be done by three main methods: point-to-point or multi lead (continuous) or hot air with solder paste method.

In summary, there are several methods for desoldering and installation of components and preparation of solder lands.

2.2 Defects Introduced by Manual Rework

As rework is most frequently a manual operation, the quality of a reworked solder joint is highly influenced by operator competence. Even the most experienced operator may produce acceptable solder joints and have variation in his work.

Microbridge Formation

The former statement is supported by a study that showed microbridge formation during solder bridge removal process (with Sn63Pb37 solder) by soldering iron and desoldering braid between gull-wing components' leads [2].

After solder bridge removal several inexplicable shortage failures were detected during electrical tests. This case study proved that during the rework process (In this case solder bridge removing.), when the solder bridge melting that copper which come from lands, component leads, and desoldering braid form Cu_6Sn_5 crystals. If the solder braid is away from the bridge (e.g., solder braid lay just at the end of the leads not on the bridge) therefore the liquid state of the solder between the leads enough for precipitation of Cu_6Sn_5 crystals that forming microbridge between the leads. The melting point of Cu_6Sn_5 intermetallic compound is 415 °C. Therefore, if the microbridge arised between leads it couldn't be remelted during the normal desoldering process because of the lower soldering temperature. The microbridge formation is shown in Fig. 1.

As for this case study, microbridges were more likely to form when the desoldering braid was applied outside edge of the lands, and when the desoldering braid was already partially soaked with solder. Microbridge did not form when the solder braid was applied directly to the solder bridge. This microbridge formation is an example of the consequence of the reworker's manual uncertainty (solder braid lay just at the edge on the lands not on the solder bridge) of the prescribed (approved) process.

Residues

In practice, another well-known source of danger to the reliable operation of the printed circuit assembly is the residue that remains after rework. It can be the cause of metal dendrite growing between non-common conductive metal surfaces in there is an electrolyte (moisture) on the PCB. These dendrites can cause catastrophic failures (electrical shortage) in the electrical circuits. The classical model of silver migration in electrical insulation was published by Kohman et al. in 1955 [3]. The electrolyte was high-purity deionized water, the model does not assume the presence of contaminating ions. This model has the following main steps, when that supply voltage is connected to the non-common conductive surfaces which start the hydrolysis of the water:

- Water dissociates under voltage (OH^- formation) at the cathode.

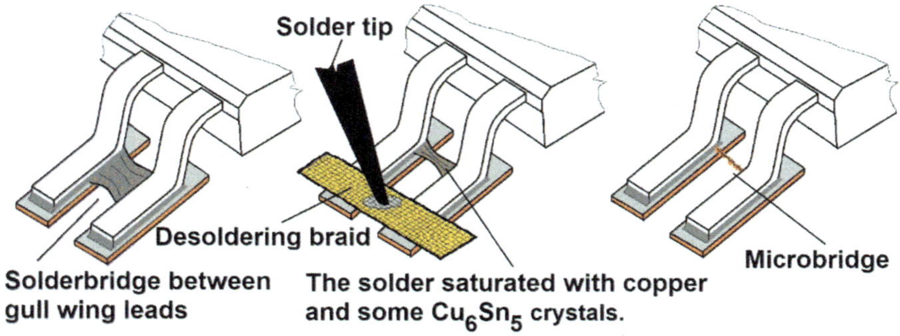

Fig. 1 Main steps of microbridge formation

Fig. 2 Example of copper dendrite formation between copper traces

- Local increase of pH value (alkalization) on the cathode.
- Alkaline and acidic ranges form in the water film. (Corrosion of metal surfaces get started.)
- Dissolution metal (cations) at the anode.
- Ion migration to the cathode due to potential difference.
- Precipitation of metallic dendrites at the cathode [4].

It is a well-known fact that flux residues have an acceleration effect of dendrite growing. In the study of Tolla et al. [5] saying "Dendrites grow approximately 10 times faster in a 0.25 M hydrochloric environment than in DI water, shorting the traces in 1 s." The test was executed on IPC B-25 test coupon where the conductor spacing between the copper traces was 0.318 mm, the supply voltage was 10 V. An example of copper dendrite formation is shown in Fig. 2.

3 Challenging the "No Rework" Rule from the Customer's Point of View

Although it is not only a technical question, but also worth devoting a few thoughts to the customer's perspective. Apart from the technical causes, I a few explanations circulates from automotive industry quality assurance specialists why their customers forbid the rework. Some of the subjective reasoning:

- "Our customer don't have any explanation, simply bans the rework."
- "It is a weapon in the hands of our customer, so that we will continuously improve our manufacturing processes, because otherwise we have to scrap the defective electronic assemblies."
- "Rework by hand soldering rework is a subjective soldering technology, the result depends on the operator's expertise. Therefore, this will not fit to our automated manufacturing technology."
- "The new is new … patched isn't new!"

Despite the former mentioned opinions, some attempts arise for reducing electronics waste where large manufacturers make contracts with professional electronics troubleshooting and repair companies. E.g., ABI Electronics Ltd. (supplier of specialist electronic test and measurement equipment in UK) works for Lockheed Martin, NASA, Continental, Ford etc. Stellantis has been operated a "Repair and Return" program for its electronics products in the car since 2011. Stellantis offers "Repair and Return" service for automatic gearboxes and complex electronic components [6]. EECONE (European ECOsystem for GreeN Electronic) project is to reduce e-waste on a European scale [7]. One of the goals of this project to provide innovative techniques for repairing, reusing, refurbishing and recycling to decrease e-waste in a new generation of electronics.

4 Reliability Aspects of Reworked Products

In this section, the reliability changes of different electronic components based on the citied studies are summarized. In these studies, the most critical test were the accelerated thermal cycling (ATC) tests. ATC tests are conducted to identify failure mechanisms caused by a mismatch in the coefficient of thermal expansion (CTE) of components and PCB and solder materials. Electronic products used in automotive application are exposed to the most extreme operating stress. The maximum and minimum temperature and the number of cycles are determined by the application of the product. Typical ramp rate during the test is 10–15 °C/min or less. In practice, the number of the thermal cycles is basically determined by customer. This varies from user to user. Table 1 shows the temperature classifications used in the automotive industry.

One of these thermal cycle tests were used in the studies described later. In the following, three different packages (chip, TSPO and BGA) lead-free experimental studies' results are summarized that shows the effect of rework on the reliability of SAC305 solder joints.

4.1 Passive Components

Chai et al. [9] found out for 2512 chip resistors (6.3 × 3.2 mm), the fatigue life of one-time reworked and three-time reworked parts experienced about a 49–79% reduction compare to adjacent components. The temperature cycle test parameters were − 55 to 125 °C with 15 min dwell time at each temperature and approx. 72 min cycle time. The total temperature cycle of the entire test was 7081. During the entire test, the resistance of each chip resistor was measured. Failure was defined as the electrical resistance increases 25 Ω above the nominal circuit resistance. SAC305 solder was used for chip termination soldering. The test board material was convectional FR4 (370HR) with OSP (Organic Solder Preservative) finish. For rework, they use hot air station for parts

Table 1 Temperature classification of automotive industry [8]

Class	Application description	Temperature ranges (°C)
I.	Passenger compartment, door interior, headliner, trunk interior	– 40 to 85
II.	Engine compartment (non-engine, -transmission mounted) Chassis mounted (not exposed to heat source) Areas exposed to direct sunlight (dashboard, rear deck, etc.)	– 40 to 105
III.	Engine compartment (engine mounted, away from exhaust manifold) Firewall mounted (not exposed to catalytic converter) Chassis-mounted (exposed to heat source)	– 40 to 125
IV.	Engine compartment (engine mounted, near exhaust manifold) Firewall mounted (exposed to catalytic converter) Transmission-/differential-mounted Internal to bell housing	– 40 to 150
V.	Brake system (exposure to oil/hydraulic lines/temperature) Throttle body, accelerator, exhaust manifold etc.	– 40 to 175 – 40 to 205

removing, copper desoldering braid for lands redressing. They use standard solder iron with 371 °C tip temperature for component installation. Alpha Telecore plus SAC305 solder wire was used for hand soldering. For chip components, two different crack paths were recognized. One of them shows multiply cracks perpendicular to the end terminal of the chip component. Then another crack path was parallel to the end termination. Here, the process variation in solder thickness under the component termination appears to be the primary source of reduced reliability. Liu et al. [10] noted that chip resistor's solder joint reliability was very sensitive to the solder thickness under the component. The three-cycle reworked SAC305 chip resistor solder thickness (approx. 19 μm) was only about half (approx. 36 μm) than the non-reworked chip component's solder thickness. Moreover, the average solder joint solder thickness difference between the two termination of 2512 resistors was approx. 12 μm. According to this study the solder crack initiated from the upper end of the end termination and continued between the terminal and the land.

As the study of Zhu et al. [11] the fatigue life of the solder joint increase with the solder thickness. The lower thickness shows bad fatigue resistance. This statement is in line with the fact mentioned above, the fatigue life of chip component solder joint influenced by the tilted or lower position of the part.

4.2 *Gull-Wing Components*

In the study of Chai et al. [9] TSOP package's SAC solder joints fatigue life was examined as well. The TSOP package had 50 leads with 0.8 mm pitch distance. The test and rework procedure were the same as the chip components described in 3.1.

Table 2 Accumulated cycles to failure for TSOP parts under − 55 to 125 °C temperature cycling [9]

Operation	Cycles to failure	
	Cycles to 10% failure	Cycles to 50% failure
SAC 3× adjacent	1789	2798
SAC 3× rework	2316	3020
SAC 1× adjacent	2021	2731
SAC 1× rework	2118	2761
SAC (no-rework)	2180	2761

According to this study an early failure was detected after 800 temperature cycle! As shown in Table 2. there isn't any difference between the failures of the 1 or 3 times reworked (once and three times reworked), adjacent and nonreworked components.

The solder joint crack initiated from the inner fillet (at the heel of the component lead) of the joint than propagated along under the lead and finished at the toe of the lead. It is well-known that the phenomenon of copper dissolution, occurs during rework process. On these test boards the copper thickness of the lands is was measured to be 64.17 ± 0.85 μm. The dissolution takes place faster in SAC solder than in SnPb, because of the higher soldering temperature and higher Sn content of SAC305 solder alloy. According to the microsectional analyses, the copper land thickness reduction from rework was approximately the same for both packages. Both passive and TSOP components' pad thickness was reduced more than 10 μm and approx. 18 μm after 1× and 3× rework cycles. Therefore, the reduction of copper pad thickness with rework cycles might increase the reliability risk and reduce the viability of rework beyond three cycles.

4.3 BGA Packages

In the study of Hongtao Ma the reliability impact of the rework process was examined [12]. The test PCB material was FR4 with OSP surface finish. The soldering material was SAC305 paste. Two FCBGA (flip-chip ball grid array) packages with SAC305 solder balls were reworked three and five times in different test boards. New FCBGA components was used for each rework. The BGA rework station had a convection spot top and a global bottom heater. A rework cycle consists of two thermal cycle: one for desoldering and one for installing the electrical component. The temperature cycle test parameters were 0–100 °C with 10 min dwell time at each temperature and the ramp rate was 10 °C/min. During the test, the electrical continuity of the joints were monitored. The experiment showed that FCBGA with microvias (U1) without rework showed no failure after 6000 thermal cycles. The other FCBGA with daisy-chain PTH (Plated Through Hole) design (U2) had barrel crack failures at much lower thermal cycles (4730) than other FCBGA with microvias. Multiple

Table 3 Characteristic life after multiply rework cycle [12]

Characteristic life	U1	U2	U4
No rework	> 6000	4730	1028
3× rework on U1	5184	1240	932
5× rework on U1	3362	–	698

reworks caused severe barrel crack failures in PTH vias. The characteristic life of two FCBGA with tree and five times rework and no-rework are summarized in Table 3. The characteristic life is defined as the thermal cycle number when 63.2% of the solder joints can be expected to have failed. As for U1 component's characteristic life, the fatigue life of three times reworked components is approx. 1.5 times of five times reworked components' life.

This study examined that situation when a CABGA (Chip Array Ball Grid Array—U4) lay on the other side of the board (so-called: clamshell design) where the rework is done. This thermal cycle degraded the component performance significantly by 10% after three times rework, and 32% after five times rework. This study also showed that the grow of intermetallic layer does not impact the temperature cycle result, the dominant failure mechanism is solder joint fatigue in bulk solder.

According to SEM (Scanning Electron Microscope) analysis this fatigue life drop of U4 component's solder joints was due to the grain coarsening of SAC solder. Because of thermal cycles the increasing grain size of the solder will cause the amount of grain boundaries to decrease. The fewer grain boundaries cause less resistance against the dislocation and therefore the soldering material of this CABGA component loses its strength. The Hall–Petch empirical equation establishes relationship between yield strength and grain size of a metal or alloy [13].

$$\sigma_y = \sigma_i + \frac{k}{\sqrt{d}} \tag{1}$$

where σ_y is the yield strength; σ_i is materials constant for the starting stress for dislocation movement; k is the strengthening coefficient. (It is the grain boundary locking term which measures the relative hardening contribution by the grain boundaries.) and d is the grain diameter. The Hall–Petch equation tells us the increasing grain size decrease the strength of the solder joint.

In addition, the previously described grain coarsening effect there is another microstructural evolution occurs as a combined effect. In the study of Hassan et al. [14] the intermetallic particles (Ag_3Sn and Cu_6Sn_5 intermetallic compounds) which surrounded the large β-Sn dendrites in the SAC305 solder grain composition were examined during aging at high (150 °C) temperature. Quantitative analysis has shown the number of intermetallic particles decreased during aging, while the average diameter of the particles increased significantly. This effect reduces their ability to block dislocation movements. It causes huge degradation in the mechanical properties as well.

5 Conclusions

In this paper, an overview is given of the reasons why the users forbid the PCA manufacturers to rework the non-complying products' solder joints and what aspects may arise when reworking is permitted. The well-known fact was highlighted, that the result of manual rework largely depends on the experience of the worker. Using examples taken from case studies, it was demonstrated that even the slightest deviation from the operating instructions of manual rework can lead to serious malfunctions (electrical shortage between components leads or lands because of microbridge or dendrite). It is likely that automation of the manual rework process is necessary for the acceptance of rework process by the user (customer). As an example, the manual land preparation (old solder removal from the lands) could be replaced by contactless solder removal solution (so-called auto scavenger). In this case, a solder removal head simply melts and removes the residual solder after the desoldering process with vacuum. The process gas can be air or nitrogen. The soldering thermal profile and the movement of the head controlled by PC to avoid the solder lands and board damages [15].

Three different reworked component packages (chip, TSOP and FCBGA) solder joints fatigue life was examined. As expected, their results were extremely varied. As for 2512 chip component, the fatigue life of the solder joint increase with the solder thickness. Between two terminals' solder joint that one will be failed where the solder thickness is smaller. One rework cycle on the chip component decreases the solder joint fatigue life almost to the half and three rework cycle almost to the one-fifth.

The TSOP rework had interesting result, because it showed no difference between the failures of the once and three times reworked, the adjacent and the non-reworked components. This result might related to the big soldered surfaces thanks to the 50 leads.

The FCBGA's operation life depends on a lot of the electrical circuit interconnection. While the microvias interconnection were good, the daisy-chain PTHs are particularly unfavorable solution. Multiple reworks caused severe barrel crack failures in PTH vias.

The FCBGA that has microvia interconnection showed degradation of its solder joint fatigue life as well but smaller extent (degradation approx. 15% after three times rework and approx. 50% after five times rework).

That situation—when a CABGA lays on the other side of the board where the rework is done—was also examined. (so-called clamshell design.) This component thermal cycle performance degraded significantly by 10% after three times rework, and 32% after five times rework due to the microstructure coarsening. On the one hand, when the solder joints course because of thermal cycles(s) of rework, the fewer β-Sn grain boundaries block less of the dislocation movements. On the other hand, the number of intermetallic particles—that surrounding the β-Sn dendrites—decrease during thermal stress while the diameters of them increase significantly.

This effect also reduces their ability to block the dislocations movements. Therefore, the strength of the solder joints are reduced.

Based on these tests, one rework cycle might be allowed by the user without significant solder joint fatigue life decreasing. A question may arise, that what is the effect of only one rework cycle on the solder joints fatigue life.

As a final note, if the customer allows rework, then this process must also be also validated similarly to the processes of new products. The cost of a new product validation is a highly expensive process. To reduce the rework validation costs, it would be advisable to examine the reworked product validation together with the new product.

References

1. IPC-T-50—Revision N—terms and definitions for interconnecting and packaging electronic circuits, Nov 2021
2. Kukelhan J (2010) Copper tin intermetallic crystals and their role in the formation of microbridges between the leads of hand reworked fine pitch components. IPC—Association Connecting Electronics Industries, Las Vegas, Nevada, USA (2010)
3. Kohman GT, Hermance HW, Downes GH (1955) Silver migration in electrical insulation. Bell Syst Tech J 34(6):1115–1147
4. Schweigart H, Meier M (2022) Influence of pH-value on the mechanism of electrochemical migration, Zestron White paper, Feb 2022
5. Tolla B, Shi Y, Wei X, Bhasvar H (2015) Chemical influences on the reliability of complex assemblies. In: Publication of SMTA international conference
6. Stellantis (2021) Corporate social responsibility report, Apr 2021. https://www.stellantis.com/content/dam/stellantis-corporate/sustainability/csr-disclosure/stellantis/2021/Stellantis_2021_CSR_Report.pdf. Last accessed 15 Oct 2021
7. https://www.eecone.com/eecone/home/. Last accessed 27 Oct 2023
8. IPC-AJ-820A assembly and joining handbook, Feb 2012, p 268
9. Chai F, Osterman M, Pecht M (2012) Reliability of gull-wing and leadless packages subjected to temperature cycling after rework. IEEE Trans Dev Mater Reliab 12(2)
10. Liu X, Lu G (2003) Effects of solder joint shape and height on thermal fatigue lifetime. IEEE Trans Compon Packag Technol 26(2):455–465
11. Zhu Y, Li X, Gao R, Wang C (2014) Investigate influence of solder thickness on the fatigue failure behavior of solder joint under high temperature. In: 2014 15th international conference on electronic packaging technology, May 2014
12. Ma H, Xie W, Subbarayan G (2012) Effects of multiple reworks on the accelerated thermal cycling and shock performance of lead-free BGA assemblies. IEEE Trans Compon Packag Manufact Technol 2(11)
13. Lumley R, Morton A, Polmear I (2006) Nanostructure control of materials, pp 219–250 (Chapter 9.2.1)
14. Hassan KMR, Alam MS, Wu J, Suhling JC, Lall P (2020) Microstructural evolution of SAC305 BGA joints during extreme high temperature aging. In: 2020 19th IEEE intersociety conference on thermal and thermomechanical phenomena in electronic systems (ITherm), Orlando, FL, USA, pp 1201–1210
15. Finetech: removal of residual solder (technical paper). https://finetech.de/technicalpaper/residual-solder-removal/. Last accessed 27 Oct 2023

Minimizing Energy Losses in the Modern E-bikes by New Integrated Strategies for Adaptive Control

Emilian Ceuca

Abstract The electric propulsion systems provide cost-effective solutions and are of great interest for companies to gain competitive edge in the market. E-bikes and electric scooters are becoming more and more attractive as they possess the advantage of high energy efficiency compared to other energy forms. This paper proposes a simulation model for design the optimal adaptive control strategy for the e-mobility. Our proposed e-bike uses a hybrid storage system solution. Based on this real equation we can propose and evaluate the real conditions and constrains. We can propose based on functional model of Li-Ion battery and supercapacitor some examples for practical setups.

Keywords E-Mobility · Adaptive control · Bicycle and vehicle specific power

1 Introduction

The optimal design of the controller is required for the power management to reduce the energy losses during acceleration and braking processes. The control unit guarantees a robust operation at different rotor speeds and at different load conditions. The proposed architecture is defined based on the connection between the components that define the energy flow routes and control strategies. In the proposed architecture the electric energy is delivered to the traction drive from a battery via a DC/DC converter.

Modern e-bikes uses 3-phase BLDC motors. Considering, the phases are shifted 120° from each other, so a convenient method for having a rotating rotor flux in the stator is the six-step commutation scheme, commutating each of the three-phase voltages 60 electrical degrees. At maximum torque and full load, the phase current should have the same waveform as the driving voltage, neglecting the inductive

E. Ceuca (✉)
Department of Informatics, Mathematics and Electronics, "1 DECEMBRIE 1918 "University of Alba Iulia, Alba Iulia, Romania
e-mail: emilian.ceuca@uab.ro

T. A. Kovács et al. (eds.), *The Impact of the Energy Dependency on Critical Infrastructure Protection*, Advanced Sciences and Technologies for Security Applications, https://doi.org/10.1007/978-3-031-78544-3_16

reactance, and the two signals need to be in-phase. In practice the load can vary in a wide range so the evaluation of road and load based on practical measurement will improve the proposed model.

The ability to recover the kinetic energy of the vehicle and store it in the battery pack is an important feature of the electric system of the e-bike. However, finding a technical solution that reduces the energy consumption is required, especially when the e-bike is driven in the urban traffic areas as the largest part of energy is drawn because the driving cycle involves frequent accelerations and decelerations. In this case, because the power demand varies according to the driving cycle, the traction system is exposed to a variable speed operation. The aim of the control unit is to minimize the power peaks upon applying the torque command. Minimizing the power peaks conducts to an improved performance against "over and under" voltages because the DC link voltage changes upon a sudden load variation. This study investigates how to reduce the current peaks during accelerating and braking operations as well as the behavior of the supercapacitor to reduce the energy loses. The current peaks can be utilized by recharging the supercapacitor until a voltage set point value is reached. Due to the time-variations of the system parameters, the control unit must take these variations into consideration. Therefore, an adaptive control method is proposed, in which the values of the controller parameters are modified during the system operation in real time to achieve good level of performances.

2 Ease of Use BMS and Energy Power Measurement Methodology

For quantify the real impacts of vehicles under on-road conditions, the Vehicle Specific Power (VSP) methodology is commonly used to perform an energy and environmental characterization of the vehicle monitored. [1–3]

This analysis provides an estimate of the power per mass unit due to a combination of vehicle dynamics (speed, acceleration, rolling and aerodynamic resistance) and road grade as well as to each point of the trip a correspondent specific power is assigned. This methodology allows comparing different technologies under similar power requirements. However, for bicycles and motorbikes only simplified approaches have been performed.[4]

Two types of bicycles were tested. One was pedal assisted with several levels of electric assistance according with the driver request (EB1) and the other was a bicycle that can be used as power on demand and pedal assist (EB2), and for the quantify the measured results a conventional bicycle (CB) was used for evaluating the difference between the electric bike and conventional one as a reference. The monitored vehicles characteristics are presented in Table 1.

Two routes in Alba Iulia town were selected to cover different types of circuits under urban context, covering a wide range of operating and driving conditions. Six volunteers between 23 and 29 years-old performed the bicycle routes.

Table 1 Bike's characterization

	Bike's characteristics		
	Bike power and gears	*Battery (kWh)*	*Bike weight (kg)*
EB1 chinese	250 W/6 gears	0,36	36
EB2 UAB	500 W/6 gears	0,36	21
CB	6 gears		12

Both vehicles were monitored several times on in each route, 10 times for each bicycle and the total distance traveled was of 114 km, as presented in Table 2.

Each trip was monitored by data collected in road and send by GSM to local server.

The GPS allows collecting speed, location, and altitude information via an integrated barometric altimeter, which allows calculating road grade based on distance and altitude, following the same reference procedures studies. [5, 6]

The GPS was adequately installed inside the backpack, to avoid pressure fluctuations due to the movement which could affect the altitude readings and an external antenna was used to avoid GPS signal losses.

Voltage probes were installed directly in the electric bicycle battery terminals, while current measurements were done on the circuit that connects the battery to the electric motor. Current measurements were calibrated before each measurement, by performing a zero adjustment. The signals provided by the probes were collected by a DAQ board installed also on the backpack. For battery voltage signal a voltage divider circuit was installed before the DAQ board to account for the 0–10 V limit of the acquisition device. Both GPS and battery data were collected in a PC using a program developed in Laboratory to integrate the different communication protocols (serial port and NMEA protocol for GPS and analog data via USB port for the voltage and current collected in the DAQ board) that allows the data synchronization.

According to the rider demands, battery data was used to determine, at each second of the trip, the power provided by the battery to the electric motor and, by integrating this information along the trip the cumulative energy spent on the tour can be obtained.

Table 2 Road's characteristic

	Table Column Head		
	Distance Km	*Average positive grade(rad)*	*Average negative grade(rad)*
R1	2.7 km	0.037	-0.029
R2	4.2 km	0.020	0,017

3 Methodology Design to Estimate Specific Power

The analysis used in this work is based on Vehicle Specific Power (VSP) to estimate the power demand by vehicles, which combines speed (v), acceleration (a) and road grade (h). This methodology allows comparing different technologies under similar power requirements. It is traditionally used on light-duty vehicles and its generic definition, which includes the forces applied to a moving body, is presented in Eq. (1). The coefficients of the equation are adjusted according to the typology of vehicle monitored ([1–3]. In this case, the coefficients used were adapted for bicycles and motorcycles based on a literature review [1].

These coefficients allowed to estimate the Bicycle Specific Power (BSP).

VSP equation for light-duty vehicles are calculated in respect to [1]

$$VSP = \frac{Power}{Mass} = \frac{\frac{d}{dt}(E_{Kinetic} - E_{Potential}) + F_{Rolling}v + F_{Aerodinamic}v}{m}$$
$$= v(a(1+\varepsilon_i) + g\theta + \frac{1}{2}\rho_a \frac{C_d A}{m}(v + v_m)^2 v \tag{1}$$

where $E_{Kinetic}$ is kinetic energy, $E_{Potential}$ is potential energy, $F_{Rolling}$ is rolling resistance force, $F_{Aerodynamic}$ is aerodynamic resistance force, $\boldsymbol{v}$ is speed, $\boldsymbol{m}$ is mass (21 or 36 kg for the bicycle and 70 kg for the rider), a is acceleration, $\boldsymbol{h}$ is the altitude, $\boldsymbol{\theta}$ is road grade, ε_i is the effect of translational mass of power train rotating components (0.01), ρ_a is air density (1.2 kg/m3), υ_w is wind speed, A is frontal area (0.50 m^2 for bicycles), g is the gravitational constant (9.81 m/s^2), C_r is the rolling coefficient (0.008 for bicycles) and C_d is the aerodynamic coefficient (1.2 for bicycles). It is worth noticing that wind speed was not accounted, it is not considered on VSP. Since it is difficult to quantify intensity and direction in a second-by-second basis, even considering an average value of wind speed would not be representative, since it could have positive or negative impacts according with relative direction with the biker.

Using the respective coefficients, the BSP equation for bicycle in (W/kg) is defined by:

$$BSP = v[1.01a + 9.81\ \sin\theta + 0.78] + 0.0041v^3 \tag{2}$$

Figure 1 presents the cumulative frequency of BSP using more than 10 h of on-road data collected with EB1 and EB2. Around 80% of collected data is between 0.86 W/kg and 2.72 W/kg, which corresponds to around 240W and is coherent with literature review [1]. The highest BSP value was found in an electric bicycle (7.45 W/kg).

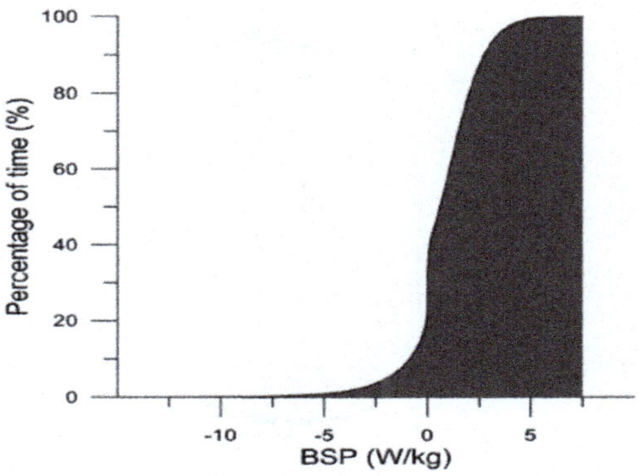

Fig.1 BSP accumulated frequency for all monitored data for EB1 and EB2

4 Characterization of Dynamics Driving and Validation of Energy Impacts

When analyzing the bicycles' data, the percentage of time spent in each BSP mode for the conventional and electric bicycles is presented in Fig. 2. For negative modes, the driving profile is very similar for both bicycles. However, on positive BSP modes, the electric bicycle presents a higher share of time spent in high BSP modes (because of higher power demands). This is a result of the electric assistance, which allows traveling at high speeds on higher slopes and combinations of higher speeds and acceleration. These two BSP profiles allowed creating a change pattern per BSP mode of shifting from CB to EB. Regarding some of the traditional difficulties associated with conventional bicycles, the factor that most influences the speed is the slope of the road, when using conventional bicycles, as it is presented in Fig. 3. In terms of conventional and electric bicycle usage comparison, an overall 16% increase in average speed was verified for the electric bicycle when compared to the conventional bicycle. This increase is mostly visible in positive road grades (37%).

Unsurprisingly, the energy rate increases with BSP mode, showing the prevalence of electric assistance on these higher modes.

To validate the analysis proposed for bicycles, a comparison between the measured and estimated electricity consumption was performed. Estimates were calculated using (2). The estimated and measured electricity consumption for the two electric bicycles and two routes, performed by 6 bikers, showing that the differences ranged from a minimum of 7.3% to 18.2%. The highest deviations found are a consequence of the different levels of electric assistance used on each route, which can be defined by the rider. On average, the differences found for both bicycles were 1.4% between estimates and measurements, while the absolute deviation was 6.0% and 6.6% for EB1 and EB2, respectively. The results indicate the adequacy and consistency of the BSP approach to quantify the energy use on specific routes.

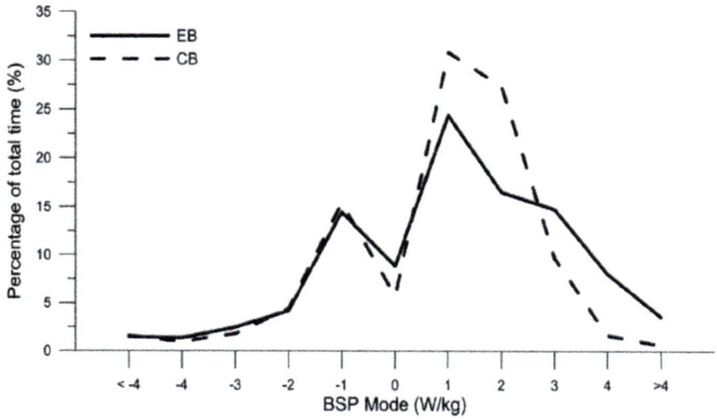

Fig. 2 Time distribution (%) per BSP mode for the two bicycles

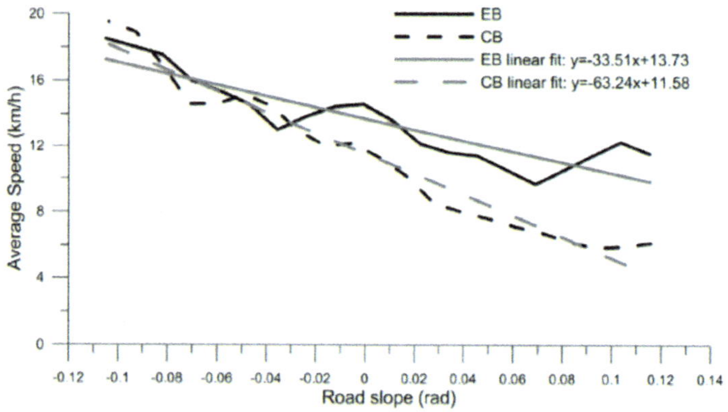

Fig. 3 Speed as a function of road grade

5 Conclusions

The concept presented in this work for on-road monitoring of bicycles and motorcycles allowed the development of the BSP methodologies to choose the strategies for charging and recovery the energy on road. These methodologies are innovative since, based on an easily collected driving cycle (speed, acceleration, and grade variables), the evaluation of the impacts of alternative technologies can be easily performed.

References

1. Magno M, Gonçalo D, Patricia B (2015) Introducing specific power to bicycles and motorcycles: application to electric mobility. Transp Res Part C 51:120–135
2. Baptista P, Silva C, Farias T, Heywood J (2012) Energy and environmental impacts of alternative pathways for the Portuguese road transportation sector. J Energy Policy 51:802–815
3. Bishop J, Doucette R, Robinson D, Mills B, McCulloch M (2011) Investigating the technical, economic and environmental performance of electric vehicles in the real-world: a case study using electric scooters. J Power Sources 196:10094–10104
4. Cherry C, Cervero R (2007) Use characteristics and mode choice behavior of electric bike users in China Transp Policy 14:247–257
5. Khayyam H, Abawajy J, Javadi B, Goscinski A, Stojcevski A, Bab-Hadiashar A (2013) Intelligent battery energy management and control for vehicle-to-grid via cloud computing network. Appl Energy 111:971–981
6. Capasso C, Veneri O (2014) Experimental analysis on the performance of lithium based batteries for road full electric and hybrid vehicles. Appl Energy 136:921–930

Analysis of Blockchain Integration from Manufacturing

Elisabeta Mihaela Ciortea and Adina Carmen Paştiu

Abstract The paper wants to present a useful method of analysis in the smart manufacturing process based on Petri nets. The method allows tracking and control of a classic smart system, which can be implemented on a supported blockchain platform and then integrated into smart system manufacturing. One of the main advantages that the method analyzes is the visualization of errors in the nodes, by following the information transfer flows. The modeling of intelligent manufacturing systems using Petri nets helps the decision-making system to detect possible errors and to be able to intervene right from the design phase. This modeling, tracking, and control system provides real-time information and can intervene in the configuration right from the design phase, to optimize the system and meet the requirements.

Keywords Manufacturing · Blockchain · Petri net · Cloud

1 Introduction

In view of the development, the manufacturing systems are in continuous development and modernization according to the market requirements, so they are systems that offer unlimited benefits, but there are many challenges regarding the centralized architecture so that all devices can be identified, authenticated, and connected through centralized servers, to meet the demands of the continuously developing market. The model proposed in this paper can be used to connect a wide range of computing devices, it will continue to support small-scale complex networks because it will not be able to provide the necessary conditions for system expansion in the future [2, 10].

E. M. Ciortea (✉) · A. C. Paştiu
"1 Decembrie 1918" University of Alba Iulia, Alba Iulia, Romania
e-mail: mciortea@uab.ro

A. C. Paştiu
e-mail: carmen.pastiu@uab.ro

T. A. Kovács et al. (eds.), *The Impact of the Energy Dependency on Critical Infrastructure Protection*, Advanced Sciences and Technologies for Security Applications, https://doi.org/10.1007/978-3-031-78544-3_17

Adopting a standardized peer-to-peer communication model for processing many transactions between devices has the potential to reduce the costs associated with setting up and maintaining large, centralized data systems and thus distribute computing and storage reports across multiple devices that define complex [2].

Blockchain's decentralized, autonomous, and non-secure capabilities make it an ideal component for its role as the building block of emerging solutions. This is why modern, enterprise technologies have become a pawn to adopt blockchain technology [2].

Establishing peer-to-peer communications presents a set of security challenges. Security means more than data protection. Thus, blockchain solutions must maintain the privacy and security of networks and use participant validation and consent for transitions to prevent hacking and theft [2, 5].

Blockchain opens up a number of scenarios that are difficult or even impossible to implement without it. The paper presents a model that will be able to deal with the exchange of information between similar devices, and they will use an intelligent industrial system that has the role of shaping the understanding between two or more parties [2].

Blockchain technologies are able to track, coordinate, transact and store information from a large number of devices, enabling the creation of applications that do not require a centralized cloud. Some companies want to apply blockchain as a technology to democratize the industrial future [2].

According to the specialized literature, so far there is an expansion of cloud computing, so we tried to adapt the manufacturing system with the principle of using and developing a fair system based on current technologies [5].

Cloud manufacturing is a globalized manufacturing framework that aims to improve the performance of manufacturing systems.

According to the specialized literature, a distributed implementation and a real-time synchronization platform are needed to enable globalized collaboration in the composition of services. This paper proposes a method of using blockchain to solve these problems. Each service composition is considered a transaction in the blockchain concept. A block includes a set of service compositions, and its validity is confirmed by a predefined mechanism [9].

The proposed model is impactful because it can increase response time using blockchain technology and improve the overall optimality of supply and demand matching in cloud manufacturing.

The efficiency of the proposed model was evaluated by investigating a service allocation problem in a cloud manufacturing system in four large-scale problems. The simulation results indicate the high quality of the proposed solution. There is a decrease in the time needed to solve possible problems, and the pace of information processing increases both in terms of the flow of information in the cloud.

The role of the cloud is highlighted by its close connections with blockchain technology. The vulnerability of this technology is the operation on trust, which is why we did not generate direct connections with the manufacturing system. The direct relationship with the manufacturing process will be resolved in further research. Since the flow of information transfer is linear, a directed graph through which information can be controlled at any time is an advantage for the decision-making system, which obtains the information at any point in real-time.

2 Description of the Integration Process

Among the advantages of adopting the blockchain connected with the manufacturing system through the cloud are the following [10]:

- all participants have the possibility to see all transactions and all nodes because each participant has his own ledger. The contents of the transaction are protected by the participant's private key, so all participants can see them but cannot directly intervene in the action and are protected. The cloud-connected manufacturing system forms a dynamic system where all connected devices can exchange information, but at the same time, the privacy of the participants is ensured.
- participants must verify transactions to approve and add them to the distributed ledger. There is no single authority that can approve transactions or set specific rules for accepting transactions. There is massive trust because most network participants must come to an agreement to validate transactions. Blockchain owns and will provide a secure platform for the cloud.
- each node has its own copy of the record containing all the transactions that have been part of the network. Blockchain is able to withstand attacks of any kind. Even if one node has been compromised, the blockchain is kept active by another node. If there is a copy of the data at each node in the process, information exchanges can be improved.
- blockchain has the ability to provide a secure network, which is necessary in dealing with modern, complex technologies with numerous and heterogeneous devices.
- a blockchain transaction is distributed across the network in minutes and will be processed at any time of the day.
- existing solutions are expensive due to high infrastructure and maintenance costs associated with centralized architecture, large servers, and network equipment. A major disadvantage is the amount of communication that needs to be managed when there are a lot of devices, which leads to a substantial increase in costs.

So far, various blockchains have been implemented and the technology seems to work well, even if most of them suffer from a lack of application of software engineering development principles.

In this paper, you use a Petri net approach in blockchain analysis with a direct impact on the manufacturing system. Using Petri nets, a single model is defined, a small data structure that highlights key information about transactions and addresses. The overall architecture and transaction schema are analyzed and implemented in a familiar and easy-to-interpret formalist.

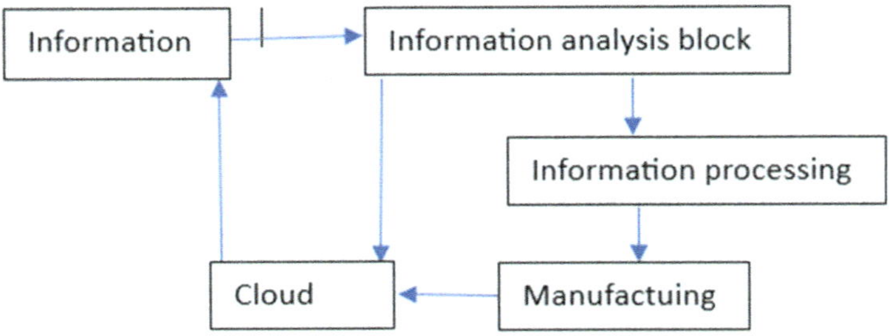

Fig. 1 The flow of information in optimal operating conditions of the system

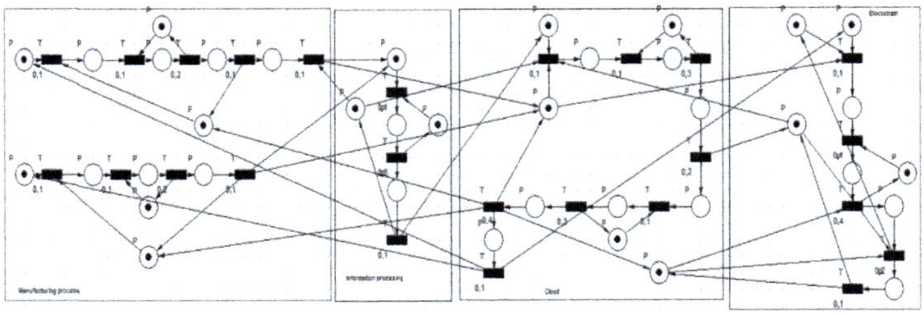

Fig. 2 Model overview

In the analysis block, the information is analyzed in a short time, what is worth noting is the variable duration of information processing. This procedure can be seen in Fig. 1.

For a real -time modeling and representative simulation, we chose a classic manufacturing system, which contains all the necessary components, Fig. 2. Thus, we outlined a general warehouse, a system that adopts blockchain technologies with analysis elements, which in some situations can be considered as external users, an information processing system, and a production system in the cloud where the information is stored for further processing. From this last section, the requested information can be accessed via the blockchain, without having direct access to the other components of the system.

3 Process Modeling and Simulation

Because within the blockchain, the variation graph of the flow of information is not linear, there are no linear errors because the technology used by the system is complete, we have only identified variations in the flow of external requests. The response received from the cloud is linear, for a complete flow of data (Figs. 3, 4, 5 and 6).

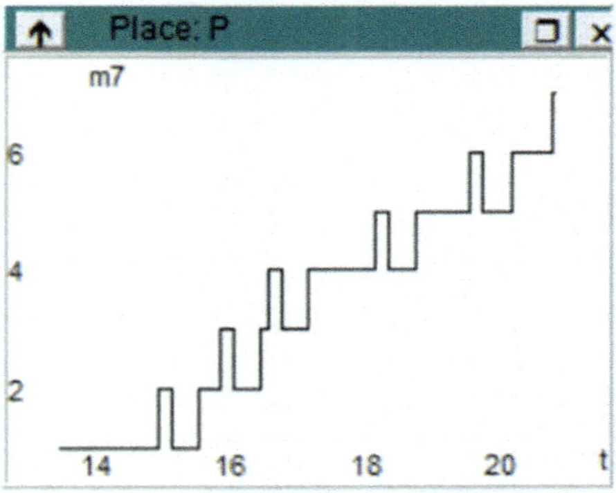

Fig. 3 The variation of the information processing flow on the manufacturing system

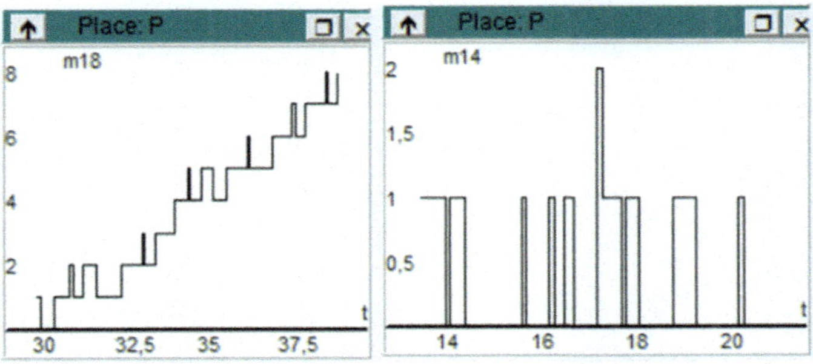

Fig. 4 The variation of the information processing flow, when preparing the information

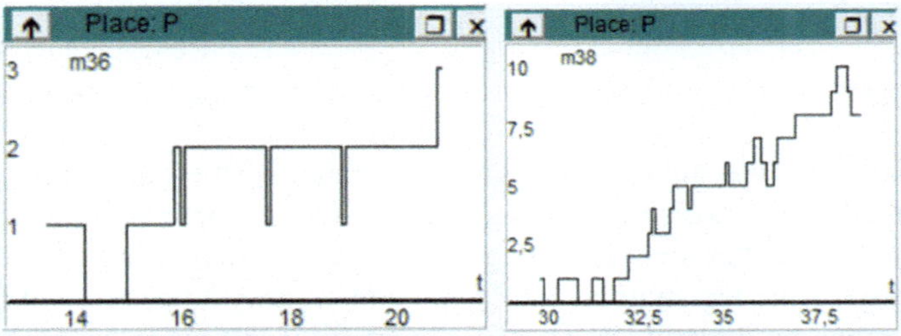

Fig. 5 The variation of the information processing flow in the cloud

4 Strategic Business Value Due to the Use of Blockchain

In accordance with McKinsay's research shows that the use of blockchain offers new business opportunities. Among the direct benefits listed are reduced compliance costs, creating more cost-effective transactions. From an economic point of view, the

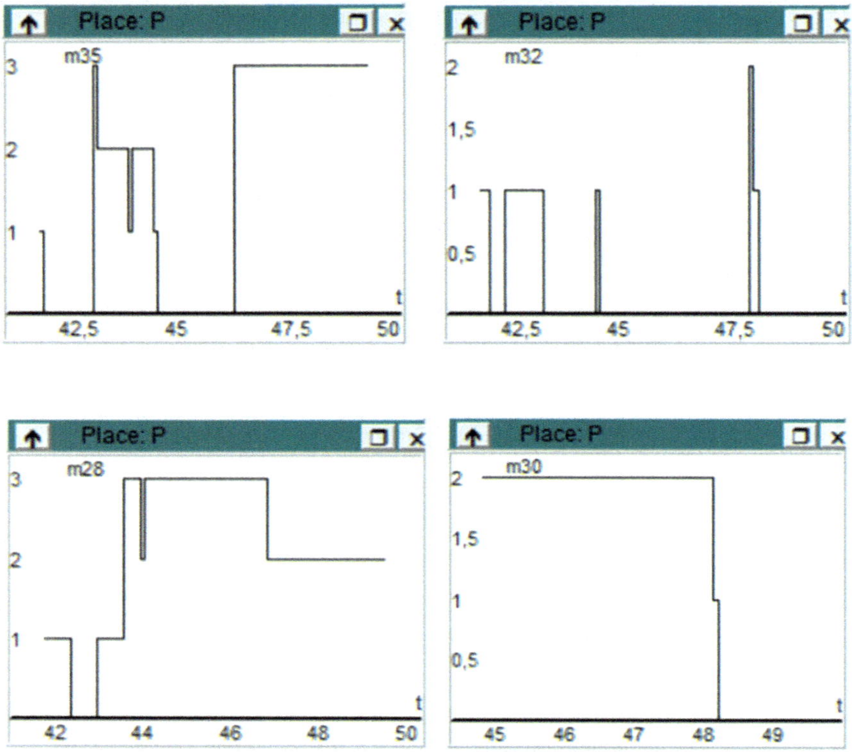

Fig. 6 Variation of information flow from blockchain users

use of Blockchain is considered a source of increasing efficiency and creating value [7, 8].

With the help of blockchain, companies can streamline the system audit procedure, they can use the records of transactions and processes that they can store and compare.

Recent research has shown the benefits of use in different industries as well as the challenges. The registration, validation, and security advantages of blockchain are accentuated by digital identity. The advantage of allocating smart contracts on public networks without the intervention of a third party. During the entire process, it can be tracked, secure, and unalterable.

Automating the agreement processes between companies and their partners and between companies and their customers will greatly reduce processing time and administrative costs.

Using Blockchain to support cloud manufacturing operations is a new manufacturing model proposed to use concepts such as the Internet of Things (IoT), service-oriented computing, and virtualization to transform manufacturing resources and operations in a set of production services that can be intelligently integrated and managed [1].

In this regard, it is proposed to use blockchain to enable secure decentralized cloud manufacturing architecture and secure knowledge sharing for manufacturing design such as design and redesign. [3, 6].

Another benefit of using blockchain is that it is used to ensure the origins of parts based on the chemical signatures of the added nanomaterials. Thus, a secure verification of the originality of the parts can be achieved by activating the supply chains in production and reducing the risks associated with the use of imitation parts. By introducing an automated process, a reduction in costs is achieved [4].

5 Conclusion

Blockchain has already demonstrated its effectiveness in various industries, but the field is in continuous evolution and development. The challenges for specialists to develop tools and processes are great.

In this paper have presented only a few aspects related to a production model that can interconnect numerous computing systems that can be configured according to production needs. The use of blockchain for cloud production is known, but we considered it necessary to present, in addition to the expectations, the configuration of the processes and the economic benefits of this system.

References

1. Al-Jaroodi J, Mohamed N (2019) Blockchain in industries: a survey in IEEE. Access 7:36500–36515. https://doi.org/10.1109/ACCESS.2019.2903554
2. Atlam HF (2018) Blochchain with internet of things: benefits, challenges and future directions. I J Intell Syst Appli 6:40–48. https://doi.org/10.5815/ijisa.2018.06.05
3. Barenji, AV, Li Z, Wang WM (2018) Blockchain cloud manufacturing: shop floor and machine level. In: Proceeding European conference on smart objects, systems and technologies, pp 1–6. VDE, Munich, Germany
4. Holland, M. Nigischer C, Stjepandi J (2017) Copyright protection in additive manufacturing with blockchain approach in transdisciplinary engineering: a paradigm shift. Amsterdam, pp 914–921. IOS Press, The Netherlands
5. Hong—Ning D, Zibin Z, Yan Z (2019) Blockchain for internet of things: a survey. IEEE Internet of Things J 6(5):8076–8094. https://doi.org/10.1109/JIOT.2019.2920987
6. Li BH et al (2010) Cloud manufacturing: a new service-oriented networked manufacturing model. Comput Integr Manuf Syst 16(1):1–7
7. Li Z, Liu L, Barenji AV, Wang W (2018) Cloud-based manufacturing blockchain: secure knowledge sharing for injection mould redesign procedia CIRP, vol 72, no 1, pp 961–966
8. McKinsay, Raport technology deep dive: DLT and blockchain. https://www.mckinsey.com/featured-insights/mckinsey-explainers/what-is-blockchain, Last Access 2023/05/30
9. Radmanesh S-A, Haji A, Fatahi Valilai O (2021) Blockchain-based cloud manufacturing platforms: a novel idea for service composition in XaaS paradigm. PeerJ Comput Sci 7:e743 https://doi.org/10.7717/peerj-cs.743

10. Tama BA, Kweka BJ, Park J, Rhee K-H (2017) A critical review of blockchain and its current applications. In: International conference on electrical engineering and computer science (ICECOS), pp. 109–113. 2017-978-1-4799-7675-1/17/2017IEE

IoT Security for Critical Infrastructure: Challenges and Best Practices

Sanja Maravić Čisar, Robert Pinter, and Petar Čisar

Abstract The rise of Internet of Things (IoT) technologies has revolutionized the way critical infrastructures operate, leading to an era of unique connectivity and automation. However, this transformation introduces a concern about the security of these interconnected systems. Ensuring the robustness of IoT security in critical infrastructure is of supreme importance, as any breach could have far-reaching consequences, affecting not only the operational integrity of these vital systems, but also potentially putting public safety and national security at risk. By examining vulnerabilities present in IoT devices, assessing risks associated with communication protocols, and delving into the various aspects of authentication and authorization mechanisms, our aim is to equip stakeholders with a comprehensive understanding of multidimensional security challenges. As IoT technologies are increasingly integrated into critical infrastructures, it is essential to meet the security requirements associated with this evolution. This collective effort will enable the establishment of a foundation that ensures the ongoing resilience and reliability of these critical systems.

Keywords IoT · Critical infrastructure · Security

S. Maravić Čisar (✉) · R. Pinter
Subotica Tech—College of Applied Sciences, Marka Oreškovića 16, 24000 Subotica, Serbia
e-mail: sanjam@vts.su.ac.rs

R. Pinter
e-mail: pinter.robert@vts.su.ac.rs

P. Čisar
University of Criminal Investigation and Police Studies, Cara Dušana 198, 11080 Zemun, Serbia
e-mail: petar.cisar@kpu.edu.rs

T. A. Kovács et al. (eds.), *The Impact of the Energy Dependency on Critical Infrastructure Protection*, Advanced Sciences and Technologies for Security Applications, https://doi.org/10.1007/978-3-031-78544-3_18

1 Introduction

The critical infrastructure serves as the fundamental framework of our modern society and contains vital components such as power generation, transportation systems, and healthcare facilities (Fig. 1). These components are increasingly relying on Internet of Things (IoT) technologies to optimize their operations and improve efficiency.

The adoption of IoT in critical infrastructure systems has been rapid. Organizations in all sectors recognize the potential of IoT to improve efficiency, reduce operational costs, and improve service delivery. As a result, IoT devices have increased in critical infrastructure, providing real-time insights and control capabilities that were previously unimaginable. However, this rapid adoption has also introduced security challenges that must be addressed to ensure the continued reliability and safety of critical infrastructure systems.

Disabling the hospital's energy supply, controlling the temperature in nuclear cooling towers, and controlling smart cars in motion are among the possible destructive scenarios. The increase in hacking poses a significant threat to important national infrastructures, which are targeted by cyber theft, spying, intimidation, disruption, and cyber terrorism [2]. A cyberattack on the infrastructure of another nation could serve as a justification for the war [3]. Consequently, protecting against cyber threats is of paramount importance for national defense. Cyberattacks have the potential to disrupt the physical systems or a nation, transfer control to external entities, deactivate critical systems, and compromise the privacy of individual data [4].

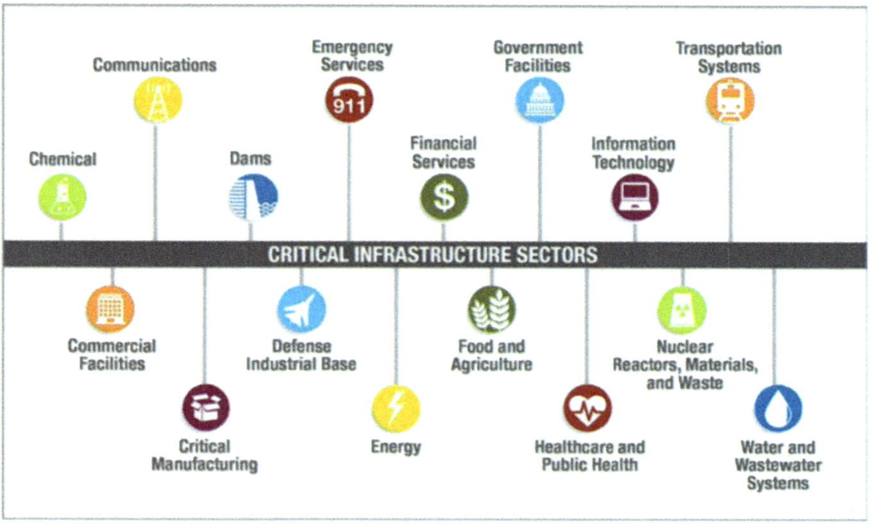

Fig. 1 Critical infrastructure sectors [1]

In [5] an analysis of cyberattacks on critical infrastructures is presented. The authors specifically investigated studies conducted between 2008 and 2019, classifying cyberattacks on various critical infrastructures and detailing the resulting damage to the systems. For example, in 2008, a tram hacking incident in Lodz, Poland, involved the manipulation of rail switches, leading to derailments and passenger injuries. In 2009, the hacking of a Texas power company disrupted its energy forecasting system for an entire day. The following year, the infamous Stuxnet attack targeted SCADA systems in an Iranian nuclear facility, causing physical damage to uranium enrichment centrifuges.

In 2011, hackers breached a water distribution system in South Houston, Texas, using stolen credentials to burn out a water pump. A similar breach occurred in 2013 when attackers accessed the SCADA system of a dam in New York, although no physical damage was reported. In 2015, a cyberattack on Ukraine's power grid caused a massive blackout, leaving 225,000 customers without electricity for hours. By 2016, a ransomware attack on San Francisco's light rail system temporarily disrupted ticketing operations, while a separate incident manipulated chemical valves in a U.S. water company.

Other notable cases include the 2017 sabotage attempt on a petrochemical plant in Saudi Arabia, where attackers aimed to trigger an explosion, and a DDoS attack on Sweden's transport network that disrupted email systems and traffic maps. In 2018, the healthcare sector was also targeted when ransomware attacked an Indiana hospital and forced it to pay a $55,000 ransom to restore system access. In 2019, a coordinated DDoS attack on Turkey's telecommunication and finance sectors led to temporary communication failures. These examples underscore the diverse and evolving nature of cyber threats targeting critical infrastructures, highlighting the urgent need for robust cybersecurity measures [5].

This paper is structured as follows: Sect. 2 discusses the adoption of IoT in critical infrastructure, emphasizing the benefits and introducing security challenges. Section 3 explores these challenges and proposes measures to address security concerns. Finally, Sect. 4 concludes by emphasizing the transformative phase introduced by IoT integration and the collective effort required to fortify critical infrastructure systems against growing cyber risks.

2 IoT Security Challenges in Critical Infrastructure

IoT devices, which can be as simple as sensors or as complex as smart machinery, enable real-time monitoring, automation, and data-driven decision-making. They improve the efficiency, safety, and reliability of critical infrastructure systems in the following ways [6]:

- Data collection: IoT devices collect data on various parameters such as temperature, humidity, traffic flow, and equipment performance, allowing continuous monitoring and analysis.

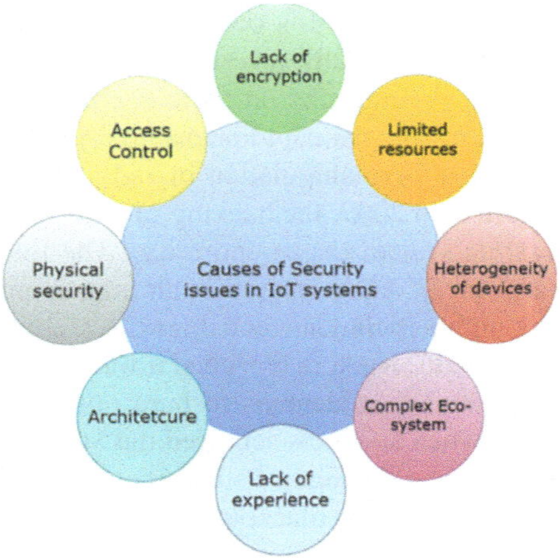

Fig. 2 IoT security factors [18]

- Automation: IoT facilitates the automation of processes, reducing human intervention and the risk of errors. For example, in energy distribution, IoT-enabled smart grids can automatically reroute power to prevent blackouts.
- Predictive maintenance: IoT devices enable predictive maintenance by detecting anomalies and wear in critical infrastructure components. This proactive approach helps prevent costly breakdowns and service interruptions.
- Resource optimization: Through data analytics, the IoT helps optimize resource allocation. For instance, in transportation, IoT can improve traffic flow and reduce congestion.

Integration of IoT devices into critical infrastructure systems presents a unique set of security challenges that require careful consideration due to the distinctive characteristics of IoT devices and the networks in which they operate (Fig. 2). These challenges include:

1. Diverse device landscape: IoT ecosystems often include a wide range of devices from different manufacturers, each with its own firmware, security protocols, and update mechanisms. This diversity can lead to compatibility issues and security gaps, making it challenging to maintain a unified security approach [7].
1. Limited computing resources: Many IoT devices have computational capabilities limited by cost and power constraints. This limitation restricts their ability to implement robust security measures, leaving them vulnerable to attacks [8].
2. Vulnerable communication channels: IoT devices rely on various communication channels, including wireless protocols and the Internet, to transmit data. These channels can be susceptible to interception, eavesdropping, and man-in-the-middle attacks if not properly secured [9].
3. Lack of standardization: The IoT lacks uniform security standards, resulting in inconsistencies in security implementations across devices and systems. This variability increases the complexity of ensuring comprehensive security [10].

4. Firmware and software vulnerabilities: IoT devices may run outdated or unpatched firmware and software due to limited maintenance or vendor support. This creates opportunities for attackers to exploit known vulnerabilities [11].
5. Inadequate authentication: Weak or absent authentication mechanisms can permit unauthorized access to IoT devices and networks, potentially enabling attackers to take control and disrupt operations [12].
6. Data privacy concerns: IoT devices collect large amounts of sensitive data, raising concerns about data privacy and unauthorized access. Mishandling or unauthorized access to these data can have severe consequences [13].
7. Supply chain risks: The global supply chain for IoT devices introduces security risks, as malicious actors can compromise devices at various points in the supply chain, compromising security from the outset [14].
8. Scale and complexity: The extensive scale and complexity of IoT deployments in critical infrastructure can overwhelm security efforts. Ensuring security in thousands or even millions of devices and their interconnected networks is a challenging task [15].

A powerful computer worm designed by U.S. and Israeli intelligence, Stuxnet targeted Iran's nuclear facilities by exploiting vulnerabilities in IoT devices, specifically Programmable Logic Controllers (PLCs). It demonstrated the potential for physical damage through cyberattacks on critical infrastructure [16].

In two separate incidents, in 2015 and 2016, hackers disrupted the Ukrainian power grid by exploiting vulnerabilities in IoT devices, leaving thousands without electricity. These attacks underscore the susceptibility of critical infrastructure to IoT-based attacks and marked the first successful cyber operations to destroy energy infrastructure [17].

These examples emphasize the urgency of addressing IoT security challenges within critical infrastructure. As the Internet of Things continues to play an integral role in modernizing and optimizing these systems, robust security measures are crucial to protect against potential threats and vulnerabilities. If left unaddressed, these weaknesses may be exploited, creating opportunities for unauthorized access or control over the very devices designed to support critical operations. Therefore, a fundamental understanding and proactive improvement of these vulnerabilities become crucial to strengthening the security of critical infrastructure.

The interconnected nature of IoT devices within critical infrastructure introduces vulnerabilities that can potentially compromise the integrity of these vital systems. The variety of communication protocols used in IoT ecosystems adds another layer of complexity. Although these protocols facilitate smooth data exchange, they also introduce potential risks. Inappropriate encryption or authentication measures within these communication channels can expose transmitted data to interception or manipulation. To ensure the confidentiality and reliability of data transmission, robust communication protocols are indispensable.

Authentication and authorization issues emerge as critical concerns in the context of IoT security within critical infrastructure. Establishing the identity of devices and ensuring they possess the appropriate permissions are pivotal in maintaining the integrity of these interconnected systems. Weaknesses in authentication mechanisms pose a direct threat, potentially leading to unauthorized access and compromising the core of critical infrastructure.

Safeguarding sensitive data is fundamental in the framework of IoT security. Inadequate encryption methods or weak privacy policies expose sensitive information to potential breaches, carrying consequences that extend far beyond initial compromises. Clearly addressing these concerns is crucial to preserving the confidentiality and integrity of data within interconnected IoT devices in critical infrastructure.

3 Addressing IoT Security Challenges in Critical Infrastructure

Effectively addressing IoT security challenges in critical infrastructure requires comprehensive vulnerability assessments, robust communication protocols, enhanced authentication measures, and secure data protection strategies (Fig. 3). By directly confronting these challenges, we fortify the foundations of our critical infrastructures, ensuring their resilience and adaptability in an ever-changing technological environment. To address the security concerns associated with IoT in critical infrastructure, we propose practical measures as follows [18].

Strengthening device security: Ensure the security of IoT devices by regularly updating firmware and managing patches. Collaboration among manufacturers, vendors, and regulators is essential to establish unified security standards.

Fig. 3 Strategies for securing IoT systems [18]

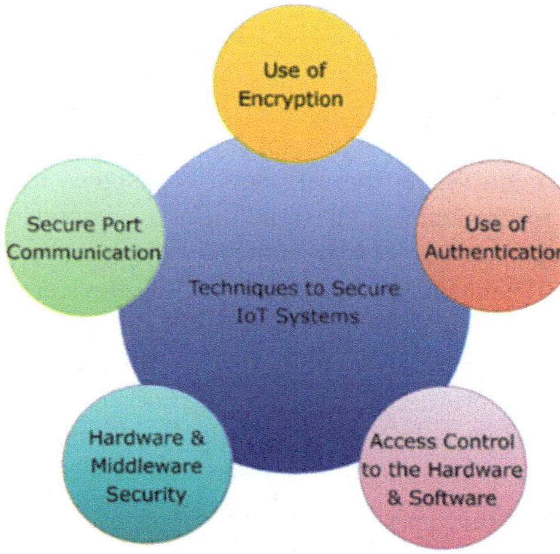

Securing communication protocols: Implement robust encryption and authentication measures for various communication channels. Adopt industry-standard protocols, conduct regular audits, and enforce encryption best practices to safeguard data transmission.

Enhancing authentication mechanisms: Strengthen authentication and authorization mechanisms through multifactor authentication, secure key management, and device identity verification. Regular evaluations and adaptive authentication strategies can strengthen defenses against unauthorized access.

Ensuring robust data protection: Emphasize data confidentiality using strong encryption methods and strict privacy policies. Regular assessments of data handling practices and compliance checks contribute to a resilient data protection framework.

Comprehensive security strategy: Adopt a comprehensive strategy that involves detailed vulnerability assessments, incident response plans, and continuous improvement. Collaboration between the public and private sectors, along with international cooperation, is crucial to address global challenges in IoT security. Regular training programs ensure that individuals are equipped to navigate evolving threats.

Secure port communication: Preventing cyber threats involves closing all open ports, with particular emphasis on closing termination ports after communication.

Role-based access control for hardware and software: Implementing role-based access ensures that only authorized individuals have access to sensitive data on IoT devices. Regular training initiatives enable individuals to stay informed about evolving threats, ensuring that they are well prepared to follow security protocols and contribute to the continuous improvement of security measures.

The adoption of these robust security measures safeguards against potential threats and vulnerabilities, ensuring the adaptability and resilience of these systems in the face of evolving cyber risks. This collective initiative to improve IoT security within critical infrastructure is crucial to sustaining the integrity, reliability, and safety of vital systems in a dynamic technological environment.

4 Conclusion

Integration of IoT technologies into critical infrastructure means a transformative phase characterized by enhanced connectivity and operational efficiency. Nevertheless, this paradigm shift is accompanied by substantial security challenges that require careful consideration.

The adoption of IoT devices within critical infrastructure has created a variety of security challenges, demanding proactive interventions. The study suggests practical strategies for enhancing device security, such as regular updates, adherence to unified standards, robust encryption, and authentication measures. It also recommends fortifying authentication mechanisms, ensuring strong data protection, and adopting a comprehensive security strategy that includes vulnerability assessments and continuous improvement.

The incorporation of secure port communication and role-based access control for hardware and software adds additional layers of protection against cyber threats. The vital role of fostering an organizational culture that prioritizes information awareness and instituting regularized training initiatives highlights the dynamic nature inherent in IoT technologies.

Addressing IoT security challenges within critical infrastructure demands a comprehensive and collaborative approach. By including the proposed practical measures and fostering a culture of awareness, we can navigate the complexities of IoT technologies and fortify critical infrastructure against the evolving landscape of cyber risks. This concerted effort is paramount for sustaining the integrity and reliability of our vital systems, ensuring they continue to operate securely and efficiently in emerging technological advancements.

References

1. Homeland Security Today. https://www.hstoday.us/featured/cyberattacks-on-critical-infrastructure-as-the-new-wmd/. Last accessed 19 Jan 2024
2. Baykara M, Das R (2015) A survey on potential applications of honeypot technology in intrusion detection systems. Int J Comput Netw Appl 2(5):203–211
3. Sagıroglu S, Arslan B (2019) Fighting with cyber terror and terrorism: threats and precautions. In: 4th international conference on computer science and engineering (UBMK), pp 239–244
4. Pacheco J, Benitez VH, Pan Z (2019) Security framework for IoT end nodes with neural networks. Int J Mach Learn Comput 9:381–386
5. Das R, Gündüz MZ (2019) Analysis of cyber-attacks in IoT-based critical infrastructures. Int J Inf Secur Sci 8(4):122–133
6. Allioui H, Mourdi Y (2023) Exploring the full potentials of IoT for better financial growth and stability: a comprehensive survey. Sensors 23(15):8015. https://doi.org/10.3390/s23198015
7. Mughal MJH, Faisal M, Munir ME (2019) Internet of things—IoT interoperability and challenges. J Mech Contin Math Sci 14(4):304–318. https://doi.org/10.26782/jmcms.2019.08.00025
8. Rainbow Secure (2014) striking a balance between security risks and benefits of IoT devices. https://www.linkedin.com/pulse/striking-balance-between-security-risks-benefits-iot-devices/. Last accessed 19 Jan 2014
9. Gerodimos A, Maglaras L, Ferrag MA, Ayres N, Kantzavelou I (2023) IoT: communication protocols and security threats. Internet Things Cyber-Phys Syst 3:1–13. ISSN 2667-3452. https://doi.org/10.1016/j.iotcps.2022.12.003
10. Al-Qaseemi SA, Almulhim HA, Almulhim MF, Chaudhry SR (2016) IoT architecture challenges and issues: lack of standardization. In: 2016 future technologies conference (FTC), San Francisco, CA, USA, pp 731–738. https://doi.org/10.1109/FTC.2016.7821686
11. Oliynyk K (2024) Analyze IoT device firmware: practical guide. https://webbylab.com/blog/analyze-the-firmware-of-iot-devices-practical-guide/. Last accessed 19 Jan 2024
12. José L (2024) Security issues of IoT: securing your IoT device in 2023. https://www.deviceauthority.com/blog/security-issues-of-iot-securing-your-iot-device-in-2023/. Last accessed 19 Jan 2024
13. Tokas Sh, Erdogan G, Stølen K (2023) Privacy-aware IoT: state-of-the-art and challenges. In: Proceedings of the 9th international conference on information systems security and privacy (ICISSP 2023), pp 450–461. https://doi.org/10.5220/0011656400003405

14. Securing the Internet of Things Supply Chain (2024) An IoTSF whitepaper. https://www.iotsecurityfoundation.org/wp-content/uploads/2022/06/RELEASE-JUNE-2022-IoTSF-supply-chain-whitepaper-v5.pdf. Last accessed 19 Jan 2024
15. Zhao B et al (2022) A large-scale empirical study on the vulnerability of deployed IoT devices. IEEE Trans Dependable Secure Comput 19(3):1826–1840. https://doi.org/10.1109/TDSC.2020.3037908
16. Fruhlinger J (2024) Stuxnet explained: the first known cyberweapon. https://www.csoonline.com/article/562691/stuxnet-explained-the-first-known-cyberweapon.html. Last accessed 19 Jan 2024
17. Council on Foreign Relations (2024) https://www.cfr.org/cyber-operations/compromise-power-grid-eastern-ukraine. Last accessed 19 Jan 2024
18. Anmulwar S, Gupta AK, Derawi M (2020) Challenges of IoT in healthcare. In: Gupta N, Paiva S (eds) IoT and ICT for healthcare applications. EAI/Springer innovations in communication and computing. Springer, Cham, pp 11–20. https://doi.org/10.1007/978-3-030-42934-8_2

Security Testing of Integrated Web Systems in the Process of Education of Software Engineers

Zlatko Čović and Zoltán Rajnai

Abstract This paper presents how engineering students become acquainted with the testing of web applications and integrated web systems during their education. The key components of integrated web systems are outlined. Integrated web system commonly includes multi-platform web applications with various levels of user access, a mobile application, and a REST API system for exchanging data among system components. To ensure that every student not only gains theoretical insights but also attains a comprehensive and robust practical understanding, it becomes imperative to seamlessly integrate laboratory exercises as an integral component of the daily educational curriculum. By engaging in hands-on activities within the controlled environment of a laboratory, students can actively apply theoretical concepts to real-world scenarios, fostering a deeper comprehension of the subject matter and enhancing their overall learning experience. This approach facilitates a well-rounded educational journey, allowing students to bridge the gap between theoretical knowledge and practical application, thereby preparing them more effectively for the challenges they may encounter in their academic and professional pursuits. Through real-life examples, students will be able to analyse security risks and apply appropriate mechanisms and protection techniques. In paper, most interesting tasks for testing the security of each component of integrated web system are described. Some of tools and services that are used by students at exercises are presented.

Keywords Security testing · Integrated web systems · Education · Software engineers

Z. Čović (✉)
Subotica Tech—College of Applied Sciences, Subotica, Serbia
e-mail: chole@vts.su.ac.rs; zlatko.covic@uni-obuda.hu

Doctoral School on Safety and Security Sciences, Obuda University, Budapest, Hungary

Z. Rajnai
Banki Donat Faculty of Mechanical and Safety Engineering, Obuda University, Budapest, Hungary
e-mail: rajnai.zoltan@bgk.uni-obuda.hu

T. A. Kovács et al. (eds.), *The Impact of the Energy Dependency on Critical Infrastructure Protection*, Advanced Sciences and Technologies for Security Applications, https://doi.org/10.1007/978-3-031-78544-3_19

1 Introduction

Most web systems created nowadays are integrated systems consisting of several components. These are often multi-platform web applications with multiple levels of user access, a mobile application, and a REST API system for data exchange between system components. When developing such complex systems, a strong emphasis should be placed on the security of each component, especially those parts through which data entry, reading, and exchange occur. At Subotica Tech—College of Applied Sciences, software engineering students engage in the development of integrated system components in several specialized subjects, while within the Integrated Web Systems course, they focus on implementing the entire system. Throughout the semester, students learn how to apply security methods and techniques in their work. They also engage in security testing of components of such systems. The first generation of the Cybersecurity program at Banki Donat Faculty of Mechanical and Safety Engineering included courses that dealt with creating web applications and potential security risks that may arise from using inadequate protection techniques and methods. The experiences gained during the engineer education at Subotica Tech in the field of security will contribute to the application of similar laboratory exercises in upcoming classes at Obuda University.

The essential operations and the microservice-based structure of mobile apps, websites, and other applications or systems are intricately linked to the indispensable role played by APIs. API stands for Application Programming Interface. It is a set of rules and tools that allows different software applications to communicate with each other. APIs define the methods and data formats that applications can use to request and exchange information. They play a crucial role in enabling the integration of different software systems, allowing them to work together seamlessly. In [1] authors addressed API security protection-related methods, and they proposed a finite state machine analysis method to enhance API security design and safety evaluation. A framework for API security auditing is introduced, focused on traffic-based API auditing, offering insights into the exploration of best practices for API protection. Ultimately, existing issues in current research are explained, and expectation is expressed for future development.

The core of computing security is secure code, yet traditional software engineering textbooks lack adequate methods for integrating security engineering into the development process. In [2] the authors emphasized the need for formal training in secure code writing. Security testing, including penetrating and destructive tests, must be added, integrating systematic approaches into software engineering curricula and development processes. Component-based development, formal methods, and automatic security checking tools are suggested for producing secure code. The authors reported some experiences applying security testing principles in their software engineering course teaching.

The Open Worldwide Application Security Project (OWASP) is a nonprofit foundation dedicated to enhancing software security. This foundation defined OWASP Top 10 as a standard awareness document for developers and web application security, encompassing a widespread consensus regarding the most crucial security risks faced by web applications. Due to the common use of APIs in various integrated projects, OWASP has proposed the TOP 10 API security risks. This document shares some similarities with the OWASP TOP 10 document in certain aspects but contains information most relevant to working with APIs [3].

This paper is organized as follows. In the continuation of the introductory part, the terms REST API and integrated web system are explained. Methods of data exchange and authentication within integrated web systems are also listed. In Sect. 2, two educational institutions are briefly presented and the importance of introducing safety education into the everyday education of engineers is described. Section 3 discusses the security risks of web applications and integrated web systems. Also, the laboratory exercises and topics are described. In Sect. 4, tools for testing are presented, as well as the way of performing laboratory exercises and assigning tasks. Several tasks from the laboratory exercises are described in detail.

1.1 REST API

Communication between different operating systems, platforms, and programming languages is mostly carried out using REST API endpoints. In integrated web systems, the security of API endpoints is of great importance. A REST API (Representational State Transfer) is an architectural style for distributed hypermedia systems. Throughout the developmental stages, API developers have the flexibility to implement REST in various manners. Like other architectural styles, REST adheres to its own set of guiding principles and constraints. To be considered RESTful, a service interface must adhere to and satisfy these principles [4].

The six guiding principles of REST are:

- Uniform Interface: A consistent and uniform way of interacting with resources (data or services) is maintained. This typically involves using standard HTTP methods (GET, POST, PUT, DELETE) to perform operations on resources.
- Client–Server Architecture: The client and server are separate entities that communicate over a network. The client is responsible for the user interface and user experience, while the server handles the application's logic, storage, and processing.
- Statelessness: Each request from a client to a server contains all the information needed to understand and process the request. The server does not store any state about the client between requests.
- Cacheable: Refers to a guideline that requires a response from a server to explicitly or implicitly indicate whether it can be cached or not. When a response is eligible

for caching, the client application is authorized to reuse the response data for equivalent requests within a specified period.
- Layered System: The layered system constraint in REST refers to the architectural principle of organizing a system into multiple layers, where each layer has a specific set of responsibilities. Each layer in the system only interacts directly with the adjacent layers, enhancing modularity and simplifying the overall system architecture.
- Code on Deman: REST permits the expansion of client functionality by enabling the download and execution of code in the form of applets or scripts.

1.2 Integrated Web System

Full-stack web development project or in other words integrated web system, commonly involves web system containing web applications, REST APIs, database servers, and mobile applications. Here's a breakdown of each element:

- Web Application: This serves as the user interface accessible through a web browser. Typically constructed using HTML, CSS, and JavaScript for front-end development. Frameworks like React, Angular, or Vue.js are frequently employed for creating interactive user interfaces. For back-end development, languages such as PHP are employed. Frameworks like Laravel or Symphony helps the development of back-end part.
- REST API: Responsible for data exchange.
- Database Server: Responsible for storing and managing data for the application. Common database types in web systems include relational databases like MySQL, PostgreSQL, or Oracle, and NoSQL databases like MongoDB.
- Mobile Application: A distinct application tailored for mobile devices like smartphones and tablets. Mobile apps can be developed natively using languages such as Swift (for iOS) or Kotlin/Java (for Android), or cross-platform using frameworks like React Native or Flutter. The mobile app often communicates with the REST API to retrieve and update data.

In integrated web systems, data exchange most commonly occurs in JSON format. JSON, or JavaScript Object Notation, is a lightweight data interchange format that is easy for humans to read and write, and easy for machines to parse and generate. JSON is based on key-value pairs and supports various data types.

Web systems grant different access levels to user groups, requiring authentication through specific information like a username/password. Upon access, systems often issue short-lived tokens for subsequent requests, establishing a session between the user and the system. Referred to as an Access Token, this token can take various forms, and its structure is determined by the system's requirements for verifying a user during the session period. Access Tokens are commonly transmitted in the header of an HTTP request.

Bearer authentication (also called token authentication) is an HTTP authentication scheme that involves security tokens called bearer tokens. The name "Bearer authentication" can be understood as "give access to the bearer of this token."

The bearer token is a cryptic string, usually generated by the server in response to a login request. The client must send this token in the Authorization header when making requests to protected resources.

2 Education and Security

Established in 1960, Subotica Tech—College of Applied Sciences stands as a revered state-run institution of higher education. With a history spanning over six decades, Subotica Tech has played a vital role in educating numerous engineers, who have subsequently contributed significantly to strengthening engineering skill. These individuals have been instrumental in shaping and revitalizing local and regional industries [5]. At Subotica Tech, emphasis is given to practical knowledge.

As of January 1st, 2010, Obuda University was established as the legal successor of Budapest Tech; however, Banki Donat Faculty of Mechanical and Safety Engineering can take pride in a history of more than 130 years. The Public Secondary Industrial School of Budapest, first legal predecessor of university, was established in 1879. After several reorganizations, it evolved into the Banki Donat Technical College of Machine Industry, where engineering specialists have been trained and welcomed in the economy since 1969. Starting from September 2023, the Cybersecurity Engineering undergraduate program has been launched at Banki Donat Faculty. This program is specifically designed to be practical, emphasizing a deep understanding of new technologies alongside theoretical classes, and providing instruction from partner institutions and industrial partners.

Students, and after graduating software engineers, need to stay alert about security concerns within an integrated web system and all its components, demonstrating the ability to identify the optimal methods or techniques to enhance the systems and components security. During their education, students at Subotica Tech can participate in software hackathons and attend conferences and workshops related to information security. Significant emphasis is placed on various aspects of security within professional courses of study. In some of these courses, protection methods and techniques are implemented through laboratory exercises, utilizing diverse approaches in testing information systems, web applications, mobile applications and other kinds of software and networks.

It is very important to give students opportunities to test their knowledge giving them tasks on laboratory exercises and homework that are related to information security.

3 Web Applications Security Risks

During their studies, students encounter security risks of web applications, as individual system or as a part of integrated web system, through various specialized subjects and curricula. Through theory and practical examples, they are introduced to the document standard defined by the OWASP foundation. All 10 key security risks are addressed, highlighting them through real-life examples. In laboratory exercises and as part of assignments, students are tasked with applying their acquired knowledge to avoid the mentioned risks.

3.1 Laboratory Exercises

Students at Subotica Tech learn theoretically and practically about following terms:

- Filtering input data from users
- Avoiding the acceptance of data that is correct in format but not in value.
- Secure file upload
- Secure user management system (register, login…)
- Secure usage of sessions and cookies
- Protection of web servers and web directories
- Cryptography and crypto algorithms
- Protection of database, SQL injection
- Protection of mobile application
- Call of restricted functions in the usage of dynamic function calls
- Creation of secure API endpoints
- Secure usage of API endpoints, testing API calls with different tools
- Token authentication in communication and data exchange.
- Avoid usage of legacy approaches and classes and methods
- Restriction of access to sensitive components of web systems
- Analysis of EXIF data in media files.

Each student has access to their virtual server and Virtualmin. Using the control panel, they can configure the web server, secure directories, create user accounts, and, importantly, have insight into error log files. By analysing the log files, they can obtain information about potential issues in their projects. It is assumed that in the development environment, error messages are displayed within the web project. However, when the project is deployed to a production server, error and warning messages must be hidden. This information is recorded in log file. Any display of error and/or warning messages can represent information that a potential attacker could exploit.

Using a web server during project work and lab exercises helps students prepare for real-life situations.

4 Tasks During Laboratory Exercises

Within the courses covering web programming and web security topics, tasks in laboratory exercises are defined to encourage students to explore security risks of web systems in various ways. These tasks also motivate them to pay more attention to security testing in their projects. The tests are diverse and depend on the type of task assigned to them. Tasks can be categorized by difficulty into several levels: beginner, intermediate, and advanced. The difficulty of tasks increases throughout the semester, adjusting to students' progress. It often happens that the simplest tasks have the most security vulnerabilities because many students focus on more complex requirements. It's also worth noting that students familiar with using certain back-end frameworks may overlook security measures in simple tasks, as these frameworks often have built-in solutions for many requirements. While helpful for more complex projects, the use of back-end frameworks becomes irrelevant for smaller projects and tasks.

In these tasks, the Web Developer Tools within the web browser are utilized. All modern web browsers come equipped with this set of tools, with Network, Console, and Storage being the most used sections. Network is used for monitoring HTTP requests and responses. By analysing this data, information about HTTP headers, request types, as well as parameters passed in the request or received in the response can be obtained. The Console section is useful for debugging JavaScript code. It sometimes happens that a programmer unintentionally outputs certain information to the console, forgetting to remove those outputs when deploying the project to a live server. Within the Storage section, there is the possibility to view cookies, and data in local or session storage can also be examined (see Fig. 1). Some tasks specifically focus on analysing data from these subparts.

For testing the security and vulnerabilities of web forms and API calls, students use the Postman tool in exercises. Postman is an API platform designed for the construction and utilization of APIs. Each step of the API lifecycle is simplified, and collaboration is streamlined by Postman, enabling the creation of better APIs at a quicker pace. Once API calls pass security checks, students use Swagger to document the API calls. When documented in an OpenAPI file, the Swagger open-source tooling enables direct interaction with the API using the Swagger UI (see Fig. 2). This

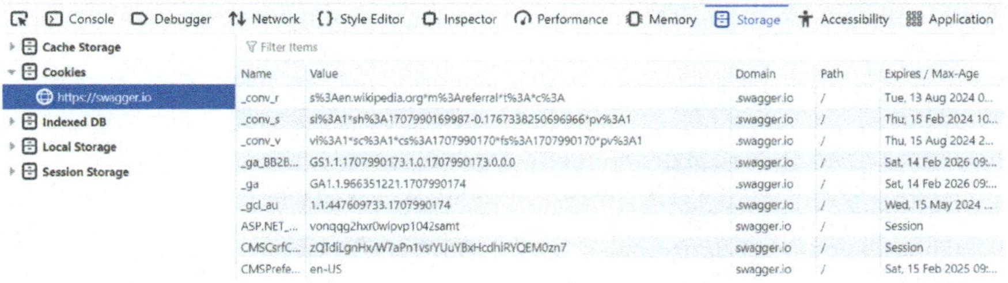

Fig. 1 Storage tab of web developer tools

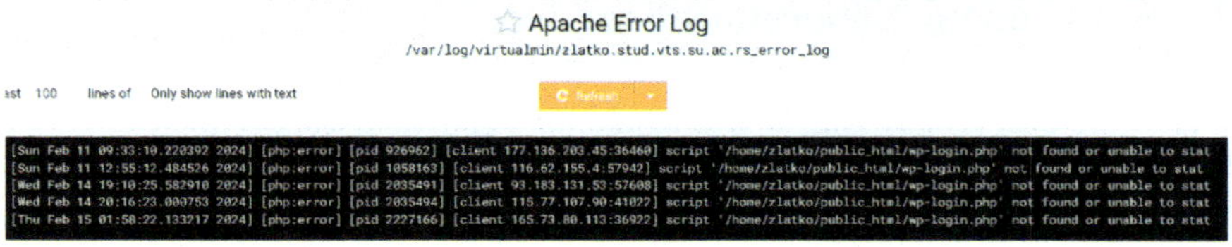

Fig. 2 Apache error log

initiative facilitates direct connections to live APIs through an interactive, HTML-based user interface. Users of the interface can make requests directly from the UI and explore the available options [6].

In addition to tasks where they work independently, implementing techniques and methods contributing to the development of secure software and/or systems, students also collaborate in teams. Team assignments represent a particular kind of challenge. Teams are divided into roles, with some defending and others attacking.

4.1 Analysing the HTML Code of Web Form, Login Page

This task is beginner level task. Students are required to review and analyse the source code of an HTML page located on the server. The web page includes a login form. By analysing the code, they can observe that the form contains three hidden fields that are not visible on the web page but are accessible in the code. Hidden fields are used within forms to include parameters necessary for the page's functionality when submitting form data. For example, on multilingual sites, they might contain the language version's name that the user sees. Hidden fields can also contain data with values of primary keys from the database.

Based on the code analysis, students were supposed to attempt to log in to the page. All attempts were stored in the database. Through analysis, students could notice that hidden fields contained a random string of characters. By copying the values from these fields and entering them into the input fields on the form, students attempted to log in. If unsuccessful on the first attempt, they tried reversing the values to log in. Interestingly, different values were recorded in the database compared to those in the hidden fields. The most common entry was the word "admin".

```
<form method="post" name="login" action="login.php">
<label for="username">Username:</label>
<input type="text" name="username" id="username"><br><br>
<label for="password">Password:</label>
<input type="password" name="password" id="password"><br><br>
<input type="hidden" name="l" value="en">
<input type="hidden" name="sgFvkwpL714rt5" value="AaA#21!3fEtV8j34">
<input         type="hidden"         name="s4rtdsds5kNmJJ043"
value="kwewe345zXm#!">
<input type="submit" name="sb" id="sb" value="login">
```

```
<input type="reset" name="rb" id="rb" value="cancel">
</form>
```

4.2 Analysing the HTML Code of Web Form with File Uploading Function

This task is similarto the previous one. Only difference is that form in this task contains file uploading function.

```
<input type="hidden" name="MAX_FILE_SIZE" value="102400">
```

Analysing the code of this page, it can be concluded that a hidden field named MAX_FILE_SIZE with a value of 102,400 is present, and only JPG images are expected in the input field for file selection. The value of MAX_FILE_SIZE is in bytes and represents the maximum file size that can be uploaded to the server using this form. This can be easily bypassed. The attacking team was tasked with attempting to avoid using this form and creating their form to try sending images larger than 100 kB and attempting to upload a different file type.

The defending team is required to protect against the input of unwanted data and files by incorporating appropriate security principles in its server-side code and on server. Also, they need to prevent getting data from different origins, to check the size and type of uploaded image. Additionally, they need to try to get information from EXIF data of image file. Exif stands for Exchangeable Image File Format. It is a standard for storing metadata in image files, including details such as camera settings, date and time, and geolocation information. This data should be recorded in database.

At the end of the class, an analysis of the success of both teams is conducted. Database records and directory contents on the server are reviewed.

4.3 Login Form, Cookies, SQL Injection and Token Usage

This task is more complex and requires the application of multiple security techniques and methods. The attacking team has been given a link to a web form located on the server. This team has information that there are users "admin" and "user1" in the database, and their data is stored in the "users" table. They also have the valid password for the user "user2." Within this task, students must complete several subtasks. Using SQL injection attacks, they need to attempt to log in to the page using the usernames "admin" and "user1" and employing an SQL injection attack instead of a password. For example:

```
SELECT * FROM users WHERE username = 'admin' OR '1'='1' -- ' AND
password = " OR '1'='1' -- '
```

In this example, the injected code: *admin' OR '1'='1'* -- is used to make the condition always true, allowing the user to log in without knowing the actual password. The double hyphen—signifies a comment in SQL, causing the rest of the original query to be ignored.

By applying SQL injection attacks, they should attempt to delete data from the "users" table.

The second task is a bit longer. It requires, after successfully logging into the page with user data for "user2," to use the Network option within the Web Developer Tools to analyse HTTP headers and data obtained for cookies and bearer tokens.

When students have extracted information from HTTP headers, as well as data from cookies and tokens, they need to use the Postman tool to attempt to send a request to the "getusers.php" page located on the web server. In the Postman tool, they should configure all the values they managed to find and retrieve for HTTP headers, cookies, and bearer tokens. This entire process needs to be repeated using the ReqBin tool. ReqBin is a REST and SOAP Online API Testing Tool that simplifies API call testing and server response validation. This tool also allows generating source code for creating requests with entered data in multiple programming languages.

The defending team needs to protect the web form from SQL injection attacks and ensure that the page receiving data checks whether the data is sent from the same origin or server. Every login attempt (successful and unsuccessful) should be stored in the database. All client-received requests should undertake validation on the server side regarding the validity of the session cookie and bearer token being passed. The token value should change periodically.

At the end of the task, an analysis is conducted together with the professor. The data in the database is reviewed, and potential vulnerabilities are discussed.

4.4 Login, Detection of Access Data, Restricted Access

The knowledge and experience gained by the students in previous laboratory exercises helped them to successfully complete this task. The attacking team has been provided with valid data for 5 user accounts existing in the database. The task of this team is to attempt to log in to the system using a VPN. Testing should be conducted across multiple web browsers and mobile devices. Within web browsers, they can use the User-Agent Switcher extension to modify the data sent in HTTP headers. In the Opera web browser, they should enable the free VPN and log into the system by changing the region. Immediately after logging in, they should log out and attempt to log in again as quickly as possible. They need to repeat this process 10 times in succession as fast as they can.

The second part of their task involves using the Postman tool to send requests by attempting to modify values for specific HTTP headers. They are required to log every message received from the login system.

The team of students tasked with defending needs to use the MobileDetect class and other appropriate techniques to detect the user agent sending the request (device type, user agent string, operating system, user's IP address, country of access, access time, and whether VPN or Thor was used in the request). External API services are used to determine VPN and/or Thor and the country of the IP address. The server-side code should record all data obtained from each request. If an IP address belongs to a VPN or Thor, access to the site is prohibited. It is also necessary to check if more than 5 requests come from one IP address within 60 s. In such a case, access to the site should be restricted as well.

At the end of the laboratory exercises, both teams present everything they have done and display messages as well as data from the database, followed by a joint analysis of the work done.

4.5 Generating Test Data, Creation, and Testing API Endpoints, Analyse of Error Logs

As one of the most crucial components of integrated systems, the API call stands out. Creating API calls requires programmers to implement security measures to ensure the use of API endpoints is secure for both parties—those using it and the web server and data on which the API endpoint resides.

Using the principles of REST architecture, a team of students needs to create valid API endpoints for manipulating data from the "movies" table. The creation of data in the "movies" table should involve using the Faker class. Each API call should be invoked in the format */api/movies/optional parameters*. It is necessary to create 4 API endpoints that satisfy basic CRUD operations. It is crucial to use the appropriate HTTP methods (POST, GET, PATCH, and DELETE) to process requests in the API call. Using the correct HTTP methods and corresponding status codes in responses is essential. All calls should be protected with some form of token, and the data format should be JSON. Students must configure*.htaccess* file in proper way for url rewriting.

The website should have a login system. After a successful login, web page shows the Bearer Token to the logged user. The attacking team needs to use the obtained Bearer Token and, using API testing tools, test the security of all endpoints. Any observed faults should be documented. Information about the API call format has been provided to the attacking-testing team. The testing team should attempt to execute API calls in a different format. If there are any errors during API calls, all such situations will be logged in the Apache Error Log file on the server (see Fig. 2). This file is accessible to students through Virtualmin. By studying the data in the log file, information about potential attacks and deficiencies in the source code can be obtained.

5 Conclusions

Most web-based information systems represent some form of integrated web system. When developing a complex integrated web system, a strong emphasis should be placed on the security of each component, especially those parts through which data entry, reading, and exchange occur.

Incorporating security subjects into the everyday education of software engineers is crucial. The tasks in laboratory exercises, where students can test the security of these components, contribute to increasing practical knowledge in information security. Some tasks are done individually, while others are performed in teams. At the end of the tasks, a discussion about requirements and results is crucial for a better understanding of potential flaws. Events like cyber challenges or hackathons provide an excellent opportunity to assess students' knowledge and gain additional experience in the realm of web security.

References

1. Sun R, Wang Q, Guo L (2022) Research towards key issues of API security. In: Lu W, Zhang Y, Wen W, Yan H, Li C (eds) Cyber security. CNCERT 2021. Communications in computer and information science, vol 1506. Springer, Singapore. https://doi.org/10.1007/978-981-16-9229-1_11
2. Wang AJA (2004) Security testing in software engineering courses. In: 34th annual frontiers in education. FIE 2004, Savannah, GA, USA, pp F1C–13. https://doi.org/10.1109/FIE.2004.1408561
3. OWASP Top 10 API Security Risks—2023. https://owasp.org/API-Security/editions/2023/en/0x11-t10/. Last accessed 2024/02/10
4. https://restfulapi.net/. Last accessed 2024/02/10
5. Čović Z (2022) Hackathon based learning in education of software engineers. In: Kovács TA, Nyikes Z, Fürstner I (eds) Security-related advanced technologies in critical infrastructure protection. NATO science for peace and security series C: environmental security. Springer, Dordrecht. https://doi.org/10.1007/978-94-024-2174-3_25
6. https://swagger.io/resources/articles/documenting-apis-with-swagger/. Last accessed 2024/02/10

Impact of Cybersecurity on Old-Type Steel Manufacturing Industrial Processes

Erzsébet Cserta, Krisztián Wizner, István E. Háber, György Fegyverneki, and László Horváth

Abstract The manufacturing industry has been one of the most targeted sectors of cyber criminals worldwide for several years. If a heavy industry company does not want to become a victim, it must be able to protect its sensitive data and industrial IT systems. Sacrifices must be made for different levels of IT protection, firewalls, log collection and analysis systems, endpoint protection, but physical security has to be increased as well. Most of these are now available as services, so a customized solution can be easily compiled according to the level of security and financial possibilities to be achieved. To find the best defence and ensure the right combination of people, process and technology, the preparation of a preliminary risk analysis is essential. For the effective realization, it is crucial to develop implementation plans and decision support tools that help owners, operators, and the vendor community asses when and how to apply cyber informed engineering in current infrastructure and technology offerings. In this paper we give an insight into the situation of an old-type steel mill company where the IT system is insufficient, or outdated, or the IT solutions implemented in technological equipment are not necessarily compatible with each other. We highlight the impact of such "semi-automated" character on cybersecurity solutions and production efficiency.

Keywords Information security · Heavy industry · Cybersecurity risks

E. Cserta (✉)
HUMDA, Budapest, Hungary
e-mail: erzsebet.cserta@humda.com

K. Wizner
University of Dunaújváros, Budapest, Hungary

I. E. Háber
Renergy Consulting Llc., Budapest, Hungary

Senergy Research and Analytics Nonprofit Llc., Budapest, Hungary

G. Fegyverneki
University of Miskolc, Budapest, Hungary

L. Horváth
HDF CIOC, Budapest, Hungary

T. A. Kovács et al. (eds.), *The Impact of the Energy Dependency on Critical Infrastructure Protection*, Advanced Sciences and Technologies for Security Applications, https://doi.org/10.1007/978-3-031-78544-3_20

1 Introduction

With the "digitalization" of human societies, as much of the technical control of our lives shifts from real, physical to cyberspace, the importance of cybercrime is also increasing exponentially. In the face of this heightened threat, it is as important to make critical infrastructure (that ensures the secure functioning of our societies) more resilient than it is to any other "type" of crime. In this white paper, we focus on the cybersecurity of a steel production plant among critical infrastructures. We chose this area of critical infrastructure because this industry field has the full spectrum of IT development. The reason for this is that this industry itself is one of the oldest technical sciences that can be found continuously throughout human history. Therefore, it did not come into existence or develop with the advent of information technology. In this study, we examine the emerging technical-IT challenges from a practical point of view, discuss the possible solutions and make suggestions.

To illustrate the tasks of cyber defense in heavy industry, we imagine and analyze an extreme situation with the help of practical examples that may arise in the case of an "old-fashion" steelmaking plant. The term "old-fashioned" refers to that era of the construction of heavy machinery plants when the IT procurements' costs were still not integrated into the necessary investment costs or only to a limited extent due to the low degree of automation of the technology at that time. This will help us shed light on the specific problems of such extreme cases. Issues related to the development of automation and thus cybersecurity will be discussed. The basic situation is taken that automation is only partially solved, and at critical technological points it is common that instead of automation, the process is handled and controlled by humans. With this, we want to illustrate the situation when the IT system is insufficient, or obsolete, or the IT solutions implemented in the technological equipment are not necessarily compatible with each other. The impact of such "semi-automated" character of the heavy industrial systems on cybersecurity and production efficiency is highlighted.

2 Effect of Cyberattacks on Steel Companies

The manufacturing industry has gone through a comprehensive transformation toward digitalization in the recent years. While these changes benefit the industry with higher productivity and performance, a big downside is the cyber security challenges arising along with the digitalized industrial processes [1]. By often overlapping and being tied to other industry sectors, the industry becomes more vulnerable to cyberattacks [1, 2]

According to the analysis of Statista [3], the global rate of cyberattacks in the manufacturing industries compared to all in 2022 was 24.8%. A considerable amount of these attacks is linked to human error and the lack of risk awareness. Based on estimations, 95% of all cyber incidents is human enabled [4]. Employees are the biggest risk when it comes to cyberattacks, and cybersecurity awareness training is proven to help reduce both system vulnerabilities and attack opportunities. As an appropriate example, phishing emails were likely to be the cause of the cyberattack that BlueScope Steel Company was subjected to, forcing production systems to be halted company-wide. For this reason, a part of the process had reverted to manual operations where possible, while some processes such as steel despatches were preserving the normal operation. Steelmaking was continuing, but the hot strip mill was temporarily disrupted at Port Kembla (Australia). However, BlueScope has not confirmed whether the attack was caused by employees opening phishing emails [5, 6]. Indeed, but evidently, vulnerability becomes much greater when working from home, taking computers outside the work environment, and connecting via VPNs and home routers [5]. Criminals often attack some remote access software to take over computers. As a basic defense, the implementation of multifactor authentication is recommended, but having a mature update process, and backing up the data are also essential [5].

IT security primarily focuses on protecting data assets stored on IT devices and, to this end, protects the security of IT systems and their components [7]. Even with the increasing rate of cyberattacks, it is rare in the manufacturing industry for hackers who operate in the digital world, to cause damage in the physical world [8]. The cyberattack on Iran's major steel maker companies in 27 June 2022 is a counter example. One of them, the Mobarakeh Steel Company [9] was forced to halt production after being hit by a cyberattack. As a direct physical effect, the heating plant started spewing molten steel and fire.

Since, high temperature machinery is operated, it could risk severe safety incidents for people if the power is lost in an uncontrolled manner. The safety must be always the first priority. Secondly, it's the concern for the environment and ensuring that no uncontrolled emissions (due to sudden machine stoppages) occur out to the air, land or water.

The question arises what kind of IT protection a similar industrial facility should have to keep its employees, equipment and business circle more or less safe. With this aid, it is always necessary to consider (1) the extent of the damage that may be caused and (2) the likelihood of their occurrence, and knowing these parameters, to decide (3) what is an appropriate cyber defense to develop. To find the best defence and ensure the right combination of people, process and technology, the preparation of a preliminary risk analysis is essential. Only if the potential risks are known can an IT security system be developed that is satisfactory for the data subject. The aim is to create closed, complete, continuous, and risk-proportionate cyber protection from the point of view of the confidentiality, integrity and availability of the IT system and the data processed in it, as well as the integrity and availability of the system elements [10].

3 Cyber Defence Aspects in the Heavy Industry

As mentioned above, the manufacturing industry has been one of the most targeted sectors of cyber criminals worldwide for several years [2]. According to current cyber defence solution trends, it would be useful to have IT systems for critical infrastructure and applications where the digital attack paths are removed or disrupted as fully as possible already by design. This is what cyber informed engineering method emphasizes (CIE) [12, 13]. With this aid, a clear focus on the risk management framework is essential to identify the most critical functions to fulfilling the organization's mission and the key points within a critical system vulnerable to a cyberattack. After that the potential consequences of a cyberattack against these functions are determined and then proven engineering strategies are to be applied to isolate and protect industry's most critical assets [11, 13]. Those operations are to be selected that must not fail and can associate attack scenarios that could bring operation processes down [12]. That would be ideal.

But, since we cannot predict where the cyberattack comes from and how the hacker thinks, the preliminary cyber planning is not satisfactory for a reliable cyber defence. Despite prior cyber-secure planning, those cyber defence technologies must be definitely incorporated into the heavy industrial IT systems and continuously developed that are suitable for averting or mitigating the consequences of a potentially successful cyberattack or restoring a normal workflow.

3 aspects of the protection are to distinguish:

Logical protection that includes all software and hardware procedures and tools that can be built into the IT system to increase its resistance.

Physical protection, whereby physical access to IT systems is restricted or completely prevented by means of locks, walls, alarm systems, manned guarding, etc.

Administrative protection refers to the introduction of procedures and rules to create and clarify the operating environment and boundaries within which stakeholders can operate. What can they do and what they can't for the security of the IT system and the data stored on it. This includes informing the affected workforce, regular safety awareness training [13–15] and back-testing, which we have already indicated above of paramount importance [16].

Companies must have the so-called Business Continuity Plan (BCP), among others, in case of an emergency. There must be described in detail how the business can continue to operate until the IT infrastructure is operational again. Relatedly, the Disaster Recovery Plan (DRP) is also an indispensable document. It discusses in detail how to recover from possible damage that was suffered during a cyber security incident, indicating tasks, deadlines, and responsible persons. Of course, during a recovery, it is very important to have backups of the system from which lost or compromised data content can be restored. It can significantly speed up and facilitate the recovery process, if a geo-redundant backup site also exists, which can be in the cloud or even in our own infrastructure located in a different location.

These documents are also listed in the statutory regulations regarding Hungary. Act No. 50/2013 [17] and BM Decree No. 41/2015 [18] regulate the mandatory operating documentation and action plan for electronic information systems (EIR) classified in different security classes. The EU is also working on multiple fronts to promote cyber resilience, fight cybercrime and enhance cyber diplomacy and defence [19].

4 "Steel Blue" Cyber Defence

The reason for the outstanding physical and material damage caused by cyberattacks is that heavy industry is a particularly dangerous area. This is especially true for steelmaking, due to the high operating temperature of equipment, toxic gases generated during technological processes, large masses, dust, etc. The price of equipment designed for these "extreme" parameters is also extreme, because in addition to material costs, it requires serious engineering, material science and IT expertise to design, manufacture and operate devices that can meet the extreme demands of the above-mentioned extreme parameters. Moreover, the necessary development of complex IT systems results in more expensive control process of the entire technology. Connectingly, coordinating individual equipment, which often has a unique IT system, result in special difficulties.

4.1 IT Challenges in Steelmaking

One of the cruces of the IT development of steel mills is that the equipment of very high value is usually not replaced, but only refurbished. In the case of a new device, Remote Terminal Units (RTUs) connected as interfaces to the controlled system might already be installed, but in the case of the "old" one, this has to be created somehow that can often be solved by local DIY. The situation is the same with the Programmable Logic Controller (PLC) of the machines.

Steel mills that were built before the year 2000 typically face the following main technical challenges:

Diversity of IT systems

(a) Due to high-value equipment, the investment is greenfield only at the start, but after that modernization typically takes place continuously, component by component. Since each equipment has its own IT system, integrating it into the plant's existing system is a major challenge (patches on each other's backs).
(b) This is especially true if the supplier of the new (modern) equipment does not allow access to its IT system, and the technological development can only be accessed to the existing IT system on special interfaces.

Lack of IT updates

(a) The security updates of the operating systems that are running on the equipment's IT devices (which are typically still simple PCs) can often be "dangerous" to the operation of the specialized software running on them. Sometimes a system update brings changes to the operating system that prevent the software that controls the equipment running on a particular computer from working properly. That is why many times the operating systems that belong to the equipment are not updated for some time.
(b) A possible solution can be, if the given PLC and the associated IT system were updated at the same time, but this in itself would entail a significant financial burden already. Furthermore, the updated system would have to be reintegrated into the company's Industrial Control System (ICS) as well. This is often neglected.
(c) There are cases where a production line has been operated by outdated IT equipment for 20–30 years (e.g. 16-bit operating system), because replacing them would be a major financial project. Savings are often necessary, and improvements are dragged out as long as possible. In such cases, plant professionals try to keep the systems together (e.g. replacing the hardware of 16-bit operating systems with a virtualized environment on a modern server), but these are only firefighting, not "optimal" solutions.

Distrust of new IT technologies

(a) The profession is constantly getting old, and old technicians often do not trust the new technology and like to take matters into their own hands.
(b) Furthermore, technicians who do manage IT devices are not IT professionals either, so they can be more easily led by a viral email or with other social engineering cyberattack methods.

Companies with smaller budgets give more space to manual process control to avoid the implementation of costly IT developments, and rely on the automatic regulation and the data collected by the IT system only at those points of the technology where it is vital already (e.g. where human no longer has access or would be harmful to human health, etc.)

4.2 Possible Physical Effects of a Cyberattack

Despite the fact that such plants are only partially automated systems and, in many places, the technological process is still under manual control, a sophisticated cyberattack can cause serious damage to the system, which can reduce productivity, lead to loss of production quality or result in scrap.

In the event of a successful cyberattack on PLCs and RTUs directly, the data of individual sensors can be compromised, not conveying the real value to ICS. Influencing the following conditions by manipulating sensors can result in harmful followers and intense losses in the operation of a steel mill. Some examples of physical harm and related consequences are given below.

Cooling outage: For the long-term operation of high-temperature equipment, it is essential to cool professionally the components exposed to high temperatures, which parameters are controlled by sensors. If the flow continuity of the refrigerant is disrupted as a result of a cyberattack, the components that need to be cooled can burn out, melt, clog and, in some cases, puncture. To estimate the magnitude of the physical damage caused in this way, practical examples are given here:

(1) In the case of an oxygen spear at a converter, if the bronze head cooled by water burns out, it must be replaced. Replacing them is relatively easy to solve, but there are only 2–3 replacement spears. If you run out of replacement spears, the head have to be mounted already, which is low in stock because it does not burn out often. *Production influencing parameter:* Until the part is purchased and installed, the production process stops.
(2) A similar problem may occur with an electro-furnace when cooling the furnace body (+ bottom + lid, etc.). There is water in the panels (pipe snakes). If the cooling water overheats the water turns into steam that will find its way somewhere due to high pressure. In the worst case, a furnace puncture may occur. *Production influencing parameter:* The steel in the furnace is drained and production stops for the time of repair.
(3) The crystallizer wall used for continuous casting is made of copper. In the event of a cooling failure, the copper plates will be punctured, thereby destroyed. *Production influencing parameter:* Production stops on the line in question and the crystallizer needs to be replaced. (Several crystallizers are ready for installation, but installation and removal can only be done when there is no casting).

Electrical power failure: It may seem less sophisticated to attack the power supply center during a cyber-attack, but the physical damage that can be achieved in this way can be very significant. This is due to the fact that there are only few systems equipped with backup power sources or alternative solutions, so in the event of a power outage, machines shut down, emergency systems are triggered, and the system is cooled down for safety. Considering that heating is the most energy-intensive process in these high-temperature installations, and cooling the system due to necessity rather than professionalism can cause unwanted material properties in the steel production process, the resulting material and asset damage can be significant.

(1) An example of a similar situation is given in relation to hydraulic and pneumatic systems: If, due to hacking of the control automatics, the turning fork on the casting tower does not turn, but the hydraulics open the cauldron and no human has access to freeze the outlet, nor does the crane have access to the cauldron to take it to the rescuer, then the steel from the tower easily makes its way down to the cellar. *Production influencing parameter:* Equipment, parts of plants melt, production stops, material damage is significant.

Incorrect display

In operations where technological parameters are set in (semi-)manual mode based on the values that are shown on the displays, the protection of data sent to the displays is even more important in addition to the human factor. Unfortunately, in these cases, the routine of experienced professionals is often the best protection against a possible cyberattack, as they mostly rely on their own, well-proven experience, treating the values appearing on the displays only as control values. Below are some critical points in steel production where the practical professional experience that are provided by humans can override the incorrect values displayed on the displays. A successful attack on these points can lead to a significant increase in costs.

(1) Wrong or not corresponding amount of alloying is added: The first and most important rule of alloying is that inserting the alloy into the steel is easy, but it cannot be removed at all. There are very expensive alloys, an overdose of which leads to serious unnecessary costs (for example, Nb, V). A misalloyed defective dose can only be sold at a low price or used as waste at a later stage. *Production influencing parameter:* Scrap production significantly increases costs and losses.
(2) Adjusting the tilt angle of the converter/furnace: The experts are quite experienced and can see how much the furnace is tilted by eye, but an incorrect tilt at the wrong time can cause steel to spill. There is a display on the adjusting table of the furnace used for controlling that also shows the tilt angle. *Production influencing parameter:* The spouting liquid steel can burn equipment in its path, e.g. the engine of a trolley or flow into equipment and passages under the furnace.
(3) Incorrect casting speed: In this case, the line may break out and liquid steel may end up in the secondary cooling zone. *Production influencing parameter:* The machine must be disassembled, cleaned, damaged parts replaced, until then production stops.

Effective IT protection should be proportionate to the potential risks, meaning that the investment in cybersecurity should be proportionate to the damage caused by a cyber-attack. Based on just a few examples above, the damage that can be caused to heavy industry as a critical infrastructure is extremely high. To successfully prevent material damage and, in the worst case, endangering human lives as a result of attacks, it is fundamental to change the attitude of both experienced technical experts and company managers. This is especially true for the "old-fashioned" steelmaking plants described above. Due to the emphasis on human control, the possibility of cyberattacks and, accordingly, the protection mechanisms that are necessary to prevent them are less widely known among employees. To strengthen effective cyber defence requires to develope implementation plans and decision support tools that help owners, operators, and the employee community asses when and how to apply cyber informed engineering (CIE) and when to implement additional IT protection in current infrastructure and technology offerings [20].

5 Conclusion

In such an "old-fashioned" steel mill, disaster is encoded in the defenseless system. Protecting a hybrid system that carries many old skeletons requires regular geo-redundant backups and appropriate redundant hardware that can be restored from backup in case of incidents. For this, of course, BCP and DRP plans must also be prepared that will surely be needed sooner or later (but rather sooner, unfortunately) in case of such a system. At the most vulnerable point of the IT systems, to strengthen the human awareness and IT expertise is crucial. Employees need to be constantly trained and skilled through regular safety awareness courses, because any system is only as secure as its weakest link.

To achieve adequate cybersecurity, it is necessary to assess the extent of the threat to the data assets to be protected, the likelihood of an attack occurring, and the amount of loss of the company in the event of a successful attack. A system of protection is required to be proportional to the risk determined in this way and shall be established to ensure safe operation.

The study stresses the importance of cybersecurity awareness to protect steel production plants from attacks and promote their cyber resilience. Through practical examples, the impact of a successful cyberattack on a steel mill is illustrated drawing attention to the magnitude of the risk.

References

1. Statista. https://www.statista.com/topics/10742/manufacturing-industry-cyber-crime/. Last accessed 25 Feb 2024
2. IBM Report. https://newsroom.ibm.com/2022-02-23-IBM-Report-Manufacturing-Felt-Brunt-of-Cyberattacks-in-2021-as-Supply-Chain-Woes-Grew. Last accessed 25 Feb 2024
3. Statista. https://www.statista.com/statistics/1315805/cyber-attacks-top-industries-worldwide/. Last accessed 25 Feb 2024
4. Siemens. https://assets.new.siemens.com/siemens/assets/api/uuid:28b9caf1-cbbf-473c-aab9-55969463d087/di-pa-dcp-metals-cybersecurity-ebook.pdf. Last accessed 25 Feb 2024
5. iTnews. https://www.itnews.com.au/news/bluescope-confirms-a-cyber-incident-is-disrupting-its-operations-548163. Last accessed 25 Feb 2024
6. ABC News. https://www.abc.net.au/news/2020-05-15/bluescope-steel-cyber-attack-shut-down-kembla-ransomware/12251316. Last accessed 25 Feb 2024
7. Time. https://time.com/6080293/norsk-hydro-ransomware-attack/. Last accessed 25 Feb 2024
8. Microsoft. https://news.microsoft.com/source/features/digital-transformation/hackers-hit-norsk-hydro-ransomware-company-responded-transparency/. Last accessed 25 Feb 2024
9. AP News. https://apnews.com/article/technology-middle-east-iran-dubai-b0404963ae23e5008439a0b607952de1. Last accessed 25 Feb 2024
10. Muha L (2008) Az informatikai biztonság egy lehetséges rendszertana. Bolyai Szemle 17(4)
11. TIME. https://time.com/6080293/norsk-hydro-ransomware-attack. Last accessed 25 Feb 2024
12. Idaho National Laboratory. Consequence-driven cyber-informed engineering. https://inl.gov/cce/. Last accessed 25 Feb 2024
13. Microsoft Security. https://www.microsoft.com/en-us/security/business/cybersecurity-awareness. Last accessed 25 Feb 2024

14. Azzani IK (2024) Enhancing awareness of cyber crime: a crucial element in confronting the challenges of hybrid warfare In Indonesia. Defense Security Stud 5:1–9
15. Fadli ZM (2022) Cyber attack awareness and prevention in network security. IJ-ICT 11(2):105–115
16. Sarker IH (2024) Multi-aspect rule-based AI: methods, taxonomy, challenges and directions towards automation, intelligence and transparent cybersecurity modeling for critical infrastructures. Internet Things 25:101110
17. Nemzeti Jogszabálytár (National Legal Regulations) Homepage (2013) évi L. törvény, https://njt.hu/jogszabaly/2013-50-00-00. Last accessed 25 Feb 2024
18. Nemzeti Jogszabálytár (National Legal Regulations) Homepage. 41/2015. (VII. 15.) BM rendelet, https://njt.hu/jogszabaly/2015-41-20-0A. Last accessed 25 Feb 2024
19. europa.eu. https://www.consilium.europa.eu/hu/policies/cybersecurity/#infrastructure. Last accessed 25 Feb 2024
20. inl.gov. National CIE Strategy. https://inl.gov/content/uploads/2023/07/FINAL-DOE-National-CIE-Strategy-June-2022_0.pdf. Last accessed 25 Feb 2024

Critical Security Points in the IT Systems of Renewable Energy Communitys

I. E. Haber, K. Haber, E. Cserta, L. Németh, and L. Horváth

Abstract Energy communities will be key members of the energy network as accounting entities within few years. They need infrastructure, communication solutions and therefore protection against cyber threats. The recent events in the world have shown that energy infrastructure is always a first row point of attack and the weakest part will be threatened when coming to opposition. In this paper, the major aspects have been collected for the proper security solution of energy communities (ECs), assuming the most obvious current technical solutions.

Keywords Energy community · Smart grid · Iot · Cybersecurity · IT infrastructure

1 Introduction

The susceptibility of crucial infrastructure to cyber threats has been amplified by the escalation in interconnectivity through increased reliance on digital technologies of the critical infrastructure. Cyber-attacks contra critical infrastructure systems are the most common tools of provocation so in military as in public level since it generates uncertainty in the population. Energy systems belong to the critical infrastructure and their level of vulnerability is a key element of a country's overall defense level. Nazare et al. [1] Since, Industry 4.0 technologies are increasingly being used to support the interoperability of these sectors as well, their security is essential to

I. E. Haber (✉) · L. Németh
Renergy Consulting Llc, Pecs, Hungary
e-mail: istvan.haber@renergy.hu

K. Haber
Senergy Research and Analytics Nonprofit Llc, Pecs, Hungary

E. Cserta
HUMDA Zrt., Budapest, Hungary

L. Horváth
HDF CIOC, Budapest, Hungary

T. A. Kovács et al. (eds.), *The Impact of the Energy Dependency on Critical Infrastructure Protection*, Advanced Sciences and Technologies for Security Applications, https://doi.org/10.1007/978-3-031-78544-3_21

ensure the continued reliability and integrity of the energy supply. Bour et al. [2] Being a crucial point of every country, it is not rare that energy systems are the main target of cyber-attacks as can be seen in the evolving situation in the Near East [3–5], Iran [6, 7], Ukraine [8], Norway [9] and several other countries worldwide [10].

At the same time, the energy sector itself requires modernization. In an age of interdependent crises (e.g. economic, health, climate change, etc.), the energy crisis showed that, in addition to technical and structural modernization, a paradigm shift on the consumer and service provider side is also vital in the energy sector. Specially, it is necessary because a significant part of our societies lives in the misconception of cheap and almost inexhaustible energy reserves, which will not be sustainable in the near future, taking into account the constantly expanding energy needs.

The energy and the utilities sectors are a crucial part of the critical national infrastructure. Their businesses that are usually preceded by private companies can be prime targets for hostile cyberattacks. Despite the major threat that is why, the importance of the energy sector's protection is often noticed only when a major attack occurs. Cybersecurity is just a "to-do-to" (do) the agenda in many cases until something big goes wrong, by which time it is often too late "2do2" anything. As an example, the distant disruption of the electricity grid could have a number of effects, e.g. causing chaos and disorder, gaining political and military power [11].

In a smart grid system, renewable energy resources such as wind, solar, and power storage units are integrated into the grid system. The smart grid employs a widespread sensor network supported by a two-way communication system for constant monitoring of the grid status. The bidirectional communication network allows the exchange of measurement data and control signals between grid entities, improving the grid and user asset monitoring and management. Moreover, to process the collected data within the required time frames, the smart grid should be supported by sufficient computational resources. The control and monitoring are conducted in a more distributed way, as the volume of the collected data is enormous, and the sensors are dispersed across the entire grid [12, 13]. In their study, Bouramdane [14] show a comprehensive exploration of cyberattacks and cybersecurity in smart grids, focusing on critical components and applications. It examines various cyberattack types and their implications on smart grids, backed by real-world case studies and quantitative models. In another comprehensive analysis made by Ding et al. [15], a global review of cyberattacks of smart grids are given. It says that the probability of ransomware intrusion through vulnerabilities is far greater than that of emails and unsafe links for the local area network of the electric power department that stores a large amount of sensitive data. In addition to demanding a huge ransom from the power company, ransomware may also cause interruption of normal user services and even serious consequences of not being able to supply electricity.

Although most large companies have cybersecurity professionals, their work is very complex and difficult to monitor from the outside and they use to be seen as a black box in board meetings. Based on a questionnaire on the cybersecurity maturity of Moroccan companies that are operating in the energy sector, the most of these companies have a high level of cybersecurity maturity, but some factors affect their effective security measures, such as lack of skilled manpower, high cost of equipment and training expenses [11, 16].

There are many efforts worldwide to make industrial areas that belong to the critical infrastructure resilient to cyberattack. For better protecting, the critical infrastructure from cyber-attacks in the US, the Biden Administration is urging organizations to build protection into systems at the design stage, rather than relying solely on IT protection. This attitude illustrates the concepts of cyber-informed design [17].

In Europe, the European Energy Information Sharing and Analysis Centre (EE-ISAC) is a cybersecurity platform for the energy sector that has been adopted by several organizations. The platform facilitates the sharing of information on cyber security assessment results and countermeasures. The applicability of cyber security assessment methods is negatively affected by a number of factors, including the level of detail of documentation, the availability of supporting tools and the continuity of support, therefore to elevate the applicability and usefulness of the platform for energy sector's stakeholders is an urging area for further development [18].

Profound research in the Indian energy sector is aimed to find the best defence against targeted attacks from cyber-physical systems. A comprehensive scoring method (BWM-I and COBRA) was used to evaluate alternatives to cyber security platforms. In collaboration with 80 energy utility managers, AI-based and end-to-end coverage methods were found to be the best [19].

The pursuit to reach high security level in industry is trackable based on the above examples. In contrast, a Renewable Energy Community (REC) is typically a citizens' initiative, therefore less knowledge, less practice and lack of security professionals characterises these projects.

The aim of the creation of energy communities (ECs) becomes exactly one of the practical means of solving the multiple crisis, which improves the living conditions of the citizens, reduces the energy bill, helps to keep financial resources in place and, indirectly, can even create new jobs of higher quality. ECs are the solution for adaptation to the natural resources without significantly changing our comfort level. It is part of the aspiration towards circular economy (CE) with the distributed energy production and it has its pros and cons against classic energy networks. The CE has the potential to harness emerging digital technologies such as big data, AI, blockchain and IoT. These technologies, combined with a good business model, can solve many of the world's problems. The paths and opportunities for CE implementation, in which IoT and AI play a key role is a highly developing field. Several articles have been published on the topic, many of which also focus on the barriers to the transition to CE, such as lack of predictability, political implications, information vulnerabilities [20].

In this study, we showcase the tangible impacts of cybersecurity on the operation of a conceptual EC, emphasizing the significance of cybersecurity measures, incident response planning, and collaboration among utilities to safeguard critical infrastructure.

2 IT Infrastructure in Energy Communities

A REC is considered to be a network of consumers and prosumers, where the community generates, consumes, stores or sells electricity from renewable energy sources. The effective management of the REC shall be carried out by the member or members whose points of connection of the consumption sites are located in the same high/medium voltage transformer station area as the points of connection of the electricity storage and power plant owned by the REC. Physically, photovoltaic systems, small wind generators, storage devices and consumption devices are connected via small voltage electricity grid. The fundamental ambition is to reach independency from the grid, but not without it, and if one entity has excess energy (a prosumer), then shares it with the other within the community and maybe with other grid consumers (Fig. 1).

Since the level of renewable energy production is uncertain [22], dependent from season/weather, it is mandatory to regulate not just the members, but consumer equipment with automatic control system (or manually). Consequently, there is a strong need for communication infrastructure within the community, which can be solved by various technologies. From security side, the best solution would be to have a dedicated wired network that have only one point in contact with WAN.

For energy network communication, the Supervisory Control and Data Acquisition (SCADA) system architecture is one that is widely used. SCADA is a robust and easily operated network, even though it is a separated network as well, it has its own weak sides [15, 23]. It is directly designed for industrial control; therefore, its usage predicts that IoT networks are more vulnerable. Moreover, SCADA needs special and expensive hardware, for this reason it is not realistic to use it in most ECs where cost efficiency and easy deployment are the top priorities. Instead, the most

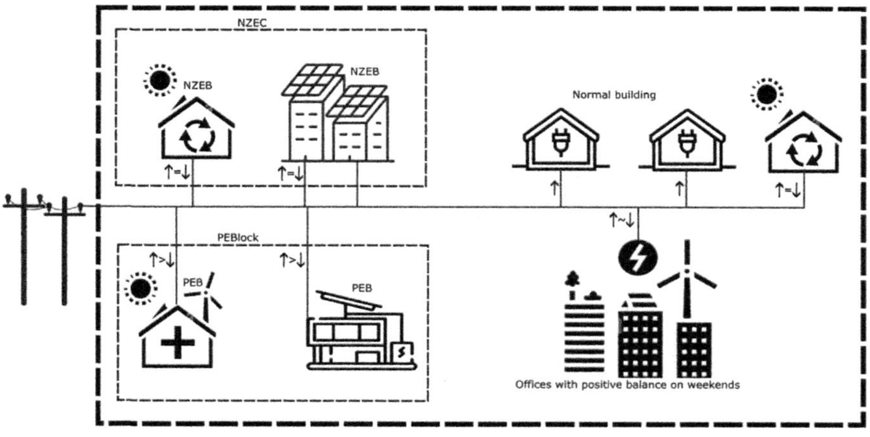

Fig. 1 A typical REC architecture (NZEB: Near Zero Energy Building, NZEC: Net Zero Energy Community, PEB: Positive Energy Building) [21]

plausible pathway is to use the already available network and inexpensive devices that include some key elements.

Key elements of an operational EC system are:

- Smart meters (IoT devices),
- IoT data management system,
- Predicting weather data-based trend maker (Prediction),
- EC software,
- EC mobile application,
- IT security, over them.

The main objective of the IT services of the EC is to provide members with information and a user-friendly environment, in addition to ensuring reliable operation. For our EC (as demonstrated in Fig. 2), the basis of the operation is identified with real-time production and consumption metering supported by smart devices with the necessary management and accounting systems in place.

As a completely new system is at stake considering both the concept and the implementation, the possibility of a consequence-driven, cyber-informed engineering design is given, obvious and recommended. The idea is to design the IT system with a 'hacker-like' approach and to develop the system until all known potential attack points have been eliminated. Consequence-driven and cyber-informed engineering (CCE and CIE) is based on securing critical infrastructure systems before their real implementations. The basic premise is that if a system is targeted to be broken, it will be. Considering this approach, a four-step process has been created that can be followed to secure the future critical infrastructure's systems. CCE looks at the whole operation of an organization, securing the essential processes and at the same time the technology itself. These frameworks also look at how a given attack affects the operation of the whole organization [24].

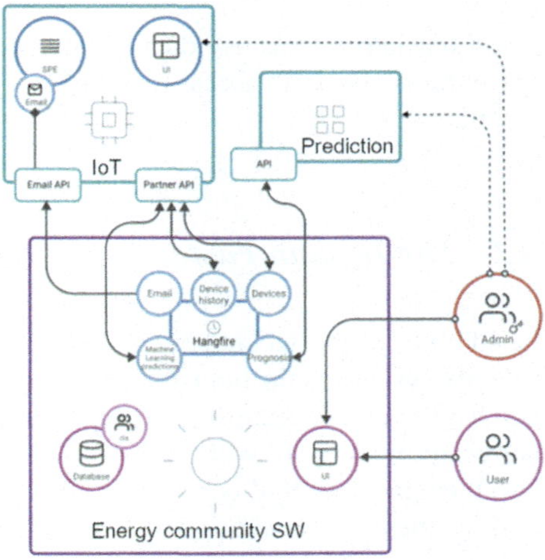

Fig. 2 An example for the IT infrastructure of an EC

3 Critical Security Points at the IT Systems of the ECs

IT security is one of the most important factors for the proper functioning of information systems supporting the EC since the vulnerability to cyber threats has been amplified by the increasing interconnectivity and digitalisation in these energy systems [25]. A well-functioning IT security is closed, comprehensive and continuous, i.e. it takes into account all relevant threats, covers all elements of the system and is implemented continuously despite changing circumstances and conditions over time. It is achieved through a combination of measures, policies, procedures and technologies. These include firewalls, anti-virus software, encryption, identification and authentication, regular application and operating system updates, monitoring and auditing, information security education and awareness. It incorporates the protection of IT systems, networks and the data they contain against attacks, phishing, malware, unauthorised access and other security threats.

3.1 Main Requirements of the ECs' IT Security

IT security aims to maintain the integrity, availability and confidentiality of the IT systems. But what does it mean in the context of the ECs?

Integrity: Data and IT systems can only be modified by those who are authorised to do so. IT security tools and measures are designed to prevent unauthorised access to data.

Availability: IT systems and the data stored in them must be made available to authorised users at the time and for the period required. IT security is designed to prevent attacks that could cause systems or networks to become inoperable, making it impossible to ensure the continuity of business operations.

Confidentiality: Sensitive data is only accessible to those persons or systems who are authorised to have access to it. IT security measures prevent unauthorised access to data.

3.2 The Human Factor

In addition to the technological aspects of cybersecurity, the weakest link of all security systems is the human factor; that is why social engineering is so successful and hard to defend against it. As humans tend to trust other humans more easily, no hardware or software will ever be sufficient. Overall, a successful strategy against social engineering will not only require the appropriate technical systems, but also stiff policies and standards and a continuous education of personnel awareness. As added value, the staff and members need to be informed and trained to avoid or at least minimize the human factor in cyber threats [26, 27].

3.3 Identifying the Critical Elements of the ECs' Support System

IoT system provides communication between the elements of the energy community support system (e.g. sensors and smart devices). This critical system supports real-time data collection and monitoring.

Physical security of devices: Devices must be located in the holes that are difficult for unauthorised persons to access and must be adequately protected to prevent theft, vandalism and physical tampering.

Authentication and access control: IoT devices and their associated systems must have appropriate authentication and access control mechanisms. Access to devices and data should be restricted to those who are authorised.

Data encryption: Data must be adequately encrypted during transmission and storage to prevent unauthorised access or data theft. The use of strong encryption algorithms and key management procedures is now unacceptable.

Updates and vulnerability management: Regular firmware updates for IoT devices are important to address security vulnerabilities and gaps. Devices should be able to receive and deploy updates in a centrally managed manner.

Communication security: This can be made possible, for example, by encrypted communication and the use of secure protocols. To avoid data loss, redundant network solutions or off-line local data storage can be used, e.g. a critical device can connect to the system in case of a network failure with a secondary mobile network solution or can store data in its own memory for an appropriate period of time and upload it back to the system later after the network failure has been resolved.

Processes and procedures: The security of IoT systems is also strengthened by appropriate processes and procedures. For example, debugging and logging can help track and learn from incidents.

Firmware and hardware security: The use of firmware or hardware components from unreliable sources should be avoided.

High availability: IoT system components are constantly handling large amounts of real-time data, so high availability should be ensured to avoid data loss. Redundancy, load balancing, restore points and data backup are some of the tools to achieve it.

Scalability: As an IoT system grows, it needs to receive and send more and more data, so it needs to be able to scale up with the requirements.

3.4 Controlling the ECs' Critical Elements

The EC subsystem, which controls each EC element allows not only to monitor and control the energy management system, but also to change its state based on the desired control strategy.

High availability is the most important cyber security aspect of control. A failure in the control of the system can lead to a physical failure of the energy system, so in addition to the traditional means of high availability (e.g. redundancy), a number of secondary solutions must be developed for device, network and software security as follows.

- Redundancy is built into the devices that provide the basic data essential for controllinhg.
- Protecting network connections and ensuring redundancy of network devices.
- Use of software development solutions that ensure high availability (e.g. microservice architecture, advanced fault management, conteneration)
- Control is implemented at multiple logical levels, e.g. control strategy is defined in the cloud environment, while execution and verification of instructions is performed locally using industrial solutions (e.g. PLC) that support fault management and protection of industrial devices.

Protection of personal and business data in the cloud. The use of cloud services has many business and IT benefits when operating an EC support system. However, in addition to traditional information security issues, new types of risks also arise. Data is out of the control of the organisation and there is no control over all layers of the cloud service. Cloud-native services face unique cybersecurity challenges due to their distributed infrastructure. They are susceptible to various threats like malware, DDoS attacks, and Man-in-the-Middle (MITM) attacks. Additionally, these services often process sensitive data that must be protected from unauthorized access. On top of that, the dynamic and scalable nature of cloud-native services makes it difficult to maintain consistent security, as deploying new instances and infrastructure introduces new vulnerabilities [28].

ECs with the right IT infrastructure and expertise can develop private cloud solutions or even hybrid cloud services. In this case, however, the increased professional demands and financial requirements of the operation must be taken into account. In a hybrid cloud solution, critical data are hosted on local servers, and geo-redundant operational and data storage solutions are to be developed. A typical example is the local storage and the archiving of system and data backups.

4 Conclusions

ECs are not always and not necessarily professional formations, therefore the first aim of them is the operability, even without protection against any cyber threats. Security always cost money, and these communities are designed even to save energy and therefore money for the members. By their very nature, the financial costs are determined to stay relatively small, that is why the economic opportunities are limited. Even in this case, the integrity, availability and confidentiality of the ECs' IT systems must be ensured through increasing the cyber awareness of the members by educating them through up-to-date information and trainings.

In the coming years, several technical (hardware + software) solutions will come out for ECs as off the shelf. These standardised solutions that may become available on the market are expected to comply with the basic and most essential principles that are collected in this paper.

References

1. Nazare T, Nardo L, Arias-Garcia J, Nepomuceno E (2023) Ensuring resilience in ocean energy power plants: a survey of cybersecurity measures. In: Proceedings of the European wave and tidal energy conference, vol 15, Sep 2023
2. Bour G, Bosco C, Ugarelli R, Jaatun MG (2023) Water-tight IoT—just add security. J Cybersecur Priv 3:76–94
3. US Condemns Iran. Issues sanctions for cyber-attacks on critical infra—infosecurity magazine (infosecurity-magazine.com). https://www.infosecurity-magazine.com/news/us-iran-sanctions-cyber-attacks/. Last accessed 02 Feb 2024
4. Israeli cyber frontlines: Inside the digital warfare of war with Hamas—The Jerusalem Post (jpost.com). https://www.jpost.com/israel-hamas-war/article-781682. Last accessed 02 Feb 2024
5. Israel launches heavy raids on southern Lebanon, while cyber-attack strickens Beirut airport—Egypt Independent. https://www.egyptindependent.com/israel-launches-heavy-raids-on-southern-lebanon-while-cyber-attack-strickens-beirut-airport/. Last accessed 02 Feb 2024
6. What to know about the cyber attack on Iran by Hackers. TIME. https://time.com/6548680/iran-hacker-gas-station-cyberattack-israel/. Last accessed 02 Feb 2024
7. Cyberattack forces Iran steel company to halt production. AP News. https://apnews.com/article/technology-middle-east-iran-dubai-b0404963ae23e5008439a0b607952de1
8. 2015 Ukraine power grid hack—Wikipedia
9. Microsoft. https://news.microsoft.com/source/features/digital-transformation/hackers-hit-norsk-hydro-ransomware-company-responded-transparency/. Last accessed 02 Feb 2024
10. Statista. https://www.statista.com/statistics/1375188/cyber-attacks-worldwide/. Last accessed 02 Feb 2024
11. Mukherjee S (2019) Implementing cybersecurity in the energy sector. Preprint. https://doi.org/10.6084/m9.figshare.9728051
12. Goudarzi A, Ghayoor F, Waseem M, Fahad S, Traore I (2022) Vulnerabilities and challenges in IoT-enabled smart grids. Retrieved from: https://encyclopedia.pub/entry/30768
13. Goudarzi A, Ghayoor F, Waseem M, Fahad S, Traore I (2022) A survey on IoT-enabled smart grids: emerging, applications, challenges, and outlook. Energies 15(19):6984
14. Bouramdane A-A (2023) Cyberattacks in smart grids: challenges and solving the multi-criteria decision-making for cybersecurity options, including ones that incorporate artificial intelligence, using an analytical hierarchy process. J Cybersecur Priv 3(4):662–705

15. Ding J, Qammar A, Zhang Z, Karim A, Ning H (2022) Cyber threats to smart grids: review, taxonomy, potential solutions, and future directions. Energies 15(18):6799
16. Raissouni K, Errabih Z (2023) Cybersecurity in the context of Moroccan energy compa-NIES. In: E3S web of conferences, vol 412, p 01038, ICIES'11 2023 (2023)
17. Virginia Wright AO (2023) Engineering cybersecurity into U.S. critical infrastructure. Cybersecurity and digital privacy. Harvard business review
18. TaniaWallis RL (2022) EE-ISAC—practical cybersecurity solution for the energy sector. Energies 15:2170
19. Rakesh Verma SK (2024) Evaluation and selection of a cybersecurity platform—case of the power sector in India. Decis Mak Appl Manag Eng 7(1):209–236
20. Chauhan C (2022) Linking circular economy and digitalisation technologies: a systematic literature review of past achievements and future promises. Technol Forecast Soc Chang 177:121508
21. Mar A, Pereira P, Martins J (2021) Energy community flexibility solutions to improve users' wellbeing. Energies 14:3403
22. Haber I, Kistelegdi I, Bötkös T, Farkas I (2013) Modeling solar irradiation data for photovoltaic energy-yield prediction. Pollack Period 8(3):27–34 (8 p)
23. Alcaraz C et al (2011) Secure SCADA framework for the protection of energy control systems. Concurr Comput Pract Exp 23:1431–1442
24. Scott Cramer RH (dátum nélk). Consequence-driven cyber-informed engineering. Idaho National Laboratory 20-50444_005_R0
25. Nazaré T, Gadelha J, Nepomuceno E, Lozi R (2023) Green computing for energy transition: a survey. IEEE Lat Am Trans 21(9):937–948
26. Papathanasiou A, Liontos G, Liagkou V, Glavas E (2023) Business email compromise (BEC) attacks: threats, vulnerabilities and countermeasures—a perspective on the Greek landscape. J Cybersecur Priv 3(3):610–637
27. Neri M, Niccolini F, Martino L (2024) Organizational cybersecurity readiness in the ICT sector: a quanti-qualitative assessment. Inf Comput Secur 32(1):38–52
28. Al lelah T, Theodorakopoulos G, Reinecke P, Javed A, Anthi E (2023) Abuse of cloud-based and public legitimate services as command-and-control (C&C) infrastructure: a systematic literature review. J Cybersecur Priv 3(3):558–590

Options for Implementing Explosion Protection in Critical Infrastructure

Norbert Daruka

Abstract When you hear the term explosion protection, everyone thinks of some kind of explosives-related activity. However, there are substances (dusts, gases, vapors, mists) that are not explosives, but are capable of triggering detonations under certain conditions. In critical infrastructure operations, especially in factories, plants, production lines, conditions and ignition sources can develop that can trigger explosions in the area. Unfortunately, there are a number of examples which, following investigations into accidents which have already occurred, most of which have resulted in fatalities, draw attention to these problem areas. Unfortunately, professionals working in the defence sector must now also take into account the fact that, in the case of deliberate damage or disturbance, they must not only keep dangerous materials (such as explosives) away from the area concerned, but also seemingly harmless materials such as combustible dusts, gases or vapors.

Keywords Ex/Hazloc · Explosion protection · Explosion safety engineering

1 Introduction

The range of challenges of the twenty-first century is vast and highly diverse. But the greatest threat, in terms of people's perceptions, is fear, which manifests itself in a wide range of challenges to our security. We fear anything (infections, threats, etc.) that directly or indirectly endangers our personal security. Today, a significant proportion of workers are still employed in industry, manufacturing and production plants and facilities. Even in these areas, safety as the most important working condition cannot be a priority.

"*All workers have the right to working conditions that respect their health, safety and dignity*" [1]. Safety in this respect is relative, since in plants where dangerous

N. Daruka (✉)
Banki Donat Faculty of Mechanical and Safety Engineering, Obuda University, Budapest, Hungary
e-mail: daruka.norbert@bgk.uni-obuda.hu

T. A. Kovács et al. (eds.), *The Impact of the Energy Dependency on Critical Infrastructure Protection*, Advanced Sciences and Technologies for Security Applications, https://doi.org/10.1007/978-3-031-78544-3_22

substances are stored, produced or processed, explosive atmospheres can be created which can endanger both human life and material goods. Explosion protection is designed to eliminate these unacceptable risks. There are, however, grey areas where those working on protection solutions are also confronted with the complexity of some elements of the legislative hierarchy, especially when they are already in doubt about the regulation of another area. There are significant differences between the regulation of explosives and the regulation of explosive atmospheres, and a clear separation between the two areas. This is not difficult, since the use of explosives in the civil sector is precisely to facilitate destruction, while in the case of explosive atmospheres the main task is to avoid explosions and destruction. I am deliberately not mentioning the law-and-order line, as it is not relevant to the explosive atmosphere.

The question arises as to what happens if the creation of an explosive atmosphere is part of sabotage or is done deliberately to eliminate an individual, a department, a factory, either partially or completely. I do not have clear evidence to justify such a claim, but it is worth reflecting on the competition for a good deal or a favorable position in certain sectors. Nor can we ignore what is clearly seen in almost every action film in the media or in the cinema, i.e. the gas valve is opened and an ignition source is used to ignite the gas spill (the explosive atmosphere), which in a quick reaction (detonation) destroys the building.

When planning the security design of an installation, it is important to consider the potential risks from this perspective, because, as they say, "the devil never sleeps".

To have the right background information, it is necessary to look in more detail at the explosive atmosphere itself and the conditions under which it is created and the risks it poses.

2 Explosion Protection and Its Legal Framework

Explosion protection is the umbrella term for the discipline that seeks to reduce the hazards of explosive materials, equipment and processes used in industrial and storage processes. In this field, explosion safety engineering tasks are carried out by specifically trained (explosion protection) professionals in order to prevent explosions and minimize the damage and injuries caused by explosions. Explosion protection covers areas such as the identification and marking of explosive atmospheres, the use of appropriate explosion protection equipment and the provision of appropriate education and training for workers in explosive atmospheres.

It almost begs the question, what is an explosive area? Well, the part of a work area where a mixture of highly flammable or explosive substances, such as combustible gases, vapours or steams, mists (aerosols) or dusts in the right proportions with air, may explode under the influence of an external ignition source. This explosion is not the same as the explosion of an explosive, as can be seen from a comparison of detonation velocities. If we are talking about explosives, detonation is a stabilized explosion process with a velocity of 1000–10,000 m/s. In this case, the detonation velocity is nothing more than the linear propagation velocity of the detonation front in the explosive [2].

If we are talking about highly flammable or explosive materials, then detonation as a concept is nothing more than: the propagation velocity of the flame front is greater than 340 m/s [3]. In this sense, the detonation front is the interface between the unconverted and the converting explosive, where the temperature and pressure change abruptly.

Based on these distinctions, a space explosion itself is nothing more than a chemical explosion in which at least one of the substances involved is gaseous, i.e. the explosive rapid oxidation of a combustible substance in an oxidizing gas or vapour. As defined in the Regulation on minimum safety requirements for workplaces in potentially explosive atmospheres:

- *explosive atmosphere*: a mixture of combustible gases, vapours, mists (aerosols) or dusts with air in which, under normal conditions, combustion spreads to the whole mixture on ignition;
- *potentially explosive atmosphere*: the part of a work area where an explosive atmosphere may be formed.

In areas where potentially explosive atmospheres may occur, the requirements of the European Union Directives, ATEX, apply. This acronym currently encompasses two directives aimed at addressing the risks from explosive atmospheres.

One is Directive 1999/92/EC—ATEX 137—which applies to workplaces in potentially explosive atmospheres. This Directive sets out minimum requirements for safety and health protection at work, i.e. it summarizes the employer's obligations. These obligations focus, for the subject of this publication, on the zoning of potentially explosive atmospheres, the assessment of explosion risks and the contrast screening of machines, tools and equipment that are properly certified for explosive atmospheres.

The second one is Directive 2014/34/EU—ATEX 114—which is relevant for machinery, tools and equipment used in potentially explosive atmospheres, i.e. for the products. The Directive harmonizes the (binding) essential health and safety obligations (EHSR—Essential Health and Safety Requirements) for manufacturers of machinery, tools and equipment that are certified (certified) for the presence of explosive atmospheres. The Directive specifies the relevant classification requirements for tools and equipment, as well as the conformity assessment procedures and obligations relating to the CE and Ex marks. The Directive is not a directly applicable legal source, it only lays down general requirements, which each Member State must transpose into separate legislation.

Hungary, like other European countries, fulfilled this obligation when it adopted Joint Decree 3/2003 (11.III.) of the Ministry of Labour and Social Affairs of the Republic of Hungary on the minimum occupational safety and health requirements for workplaces in potentially explosive atmospheres [4]. In relation to Directive 2014/34/EU, the Ministry of National Economy issued Decree 35/2016 (27.IX.2016) on the testing and certification of equipment and protective systems intended for use in potentially explosive atmospheres [5].

The knowledge of the international and relevant national legal environment of explosion protection and its changes is essential to maintain an adequate level of knowledge in order to be able to perform competently in this field. In addition to keeping abreast of ever-changing legal sources, the technological state of the art is constantly evolving, with new manufacturers, new equipment and new procedures making it more difficult for professionals working on protection solutions.

As in all industries, occupational health and safety is a key area of specialization, so it is the starting point for almost all European countries' legislation in other areas of regulation focusing on safety solutions. Regulation 3/2003 deals in detail with the general duties of the employer, the basic obligations of the employer in the field of occupational safety and health, the obligation to carry out risk assessments and the coordination duties. Perhaps the most relevant part of the joint regulation, in terms of the subject of this publication, is the provisions on explosion risk areas and explosion protection documentation.

As regards the content of the explosion protection documentation, I do not wish to go into detail and explain it, as it is explained in detail in the Regulation. What is important to me now is the zoning of workplaces where explosive atmospheres may occur:

On this basis, the zoning is as follows:

- For combustible gases, vapours or mists (aerosols)
  Zone 0—where explosive atmospheres are present permanently, for long periods or frequently;
  Zone 1—where explosive atmospheres may develop;
  Zone 2—where the explosive atmosphere is present infrequently and for short periods.
- For airborne combustible dusts
  Zone 20—explosive atmospheres are present permanently, for long periods or frequently present;
  Zone 21—explosive atmosphere may occur;
  Zone 22—the explosive atmosphere is present infrequently and for short periods [2].

Since zoning requirements can be formulated for different installations, the next step is to consider the machines that will be placed in these zones.

3 Equipment and Protective Systems

With regard to equipment and protective systems intended for use in potentially explosive atmospheres, a primary consideration is the need to clarify the criteria determining the categorization of application groups. It makes a difference in which area a piece of protective equipment is intended to be used, for example above ground or buried below ground level. These factors may generate additional risk factors. The following classifications should be used, taking into account the requirements of the relevant legislation:

- Application Group I: relevant for deep mines;
- Application Group II: relevant for above-ground disposal.

  Category 1: equipment designed for use in spaces where explosive atmospheres caused by the mixing of air with gases, vapours, mists or airborne particles occur continuously, for long periods or frequently;

  Category 2: Equipment designed for use in areas where explosive atmospheres caused by gases, vapours or mists or mixtures of air and dust are likely to occur occasionally;

  Category 3: equipment designed for use in spaces where explosive atmospheres caused by air and gases, vapours, mists or air-dust mixtures are less likely to occur or, if they do occur, are infrequent and of short duration [5].

There are also minimum requirements for the design and manufacture of different equipment that can be crucial in our case. Equipment that may emit flammable gases or dust should be designed with enclosed spaces as far as possible. It is also important that if these enclosed spaces have openings or unsealed joints, they should be designed in such a way that the evolving gases or dusts cannot lead to an explosive environment outside the equipment.

And this is the point where you have to go to the facilities where you want to use these devices. For infrastructure, there are a myriad of classification options. In our case, I have focused on those infrastructures that are critical elements in a given region, country or area. By critical element, I mean a key element whose partial or total loss could have a fundamental impact on the maintenance of the normal functioning of a region.

Figure 1 clearly shows the vast extent of different areas (such as transport) as infrastructure across different sectors and industries. Potentially explosive industrial areas, such as chemical plants, gas pipeline networks and gas distributors, airports, power plants, ports, painting plants, oil refineries, cement plants, food processing plants, woodworking and last but not least plastic granulate plants.

Of course, there are factors in this list that are difficult to imagine as a critical infrastructure element, but could be part of those where an explosive environment could develop at any time.

4 Conditions for the Creation of a Potentially Explosive Atmosphere

Combustion, as we have already learned in our elementary studies, requires oxygen and combustible material, and an ignition source to ignite the combustible material. This is not so simple in the case of explosive atmospheres, since the combustible

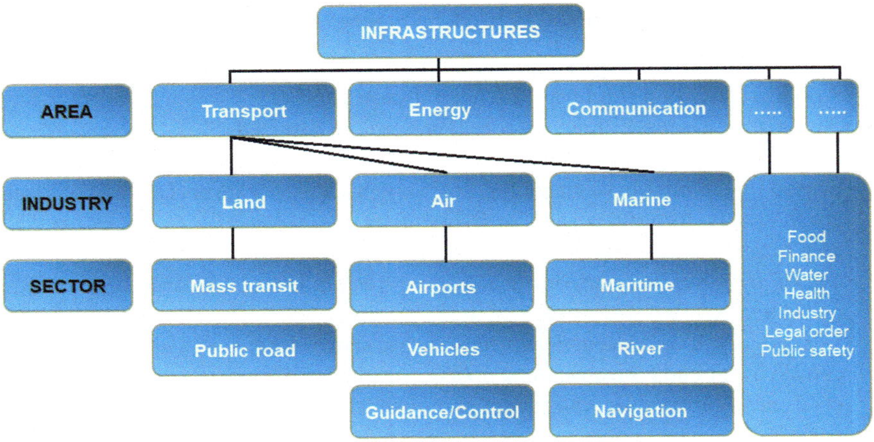

Fig. 1 Distribution of infrastructures [6] (by the author)

material is provided by the mixture of dusts, gases, vapours and mists with the appropriate air.

In order for an explosive atmosphere to exist, at least seven conditions must be met simultaneously for dusts. The properties of dusts differ significantly from those of the solids from which they are formed. The basic reason for this is that the surface area is larger—by orders of magnitude—for the same mass, and therefore the intensity of oxidation is greater. This process, when it turns into a chain reaction, is called a dust explosion. For a dust explosion to occur, the following conditions must be fulfilled at a given time:

- combustible material (dust, particles) must be present;
- the oxidizing agent (e.g. oxygen in the air) necessary for combustion to occur;
- the dust particles must be in a dispersed (suspended) state in the given air space;
- dust particles must have a particle size of less than 0.5 mm;
- the dust cloud must be in an enclosed or partially enclosed space;
- the dust concentration must be between the lower flammability limit value (LFL—Lower Frammable Limit) and the upper flammability limit value (UFL—Upper Frammable Limit);
- there must be an ignition source of sufficient power to ignite the dust cloud [7].

Flammable substances mixed with air have lower and upper explosive limits.

The range between these two values is the range in which an explosion may occur under the influence of an external ignition source.

Highly flammable and explosive substances in a combustible medium can initiate a high-speed combustion process when exposed to an ignition source. Consequently, the phenomenon of no explosion below and above a certain concentration also applies to this type of combustion.

For gases, vapours and mists, these lower and upper explosion limits are defined in % by volume, while for dusts they are defined in g/m^3. The lower explosion limit is taken into account, for example, when determining the safety of workplaces. In terms of safety, a value less than 80% of the lower explosive limit (LEL 20%) is considered safe. This is also related to explosion protection, which aims to replace combustible material or reduce the amount of oxygen to a level where there is no risk of explosion.

It is also important to note that the explosive state can only be described in dispersed form, as the settled dust behaves as a solid until it is stirred. It is important to know that fine dust mixed with air is very similar to a mixture of combustible gases with air, but it is also much more dangerous, because once the gas mixture has exploded, there is no longer any explosion hazard, so to speak. A dust explosion, on the other hand, does not behave in this way, as it can cause further so-called series explosions by stirring up (dispersing in air) the amount of dust deposited in its vicinity. This is only possible if the combustible material has combustion propagation and the stirred-up dust is mixed with the air in the area in the right proportion. If this condition does not occur, i.e. the right proportion is not present, then on the one hand the heating up to the ignition temperature between the particles is not ensured (there is little combustible material in the air) and on the other hand the burnt particles draw off the available oxygen from the particles that have not yet ignited and cool the gases that have been produced (there is a lot of combustible material in the air). In this respect, it is important to know the upper and lower explosion limits for a given material. Generally speaking, the ignition temperature of dust-air mixtures is in the range 300–600 °C. Only very few materials ignite at temperatures below 300 °C [2].

The maximum rate of explosion pressure rise depends on the values of the following parameters:

- the characteristics of the highly flammable and explosive material (ignition temperature, ignition energy, etc.);
- the size of the dust particle;
- the distribution of the dust particle;
- turbulence;
- the shape of the chamber or containment vessel;
- the characteristics of the ignition source.

If we look at these characteristics, we can conclude that even small changes in the environment, whether reduced or increased, will have very wide variations. It follows that the size and design of the drum, mixing vessel or kettle used for mixing explosives and highly flammable and explosive materials can have a significant influence on the conditions under which an explosive atmosphere can be created.

We also need to look at the risk areas of gases and vapours. It is important to remember that mixtures of flammable gases or vapours with air can only ignite at the right concentration. This limit, as in the case of the dusts mentioned above, will fall between the LFL and the UFL. The lower explosion limit is defined as the lowest concentration of gas or vapour that can be experimentally verified to be at a level at which flame propagation can still be observed after ignition. For the upper explosive limit, this means the point where the flames are still spreading [2].

If these aspects can be excluded, or if the quantity or mixing ratio of the combustible material can be brought below or above the limit value, an explosion can be avoided.

5 Protection Methods and Options

Explosion protection basically considers three distinct methods of protection to be acceptable to avoid the risk posed by an explosive atmosphere and the harmful effects of detonation during an explosion.

The first is to prevent the various dusts, gases, vapours and mists from entering the atmosphere in question, i.e. to prevent them from reaching the quantity necessary for an explosion to occur, once they have entered the atmosphere, by either displacing or excluding the explosive substance(s), i.e. by limiting the quantity of the medium feeding the combustion.

These proposed solutions can also be implemented in practice by, for example, using closed technology, thus not allowing the right conditions to develop. The other option is to replace hazardous substances, i.e. to try to replace flammable and explosive substances with substances which, if possible, are not capable of igniting or which are less hazardous in the technological processes concerned. Another method is to remove oxygen or the air containing it from the hazardous areas of the technology or to increase/decrease its concentration to the point where the parameters for combustion are no longer optimal.

Secondarily, in the case of the formation of an explosive medium, the medium is potential ignition sources are excluded from the explosive atmosphere. This requires the identification of all ignition sources that could cause combustion and subsequent explosion in the explosive atmosphere for the given technology. Possible ignition sources include:

- Hot surfaces;
- Flames and hot gases (including hot particles);
- Mechanically generated sparks;
- Electrical source;
- Stray electric currents, cathodic corrosion protection;
- Static electricity;
- Lightning;
- Radio frequency (RF) electromagnetic waves from 10^4 Hz to 3×10^{12} Hz;
- Electromagnetic waves including optical radiation from 3×10^{11} Hz to 3×10^{15} Hz;
- Ionizing radiation;
- Ultrasonics;
- Adiabatic compression and shock waves;

- Exothermic reactions, including self-ignition of dusts [2].

Secondarily, in the case of the formation of an explosive medium, the medium is potential ignition sources are excluded from the explosive atmosphere. This requires the identification of all ignition sources that could cause combustion and subsequent explosion in the explosive atmosphere for the given technology.

For the listed ignition sources, space limitations do not allow me to analyze all ignition sources in detail, but it is important to note that there are companies specialized in ignition source analysis that have the necessary expertise and thoroughness to assess the potential hazards. In order to properly identify the ignition sources that are relevant and relevant to a given technology, the zoning of the technology's environment is the most important step. Zoning can also be carried out by a person with appropriate expertise and practical experience.

Thirdly, the effects of an expected explosion should be kept under control and, where appropriate, reduced. Several methods are known to achieve this, such as isolation, suppression and deflagration. The most common is the use of slotted openings, which is not only practical but also economical. The only difficulty is perhaps that it is not easy to size such surfaces.

6 Intentionally Created Explosive Atmosphere, Sabotage

In the foregoing, I have described the conditions necessary to create an explosive atmosphere and the ignition sources required to ignite it. There are infrastructures where these conditions can arise on a daily basis and where the security professionals and the various security devices work together to prevent the explosion itself from occurring.

It should be mentioned, however, that the deliberate creation of explosive atmospheres is also a feasible activity, although fortunately we only see examples of this in various action films. But what happens if the disaster in the film suddenly becomes reality. There are many examples in history of attacks on critical infrastructure, usually with explosives [8]. Attacks on structures with explosive devices [9] are also quite frequent and strategic attacks on infrastructures have been observed in recent years.

But what many forget is that cutting-edge technology solutions make it much harder to sabotage infrastructures. The potential for malicious attacks (not caused by human negligence) will always be present, as critical infrastructures by their very nature can have a huge influence or strategic importance.

It is therefore necessary to explore the possibility of how, in the case of a strategically important critical infrastructure, insider offenders can be filtered out and the conditions for creating an explosive atmosphere minimized or eliminated. Think of the meticulous measures that would be needed to prevent an insider (potential bomber) from gaining access to, for example, gas lines, electrical wiring, motors, machinery, switch cabinets. I could also mention here the various chemicals and cleaning agents that are basically found in all facilities and are usually accessible to anyone. In my opinion, these risks can only be addressed by thorough, systematic and thorough investigations from a safety point of view and by involving the most qualified professionals. Perhaps it is not surprising that economic considerations should not be the most important factor in critical infrastructure [10]. An individual with authorized access to associated facilities or associated activities or to sensitive information or sensitive information assets, who could commit, or facilitate the commission of criminal or intentional unauthorized acts involving or directed at nuclear material, other radioactive material, associated facilities or associated activities or other acts determined by the State to have an adverse impact on nuclear security [11].

The question may arise: who could be an insider? In order to understand the scale of the threat, let us clarify the persons, who can become insiders. It can be anybody who has permission to enter the site and/or authorized access to the systems of the facility, thus in particular, but not exclusively: officials of the management of the facility; employees of the facility; security personnel, guards; system administrators of the IT system; external contractors, partners; maintenance personnel; official persons; employees of public utilities (electricity, gas, water, sewer, Internet, waste management); vendors, courier; visitor [10].

All of these persons could be potential attackers if their knowledge allows them to create a situation (explosive atmosphere) that, given the right ignition source, could shut down or destroy a particular machine, sector, section or the entire facility and thus the entire critical infrastructure.

7 Conclusion

As technology develops and industries become more and more diverse, the issue of security is becoming increasingly important. I have shown that the creation of explosive atmospheres is a major risk factor in various infrastructures. The extent of this airspace can pose a threat to the technology, the department and ultimately the whole facility. If a critical infrastructure is breached, it can contribute to a change in the normal functioning of an entire community (county, country, region), with unforeseen consequences.

Today, there are still ways to prevent explosive atmospheres from forming or igniting. In fact, there are also scientifically proven solutions for the controlled management of explosive atmospheres and the mitigation of their effects. As all technologies evolve, so does this field, and we must keep pace with scientific advances and make the best possible use of them.

In my opinion, explosion protection in practice is not the same as preventing the formation of explosive atmospheres, or controlling their reaction. We also have to consider that insider offenders can be as dangerous as an improperly identified ignition source. In facilities where the level of threat to critical infrastructure is high—dams, nuclear facilities, chemical plants and so on—the feasibility of sabotage must be assessed and an internal employee climate must be established and controlled with appropriate controls. Otherwise, we create the potential for an accidental or deliberate explosion to endanger our lives.

The possibility of sabotage arises in connection with rapidly advancing technologies today, such as the increased scrutiny of 3D printing. This is a potential avenue for further research, [12] exploring its potential connections to explosive techniques and how it might impact critical infrastructure in the future.

References

1. Fundamental Law of Hungary. Dordrecht. https://net.jogtar.hu/jogszabaly?docid=a1100425.atv
2. Norbert D (2023) Current issues and implementation options for explosion protection in relevant areas of explosives production. Obuda University, Diploma thesis 2023
3. Technical Directive on Fire Safety—TvMI 13.3:2022.06.13. Explosion protection. Ministry of the Interior—National Directorate General for Disaster Management, p 60, 2022. Online https://www.katasztrofavedelem.hu/application/uploads/documents/2022-04/78650.pdf
4. Joint Decree No. 3/2003 (11.III.) of the Ministry of Health and Social Affairs on the minimum requirements for the protection of workers in potentially explosive atmospheres. Online: https://net.jogtar.hu/jogszabaly?docid=a0300003.fmm
5. 35/2016 (IX.27.) Decree of the Ministry of National Economy on the testing and certification of equipment and protective systems intended for use in potentially explosive atmospheres. Online: https://net.jogtar.hu/jogszabaly?docid=A1600035.NFM&txtreferer=A1100187.TV
6. Daruka N (2022) Critical infrastructure protection in the production and use of explosives industry. Products. In: Kovács TA, Nyikes Z, Fürstner I (eds) Security-related advanced technologies in critical infrastructure protection. NATO Science for Peace and Security Series C: Environmental Security. Springer, Dordrecht, pp 297–313. https://doi.org/10.1007/978-94-024-2174-3_26. Online ISBN 978-94-024-2174-3
7. Ervin P (2019) Explosion safety engineering VII—what you need to know about explosive dusts. 170. Electricians J Build Electr Eng XVIII(9):6–10
8. Lukács L, Balogh Zs (2018) Protection of critical infrastructure facilities against bombing attacks. In: Kiss PA (szerk) National resilience: opportunities and challenges in a changing security environment, Budapest, Hungary. Hungarian Defence Forces (HDF) General Staff Scientific Research Centre, vol 327, pp 183–204, 22 p
9. Zoltán K (2016) Explosion of improvised explosive device on structures. Hadmérnök 1788-1919. 11(1):56–63. http://www.hadmernok.hu/161_06_kovacsz.pdf
10. Bunyitai Á (2023) Insider threat mitigation in high security facilities. Nemzetbiztonsági Szemle, 11. évfolyam (2023) 1. szám 49–61. Online https://folyoirat.ludovika.hu/index.php/nbsz/article/view/6464/5371
11. IAEA—International Atomic Energy Agency (2013) Objective and essential elements of a state's nuclear security regime. NSS-20. International Atomic Energy Agency, Vienna
12. Ember I (2022) Modern kumulatív töltetek hatékonyságának vizsgálata. Haditechnika 56(6):15–20

The Impact of Polluting Chemical Compounds Resulting from the Gases Exhausted by Road Vehicles in Urban Environment

Levente Dimén and Ruscă Marcel

Abstract At the national level, the level of air pollution (especially in the big cities) is above the permissible limits. This undesirable reality affects the health of the inhabitants and the state of the environment—furthermore, the high level of atmospheric pollution causes a series of respiratory and cardiovascular diseases. It is important to pay more attention to the actions taken to identify sources of air pollution and their chemical composition, as well as to establish measures to monitor and reduce these pollutants. The city of Alba Iulia was chosen for the evaluation of air quality in urban agglomerations due to being one of the cities in Romania with a rich and ancient history that has undergone an extraordinary economic, social, and cultural transformation in the past 15 years. To carry out the determinations, points were established in representative areas of the city in terms of traffic intensity, large flow of children and adults to and from schools, kindergartens, and hospitals. Measurements took place in two representative periods of the year, spring, and autumn, resulting in the determination of the amount of fine material particles in the atmosphere, as well as a series of gaseous compounds emitted mainly by motor vehicles. In the period 2016–2023 it appears that no studies of this scope were carried out in Alba-Iulia, regarding the level of pollutant emissions from road vehicles. Our contribution covers the research gap on the polluting factors in the city and can help the competent authorities in the implementation of the necessary measures. Finally, some current and future directions of action were evoked in order to reduce the level of pollution caused by emissions.

Keywords Pollution · Catalytic reduction systems · Pollution sources · Fine particles

L. Dimén · R. Marcel (✉)
Faculty of Exact Sciences and Engineering, University „1 Decembrie 1918" of Din Alba Iulia, Alba Iulia, Romania
e-mail: rusca.marcel@gmail.com

T. A. Kovács et al. (eds.), *The Impact of the Energy Dependency on Critical Infrastructure Protection*, Advanced Sciences and Technologies for Security Applications, https://doi.org/10.1007/978-3-031-78544-3_23

1 Introduction

The atmosphere can be likened to a large chemical reactor in which there are a multitude of chemical substances originating from natural and anthropogenic phenomena. Among the anthropogenic activities, encountered mainly in the urban environment, among the activities that bring a significant contribution to the pollution of the atmosphere are the industrial activities, the activities of means of transport, thermal power plants, etc. These emit considerable amounts of pollutants into the atmosphere such as: suspended dust, oxides of nitrogen, carbon oxides, sulfur dioxides. Dry air represents 99.964%. If water vapor, nitrogen, oxygen, argon was removed from the air, the remaining air of 0.036% represents the standard composition of the lower atmosphere, being made up of carbon dioxide (0.0325%), neon (0.00182%), helium (0.000524%), methane (0.00015%), krypton (0.000114%), hydrogen (0.00005%). The other three dominant gases in the atmosphere are—nitrogen (78.84%), oxygen (20.946%), argon (0.934%) [1].

Today, the subject of pollution is a great interest, due to the problems related to urbanization and extreme industrialization, which negatively affect all living creatures. In the atmosphere, a substance undergoes changes as a result of chemical and photochemical reactions through the influence of sunlight, forming new substances that can act as secondary pollutants. Chemical pollution is considered to be the most dangerous pollution for human health through the direct action of pollutants [2, 3]. In the area of urban agglomerations, due to various human activities, air quality is strongly influenced by industry, transport and service activities, which generate both diffuse and concentrated pollution. The impact of these organic compounds is increasing sharply due to polluting vehicles (petrol and diesel) and due to the increase in the influx of travelers in most congested areas, as well as from the atmospheric pollution due to industrial and agricultural activities.

In Romanian cities, big problems are generated by insufficient green spaces and intense car traffic, the latter generating linear pollution along traffic routes that are very difficult to manage. We consider it necessary to reduce the use of fossil fuels, improve energy efficiency and use renewable energies.

Our novel contribution fills the research gap regarding the level of chemical pollutants of road vehicles from the city of Alba-Iulia, which is of major interest due to their impact [2].

For the evaluation of air quality in urban agglomerations, the city of Alba Iulia was chosen because it is one of the Romanian cities with a rich and ancient history that has undergone an extraordinary economic-social transformation in the last 15 years (Fig. 1).

The city of Alba Iulia is the center of residence of Alba County, located at an altitude of 270 m, has a total area of 103.65 km^2, a population of 63,536 inhabitants (as of 2011) and a density of 566 inhabitants/km^2. It is located at a distance of 340 km from Bucharest, 100 km from Cluj, 70 km from Sibiu. The city is located in the perimeter formed by the rivers Ampoi, Sebeş, Mureş and the ridges of the Apuseni mountains [4, 5].

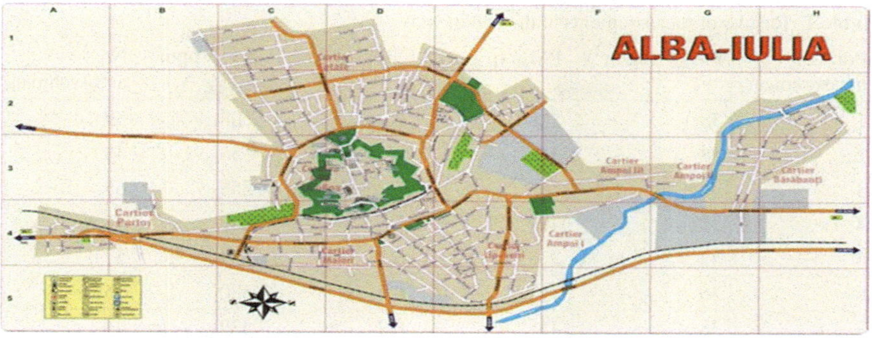

Fig. 1 Alba Iulia city map

1.1 Results and Discussion

The results of the measurements from the two-time intervals are presented below.

It is worth mentioning that according to Law 104/2011 the daily limit value for PM10 is 50 μg/m^3, provided that this value is not exceeded more than 35 times in a calendar year, the annual limit value being 40 μg/m^3. For $PM_{2.5}$, the annual limit value is 25 μg/m^3. As it can be seen from Tables 1 and 2, in the first stage of measurement, the limit values were exceeded in a few measurement points, whereas in the second stage of measurement, almost all measurement points exceeded the limit values. These values are compared with the values from the automatic city station AB1, finding that the maximum allowed values are generally exceeded 15–20 times in a year. Regarding $PM_{2.5}$ values, we observe that both on-site measurements and average calculated were higher than the specified limit.

At the European level, rigorous control over atmospheric pollution is mandatory, especially with regard to certain polluting substances such as: sulphur dioxide (SO_2), carbon dioxide (CO_2), carbon monoxide (CO), benzene (C_6H_6) and lead (Pb), suspended dust.

As expected, the intensively trafficked areas, especially with heavy means of transport, were the most polluted. The areas around the county hospital, and schools were heavily polluted with fine material particles ($PM_{2.5}$). According to the measurements in the first period of time, the average of the $PM_{2.5}$ values measured—38.03 μg/m^3 is higher than 20, the limit for this indicator, according to the legislation. At the opposite pole, in the first period, the average of the measured PM_{10} values—38.80 μg/m^3 is lower than the legal limit of 50 μg/m^3. According to the measurements in the second period of time, the average of the $PM_{2.5}$ values measured—53.73 μg/m^3 is higher than the 20 limit, whereas the average of the measured PM_{10} values—65.857 μg/m^3 is also higher than the legal limit of 50 μg/m^3. Between 2016 and 2023, it seems that no research studies of this scope were conducted in Alba-Iulia, focusing on the extent of pollution emissions from road transportation. This current paper addresses the research void concerning the air pollutants in the city and may assist relevant

Table 1 Results of measurements at the end of May

Point of measurement	$PM_{2.5}$ (μg/m³)	PM_{10} (μg/m)³	VOC (μg/m³)	CO_2 (ppm(	Nr. auto-vehicles
1	38	36	6.8	608	59
2	28	38	0.5	556	48
3	49	55	1.2	558	65
4	36	48	1.6	556	46
5	32	33	0.2	550	57
6	36	31	0.4	580	55
7	38	41	0.2	539	44
8	36	40	0.4	560	57
9	42	45	0.2	552	48
10	46	30	0.4	638	50
11	34	32	0.2	525	44
12	38	36	0.2	506	44
13	32	45	3.1	562	38
14	34	43	1.3	585	44
15	33	39	0.5	600	42
16	47	33	1.2	516	54
17	36	44	3.1	562	46
18	53	49	1.3	585	63
19	46	44	0.5	600	54
Average	38.63	40.21	1.436	561.53	50

officials in enacting appropriate actions. For a more realistic comparison and interpretation of the results, we refer to the values determined by the local station in Alba Iulia, where the values of PM_{10} and $PM_{2.5}$ particles for each year are presented in the annual reports [5] (period 2016–2013). It was found that the average annual values are lower than the values determined in this paper.

Lowering pollutants emitted by vehicles.

The main measures to reduce pollution due to road traffic in urban centres could be the following: reduction of road traffic through the city; the control of industrial pollutant emissions, as well as the establishment of clear rules regarding the accepted level of emissions from motor vehicles; elaboration and updating of local and regional action plans in which the established conditions to be taken to improve air quality are explained; use of electric vehicles; the use of catalytic converters; keeping the streets clean [6, 7]. One way to reduce the emissions from fuel combustion is represented by the catalytic systems associated with vehicle engines. These are devices with the role of converting the emissions produced by burning the fuel into less dangerous gases before they are released into the environment. We present a catalyst model called the SCR (Selective Catalytic Reduction) system, which significantly reduces

Table 2 Results of measurements at the end of September

Point of measurement	$PM_{2.5}$ ($\mu g/m^3$)	PM_{10} ($\mu g/m^3$)	VOC ($\mu g/m^3$)	CO_2 (ppm)	Nr. autovehicles
1	70	95	6.4	608	74
2	70	93	9.0	686	73
3	68	90	4.1	448	68
4	63	72	3.1	448	64
5	35	42	0.2	445	48
6	30	36	0.2	597	–
7	35	41	0.2	399	–
8	40	44	1.6	415	48
9	35	40	0.2	405	–
10	68	87	1.4	562	74
11	49	64	1.1	439	–
12	46	56	2.5	429	63
13	32	51	0.5	445	–
14	43	50	0.3	415	–
15	66	78	0.6	463	67
16	48	56	0.5	460	–
17	44	54	0.2	426	–
18	84	92	2.5	469	74
19	95	105	0.2	495	–
Average	53.73	65.57	1.82	455.44	65.3

the nitrogen oxides emitted. The system allows the use of a trivalent catalyst that ensures the simultaneous reduction of the resulting pollutants: through oxidation reactions, it reduces carbon monoxide (CO) and gaseous hydrocarbons (HC). Also, through reduction reactions it reduces nitrogen oxides (NOx). In order to obtain that, selective catalytic reduction (SCR) uses the injection of a reducing agent (ammonia or urea) into the exhaust gases [8].

The reactions to reduce nitrogen oxides to nitrogen by adding ammonia to the catalytic device are presented:

$$4\,NO + 4NH_3 + O_2 \Rightarrow 4N_2 + 6H_2O \tag{1}$$

$$6NO + 4NH_3 \Rightarrow 5N_2 + 6H_2O \tag{2}$$

$$2NO_2 + 4NH_3 + O_2 \Rightarrow 3N_2 + 6H_2O \tag{3}$$

$$6NO_2 + 8NH_3 \Rightarrow 7N_2 + 12H_2O \tag{4}$$

By using the SCR system, nitrogen oxides are reduced, the temperature of the burnt gases having the most significant influence on the amount of NO_x produced [3, 7, 9, 10].

2 Conclusions

The values established for reducing the level of polluting emissions provide for a reduction of approximately 40% by the year 2030 of polluting emissions and the use of green energies in a percentage of at least 30%.

Following the research carried out, a series of recommendations can be made regarding the location of some social objectives (housing, schools, kindergartens, playgrounds for children, etc.) in areas with low pollution. Also, sensitive people or people with lung conditions can choose their residence in a less polluted area.

Some examples of measures needed to reduce the level of pollution involve the use of dust-absorbing green curtains and catalytic NO_x reduction systems, modifying some organizations procedures, or by using alternative routes.

The SCR system is an efficient method for reducing nitrogen oxides which are important constituents of atmospheric pollution.

2.1 Experimental Section

Based on preliminary observations made over a period of several months, it was decided to carry out two measurement campaigns in the summer and autumn period because the activities and traffic in the city are strongly influenced by social activities, such as: school and university activities, cultural activities, tourism etc.

The areas are the following:

- Zone 1: Str. Tudor Vladimirescu, Calea Motilor, Str. Tudor Vladimirescu,
- Zone 2: B-dul Ferdinand (Entrance/exit from Sebeș)
- Zone 3: Piața Garii—Bus station/CFR train station (Entry/exit from Sebes
- Zone 4: Cetate Hotel, Cetate Park
- Zone 5: Horea Boulevard, Str. Unirii, Str. Cloșca (Exit to Cluj)
- Zone 6: Horea Boulevard, (Horea Cloșca and Crișan National College)
- Zone 7: Municipal hospital
- Zone 8: Agras Market

The measurement areas were established based on heavy traffic criteria and on the importance of air quality in certain special areas, such as schools or hospitals. The two data collection sessions took place at the end of May and September and were carried out during the week. The measurements focused on air pollutants resulting mainly from car traffic, on the noise level and the number of vehicles transiting the area, as well as their type. For atmospheric pollutants, continuous measurements are required, and for the average values obtained related to the measurement time interval, discrete measurements are used.

In order to measure the air quality, we have used the UNI-T UT33C model portable devices (Figs. 2 and 3) [11] which can determine the fine particle size, distribution and composition. These models of equipment have the advantage that they can be used in any polluted area, being easy to place and the determination of values of fine particles are done easily.

The combustion of exhaust gases generates polluting emissions in the atmosphere, of which nitrogen oxides ($NO_x = NO + NO_2$), carbon monoxide (CO), gaseous hydrocarbons (HC) and PM particles (particulate matter) are limited by regulations.

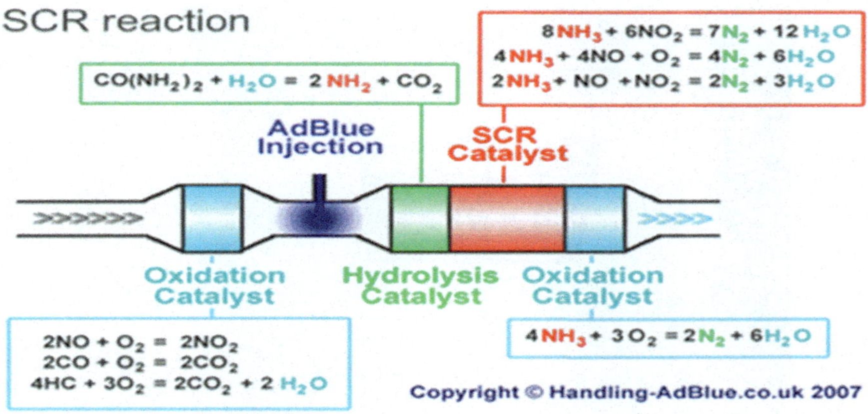

Fig. 2 Catalytic converter to reduce pollution due to exhaust gases—complete chemical reactions of the SCR system

Fig. 3 Location of the areas where the measurements will be made [5]

PM are made up of solid or liquid particles found in the air. Coarse (bigger) particles, called PM10 can cause eye, nose or throat irritation. Fine (smaller) particles, called $PM_{2.5}$ are even more dangerous because they can get deeper in the lungs.

For the measurements made in order to obtain the values of PM_{10} and $PM_{2.5}$, we used a high-performance suspended dust analyzer (Portable air quality meter UT338C Aerosol Monitor DUST TRAK DRX), as shown in Fig. 4. This device combines the functions of a photometer with an optical particle counter, allowing the intake of an air volume of 3 l/min. This measuring instrument is digital and has a lot of advantages, such as: high measurement speed, elimination of reading errors, high precision and low sensitivity to disturbances [12].

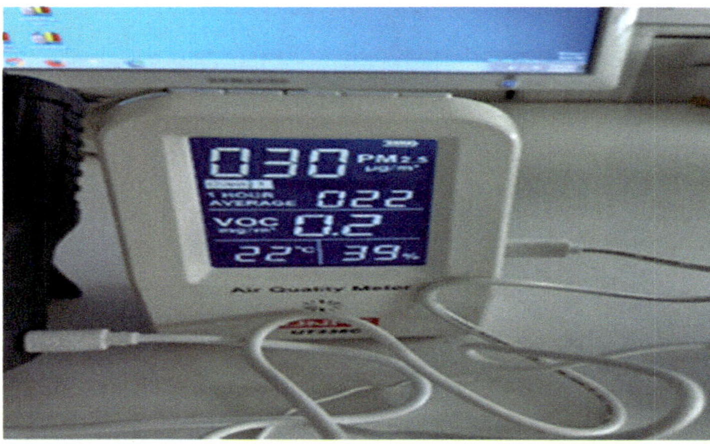

Fig. 4 Portable air quality meter UT338C [12, 13]

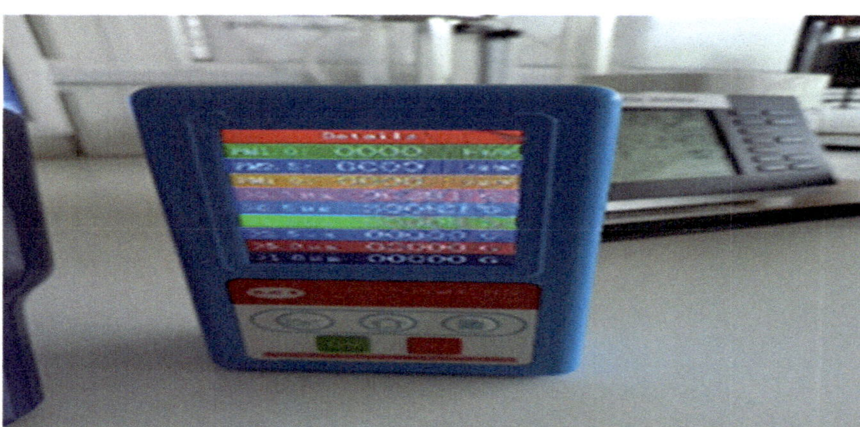

Fig. 5 Portable air quality meter UT338C [12, 13]

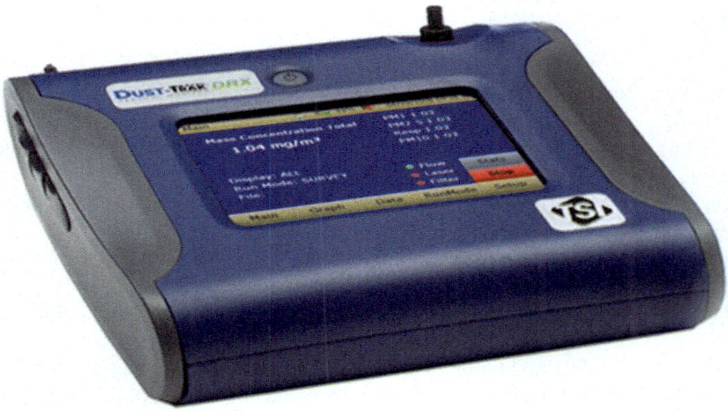

Fig. 6 DustTrak DRX aerosol monitor 8533 [12]

In the present article, the measurement method was direct, in real time. The nozzle of the flow meter classifies PM fractions according to size (Fig. 6).

Atmospheric particles are picked up by a suction pump and passed through a HEPA filter that stops solid particles, generating aerosols. These reach the optical chamber, where a laser light stream passes (with a wavelength of 655 nm), which helps to identify the chemical substances present in the studied air. The diffuse light in the dispersion angle of $90° \pm 62°$ is captured by a layer of spherical mirrors and focused on a photodetector.

The photodetector converts the light into a voltage in order to determine the mass concentrations of the aerosols (mg/m^3). The photometer measures particle sizes from 0.1 to 10 μm, with concentrations ranging from 0.1 to 100 mg/m^3 or more [12, 13].

In each area, several measurements were made in the time interval 9.30–18.30 in previously established areas (19 collection points). The measurements were made in the two specified time periods (summer–autumn). Through mathematical calculation, the average value for each point of measurement was established. The values of PM_{10}, $PM_{2.5}$, VOC (volatile organic compounds), CO_2 were determined. Volatile organic compounds (VOCs) are present in the atmosphere in a gaseous state and are at the concentration level of 0.1–10 $\mu g/m^3$ air. The emission of VOCs from natural sources is 5.5 times higher than that caused by human activities [5].

Also, the intensity of road traffic was tracked by counting vehicles in the established time unit of 15 min. Therefore, in a first stage of measurements, the average values for each chosen point were entered in the table below. Air quality, as well as the monitoring of quality indicators, is regulated at the national level by Law no. 104/2011 [14].

References

1. Horizon 2020 Report—European Union framework program for research and innovation (2014–2020)
2. WHO (2016) Ambient air pollution report: a global assessment of exposure and burden of disease
3. Lăzăroiu G (2006) Modern air pollution solutions. Publishing House, Agir, Bucharest, Romania
4. Preliminary reports on ambient air quality for 2015–2018 in Alba-Iulia
5. Environmental protection agency annual reports regarding the state of the environment 2017–2019. http://www.anpm.ro/web/apm-Alba/rapoarte-anuale
6. Dascăl IF (2014) Control and reduction of pollution from road vehicles. Pim Publishing House, Iaşi
7. Voss KE (1994) Catalytic oxidation of diesel particulates with base met-al oxides. In: Proceedings of the third international congress on catalysis and automotive pollution control, Brussels, p 497
8. Granger P, Pârvulescu VI (2011) Catalytic NOx abatement systems for mobile sources: from three-way to lean burn after-treatment technologies. Chem Rev 111:3155–3207
9. Sandu V (1998) Chemical and acoustic treatment of exhaust gases emit-ted by diesel engines. PhD Thesis, Transilvania University Braşov
10. Foerter D, Jozewicz W (2001) Cost of selective catalytic reduction (SCR). U.S. Environmental Protection Agency Office of Research and Development, Washington
11. Rusu T, Teodorof L (2009) Environmental quality analysis and assessment tools. Publishing House: UTPRESS, Cluj-Napoca. ISBN-978-073-662-436-0
12. Rusu T, Vida C, Suciu A, Rusu TA (2014) Methodology for determining the concentrations of material particles in the atmosphere of Cluj-Napoca. In: The 14th international multidisciplinary conference "Professor Dorin Pavel—the founder of Romanian hydropower". Sebes
13. Suciu A (2015) Studies and research on the determination of material particles in suspension in the urban environment. PhD Thesis, Cluj-Napoca Technical University
14. Law 104/15.06.2011 on ambient air quality published in official Monitor part I 452 of (28.06.2011)

The Possible Impact of the Proliferation of 3D Printers on the Protection of Critical Infrastructure

István Ember

Abstract 3D printing is one of the rapidly advancing and spreading technologies of our time. It is not uncommon for households to have such devices, often used for creating plastic components. These widely accessible 3D printers are primarily suitable for manufacturing plastic parts. One might think that low-density materials have little significance in explosives technology, but this notion can be convincingly challenged. Shaped and cutting charges made through additive manufacturing can have an impact on the protection of critical infrastructure, as they can be crafted anywhere in the world with some expertise and practice.

Keywords 3D Printing · Additive · Charge · Blasting

1 Introduction

3D printing is one of the rapidly advancing technologies in our present time. Various technically capable devices can produce objects, components, and prototypes from a wide range of materials. Previously, it was thought that this technology could serve as a fundamental element in rapid prototype manufacturing. However, due to its rapid development, it no longer holds true in today's context. For plastic products, a manufacturing axiom is that injection moulding is the most cost-effective solution when producing them in quantities reaching the hundreds of thousands. This is undeniable; however, certain 3D printers are now capable of producing not only small batches but also marketable quantities at competitive prices, even in the range of several thousand units.

The evolving technology in this field has resulted in an exceptionally broad range of manufacturing processes. Several procedures are available, each with its unique capabilities, advantages, and disadvantages. Therefore, it is not advisable to rank

I. Ember (✉)
Faculty of Military Science and Officer Training, Ludovika University of Public Service, Budapest, Hungary
e-mail: ember.istvan@uni-nke.hu

T. A. Kovács et al. (eds.), *The Impact of the Energy Dependency on Critical Infrastructure Protection*, Advanced Sciences and Technologies for Security Applications, https://doi.org/10.1007/978-3-031-78544-3_24

these solutions. To provide a comprehensive overview of the field, I will introduce some technologies in the following.

It is my assumption that the application of 3D printing in explosive devices not only offers new possibilities for legally operating professionals in the field of demolition but may also facilitate the work of terrorists in the realm of criminally motivated explosions. This, in turn, could enhance the efficiency of attacks against critical infrastructures.

There will likely be additional areas where additive manufacturing and the protection of critical infrastructures intersect. However, in this writing, I will exclusively analyse the connections between criminally motivated explosions and explosive technology.

2 3D printing

The Fused Deposition Modelling (FDM) method is the first one I would like to introduce. FDM printers primarily use a thermoplastic polymer as the raw material. This technology is currently one of the most widespread in the world of 3D printing. The threads, known as "filament," can be obtained by winding, and there is a wide variety available. Among the most common are polylactic acid (PLA), polyethylene terephthalate glycol (PET-G), acrylonitrile styrene acrylate (ASA), and acrylonitrile butadiene styrene (ABS). There are also variations of the process where continuous fíber reinforcement can be used to improve the material properties of objects. The filament is fed into the "hotend" which is a component that melts it at the appropriate temperature, using the stepper motor. Due to the continuous material flow, the melted material is extruded through the nozzle onto the build platform in the build chamber. The build platform is usually heated, and its temperature is customized for the specific polymer. Typically, the movement of the "hotend" provides the horizontal dimensions, while the build platform moves vertically. However, practically any combination of movements can be found. The layers adhere to each other as the melted material flows out of the nozzle and cools, forming the desired object [1].

The second type is the so-called Stereolithography (SLA) additive manufacturing. In this process, the final object is created using photopolymers. The build platform in this technology can only perform vertical movement to ensure precise layer-by-layer construction. A laser illuminates and solidifies the appropriate points in the liquid material, which is submerged one layer deep on the build platform. Once all the necessary points of the first layer are solidified, the build platform moves one layer, and the laser phase repeats. At the end of the process, the finished object needs to be cleaned of residual material in isopropyl alcohol and then exposed to ultraviolet (UV) radiation to achieve its final surface quality. The end product is characterized by its outstanding surface quality [1].

Selective Laser Sintering (SLS) is the next fundamental technology. Objects are made from powdered material, most commonly some form of polyamide. The device spreads a layer of powder onto the build platform, and similar to the SLA process, laser beams illuminate the appropriate points to solidify the material. The process continues as each new layer of powder is added, followed by laser solidification. The laser's energy doesn't completely melt the material, but a chemical bond forms between the particles. A distinctive feature of this process is that the spread but unused powder can be reclaimed and, with the right mixing ratio, reused for new projects [1].

In Laser Melting (SLM) and Electron Beam Melting (EBM), the material undergoes melting as part of the process. The technology of these processes is similar to the SLS system, but the final product is of higher quality. This makes these technologies ideal for the production of metal components.

The Photopolymer Jetting (PJ) process significantly differs from those presented above. In this method, the build chamber is continuously exposed to UV light, and the material is deposited through droplets. The drops solidify under the influence of light, slowly building up the final object. In Binder Jetting (BJ) technology, a binding agent is applied to the layers of powder, constructing the desired product. High-quality objects can be produced using both methods [1].

The last presented method is the Laminated Object Manufacturing (LOM) based on the principle of lamination. The technological process begins with the introduction of sheets or layers. These layers are cut along the desired contour lines using a laser. The cut sections are then fused together by a heated roller. This process is repeated iteratively to build up the final object [1].

From the standpoint of my research, a significant portion of the aforementioned technologies can be considered as methods for manufacturing components of explosive devices for criminal purposes. However, I believe that a terrorist cell is unlikely to acquire high-value or conspicuous devices, some of which require specialized knowledge to operate. Therefore, ease of operation, relatively low acquisition costs, and material expenses become the primary considerations. Based on these factors, I believe that FDM, or possibly SLA technology, will be selected. In the following, I will examine the FDM type as a potential method.

It is possible to 3D print metal components using various processes, significantly enhancing the efficiency of the charges mentioned below. However, acquiring these capabilities involves significant financial investment, and their application requires a substantial technical background and, at times, a specially designed facility. Manufacturers often permit commissioning only under continuously monitored installations, and the blueprints of the produced components are typically transmitted to the printer through a cloud-based system. None of these aspects support the acquisition of such equipment for criminal purposes.

3 Shaped and cutting charges

Of the three main types, the condensed-phase explosion is the one we need to investigate in this case. The chemical transformation of explosives is an extremely vigorous process. These energy carriers undergo complete chemical transformations in a matter of microseconds, resulting in extreme heat and pressure. The former fundamental effects may be complemented, in some cases, by significant light and sound effects, shockwaves, and seismological impacts [2].

However, the tremendous, released energy is not the only aspect to consider, as there are methods to control and concentrate it. In practice, this can be achieved with shaped charges, which can have a focused piercing effect, an extended cutting effect, or form-fitting charges. These specialized technologies provide significant assistance to explosive experts, and their applications are possible in both civilian and military domains, although there is a greater range of possibilities for military forces.

The discovery of the cumulative effect is attributed to chance, as is the case with many similar phenomena in the world of science. Several researchers have studied this effect, and one widely accepted term for the cohesive flow of gas products generated during the explosion of hollow charges is the Munroe Effect, named after Charles Edward Munroe.

Munroe examined variations of the hollow charge with metal linings, but it was not until 44 years later that comprehensive experimentation on the topic led to the standardized cumulative charge. The so-called Neumann Effect, named after the research of M. Neumann and Egon von Neumann, is considered the basis for the cumulative charges currently in use today [3].

Subsequently, the development field expanded immensely. Researchers were able to determine the capabilities of the process by detonating various metals and insert shapes. Professionals extended this line of thought when they designed shaped charges optimized for cutting metals [3].

As armoured vehicles nearly impervious to conventional weapons emerged on the battlefield during this time, the accumulated knowledge was applied to the development of weapons. Since shaped charges were fundamentally designed for penetrating metals, it seemed like a logical solution to employ this technique for military explosive devices aimed at destroying armoured targets. Between the two World Wars, significant efforts were thus initiated to develop weapons based on the cumulative principle [3].

Nowadays, modern anti-tank weapons are used in contemporary warfare, yet the cumulative effect remains present in all of them. Currently, there is no alternative that proves more effective on the battlefield, although modern munitions significantly differ from their initial designs.

Linear cutting charges and cumulative-shaped charges can fundamentally be applied for cutting metals. Naturally, there is an overlap between civilian and military applications in this regard. Initially, military development and use were prevalent, later adopted by the civilian sector. This has blurred the boundary in applications, especially true in the field of explosive demolition of structures. While the methods may differ in their end goals for civilian and military tasks, the materials and tools employed often align.

The demolition of steel structures is a typical application area for cutting charges. This may arise, for instance, during the explosive dismantling of bridges. In such a task, cutting charges can play a significant role, aiding in the optimal use of explosives and initiators, as well as reducing the adverse effects of the explosion. They can also be advantageous due to quick deployment, exceptional efficiency, and promoting safe execution. Another non-negligible consideration is the cost aspect in debris removal, where it is also feasible to handle optimally sized, fragmented metal structures [4].

Among exclusively military explosive demolition tasks, it is noteworthy that military engineering applications have a history spanning several decades. The sizing of extended shaped charges is a task that can be performed for Hungarian soldiers with relatively straightforward rules [5].

It is already evident that with such specialized charges, highly efficient explosive demolition work can be carried out. It raises the assumption that this type of procedure could also appear in certain criminally motivated explosions. In the following, I will examine how modern technological tools can create synergy with classical explosive demolition procedures.

4 Impact on critical infrastructure

In my earlier research, I have already demonstrated that components of shaped charges, including both linear and shaped versions, created with a 3D printer can be effective even when made solely from low-density materials. Not only can they be effective, but they may also possess advantageous properties in certain explosive ordnance disposal tasks and neutralization processes [6–8].

However, these results primarily support legal, mostly military applications. The potential for criminal use may arise in the hands of terrorists, as these individuals exploit every opportunity to cause destruction and disruption [9]. Terrorist cells often obscure their finances in an unidentifiable manner, [10] which may be a fundamental reason for avoiding high-profile investments. Therefore, cheap yet relatively reliable 3D printers could be the tools of choice for such tasks. These can be easily purchased, and materials can be readily obtained, even with cash transactions, without leaving a trace of the purchaser.

The concentrated cumulative charges presented above, in terms of their effectiveness, can easily penetrate up to 15 mm thick of steel, all while maintaining relatively simple geometry and small size. The charge depicted in Fig. 1 was created using the freely downloadable FreeCAD application. The design does not require significant technical expertise; the challenge lies in determining the parameters, but for an explosives-savvy professional, this may not be a substantial hurdle. After the design phase, the manufacturing was carried out using an FDM 3D printer, practically under default settings. The performance speaks for itself, as the entirely non-metallic charge components successfully met the expectations.

Furthermore, I examined linear cutting charges (Fig. 2). In their case, I successfully tested the distance for the initiation shockwave to ramp up and determined the effective operational phase of the charge. The process completely matched the previous one in terms of design and manufacturing.

I have also conducted successful tests for profile shaped charges. I performed the sizing, design, and printing using the already established procedure. The charge

Fig. 1 A concentrated cumulative charge

Fig. 2 A linear cumulative charge

shown in Fig. 3 met the expectations. However, it is worth noting that due to the peculiarities of the manufacturing technology, some difficulties arose, which could only be resolved with a certain level of 3D printing knowledge. These challenges did not make the production of the components impossible but did influence the design

Fig. 3 A profile shaped charge

process. This implies that some engineering foresight is necessary when selecting dimensions and shapes in the case of additive manufacturing processes.

Since the process is efficient, there is a possibility that terrorists could manufacture components for improvised explosive devices from these charges. It is widely acknowledged within professional circles that the design of these devices is limited only by imagination, [11] and there are numerous types that armed forces have to confront [12, 13]. Terrorists do not recognize insurmountable obstacles and taboos if the target is a critical infrastructure [14]. The 3D printer can facilitate this reality, and various plastic components can be produced on a desktop.

Embarking on this path merely as a feasibility demonstration, I have created some specialized charges that could potentially be used for criminal purposes to attack certain elements of critical infrastructure. In Fig. 4, a semi-circular cutting charge is displayed, which could be excellently suited for cutting and damaging pipelines. In today's ever-changing climate, serious challenges arise, with one of the particularly vulnerable areas being related to freshwater and water supply [15]. In this regard, the supply and the pipeline network may become increasingly valuable targets that can be significantly attacked with fewer explosives [16].

The next type is a right-angled bent cutting charge, which could be capable of damaging protected distribution cabinets and sheet cabinets, thereby causing systemic damage within a facility. Here, I would like to plant a thought regarding explosion-resistant cabinets in connection with armoured cabinets. It would be worthwhile to examine these with such cutting charges, as they may be readily available to perpetrators (Fig. 5).

I do not share the implementation parameters of these aforementioned charges publicly, to avoid easy replication. I also keep the details of sizing within a closed circle, as it is a fundamental element of their effective operation. Without these details, only similar but not particularly effective components can be produced. However, it is essential to keep in mind that a determined terrorist could eventually find a solution to create a functional charge. This can happen because there are numerous advantageous aspects for them:

- inexpensive plastic components can be produced;

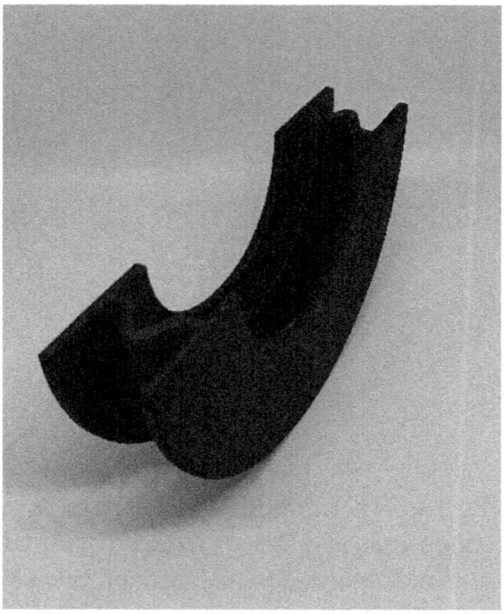

Fig. 4 A semi-circular cutting charge

Fig. 5 A right-angled bent cutting charge

- custom shapes can be designed;
- the buyer is not easily traceable or traceable at all;
- much less explosive material is needed for the destruction of metal objects;
- the finished components are difficult to identify even for a trained eye;
- tools can be made in small pieces, facilitating their transport and smuggling.

However, alongside the numerous advantages, there are also disadvantages:

- design and execution require some degree of expertise.
- affordable components can only be manufactured from plastic materials.
- highly complex shapes can only be produced in pieces.
- there are unique limitations to manufacturing technologies.

5 Conclusion

It is somewhat uncomfortable to state that a widely used technical device, found in an increasing number of households, can be used for criminal purposes. However, based on the above analysis, FDM-based 3D printers, confirming my assumption, are suitable for producing components of charges for illicit purposes that could be used in attacks against critical infrastructure.

At the moment, this doesn't seem to be a widespread method or an established practice for making explosive devices. However, it's only a matter of time before these changes. There is an opportunity now to prepare for this situation by controlling and documenting purchases and acquisitions. Additionally, it would be forward-thinking to train security personnel to visually identify such plastic components. This could be complemented by enhanced explosive detection capabilities for critical infrastructure, as plastic components are essentially worthless without explosive material and an initiator. Although, for a specific purpose, significantly less explosive material may be sufficient when combined with them.

References

1. Gál B, Németh A (2019) Additív gyártástechnológiák katonai alkalmazásának vizsgálata, különös tekintettel a katonai elektronika területére. Hadmérnök 14(1):231–249
2. Lukács L (2017) Szemelvények a magyar robbantástechnika fejlődéstörténetéből – Különös tekintettel a továbbfejlesztés várható irányaira és a kor új kihívásaira. Dialóg Campus Kiadó, Budapest
3. Lukács L (2010) A kumulatív töltetek és gyakorlati alkalmazásuk. Műszaki Katonai Közlöny 20(1–4):178–179
4. Duan B et al (2017) Blasting demolition of steel structure using linear cumulative cutting technology. Adv Mech Eng 9(11):1–11
5. Lukács L (1992) A kumulatív hatás és a kumulatív töltetek méretezése – Jegyzet a Szárazföldi Haderőnemi Fakultás műszaki hallgatói számára. Magyar Honvédség, Zrínyi Miklós Katonai Akadémia, Műszaki tanszék
6. Ember I (2022) Modern kumulatív töltetek hatékonyságának vizsgálata. Haditechnika 56(6):15–20
7. Ember I (2024) Additív eljárással készült lineáris vágótöltetek működésének vizsgálata. Hadmérnök 18(3):5–17
8. Ember I (2023) 3D nyomtatott kumulatív idomtöltetek tesztrobbantása. Műszaki Katonai Közlöny 33(3):29–40
9. Daruka N (2012) Bűnös célú robbanószerkezetek alkalmazásának és hatástalanításának sajátosságai. In: Bohus G (ed) A fúrás-robbantástechnika 2012 c. konferencia előadásai. Országos Magyar Bányászati és Kohászati Egyesület Robbantástechnikai Szakbizottság, Miskolc, pp 109–118
10. Tomolya J, Padányi J (2012) A terrorizmus jelntette kihívások. Hadtudomány 22(3–4):34–67
11. Daruka N (2018) A jövő háborúi az improvizált robbanószerkezetek alkalmazásának tekintetében. Seregszemle 16(2):7–22
12. Kovács Z (2012) Az improvizált robbanóeszközök főbb típusai. Műszaki Katonai Közlöny 22(2):37–52
13. Kovács Z (2012) Fontos létesítmények IED elleni védelme. Műszaki Katonai Közlöny 22(ksz.), pp 35–44

14. Daruka N (2010) A bűnös célú/terrorista robbantások és az ellenük való védekezés lehetőségei. Műszaki Katonai Közlöny 20(1–4):229–241
15. Padányi J (2022) Kihívások, kockázatok, válaszok—Az éghajlatváltozás okozta kihívások és azok hatása a katonai erőre. Ludovika Egyetemi Kiadó, Budapest
16. Földi L, Berek T, Padányi J (2021) Hungary's energy and water security countermeasures as answers to the challenges of global climate change. AARMS—Acad Appl Res MilY Public Manag Sci 20(2):87–96

Revolutionizing Industries Through Virtual Reality: Current State, Applications, and Future Directions

Dániel Frankl

Abstract Virtual Reality (VR) as an immersive technology, has revolutionized the way interactive content is consumed, spanning from the entertainment industry through education, healthcare, and engineering design to manufacturing. Highlighting the educational and training potential of VR, the study points out that the technology enables students and professionals in the field of health-care, engineering or medical science to gain practical experiences in a risk-free environment. The study concludes that while it is advisable for industries and researchers alike to invest in fully leveraging the potential of VR, further research and development of VR technology are crucial for enhancing efficiency, fostering innovation, and enriching human experience.

Keywords Virtual reality · Industry applications · Immersive technology · Innovation

1 Introduction

Over the past decade, Virtual Reality (VR) technology has undergone significant development, transforming the way we interact with digital content. Initially emerging as a revolutionary innovation within the entertainment industry, it soon became evident that VR's potential extends far beyond gaming and media consumption. Today, the application of VR technology spans education, healthcare, engineering design, manufacturing, and many other areas, allowing us to approach complex problems and tasks in new and innovative ways.

This technology creates immersive, 3D environments that are interactive and lifelike, enabling users to fully immerse themselves in a virtual world. The depth of experience and level of user interaction offered by VR opens new opportunities for

D. Frankl (✉)
Doctoral School on Safety and Security Sciences, Obuda University, Budapest, Hungary
e-mail: frankl.daniel@stud.uni-obuda.hu

T. A. Kovács et al. (eds.), *The Impact of the Energy Dependency on Critical Infrastructure Protection*, Advanced Sciences and Technologies for Security Applications, https://doi.org/10.1007/978-3-031-78544-3_25

developing practical skills in education, practicing risk-free surgical procedures in healthcare, product designs and testing in the industry.

The further development and application of VR technology in various industrial sectors not only create new opportunities across a broad spectrum of human activity and interactions but also present challenges for designers, developers, and researchers. These challenges include overcoming hardware and software limitations, enhancing user experience and addressing ethical and privacy issues.

In the following pages, we will examine the evolution, current state, possibilities, and prospects of VR technology. Based on past experiences and literature references, we will demonstrate how VR can change the future ways of working and living, facilitating efficiency improvement, innovation, and enrichment of human experience.

2 The Origins and Evolution of Virtual Reality Technology

The journey of Virtual Reality (VR) from a conceptual idea to a cornerstone of technological innovation spans several decades, tracing back to the 1960s. Initially conceived to fully immerse users in a digitally created world that mimics reality, VR's early days were marked by rudimentary systems constrained by the technological limits of the era. Despite these initial limitations, the ambition to create an immersive interactive environment was clear and unwavering, laying the foundational stone for the future evolution of VR technology.

One of the seminal moments in the history of VR was the development of "The Sword of Damocles" by Ivan Sutherland in 1968. This was not the first ever Head-Mounted Display (HMD), for instance, a collection of images compiled by Stephen R. Ellis of NASA Ames includes one dating back to 1613 [1]. Regarded as the first operational VR system, it was primitive by today's standards but revolutionary for its time. It introduced the concept of a head-mounted display (HMD), offering users real-time 3D graphics that, although simple, opened new possibilities for human–computer interaction. This pioneering system demonstrated the feasibility of VR, providing a glimpse into its potential to transform the way we engage with digital environments.

Throughout the 1980s and 1990s, VR technology witnessed further development, primarily within the realms of entertainment and military training [2]. This era marked the arrival of the first commercial VR systems. While still expensive and relatively limited in capabilities, these systems managed to deliver immersive experiences, showcasing VR's potential to captivate and engage users in novel ways. The advancements during this period were instrumental in demonstrating VR's applicability beyond mere conceptual exploration, suggesting its utility in practical and recreational domains alike.

The early twenty-first century marked a renaissance in VR technology, fueled by significant advancements in hardware and the concerted efforts of entities. The unveiling of the Oculus Rift prototype in 2012 signified a pivotal shift, heralding the modern era of VR [3]. This milestone made VR technology more accessible and sophisticated, significantly expanding its potential applications to the public. The Oculus Rift and similar subsequent innovations have played a crucial role in democratizing VR, making it a viable tool for a broad array of applications ranging from education and healthcare to entertainment and beyond.

Today, VR stands as a beacon of digital innovation, with its applications extending across various sectors. The technology enables immersive learning experiences in education, realistic simulations for surgical training in healthcare, and innovative design and prototyping tools in engineering, among others. The evolution of VR is a testament to the enduring human quest to blend the digital and physical realms, enhancing our interaction with and understanding of both. As VR technology continues to advance, it promises to further bridge the gap between digital experiences and physical reality, offering even more immersive and interactive opportunities for users worldwide.

3 Physical Limitations

Despite the rapid development of Virtual Reality (VR) technology, it faces numerous physical limitations that affect its usability, as well as its mechanical and electronic components. These limitations can impact the quality of the VR experience, user comfort, and the ability to expand application areas. This section will outline the physical limitations of VR.

3.1 Cybersickness

The discrepancies between virtual and real motion can induce motion sickness in some users, leading to nausea, dizziness, and loss of balance. This phenomenon occurs due to the brain receiving conflicting signals from the visual system, which perceives motion in the virtual environment, and the vestibular system, which does not detect corresponding physical movement [4]. This sensory mismatch can lead to a disorientation like motion sickness, but it is experienced within the context of VR.

The causes of cybersickness are multifaceted, encompassing factors such as the quality of the VR experience, including resolution and refresh rates, the user's movements within the virtual environment, and individual susceptibility to motion sickness. For instance, rapid movements or extensive navigation in the VR space without corresponding physical motion can exacerbate the feeling of disorientation and discomfort.

Ultimately, minimizing cybersickness is crucial for enhancing user comfort and expanding the applicability of VR technology across various domains, from entertainment and training to healthcare and education. Ongoing research and development in VR system design and user experience optimization remain key to addressing the challenges of cybersickness and unlocking the full potential of VR as a transformative technology.

3.2 Discomfort and Fatigue

Mechanical limitations of Virtual Reality (VR) technology, particularly in the context of head-mounted displays (HMDs), can significantly impact user comfort and overall experience. Discomfort and fatigue are common issues reported by users of VR systems, primarily attributed to the physical design and ergonomics of HMDs. The weight of these devices and their tight fit around the head and face can exert undue pressure, leading to discomfort during extended use [5].

This physical strain not only detracts from the immersive experience but can also discourage prolonged interaction with VR content, limiting the technology's potential for applications requiring longer usage periods, such as education, training, and certain types of entertainment.

Addressing these mechanical limitations involves a multi-faceted approach focusing on the design and engineering of VR hardware. Efforts to reduce the weight of HMDs using lighter materials and more compact electronics are crucial. Additionally, improving the adjustability and fit of these devices can help accommodate a wider range of head sizes and shapes, thereby reducing pressure points and enhancing comfort. Ergonomic design principles that distribute the weight of HMDs more evenly can also alleviate the strain on specific areas of the head and face.

Advancements in materials science and engineering may offer novel solutions, such as flexible displays and adaptive materials that conform more naturally to the user's head. Furthermore, the development of wireless HMDs can eliminate the discomfort and restrictions associated with tethered cables, offering users greater freedom of movement, and reducing the risk of fatigue associated with navigating physical constraints.

As VR continues to evolve, ongoing research and innovation in HMD design will play a pivotal role in enhancing the usability and appeal of VR experiences. By prioritizing user comfort and ergonomic design, developers can ensure that VR remains an engaging and accessible technology for a broad audience.

3.3 Resolution and Field of View

The limitations in resolution and field of view (FoV) inherent in Virtual Reality (VR) devices are critical factors that can significantly impact the quality of the user experience, diminishing the realism of the virtual environment. High resolution is essential for creating clear, detailed virtual scenes, while an expansive field of view contributes to a user's sense of immersion, making the virtual world feel more expansive and real. When either of these aspects is lacking, it can result in a less convincing virtual experience, potentially leading to a disconnect between the user and the virtual environment. This reduction in realism can hinder the effectiveness of VR applications, whether they are intended for entertainment, education, or professional training purposes.

Improving the geometric calibration of head-mounted displays (HMDs) is one approach to addressing these limitations. Through precise calibration techniques, it is possible to enhance the accuracy of distance estimation within the virtual environment, thereby improving the overall user experience. Such improvements can make virtual objects and spaces appear more true-to-life, enhancing depth perception and spatial awareness for the user. As the technology continues to evolve, advancements in display resolution and FoV will play a pivotal role in overcoming current electronic limitations, pushing the boundaries of what is possible in virtual reality and expanding its application across various fields [6].

3.4 Wireless Data Transmission

In case of not using a wired equipment, limitations of wireless technology, such as low data transmission speed and latency, play a critical role in shaping the quality of the Virtual Reality (VR) experience. These limitations can hinder the seamless integration of real-time user actions with the virtual environment, a cornerstone for achieving a truly immersive VR experience. Low data transmission speeds can cause delays in rendering graphics and processing user inputs, while high latency can disrupt the synchronization between the user's physical movements and the virtual response. This lack of immediacy and responsiveness can significantly diminish the user's sense of presence within the virtual world, one of the most valued aspects of VR technology.

The challenge of overcoming these wireless data transmission limitations is multifaceted. It involves not only advancements in hardware, such as the development of more efficient wireless communication protocols and the integration of higher bandwidth capabilities, but also innovations in software algorithms that can better predict and compensate for potential lags in data transmission.

Efforts to mitigate the impact of wireless limitations on VR also include the optimization of VR content to require less bandwidth without compromising quality. Techniques such as data compression, efficient encoding schemes, and the intelligent prioritization of data packets can help ensure that critical information is transmitted with minimal delay. Additionally, local processing and the use of edge computing can reduce the reliance on distant servers, thereby decreasing the potential for latency.

The ongoing research and development in this area are vital, as highlighted by Sra and Schmandt's work on MetaSpace II, which explores object and full-body tracking for interaction and navigation in VR [7]. By addressing the constraints of wireless data transmission, future VR systems can offer users smoother, more responsive experiences that are closer to real-world interactions, further blurring the lines between physical and virtual realities.

3.5 Limited Physical Interaction

One of the prevailing challenges within Virtual Reality (VR) technology is the limitation in facilitating comprehensive physical interaction, particularly through haptic feedback. This form of feedback is crucial for simulating the sense of touch, allowing users to experience and manipulate virtual objects as if they were real. The current constraints in delivering detailed haptic feedback significantly impact the user's ability to fully engage with the virtual environment, thus diminishing the overall depth of immersion. Without the tactile response that confirms interactions with virtual elements, the boundary between the virtual and the real remains distinctly noticeable, preventing users from achieving a complete sense of presence within the virtual space, especially when using hand tracking features.

Addressing the limitations in haptic feedback requires innovative approaches to simulate physical touch and texture in a manner that convincingly replicates real-world interactions. Developments in this area, such as the work by Zenner and Krüger on "Shifty: A Weight-Shifting Dynamic Passive Haptic Proxy to Enhance Object Perception in Virtual Reality," represent significant strides towards overcoming these challenges [8]. Their research explores the use of dynamic passive haptic proxies that can shift their weight in response to virtual interactions, offering users a more tangible connection to virtual objects without the need for complex and costly active haptic devices.

Future advancements in haptic technology aim to create more sophisticated systems capable of delivering a wide range of tactile sensations, from the texture of surfaces to the resistance of objects. These technologies include wearable haptic suits, gloves equipped with force feedback mechanisms, and localized vibration feedback devices, all designed to enhance the realism of virtual experiences. By integrating these advanced haptic feedback mechanisms, VR can transcend its current limitations, offering users not only visually immersive experiences but also physically interactive environments where they can truly feel the virtual world around them.

The integration of enhanced haptic feedback mechanisms will play a pivotal role in bridging the gap between virtual and physical realities, paving the way for more immersive and interactive virtual experiences. The ongoing research and development in this field are crucial for realizing the full potential of VR as a transformative medium for gaming, education, training, and beyond, while also making these technologies more affordable.

4 Software Limitations

Despite the rapid advancement of Virtual Reality (VR) technology, it faces several software limitations that present significant challenges in technology development and application. These limitations can affect the availability, performance, and user experience of VR applications. This section details the software limitations impacting both development and usage, including the complexity of content development, compatibility issues, user interaction constraints, and the requirements for real-time performance.

4.1 Complexity of Content Development

The creation of Virtual Reality (VR) content represents a significant challenge within the field, characterized by its demand for extensive time, resources, and specialized skills. The intricacies of developing immersive VR experiences stem from the necessity for expertise in several key areas: 3D modeling, user interaction design, and the optimization of virtual environments. Each of these components plays a vital role in creating VR content that is not only visually compelling but also interactive and engaging for the user.

3D modeling forms the backbone of VR content, requiring detailed creation and rendering of objects and environments that users will explore. This process demands not only technical skills in software and graphic design but also a creative understanding of spatial composition and aesthetic appeal.

User interaction design in VR goes beyond traditional interface design, encompassing the development of intuitive controls and mechanisms that allow users to naturally navigate and interact with the virtual world. This aspect of VR content development requires a deep understanding of human–computer interaction principles, ensuring that users can perform actions within the VR environment with ease and precision.

Optimization of virtual environments is critical to ensuring that VR experiences run smoothly across different hardware setups. Developers must balance high-quality graphics and complex interactions with the technical limitations of VR systems, such as processing power and memory constraints. This optimization process is essential for maintaining immersion and preventing issues like latency or frame rate drops, which can break the sense of presence in VR and potentially lead to discomfort or motion sickness.

The comprehensive insight into the foundational methods and challenges of VR content development, as discussed in the work by Dörner, Broll, and Grimm, underscores the multifaceted nature of creating engaging and effective VR experiences [9]. As the field continues to evolve, the development of tools and techniques that streamline these processes and lower the barriers to entry will be crucial for expanding the reach and impact of VR technology.

4.2 Compatibility Issues

The development of Virtual Reality (VR) applications faces significant challenges due to the diversity of hardware platforms, including various head-mounted displays and input devices. Each hardware platform often requires unique optimizations to accommodate its specific specifications, such as screen resolution, field of view, and processing capabilities. This necessity for tailored development can lead to compatibility issues, where an application optimized for one device may not perform adequately on another, potentially degrading the user experience.

To mitigate these issues, developers may need to produce multiple versions of their VR applications, each optimized for different devices. This process increases the complexity and cost of VR content development. Industry-wide standards and cross-platform tools are emerging solutions to ease these compatibility challenges, aiming to streamline the development process and ensure a more uniform VR experience across various devices. Despite these efforts, achieving universal compatibility remains a complex task that underscores the need for ongoing collaboration within the VR industry.

4.3 Dynamic Evolution of VR Development Environments

The development environments within the realm of Virtual Reality (VR) technology are marked by significant and rapid evolution. This constant state of flux is driven by the continuous advancements in VR hardware and software, as well as the growing demands of applications that seek to push the boundaries of what is currently possible. For developers, this presents both an exciting opportunity and a formidable challenge.

As VR technology progresses, developers must adapt to new tools, platforms, and programming paradigms to leverage the latest advancements. This might include transitioning to more powerful engines capable of rendering more complex virtual environments, adopting new programming languages better suited for VR development, or integrating novel interaction techniques that enhance user immersion and experience.

The fast pace of change in the development environments necessitates a commitment to lifelong learning and flexibility among VR professionals. Developers often find themselves on the cutting edge of technology, experimenting with untested software or hardware to create experiences that were unimaginable just a few years prior. This environment fosters innovation but also requires a willingness to navigate the uncertainties and challenges inherent in working with emerging technologies.

Moreover, the rapid evolution of VR development tools can lead to fragmentation, with a diverse array of platforms and standards complicating the development process. To mitigate these challenges, the VR community is increasingly focusing on creating more unified and stable development ecosystems. Initiatives aimed at standardizing VR development practices and fostering interoperability between different VR systems are crucial for simplifying the development process and ensuring a wider adoption of VR technology.

4.4 Real-Time Performance Requirements

For Virtual Reality (VR) applications to achieve the goal of immersing users in a believable alternate reality, they must be capable of responding to user inputs and interactions without perceptible delays. This requirement for instantaneous feedback places substantial demands on both the software and hardware running VR applications. The essence of a seamless VR experience lies in its ability to mirror the immediacy of the real world, where actions and reactions occur in real-time.

Delays in processing or rendering not only disrupt the immersive experience but can also contribute to motion sickness, a phenomenon exacerbated by the disconnect between what users see in the VR environment and how their bodies perceive motion in the physical world as mentioned before. To mitigate these issues, VR systems must be equipped with high-performance hardware capable of rendering complex 3D environments at high frame rates. Similarly, VR software must be optimized for efficiency, minimizing the time between user actions and on-screen responses.

The challenges of achieving real-time performance in VR are multifaceted, involving advances in graphical processing, data throughput, and latency reduction. Developers and researchers continuously seek innovative solutions to these challenges, such as more powerful processing units, optimized rendering techniques, and algorithms that predict user movements to pre-render scenes.

As VR continues to evolve, the ability to meet these performance demands will be a key factor in determining the success and user acceptance of VR applications across various domains, from gaming and entertainment to education and training.

5 Deployment of Virtual Reality Across Industries

Virtual Reality (VR) technology is reshaping a wide range of industries by offering immersive experiences that bridge the gap between virtual simulations and real-world scenarios. In mental health, VR serves as a powerful tool for therapists to treat conditions like phobias, anxiety disorders, and PTSD through controlled exposure therapy. The sports sector benefits from VR's ability to simulate specific training scenarios, enhancing athletes' performance without the risk of injury. The retail industry is witnessing a transformation with VR shopping experiences allowing customers to try on clothes and explore products virtually, enhancing the convenience and enjoyment of online shopping.

Event planning and real estate have also embraced VR for virtual tours and previews, enabling clients to visualize spaces and properties before they are physically realized, thereby streamlining decision-making processes. The automotive industry leverages VR for both design and customer engagement, offering virtual test drives and safety simulations. In entertainment and media production, VR introduces immersive storytelling, allowing viewers to actively explore and interact with cinematic worlds.

The rise of remote work has spurred the development of VR-enhanced collaborative workspaces, creating realistic environments for meetings and teamwork. Environmental and urban planning professionals use VR to visualize and assess the impact of projects, improving stakeholder engagement and decision-making. Education benefits from VR's ability to take students on virtual field trips, making learning more accessible and engaging by transcending geographical limitations.

Furthermore, VR's application in engineering design, healthcare training, manufacturing, and military training exemplifies its role in enhancing learning, design, and operational efficiencies. By simulating complex procedures and scenarios, VR technology allows for hands-on experience in a risk-free environment, thereby enhancing skills, reducing development cycles, and ensuring safety.

Across these diverse applications, VR technology stands out as a transformative tool that not only enhances current practices but also opens new possibilities for innovation and engagement in various sectors. As VR technology continues to evolve, its potential to revolutionize how we learn, work, design, and interact within both virtual and physical realms becomes increasingly evident, promising a future where VR applications continue to expand and enrich our experiences.

6 Concluding Reflections

The rapid development and versatile application of Virtual Reality (VR) technology across various industrial sectors have opened new dimensions in training, design, manufacturing, healthcare, and many other areas. VR's ability to create immersive, interactive environments allows users to gain realistic experiences without being physically present in a specific situation or location. This is particularly important in the fields of education and training, where VR enables the practice of complex operations within a safe and controlled environment, thereby increasing the efficiency of the learning process and reducing the risk of errors in a cost-effective manner.

In the coming years, VR technology is expected to continue evolving, with the introduction of new tools and software solutions that further exploit this technology's potential. Through the expansion of VR applications, ongoing research and development aimed at overcoming technological limitations, VR is set to become an integral part of industrial and societal applications even more widely, promoting efficiency enhancement, innovation, and the enrichment of human experience.

Considering these developments, Virtual Reality is not just a transient technological trend but a profound and lasting tool that is shaping the future ways of working and living. Industries and researchers alike should invest in further research and development of VR technology to maximize its benefits and possibilities for society, as its full potential is yet to be realized.

References

1. Slater M, Sanchez-Vives MV (2016) Enhancing our lives with immersive virtual reality. Front Robot AI 3:74
2. Lanier J (2017) Dawn of the new everything: encounters with reality and virtual reality. Henry Holt and Company
3. Gleasure R, Feller J (2016) A rift in the ground: theorizing the evolution of anchor values in crowdfunding communities through the oculus rift case study. J Assoc Inf Syst 17(10):1
4. Porcino TM, Clua E, Trevisan D, Vasconcelos CN, Valente L (2017) Minimizing cyber sickness in head mounted display systems: design guidelines and applications. In: 2017 IEEE 5th international conference on serious games and applications for health (SeGAH). IEEE, pp 1–6
5. Lee J, Kim M, Kim J (2017) A study on immersion and VR sickness in walking interaction for immersive virtual reality applications. Symmetry 9(5):78
6. Kellner F, Bolte B, Bruder G, Rautenberg U, Steinicke F, Lappe M, Koch R (2012) Geometric calibration of head-mounted displays and its effects on distance estimation. IEEE Trans Visual Comput Graphics 18(4):589–596
7. Sra M, Schmandt C (2015) Metaspace ii: object and full-body tracking for interaction and navigation in social vr. arXiv preprint arXiv:1512.02922
8. Zenner A, Krüger A (2017) Shifty: a weight-shifting dynamic passive haptic proxy to enhance object perception in virtual reality. IEEE Trans Visual Comput Graphics 23(4):1285–1294
9. Doerner R, Broll W, Grimm P, Jung B (eds) (2022) Virtual and augmented reality (VR/AR): foundations and methods of extended realities (XR). Springer Nature

Embracing the Sharing Economy: A Global Imperative

Tobias Gölling

Abstract The Sharing Economy, a multifaceted phenomenon, has garnered significant attention globally for its transformative potential in reshaping economic, social, and environmental dynamics. This abstract provides a succinct overview of key aspects explored in the paper regarding the Sharing Economy's implications on a global scale. Beginning with a definition, the paper elucidates the concept of the Sharing Economy, highlighting its diverse manifestations across various sectors and its underlying principles of collaborative consumption and resource optimization. Drawing upon theoretical frameworks and empirical research, the paper delves into the core components of sharing practices, encompassing involved actors, compensation schemes, participation motives, and ownership transfer. Furthermore, the paper examines strategies for competing in the Sharing Economy, focusing on differentiation approaches such as technology, partnership, and user-experience strategies. It also discusses the challenges and opportunities inherent in sharing practices, including issues related to pricing, trust, and resource allocation, as well as the potential for enhanced resource utilization and decentralized value creation. Finally, the paper explores implications for both practice and theory, emphasizing the importance of continuous adaptation and innovation in navigating the complexities of the Sharing Economy. By understanding the drivers and dynamics of sharing practices, stakeholders can harness its transformative potential to foster sustainability, inclusivity, and economic resilience in an increasingly interconnected world.

Keywords Sharing Economy · Sharing practices · Business models

T. Gölling (✉)
Obuda University, Budapest, Hungary
e-mail: tobiasgoe@googlemail.com

T. A. Kovács et al. (eds.), *The Impact of the Energy Dependency on Critical Infrastructure Protection*, Advanced Sciences and Technologies for Security Applications, https://doi.org/10.1007/978-3-031-78544-3_26

1 Introduction

In an era characterized by rapid technological advancements, shifting consumer behaviors, and growing environmental concerns, the concept of the Sharing Economy has emerged as a transformative force with profound implications for societies worldwide. The Sharing Economy, often hailed as a paradigm shift in economic and social relations, revolves around the collaborative consumption and utilization of resources, facilitated by digital platforms and peer-to-peer networks. This phenomenon encompasses a diverse array of sectors, from transportation and accommodation to goods and services, challenging traditional ownership models and fostering a culture of resource optimization and communal exchange.

The significance of the Sharing Economy extends far beyond its economic implications; it embodies a fundamental shift in values, emphasizing access over ownership, sustainability over excess, and community over individualism. As societies grapple with pressing issues such as climate change, resource scarcity, and socioeconomic inequality, the Sharing Economy offers a promising alternative framework that promotes efficiency, inclusivity, and environmental stewardship.

Against this backdrop, this paper seeks to explore the importance of the Sharing Economy on a global scale. By examining its impact across different regions, industries, and demographic groups, its aim is to shed light on the multifaceted benefits and challenges associated with embracing sharing practices. Through an analysis of case studies, empirical research, and theoretical frameworks, the transformative potential of the Sharing Economy will be elucidated in addressing pressing global challenges and shaping a more sustainable and equitable future for all.

Drawing upon insights from economics, sociology, environmental studies, and technology, this paper will provide a comprehensive overview of the Sharing Economy's relevance in today's interconnected world. By understanding the drivers, dynamics, and implications of sharing practices, policymakers, businesses, and individuals can harness the full potential of the Sharing Economy to foster innovation, promote social cohesion, and build resilient communities in an increasingly interconnected and resource-constrained world.

2 Sharing Economy

2.1 Definition

The phenomenon known as the Sharing Economy encompasses various terms such as Access-Based Consumption [1], Collaborative Consumption [2], or Collaborative Economy [3], leading to challenges in defining clear conceptual boundaries [4].

The concept of ownership and sharing can be illustrated in a diagram [5], which is depicted in Fig. 1.

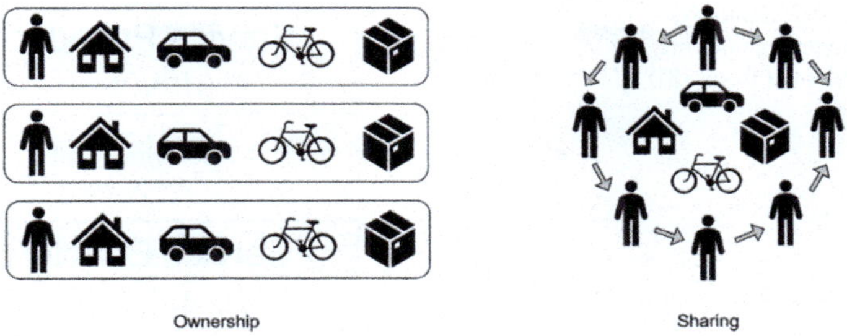

Fig. 1 Depiction of ownership and sharing (adopted from: [5])

While ownership entails individual possession and use of belongings, sharing involves a collective utilization of resources among a group of individuals.

The Sharing Economy involves the exchange of underutilized goods and services for both monetary and non-monetary gains [3], either through peer-to-peer or business-to-customer models. It can also be seen as a community where consumers grant temporary access to their underutilized or unused physical goods, often for a fee [6]. Another perspective defines it as a socioeconomic system facilitating the exchange of resources between individuals and companies through intermediaries, aimed at optimizing resource efficiency. Alternatively, it may serve as an overarching term encompassing various related concepts, aiding in understanding and implementing new societal behavioral patterns [7].

In summary, the Sharing Economy involves individuals sharing underused products or services with one another for socio-economic benefits, typically for a limited duration and in exchange for some form of compensation.

2.2 *Concepts and Theories*

The authors [8] propose a framework illustrating the functioning of Sharing Economy companies. Their model comprises three key elements: a customer seeking a service, a service provider delivering the service (e.g., a driver), and a service enabler who facilitates the connection between the customer and the service provider through their platform. Feedback from both the customer and the service provider is solicited by the service enabler following each interaction. This model, exemplified through car sharing in Fig. 2, elucidates the socioeconomic transactional dynamics involved. These three roles are similarly delineated using the example of Airbnb [9]. Offers depiction an initial understanding of the Sharing Economy dynamics in a broader context [8].

In the following subsection, a more detailed framework outlining key concepts within the Sharing Economy allows distinct perspectives on this subject matter.

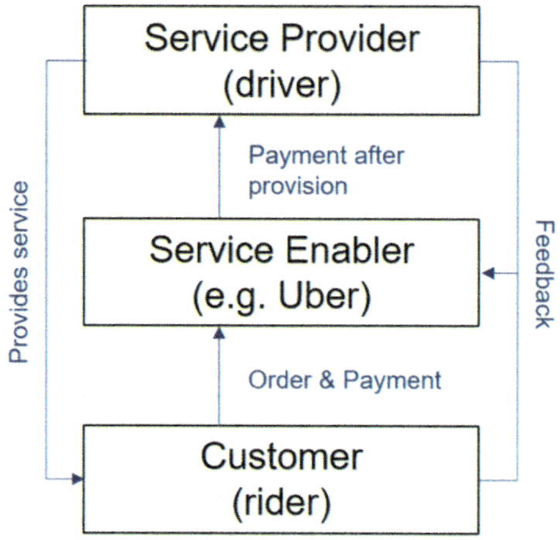

Fig. 2 Sharing Economy business model, car sharing example (adopted from: [8])

2.3 Core Components of Sharing Practices

Four fundamental components can be delineated pivotal in structuring sharing practices, namely the involved actors, compensation schemes, participation motives, and ownership transfer [10]. These components will now be elaborated upon individually.

The term 'involved actors' pertains to the participants engaged in sharing practices, encompassing consumers, resource owners, and businesses. Such practices can be categorized based on the ownership of shared resources, with distinctions drawn between Consumer-to-Consumer (C2C), Business-to-Customer (B2C), and Government-to-Customer (G2C) models [10–12].

'Compensation schemes' denote the methods through which the utilization of sharing practices is reciprocated. These schemes include direct monetary compensation, such as transaction fees or membership dues, as well as direct and indirect non-monetary compensation, involving bartering or gifting [1, 13].

'Participation motives' elucidate the underlying reasons for individuals and resource owners to engage in sharing practices, spanning economic benefits, environmental consciousness, social connections, cultural inclinations, and dissatisfaction with traditional markets [1, 8, 13].

Lastly, 'ownership transfer' distinguishes whether there is a transfer of property between individuals. While some researchers assert that sharing economy transactions entail non-ownership access to resources, others allow for varying degrees of ownership transfer, encompassing actions like donating, swapping, or trading [6, 14]. It is widely acknowledged that both physical properties (e.g., cars, apartments) [1, 11] and immaterial goods (e.g., knowledge, skills) are integral to sharing practices [2, 15].

2.4 Importance of Sharing Economy

Several reasons can be provided for the question, why the Sharing Economy is such an important part of todays world. Different approaches are stated in the following:

Resource Optimization: With growing concerns about sustainability and resource scarcity, the sharing economy enables more efficient use of existing resources. By sharing goods, services, and assets, fewer resources are wasted, contributing to a more sustainable future.

Cost Savings: In an era of economic uncertainty and fluctuating market conditions, the sharing economy offers cost-effective alternatives to traditional ownership. Consumers and businesses can access what they need without the upfront costs and ongoing expenses associated with ownership.

Flexibility and Convenience: The sharing economy provides flexibility and convenience, allowing users to access goods and services on-demand and often through digital platforms. This aligns with changing consumer preferences for experiences over ownership and the desire for instant gratification.

Entrepreneurship and Employment: The sharing economy fosters entrepreneurship by enabling individuals to monetize their assets or skills. Platforms like Airbnb, Uber, and TaskRabbit allow people to earn income by renting out their properties, providing transportation services, or offering freelance services. This can create opportunities for income generation, especially in regions with high unemployment rates or limited job opportunities.

Community Building: Sharing economy platforms often foster a sense of community by connecting individuals with shared interests or needs. This can lead to social interactions and collaborations that strengthen social bonds and build trust among participants.

Technological Advancements: Advances in technology, such as mobile applications, GPS tracking, and online payment systems, have facilitated the growth of the sharing economy. These technological innovations make it easier for users to participate in sharing economy transactions, driving its widespread adoption.

Disruption and Innovation: The sharing economy has disrupted traditional industries and business models, forcing companies to innovate and adapt to changing consumer preferences. This drives competition and encourages businesses to improve their offerings, leading to better products, services, and experiences for consumers.

Regulatory and Policy Implications: The rise of the sharing economy has prompted governments to reevaluate existing regulations and policies to accommodate these new business models. This ongoing dialogue between policymakers and industry stakeholders is shaping the future of the sharing economy and its impact on society.

Overall, the sharing economy addresses pressing societal challenges while also driving innovation. Its importance today lies in its ability to transform industries, empower individuals, and create more sustainable and inclusive communities.

2.5 Strategies for Competing in the Sharing Economy

In addition to the discourse on the Sharing Economy, there exist strategies that companies can employ to position themselves effectively in a competitive peer-to-peer market and leverage the benefits of sharing practices [16]. The authors research identifies three distinct differentiation strategies, each illustrated with relevant case studies.

The first strategy, termed the technology differentiation strategy, is ideal for companies leveraging technology to gain a competitive edge. For instance, Uber, a prominent car-sharing provider, harnesses technology to identify and capitalize on new opportunities, striving for superior efficiency compared to rivals.

The second strategy is the partnership differentiation approach, wherein companies forge alliances with resource and infrastructure suppliers to gain competitive advantages. An illustration of this is SWA Carsharing, a regional car-sharing service, which strengthens its position in the Augsburg area through simple technologies and partnerships with other car-sharing entities.

Lastly, the user-experience differentiation strategy focuses on delivering a unique user experience, prioritizing social interactions. For instance, BlaBlaCar enhances user engagement by offering driver profiles on its platform, allowing customers to personalize their carpooling experiences based on individual preferences and interests.

It is imperative for companies to reflect on their identity, competition, and the shared product to determine the most effective differentiation strategy. While these strategies are distinct, they can also be complementary when used in conjunction.

2.6 Difficulties and Opportunities

Finally, the potential threats and opportunities associated with sharing practices from various viewpoints should be depicted [8].

The Sharing Economy model encounters both internal and external challenges. Service enablers must navigate factors such as competition-driven price undercutting while ensuring fair compensation for service providers. Failure to meet customer expectations risks eroding trust and customer migration to competitors. Retaining customers amidst a landscape of shifting preferences poses a significant concern for service enablers, who must balance the needs of both providers and customers while maintaining service capacity and quality.

From the perspective of service providers, autonomy in setting prices for their services is highly valued. However, aggressive pricing strategies employed by service enablers may lead to dissatisfaction among providers regarding compensation. Additionally, limited interaction with enablers hinders the resolution of provider concerns, and providers must also arrange their own insurance coverage.

Customers primarily evaluate sharing services based on charges and service quality, directly impacting demand. Dissatisfaction stemming from inadequate information and communication channels can prompt customers to seek alternatives.

To address these challenges, various strategies can be employed. Implementing performance-based remuneration can incentivize service quality improvement and adherence to company standards. Service enablers can mitigate supply gaps and customer/provider turnover by fostering stronger company loyalty. Offering unique value propositions can justify higher prices for cost-conscious customers, provided the overall service experience meets expectations.

For service providers, adopting performance-based compensation schemes and establishing direct communication channels with enablers can enhance satisfaction and engagement. Integrating insurance options into business offerings can further incentivize provider retention.

From the customer perspective, enhancing navigation tools and providing training for ridesharing services can improve the overall experience. Adjusting rating systems can also contribute to a more positive customer experience.

Overall, addressing these challenges and capitalizing on opportunities requires a multi-faceted approach that prioritizes alignment between providers, enablers, and customers.

The significant influence of sharing economies on businesses spreads across various sectors and factors. Some of the key impacts are stated below:

Disruption: Traditional business models are disrupted by sharing economy platforms that enable peer-to-peer transactions. This challenges established industries such as transportation, accommodation, and retail by offering alternative ways for consumers to access goods and services.

Utilization of Assets: Sharing economy platforms allow businesses to optimize the use of their assets. For example, companies can rent out excess office space or equipment during idle periods, generating additional revenue streams.

Cost Reduction: Businesses can save costs by utilizing sharing economy services instead of purchasing or maintaining assets. For instance, instead of maintaining a fleet of vehicles, a company can use ride-sharing services for employee transportation.

Access to Talent: Freelance and gig economy platforms provide businesses with access to a broader pool of talent. This allows companies to find specialized skills on-demand without the need for full-time hires, reducing overhead costs.

Customer Experience: Sharing economy platforms often prioritize user experience and convenience, which raises customer expectations for all businesses. Companies need to adapt by offering more personalized and flexible services to remain competitive.

Regulatory Challenges: The rise of the sharing economy has led to regulatory challenges as traditional regulations may not adequately address these new business models. Businesses operating in the sharing economy must navigate complex legal and regulatory frameworks, which can impact their operations and profitability.

Sustainability: Sharing economy models can contribute to sustainability by promoting resource sharing and reducing waste. Businesses can benefit from aligning with these principles, attracting environmentally conscious consumers, and improving their corporate social responsibility image.

Competition and Collaboration: Traditional businesses face increased competition from sharing economy startups. However, there are also opportunities for collaboration, such as partnering with sharing economy platforms to reach new customers or integrating sharing economy principles into existing business models.

Overall, the sharing economy has reshaped the business landscape, presenting both challenges and opportunities for companies across various industries. Adaptation and innovation are essential for businesses to thrive in this evolving environment.

3 Implications for Practice and Theory

Given the absence of a precise definition for the Sharing Economy, the necessity for a new framework to establish clear distinctions among interchangeable terms persists [10]. Moreover, researchers should seize the opportunities presented by the Sharing Economy and propose strategies for its enhancement, aiming for more efficient resource utilization and decentralized value creation [17].

Practical insights for service enablers highlight the importance of consistently acquiring new customers and service providers while effectively allocating resources to thrive in the Sharing Economy [8]. Firstly, service enablers should assess their existing customer and provider base to pinpoint and rectify any weaknesses or gaps. Secondly, efforts should be made to re-engage lost customers, probing into reasons for their departure. Lastly, service enablers must incentivize providers to sustain their services and align with enabler demands. The sustained involvement of both new and existing customers and providers is crucial for achieving profitability from the perspective of the service enabler.

The implications of the sharing economy offer valuable insights for businesses, policymakers, and scholars alike.

From a practical perspective, businesses must adapt to the changing landscape of consumer behavior and preferences driven by the sharing economy. This entails rethinking traditional business models to embrace collaborative consumption, access-based approaches, and platforms that facilitate peer-to-peer exchanges. Companies that successfully integrate sharing economy principles into their operations can unlock new revenue streams, enhance customer engagement, and foster greater sustainability.

Moreover, the sharing economy presents opportunities for businesses to innovate and differentiate themselves in the marketplace. By leveraging digital platforms, data analytics, and user-centric design, companies can create seamless and personalized sharing experiences that meet the evolving needs of consumers. This requires a shift towards a more agile and adaptive organizational culture that prioritizes experimentation, collaboration, and customer feedback.

On a theoretical level, the sharing economy challenges conventional economic paradigms and offers new perspectives on consumption, ownership, and value creation. Scholars across disciplines such as economics, sociology, and business studies are exploring the implications of the sharing economy for theories of market behavior, social capital, and sustainability. This research contributes to our understanding of how collaborative consumption practices shape societal norms, economic systems, and environmental outcomes.

Furthermore, the sharing economy raises important questions about regulation, governance, and ethics in the digital age. Policymakers face the challenge of balancing innovation and consumer protection while ensuring fair competition and addressing concerns related to labor rights, data privacy, and market concentration. By engaging with stakeholders and conducting empirical research, policymakers can develop regulatory frameworks that support the growth of the sharing economy while safeguarding the public interest.

4 Conclusion

In conclusion, the Sharing Economy represents a pivotal shift in societal norms and economic paradigms, offering a compelling vision for a more sustainable, inclusive, and resilient world. Through the collaborative consumption and utilization of resources, facilitated by digital platforms and peer-to-peer networks, the Sharing Economy fosters efficiency, social cohesion, and environmental stewardship on a global scale.

The importance of the Sharing Economy lies in its ability to address pressing global challenges, ranging from climate change and resource scarcity to socioeconomic inequality. By promoting access over ownership, sustainability over excess, and community over individualism, sharing practices offer practical solutions to some of the most complex issues facing humanity.

Moreover, the Sharing Economy holds immense potential for economic growth and innovation, unlocking new opportunities for entrepreneurship, job creation, and value generation. By embracing sharing practices, individuals, businesses, and governments can harness the power of collaboration to drive positive change and build a more resilient and equitable future for all.

In essence, the Sharing Economy is not just a trend or a business model; it is a fundamental reimagining of how we interact with each other and our environment. As we continue to navigate the challenges of the twenty-first century, the Sharing Economy offers a compelling blueprint for a more sustainable and prosperous world, where resources are shared, opportunities are accessible to all, and collective well-being takes precedence over individual gain. Embracing the Sharing Economy is not only important; it is imperative for the future of our planet and humanity.

References

1. Bardhi F, Eckhardt GM (2012) Access-based consumption: the case of car sharing. J Consum Res 39(4):881–898
2. Möhlmann M (2015) Collaborative consumption: determinants of satisfaction and the likelihood of using a sharing economy option again. J Consum Behav 14(3):193–207
3. Botsman R (2013) The sharing economy lacks a shared definition. Fast company website
4. Acquier A, Daudigeos T, Pinkse J (2017) Promises and paradoxes of the sharing economy: an organizing framework. Technol Forecast Soc Chang
5. Matzner M, Chasin F, Todenhöfer L (2015) To share or not to share: towards understanding the antecedents of participation in IT-enabled sharing services. In: Twenty-third European conference on information systems
6. Frenken K, Schor J (2017) Putting the sharing economy into perspective. Environ Innov Soc Trans 23:3–10
7. Muñoz P, Cohen B (2017) Mapping out the sharing economy: a configurational approach to sharing business modeling. Technol Forecast Soc Chang 125:21–37
8. Kumar V, Lahiri A, Dogan OB (2018) A strategic framework for a profitable business model in the sharing economy. Ind Mark Manage 147–160
9. Benoit S, Baker TL, Bolton RN, Gruber T, Kandampully J (2017) A triadic framework for collaborative consumption (CC): motives, activities and resources & capabilities of actors. J Bus Res 79:219–227
10. Trenz M, Frey A, Veit D (2018) Disentangling the facets of sharing: a categorization of what we know and don't know about the sharing economy. Internet Res 28(4):888–925
11. Cohen B, Kietzmann J (2014) Ride on! mobility business models for the sharing economy. Organ Environ 27(3):279–296
12. Andersson M, Hjalmarsson A, Avital M (2013) Peer-to-peer service sharing platforms: driving share and share alike on a mass-scale. In: Thirty fourth international conference on information systems
13. Lamberton CP, Rose RL (2012) When is ours better than mine? A framework for understanding and altering participation in commercial sharing systems. J Mark 76(4):109–125
14. Barnes SJ, Mattsson J (2016) Understanding current and future issues in collaborative consumption: a four-stage Delphi study. Technol Forecast Soc Chang 104:200–211
15. Gargiulo E, Giannantonio R, Guercio E, Borean C, Zenezini G (2015) Dynamic ride sharing service: are users ready to adopt it? Proc Manuf 3:777–784
16. Frey A, Trenz M, Veit D (2019) Three differentiation strategies for competing in the sharing economy. In: MIS quarterly executive, pp 143–156
17. Heinrichs H (2013) Sharing economy: a potential new pathway to sustainability. GAIA Ecol Perspect Sci Soc 22(4):228–231

Investigation of the Tightening Torque of Electrical Connectors

Ferenc Haraszti

Abstract Some electrical connectors are prone to corrosion. When metals of different potentials come into contact, galvanic corrosion processes may start. This adversely affects the electrical resistance and mechanical properties of the connection. Heat is generated in parts with an electric current, impairing the parameters of electrical conduction. These heat effects are imperceptible for a long time. With the use of a thermal camera, the defects of the connection and problems resulting from an insufficient tightening torque can be detected in time.

Keywords Galvanic corrosion · Resistance · Thermal imager

1 Introduction

Electrical connectors are indispensable components of the electrical and energy industry. These are usually connectors consisting of two or more elements, whose task is to conduct electrical current and ensure mechanical strength. They can be made from various materials. Copper, steel, aluminium and their alloys are most commonly used. Due to economic considerations, usually, two different metals come into close contact in a connector. They are exposed to contact (galvanic) corrosion due to the influence of the environment and the current, but mainly due to the difference in the electrode potential of the metals. The resulting corrosion reaction products deteriorate contact mechanics therefore the joints heat up due to the increased resistance in the connector. The temperature of the connection further increases if the torque of tightening is not appropriate. As a result, the resistance of the connector changes and the temperature of the connector may increase further.

F. Haraszti (✉)
Banki Donat Faculty of Mechanical and Safety Engineering, Obuda University, Budapest, Hungary
e-mail: haraszti.ferenc@bgk.uni-obuda.hu

T. A. Kovács et al. (eds.), *The Impact of the Energy Dependency on Critical Infrastructure Protection*, Advanced Sciences and Technologies for Security Applications, https://doi.org/10.1007/978-3-031-78544-3_27

2 Examination of Tightening Torque Without Corrosion

The current that on electrical connector can conduct does not only depend on the roughness, lack of oxides, shape, temperature, and resistance of the contact surfaces. It is also influenced by the force that compresses the surfaces together. The resistance of the connector changes depending on the tightening torque of the nut or the angle of rotation of the nut proportional to it. A clamping force set to the prescribed value reduces the resistance between the conducting surfaces. The number of surface roughness peaks decreases, and the oxide layer that reduces conductivity may break. In the following figures, we can see the surface and cross-sectional images of the galvanized screw and the copper terminal before corrosion, produced with an atomic force microscope (Figs. 1 and 2).

Galv. screw (alone)	Cross-section
	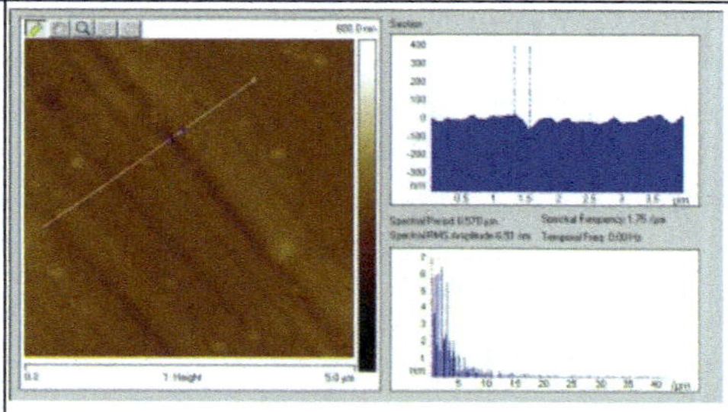

Fig. 1 AMF image of a galvanized screw before corrosion

Terminal (alone)	Cross-section
	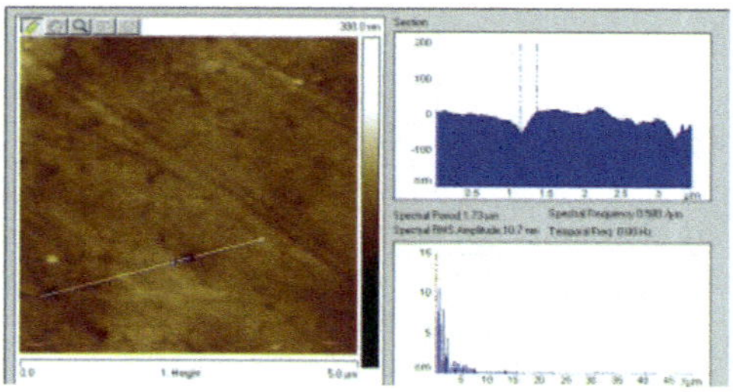

Fig. 2 AMF image of the terminal before corrosion

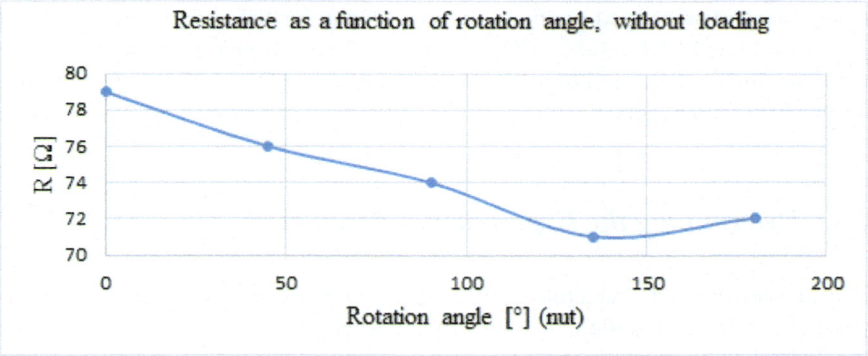

Fig. 3 Resistance as a function of the rotation angle of the nut

The high-resolution microscopic images show the morphology of the examined surfaces. The images show that the surfaces are not perfectly smooth even immediately after production, which affects the resistance of the electrical connection and the resulting heating.

Figure 3 shows the change in resistance as a function of rotation angle (torque).

Resistance is high in the zero-degree (loose) "non-rotation" of the nut. As more and more torque is applied, resistance decreases until the conductive layers are smoothed and the protruding edges disappear. The curve can be approximated with a linear function.

2.1 Investigation of the Galvanic Corrosion of Electrical Connectors Under Laboratory Conditions.

My goal was to produce an accelerated test, with which the behaviour of the material can be observed in a corrosive environment. If its size allows it, the part is examined directly, if it is not possible, a sample is cut from the structural material to be tested and the sample is examined. Laboratory methods can only approximate operational (atmospheric, soil, etc.) conditions, but they can approximate real conditions very well. There are standards for the examination of galvanic corrosion, with which the expected corrosion damage can be predicted. Such standards were, for example, the MSZ 4380-86 standard, which examined the corrosion of metals in contact under natural atmospheric conditions, or the ISO 8044:2020 standard, which estimated galvanic corrosion based on measured potential difference. There is also a corrosion standard showing what metals can be used together [1, 2]. Unfortunately, these standards have been withdrawn and replaced with the MSZ EN ISO 7441:2015 standard [3]. With this standard, galvanic corrosion damage resulting from the use of different metals can be prevented. Currently, there are test methods for measuring corrosion damage and standardized methods and procedures for the prevention of galvanic corrosion (ASTM G82, G71, ISO 23226). My research goal is to develop a new method which can be used to model the galvanic corrosion of electrical connectors in the electrical industry.

2.2 Determining the Experimental Parameters

The aforementioned accelerated process can be produced by influencing several parameters. As a rule, the corrosion process is accelerated if the concentration of the created environment is increased or the composition of the medium is changed appropriately. It also accelerates as temperature rises. The result changes if the relative speed of reactions between the workpiece and the surrounding medium changes. The final result may be very different from the expected result if the effect of the parameters of the corrosion experiment is not sufficiently known.

Main aspects of the examination of the galvanic corrosion of electrical connectors.

The liquid immersion test is a widely used method for the corrosion testing of metals of different types and compositions. In this test, the various parameters and their change can be followed accurately. In order to produce the best possible result, the following factors must be taken into account:

- Workpieces can be obtained by disassembling or cutting from a non-operational circuit, or an original, unused part can be used as well. We need to know their chemical and metallurgical composition, and what metalworking was performed on them. Within a series, homogeneity, or uniform characteristics is of paramount importance. Some test methods require various forms of surface preparation. In our case, we use the original surface or a polished surface, depending on the test method.
- The temperature, air pressure, and relative humidity of the testing room should be strictly constant. These variables can significantly influence the results. If fluctuation occurs, it must be recorded in all cases.
- The physical and chemical parameters variables of the solution surrounding the test piece must also be examined and monitored, including pH, temperature, composition, and relative speed. Even a small change in these will considerably change the measurement results. If the corrosive solution tends to evaporate, it must be replenished at appropriate intervals.
- It is advisable to separate the workpieces. In the case of quick tests, where high accuracy is not necessary, it may be permissible to test several samples simultaneously in a common medium. In the case of precise, scientific tests, however, this is not possible. The surface-to-volume ratio of the workpiece and the test medium must be kept constant. Based on my experience, at least 100 ml of solution is necessary for 1 cm^2.
- The homogeneity of the corrosion solution is very important. There can be no concentration differences in different places. Not even a minimal temperature

fluctuation or flow is permissible. The best results can therefore be achieved in a constantly mixed liquid. If mixing speed is optimal, we can prevent the formation of corrosion products in undesirable areas of the tested workpiece. Too high a mixing speed should also be avoided.

- When dual metals are tested, they should always be fully immersed in liquid. The sample should not be immersed only partly, because the oxygen supply at the air-metal-solution boundary is always far higher than inside the liquid, so a differential concentration element can form, which can falsify the results.
- On the surface of the special screw–terminal joints, it is necessary to ensure that resistance can be measured undisturbed. Corrosion must not affect the entire surface of the system of two metals, so it is necessary to ensure proper contact with the clamping structure. Stainless steel is preferred because of its durability. In general, contact with other test pieces should be avoided, here good contact between the systems of two metals is the goal.

2.3 The Effect of Galvanic Corrosion on the Most Important Operating Characteristics of Electrical Connectors

The measurement methods and standards described in the previous section can characterize the degree of corrosion from many aspects. However, they do not show the galvanic corrosion of different metals clearly. The connection of two or more metals significantly limits the available methods for determining their rate of corrosion. Most of the tests described so far cannot be used. Connectors tightened with a predetermined torque only partially enable visual and microscopic examination, graviometry, etc. We must choose a physical parameter that changes proportionally to the degree of corrosion and can accurately express the degree of galvanic corrosion.

It is often very difficult or impossible to test the corrosion of parts under natural conditions. We can only model the behaviour of electrical connectors in a circuit in a laboratory environment. Here, corrosion in an industrial or other environments can be safely modelled without high or low voltage. Among the many physical properties of metals, the change in electrical resistance is most suitable for accurately determining the corrosion rate of a closed system. The accelerated corrosion process and simultaneous high-precision resistance measurement accurately shows the state of the connection between the two metals.

The goal is to design and build a device that shows the progress of galvanic corrosion. In such a device, different pairs of metals can be tested at the same time, different anode–cathode ratios can be used, and if the device is placed in a closed space, temperature and pressure can be changed, which can also be an influencing factor in a corrosion process.

The use of the test equipment enables immediate monitoring during the entire period. Samples can be replaced independently of the others—it does not affect their results. With computer tools and software, we can create graphs and figures and draw immediate conclusions from the results. We can analyse the effect of various anti-corrosion materials and inhibitors using the corrosion rate. The number of samples that can be tested at the same time depends only on the size of the device, which in our case is similar to that of desktop instruments. The material of the clamping device is made from Cu E 99.9 for good electrical conductivity, and copper-plated steel nuts are used for fastening. The part of the sample holder in contact with steel (aluminium) is treated with a protective paste, which prevents the formation of oxides and avoids the galvanic effect during the measurement. The electrical resistance characterizing the kinetics of contact corrosion is measured with a highly sensitive Wheatstone bridge with a range from 10^{-4} to 10^{6} Ω and is powered by a stabilized direct current power supply. The sensitivity of the associated galvanometer is 0.095×10^{-8} A. The test tank is made of thick-walled heat-resistant borosilicate glass, which can be heated up to 155 °C with a heat distributor. For a homogeneous temperature and corrosion solution (3 m /m %), constant mixing is necessary, which is most ideal at 300 rpm, based on my tests. The built-in mixer can be heated up to 380 °C, the number of samples that can be tested at the same time depends only on the dimensions of the device. The monitored cell and specimen can be selected with the switches. For the tests, I had to build a new device (Fig. 4), for which I used the measuring devices described in the literature [1, 4, 5] and equipped it with new measuring elements for my own tests. In this way, I created a previously unused measurement setup.

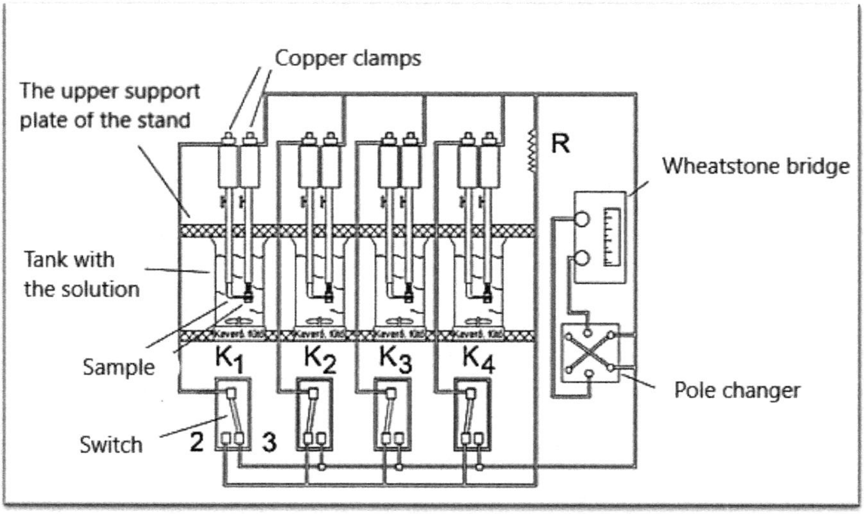

Fig. 4 The equipment for the examination of galvanic corrosion (the author)

2.4 Tightening Torque Test with Corrosion

The device makes it possible to carry out examinations lasting days or weeks. The most important task of an electrical connector consisting of several parts is the conduction of current between the individual parts with as little loss as possible. With the device outlined above, the changes in the most important parameters can be followed. The figure below shows the atomic force microscope images of the various surfaces of the fastener after being exposed to the corrosive solution. On Fig. 5 we can observe the increase in resistance caused by corrosion (Fig. 6).

3 Thermographic Investigation of the Galvanic Corrosion of Electrical Connecting Elements as a Function of Torque

The traditional testing of connectors used in the electrical industry is difficult and dangerous from a safety and operational point of view. However, with the spread of thermography, these tests can be performed more simply and safely, and faster.

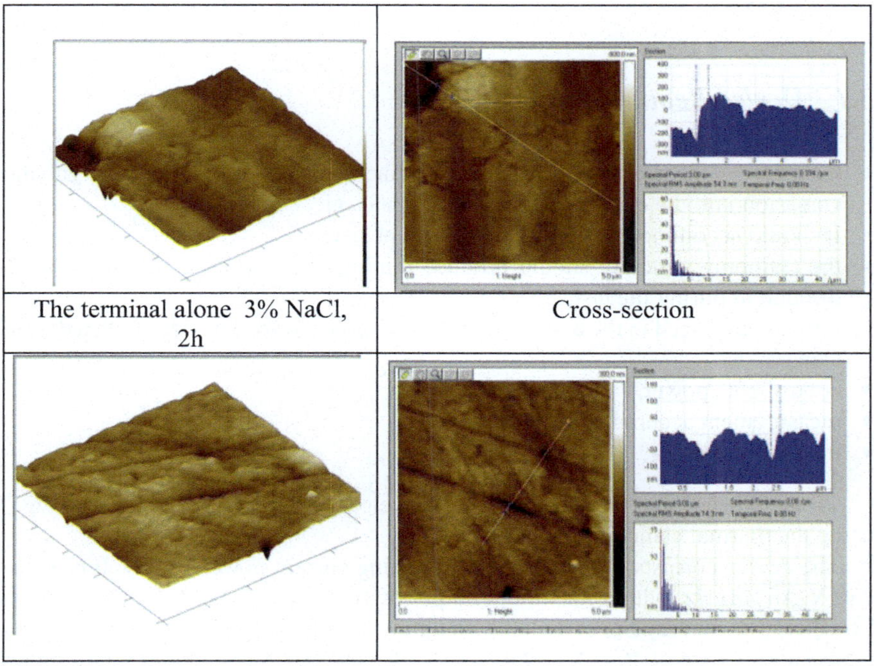

Fig. 5 AMF image of a terminal and galvanized nut after corrosion

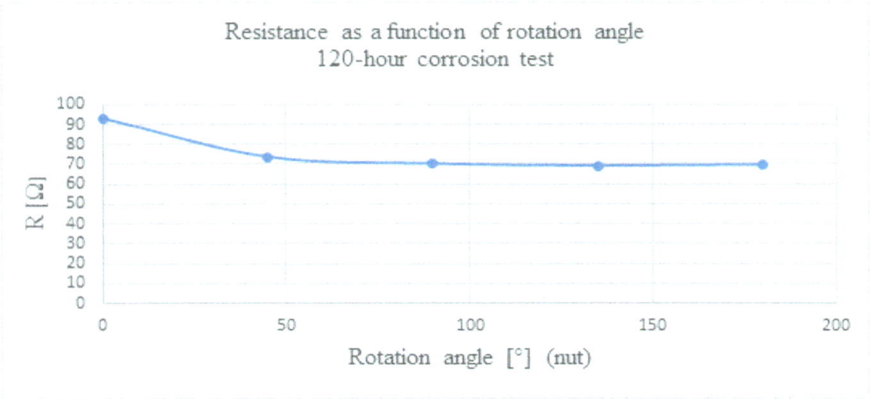

Fig. 6 Resistance as a function of the rotation angle of tightening, after a 120 h corrosion test

In industrial and domestic environments, a properly selected thermal camera can therefore be of great help. This is no different even in laboratory conditions, where this modern diagnostic tool can supplement our arsenal of measuring equipment.

My goal is to design and build a device that enables the circuit monitoring and temperature detection of the tested sample by thermography.

3.1 Device Requirements

- Connecting the electrical connecting element previously exposed to galvanic corrosion into the circuit should be quick and safe.
- It should accommodate samples of different sizes,
- the clamping structure should produce minimal heat when a current is passed through id during the test,
- the sample placed in the device can be observed from several sides with a thermal camera,
- controllable power,
- measurement accuracy,
- in-situ monitoring.

Taking into account the requirements, I developed the measuring system in Fig. 7, using the thermal camera in Fig. 8.

The device consists of a stabilized alternating current power supply. It produces 230 V AC, 50 Hz, with a maximum output of 10 A, which can be adjusted freely. The measuring arrangement includes an ammeter with a sampling option of 2.5 x/s, a measuring range of 50 mA/500 mA/10A AC, and ± 1.5% + 2 digit measurement

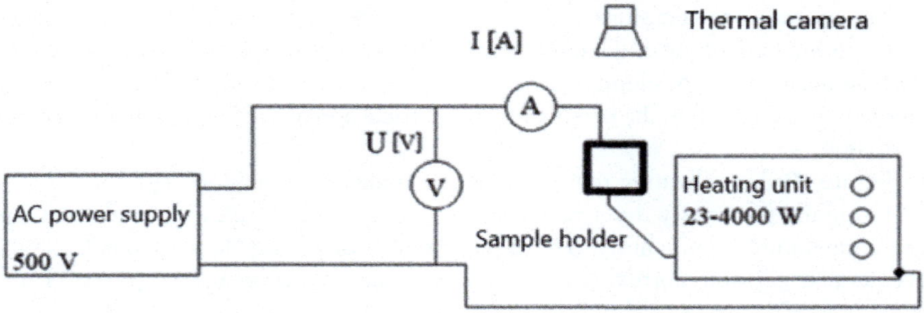

Fig. 7 Schematic diagram of the measurement setup with the thermal imaging camera (with measuring equipment) (the author)

Fig. 8 Thermal camera

accuracy. A voltmeter unit has also been installed, which also has a sampling capability of 2.5 x/s, an accuracy of ± 1.5% + 2 digits and a measuring range of 5 V/50 V/500 V/750 V.

The arrangement and material of the electrical connector holder is the same as the sample holder in the corrosion testing equipment. The power of the heating unit can be adjusted between 23 and 4000 W continuously, without steps. Nominal current consumption can be regulated up to a maximum of 16 A. Protection class is IP 20. Heat is removed by a single-phase fan equipped with a thermostat. The focus distance of the thermal camera can be adjusted between 100 and 300 mm, the axis of symmetry of the lens is perpendicular to the surface of the sample. The schematic diagram of the equipment is shown below.

The operating temperature of the electrical connection is significantly influenced by the tightening torque of the nut fastening the connection (or its rotation angle). As we have seen above, the clamping force affects the resistance of electrical connection element. Figure 9 shows the heating of the electrical connector in normal mode before corrosion.

Figure 10 shows heating in the circuit as a function of rotation angle.

Using the previously described corrosion equipment changes the resistance and the temperature significantly. The figure below shows the thermal image of the accelerated galvanic corrosion of the copper–steel screw connector (Figs. 11 and 12).

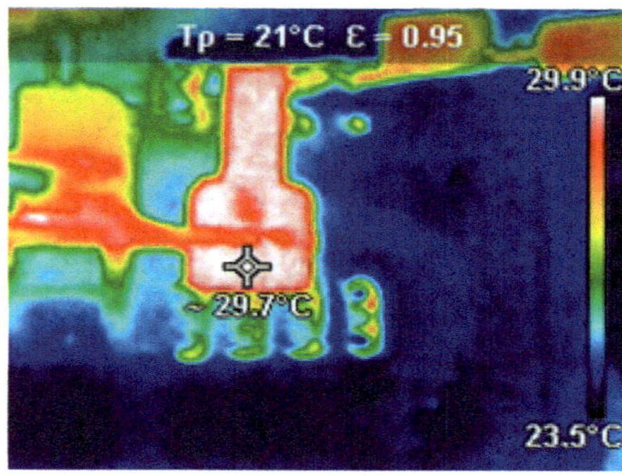

Fig. 9 The thermal image of a copper terminal—coated steel screw before corrosion

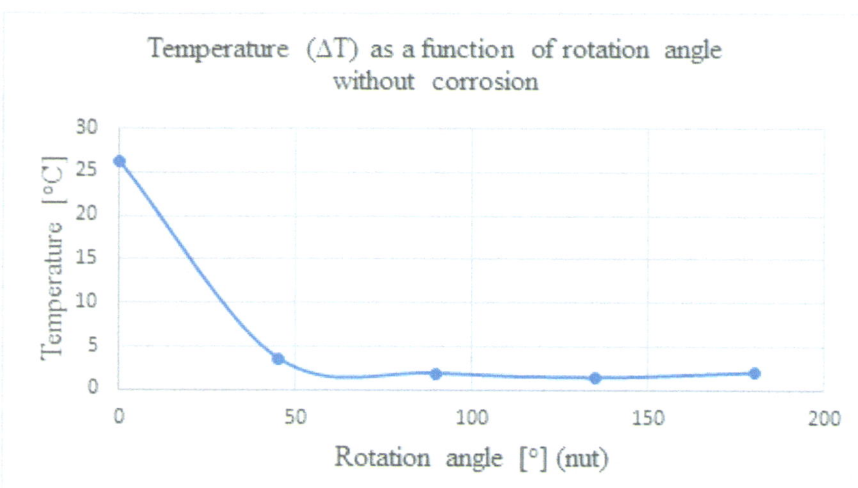

Fig. 10 Temperature as a function of tightening torque, which is proportional to the rotation angle of the nut

Fig. 11 Thermal image of copper terminal and coated steel screw after corrosion

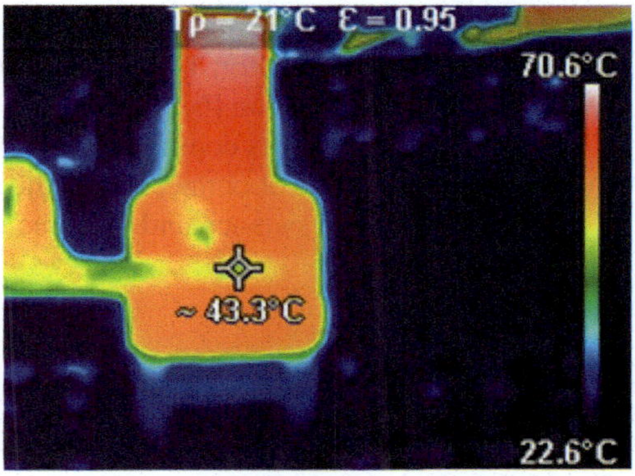

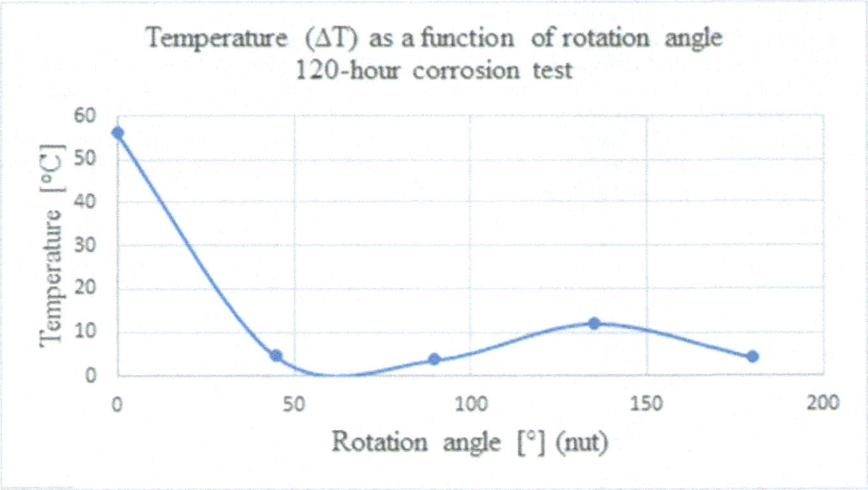

Fig. 12 Temperature as a function of tightening torque, which is proportional to rotational angle

4 Summary

The graphs clearly show the effect of the tightening torque (rotation angle of the nut) on the temperature of the connection. A loose joint tightened with insufficient torque heats more. A rotation angle of forty-five to fifty degrees is necessary for heating to decrease to a permissible level. If the environment causes galvanic corrosion on the connection, it generates twice as much heat, which causes increased electrical losses, faster deterioration, and safety problems. It is not the subject of this article, but the corrosion products have a reducing effect on clean metallic contact surfaces. The effects mentioned above cause resistance to increase. The increasing resistance causes local heating. It has been proved that the tightening torque of the screw connection (which is proportional to the rotation angle) is inversely proportional to the resistance of the connector. The change in resistance over time is proportional to the rate of galvanic corrosion [6, 7].

The measurement results confirm the assumption that in the case of the copper–zinc coated steel electrical connection without corrosion (Fig. 3) and after 120 h of corrosion (Fig. 6) resistance and galvanic corrosion is reduced as a function of tightening torque.

References

1. MSZ 4380-86 Corrosion in natural atmospheric conditions
2. ISO 8044:2020 Corrosion of metals and alloys
3. ISO 7441:2015 Corrosion of metals and alloys
4. Holló M (1964) Korróziós vizsgálatok. (Corrosion studies.) Műszaki Könyvkiadó
5. During EDD (2018) Corrosion atlas: a collection of illustrated case histories. 3rd edn. Elsevier
6. Tóth L, Haraszti F, Kovács T. Surface roughness effect in the case of welded stainless steel corrosion resistance. https://doi.org/10.2478/amt-2018-0018
7. Haraszti F et al (2017) Plastic deformation effect of the corrosion resistance in case of austenitic stainless steel. IOP conference series: materials science and engineering, vol 175 p 012048. https://doi.org/10.1088/1757-899X/175/1/012048

Tools for a Root Cause Analysis for Safety–Critical Components, A Review

Csenge Huszák, Péter Pinke, and Tünde Anna Kovács

Abstract In our paper, we introduce the Safety–Critical Components, then in the following, we highlight the importance of Root Cause Analysis and then we introduce classical tools for Root Cause Analysis such as 5W, Fishbone, Fault Tree Analysis, and the Event Tree Analysis. Additionally, we discuss the PDCA cycle, FMEA and the 8D which can be employed to eliminate potential failures. Towards the end of our article, we are reviewing some of the state-of-the-art techniques and prospects of the Root Cause Analysis.

Keywords 5W · Fault tree analysis · Fishbone · FMEA · Root cause analysis · Safety–critical component

1 Introduction, the Relationship Between Safety–Critical Components and Root Cause Analysis

A Safety–Critical Component is a specific part or element within a system whose failure or malfunction could lead to significant harm to people, environmental damage, or financial loss. These components are essential for the safe operation of various systems and industries, ranging from vehicles (such as aircraft, automobiles, and trains) to medical devices, nuclear facilities, and industrial machinery [1, 2].

The significance of Safety–Critical Components lies in their ability to maintain the overall integrity and functionality of a system. Failure of these components can

C. Huszák (✉)
Doctoral School on Safety and Security Sciences, Obuda University, Budapest, Hungary
e-mail: huszak.csenge@bgk.uni-obuda.hu

P. Pinke · T. A. Kovács
Banki Donat Faculty of Mechanical and Safety Engineering, Obuda University, Budapest, Hungary
e-mail: pinke.peter@bgk.uni-obuda.hu

T. A. Kovács
e-mail: kovacs.tunde@bgk.uni-obuda.hu

T. A. Kovács et al. (eds.), *The Impact of the Energy Dependency on Critical Infrastructure Protection*, Advanced Sciences and Technologies for Security Applications, https://doi.org/10.1007/978-3-031-78544-3_28

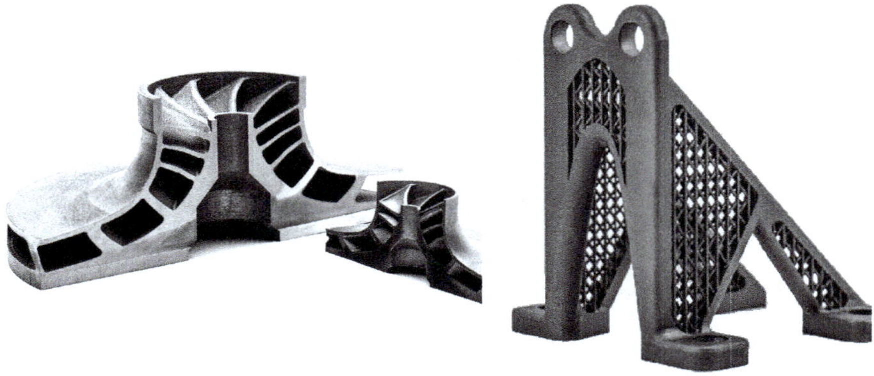

Fig. 1 3D printed components for aerospace [15, 16]

result in accidents, injuries, loss of life, environmental disasters, or financial implications. Consequently, the design, manufacturing, and maintenance of Safety–Critical Components are subjected to rigorous standards, regulations, and quality control measures to ensure their reliability and performance under various conditions. These components are carefully engineered to perform vital functions, often involving intricate processes and cutting-edge technologies such as 3D printing [3–6], sensor technology with big data, data analysis with AI technologies [7–13].

In the aerospace industry, Safety–Critical Components include parts of the aircraft's engine, avionics systems, structural elements etc. For example, a failure in an aircraft's engine turbine blades or flight control systems could lead to a catastrophic accident [14]. In Fig. 1, there are examples of 3D printed aerospace components.

In the healthcare sector, medical devices like pacemakers, breathing machines, and infusion pumps contain Safety–Critical Components. Malfunctions in these devices can directly impact patients' health and safety, making it imperative to ensure their flawless operation [17]. In Fig. 2 there is an example of a medical device.

In the nuclear industry, Safety–Critical Components include among others reactor control systems, cooling systems, and containment structures. Failures in these components can lead to nuclear accidents, posing severe environmental and public health risks [19]. Other power plans, like a solar or wind turbine plan, can have a failure [20], therefore in the urban zone which depends on the power plan during a long power outage the power shortage can negatively impact the life of the habitant [21, 22]. In Fig. 3, there is a burning wind turbine in a wind farm.

The railway has Safety–Critical Components and systems too. Its operation is associated with potential risks to passenger and public safety, environmental impact, and economic consequences. A railway failure could be caused by high speed and heavy load, a small error in the railway's complex systems which could generate a bigger failure in another system, human error, high traffic density, ageing infrastructure, or unpredictable environmental factors. Ensuring railway safety requires rigorous standards, regular maintenance, advanced signalling systems, staff training,

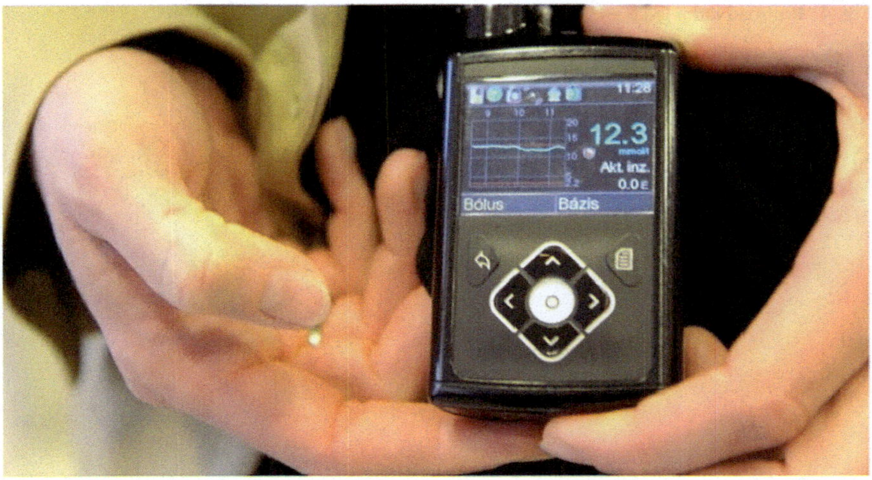

Fig. 2 Insulin pump [18]

Fig. 3 Wind turbine on fire [23]

and thorough risk assessments. Safety protocols and Root Cause Analysis techniques are essential in identifying and mitigating risks, making railways a critical component of modern transportation [7, 24–30]. Similarly, in the automotive industry, components such as airbags, anti-lock braking systems, and electronic stability control systems are safety critical. Any malfunction in these or other Safety–Critical Components can result in accidents and injuries to drivers and passengers. In Fig. 4, there is a railroad lamp like an example of a railway safety component.

To ensure the reliability of Safety–Critical Components, thorough risk assessments, stringent testing protocols, and continuous monitoring are conducted throughout their lifecycle. Moreover, Root Cause Analysis techniques are employed to identify and rectify any issues promptly, minimizing the potential for failures and ensuring the safety of individuals and the environment [29, 32].

Fig. 4 Railroad lamp [31]

Root Cause Analysis plays a central role in safety management and risk assessment across various industries. It is a systematic approach used to identify the underlying causes of problems, incidents, or failures. In the context of safety management, Root Cause Analysis serves as a fundamental tool for understanding the factors that contribute to accidents, errors, or equipment failures. By employing Root Cause Analysis techniques, organizations can delve deep into incidents, moving beyond immediate symptoms to uncover the root causes. This understanding is crucial for developing effective preventive measures, thereby mitigating the risk of similar incidents in the future. Safety management systems rely on Root Cause Analysis to analyze past incidents comprehensively, allowing organizations to learn from failures and continuously enhance safety protocols [33–35].

Identifying root causes through Root Cause Analysis leads to more targeted and effective corrective actions. These are the following (we will introduce them later in this article):

1. Precision in Problem Solving
2. Preventing Recurrence
3. Proactive Risk Mitigation
4. Resource Optimization
5. Continuous Improvement.

In summary, any part of a system, or even the whole system, can be a critical component from the point of view of security. Safety–Critical Components are crucial because their failure can endanger human life and the environment. While the Root Cause Analysis is the cornerstone of effective safety management and risk assessment. It provides the insights necessary for organizations to learn from incidents, prevent their recurrence, and continuously enhance their safety measures. By focusing on the root causes, organizations can implement precise, targeted, and efficient corrective actions, ensuring a safer and more secure operational environment [33, 34].

Classical Root Cause Analysis Methods

The Root Cause Analysis in each case is a unique process for finding why a component, machine or equipment had a failure. This component-specific solution has some points, which can be common with other root cause analyses. This problem-solving method generally contains the following steps [35, 36]:

1. Problem description:

   Clearly state the problem or issue which you would like to eliminate.
2. Data collection:

   Gathering relevant data and information which are related to the problem or issue. This can include reports, performance metrics, observations, measurements, and any other relevant data sources.
3. Problem analysis:

   Using the collected data, we must strip down the issue to the underlying causes of the problem. A classical tool for this step is the Five Whys, where we start with the initial problem and repeatedly ask "Why" until we can identify the root cause or causes of the problem or issue.
4. Visually representation of the potential problems:

   The fourth step of problem-solving is a summary of the collected potential failures. A classical tool for this step is the Fishbone Diagram.
5. Prioritize Contributing Factors:

   Assess the potential contributing factors identified in the previous step (e.g., with the Fishbone Diagram) and prioritize them based on their significance and impact on the problem.
6. Data Analysis:

   Analyse the data which was collected in 2nd step to validate the potential contributing factors and identify correlations or patterns.
7. Root Cause Identification:

   Based on the data analysis and Fishbone Diagram (or other problem representation diagram or visualization tool), identify the root cause, or causes that are most likely responsible for the failure or problem.
8. Corrective Actions:

   Propose and document corrective actions to address each root cause. These actions should be specific, measurable, relevant, and time-bound.
9. Implementation Plan:

   Develop a plan for implementing the corrective actions, including assigning responsibilities, setting timelines, and allocating resources.
10. Monitoring and Evaluation:

    Define a method for monitoring the effectiveness of the corrective actions over time. Set up review points to evaluate progress and adjust as needed.

11. Preventive Measures:

    Suggest preventive measures to avoid similar problems in the future.
12. Documentation and Communication:

    Document the entire Root Cause Analysis process and share the findings, corrective actions, and preventive measures with relevant stakeholders.
13. Redo the previous steps (PDCA cycle):

    We are not able to eliminate all the potential failures, therefore we should repeat the previous steps to find more potential problems and failures. We can do it immediately after the previous steps or after some time, while we collect more experience from the machine, component, or equipment.

In the following subsections, we will review some of the classical Root Cause Analysis tools and methods.

5 Whys (5W, Five Whys)

The 5 Whys is a method with which we can detect the cause and effect of a failure or an error. This technique is an iterative method where we ask in each case why the failure happens, and we repeat the question until we can deduct the cause of the failure. In general, 5 questions can produce the root cause of the failure, but less and more may also occur [36–38].

In the following, we can see an example of this Root Cause Analysis using a coffee machine, in which the water heater element was burned out.

1. Why did the water heating element burn out?

   The water heating element burned out due to overheating.
2. Why did the water heating element overheat?

   The thermostat malfunctioned, causing the heating element to operate at excessively high temperatures.
3. Why did the thermostat malfunction?

   The thermostat malfunctioned because of wear and tear over time and a lack of regular maintenance.
4. Why was there a lack of regular maintenance?

   The coffee machine users were not aware of the necessity for regular maintenance, and there were no clear instructions provided regarding maintenance procedures.
5. Why were users not aware, and why were there no clear instructions?

   The user manual provided with the coffee machine did not emphasize the importance of regular maintenance, leading to user negligence. Additionally, the manufacturer did not include detailed maintenance instructions, assuming users' familiarity with general appliance care.

The 5W is a basic tool for Root Cause Analysis. It was introduced more than 100 years ago. Today we have more sophisticated Root Cause Analysis techniques, therefore the 5W is used as a tool whose output is an input for a newer Root Cause Analysis tool. The 5W has received some criticism since its creation [39].

1. Investigators often stop at identifying surface-level symptoms instead of probing deeper to uncover underlying root causes.
2. The 5 Whys method is constrained by the investigator's existing knowledge, making it challenging to uncover causes beyond their current understanding.
3. It lacks adequate guidance and support to assist investigators in arriving at accurate answers to the successive "why" questions.
4. The results obtained through the Five Whys technique tend to lack consistency and repeatability, with different individuals using the method arriving at varying causes for the same problem.
5. The technique has a predisposition to pinpointing a single root cause, overlooking the potential for multiple root causes to emerge in response to each "why" question posed.

Fishbone Diagram

The Fishbone Diagram (also known as the Ishikawa Diagram or Cause-and-Effect Diagram), is a visual tool used for Root Cause Analysis and problem-solving. It was developed by Japanese quality control expert Kaoru Ishikawa in the 1960s. This diagram is particularly useful when trying to identify the root causes of a problem or when brainstorming potential causes of a particular issue [40–42].

The Fishbone Diagram is typically structured by the following [40]:

1. **Problem Statement**

   At the head of the diagram, the problem or issue which we want to investigate is written down.
2. **Spine**

   Which is a horizontal line extending from the problem statement, resembling the spine of a fish.
3. **Categories**

   Branching off the spine, which are several diagonal lines like the bones of a fish. At the end of each line, there is a category that could contribute to the problem.
4. **Causes**

   Under each category, there are additional lines as branches. On these branches, specific factors or causes are listed, which could be related to the category. These factors are often brainstormed with a team of individuals who know the problem or issue.
5. **Analysis**

   Once all the potential causes are identified and added to the diagram, the team can start analyzing each cause to determine if it plays a role in the problem. This involves discussing and investigating each cause thoroughly.

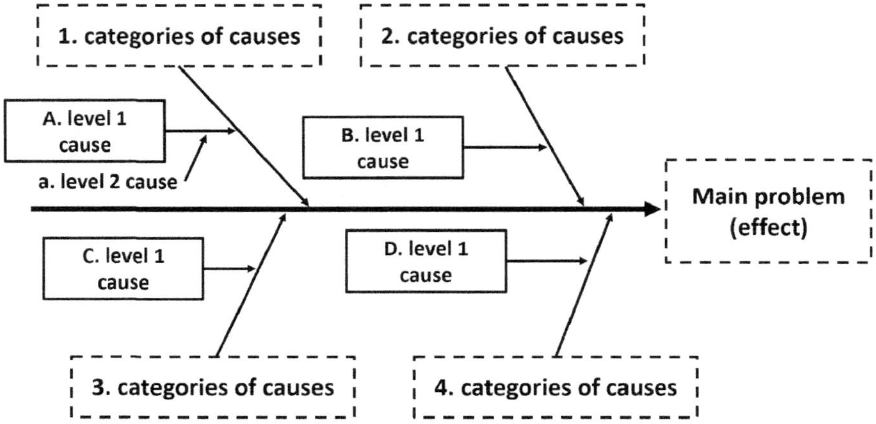

Fig. 5 The basic structure of an Ishikawa diagram [41]

6. **Root Cause Identification**

   The goal is to identify the root cause or causes of the problem. This might involve further analysis, data collection, or testing to confirm which causes have the most significant impact.

The Fishbone Diagram is a valuable tool because it encourages a structured and systematic approach to problem-solving. It helps teams look beyond the obvious symptoms of a problem and dig deeper to find the underlying causes. By categorizing causes into different aspects of the problem, it becomes easier to identify which areas require improvement. You can find the basic structure of a Fishbone Diagram in Fig. 5.

The Fishbone Diagram is a powerful tool for Root Cause Analysis, it's most effective when used in conjunction with other problem-solving techniques and data analysis methods to validate the identified causes and develop effective solutions [41, 42]. There are several applications which are shown in the Fishbone Diagram's usability in various places like for centrifugal pumps [42], cardiovascular loads [43], analyzing the ergonomic risk in onshore oil rig operations [44], decision making [45], manufacturing [46] etc.

In Fig. 6, there is a Fishbone Diagram for a Coffee taste issue.

Fault Tree Analysis

Fault Tree Analysis (FTA) is a systematic method used to analyse the causes of a specific undesirable event or failure. The Fault Tree Analysis was first introduced in 1962 by the Bell Telephone Laboratories then it was developed during the next decades [47–49].

This analysis is a visual tool for representing a problem and its presumed root causes. The Fault Tree Analysis looks like a tree where the main problem is on the top of the tree, then the possible root causes are listed below the top event. After the first level, root causes other levels could be listed till the bottom, where the final

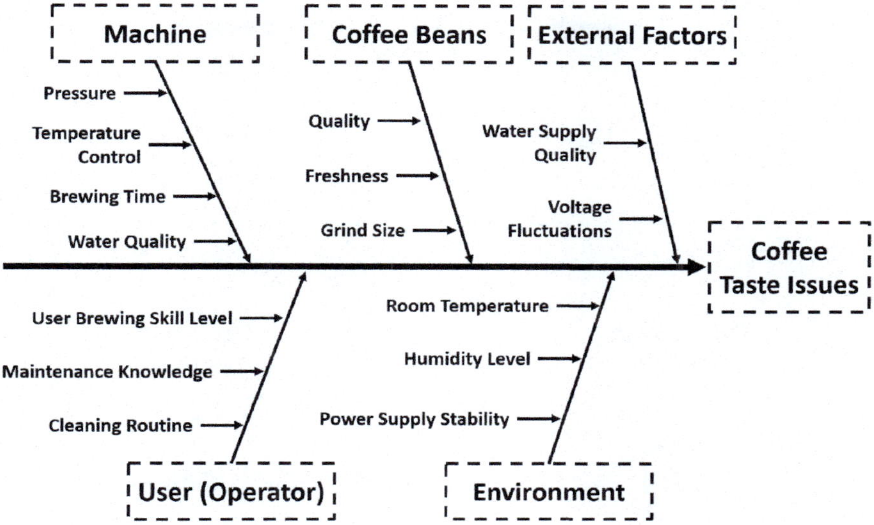

Fig. 6 Fishbone diagram of a centrifugal pump (*Source* Author)

root causes are. The FTA is built from graphical symbols, like event symbols (Basic, External, Conditional, Intermediate, and Undeveloped events), logical gate symbols (AND, OR, Exclusive OR, Priority And, and Inhibit gate), and Transfer symbols (Transfer in and Transfer out symbols) [47, 48]. For the previous coffee machine example the FTA is shown in Fig. 7.

To create a Fault Tree Analysis, we must follow the following five steps [47, 49]:

1. Define the Top Event
2. Identify Basic Events
3. Identify Intermediate Events
4. Construct the Fault Tree Diagram
5. Analyze the result and control the identified hazards.

The Fault Tree Analysis is a top-down method, which systematically breaks down the causes of the undesirable event, a fault tree analysis helps identify the critical factors contributing to the problem and allows for proactive risk mitigation and problem-solving.

Event Tree Analysis

The Event Tree Analysis (ETA) was developed in the beginning of 1970. This tool is a systematic and graphical tool which can be used in risk assessment and decision-making to analyze the potential outcomes of a single specific event or a series of events. In contrast to the Failure Tree Analysis (FTA), the ETA contains not only the failure outcomes but the success outcomes too. For each outcome the Event Tree Analysis provide a probability, therefore the possibility of an outcome is statistically comparable with the other outcomes [50, 51].

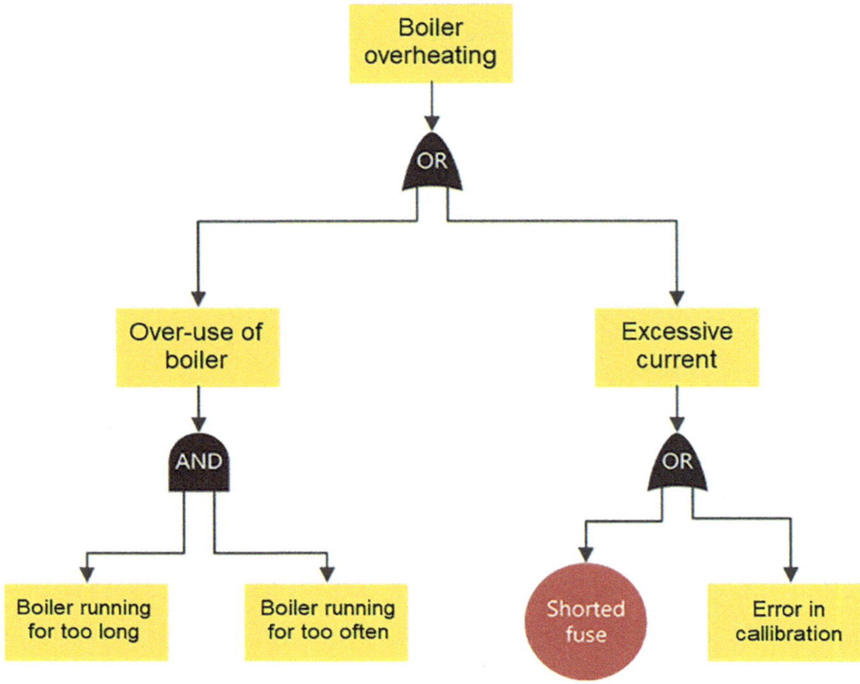

Fig. 7 Fault tree analysis of an overheating boiler (*Source* Author)

The Event Tree Analysis has the following steps:

1. Initiating Event

   Identify the starting event.
2. Branching

   Create branches for potential outcomes.
3. Probability Assessment

   Assign probabilities to each branch.
4. Consequence Analysis

   Evaluate consequences for each branch.
5. Decision-making

   Use results for informed decisions.
6. Sensitivity Analysis

   Assess the impact of parameter changes.
7. Documentation

   Present results in a visual tree diagram.

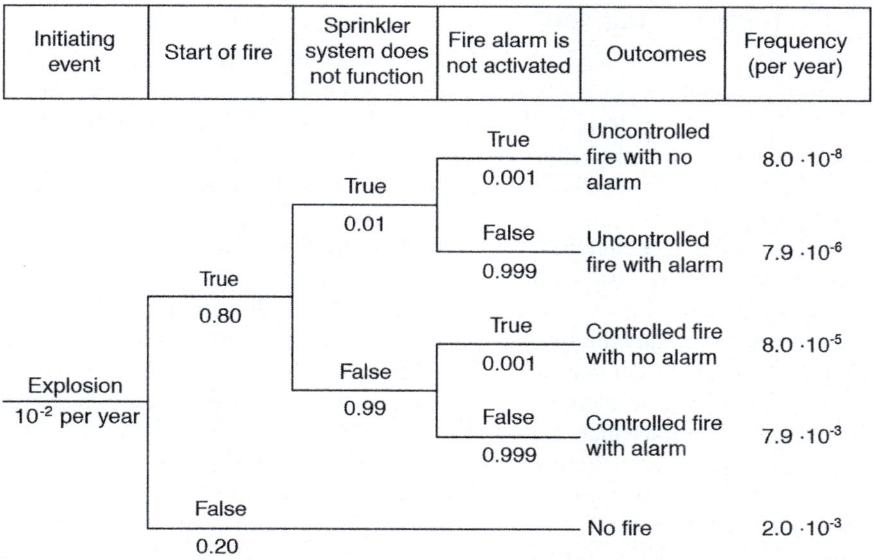

Fig. 8 "Could an explosion start a fire?", an example of the Event Tree Analysis [51]

An example of the ETA is shown in Fig. 8.

The ETA has some advantages, it is a structured analysis, therefore it provides an organized framework for evaluating complex events with the highlights of the potential outcomes. It helps identify the critical paths and events within a sequence. The Event Tree Analysis helps the decision-makers who can make more informed choices based on quantified risks and consequences, ultimately leading to better risk management strategies.

Where there are advantages, there are also disadvantages and limitations too. The Event Tree Analysis has examined the following limitations. The quality of the ETA heavily relies on the availability of accurate data and expert input for probability and consequence assessments, inaccurate data can lead to unreliable results. ETA assumes that events are independent of each other, which may not always hold in complex systems. This simplification can lead to an incomplete understanding of risk. For large and intricate systems, ETA can become complex and time-consuming to develop and analyse, potentially requiring significant resources. The quality of ETA analysis can vary based on the expertise of the analysts involved. Different analysts may produce different results, impacting the reliability of the assessment.

In summary, while Event Tree Analysis is a valuable tool for risk assessment and decision-making, it is essential to consider its limitations and ensure the availability of accurate data and expert input to maximize its effectiveness in assessing complex event sequences.

2 Problem-Solving Tools

If there is a possibility that a component or a machine has a failure, or the failure already happened, the Root Cause Analysis tools and methods could be used to determine the cause of the failure. When we know the cause, we are fixing it due to technical, economic and safety interests. This is where problem-solving techniques play a pivotal role.

Problem-solving techniques are systematic approaches designed to identify, analyze, and address issues for the failures of a component or a system. They provide a structured framework for tackling problems, making informed decisions, and driving continuous improvement. These techniques empower individuals and teams to navigate the complexities of problem-solving with precision and purpose.

There are a lot of problem-solving techniques and methodologies which already exist. There are some which can be used generally [33, 34, 52], but there are some component, system, or failure specific tools too [3, 51]. In the following, we will introduce shortly three problem-solving methodologies, which are often utilized across diverse industries. These three tools are PDCA (Plan-Do-Check-Act), FMEA (Failure Mode and Effects Analysis), and 8D (Eight Disciplines). Each technique offers its unique strengths and applications, enabling organizations to address a wide range of challenges, from quality control and process optimization to risk management and Root Cause Analysis [40, 51].

PDCA (Plan-Do-Check-Act)

The PDCA (Plan-Do-Check-Act) cycle is a fundamental and probably the most important component of each problem-solving process. The PDCA is a cycle, therefore it must be used repeatedly for continuous improvement. If the methodology does not contain directly the PDCA cycle common sense is required [40, 50]. The representation of the PDCA cycle is shown in Fig. 9.

The first part of the PDCA cycle is the Plan. In this initial phase, there is a suspected or occurred failure that the troubleshooting team wants to fix. This step

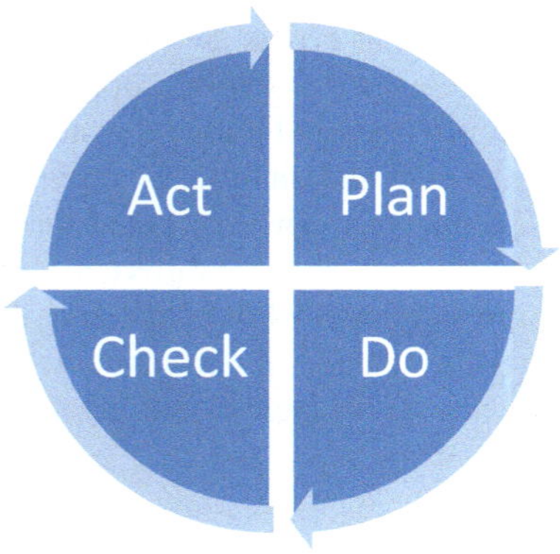

Fig. 9 PDCA cycle (*Source* Author)

involves gathering data, identifying root causes, and devising a strategy to address the issue. Planning is a critical step that sets the direction for the entire PDCA process.

The second part of the cycle is the Do step, where the problem-solving team implements the proposed changes or improvements.

After the plan is executed the third part of the cycle is next, which is the Check. In this phase, the team assesses and evaluates the results of the actions taken in the "Do" phase. This involves comparing the actual outcomes with the expected results outlined in the planning phase. Data analysis and performance measurements are key activities to determine the effectiveness of the implemented changes.

The next step is the Act, where based on the findings and insights from the previous steps, the organization makes decisions and acts. If the results meet the objectives and expectations, the changes are standardized, and the process is improved accordingly. If the outcomes do not align with the desired goals, further adjustments and refinements are made, and the PDCA cycle continues iteratively.

The PDCA is a cycle, and it is used for continuous improvement, therefore after the fourth step of the PDCA cycle, the cycle continues, with the fifth step which is the Plan. Using this fundamental tool, we can eliminate failure causes for a Safety–Critical Component during its design phase and in the implemented status too. The PDCA is a general tool, therefore it can be used for a wide range of applications [40].

FMEA—Failure Mode and Effects Analysis

Failure Mode and Effects Analysis (FMEA) is a structured and systematic approach used to proactively identify and assess potential failure modes, their causes, and their consequences within a system, process, or product. FMEA is a valuable tool for risk management, quality improvement, and enhancing the reliability and safety of various operations across multiple industries [40, 50].

FMEA was initially developed by NASA in the early 1960s. Thanks to a systematic approach each failure mode and its effect is listed and this nonnumerical data is converted to numeric information which is easily comparable. After the first success of the FMEA other industries also started to use, during the 1960s the aerospace industry, while by the mid-1970s, the FMEA was used in the nuclear and the automotive industries as well. Over the years, FMEA has evolved and is now widely employed in various industries, including electronics and software, as a standard practice for risk assessment and mitigation [34, 40, 51].

FMEA can be categorized into several types, each tailored to address specific aspects of the product or process being analyzed. In the following, we present 5 types of the FMEAs:

1. Design FMEA (DFMEA): DFMEA specifically assesses the design phase of a product or system. It aims to identify potential design-related failure modes and their consequences, to improve the product's design for reliability and safety.
2. Process FMEA (PFMEA): PFMEA concentrates on the manufacturing or production process itself. It seeks to identify potential process-related failure modes and their impacts on product quality, aiming to enhance the manufacturing process to prevent defects.

3. Process Design FMEA (PD-FMEA): PD-FMEA combines aspects of both DFMEA and PFMEA to assess the interaction between the product design and the manufacturing or production process. It aims to optimize the design with manufacturing considerations in mind.
4. Functional FMEA (FFMEA): FFMEA assesses the functional aspects of a system or product, concentrating on how well it performs its intended functions. It identifies failure modes that could hinder the primary functions and user requirements.
5. Environmental FMEA (EFMEA): EFMEA is used to assess the environmental impact and risks associated with a product or process. It helps organizations identify and mitigate environmental hazards and comply with regulations.

The choice of FMEA type depends on the specific context and objectives of the analysis. Different types of FMEA can be used in combination to comprehensively assess and mitigate risks throughout the product lifecycle or process development. The previous list is not full, other types of FMEA exist [53, 54]

The Failure Mode and Effects Analysis starts with the scope definition, where the system, process, product, or the scope of the analysis is clearly defined. Then a team assembles, preferably a multidisciplinary team with members who are experienced. Then the team identifies failure modes and determines its causes. The team for each failure assess the severity, the frequency of occurrence and the detectability. The severity (S), the occurrence (O) and the detectability (D) are classified by number (e.g., from 1 to 10), and then the team multiplies these three numbers. The result of this multiplication is the Risk Priority Number ($RPN = S \times O \times D$). The failures are prioritized by the RPN, then the team creates action plans to reduce the identified risks from high RPN to low. The last step of an FMEA is the PDCA cycle where the team monitors and reviews the operation of the subject of the FMEA [40].

In Table 1, there is an example for FMEA, where a coffee maker's brewing process, including water heating, coffee ground dispensing, and brewing was analyzed.

As it was shown in the previous figure the FMEA by systematically analysing and addressing failure modes and their effects, FMEA empowers organizations to make informed decisions, prioritize resources, and ultimately deliver higher-quality products and services to customers while minimizing risks and liabilities.

8D (8 Disciplines)

The 8D (8 Disciplines) problem-solving methodology is a systematic approach used in various industries to identify, correct, and prevent recurring problems and defects. The 8D was developed by Ford Motor Company in 1986. Originating from the automotive industry but the 8D can be employed in various industries like maintenance, manufacturing, aerospace, healthcare, IT, customer services, and where quality, safety and customer satisfaction are important. 8D provides a structured process for addressing complex issues and improving processes to prevent their recurrence. Each "D" represents a step or discipline in the process. The disciplines of the 8D are the following [55]:

1. D1—Form a Team:

Table 1 Coffee machine example for FMEA (*Source* Author)

Failure modes	Possible causes	Severity (S)	Occurrence (O)	Detection (D)	RPN	Prioritize actions	Action taken
The water heating element burns out	Overheating due to a malfunctioning thermostat	Sever The user may suffer burns and there is no coffee	Low The heating element rarely burns out	High Easily detected by temperature sensors	9 × 3 × 10 = 270	1st	Improve the heating element's design and quality
Coffee grounds dispenser jams	Coffee grounds are not ground to the correct consistency	Moderate Coffee won't brew, but no safety risk	Moderate The dispenser jams occasionally	Moderate It May require user observation	4 × 5 × 5 = 100	3rd	Implement a better grinder mechanism
Water reservoir leaks	Reservoir crack caused by pressure build-up	Moderate Water leakage, potential electrical issues	Low Leaks are infrequent	High Leaks are visible	6 × 3 × 8 = 144	2nd	Enhance the reservoir material

Assemble a cross-functional team with members who have expertise related to the problem. The team will be responsible for solving the issue.

2. D2—Define the Problem:

   Clearly define the problem, specifying with the 5W + 2H questions which are what, where, when, why, who, how and how many. Use data and facts to accurately describe the issue.

3. D3—Develop Interim Containment Actions:

   Implement temporary measures to contain the problem and prevent it from escalating. These actions are short-term solutions while the team investigates the root cause.

4. D4—Identify and Verify Root Causes:

   Use tools like Fishbone Diagrams, 5 Whys, or Fault Tree Analysis to identify the root causes of the problem. Verify these causes through data analysis and testing.

5. D5—Choose and Verify Permanent Corrective Actions:

   Develop and implement permanent solutions to address the root causes. These actions should prevent the problem from recurring. Verify their effectiveness through testing and analysis.

6. D6—Implement and Validate Corrective Actions:

   Put the corrective actions into practice and ensure they are effectively addressing the problem. Validate their success through monitoring and data analysis.

7. D7—Prevent Recurrence:

   Modify processes, systems, or training to prevent the problem from recurring in the future. Implement preventive measures to ensure the issue does not happen again.
8. D8—Recognize the Team:

   Acknowledge and appreciate the efforts of the team members. Document the lessons learned and share the knowledge gained from the problem-solving process.

In the following, we present an 8D analysis where the issue is a quality problem, where a coffee machine produces weak coffee due to inconsistent brewing temperatures. The 8D process is the following:

1. D1—Form a Team:

   Assemble a team comprising coffee machine engineers, experienced baristas, and quality control specialists.
2. D2—Define the Problem:

   Clearly define the problem: "Inconsistent brewing temperatures leading to weak coffee, impacting customer satisfaction and product quality."
3. D3—Develop Interim Containment Actions:

   Implement temporary measures: Increase temperature monitoring frequency, manually adjust brewing temperatures during production, and conduct additional quality checks on brewed coffee.
4. D4—Identify and Verify Root Causes:

   Conduct a Root Cause Analysis: Use for example Fishbone Diagram or 5 Whys to identify root causes. Determine that the inconsistent brewing temperatures are caused by a malfunctioning heating element regulator.
5. D5—Choose and Verify Permanent Corrective Actions:

   Develop permanent solutions: Replace the malfunctioning heating element regulator, implement more precise real-time temperature sensors for automatic adjustments, and conduct regular equipment maintenance. Verify effectiveness through extensive testing and monitoring.
6. D6—Implement and Validate Corrective Actions:

   Implement changes in the production process: Train staff on the new procedures, update the machine's firmware to integrate automatic temperature adjustments, and establish a routine maintenance schedule. Validate effectiveness through continuous temperature monitoring and customer feedback.
7. D7—Prevent Recurrence:

Prevent future inconsistencies: Implement a proactive maintenance program for all coffee machines, conduct regular staff training sessions on machine maintenance, and establish a quality control team to monitor coffee quality. Create a standardized operating procedure to ensure consistent brewing temperatures.

8. D8—Recognize the Team:

   Acknowledge the efforts: Recognize and appreciate the team's dedication. Document lessons learned, update the training curriculum, and share the experience across the company to improve quality control processes.

Through the previous example of the 8D process, we introduced an issue, where a coffee machine's inconsistent brewing temperatures were identified and resolved. The implementation of automatic temperature adjustments and a proactive maintenance program ensured consistent coffee quality, leading to increased customer satisfaction and improved product quality. The collaborative efforts of the cross-functional team resulted in a successful resolution to the problem [55].

In summary, the 8D methodology emphasizes a team-based approach, data-driven decision-making, and a focus on preventing problem recurrence. It is widely used in manufacturing, maintenance, engineering, and various other industries to improve quality, enhance processes, and address customer concerns effectively.

3 State of Art and Future Approaches for Root Cause Analysis

As of the current state, Root Cause Analysis has evolved significantly in recent years, thanks to advancements in technology, data analytics, and methodologies. Traditional methods which were listed before, like 5 Whys, Fishbone Diagrams, and FMEA continue to be widely used. However, organizations are increasingly integrating more sophisticated tools which can help the problem-solving team to track down and count the possible root causes [11, 56, 57]. This tool contains one or more tools from the following:

1. Data Analytics

   Advanced data analytics techniques, including machine learning and artificial intelligence, are being utilized to process large datasets quickly. This helps in identifying complex patterns and hidden correlations, enabling organizations to pinpoint root causes more accurately [7, 11–13, 56–60].

2. Big Data Integration

   With the rise of big data technologies, organizations can now analyze vast amounts of data from various sources, allowing for a more comprehensive understanding of the factors contributing to problems or failures [12, 13, 56].

3. Simulation and Modelling

Simulation tools allow organizations to create virtual models of systems and processes. By simulating different scenarios, they can identify potential root causes and test the effectiveness of corrective actions without real-world consequences [61–63].

4. Collaborative Tools

   Online collaboration platforms and cloud-based software facilitate real-time collaboration among team members, enabling geographically dispersed teams to work together seamlessly in identifying and addressing root causes [64].

5. Integration with IoT

   Internet of Things (IoT) devices provide real-time data from equipment and processes. Analyzing this data can help in identifying anomalies, predicting failures, and understanding the root causes of issues in real time [59, 64].

The Root Cause Analysis tools which we use nowadays are a high-end approach compared to the traditional Root Cause Analysis tools, but the future could contain more accurate and precise tools with automation. These future tools will use methods and tools that are already being outlined. Some of these methods are the following:

1. Predictive Analytics

   The future of Root Cause Analysis lies in predictive analytics, where algorithms can predict potential issues before they occur. This proactive approach helps organizations prevent problems, improving overall efficiency and reducing downtime [58–60].

2. Explainable AI

   As AI techniques become more advanced, the ability to explain the reasoning behind AI-driven Root Cause Analysis will be crucial. Explainable AI ensures that decisions are transparent and understandable, especially in critical applications [11–13].

3. Augmented Reality (AR) and Virtual Reality (VR)

   AR and VR technologies can create immersive environments for Root Cause Analysis, allowing experts to visualize complex systems, identify potential failures, and test solutions virtually [64–67].

4. Advanced Sensor Technologies

   Continued advancements in sensor technologies will provide even more detailed and accurate data, enabling organizations to perform Root Cause Analysis at a finer granularity and identify subtle issues that might be missed otherwise [8–10].

5. Cognitive Computing

   Cognitive computing systems, designed to mimic human thought processes, will be employed in Root Cause Analysis. These systems can understand, reason, and learn from vast amounts of data, enhancing the accuracy and efficiency of Root Cause Analysis [58, 59].

6. Ethical and Social Implications

As Root Cause Analysis becomes more automated and reliant on AI, ethical considerations surrounding bias, privacy, and accountability will be critical. Future advancements will need to address these ethical and social implications to ensure the responsible use of Root Cause Analysis technologies [11, 12, 59].

4 Summary

In the first chapter, we presented the Safety–Critical Components with some examples. In the second chapter, we introduced some classical Root Cause Analysis tools, the 5W, the Fishbone Diagram, the FTA, and the ETA. After that, we presented three problem-solving tools which were the PDCA, FMEA and the 8D. In the last chapter of our article, we summarized that the future of Root Cause Analysis is exciting and is driven by the development of many new technologies, such as data science, artificial intelligence, IoT and collaborative technologies. These developments allow organizations to not only identify and manage problems more effectively, but also to proactively prevent problems, which leads to more reliable, efficient, and sustainable operations not only in the area of Safety–Critical Components but also for other components.

References

1. Maurya A, Kumar D (2020) Reliability of safety-critical systems: a state-of-the-artreview. Qual Reliab Eng 36(7):2547–2568. https://doi.org/10.1002/qre.2715
2. Knight JC (2022) Safety critical systems: challenges and directions. In: Proceedings of the 24th international conference on software engineering, pp 547–550
3. Ficzere P, Lukacs N, Borbas L (2021) The investigation of interlaminar failures caused by production parameters in case of additive manufactured polymers. Polymers 13(4):556. https://doi.org/10.3390/polym13040556
4. Kónya G, Ficzere P (2023) The effect of layer thickness and orientation of the workpiece on the micro- and macrogeometric properties and the machining time of the part during 3D printing. Periodi Polytech Mech Eng 67(2):143–150. https://doi.org/10.3311/PPme.21473
5. Alzyod H, Ficzere PP (2022) The influence of the layer orientation on ultimate tensile strength of 3D printed poly-lactic acid. Jordan J Mech Ind Eng 16(3):361–367
6. Alzyod H, Ficzere PP (2023) Material-dependent effect of common printing parameters on residual stress and warpage deformation in 3D printing: a comprehensive finite element analysis study. Polymers 15(13):2893. https://doi.org/10.3390/polym15132893
7. Lu S, Wang J, Jing G, Qiang W, Rad MM (2022) Rail defect classification with deep learning method. Acta Polytech Hung 19(6):225–241. https://doi.org/10.12700/APH.19.6.2022.6.16
8. Sekhar R, Solke N, Shah P (2023) Lean manufacturing soft sensors for automotive industries. Appl Syst Innov 6(1):1–30
9. Simon J, Santa R (2023) Energy efficient smart home heating system using renewable energy source with fuzzy control design. Decis Mak Appl Manag Eng 6(2):948–948. https://doi.org/10.31181/dmame622023825
10. Gogolák L, Fuerstner I (2020) Wireless sensor network aided assembly line monitoring according to expectations of industry 4.0. Appl Sci 11(1):25. https://doi.org/10.3390/app11010025

11. Sharma AK, Pinca-Bretonean C, Sharma S (2023) Artificial intelligence in lean manufacturing paradigm: a review. In: E3S web of conferences, 4th international conference on design and manufacturing aspects for sustainable energy (ICMED-ICMPC 2023), vol 391, p 01163. https://doi.org/10.1051/e3sconf/202339101163
12. Ma Q, Li H, Thorstenson A (2021) A big data-driven root cause analysis system: application of machine learning in quality problem solving. Comput Ind Eng 160(1):107580. https://doi.org/10.1016/j.cie.2021.107580
13. Zhou H, Li Y, Yang JJ, Li W (2018) BigRoots: an effective approach for root-cause analysis of stragglers in big data system. IEEE Access 6:41966–41977. https://doi.org/10.1109/ACCESS.2018.2859826
14. Nalçacıgil E, Kaçar B (2021) Safety culture and safety management system in aviation. In: Karakoc TH, Atipan S, Dalkiran A, Ercan AH, Kongsamutr N, Sripawadkul V (eds) Research developments in sustainable aviation. Springer. https://doi.org/10.1007/978-3-031-37943-7_26
15. GoEngineer, Aerospace 3D printers. GoEngineer [Online]. Available: https://www.goengineer.com/3d-printing/aerospace. Accessed 08 Nov 2023
16. Materialise (2023) Case study: a 63% lighter titanium aerospace part. Materialise. [Online]. Available: https://www.materialise.com/en/inspiration/cases/3d-printed-lighweight-titanium-aerospace-part. Accessed 08 Nov 2023
17. Stauder L, Knott A, Schmitt RH, Bergs T (2022) Methodology for the integrative adaption of manufacturing process and inspection sequences to component changes of safety–critical medical products. Prod Eng Res Devel 16(4):535–544. https://doi.org/10.1007/s11740-022-01114-1
18. Házipatika (2022) Cukorbetegség: itt az automata inzulinpumpa. Házipatika, 10 Aug 2022 [Online]. Available: https://www.hazipatika.com/napi_egeszseg/cukorbetegseg/cikkek/cukorbetegseg_itt_az_automata_inzulinpumpa. Accessed 08 Nov 2023
19. National Academies of Sciences, Engineering, and Medicine (2016) Lessons learned from the Fukushima nuclear accident for improving safety and security of U.S. Nuclear plants: phase 2, The National Academies Press, Washington, DC. https://doi.org/10.17226/21874
20. Hetyei C, Nagy R (2021) Review of wind turbine failures, highlighting fire accidents. Műszaki Katonai Közlöny 30(2):43–56. https://doi.org/10.32562/mkk.2020.2.4
21. Ahmed I, Parikh P, Munezero P, Sianjase G, Coffman D (2023) The impact of power outages on households in Zambia. Econ Politica 40(8):835–867. https://doi.org/10.1007/s40888-023-00311-0
22. Hinze J (2005) Practices that influence safety performance on power plant outages. Pract Period Struct Des Constr 10(3):190–194. https://doi.org/10.1061/(ASCE)1084-0680(2005)10:3(190)
23. M Live Media Group, Wind turbine catches fire in Huron County [Online]. Available: https://www.mlive.com/news/saginaw-bay-city/2019/04/wind-turbine-catches-fire-in-huron-county.html. Accessed 09 Nov 2023
24. Kliuchnyk S, Ovchynnykov P (2022) Study of railway steel bridges' behaviour in order to identify the causes of their defects. Acta Polytech Hung 19(3):77–88. https://doi.org/10.12700/APH.19.3.2022.3.7
25. Macura D, Laketić M, Pamučar D, Marinković D (2022) Risk analysis model with interval type-2 fuzzy FMEA—case study of railway infrastructure projects in the republic of Serbia. Acta Polytech Hung 19(3):103–118. https://doi.org/10.12700/APH.19.3.2022.3.9
26. Ficzere P (2023) The role of artificial intelligence in the development of railway transportation. Des Mach Struct 13(1):58–63. https://doi.org/10.32972/dms.2023.005
27. Alzyod H, Ficzere P (2022) Industry, failures/breakdowns due to residual stresses in the vehicle. Int J Traffic Transp Eng 12(2):291–301. https://doi.org/10.7708/ijtte2022.12(2).10
28. Li Y, Liu H, Wang S, Jiang B, Fischer S (2023) Method of railway subgrade diseases (defects) inspection, based on ground penetrating radar. Acta Polytech Hung 20(1):199–211. https://doi.org/10.12700/APH.20.1.2023.20.14
29. Kurhan M, Kurhan D, Hmelevska N (2022) Maintenance reliability of railway curves using their design parameters. Acta Polytech Hung 19(6):115–127. https://doi.org/10.12700/APH.19.6.2022.6.9

30. Gawlak K (2022) Safety critical components (SCC) in the maintenance management system for railway vehicle. Diagnostyka 23(4):1–9. https://doi.org/10.29354/diag/156166
31. Beyond Silence (2006) Railroad lamp-2.jpg. Wikimedia Common [Online]. Available: https://commons.wikimedia.org/wiki/File:Railroad_lamp-2.jpg. Accessed 08 Nov 2023
32. Pletikoszity Á, Fürstner I, Gogolák L, Tarján L (2022) Interconnected drone systems for monitoring regional safety issues in field agriculture. In: Security-related advanced technologies in critical infrastructure protection. Springer. https://doi.org/10.1007/978-94-024-2174-3_21
33. Jianjun G, Yinhui C, Lei M (2017) Incident investigation and root cause analysis method and technology review. In: Proceedings of the 2017 25th international conference on nuclear engineering. Shanghai, China, 2–6 July 2017, p V001T01A045. https://doi.org/10.1115/ICONE25-67834
34. Alinezhad HS, Roohi MH, Chen T (2022) A review of alarm root cause analysis in process industries: common methods, recent research status and challenges. Chem Eng Res Des 180(9):846–860. https://doi.org/10.1016/j.cherd.2022.10.041
35. Sahoo T (2021) Root cause failure analysis: a guide to improve plant reliability. Wiley. https://doi.org/10.1002/9781119615606
36. Gangidi P (2019) A systematic approach to root cause analysis using 3 × 5 why's technique. Int J Lean Six Sigma 10(1):295–310
37. Ohno T (1988) Toyota production system: beyond large-scale production. Productivity Press, Cambridge
38. Rahmana A, Fauzy M, Suyono AM (2021) 5 why analysis implementation to detect root cause of rejected product (study at aerospace industry). Turk J Comput Math Educ 12(8):1691–1965
39. Minoura T (2003) The "thinking" production system: TPS as a winning strategy for developing people in the global manufacturing environment. Public Affairs Division, Toyota Motor Corporation
40. Ászity S (2018) Járműipari gyártási folyamatok minőségbiztosítása (Quality assurance of automotive production processes). Akadémia Kiadó, Budapest. https://doi.org/10.1556/9789634543015
41. Fabiś-Domagała J (2017) Application of ishikawa diagram for failure analysis of a car. Tech Trans 114(8):193–198. https://doi.org/10.4467/2353737XCT.17.141.6892
42. Ker JI, Khan ME (2017) Vertical integration of engineering courses for effective learning of continuous quality improvement concepts. In: 2017 ASEE Gulf—Southwest section annual conference, Dallas, Texas, USA. https://doi.org/10.18260/1-2-1153-33831
43. Purnawan AP, Hamali S (2023) Workload analysis using cardiovascular load, Nasa-TLX, and fishbone diagrams. Int J Soc Sci Human Res 6(8):4761–4767. https://doi.org/10.47191/ijsshr/v6-i8-19
44. Bernard A, Varaprasad G (2022) Fishbone diagram analysis for assessing ergonomic risks in onshore oil rig operations. In: Ergonomics for design and innovation. Springer. https://doi.org/10.1007/978-3-030-94277-9_117
45. Sakdiyah SH, Eltivia N, Afandi A (2022) Root cause analysis using fishbone diagram: company management decision making. J Appl Bus Tax Econ Res 1(6):566–576. https://doi.org/10.54408/jabter.v1i6.103
46. Subandi, Darunato D, Sumirahwati, Wahyuni RE, Saidah D, Binarti I (2023) Implementation of the lean manufacturing concept to reduce waste in spare part repair activities at PT. X. Int J Sci Soc 5(2):10–18
47. Baig AA, Ruzli R, Buang AB (2013) Reliability analysis using fault tree analysis: a review. Int J Chem Eng Appl 3(169–173):4. https://doi.org/10.7763/IJCEA.2013.V4.287
48. Patil RB, Waghmode LY, Chikali PB, Mulla TS (2009) An overview of fault tree analysis (FTA) method for reliability analysis & life cycle cost (LCC) management. In: Proceedings of the second international conference on emerging trends in engineering (SICETE). Nagpur, India
49. Halme J, Antti Aikala A (2012) Fault tree analysis for maintenance needs. J Phys Conf Ser 364:012102. https://doi.org/10.1088/1742-6596/364/1/012102
50. Szabó G (2001) Nagy biztonságú rendszerek megbízhatósági analízise és közlekedési alkalmazásai (Reliability analysis of high-security systems and their traffic applications). Ph.D. disszertáció (Ph.D. dissertation). BME Közlekedésautomatikai Tanszék, Budapest

51. Rausand M, Barros A, Høyland A (2021) System reliability theory—models, statistical methods, and applications. Wiley, Hoboken. https://doi.org/10.1002/9781119373940
52. Hastuti EP, Purwadi P, Sudarmono S, Rohanda A, Permana S (2023) Evaluation on structure, system, and component qualification in the periodic safety review of RSG-GAS. Adv Sci Technol 126:29–39. https://doi.org/10.4028/p-rbyb74
53. Del Rosso Calachea LD, Zanona LG, Munhoz Arantesa RF, Osirob L, Ribeiro Carpinettia LC (2021) Risk prioritization based on the combination of FMEA and dual hesitant fuzzy sets method. Production 31(9):e20200081. https://doi.org/10.1590/0103-6513.20200081
54. Mascia A, Cirafici AM, Bongiovanni A, Kisslinger A (2020) A failure mode and effect analysis (FMEA)-based approach for risk assessment of scientific processes in non-regulated research laboratories. Accred Qual Assur 25(8):311–321. https://doi.org/10.1007/s00769-020-01441-9
55. Divanoğlu SU, Taş Ü (2022) Application of 8D methodology: an approach to reduce failures in automotive industry. Eng Fail Anal 134(1):106019. https://doi.org/10.1016/j.engfailanal.2021.106019
56. Papageorgiou K, Theodosiou T, Rapti A, Papageorgiou EI, Dimitriou N, Tzovaras D, Margetis G (2022) A systematic review on machine learning methods for root cause analysis towards zero-defect manufacturing. Front Manufact Technol 2:972712. https://doi.org/10.3389/fmtec.2022.972712
57. Barsalou M (2023) Root cause analysis in quality 4.0: a scoping review of current state and perspectives. TEM J 12(1):73–79. https://doi.org/10.18421/TEM121-10
58. Bulla C, Birje M (2022) Improved data-driven root cause analysis in a Fog computing environment. Int J Intell Inf Technol 18(1):1–28. https://doi.org/10.4018/IJIIT.296238
59. Bulla C, Mahantesh Birje M (2021) Improved data-driven root cause analysis in Fog computing environment. J Reliable Intell Environ 8(3):359–377. https://doi.org/10.1007/s40860-021-00158-x
60. Gao Z, Xiang L, Fekete G, Baker JS, Mao Z, Gu Y (2023) A data-driven approach for fatigue detection during running using pedobarographic measurements. Appl Bionics Biomech 3(1):1–11. https://doi.org/10.1155/2023/7022513
61. Yang Y, Mohr WC (2015) Finite element creep-fatigue analysis of a welded furnace roll for identifying failure root cause. J Mater Eng Perform 24(11):4388–4399. https://doi.org/10.1007/s11665-015-1746-0
62. Dondapati S, Trivedi M, Dondapati RS, Chandra D (2017) Investigation on the mechanical stresses in a muffler mounting bracket using Root Cause Failure Analysis (RCFA), finite element analysis and experimental validation. Eng Fail Anal 81:145–154. https://doi.org/10.1016/j.engfailanal.2017.08.010
63. Simms ER, Slakey D, Garstka ME, Tersigni S, Korndorffer JR (2012) Can simulation improve the traditional method of root cause analysis: a preliminary investigation. Surgery 152(3):489–497. https://doi.org/10.1016/j.surg.2012.07.029
64. Ito A, Hagström MH, Bokrantz J, Skoogh A, Nawcki M, Gandhi K, Bergsjö D, Bärring M (2022) Improved root cause analysis supporting resilient production systems. J Manuf Syst 64(18):468–478. https://doi.org/10.1016/j.jmsy.2022.07.015
65. Jia J, Chan TT, Lian T, Rio KW (2023) Local pupil swim in virtual- and augmented-reality: root cause and perception model. J Soc Inform Display 31(5):230–240. https://doi.org/10.1002/jsid.1210
66. Jia J, Chan TT, Lian T, Rio KW (2023) 20-2: distinguished paper: local pupil swim in VR/AR: root cause and perception model. SID Sympos Dig Tech Pap 54(1):255–258. https://doi.org/10.1002/sdtp.16539
67. Simon J, Gogolák L, Sárosi J, Fuerstner I (2023) Augmented reality based distant maintenance approach. Actuators 12(7):302. https://doi.org/10.3390/act12070302

Enhancing Computer Forensics: Standardisation of Forensic Information Technology Expert Assignments

Zsolt Illési, Vid Sebestyén Honfi, and Pál Bárkányi

Abstract The advancement of information technology (IT) has significantly impacted forensic science, necessitating the development of standardised methods within computer forensics to improve investigation quality and reliability. This paper aims to address the challenges faced in forensic information technology examinations by proposing a structured framework for categorising expert activities and formulating guiding questions for forensic experts. Employing a detailed analysis of current practices and identifying common pitfalls, this study introduces a two-dimensional classification of forensic expert activities into direct, indirect, and elementary and complex tasks. This categorisation facilitates a nuanced understanding of expert roles and the intricate nature of forensic investigations. The paper outlines key activity categories, including professional interpretation, compliance testing, digital evidence acquisition and management, laboratory examinations, and evidence presentation. The findings suggest that employing standardised methods and specific guiding forensic tasks or assignments that competent authorities can utilise can significantly enhance forensic investigations' consistency, efficiency, and effectiveness. Ultimately, the paper concludes that the proposed solutions offer a promising approach to improving the practice of forensic information technology expertise, thereby supporting the pursuit of justice in the digital age.

Keywords Computer forensics · Digital evidence · Forensic standardization · Computer forensics assignments

Z. Illési · V. S. Honfi (✉) · P. Bárkányi
Milton Friedman University, Budapest, Hungary
e-mail: honfi@uni-milton.hu

P. Bárkányi
Obuda University, Budapest, Hungary

T. A. Kovács et al. (eds.), *The Impact of the Energy Dependency on Critical Infrastructure Protection*, Advanced Sciences and Technologies for Security Applications, https://doi.org/10.1007/978-3-031-78544-3_29

1 Introduction

During criminal proceedings, it is necessary to examine information technology and information communication devices, systems, and their elements from various perspectives. These systems and elements can serve as targets of criminal activities, objects or environments of implementation or commission, facilitators of commission, symbols of commission, or witnesses to commission.

The tools and methods applicable during the investigation can vary significantly depending on the roles of these examined elements. In our country, the term "forensic sciences," commonly used in Anglo-Saxon literature, is also prevalent for denoting expert fields. Therefore, in this study, the authors consider forensic examinations aimed at detecting criminal activities involving IT devices as part of forensic information technology examinations.

The emergence and development of forensic sciences can be attributed to the societal demand to identify the perpetrator of a crime and incontrovertibly prove their guilt. Edmond Locard was the first to articulate the scientific principle in evidence gathering: every contact leaves a trace. Accordingly, according to the theory, every interaction between different objects results in trace formation; whatever activities the perpetrator engages in at the scene will leave a trace. Thus, the authors can establish a connection between the perpetrator, the crime scene, and the victim. Traces can include fingerprints, material residues, and data remnants. Forensic science refers to material changes in a crime's commission as traces or material residues. It considers traces of those changes that reflect the shape and surface characteristics of the object created. In contrast, material residues refer to materials derived naturally or as a result of physical force from individuals or objects involved in the event under investigation. These materials can be understood as originating from a single, random sampling, thereby sharing characteristics with the object from which they originated.

An essential step in ensuring the reliability and acceptance of forensic examinations came in 1993 with the decision of the Supreme Court of the United States in the case Daubert v. Dow Pharmaceuticals. The court ruled that, regarding the reliability of expert opinions, the judge must examine whether:

- the expert technique or the underlying theory has been tested or is testable;
- the method or underlying theory has been published in a peer-reviewed journal;
- the error rate of the technique or method has been determined during its application, and whether there are standards for controlling the technical process;
- the acceptance of the field by the scientific community.

2 Typical Errors Related to Forensic Investigations

During forensic expert examination, experts can commit various errors, violating the rights of one or more parties. Such infringements may occur when:

- the right of the state to protect society is violated,

- the right of the accused to a fair trial is infringed upon,
- the right of the victim or aggrieved party to restitution or compensation is violated.

A faulty expert opinion can thwart or limit legal remedies. The successful redress, compensation can be hindered by an inadequate approach, resulting in:

- insufficient quantity and quality of evidence being gathered,
- inadequate documentation of forensic examinations,
- non-repeatability of examinations due to improper evidence handling,
- errors occurring during analysis or interpretation [1, 2].

Next, the authors summarise the most common errors the appointing authority or the appointed expert may commit. Section 3 (1) of Act XXIX of 2016 on forensic experts [3] (from now on referred to as the Expert Act) prescribes that forensic experts prepare their opinions utilising the results of science and technological advancements.

2.1 Violation of Competence

One of the errors that appointed forensic computer experts can commit is when they expert exceed the boundaries outlined in their appointment. Such an error occurs particularly when:

- the expert provides an opinion on matters of law,
- the expert responds to a question not posed in the appointing order,
- the expert provides an opinion based on evidence from a source outside the forensic report [1].

Section 47 (6) of the Expert Act prohibits experts from expressing opinions on legal matters by force of law. Similarly, it constitutes an error if the expert, exceeding their authority, expresses opinions on matters not raised by the appointing authority. Thus, it is not uncommon for a poorly formulated question to elicit an incorrect response.

2.2 Substantive Lack of Foundation

Act XC of 2017 on Criminal Procedure [4] (from now on referred to as the Criminal Procedure Act) Section 197 (1) specifies the cases in which an expert opinion cannot be accepted due to some deficiency. Beyond formal requirements, the law sets the expectation that the expert opinion be clear and consistent with itself and the data provided to the expert and leaves no doubt about its correctness. From a practical standpoint, the expert opinion is not acceptable if the expert opinion is:

- inconsistent with the documents,
- contains erroneous conclusions,

- lacks essential factual findings, or
- contains unresolved issues in the questions posed [1].

In such cases, upon request from the court, the prosecutor, or the investigating authority, the expert must provide clarification or supplement the expert opinion. Avoiding this situation requires that the expert, irrespective of the number and quality of the issues to be examined, conducts the examination as thoroughly and meticulously as possible within the available time frame.

2.3 Lack of Methodological Foundation

According to Section 3 (1) of the Expert Act, the expert must conduct the examination forming the basis of the expert opinion using tools, procedures, and methods consistent with the state of science and contemporary professional knowledge. Unfortunately, the legislator does not provide clear guidance on the exact definition of these concepts. Therefore, the authors consider it a methodological lack of foundation when the expert fails to present or superficially identify the examination methods when the expert needs to substantiate the validity of the applied method or superficially identifies it [1, 2].

From a practical perspective, this translates into the requirement that the expert should scientifically substantiate every method they employ, continuously monitor the advancement of science, and apply its latest developments. However, whether the examined field can meet this criterion remains to be seen.

2.4 Issues Regarding the Involved Experts

Regarding the involvement of experts in some instances, Section 188 (1) of the Criminal Procedure Act stipulates that if particular expertise is required to establish or assess a fact, an expert must be employed. The law also enables the determination of these areas for other regulations.

However, there may be extreme cases where the necessity is incorrectly determined, resulting in the omission of expert examination, failure to appoint an expert, or failure to request one.

Although Section 5 of the Expert Act defines, who can be a forensic expert, becoming an expert only requires formal proof of expertise. This fact alone may raise concerns about whether experts with the necessary competencies have been appointed for every case. This is particularly risky in information technology, considering the significant advancements in IT solutions since 2006, not to mention the rapidly evolving field of artificial intelligence solutions and their vulnerabilities.

Significant risks are associated with appointing multiple experts consecutively for the same case. Although Section 192 (3) states that "if an expert examines tangible evidence or electronic data that may change or be destroyed during the examination, the expert must ideally preserve part of it in its original state so that its identity and origin can be determined," can the authors be assured that subsequently appointed experts will examine the same evidence if the initial expert did not exercise due diligence? This situation is somewhat mitigated by the same provision, which states that "if multiple experts conduct examinations in criminal proceedings, they must inform each other about the forensic examination they intend to conduct. The notified expert may be present during the examination conducted by the other expert." It is easy to see that in these cases, the interests of any party may easily be compromised.

The professionalism of expert work is also questioned by the fact that there is an opportunity for ad hoc expert appointments in procedures. Section 2 (4) of the Expert Act defines an ad hoc expert as "a natural or legal person with appropriate expertise to establish or assess significant facts or other circumstances to be established or assessed in the procedure, who is not a forensic expert, and a forensic expert who provides expert opinions in fields of forensic expertise and in fields not specified in the regulation on forensic expertise." To what extent does applying contemporary professional knowledge in line with the state of science apply in these cases? It can be unequivocally stated that if a forensic expert is available in the relevant field, appointing an ad hoc expert is incorrect. Of course, the immediate question arises whether, based on the applied terminology, it can be determined whether there is an expert whose task and competence are to perform the given task. This brings us to the next problem area, which is the definition of fields of expertise.

2.5 Issues Related to Fields of Expertise

In our assessment, ensuring professionalism in expert work poses a concern in the delineation of expertise in the field of information technology. The latest attempt to define the scope of expertise in this field was made by Regulation 9/2006 (II. 27.) of the Ministry of the Interior on forensic expertise fields and related qualifications and other professional requirements [5].

After the authors compared this regulation to the U.S. Department of Justice, FBI, Laboratory Division: Handbook of Forensic Services [6], it is easily understood that this legislation needs to be updated (Table 1).

The authors agree with Nogel's observations [7] that it is increasingly urgent to reclassify fields of expertise and reconsider qualification requirements.

2.6 Legal Status of the Experts

A common characteristic of expert appointments in Hungary is the excessive trust placed in the individual expertise of the expert. The most crucial criterion for the

Table 1 Fields of forensic computer expertise: [5, 6] based on own editing

Information technology fields	Computer forensics tasks
• Computer equipment, computers, peripherals, and local networks (hardware) • Information security • Design and organisation of information systems • Activities related to studio technology and multimedia • Computer databases, data structures	• Content analysis • Comparison • Establishing the timeline of transactions/events • Data file recovery • Recovery of deleted data files • Format conversion • Keyword search • Password recovery • Source code analysis

adequacy of the expert opinion is the expert's specialised knowledge and competence. In practice, since there are no fixed, standardised methodologies, the expert's approach ensures the adequacy of the conducted examinations. The problem is that since the methodology is not uniform, each expert decides how to examine a particular case. This broad independence in approaches leads to the consequence that the result depends on the diligence of the expert. Another issue arises when evaluating the expert opinion, as the objective criteria for evaluation still need to be fully met since both the judge and the appointing authority can be considered laypersons in the specific issues.

The expert's work is only questioned when the opposing party or its legal representative can undermine the credibility of the expert opinion and support it with appropriate arguments. This practice in Hungary is contrary to international practices, as while in our country those selected can be experts and conduct the work as they wish, in other countries, the examination methodology is the decisive factor, and anyone who can perform the examinations according to the standardised methodology can be an expert. In the Hungarian practice, excessive trust surrounds the individual expert because they ensure credibility, not the applied methodology.

This situation is exacerbated by the fact that experts do not receive adequate support for their work. Paragraph 15 (1) of the Expert Act makes participation in legal education and passing the legal exam mandatory. The same legal regulation, in paragraph 16 (1), requires participation in regular professional training sessions organised by the Chamber or the Minister for the practice of expert activities. However, these methodological training and professional consultation forums still need to be improved; thus, ensuring the expected professionalism in the applied methodologies is not guaranteed.

3 Proposed Solutions

The authors formulate proposals to address the identified problems in forensic information technology expertise in the following sections. They emphasise expert activity standardisation in this regard. Parallel to classifying forensic information technology expert activities and defining sub-activities, the authors formulate typical questions for which experts can be commissioned.

3.1 Forensic Expert Activities

The activities performed by forensic experts can be classified from several perspectives. Accordingly, distinctions can be made between tasks based on their immediacy and complexity.

The authors suggest the following two-dimensional categorisation:

- ***direct*** (usually specified in the appointment order)
- ***indirect*** (typically not specified but necessary to conduct the forensic examination),

and

- ***elementary*** (related to individual activities)
- ***complex*** (requiring the sequential execution of multiple forensic activities)

Direct activities are those typically included in the expert appointment order (e.g., Determine if the investigated system contains images of naked individuals!). Indirect activities are those not included in the expert appointment order but necessary to achieve professional results. An example of such an activity is verifying the accuracy of system time, which is a prerequisite for investigations involving any time-related data requested by the appointing authority.

Elementary activities include discrete tasks which can be performed independently of other tasks, and their output might serve as input for other elementary or complex forensics activities. Backing up a hard drive is an excellent example of an elementary activity. ***Complex*** activities involve a group of elementary or complex tasks. Backing up a mobile phone can serve as a sample for a complex task. It includes backing up data from the device, saving data from the SIM card and the memory card, and, if necessary, cloning the SIM card without proper security data.

Table 2 Representative or sample assignment of professional interpretation, based on own editing

Forensic information technology expert activity	Representative or sample assignment
Presentation	• Describe what an [**assignment:** IT method or technology] is and how it is used! • Explain under what conditions [**assignment:** IT event] can result in [**assignment:** consequence]!

3.2 Classification of Forensic Information Technology Expert Activities

Next, the authors will describe sub-activities falling under the main types of activities and representative or sample assignments prescribed by the competent authorities that experts should perform in answering questions relevant to the case.

The main types of activities performed by forensic information technology experts include:

- professional interpretation,
- matching, compliance testing,
- digital evidence acquisition and management,
- laboratory examinations,
- report presentation.

3.2.1 Professional Interpretation

Tasks falling under this category involve answering general questions requiring specialised, in this case, information technology expertise. This includes providing layman's explanations of concepts or phenomena or conducting a situation analysis to make it understandable to stakeholders whether "a particular event could have happened this way" or "what was necessary for…" (Table 2).

3.2.2 Matching, Compliance Testing

This category encompasses questions regarding hardware or software's compliance with the following:

- standards, recommendations,
- general, best, or Hungarian practice,
- tender or contract terms.

The expert's task is to compare the operation of the questioned hardware or software device with the requirements of a standard, recommendation, tender specification, or technical annex of a contract and determine its compliance or degree of non-compliance. The related sample questions are included in Table 3.

Table 3 Representative or sample matching and compliance testing assignment, based on own editing

Forensic information technology expert activity	Representative or sample assignment
Matching, compliance testing	• Determine if the developed [**assignment:** system] system meets the requirements defined in the [**assignment:** documented user specification] specification! • Determine if the project [**assignment:** case the project documentation] complies with the [**assignment:** standard or guideline] documentation!

In such cases, the expert makes technical concepts understandable to laypeople, elucidates their content, and outlines choices or professional analyses of a phenomenon or event.

Professional interpretation usually requires direct and complex activities. (For example, the question defines the full scope of interpretation.) However, the judicial expert must gather underlying concepts, techniques, and methodologies and comprehensively assess them.

3.2.3 Digital Evidence Capture and Management

In this task group, the authors classify forensic information technology expert activities related to the secure recording and authentication of data sources used for analysis. This includes activities preceding data capture, ensuring that the captured data remains separate from other data (e.g., from previous cases), and restricting access to data managed by experts. Activities for retrieving existing data are also included in this group.

This leading group encompasses all activities related to the acquisition, preliminary assessment, and storage (and possibly deletion) of digital evidence, including:

- criminal evidence and data handling
- data carriers ("dead" system) backup
- "live" system data backup
- network data traffic backup
- duplicate creation.

Criminal evidence and data carrier handling typically involve activities not explicitly specified in the expert appointment order. However, specific regulations may partially cover these activities (e.g., rules for document management). These activities primarily involve identifying and labelling target data carriers, sterilising target data carriers (overwriting with hex 0 values bit-by-bit), post-examination closure and dispatch, handling incoming criminal evidence, and storage. There are no representative or sample assignments within this activity scope.

The most common method of evidence preservation for a forensic information technology expert is the bit-by-bit identical backup of data content from data carriers. When performing this task, the data carrier under investigation must be backed up to ensure the copied data content is an accurate and authentic replica of the original. Examining a "dead" system involves examining data carriers of shutdown computers using one's reliable tools in a reliable environment. Although this approach yields adequate results, it may not always be feasible in practical scenarios, such as when examining unique tools and prototypes, lacking a specialised examination environment and tools, or requiring rapid, immediate on-site analysis. During backup, the expert must also preserve data that is not obvious to laypersons (e.g., hard disk DCO, HPA, and SMART data). Common data carriers include hard drives, individual hard drives, RAID volumes, SIM cards, optical storage CDs/DVDs, SSDs, memory cards, and USB data storage devices.

Backing up "live" system data involves examining the computer or device under analysis using its operating system, user interface, and utility programs. During backup, the expert must preserve hidden data and minimise the environment's impact on the backed-up data. Backing up "live" system data may include backing up server/workstation memory, server/workstation data carriers, virtualised systems, network data storage, mobile phones, tablets, and PDAs.

Backing up network data traffic—mobile phone data traffic or other personal data cases—can only be performed with special authorisation. If the goal is to preserve the technical characteristics of the traffic for later analysis, a forensic expert can perform the examination. However, it should be noted that data recording performed simultaneously with the offence constitutes covert data collection according to Section 214 of the Criminal Code.

Duplicate creation refers to creating a bit-by-bit copy of already recorded evidence. Cloning SIM cards is considered an indirect activity if copying data from a mobile phone requires circumventing SIM protection data (PIN, PUK). Table 4 includes the most representative or sample assignment encountered during digital evidence capture and management.

Table 4 Representative or sample assignment of digital evidence capture and management, based on own editing

Forensic IT expert activity	Representative or sample question
Backup data storage ("dead" system)	• Create a copy of the examined [**assignment**: data carrier] data carrier!
"Live" system data backup	• Create a copy of the examined [**assignment**: system or computer]'s [**assignment**: memory or storage]!
Network data traffic backup	• Save the data of the examined [**assignment**: computer network or network segment] network!
Duplicate creation	• Create [**assignment**: 1...n] copies of the saved data!

3.2.4 Laboratory Examinations

Laboratory examinations encompass activities performed on preserved data in an informatics laboratory. Such activities include:

- search,
- pattern search,
- feature search,
- cryptanalysis,
- transaction analysis,
- program analysis.

Under search, the authors understand that forensic investigations aimed at analysing data collected at the physical layer, identifying and segregating primary data containers (digital objects) following the recording of data passing through the network, stored in memory or on other data carriers by the processor. These are not files because not all data have file entries. Search objectives may include file system data, unpartitioned space, partitioned space, and primary digital objects. These are indirect tasks; typically, they do not involve representative or sample assignments.

Another group of searches includes file (deleted, regular) searches, secondary digital object searches (embedded digital objects, partially deleted files), and network data searches (data frames, data packets, control data, protocol hierarchy, embedded protocols). These searches include primary activities (as described in the appointment order).

In the context of pattern searches, the authors refer to keyword searches, pattern matching, and searching for hidden correlations (data mining).

In keyword searches, the expert searches for words or word fragments relevant to the case. Searches can be conducted directly on digital objects or in an index database prepared beforehand to increase efficiency and speed by indexing all interpretable data portions found in the files involved in the investigation. This is particularly applicable when numerous words or fragments are expected to be searched or when it is unclear what exactly needs to be searched for during the investigation. In such cases, the examiner must meticulously identify traces of the offence using a comprehensive keyword database.

During pattern-matching searches, unknown data content is sought based on a known structure, such as a date, bank card number, email address, etc. Regular expressions (RegEx) are used to describe the characteristics of the searched pattern.

Data mining involves the semi-automatic discovery of information hidden in large quantities of data using various algorithms. Under data mining, the authors understand the search for novel, valid, non-trivial, and likely valuable and explainable correlations in large datasets (Table 5.).

Table 5 Representative or sample assignment for laboratory examinations, based on own editing

Forensic IT expert activity	Representative or sample assignment
File search	• Determine if the [**assignment:** data] contains the following keywords [list: keywords]!
Pattern search	• Determine if the [**assignment:** data] contains data referring to [**assignment:** keyword list]! • Determine if the [**assignment:** data] contains [email addresses!
Cryptanalysis	• Identify encrypted data in the [**assignment:** target data]!
Transaction analysis	• Determine who used the [**assignment:** target system]! • Determine if there was a backup in the [**assignment:** target system] between [**assignment:** date/time] and [**assignment:** date/time]!
Program analysis	• Determine if the [**assignment:** analysed program] contains programmed/ malicious code snippets! • Determine if the [**assignment:** analysed program] transmits data to external systems during operation! If so, identify the transmitted data and the target computer! • Back up the [**assignment:** data] data stored in the [**assignment:** analysed program]!

3.2.5 Evidence Presentation

While preserving digital objects, the expert copies the data (digital objects) identified and retrieved in previous phases onto a data carrier serving as an electronic annex to the expert opinion. This is done to make these data accessible and analysable to the appointing authority or non-IT experts conducting further examinations. During the transformation, the forensic IT expert converts the digital objects into an understandable and manageable form for the appointing authority. Additionally, the expert ensures the encoding of identified (possibly hidden or indirectly interpretable) data and their visualisation (e.g., representing timelines in activity diagrams or other formats). The visualisation is typically not included in the appointment order. Still, it is prepared by the expert as an indirect task to support their opinion and enhance the understanding of the appointing authority (Table 6).

Table 6 Representative or sample assignment related to evidence presentation, based on own editing

Forensic IT Expert Activity	Representative or sample assignment
Evidence Presentation	• Save the identified [**assignment:** digital evidence] files! • Transform the identified [**assignment:** digital evidence] to [**assignment:** format] format!

4 Conclusions

The detailed analysis demonstrates that many specific types of information technology are used for forensic purposes. Standardised methods are essential in this area. The proposed solutions address the challenges by defining expert activities, sub-activities, and questions that will guide forensic information technology experts in their work.

The categorisation of forensic expert activities into direct and indirect, elementary and complex, can provide a structured framework for understanding the diverse tasks encountered in forensic examinations. This classification facilitates a nuanced comprehension of the expert's role by distinguishing between direct activities specified in appointment orders and indirect activities essential for professional outcomes. Similarly, the differentiation between elementary tasks that can be performed independently and complex tasks requiring sequential execution underscores the intricacies involved in forensic investigations. It emphasises the need for meticulous planning and execution.

Professional interpretation, matching, compliance testing, digital evidence acquisition and management, laboratory examinations, and evidence presentation are some categories of forensic information technology expert activities. This reveals the wide range and depth of knowledge required in this area. Each activity type encompasses a range of tasks that demand specialised knowledge and skills, from providing layman's explanations of complex concepts to conducting intricate analyses of digital evidence.

The authorities and forensic information technology experts use the questions to guide their work and ensure their examinations are thorough and rigorous. These questions explain the main parts of each activity, like how professionals interpret, check if something is done right, get data, analyse it, and present it. These representative or sample assignments help to make forensic examinations more consistent and reliable, improving the quality and credibility of expert opinions. The solutions proposed in this study are an excellent first step toward improving the practice of forensic information technology expertise. This approach makes forensic investigations more transparent, efficient, and effective by setting up a system for categorising expert activities and formulating questions. In an ever-evolving digital world, pursuing justice requires further standardisation of computer forensics techniques and implementing these methods to improve the quality and capabilities of computer forensics.

References

1. Tremmel F (2006) Bizonyítékok a büntetőeljárásban (Evidence in criminal proceedings); Dialóg Campus Kiadó, Budapest-Pécs. ISBN 963-7296-72-7
2. Illési Z (2013) Cyberterrorism from IT Forensics Perspektive. Magyar Rendészet 13:55–62, 8 p
3. 2016. year XXIX. Law on forensic experts
4. 2017. yearXC. Law on criminal procedure

5. 9/2006. (II. 27.) IM Regulation on the specialisation of forensic experts and the qualifications and other professional conditions related thereto
6. U.S. Department of Justice, FBI, Laboratory Division: Handbook of Forensics Services. https://www.fbi.gov/file-repository/handbook-of-forensic-services-pdf.pdf/view (2023 Nov 20)
7. Nogel N (2018) Igazságügyi szakértői vélemények hiteltérdemlősége a büntetőeljárásban (The discrediting of forensic expert opinions in criminal proceedings). Ph.D. értekezés, Pécsi Tudományegyetem Állam- és Jogtudományi Kar Doktori Iskola. https://ajk.pte.hu/files/file/doktori-iskola/nogel-monika/nogel-monika-muhelyvita-tezisek.pdf (10 Dec 2023)

Estimation of the Thermal Influence Size Zones of Oil Spill Fire

Yaroslav Kalchenko, Kostiantyn Afanasenko, Natalia Minska, and Valentin Kerdyvar

Abstract The study assessed the consequences of emergencies at oil refineries. Three scenarios for the development of an emergency situation were considered. Local depressurization of the tank with formation of oil spill fire in the dike. Full depressurization of the oil tank with petroleum products without a breakthrough wave with the formation of a fire in the dike. Complete depressurization of the petroleum product tank with formation of a breakthrough wave on the spill fire. Graphs of the dependence of the intensity of thermal radiation on the distance to the center of the spill for oil tanks of different volumes were constructed. The numerical values of distances from the oil spill fire for different degrees of human injury, which are presented in the form of a table, were determined.

Keywords Oil spill · Oil tank · Spill fire

1 Introduction

The fire hazard of petrochemical industry facilities must meet the requirements of fire, energy, economic, and environmental safety. The issues of ensuring fire and explosion safety of enterprises of the oil refining industry are very urgent. This is explained by: the presence of potential dangers causing material and human losses; the concentration of chemical energy carriers, oil and oil products, their ability to burn, explode and pollute the atmosphere [1] and underground and river waters due to the seepage of liquid deep into the soil [2] by dangerous emissions, extremely high energy saturation of oil refining industry facilities. Spreading over considerable distances, these significantly affect the state of the air [3].

About 60% emergency situations with fires are fires involving flammable liquids [4]. Accidents of vertical steel tanks, as a rule, are accompanied by oil spill and fires

Y. Kalchenko (✉) · K. Afanasenko · N. Minska · V. Kerdyvar
National University of Civil Defence of Ukraine, Kharkiv, Ukraine
e-mail: kalchenko_yaroslav@nuczu.edu.ua

T. A. Kovács et al. (eds.), *The Impact of the Energy Dependency on Critical Infrastructure Protection*, Advanced Sciences and Technologies for Security Applications, https://doi.org/10.1007/978-3-031-78544-3_30

caused by it, which leads to man-made disasters, violations of normal operation and significant environmental pollution [5].

In the case of fires and explosions that occur in the event of an accident of oil tanks, the zone affected by open fire and thermal radiation should be determined. The size of the zone of impression by open fire is determined by the area where its appearance is possible.

One of the causes of accidents occurring in oil tanks is damage to the tank casings due to many reasons [6]: violations of the technological conditions of operation, defects and violations during installation, damage by debris and shock waves during shelling of oil tanks, corrosion, etc.

In [7], the thermal effect of a fire in a tank with petroleum products on a neighboring tank is investigated. Using the thermal conductivity equation, the temperature distribution inside the tank wall is determined. However, this research paper does not consider a spilled liquid fire. Using the thermal conductivity equation, the temperature distribution inside the tank wall is determined. In [8], a model of the thermal effect of a flame on a container with oil products located nearby is proposed, but this model does not take into account the different temperatures on the outer and inner surfaces of the tank. In [9], the authors analyse the effect of a fire in a nearby tank and a fire-in-collapse on a oil product-containing tank.

In [10], a model of combustion of an oil spill on a horizontal surface in the shape of a circle is presented. The disadvantage of this model is that the slope of the surface is not taken into account, which leads to the stretching of the spill area along the direction of the slope [11]. In [12], a procedure for experimentally determining the parameters of infiltration into the soil is given, but the parameters of the flame above the burning and propagating liquid are not taken into account. In [13], a model of the convection component of the heat flow from a spill fire of arbitrary shape is described. At the same time, these works do not consider the consequences of accidents caused by oil spills caused by fire, namely, the sizes of thermal impact zones and their impact on people.

2 Research Materials and Methods

In the study were estimated the consequences of emergency situations at oil depots. For research were chosen three possible scenarios of fire occurrence and development:

- local depressurization with the occurrence of an oil spill fire in the diking;
- full oil tank depressurization without wave with the occurrence of an oil spill fire throughout all the dike area;
- full oil tank depressurization with wave which burns and goes through the dike area.

The data were used for the studies presented in Tables 1 and 2. For research were used 5000, 10,000 and 20,000 m^3 oil tanks were, as are the most common in Ukraine.

Table 1 Technical characteristics of the tanks used in the calculations

Oil tank V = 5000 m^3		Oil tank V = 10,000 m^3		Oil tank V = 20,000 m^3	
Height, m	12.0	Height, m	12.0	Height, m	18.0
Diameter, m	22.8	Diameter, m	34.2	Diameter, m	40.0

Table 2 Initial data

Parameter	Value
Oil density, kg/m^3	850
Coefficient of spillage during spreading, f_p, m^{-1}	20
Oil temperature, °C	39
Lower flame propagation concentration limit, %	0.9
Molecular weight, kg/kmol	50
Acceleration of gravity, m/s^2	9.8
Air pressure, kPa	101.325
Degree of filling of oil tank	0.9
Block time, s	3600
Saturated oil vapor pressure, kPa	60
Saturated oil vapor concentration, %	59.215
Diameter of hole formed during depressurization, m	0.05

Calculations will be carried out by computing equipment using the Maple program.

3 Mathematical Model of the Oil Spill Fire Thermal Influence Size Zones Estimation

Thermal radiation intensity is determined by the formula [14]

$$q = E_f F_q \tau, \tag{1}$$

where $E_f = 60$ kW/m^2—average speed of thermal radiation intensity, τ—atmospheric transmission coefficient is determined by the formula

$$\tau = \exp\left[-7.0 \times 10^{-4}\left(\sqrt{r - 0.5d}\right)\right], \tag{2}$$

where *r*—the distance from the geometric center of the spill fire to the object of exposure.

The effective diameter of the spill is determined by the formula

$$d = \sqrt{\frac{4F}{\pi}}, \tag{3}$$

where F—oil spill square which determined by the formula

$$F = fV_w. \tag{4}$$

The volume of oil flowing out of the tank from the moment of the accident to the blocking of the hole is described by the expression

$$V_w = \int_0^{\tau_{approx.}} q(\tau)d\tau, \tag{5}$$

where $q(\tau)$—change of oil flow rate at its free outflow from the hole can be described by the formula

$$q(\tau) = \mu F_{open} \nu(\tau), \tag{6}$$

where $\nu(\tau)$—change in the rate of oil outflow with its free outflow from the hole is described by the formula

$$\nu(\tau) = \sqrt{2gh(\tau)}. \tag{7}$$

Change of oil head at its free outflow from the hole can be determined from the formula

$$h(\tau) = \left(\frac{2\sqrt{\varepsilon H_p - h_{open}} - \sqrt{2g}\frac{\mu F_{open}}{f_p}\tau}{2} \right)^2. \tag{8}$$

For a scenario with full depressurization of the tank without creating a breakthrough wave, the volume of oil flowing out is determined by the formula

$$V_w = h_{obv}\frac{\pi D_p^2}{4}. \tag{9}$$

For a scenario with complete depressurization of the tank with the formation of a breakthrough wave, the volume of oil spilled will be determined by the formula

$$V_w = \varepsilon H_p\frac{\pi D_p^2}{4}. \tag{10}$$

The angular irradiance coefficient is determined by the formula

$$F_q = \sqrt{F_v^2 + F_H^2}, \tag{11}$$

where

$$F_v = \frac{1}{\pi}\left[\begin{array}{l} \frac{1}{S}\operatorname{arctg}\left(\frac{h}{\sqrt{S^2-1}}\right) - \frac{h}{S} \\ \cdot\left\{\operatorname{arctg}\left(\sqrt{\frac{S-1}{S+1}}\right) - \frac{A}{\sqrt{A^2-1}}\operatorname{arctg}\left(\sqrt{\frac{(A+1)\cdot(S-1)}{(A-1)\cdot(S+1)}}\right)\right\} \end{array}\right]; \tag{12}$$

$$F_H = \frac{1}{\pi}\left[\frac{B-1/S}{\sqrt{B^2-1}}\operatorname{arctg}\left(\sqrt{\frac{(B+1)(S-1)}{(B-1)(S+1)}}\right) - \frac{(A-1/S)}{\sqrt{A^2-1}}\operatorname{arctg}\left(\sqrt{\frac{(A+1)\cdot(S-1)}{(A-1)\cdot(S+1)}}\right)\right]. \tag{13}$$

Dimensions h, S, A, B are determined from expressions

$$\begin{gathered} h = 2H/d;\ S = 2r/d; \\ A = \left(h^2 + S^2 + 1\right)/2S;\ B = \left(1 + S^2\right)/2S. \end{gathered} \tag{14}$$

Flame height is determined by the formula

$$H = 42d\left(\frac{m_y}{\rho_B\sqrt{gd}}\right)^{0.61}, \tag{15}$$

Where—$m_y = 0.06$ kg/(m^{2}s) specific mass rate of oil burning; ρ_B—air density.

The area of the oil spill for various emergency scenarios, which was calculated based on the data in Table 1, is presented in Table 3.

Table 3 Oil spill area for different emergency scenarios

Oil tank V = 5000 m^3		Oil tank V = 10,000 m^3		Oil tank V = 20,000 m^3	
Local depressurization					
Oil spill area, m^2	1264	Oil spill area, m^2	1267	Oil spill area, m^2	1555
Full oil tank depressurization without wave					
Oil spill area, m^2	12,248	Oil spill area, m^2	17,564	Oil spill area, m^2	17,564
Full oil tank depressurization with wave					
Oil spill area, m^2	132,283	Oil spill area, m^2	297,637	Oil spill area, m^2	407,150

4 Results of Estimation of the Oil Spill Fire Thermal Influence Size Zones

According to formula (1), graphs of the dependence of the amount of thermal radiation during the burning of an oil spill in the case of local depressurization, full depressurization without the formation of a spill wave, and full depressurization with the formation of a spill wave were constructed, which are presented in Figs. 1, 2 and 3.

From the analysis of the resulting graphs, follows that the magnitude of the intensity of thermal radiation decreases exponentially. The dependence of the intensity of thermal radiation on the distance for the scenarios of local depressurization and full depressurization without wave formation differs slightly. This is due to the fact that for both cases the fire area will be limited to the diking area, but the volume of oil spill during the accident will be different.

For the scenario of full depressurization of the oil tank with the formation of a breakthrough wave, the spill fire area will increase significantly. Accordingly, the distance from the accident site at which the thermal radiation intensity will be dangerous will also increase.

Table 4 shows the maximum permissible intensity of thermal radiation from an oil spill fire.

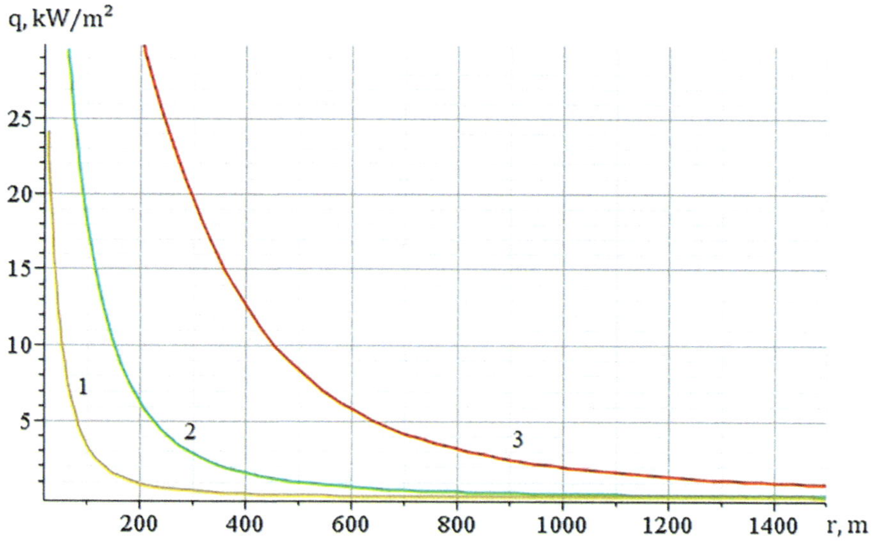

Fig. 1 Dependence of thermal radiation intensity for the oil tank (V = 5000 m^3): 1—local depressurization; 2—full depressurization, without wave; 3—full depressurization, with wave

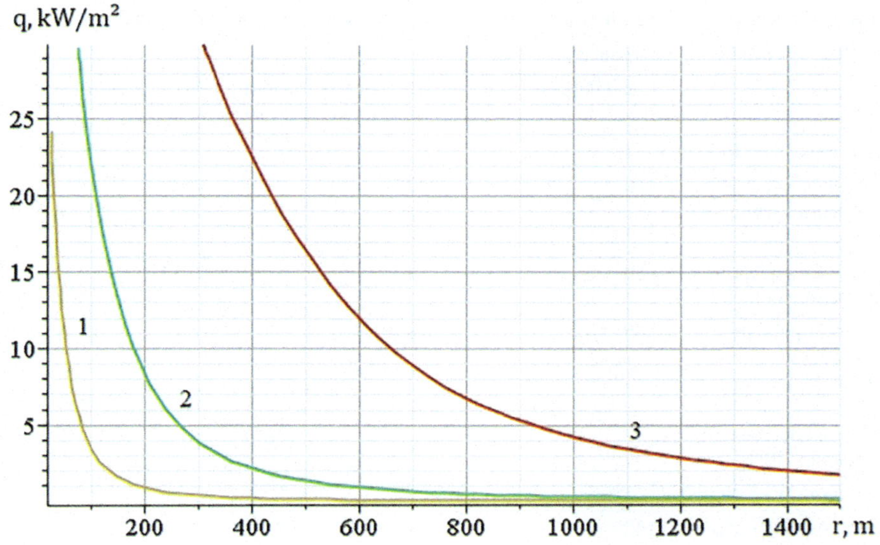

Fig. 2 Dependence of thermal radiation intensity on the radius for the oil tank (V = 10,000 m^3): 1—local depressurization; 2—full depressurization, without wave; 3—full depressurization, with wave

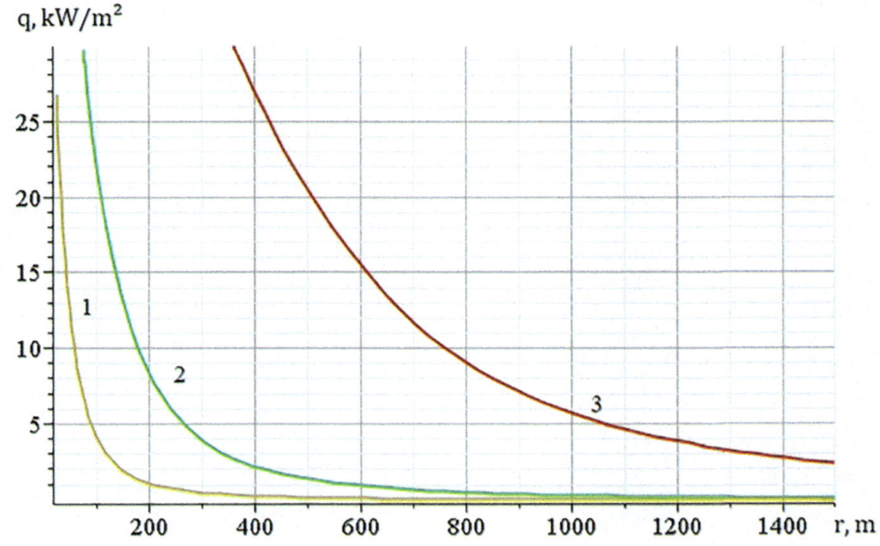

Fig. 3 Dependence of thermal radiation intensity on the radius for the oil tank (V = 20,000 m^3): 1—local depressurization; 2—full depressurization, without wave; 3—full depressurization, with wave

Table 4 The maximum permissible intensity of thermal radiation from an oil spill fire

Affections degree	Distance, m		
	Local depressurization	Full depressurization, without wave	Full depressurization, with wave
Oil tank (V = 5000 m³)			
Without any negative affections while long time	160	425	1180
Save for people in tarpaulin clothes	80	240	700
Violent pain after 20…30 s First degree burn after 15…20 s Second degree burn after 30…40 c	65	185	545
Oil tank (V = 10,000 m³)			
Without any negative affections while long time	165	480	1670
Save for people in tarpaulin clothes	95	280	1000
Violent pain after 20…30 s First degree burn after 15…20 s Second degree burn after 30…40 c	67	220	785
Oil tank (V = 20,000 m³)			
Without any negative affections while long time	180	495	1930
Save for people in tarpaulin clothes	100	290	1150
Violent pain after 20…30 s First degree burn after 15…20 s Second degree burn after 30…40 c	73	240	910

5 Discussion of the Results

The assessment of the consequences of an accident at petrochemical facilities, namely an oil spill fire, was investigated for the case of their spreading without the formation of an explosive vapor-air mixture. In this case, the spread of oil occurred on a flat asphalt surface.

From Table 3 follows that for the case of complete depressurization with the formation of a breakthrough wave, the fire area will increase in 10.8 times for a 5000 m^3 tank, in 16.85 times for a 10,000 m^3 tank and in 23.18 times for a 20,000 m^3 tank. From Table 4 follows that the safe distance from the oil spill fire source for the scenario of full depressurization of the tank with the formation of a breakthrough wave compared to the scenario of full depressurization without the formation of a breakthrough wave, in which the fire develops only in the diking area, increases in 2.77 times for a tank with a volume of 5000 m^3, in 3.47 times for a 10,000 m^3 tank and in 3.89 times for a 20,000 m^3 tank. Based on this, it can be concluded that the correct design and operation of the dike around the tanks can significantly reduce the fire area and, accordingly, reduce the consequences of accidents at petrochemical facilities.

From the analysis of the results which presented in Table 3 follows that even when spill fire due to a local depressurization of a 5000 m^3 oil tank the safe distance for a person without special thermal protective clothing is 160 m, for a 10,000 m^3 oil tank—165 m, and for a 20,000 m^3 tank—180 m.

At the same time, for a 5000 m^3 oil tank the presence of a person on a distance less than 65 m, for a 10,000 m^3 oil tank at a distance of 67 m and for a 20,000 m^3 oil tank on a distance less than 73 m can lead to burns of the 1st degree in 15–20 s.

For the case of full depressurization with wave 5000 m^3 oil tank the safe distance for a person is at least 1180 m, for a 10,000 m^3 oil tank—1670 m, and for a 20,000 m^3 oil tank—1930 m. For firefighters in special protective tarpaulin clothing, this distance will be 700, 1000 and 1150 m, respectively.

The obtained results can be used to predict and assess the consequences of accidents at objects of the petrochemical industry, as well as fire departments in the preparation of operational fire extinguishing plans and the development of tactical exercises to extinguish such objects.

6 Conclusions

An estimation of the size of heat-affected zones in the event of a fire spilling petroleum products from tanks for various scenarios of the development of accidents were carried out. Distances at which a person will be affected by the thermal radiation of an oil spill fire were determined.

Safe distance from the oil spill fire for people without special cloth for 5000 m^3 oil tank is 160 m—for local depressurization, 425 m—for full depressurization, without wave and 1180 m—for full depressurization, with wave.

Safe distance from the oil spill fire for people without special cloth for 10,000 m^3 oil tank is 165 m—for local depressurization, 480 m—for full depressurization, without wave and 1670 m—for full depressurization, with wave.

Safe distance from the oil spill fire for people without special cloth for 20,000 m^3 oil tank is 180 m—for local depressurization, 495 m—for full depressurization, without wave and 1930 m—for full depressurization, with wave.

References

1. Kustov M, Kalugin V, Tutunik V, Tarakhno E (2019) Physicochemical principles of the technology of modified pyrotechnic compositions to reduce the chemical pollution of the atmosphere. Voprosy khimii i khimicheskoi tekhnologii 1:92–99
2. Loboichenko V, Strelec V (2018) The natural waters and aqueous solutions express-identification as element of determination of possible emergency situation. Water Energy Int 61(9):43–50
3. Popov O, Iatsyshyn A, Kovach V, Artemchuk V, Kameneva I, Taraduda D, Sobyna V, Sokolov D, Dement M, Yatsyshyn T (2020) Risk assessment for the population of Kyiv, Ukraine as a result of atmospheric air pollution. J Health Pollut 10:25
4. Fabiano B, Caviglione C, Reverberi AP, Palazzi E (2016) Multicomponent hydrocarbon pool fire: analytical modelling and field application. Chem Eng Trans 48:187–192
5. Raja S, Tauseef SM, Abbasi T (2018) Risk of fuel spills and the transient models of spill area forecasting. J Fail Anal Prev 18:445–455
6. Vasilchenko A, Otrosh Y, Adamenko N, Doronin E, Kovalov A (2018) Feature of fire resistance calculation of steel structures with intumescent coating. MATEC Web Conf 230:02036
7. Semerak M, Pozdeev S, Yakovchuk R, Nekora O, Sviatkevich O (2018) Mathematical modeling of thermal fire effect on tanks with oil products. MATEC Web Conf **247**(00040)
8. Espinosa SN, Jaca RC, Godoy LA (2019) Thermal effects of fire on a nearby fuel storage tank. J. Loss Prevent. Process Ind. **62**(103990)
9. Ahmadi O, Mortazavi SB, Pasdarshahri H, Mohabadi HA (2019) Consequence analysis of large-scale pool fire in oil storage terminal based on computational fluid dynamic (CFD). Process Saf Environ Prot 123:379–389
10. Abramov Y, Basmanov O, Krivtsova V, Salamov J (2019) Modeling of spilling and extinguishing of burning fuel on horizontal surface. Naukovyi Visnyk NHU 4:86–90
11. Oliinik V, Basmanov O (2023) Model of spreading and burning the liquid on the soil. Probl Emerg Situat 1(37):18–30
12. Abramov Y, Basmanov O, Khmyrov I, Oliinik V (2022) Justifying the experimental method for determining the parameters of liquid infiltration in bulk material. Eastern-Eur J Enterp Technol 4/10(118):24–29
13. Abramov Y, Basmanov O, Oliinik V, Khmyrov I, Khmyrova A (2022) Modeling the convective component of the heat flow from a spill fire at railway accidence. EUREKA: Phys Eng 6:128–138
14. ДСТУ Б В.1.1-36:2016 Визначення категорій приміщень, будинків та зовнішніх установок за вибухопожежною та пожежною небезпекою. https://online.budstandart.com/ua/catalog/doc-page.html?id_doc=65419

The Possibility of Energy Enterprises Technological Units' Protection from the Fragmentation Effect of Ammunition

Kostiantyn Afanasenko, Volodymyr Lypovyi, Yaroslav Kalchenko, and Oleksandr Hryhorenko

Abstract The analysis of rocket and artillery strikes on the energy complex of Ukraine since February 24, 2022 has been carried out. It is shown that most of the country's energy infrastructure has received some kind of damage. Based on this, an analysis of the ammunition used for strikes was carried out, and their main characteristics were given. The analysis showed that most of the weapons have a high-explosive and fragmentation effect. The main structures for protecting open technological installations from fragments are considered. The main calculation methods for determining the probability of object hitting (an open technological installation) by a fragmentation field are considered. On the basis of the analysis, calculations of the damage probability and the possibility of protecting technological installations from frags were made. It is shown that the use of a protective structure makes it possible to reduce the probability of hitting an object at extreme distances up to 80% with a protection area of 10% of the vulnerable area to 99% with 90% of the vulnerable area.

Keywords Energy enterprises · Technological units' protection · Fragmentation effect · Fragmentation ammunition

1 Introduction

In the condition of Russia's military and aggression against Ukraine, in addition to military and civilian objects, critical infrastructure objects, including objects of the state's energy complex, namely, heat and power supply objects, oil refineries, and their structural elements were also attacked.

According to various sources, since the beginning of the armed aggression, all oil refineries, from 40 to 50 small and large storages of oil and fuel, have been partially

K. Afanasenko (✉) · V. Lypovyi · Y. Kalchenko · O. Hryhorenko
National University of Civil Defence of Ukraine, Kyiv, Ukraine
e-mail: armfree0@gmail.com

T. A. Kovács et al. (eds.), *The Impact of the Energy Dependency on Critical Infrastructure Protection*, Advanced Sciences and Technologies for Security Applications, https://doi.org/10.1007/978-3-031-78544-3_31

or completely destroyed or damaged, which led to huge financial losses and in the spring of 2022 caused a serious crisis with the provision of automotive fuel for both the civilian and military infrastructure [1] (Fig. 1).

The next mass missile attacks, which started in October 2022, damaged Ukraine's electricity and heat production and distribution capabilities. Currently, of course, information on the extent of the damage of power equipment is mostly classified, but from open sources and indirect evidence, we can state that almost all generating and distribution facilities have received damage of varying extent of severity [2] (Fig. 2).

Fig. 1 Examples of damage to energy sector equipment after being hit by ammunition

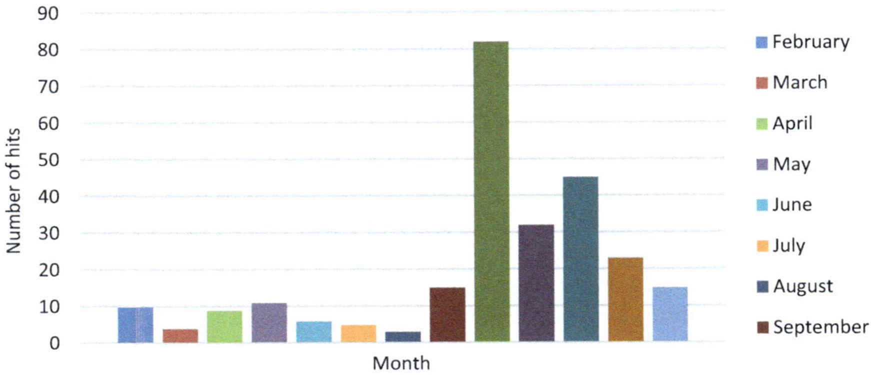

Fig. 2 Statistics of massive missile strikes on the Ukraine energy structure during 2022–2023

At the same time, the initial occurrence of fires during shelling repeatedly occurred on several open technological elements of the energy complex at the same time due to their open and tight location on the territory of critical infrastructure objects.

Taking into account the above, the paper analyzes the possibility of protecting external installations of energy facilities from the impact of ammunition.

2 Analysis of the Means of Destruction Used Against Energy Facilities of Ukraine

2.1 Types of Ammunition and Characteristics of Their Damaging Effect

In order to ensure the hit of targets, it is necessary to exert some kind of effect on them. It is customary to single out [3]:

- the effect of the fragmentation flow;
- penetrating effect;
- cumulative effect;
- the effect of the shock wave and explosion products;
- incendiary effect;
- the effect of penetrating radiation and electromagnetic radiation of a nuclear explosion.

In accordance with what damaging factor determines the main effect of a particular ammunition, all ammunition is divided into the following types:

- fragmentation ammunition (munitions of natural, specified crushing and with ready-made submunitions, rod warheads, powder and explosive shrapnel, buckshot);
- penetrating ammunition (small arms ammunition, armor-piercing and concrete-piercing);
- cumulative ammunition (cumulative ammunition and warheads with one cumulative unit and tandem, ammunition with shell-forming charges);
- high-explosive munitions, including explosive munitions;
- incendiary ammunition;
- nuclear weapons (based on fission reaction);
- unified in action (high-explosive fragmentation, armor-piercing incendiary, cumulative fragmentation, etc.).

Despite the huge variety of designs of weapons and ammunition, as we can see, there are relatively few ways to effect targets. Thus, there is a need to consider the characteristics of munitions and the negative effects they have on critical infrastructure.

2.2 The Ammunition and Characteristics of Their Damaging Effect

To date, it is known about the use of ammunition during missile attacks, data on which are given in Table 1 [4].

In addition, the Russian Federation, according to the UN data, uses about 20,000 artillery shells of various types daily, which is 99.9% of the fire damage to troops, equipment of military and civilian facilities. This number includes high-explosive fragmentation ammunition (Tables 2 and 3).

Table 1 Used in strikes against Ukraine's critical infrastructure missiles

Missile classification (by Russian Federation)	Missile classification (by NATO)	The mass of the warhead (kg)	Operational range (km)
Kh-22	AS-4 "Kitchen"	960	600
OTR-21 "Tochka U"	SS-21 "Scarab"	480	120
Kh-59	AS-13 "Kingbolt"	310	290
"Iskander"	SS-26 "Stone"	480	500–2500
Kh-55/555	AS-15 "Kent"	450	2000–3500
3M-14 "Kalibr"	SS-N-30A	200–450	2600
Kh-47 "Dagger"	"Killjoy"	500	3000
Kh-101/102	Kh-101	430	5000
S-300 (Modified)	SA-10 "Grumble"	130	120
P-800 "Onyx"	SS-N-26 "Strobile"	200–300	300–800

Table 2 Towed guns and mortars

Weapon	Maximum firing range (km)	The mass of the shell (kg)	The mass of weapon (tonne)
82 mm mortar 2B14-1 "Tray"	4.7	3.1	0.039
120-mm portable mortar 2S12 "Sani"	7.1	16	0.21
120-mm gun 2B16 "NONA-K"	8.8	17.3	1.2
122-mm howitzer D-30A	15.3	21.8	3.2
152 mm howitzer gun D-20	17.4	43.56	5.6
152-mm howitzer 2A65 "Msta-B"	28.5	43.56	7.0
152-mm gun 2A36 "Hyacinth-B"	33.0	46.0	9.8

Table 3 Self-propelled guns and mortars

Weapon	Maximum firing range (km)	The mass of the shell (kg)	The mass of weapon (tonne)
120-mm gun 2S23 "NONA-SVK"	12.8	17.3	14.5
122-mm self-propelled howitzer 2S1 "Gvozdika"	15.2	21.8	15.7
152-mm self-propelled howitzer 2S3 "Acacia"	17.3	43.56	27.5
152-mm self-propelled gun 2S5 "Hyacinth-S"	33.0	46	28.2
152-mm self-propelled howitzer 2S19 "Msta-S"	28.5	43.56	42.5
203-mm self-propelled gun 2S7 (2S7M) "Pion" ("Malka")	47.5	110.0	46.0
240-mm self-propelled mortar 2S4 "Tulip"	19.0	130.7	27.5

Based on the data given in Tables 1, 2 and 3, the main weapons inflicting fire damage on energy infrastructure facilities are fragmentation and high-explosive missiles and ammunition.

However, if protection against high-explosive impact is practically impossible except for a stationary powerful shelter, then the analysis of the issue of protection of open technological objects is an urgent task.

2.3 Engineering Structures and Means of Protection Against the Fragmentation Action of Ammunition

To substantiate protection against the impact of fragmentation of ammunition, prefabricated shelters and fortifications are considered in the work: protection from concrete blocks and military gabions, which are already used in Ukraine (Fig. 3).

The analysis of the literature [5] preference was given to gabions due to their easier use and the possibility of ricochets from concrete blocks and hitting the target with ricochet frags.

a) b)

Fig. 3 Examples of protecting objects from hitting by frags: **a** concrete shelter; **b** a wall of military gabions

3 Calculation of Open Technological Units' Protection Possibility from the Impact of Fragmentation of Ammunition

3.1 Fragments Dispersion Fields, Methods of Dispersion Field Management

When a fragmentation munition explodes, a fragment field is formed—a stream of fragments, characterized by the direction and speed of movement, and even density, that is, the number of fragments per unit area that they cross. The density and speed of the stream of fragments are the most important characteristics that determine the possibility of impacting the target. Since the target itself, in general, can be at an absolutely arbitrary angle in relation to the axis of the ammunition, when solving the problem of determining the probability of fragments hitting the target, it is first necessary to know how many fragments fly in a given direction. The answer to this question is given by the so-called law of fragmentation. It is the dependence of the relative number and speed of fragments flying in this direction relative to the axis of the ammunition. Usually, this direction is specified in the spherical coordinate system by two angles—an angle θ in the equatorial and at an angle φ in the meridional planes. The angle φ is counted from the axis of the warhead and can vary from 0 to π. The angle θ varies from 0 to 2π.

According to the shape of the fragment field, two main classes are distinguished: axisymmetric fields; non-axisymmetric fields. The first class is the largest and includes the fields of all barrel munitions and aerial bombs, most of the fragmentation warhead of missiles and most of the engineering fragmentation mines. According to the configuration of the fragment fields, three main classes can be distinguished.

We introduce the classification of axisymmetric fragment fields depending on the angle of dispersion $\Delta\varphi = \varphi_2 - \varphi_1$, which includes 80% or 90% of all fragments. For cylindrical shells with elongation $L_0/d_0 = 1.5-2.5$ with point initiation from the end of the charge and from the center of the charge, the spread angle is, respectively, about 15° and 25°. Small launch angles $\Delta\varphi = 5°-10°$—fragmentation fields of the "cutting disc" type) are implemented using a shell with a concave forming, two-point initiation on the ends of the charge, multi-point synchronous initiation on the axis of the charge.

3.2 *Methods of Destruction Zones Calculation of Objects by Fragmentation*

The analysis of the literature [6, 7] shows that all munitions when assessing the effectiveness of the striking effect can be divided into impact or contact munitions capable of hitting a target only when directly hitting it (cumulative, armor-piercing, concrete-piercing), and remote-action munitions capable of hitting a target when it explodes at some distance (fragmentary, high-explosive).

Damage to the target by remote munitions is carried out either due to the effect on the target of the explosion products and shock wave (high-explosive munitions), or due to the damage by frags of individual vulnerable units of the target (fragmentation munitions), or by a combination of both factors.

The probability of hitting a target with fragments is determined by the relative position of the fragmentation point of the fragmentation munition and the target. In the system of coordinates related to the point of rupture, the coordinate law G of hitting a target is the probability of hitting a target located on the surface of the earth at a point with certain coordinates. In this case, it is natural to accept flat polar coordinates r, x, that is, to set $G = G(r, x)$.

The probability of hitting the target [8, 9] P is the probability of hitting it with at least one striking fragment.

$$P = 1 - e^{-nP_0} \tag{1}$$

where n is the mathematical expectation of the number of fragments falling into the full (average course) area of the target; S_0, P_0—the probability of hitting the target with one fragment.

The product S_0P_0 has the dimension of the area, is the vulnerable area S_{vuln} of the target, whence

$$G = 1 - e^{-F_d S_{vuln}}. \tag{2}$$

To determine the dynamic field density, we rebuild the static spread diagram into a dynamic one. The shrapnel field is assumed to be single-zone, that is, it is located within one angular zone with a constant shrapnel density and speed.

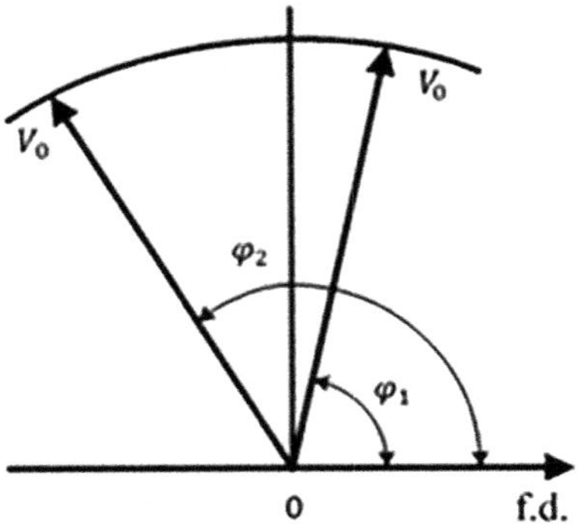

Fig. 4 Angular fragmentation zone for a stationary shell

The view of the corner zone for a stationary shell (in statics) is shown in Fig. 4: where V_0—the initial speed of fragments; φ_1—front angle of the field in statics; φ_2—back corner of the field in statics; $\Delta\varphi = \varphi_2 - \varphi_1$.

If the frag in statics moved at an angle φ to the direction of movement of the shell V_{sh}, then as a result of considering its velocity of the shell, i.e., the vector addition of velocities V_0, V_{sh}, we obtain (Fig. 5):

$$V_d = \sqrt{V_0^2 + 2 \cdot V_0 \cdot V_{sh} \cdot \text{Cos}\,\varphi + V_{sh}^2}, \quad (3)$$

where γ—the frag movement angle in dynamics; V_d—the dynamic frag speed.

These ratios determine the front and rear dynamic angles of the field γ_1, γ_2 and corresponding speeds V_{d1}, V_{d2}. The dynamic field density at different distances from the detonation point is obviously defined as F_s

$$F_s = \frac{N}{S_{\text{sb}}} \quad (4)$$

where N—the number of striking elements; S_{sb}—the surface area of the spherical belt.

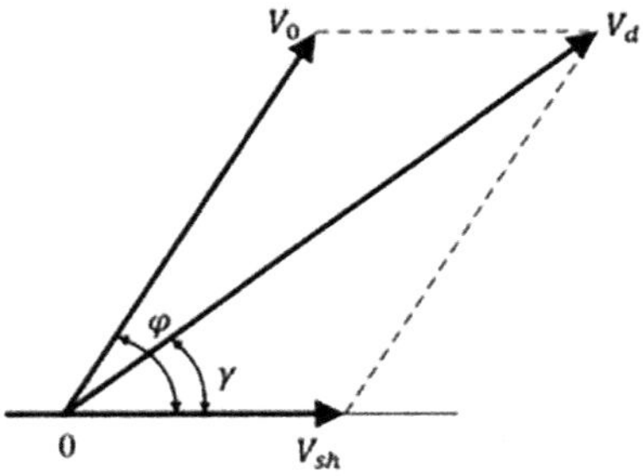

Fig. 5 Vector summation of the velocities of the frags and the velocity of the shell directly

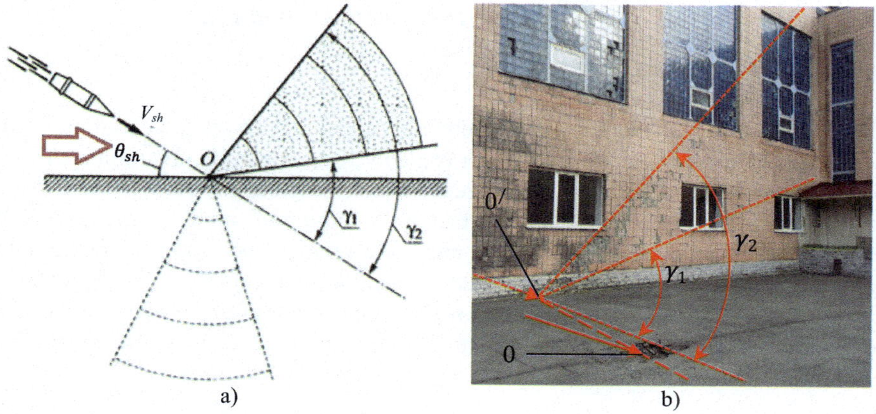

Fig. 6 Sectors of the affected zone in the area during ground detonation of the high-explosive shell: **a** illustration; **b** hitting the territory of National University of Civil Defence of Ukraine

Taking into account that $S_{sb} = 2\pi rh$, where: $h = r(\text{Cos}\,\gamma_1 - \text{Cos}\,\gamma_2)$ (Fig. 5), we get $S_{sb} = 2\pi r^2(\text{Cos}\,\gamma_1 - \text{Cos}\,\gamma_2)$ and, respectively:

$$S_{sb} = \frac{N}{2\pi r^2(\text{Cos}\,\gamma_1 - \text{Cos}\,\gamma_2)} \tag{5}$$

The formation of affected sectors during a ground rupture is shown in Fig. 6.

The shrapnel field emanates from a point located in the plane of the earth. The boundaries of the affected sectors are the traces of the intersection of two conical surfaces by the ground plane passing through their common apex. The angular boundaries of the affected sector on the terrain are determined by the expressions:

$$\xi_1 = \arccos\frac{\text{Cos}\,\gamma_1}{\text{Cos}\,\theta_{sh}} \tag{6}$$

$$\xi_2 = \arccos\frac{\text{Cos}\,\gamma_2}{\text{Cos}\,\theta_{sh}} \tag{7}$$

Summarizing all of the above, we obtain for a fixed vulnerable area S_{vuln} of the target and constant mass frags the coordinate law of damage in the sector $\xi_1 - \xi_2$ in the form

$$\begin{cases} G = 1 - e^{-\frac{NS_{vuln}}{2\pi r^2(\text{Cos}\,\gamma_1 - \text{Cos}\,\gamma_2)}} & with\ r \le I_{\text{kill}} \\ G = 0 & with\ r \ge I_{\text{kill}} \end{cases} \tag{8}$$

3.3 Calculation of Possibility of Objects Protection from Hitting by Frags

In the paper, the calculation of the probability of hitting of an open technological object in the event of a rupture of a high-explosive fragmentation munition is carried out. As initial data, typical areas of large open technological installations were taken as the affected area. As an influencing factor, the minimum and maximum values of the number of frags from ammunition and the average value of the killing interval of a fragment were used. The performed calculations are shown in Figs. 7 and 8.

It has been established (Fig. 7) that for open installations at the maximum range of the killing interval and the maximum fragmentation field density, the probability

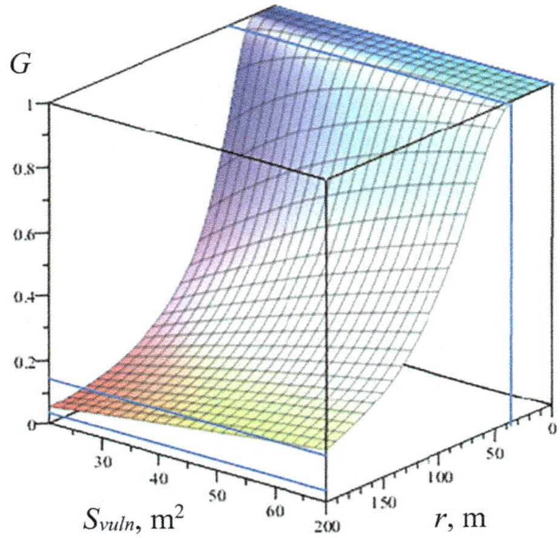

Fig. 7 The probability of object hitting, depending on the distance to the ammunition explosion and the vulnerable area

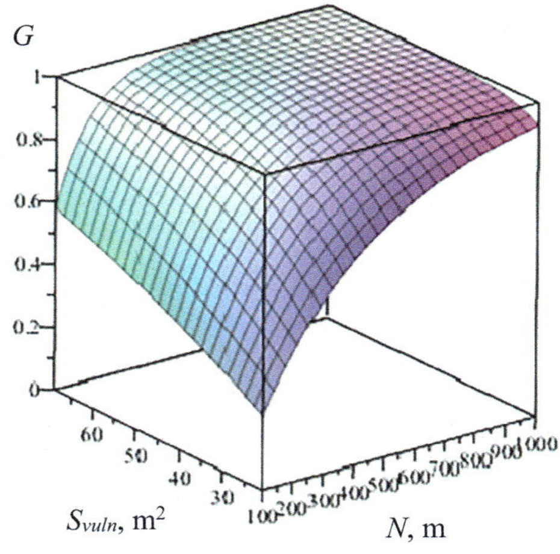

Fig. 8 The probability of object hitting depending on the fragmentation field density and the vulnerable area

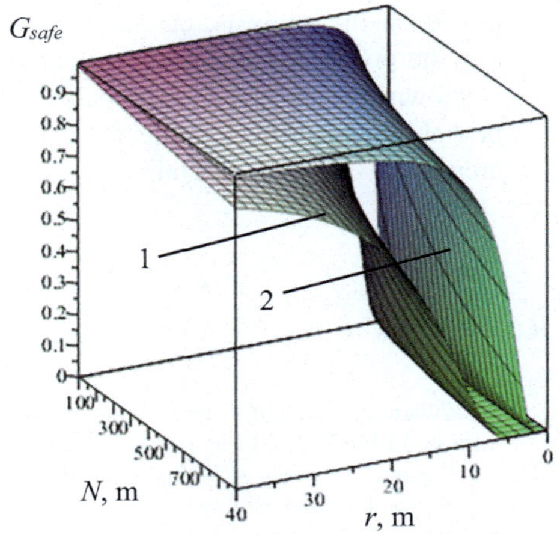

Fig. 9 Probability of protecting an object at different areas of the protective structure, depending on the distance to the fall of the ammunition and the number of frag: 1—the area of protection is 10% of the vulnerable area; 2—protection area 90% of the area of the vulnerable area

of being hit by at least one frag is from 5 to 17%. At the same time, data analysis shows that at distances of 40 m, the probability of hitting is 95–100%.

Analysis of Fig. 8 shows that varying the vulnerable area and the number of frags at the minimal distance from the shell burst, the minimum probability of hitting an object is 23%.

As a result, the probability of avoiding damage to the object by frags was calculated in the work (Fig. 9). As an assumption, it is accepted that the protective structure is impenetrable.

Analysis of Fig. 9 data shows that the use of a protective structure makes it possible to reduce the probability of hitting an object at extreme distances up to 80% with a protection area of 10% of the vulnerable area to 99% with 90% of the vulnerable area.

4 Conclusions

1. The paper analyzes the attacks on the energy complex of Ukraine in the period from February 24, 2022. It is shown that most of the country's energy infrastructure has received some damage.
2. An analysis of the ammunition used for strikes was carried out, their main characteristics are given. It is concluded that most of the weapons have a high-explosive fragmentation action. The main structures for protecting open technological installations from fragments are considered.
3. The main calculation methods for determining the probability of hitting an object (an open technological installation) by a fragmentation field are considered.

4. Based on the analysis, calculations were made of the probability of damage and the possibility of protecting technological installations from fragments. It is shown that the use of a protective structure makes it possible to reduce the probability of hitting an object at extreme distances up to 80% with a protection area of 10% of the vulnerable area to 99% with 90% of the vulnerable area.

References

1. https://www.energycharter.org/. Ukrainian energy sector evaluation and damage assessment—VI (as of 24 Jan 2023). Last accessed 20 March 2023
2. https://strana.today/news/427120-skolko-udarov-rf-nanesla-po-enerhoobektam-ukrainy-za-hod-infohrafika-enerhoatoma.html. Last accessed 1 March 2023
3. Xin T et al (2014) Research on system planning of retired or scrap weaponry and equipment recycling. Adv Mater Res 1010–1012:979–983
4. Gordon Y (2004) Soviet/Russian aircraft weapons: since World War Two. Midland Publishing, Hinckley, England
5. Surianinov M, Andronov V, Otrosh Y, Makovkina T, Vasiukov S (2020) Concrete and fiber concrete impact strength. Mater Sci Forum 1006:101–106
6. Novik GP (2022) Analysis of samples of high explosives extracted from explosive remnants of war. Sci Total Environ 842
7. Elshenawy T, Zaky MG, Elbeih A (2022) Experimental and numerical studies of fragmentation shells filled with advanced HMX-plastic explosive compared to various explosive charges. Braz J Chem Eng (2022)
8. Ivanov AI, Syrunin MA, Fedorenko AG et al (2001) Fragmentation of spherical shells under internal blast loading. Strength Mater 33:150–156
9. Fan SC, Lee CK, Kang KW, Wu ZJ (2015) Validation of a flight model for predicting shrapnel trajectory from the explosion of an ammunition storage magazine. J Wind Eng Ind Aerodyn 136:114–126

Genetic Algorithm Optimization of PID and LQR Controllers for Enhanced Steer-by-Wire (SBW) System Performance

Karrar Y. A. Al-bayati, Ali Mahmood, and Róbert Szabolcsi

Abstract One of the most significant demands of autonomous vehicles is the precise steering vehicle orientation. This work deals with two control techniques of a steer-by-wire (SBW) system based on a Genetic Algorithm GA methodology. The PID and LQR controllers are implemented and the PID parameters and LQR matrices are optimized using the GA to improve the dynamic performance and stability of the SBW system. A comparative analysis study of both PID and LQR controllers is presented using MATLAB simulation under different fitness function scenarios (MSE, ISE, IAE, and ITAE). That demonstrates the effective SBW system response based on the GA impact to minimize steering angle error and approach to the desired vehicle orientation. Four fitness functions are applied to contribute to the control enhancement of the steer-by-wire system that sheds light on different time response specifications regarding the fitness function of PID and LQR controllers.

Keywords Autonomous vehicles · Steer-by-Wire (SBW) · Genetic algorithm · PID · LQR · MSE · ISE · IAE · ITAE

K. Y. A. Al-bayati (✉) · A. Mahmood
Doctoral School On Safety and Security Sciences, Obuda University, Budapest, Hungary
e-mail: karrar.albayati@uni-obuda.hu

A. Mahmood
e-mail: ali.mahmood@uni-obuda.hu

K. Y. A. Al-bayati
Electronic and Communication Engineering Department, University of Kufa, Najaf, Iraq

A. Mahmood
Systems and Control Engineering Department, Ninevah University, Mosul, Iraq

R. Szabolcsi
Banki Donat Faculty of Mechanical and Safety Engineering, Obuda University, Budapest, Hungary
e-mail: szabolcsi.robert@kvk.uni-obuda.hu

T. A. Kovács et al. (eds.), *The Impact of the Energy Dependency on Critical Infrastructure Protection*, Advanced Sciences and Technologies for Security Applications, https://doi.org/10.1007/978-3-031-78544-3_32

1 Introduction

In recent years, the field of autonomous vehicles has evolved with numerous innovations to improve the performance, accuracy, and safety of vehicles [1–4]. The automotive steering system is considered one of these innovations, which utilizes a feedback control system using sensors and actuators to optimize the vehicle steering angle instead of the traditional mechanism with mechanical units. The Steering-By-Wire (SBW) system shows the best steering response regarding the traditional steering system in terms of stability, safety, and comfortability, as well as power consumption [5, 6]. Therefore, many vehicle companies have proposed a lot of research about the SBW system with various control techniques to minimize the steering angle error and track the desired orientation [7, 8].

Different control approaches have been adopted to obtain the optimal response of the front wheel steering angle regarding the desired handle wheel driver via the SBW system. Lei-yan and Ping-li [9] have applied a PID controller via the SBW system to meet the actual steering angle based on the two steering sensors' feedback signals while Xiuwei at el. [10] have been used PID controller with partition integration to ensure the steering wheel returns to the center and gets more stability compared with conventional PID. Zhai at el. [11] have introduced bilateral control loops using two PID controllers for both motors that are mounted in the steering wheel and the front wheel with take in count the road feeling reflection and changing vehicle speed dynamically. Zhang et al. [12] have presented a combination of PD controllers with feed-forward compensation while Sahboun and Emhemed [7] have proposed PSO-PID and ILC to obtain the effectiveness of tracking actual hand wheel angle.

From the other researchers point view, the other control approaches have employed the sideslip angle strategy with a yaw rate to improve the stability of the SBW [13]. Shi et al. [14] have implemented the sliding mode control strategy to tackle the yaw rate error and to calculate the sideslip precisely during the steering torque and disturbances. Sung et al. [15] utilized the sliding mode control also to detect the sensor error for SBW vehicles while Sun [16] addressed the uncertainties and the self-aligned torque of the tire by using the adaptive sliding mode control.

Furthermore, several papers have proposed fuzzy logic control for the nonlinear behavior of SBW vehicles. It shows robust SBW control with a fast response and the ability to reject the disturbances of tire friction and aligned torque of sideslip [17–19]. The linear quadratic regulator (LQR) algorithm has been examined by researchers also to optimize the performance and robust control of the SBW system. Fahami et al. [20] have proposed the LQR controller to represent the steering wheel reaction due to the tire torque alignment which led to increased system stability and hand wheel driver feel. Zhang et al. [21] employed the LQR controller with a compensator to synchronize the front axle wheel angle with the desired hand wheel drive by eliminating the tracking angle error against external disturbances such as friction and self-aligned torque. In this paper, a comparative study of the PID and LQR controllers of SBW vehicles is proposed using a genetic algorithm to specify the optimal parameters of both controllers to stabilize the system, reduce the steady state error, and minimize the time response dynamically.

This paper organized as following: After the introduction section, the electronic steering by wire system of vehicle model is presented as well as, the PID and LQR controllers are designed, followed by optimization of both controllers' parameters via the genetic algorithm in section II. Then, simulation results are presented using Matlab/Simulink to demonstrates the validation of tuning controllers' parameters using genetic algorithm with applying a various fitness function in section III. Finally, the conclusion summarizes the compersim result of PID and LQR controllers in the last section.

2 Methodology

This section contains the required mathematical background to firstly describe the model of the Steering-by-wire system then it will describe the GA optimization method. Finally, it will end with the description of the controllers.

2.1 Steering-by-Wire System Model

The Steering-by-wire system of the vehicle shown in Fig. 1 consists of two blocks regarding the steering wheel part and the front axle wheel part. There are two sensors mounted in the steering wheel and front wheel blocks to provide the desired steering angle and the actual pinion angle of the front wheels respectively [22]. In addition to the two sensors, two motors are employed to estimate the torque on the wheel hand driver as a feedback torque of the front axle tire with road interactions, and the other motor is used to generate torque for directing the front wheels to the desired orientation.

According to [22] a simplified second-order SBW system model of the front wheel steering angle (x) output related to the input voltage of the steering motor (u) that focuses on the steering front wheel control part is:

$$J\ddot{x} + C\dot{x} + \rho \text{sign}(\dot{x}) + \xi \tanh(x) = bu \quad (1)$$

where:

J	compound moment of inertia of front wheels and steering motor
C	compound viscous friction front wheels and steering motor
R	the constant of Coulomb friction
sign ()	the standard signum function
b	the scaling factor
ξ	the self-aligning torque coefficient.

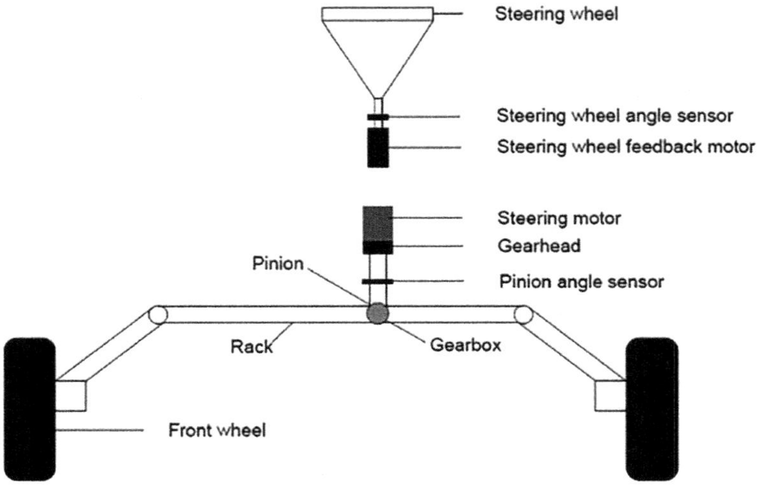

Fig. 1 The SBW system structure

The term $\xi \tanh(x)$ represents the self-aligning torque of the tire on the road which plays a violet role in the SBW dynamic equation. According to a simplification of a second-order differential equation of SBW by neglecting the external torque disturbances temporarily [22], the transfer function of SBW of the front wheel would be:

$$G(s) = \frac{X(s)}{U(s)} = \frac{b}{J_o s^2 + c_o s} \tag{2}$$

2.2 *Optimization Method GA*

The genetic algorithm (GA) is one of the most used methods to find an optimization solution for the controllers in the process of design. Its rule is to find the optimal values for the controllers' parameters.

GA was first proposed by Holland [23, 24] for the quality of computational procedures enhancement. Its idea is to use the natural development process of the chromosome where the first step is the representation of the chromosome's structure [25, 26]. This structure consists of all the elements that need to be optimized. For the PID controller, the design procedure requires the selection of its three gains (K_p, K_i, and K_d) while for the LQR controller, it requires the selection of the Q and R matrices [27, 28]. So, after forming the structure of the chromosome, the GA optimization method will start where the first step is the selection of the chromosome population (number of the chromosomes) in a random way. In this step, the fitness function will be used

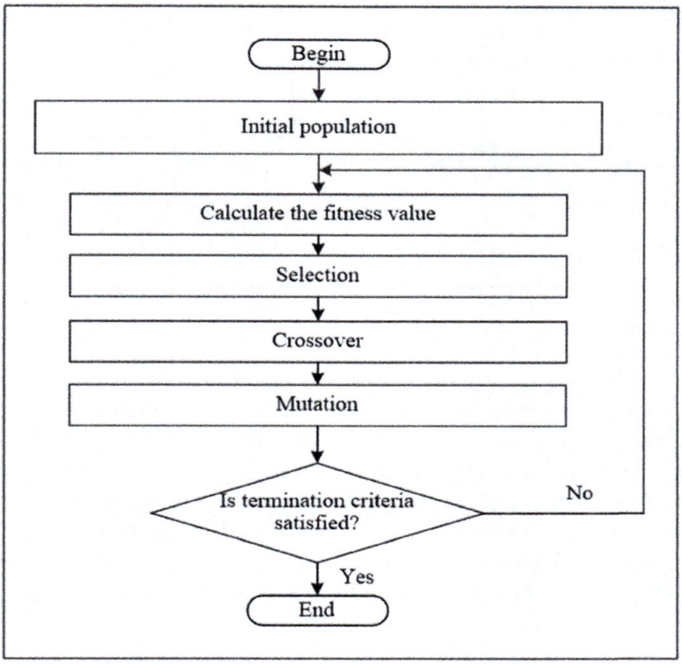

Fig. 2 GA flow chart

to measure the performance of each chromosome. Following is the crossover that happens between a pair of individuals to exchange information between two chromosomes to create new chromosomes. The next step is mutation which will happen to some of the individuals in the population; this involves a random change to these individuals to bring diversity into the population.

Figure 2 shows the flowchart of the GA.

2.3 Controllers

This section will discuss the two controllers that will be designed, the first part will be for the PID controller then in the second part, LQR will be discussed.

PID controller. The first controller is the traditional PID controller. This controller is still the most used because of its simplicity and its ability to deal with enormous systems. This controller consists of three parts and their combination creates the control strategy that controls the angle of the SBW system. These parts as it is known are the proportional, integral, and derivative and the gains for these parts are the K_p, K_i, and K_d respectively. Each of these parts has its own effect on the response of the

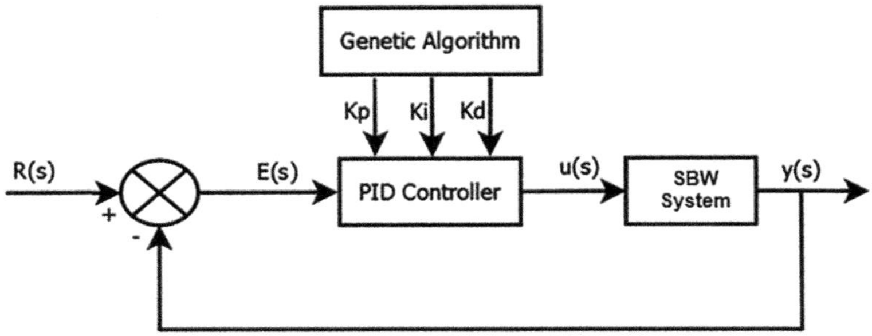

Fig. 3 PID with GA

system. For the proportional part, increasing the K_p values will increase the response speed and decrease the steady state error but at the same time, it will increase the overshoot. For the derivative part, increasing the K_d value will mainly reduce the overshoot of the system and any sudden change in the response, this will decrease the time needed for the response to settle. Adding the integral part will diminish the steady state error and make it equal to zero even though it increases the overshoot.

The control signal of the PID controller u(t) will be presented in Eq. (3)

$$u(t) = K_p.e(t) + K_i \int e(t)dt + K_d \frac{de(t)}{dt} \tag{3}$$

To see the use of the GA optimization method with the PID controller and the SBW system, see Fig. 3.

LQR controller. Linear Quadratic Regulator (LQR) is one of the commonly used control strategies in the field of control theory due to its simplicity and effectiveness. It is designed to minimize the quadratic cost function by optimizing the weighting matrices.

The LQR controller shown in Fig. 4 is a state feedback control technique with a quadratic cost function that has been used to determine the optimal control input that could be used to minimize the system's performance index. The feedback loop will measure the state variables of the system then the optimal control inputs will be calculated.

The state space representation of the LQR controller for a continuous-time linear time-invariant system:

$$\dot{x}(t) = Ax(t) + Bu(t) \tag{4}$$

where:

x(t) is the state vector,
u(t) is the control input,
A is the state matrix,

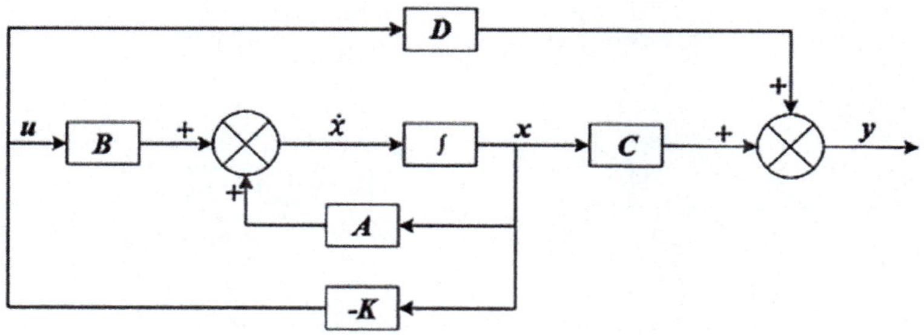

Fig. 4 The system block diagram with LQR controller

B is the input matrix.

The quadratic cost function to be minimized is given by:

$$J = \int_{0}^{\infty} \left(x^T(t)Qx(t) + u^T(t)Ru(t)\right)dt \tag{5}$$

where

Q is the state cost matrix, R is the control cost matrix.

By finding the optimal values of the Q and R matrices using the GA optimization algorithm, the value of the controller gains K can be computed by solving the continuous-time algebraic Riccati equation:

$$A^T P + PA - PBR^{-1}B^T P + Q = 0 \tag{6}$$

Once P is obtained, the control input is given by:

$$u(t) = -R^{-1}B^T Px(t) \tag{7}$$

The state feedback matrix K is then given by:

$$K = R^{-1}B^T P \tag{8}$$

Figure 5 shows its block diagram of the LQR controller optimized by GA optimization method.

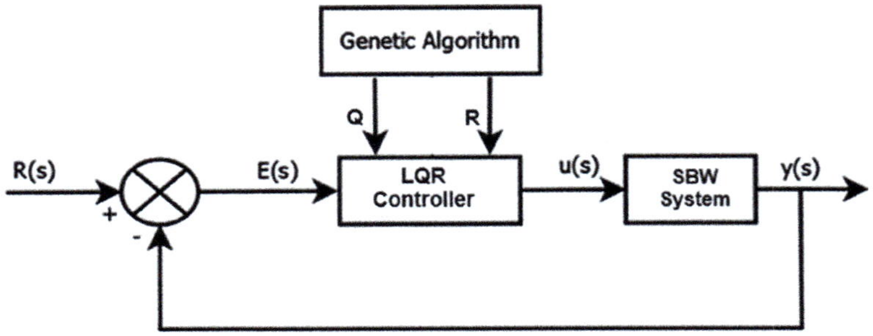

Fig. 5 LQR controller with GA

3 Results and Discussion

In this section, the obtained results will be presented and discussed in detail. Firstly, the PID controller with GA optimization method will be discussed for the four fitness functions which are the MSE, ISE, IAE and ITAE as shown in Fig. 6.

After that, the same procedure will be repeated for the LQR with GA optimization method for the same fitness function. Figure 7 shows its step response.

To compare the response of the fitness functions of the PID controller, Table 1 shows the transient and the steady state response of them.

For LQR controller, Table 2 shows the time response specifications for the fitness function.

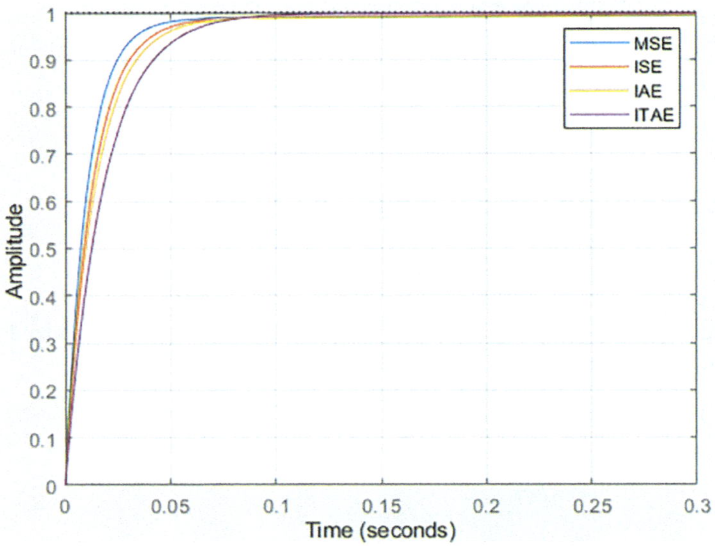

Fig. 6 Step response of the PID with GA

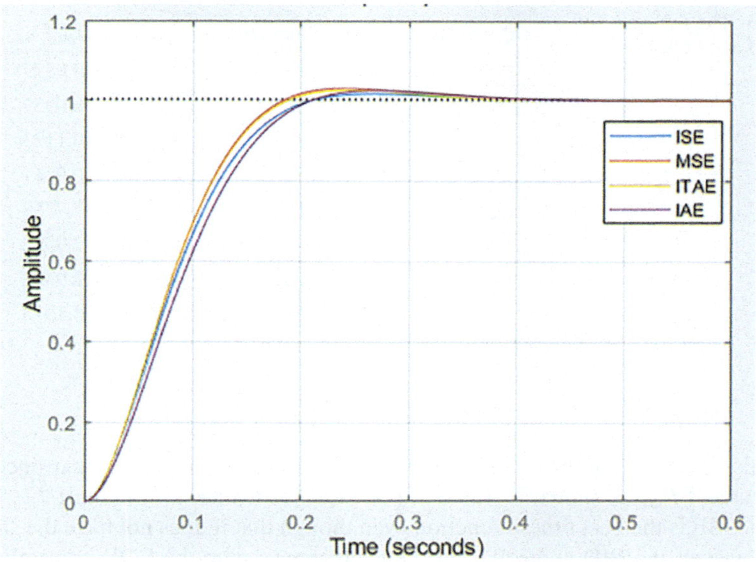

Fig. 7 Step response of the LQR with GA

Table 1 Gains' values and time response specification for the PID Controller

Fitness function	K_p	K_i	K_d	%OS	T_r	T_s	E_{ss}
ISE	48	8	28	0	0.0292	0.0585	0
IAE	48	8	22	0	0.0325	0.0634	0
ITAE	48	8	20	0	0.0399	0.0724	0
MSE	50	10	35	0	0.0234	0.0477	0

Table 2 Gains' values and time response specification for the LQR Controller

Fitness function	Q_{11}	Q_{22}	R	%OS	T_r	T_s	E_{ss}
ISE	2.3	1500	0.001	1.4077	0.1259	0.1922	0
IAE	1	1000	0.001	2.5088	0.1302	0.3148	0
ITAE	1.2	1500	0.001	2.5374	0.1176	0.2853	0
MSE	1	1500	0.001	2.7916	0.1161	0.2901	0

As it is hard to find which fitness function is the best for the controllers, especially for the LQR controller, one solution is to select depending on the most important characteristic like the rise time. Another solution which is more accurate is to sum the response specifications together as shown in Eq. (9) and the fitness function of the smallest J_{min} would be the best.

Table 3 The J_{min} value of the PID and LQR controllers

Controller	Fitness function	J_{min}
PID	ISE	0.0877
	IAE	0.0959
	ITAE	0.1123
	MSE	0.0711
LQR	ISE	1.7258
	IAE	2.9538
	ITAE	2.9403
	MSE	3.1978

$$J_{\min} = \sum \%OS + T_r + T_s \tag{9}$$

In this case, for the PID controller, the MSE fitness function is the best since it has the smallest $J_{\min}$ as shown in Table 3. For LQR controller, the smallest $J_{\min}$ shows that the ISE is the best fitness function even though that it does not have the smallest T_{r}. However, the PID controller has better response than the LQR controller even though they both have acceptable response which is fast and very small overshoot.

Figure 8 shows the time response of the best fitness functions of the PID and LQR controllers.

To sum up the results above, the best controller with the best fitness function to be used is the PID controller with the MSE fitness function since it has the best response

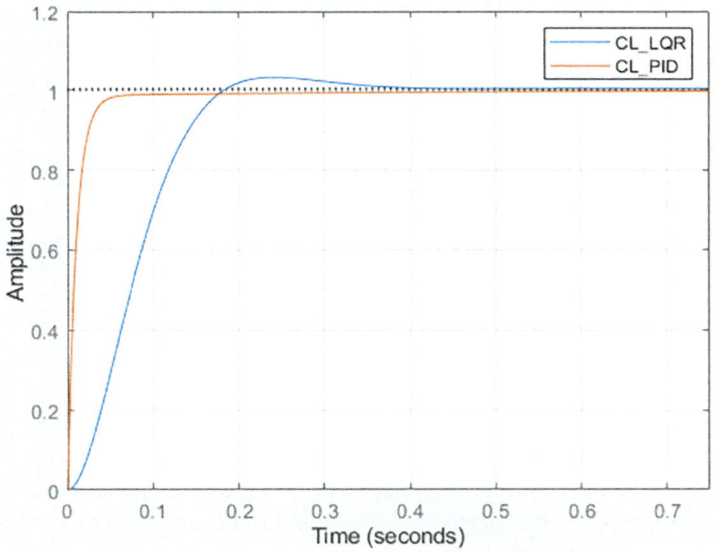

Fig. 8 Time response of the Best

specification in terms of %OS, T_r, T_s and E_{ss}. This will lead to having the minimum J_{min}.

4 Conclusion

This study presents a comprehensive analysis of the Steer-by-Wire System (SBW), evaluating the effectiveness of Genetic Algorithm (GA)-tuned Proportional-Integral-Derivative (PID) and Linear Quadratic Regulator (LQR) controllers for the Steer-by-Wire (SBW) System. The designed controllers have been optimized using the Genetic Algorithm (GA) optimization method. The performance of the two controllers has been assessed based on the transient and steady state specifications which are the rise time, percent overshoot, settling time and the steady state error. For deep comparison and to further assess the efficiency of the two designed controllers, four fitness functions have been utilized which are the Integral Squared Error, Integral Absolute Error, Integral of Time Absolute Error in addition to the Mean Squared Error. After running the optimization code, analyze the results and comparing them, it has been indicated that the PID controller utilizing the Mean Squared Error outperforms the LQR controller regarding the time response specifications with the minimum cost function (J_{min}).

References

1. Mortazavizadeh SA, Ghaderi A, Ebrahimi M, Hajian M (2020) Recent developments in the vehicle steer-by-wire system. IEEE Trans Transp Electrif 6(3):1226–1235. https://doi.org/10.1109/TTE.2020.3004694
2. Zhao W, Wang A, Zou S, Zhang H (2021) Individual auxiliary and fault-tolerant control of steer-by-wire system considering different drivers steering characteristics. IEEE/ASME Trans Mechatron 26(3):1558–1569. https://doi.org/10.1109/TMECH.2020.3023947
3. Huang, C., Li, L.: Architectural design and analysis of a steer-by-wire system in view of functional safety concept. Reliab Eng Syst Saf 198(April 2019):106822. https://doi.org/10.1016/j.ress.2020.106822
4. Mahmood A, Almaged M, Noaman MN, Alnema YHS (2021) Adaptive cruise control of a Simscape driveline vehicle model using PID controller. J Eng Sci Technol 16(1):681–695
5. Setlur P, Wagner JR, Dawson DM, Braganza D (2006) A trajectory tracking steer-by-wire control system for ground vehicles. IEEE Trans Veh Technol 55(1):76–85. https://doi.org/10.1109/TVT.2005.861189
6. Zhang L, Zhang Z, Wang Z, Deng J, Dorrell DG (2021) Chassis coordinated control for full X-by-wire vehicles—a review. Chinese J Mech Eng (English Ed) 34(1). https://doi.org/10.1186/s10033-021-00555-6
7. Sahboun SM, Emhemed AAA (2022) Controller design for steer-by-wire system. J Mechatron Robot 6(1):1–6. https://doi.org/10.3844/jmrsp.2022.1.6
8. Arango JF et al (2020) Drive-by-wire development process based on ROS for an autonomous electric vehicle. Sensors (Switzerland) 20(21):1–31. https://doi.org/10.3390/s20216121

9. Yu LY, Yun PL (2009) Motor motion control of automobile steer-by-wire system in electric vehicles. In: 5th IEEE vehicle power and propulsion conference (VPPC'09), pp 1502–1505. https://doi.org/10.1109/VPPC.2009.5289552
10. Fu X, Kong F, Fu L (2009) Research of automotive steer-by-wire control based on integral partition PID control. In: 3rd international conference on genetic and evolutionary computing (WGEC 2009), pp 561–564. https://doi.org/10.1109/WGEC.2009.21
11. Zhai P, Du H, Li Z (2013) Bilateral control of vehicle steer-by-wire system with variable gear-ratio. In: Proceedings of the 2013 IEEE 8th conference on industrial electronics and applications (ICIEA 2013), pp 811–815. https://doi.org/10.1109/ICIEA.2013.6566478
12. Zhang X, Kong H, Wang H (2015) A composite control scheme for automotive steer-by-wire system. In: Proceedings of the 2014 international conference on modelling, identification & control (ICMIC 2014), pp 237–242. https://doi.org/10.1109/ICMIC.2014.7020758
13. Duan J, Wang R, Yu Y (2010) Research on control strategies of steer-by-wire system. In: 2010 international conference on intelligent computation technology and automation (ICICTA 2010), vol 2, pp 1122–1125. https://doi.org/10.1109/ICICTA.2010.238
14. Shi L, Wang H, Huang Y, Jin X, Yang S (2018) A novel integral terminal sliding mode control of yaw stability for steer-by-wire vehicles. In: Chinese control conference (CCC), vol 2018-July, pp 7787–7792. https://doi.org/10.23919/ChiCC.2018.8483780
15. Im JS, Ozaki F, Matsunaga N, Kawaji S (2006) Bilateral control for steer-by-wire vehicles. In: 2006 SICE-ICASE international joint conference, pp 528–533. https://doi.org/10.1109/SICE.2006.315541
16. Sun Z, Zheng J, Man Z (2016) Adaptive sliding mode control for a vehicle steer-by-wire system. In: Proceedings of 2016 2nd international conference on control science and systems engineering (ICCSSE 2016), pp 77–81. https://doi.org/10.1109/CCSSE.2016.7784357
17. Zhou Y, Liu MS, Hu LS, Da Zhang Z (2010) Application of fuzzy five states control system base on phase plane in steer by wire system of loader. In: Proceedings—2010 seventh international conference on fuzzy systems and knowledge discovery (FSKD 2010), vol 2, no Fskd, pp 649–653. https://doi.org/10.1109/FSKD.2010.5569427
18. Qu AQ, Wang TJ, Liu XH (2010) Study on hybrid-fuzzy control algorithm of wheel loader steer by wire system. In: 2010 2nd international conference on industrial and information systems (IIS 2010), vol 2, pp 398–400. https://doi.org/10.1109/INDUSIS.2010.5565736
19. Huang C, Du H, Naghdy F, Li W (2015) Robust fuzzy tracking control of uncertain steer-by-wire systems with network time delays. In: 2015 Australasian universities power engineering conference: challenges for future grids (AUPEC 2015), pp 1–5. https://doi.org/10.1109/AUPEC.2015.7324795
20. Fahami SMH, Zamzuri H, Mazlan SA (2015) Development of estimation force feedback torque control algorithm for driver steering feel in vehicle steer by wire System: hardware in the loop. Int J Veh Technol 2015. https://doi.org/10.1155/2015/314597
21. Zhang G, Wang X, Li L, Shao W (2021) Design and applications of steering angle tracking control with robust compensator for steer-by-wire system of intelligent vehicle. IFAC-PapersOnLine 54(10):221–227. https://doi.org/10.1016/j.ifacol.2021.10.167
22. Sun Z (2017) Robust control design for vehicle steer-by-wire systems. [Online]. Available: https://researchbank.swinburne.edu.au/items/d86bd8ec-9593-47c1-990b-3ea4bbb8c3d4/1/Zhe%20Sun%20Thesis.pdf
23. Anandan R (2022) Adaptive heuristic—genetic algorithms. In: Mathematics in computational science and engineering, pp 329–342. https://doi.org/10.1002/9781119777557.CH15
24. Rodkaew Y (2021) A genetic algorithm as a classifier. In: ECTI-CON 2021—2021 18th International Conference on Electrical Engineering/Electronics, Computer, Telecommunications and Information Technology: Smart Electrical System & Technology, pp. 254–257, May 2021. https://doi.org/10.1109/ECTI-CON51831.2021.9454722
25. Alam T, Qamar S, Dixit A, Benaida M (2020) Genetic algorithm: reviews, implementations, and applications. Int J Eng Pedagog 10(6):57–77. https://doi.org/10.3991/IJEP.V10I6.14567
26. Ye F, Chen J, Tian Y, Jiang T (2020) Cooperative task assignment of a heterogeneous multi-UAV system using an adaptive genetic algorithm. Electronics 9(4):687. https://doi.org/10.3390/ELECTRONICS9040687

27. Abdullah AI, Mahmood A, Thanoon MA (2022) Design of a linear quadratic regulator based on genetic model reference adaptive control. J Autom Mob Robot Intell Syst 16(3):75–81. https://doi.org/10.14313/jamris/3-2022/26
28. Dawood YS, Mahmood AK, Ibrahim MA (2018) Comparison of PID, GA and fuzzy logic controllers for cruise control system. Int J Comput Digit Syst 7(5)

How to Motivate Students of Critical Infrastructure Protection, Security Technology Engineering and Law Enforcement in ESP Classes?

Éva Kovács, Gabriella Ürmösné Simon, and László Hanka

Abstract The aim of the article is to demonstrate to the reader the most motivating innovative teaching methodological approaches that are applied with English for Special Purposes (ESP) students during the course of both material design and classroom solutions. Our students represent the special segment of the future professions in critical infrastructure protection, security technology engineering, private security and law enforcement. What connects those lines of expertise is their technology-rich working environment as well as the prompt practical implementation of theoretical knowledge having gained during undergraduate studies. Therefore, it is imperative to us, professors, tutors, instructors that we engage in and absorb the most recent advancement of the professions concerned while synthesizing them with an array of state-of-the-art teaching methods that are applied in the learning environment. As a reward, our recent surveys, studies and applied methods have led to the recognition of our work in the form of various awards.

Keywords Security Technology English · English for law enforcement · Mission-oriented preparation · Flow · Creativity · Augmented reality · HY-DE model

1 Introduction

In the academic world, if English as a foreign language is taught at any training level, it is mostly done to university students or people already employed in the line of profession of their chosen field of study. Therefore, the focus of teaching shifts to the particular vocabulary and skill set they will be using in their occupation or in the

É. Kovács · G. Ü. Simon (✉) · L. Hanka
Ludovika University of Public Service, Budapest, Hungary
e-mail: simon.gabriella@uni-nke.hu

É. Kovács · L. Hanka
Obuda University, Budapest, Hungary

T. A. Kovács et al. (eds.), *The Impact of the Energy Dependency on Critical Infrastructure Protection*, Advanced Sciences and Technologies for Security Applications, https://doi.org/10.1007/978-3-031-78544-3_33

academic world itself. ESP, as a subcategory of English as a Second Language (ESL) targets exactly those learners, with a lot of attention to up-to-date content in its course design [1]. The authors of this article have been dealing with learners that already are or will later be employed in future positions in security technology engineering, critical infrastructure protection or various fields of law enforcement such as private security, criminal investigation or intelligence services at Obuda University (OU) and Ludovika University of Public Service (LUPS), Hungary. Typically, we deal with young adults or adults whose proficiency in the target language reaches or goes beyond intermediate level. With a number of them already having obtained some experience in the discipline they are also learners of, when choosing an ESP class, they are well aware that the program offered will be relevant to them and are also able to assess the suitability of material and the congruity of teaching methods as well.

2 Researches Results Show Needs and Prompt Changes

2.1 The Catalyst of Changes: Online Education During COVID-19 and a Relevant Research

Despite the continuous application and deeper inclusion of technology and online platforms in our classroom activities, we could not have predicted the nature and scale of the abrupt change we needed to adjust to due to emergency online teaching during the pandemic years of COVID-19. Nationwide, teaching practices shifted towards technology-laden methods and solutions, maintaining distant online learning through semesters in Hungarian higher education [2]. This meant a unanimous application of online tools, the use of which had formerly been lesser-known and applied by students and professors equally. Suddenly, the platforms for teaching were no longer called auditoriums, classrooms or laboratories, instead, the participants logged on to videoconferences and online meetings on MS Teams, BigBlueButton, along with others. Managing the technological environment including constant Internet provision, the flawless operation of web cameras, built-in microphones, headsets or creating an undisturbed learning environment posed several challenges to all parties. At LUPS, the authors conducted a research to map the attitudes to the altered learning and teaching circumstances, methodologies and material. We analysed the textual and numeric data derived from the answers of two major segments of academic population, the professors and students. We compared their answers using statistical methods, applying a series of tests such as a Kolmogorov-test to analyse the normality of data set, a two-sample t-test to verify the equality of standard deviation, and, finally, *Welch*-test, to confirm the conclusion of the previous step, and to be applied in case standard deviations were not equal. Thus, the meticulous analyses of statistical results enabled the confirmation or rejection of our assumptions. Our presumptions about the perception of various aspects of online teaching and learning by students and professors were formulated in hypotheses, the results of which are demonstrated in Tables 1, 2 and 3.

In case of the comparison between the first three questions and results, there is no difference in the statistical values gained as conclusions. We must therefore state that both students and professors perceived the mentioned topics the same way, having no alteration in their opinions. Demonstrated in Tables 4, 5 and 6 are further research questions and testing findings with differing attitudes.

Table 1 The results of the hypothesis testing procedure for question 1

Hypothesis testing for Question 1. In your view, were you able to acquire/pass on the same knowledge in the same depth as in the case of contact teaching?					
F-test		*t*-test		*Welch*-test	
Test statistics	1.53	Test statistics	0.27	Test statistics	0.31
Critical value	1.39	Critical value	1.96	Critical value	1.98
Conclusion	$\sigma_1 \neq \sigma_2$	Conclusion	$\mu_1 = \mu_2$	Conclusion	$\mu_1 = \mu_2$

Table 2 The results of the hypothesis testing procedure for question 2

Hypothesis testing for Question 2. If you had the opportunity to continue studying/teaching online, would you prefer to do so?					
F-test		*t*-test		*Welch*-test	
Test statistics	1.07	Test statistics	–0.67	Test statistics	–0.66
Critical value	1.35	Critical value	1.96	Critical value	1.98
Conclusion	$\sigma_1 = \sigma_2$	Conclusion	$\mu_1 = \mu_2$	Conclusion	$\mu_1 = \mu_2$

Table 3 The results of the hypothesis testing procedure for question 3

Hypothesis testing for Question 3. Have you received/provided all possible help within the framework of online education?					
F-test		*t*-test		*Welch*-test	
Test statistics	1.34	Test statistics	1.67	Test statistics	1.56
Critical value	1.35	Critical value	1.96	Critical value	1.98
Conclusion	$\sigma_1 = \sigma_2$	Conclusion	$\mu_1 = \mu_2$	Conclusion	$\mu_1 = \mu_2$

Table 4 The results of the hypothesis testing procedure for question 4

Hypothesis testing for Question 4. Have you/your students been motivated enough within the framework of the online education?					
F-test		*t*-test		*Welch*-test	
Test statistics	1.43	Test statistics	2.03	Test statistics	2.22
Critical value	1.39	Critical value	1.96	Critical value	1.98
Conclusion	$\sigma_1 \neq \sigma_2$	Conclusion	$\mu_1 \neq \mu_2$	Conclusion	$\mu_1 \neq \mu_2$

Table 5 The results of the hypothesis testing procedure for question 5

Hypothesis testing for Question 5. Have you/your students been able to concentrate the same way both during online and offline education?				
F-test		*t*-test	*Welch*-test	
Test statistics	2.03	Does not work	Test statistics	2.30
Critical value	1.39		Critical value	1.98
Conclusion	$\sigma_1 \neq \sigma_2$		Conclusion	$\mu_1 \neq \mu_2$

Table 6 The results of the hypothesis testing procedure for question 6

Hypothesis testing for Question 6. Have you had disturbing factors during online education?				
F-test		*t*-test	*Welch*-test	
Test statistics	2.47	Does not work	Test statistics	5.89
Critical value	1.39		Critical value	1.98
Conclusion	$\sigma_1 \neq \sigma_2$		Conclusion	$\mu_1 \neq \mu_2$

Regarding the comparison between the responses given to questions 4–6, there is significant difference in the statistical values gained as conclusions. We must therefore state that students and professors perceived the mentioned topics differently, with professors judging the situation less favourably than students [3].

2.2 *Lessons Learned and Follow-Up Suggestions*

Furthermore, we also provided a number of conclusions and recommendations for our subsequent teaching and learning practices. Among those, we had to conclude and propose that online teaching and learning should partially and gradually be introduced into the academic curriculum, not only in order to bridge gaps of distance but also to keep up and develop digital literacy and a variation of methods that the academic population will benefit from. A further reason for the above is that countrywide, students prefer the autonomy individual online learning provided for them, an asset which must deserve higher scores in the credit system of institutions, as other domestic surveys into the subject have shown [4].

Last but not least, our proposals also include that the initiation of online education shall only be in case of lectures, correspondence education, and theoretical modules, whereas practice sessions should remain in contact form.

2.3 The Impact of Digital Learning During COVID-19 on Our Subsequent Teaching Practices

It was evident to all of us engaged in ESL teaching, that after conducting teaching online through semesters, using a variety of platforms, programs and resources to make our lessons more intriguing and colourful, teaching 'normally' in offline classrooms may not revert to the situation before as if online teaching had never occurred. We understood that with the abundance of digital material and academic teaching resources, the application of technology in offline classroom environment is not only desirable but it is becoming an unavoidable, tight-woven thread in the fabric of our everyday classroom activities. Our ESP lessons and general academic teaching practices are enriched with all those new resources which had been discovered and were made to apply distantly.

Therefore, we put special emphasis on course design. This being a priority to cater for students' needs and creating an active learning environment, we add diverse tools and multiple forums where students can create knowledge as well. Needless to say, the process involves bringing technology into classroom, and as we do so, there are some prerequisites. First and foremost, the technology must be present in the classroom. This normally involves at least one central computer, in some cases, many other individual desktops available for the learners. Also, the projection of the visual content of the central computer is normally solved by using a ceiling-mounted projector and a screen or a connected television monitor.

3 Course Design and Textbook Samples for Students of Security Technology English Based on Most Recent Research Findings

3.1 Mission-Oriented Preparation and Security Technology English

At OU, within the framework of Mission-Oriented Preparation (MOP) [5], a unique ESP course has been offered for semesters mainly for students of the relevant security technology engineering major. Yet, other keen students from various faculties of the university have also been accepted, a lot of them having foreign origins staying in Hungary on a Stipendium Hungaricum scholarship. The course boasts being the exclusive representative of the technical language of security technology, and as such, it is heavily filled with topics that are align with the professional subjects of the security technology major. To achieve the highest level of professionalism in combination with advanced teaching methods and pedagogical approaches, course design is continuous. The inclusion of areas such as disaster management, fire protection, occupational and health safety, biometrics or access control systems guarantee validity and authenticity, as well as close collaboration with professors of various subjects. The methodological freshness is represented in handouts, exercises and projects all contributing to a textbook, glossary and dictionary of this field of ESP.

3.2 Language Proficiency Levels: Input and Output

Even with the unstable and ever-changing Hungarian preconditions regarding the attainment of language examination certificates for higher lever education admission, our best supposition suggests that incoming students possess English proficiency at B2 level. In each individual case, however, admission to the course may be granted or denied based solely on the level of students' capability to perform academic tasks in the target language. During the course, we aim at attaining proficiency for students as described in the Common European Framework for Reference (CERF):

> C1: Can understand a wide range of demanding, longer texts, and recognise implicit meaning. Can express him/herself fluently and spontaneously without much obvious searching for expressions. Can use language flexibly and effectively for social, academic and professional purposes. Can produce clear, well-structured, detailed text on complex subjects, showing controlled use of organisational patterns, connectors and cohesive devices [6].

3.3 Tailored Course-Design and Textbook

While compiling the multitude of tasks targeting special language vocabulary enrichment, one must do so relying on the latest research result of ESL teaching or general pedagogy. A meta-analyses of numerous studies into the skill-based teaching techniques concludes the relative ineffectiveness of separated skills-development and points to integrated skills development instead, as a teaching approach that yields significantly better results during testing [7]. This finding impacted the layout of the planned course book together with the focus shifted to complex skills-development. The integrated skills-development approach is well reflected in several excerpts.

Here, a combination of passive and active skills practice may be traced in connection with the same topic, with the potential to further conversion to a writing task. Also, the tasks above demonstrate another recent approach that reflects the youngest generations' attention span. The idea of 'nano-learning' takes into consideration all the constituents that should be factored in that result in current students' reduced attention span [8]. One such component is the increased video-content that those generations consume through various social media feeds. Therefore, the course material must be bold enough to employ video resources. Also, with regard to the astonishingly short focus time of 1.3 s in the case of Generation Z [9], we must economize the material and create bite-size task chunks and video length for every learner to be able to benefit from any given material and lesson.

There are other areas of language skills development that are emphasized during the course of Security Technology English. However, CERF has been present governing foreign language teaching and testing, it had also been criticized for long [10]. It had been lacking level and subskill descriptors for mediation. Not only the descriptors were missing, but also mediation skill testing gradually disappeared from the assortment of the available accredited language examinations. Yet, as a Hungarian scholar pointed out, the omission of mediation in written language examination decreases the validity of the whole test paper, therefore the theory and practise of Hungarian language testing should be rethought [11]. However, Security Technology English, in accordance with CERF sub-skill descriptions and other recommendations of studies aims to cease that particular shortcoming in its classroom practises. We believe that similarly to other general and detailed descriptions, the ones regarding mediation will also exert monumental influence on ESL teaching and testing, the subskill description will eventually affect the methods and official examinations too. The Security Technology English course has preceded that foreshadowed trend by having incorporated mediation into its curriculum. The skill has facilitated better understanding of professional terminology and thus, its frequent inclusion in classroom activities significantly improves the chosen field of study. As an additional benefit, students play a part in coining terms in their native language for objects or concepts that exist in English but lack a proper or wide-spread Hungarian translation due to the rarity or the novelty of a scientific term. The excerpt from the widely available online material for learners at LUPS [12] makes students apply mediation skills that coursebooks on General English normally disregard.

You are watching the news during your break at police headquarters. You have a colleague from abroad who you are cooperating with. He becomes interested and asks you to interpret the main ideas. Help him understand what you are watching by telling him the most important facts.

4 Innovative Solutions, Their Rewards and Awards

4.1 The Most Innovative Department

In 2021, the Department of Foreign Languages for Specific Purposes (hereinafter: Language Department) at LUPS was awarded the 1st place for "the Most Innovative Department" initiated by the Creative Learning Programme. One of the innovative features of the Language Department highlights that ICT-supported seminars are highly motivating for students [13]. The aim is to propose what elements are indispensable to make the students' individual flow experience visible in higher education, and thus, to implement the flow-based pedagogical model in education [14]. Students are expected to be able to constantly be attentive, independent and creative. Our research confirms that students enjoy the digital environment because they are challenged to think, problem-solve, and collaborate, thereby they develop their competences and creativity [15]. In order to enhance the students' individual flow state and knowledge, a pilot research was carried out to measure the application of Augmented Reality [16] and the HY-DE model [17]. The research was conducted in the framework of English for law enforcement in the academic year of 2021/2022. At the end of the test seminars, a Flow State Questionnaire was filled in by the students and their vocabulary knowledge was measured with a self-made test. It is assumed that the use of digital technology in the classroom has a positive effect on the development of the students' motivation and language skills. According to the initial results, there is significant difference between the students' individual flow state in the study and the control group. It means that students in the study groups, where the teaching material was transferred by the use of ICT tools, perceived learning as an experience and a challenge.

4.2 High Expectations

The expectations of the Creative Learning Programme (CLP) was initiated by the Vice Rector of the Ludovika University of Public Service for the nomination of the most innovative department [13]. The expectations were as follows: fostering and attracting talents, conveying knowledge transfer, collaboration with other departments, the promotion of learning, innovation and creativity, inspiration, and future orientation. The content of the call for the competition of the 'Most Innovative Department' covered a wide range of students' expectations. These are the promotion of the individual learning, motivation of students and professors to develop skills. Moreover, the presence of collaboration and the exploitation of collective knowledge was also highlighted in the project as they were stressed in relevant works [18, 19]. Taking into account the key competences and the concepts of the CLP, openness to innovations, exploitation of internal and external partnerships and networks were also emphasized amongst the expectations we had to meet. The following issues were also enhanced by the committee of the CLP: the measurement of successfulness and motivation in education; the improvement towards the sub-optimal areas, mutual values, clear objectives, well-structured methods for creativity of the department, best methods, the implementation of good practices. In addition, in order to have a chance to be awarded we had to give a real insight into how the colleagues inspire each other and the students as well, how they increase the prestige of the department. Furthermore, the competition involved questions related to how the departments represent collective pedagogical values, educational policy and future perspectives. Finally, the applicants had to prove that the colleagues of the departments could make decisions flexibly, learn fast, how the organizational culture promotes new ideas and how they can make the performance of education visible for the whole faculty and for the students as well. It is evident that as teachers we have to face Generations Y and Z, who want the maximum benefit with the least effort and whose attention is difficult to capture. Unavoidably, we are made to turn to the opportunities offered by the digital world, for example, learning and teaching e-investigation skills as well [19]. Overall, it is indispensable to apply new methodologies to capture the students' attention in both seminars and lectures today [20].

4.3 Applied Methodologies and Building Networks

The colleagues of the Language Department use a wide range of ICT platforms to create competitive spirit, and enhance motivation such as Mentimeter, Quizlet, Quizzes, Kahoot, Redmenta, Padlet, Wordwall, BigBlue Button, Moodle, Classroomscreen, Quiz Show, Mindomo (mindmap), Edpuzzle, Genially, Twee, Word Art and so on [20]. There are three more methods we started to test in the framework of classroom research in order to enhance the students' motivation: (1) the application of Augmented Reality (AR), (2) the HY-DE model, which influences the hyper and deep attention of the students by multitasking effects such as images, sounds, videos, power point presentations and so on. We also apply (3) the method of MOP in strong cooperation with professionals in a situational linguistic environment, targeting the students of private security. Their innovative teaching method combines methodological and pedagogical approaches and applies them in a practical way. Its objective and mission is to prepare students to be fit for their future task in the target language.

In order to comply with the above-mentioned criteria devised by the CLP, various events are organised by the Language Department: we have two compulsory workshops annually in order to train the teaching methods. One of them is the *Innovative methods in language education for special purposes*, where new trends in language teaching, ICT platforms and good practices are introduced such as the application of AR in the classroom, new methods in language teaching, gamification or escape rooms. The other one is the *Law Enforcement Workshop*, where we have opportunities to keep up with the latest information concerning law enforcement topics like cyber criminality, dark web, terrorism, new psychoactive materials, biometrics, organised crime, security technology and so on. We have participated in workshops where we were trained in the topics of forensic linguistics, crime scene investigation, terrorism, cyber criminality, new psychoactive materials and profiling, since we teach English, German and Russian for law enforcement at the Faculty of Law Enforcement (FLE).

Additionally, we regularly participate in different projects or competitions and three of them have already been awarded: 1st prize was given by the Internal Affairs Scientific Committee in 2020, then we were the recipient of the 1st prize winner of "the Most Innovative Department Prize" in 2021, as mentioned before. The next year the 3rd prize was received from the Disaster Management Scientific Committee. Last but not least, our colleague and fellow researchers took part in the Profformance Higher Education Teacher Award Call in 2022, where the 2nd prize was awarded based on project work of the AR-supported good practice. Furthermore, other calls for applications such as Hanns Seidel and the European Educational Research Association Award were also received. Besides projects and competitions, we take part in the publication of Q studies as well. We participate in conferences and congresses in the scientific fields of psycholinguistics, languages for special purposes, pedagogy and private security as presenters, as we take part in conferences on law enforcement issues as an audience.

As educators in the academic world, we also place special importance on the constant development of training materials, since they function as textbooks or course books during lessons while keeping updated with the latest scientific result of our various domains. In the last 4 years, 2 bilingual English and Hungarian dictionaries for law enforcement, border policing, and immigration policing, another two German and two English coursebooks for law enforcement were published by the Language Department. In process, we have a new coursebook in the field of law enforcement, customs administration, disaster management and private security with glossary as well.

Research activities are constant and they are also carried out by the Language Department in the fields of psycholinguistics [21], Greek-Hungarian bilinguals [22], sociolinguistics [23], law enforcement topics such as special penal institutions [24], serial killers [25], law enforcement lexemes, forensic linguistics [26] and the effectiveness of online education at the FLE [2]. The technical language of customs administration, border policy [27], immigration, disaster management, private security [28], biometrics [29], psychoactive substances [30], innovative language teaching methods, e-learning [31], the integration of ICT platforms, the MOP [32], intercultural competence [33], comparative literature [34], and the Russian literature are also research topics at the Language Department. As we have observed the expectations of students, the use of language in the period of COVID-19 pandemic [35], the effect of online education and the usefulness of education maintained by the principle of Computer-Assisted Language Learning (CALL) and Mobile Assisted Language Learning (MALL) were also measured at the FLE. In addition to the preparation for the language test of Language for Services (LforS), general intermediate and advanced language examinations, consultations, competitions, scholarships, the building of foreign networks also highlighted in the project. We upgrade the students who do not possess B2 language exam, and we mentor them for the LforS ESP language examination. Earlier we had an opportunity to prepare the students for the Chinese scholarship to the Zhongnan University and cyber competition as well.

Within the framework of FRONTEX and the international work team collaborating, we have Italian, German, Romanian and Slovakian connections and due to the CEPOL training programme, we also have Estonian and Cretan collaborations. We foster Irish relationships too, with the University of Limerick. In the framework of the series of the Summer School of Psycholinguistics, we cherish international connections. Regarding our memberships in national organisations, we can enlist the Hungarian Association of Teachers and Researchers of Languages for Specific Purposes, the Hungarian Association of Applied Linguists and Language Teachers, the Summer School of Psycholinguistics, Fédération Internationale des Professeurs de Langues Vivantes and the ICT MasterMinds research teams. We keep in touch with the experts of the other departments such as the National Tax and Customs Administration, the Border Policing, the authorities that let us participate in field works in authentic environment. Our relationship with a forensic linguist and a crime scene investigator is also maintained. In the framework of the project, *the international connections via Erasmus + mobility* and the cooperation with different universities were highlighted such as the University of Nicosia in Cyprus, the Istanbul Kültür University, the Estonian Academy of Security Sciences in Tallin, the University of Zilina, the Police Academy in Szczytno, and in Appeldorn. We used to teach in the

Academy of Police Force (Bratislava), and in Rothenburg (Hochschule der deutschen Polizei) as well.

4.4 Augmented Reality

Augmented Reality is an increasingly widespread technology that allows the computer to generate virtual elements into the physical world such as a 3D model, an animation or a video [36]. According to the most widely accepted definition of AR, it combines the real and virtual world interactively, all registered in 3D [37]. The augmented system takes place between the real and virtual worlds [38]. There are several types of AR: markerless-, location, marker or superimposition-based AR [39]. It can be adaptable for military purposes, entertainment industry, marketing or medical purposes as well. There are a number of benefits why the educational uses must be in the centre of our attention. Firstly, it provides a motivational basis in the classroom, and supports both project and team work, too. Secondly, it is useful for a wide range of classroom activities such as cooperative tasks, simulation, discussion, debate, presentation, explanation and demonstration as well [40]. Either a smart phone or a tablet is required for the visual image [39]. The application of technologies like Artificial Intelligence (AI), Virtual Reality (VR), Augmented Virtuality (AV) and Augmented Reality (AR) has appeared and is expected on the labour market. Due to the cooperation with the Stiefel Interactive Ltd., the developer of the eduARdo software in Hungary, the students of the FLE were able to test-run it for for content consumption and creation. In addition, we could measure the added value of the application of software for the goal of improved learning and teaching processes to develop students' vocabulary knowledge at LUPS [41].

5 Conclusion

As ESP teachers, we understand the significance of the professional content of the syllabus for our future critical infrastructure, security technology, private security and law enforcement experts in the labour market who must keep pace with topical issues. We address them in the target language coupled with innovative pedagogical and teaching methods that not only have been researched well and proven to be highly effective, but also initiate and deploy various technological innovations in the classroom. The use of the above detailed platforms, methods and devices maintain the learners' interest and motivation to continuously develop their skills in the foreign language. This happens primarily because they can immediately become more engaged in classroom activities and harvest better results, making the process of learning effortless and meaningful. Last but not least, the researches we conducted and the conclusion that we drew help to refine the aforementioned techniques and fructify valuable and renowned academic awards.

References

1. Basturkmen H (2010) Developing courses in English for specific purposes. Palgrave MacMillan, Great Britain
2. Simon ÜG, Kovács É (2022) Az online oktatás hatékonysága a Rendészettudományi Karon az oktatók és a hallgatók szemüvegén keresztül – egy hibrid kutatás eredményei. Védelem Tudomány 7(1):191–219
3. Ürmösné Simon G, Kovács É, Hanka L (2023) How did the professors and the students of the Faculty of Law Enforcement manage online education deriving from the proceed of the pandemic? In: I. Ludovika International Law Enforcement Research Symposium Conference Proceedings, pp. 169–171
4. Hallgatói Önkormányzatok Országos Konferenciája: Távoktatás jelentés. https://hook.hu/hu/felsooktatas/tavoktatas-jelentes-2851. Last accessed 14 March 2024
5. Kovács É (2021) The experiences of running 'Advanced Technical English' courses at Donat Banki Faculty of Mechanical and Safety Engineering, Obuda University. Biztonságtudományi Szemle 3(1):65–78
6. Council of Europe (2001) Common European framework of reference for languages: learning, teaching, assessment—structured overview of all CERF scales 5–6
7. Conway B (2023) Emerging stronger. Kappan 105(1):24–28
8. Debétaz E (2024) Top education trends to watch in 2024. https://hospitalityinsights.ehl.edu/education-trends. Last accessed: 14 March 2024
9. Lebow S (2024) Gen-Z has a 1-second attention span. That can work to marketer's advantage. https://www.insiderintelligence.com/content/gen-z-has-1-second-attention-span-work-marketers-advantage. Last accessed: 14 March 2024
10. Piccardo E (2024) The common European framework of reference in language education: past, present and future. TIRF Language Education in Review Series. https://www.tirfonline.org/wp-content/uploads/2021/10/LEiR_CEFR.pdf. Last accessed: 14 March 2024
11. Dévény Á (2008) Az idegen nyelvi közvetítés feladat helye, szerepe a kritériumfüggő nyelvvizsgán
12. Acsai G, Borszéki J, Gyura V, Hársia B, Kovács É, Kudar M (2024) Crime and law enforcement. https://rtk.uni-nke.hu/document/rtk-uni-nke-hu/crime_and_law_enforcement.pdf Last accessed: 14 March 2024
13. Ürmösné Simon G, Barnucz N, Kudar M (2022) Út az Innovatív Tanszék első helyezett díja felé'. Belügyi Szemle 70(10):2113–2133
14. Dominek DL (2022) On a Flow-based pedagogical model the emergence of experience and creativity in education. Eruditio-Educatio 17(3):72–82
15. Dominek DL (2021) Élményalapú környezeti nevelés a természettudományi múzeumokban. Belvedere Meridionale, Szegedi Egyetemi Kiadó
16. Czékmán B (2017) Mobiltechnológia a tanórán: oktatási tartalmak, oktatást segítő digitális megoldások. Mobil–Világ–Iskola: Válogatott tanulmányok az I. Mobil eszközök az oktatásban konferenciáról. Debreceni Egyetemi Kiadó
17. Dani E (2014) A kétfázisú HY-DE-modell: A hiper- és mélyfigyelem fázisváltásai a katedrától a hallgatói önfejlesztéséig. Informatika a felsőoktatásban konferencia
18. Mátyás S (2020) A Kárpát-medence talentumföldrajza. Mátyás Szabolcs és Társa. Debrecen
19. Nyitrai E (2022) A magyar nemzeti adatvagyon jelentősége a bűnüldöző szervek munkája során. Információs Társadalom 1:67–80
20. Ürmösné SG, Borszéki J, Barnucz N, Uricska E (2021) A rendészeti szaknyelv szerepe a felsőoktatásban és az új nyelvi stratégia bevezetése. Porta Lingua 1:185–195
21. Ürmösné SG (2012) Hungarian-Greek communicative strategies in respect of gender. Mentális folyamatok a nyelvi feldolgozásban. Budapest.Tinta Könyvkiadó, 280–290
22. Ürmösné SG (2017) A case study based on a spontaneous discourse of Greek-Hungarian bilinguals in respect of interjections, swear words and syntactical mistakes, as regards gender. Magyar Rendészet 17(4):193–210

23. Ürmösné SG (2014) Gender, szubkultúrák és genderpragmatika. Transzdiszciplináris üdvözletek: Lengyel Zsolt számára. Budapest, Gondolat Kiadó, 210–218
24. Ürmösné SG (2019) Amazing penal institutions and dwelling circumstances of inmates. Internal Secur 11(2):179–190
25. Ürmösné SG (2020) The portrayal and the attributes of serial killers and some of the most notorious ones. Internal Secur 12(2):261–273
26. Ürmösné SG, Nyitrai E (2021) The phenomena of epidemic crime, deepfakes, fake news, and the role of forensic linguistics. Információs Társadalom 4:86–101
27. Borszéki J (2021) The role of domain experts in the development of the course English for Border and Coast Guards. Rendőrségi Tanulmányok, 6–23
28. Christián L, Kovács É (2018) Private security and municipal policing in academic education—an international overview. Magyar Rendészet 18(4):49–60
29. Kovács É, Kovács T (2022) Identification and authentication potentials based on limited biometric data. In: Critical infrastructure protection: theoretical and practical approach. Springer, Heidelberg, Netherlands, pp 391–404 (2022)
30. Bacsó B (2023) A kábítószerekkel összefüggő bűnözés elleni küzdelem és a főbb térbeli jellemzői a kábítószer-bűnözésnek a német nyelvű országokban. Bűnözésföldrajzi Közlemények 4(3–4):9–18
31. Borszéki J (2021) The development of e-learning tools used for teaching English for Law Enforcement. Internal Secur 13(1):181–200
32. Szabó A, Kovács É (2021) A személy- és vagyonőrök objektumspecifikus fejlesztési terve (IDPS) MOP-kiterjesztéssel. Magyar Rendészet 21(3):139–154
33. Nagy G (2022) A rendészeti hallgató gondolkodásmódja és az interkulturális kompetencia fejlesztése. Magyar Rendészet 22(4):153–169
34. Nagy É (2022) A turgenyevi témák és I. Sz. Turgenyev "alakjának" problémái F. M. Dosztojevszkij "Ördögök" című regényében. Műértelmezés – fordításkritika – transzmedialitás. Budapest. ELTE Orosz Irodalom és Irodalomkutatás - Összehasonlító Tanulmányok Doktori Program, pp 119–131
35. Uricska E (2021) COVIDictionary new words and phrases related to the global pandemic. Rejtjel Kiadó, Budapest
36. Matuszka T (2012) Kiterjesztett valóság alkalmazások fejlesztése, elemzése és a fejlesztőeszközök összehasonlítása. ELTE IK. http://people.inf.elte.hu/tomintt/tdk.pdf Last accessed: 4 March 2024
37. Azuma R T (1997) A survey of augmented reality. Presence Teleoperators Virtual Environ 6(4):355–385
38. Aknai DO, Czékmán B, Fehér P (2021) Kiterjesztett valóság (AR) alkalmazások használata és készítése az iskolában. Iskola a társadalmi térben és időben VII. Pécs: PTE BTK "Oktatás és Társadalom" Neveléstudományi Doktori Iskola, pp 150–156
39. Barnucz N (2020) A kiterjesztett valóság alkalmazása a nyelvoktatásban különös tekintettel a rendészeti szaknyelvre. Educatio 29(4):644–652
40. Ibañez MB, Uriarte AP, Zatarain RC, Barrón ML (2020) Impact of augmented reality technology on academic achievement and motivation of students from public and private Mexican schools. A case study in a middle-school geometry course. Comput Educ 145:103734
41. Dominek DL, Ceglédi S, Barnucz N (2024) The study of instructors' digital competence in higher education – comparative analysis. Információs Társadalom. (under release)

Microstructural Analysis of High-Strength Steel Post Gleeble Modelling

Lama Mkanna

Abstract High-strength steel alloys are widely used in critical engineering applications due to their exceptional mechanical properties. To ensure their reliability and performance under extreme conditions comprehensive understanding of their microstructural changes during testing and processing is crucial. This study investigates the microstructural evolution of high-strength steel samples subjected to Gleeble modelling, a thermomechanical simulation technique that replicates real-world conditions. The research was conducted on a series of high-strength steel specimens, where varying combinations temperature, strain rate, and applied stress were employed to simulate a range of operational scenarios. The microstructural analysis was carried out using advanced microscopy techniques, including optical microscopy, scanning electron microscopy (SEM). Our findings reveal intriguing insights into the dynamic changes occurring within the steel microstructure during Gleeble modelling. At elevated temperatures, grain growth was observed, affecting both the grainsize and distribution. The effect transformation and recrystallization were also examined. This research not only contributes to a deeper strain rate variations on phase comprehension of the microstructural alterations in high-strength steel during Gleeble modelling but also offers valuable information for the optimization of steel processing and heat treatment strategies. Such insights are paramount for engineers and metallurgists working in fields requiring high-strength steel. Such as aerospace, automotive, and structural engineering.

Keywords High-strength steel · Microstructural analysis · Gleeble modelling · Grain growth · Phase transformation

L. Mkanna (✉)
University of Dunaújváros, Dunaújváros, Hungary
e-mail: Lamamkanna@gmail.com

Multidisciplinary Doctoral School of Engineering Sciences, Széchenyi István University, Győr, Hungary

T. A. Kovács et al. (eds.), *The Impact of the Energy Dependency on Critical Infrastructure Protection*, Advanced Sciences and Technologies for Security Applications, https://doi.org/10.1007/978-3-031-78544-3_34

1 Introduction

High-strength steel alloys are widely used in critical engineering applications due to their exceptional mechanical properties. To ensure their reliability and performance under extreme conditions, a comprehensive understanding of their microstructural changes during testing and processing is crucial. This study investigates the microstructural evolution of high-strength steel samples subjected to Gleeble modelling, a thermomechanical simulation technique that replicates real-world conditions [1].

2 High-Strength Steels Are Highly Susceptible to Cold Cracking and How to Avoid It

The inherent strength of high-strength steels makes them indispensable in various industrial applications; however, their susceptibility to cold cracking poses a significant challenge during welding and heat treatment processes. Cold cracking, also known as hydrogen-induced cracking, arises from the interaction of hydrogen with susceptible microstructures within the steel, leading to the formation of cracks. To circumvent this issue, meticulous precautions must be taken. Employing controlled cooling procedures, such as preheating the steel before welding and implementing post-weld heat treatment, is a proven strategy to mitigate the risk. Additionally, the use of low hydrogen electrodes, maintaining a dry welding environment, and adhering to proper welding techniques are crucial in preventing the introduction of hydrogen, a key contributor to cold cracking. Material selection plays a pivotal role and choosing high-strength steels with enhanced resistance to cold cracking, often achieved through alloying elements, is fundamental. By integrating these preventive measures into welding and material selection processes, industries can effectively navigate the challenges posed by cold cracking in high-strength steels, ensuring the integrity and durability of critical components [2].

3 The Critical Cooling Time Can Be Calculated Based on Heat Process Modeling

In the realm of materials engineering, understanding and mitigating cold cracking risks is paramount, particularly when dealing with high-strength steels subjected to welding or heat treatment processes. The calculation of critical cooling time, a pivotal factor in preventing issues like cold cracking, involves a nuanced approach rooted in heat process modeling. This intricate process commences with the comprehensive assessment of material properties, such as thermal conductivity and phase transformation characteristics, and extends to the meticulous consideration of heat input during the specific operation, whether it be welding or heat treatment. Applying heat transfer equations to model temperature distributions within the material, especially accounting for phase transformations like austenite-to-martensite, enables the calculation of cooling rates at various points. The critical cooling time, essential for avoiding cold cracking, emerges as the duration required for the material to cool through a temperature range critical for preventing detrimental effects. Engineers and researchers often leverage computational tools, finite element analysis, and numerical simulations to refine and optimize cooling rates, ensuring the robustness of materials in demanding industrial applications [3].

Three-dimensional heat dissipation (3D):

$$\Delta t_{T_1-T_2} = \frac{(q/v)_{eff}}{2\pi\lambda}\left(\frac{1}{T_2 - T_0} - \frac{1}{T_1 - T_0}\right) \tag{1}$$

Two-dimensional heat dissipation (2D):

$$\Delta t_{T_1-T_2} = \frac{(q/v)^2_{eff}}{4\pi\lambda\rho c s^2}\left(\frac{1}{(T_2 - T_0)^2} - \frac{1}{(T_1 - T_0)^2}\right) \tag{2}$$

$$(q/v)_{eff} = \frac{UI\eta_{eff}}{v_{heg}} \tag{3}$$

4 Materials Used in the Research

Embarking on a meticulous exploration of materials within the realm of my research, the spotlight converges on a series of high-strength steels that have played a central role in shaping the narrative. The selection includes a spectrum of alloys—S355MC, S500MC, S700MC, S960MC, and S1100MC—each possessing unique mechanical characteristics and tailored for specific applications. These materials, known for their exceptional strength and resilience, encapsulate the essence of the study, embodying a crucial amalgamation of performance and versatility. As we journey through the intricacies of their properties, the nuanced tale of these alloys unfolds, offering insights into their collective impact on the broader canvas of materials engineering. See Fig. 1.

5 Modelling in the Gleeble 3500

Embedded in the dynamic landscape of materials engineering, our scrutiny of high-strength steels—S355MC, S500MC, S700MC, S960MC, and S1100MC—takes a profound leap forward through meticulous modeling on the Gleeble 3500 system.

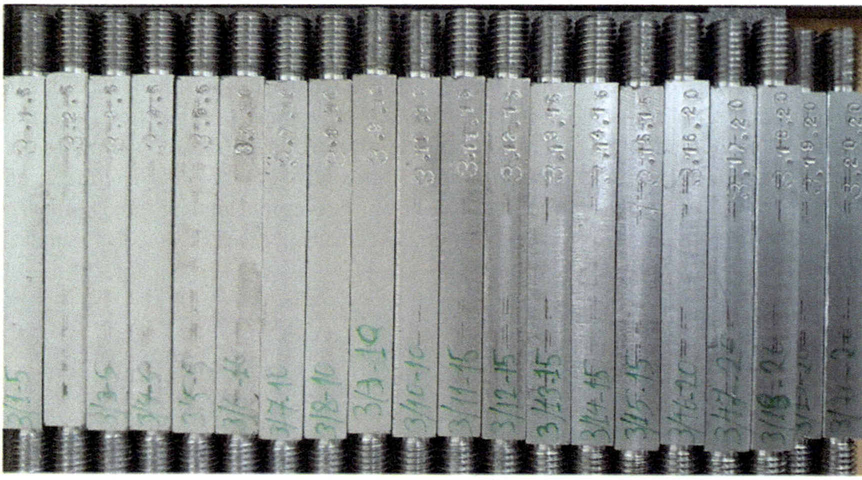

Fig. 1 Some of the cutted samples before the Gleeble modelling

This advanced apparatus, renowned for its precision and versatility, serves as the crucible for subjecting these alloys to controlled thermal regimes, mimicking real-world scenarios such as welding and heat treatment. What sets this investigation apart is the deliberate integration of varied cooling times—5 s, 10 s, 15 s, and 20 s—each representing a crucial parameter in the thermomechanical behavior of these materials. As we navigate the virtual landscape of the Gleeble 3500, this study not only unravels the thermal responses of high-strength steels but also systematically explores the impact of different cooling times, providing a nuanced understanding of their performance characteristics. This pivotal juncture in materials research sheds light on the suitability of S355MC, S500MC, S700MC, S960MC, and S1100MC for critical engineering applications, ushering in a new era of informed materials selection and application. See Figs. 2 and 3.

6 Heat Cycle Simulation

The crux of our materials engineering investigation lies in the intricate heat cycle simulations conducted using the cutting-edge Gleeble 3500 system. This sophisticated apparatus stands as a linchpin in our research, offering a controlled environment to meticulously replicate the thermal conditions that high-strength steels (S355MC, S500MC, S700MC, S960MC, and S1100MC) encounter during critical processes like welding and heat treatment. The Gleeble 3500's precision and adaptability enable us to subject these materials to varying thermal cycles, crucially encompassing heating and cooling phases. From rapid heating to dwell times and the deliberate introduction of different cooling times (5 s, 10 s, 15 s, and 20 s) this simulation

Fig. 2 Figure Gleeble3500, welding the thermal Caples before inert it to the machine

Fig. 3 Figure Impactor before simulation and during simulation

captures the dynamic interplay between temperature and time. As we delve into this heat cycle simulation, our aim is to unravel the intricate thermomechanical responses of these high-strength steels, providing valuable insights that transcend theoretical boundaries and inform practical applications in the realm of materials engineering. This pivotal facet of our research underscores the significance of controlled heat cycle simulations in understanding the nuanced behavior of materials under diverse thermal conditions. See Fig. 4 [4].

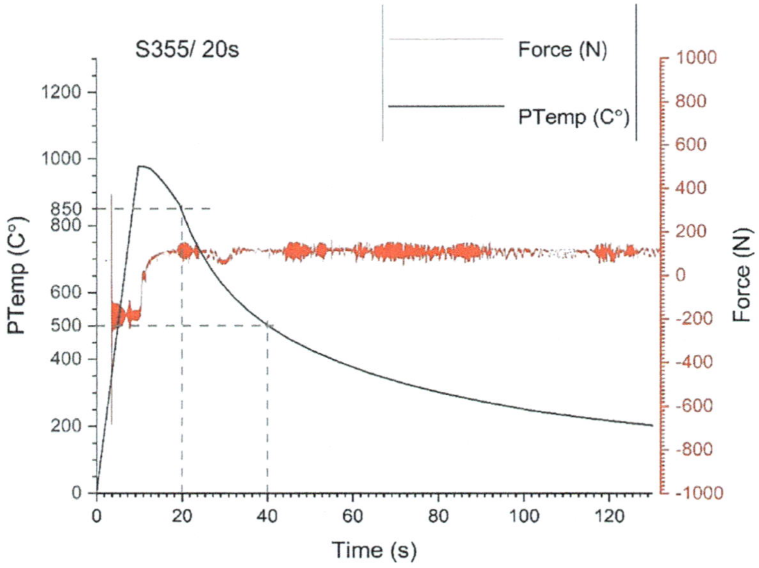

Fig. 4 Figure example of the heat cycle in case of S355/20 s

7 Hardness Testing

Complementing our in-depth heat cycle simulations on the Gleeble 3500, the investigation delves into the mechanical characteristics of high-strength steels (S355MC, S500MC, S700MC, S960MC, and S1100MC) through meticulous hardness testing. Harnessing advanced testing methodologies, this phase of the study aims to quantify the materials' resistance to deformation and ascertain their mechanical strength under different thermal conditions [5]. Employing standardized hardness tests such as Vickers, our examination provides a quantitative measure of the steels' hardness, offering a tangible representation of their structural integrity. The hardness testing phase serves as a critical bridge between theoretical thermomechanical simulations and practical material behavior, enhancing our understanding of how these alloys respond to diverse thermal cycles. This integrated approach, combining heat cycle simulations and hardness testing, marks a comprehensive exploration of the high-strength steels, paving the way for informed materials selection and robust engineering applications. See Fig. 5 and 6.

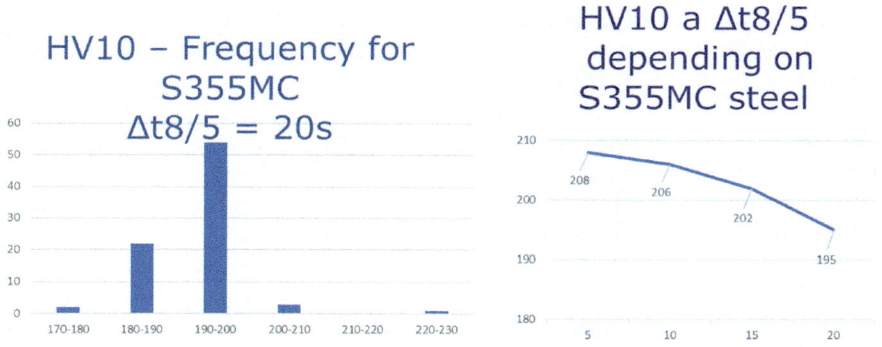

Fig. 5 Figure Hardness results in case of S355MC/20 s and the curve for the relation between the hardness and the cooling time

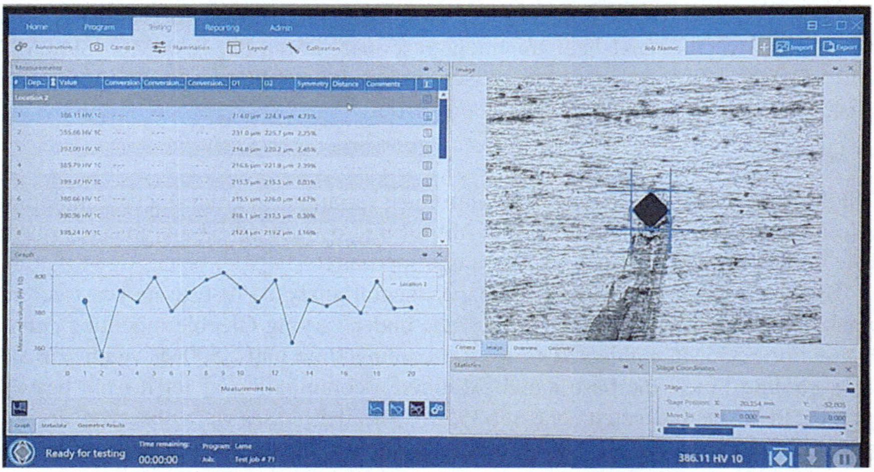

Fig. 6 Figure photo from the program during the hardness test

8 Microscope Testing and SEM

The application of the Gleeble modeling under different cooling times (5 s, 10 s, 15 s, and 20 s) introduces a time-dependent aspect to the thermo-mechanical treatment. This variable cooling rate can significantly influence the resulting phases and microstructures in the steel samples.

Ferrite Phase

S355MC: Varying cooling times can affect the ferrite phase, influencing grain size and distribution. Shorter cooling times might lead to finer ferrite grains. S500MC,

S700MC, S960MC, and S1100MC: The cooling rate can impact the formation of ferrite, affecting the overall strength and toughness of these high-strength steels.

Pearlite Phase

S355MC: The cooling times during Gleeble modeling could alter the pearlite phase, influencing hardness and strength. S500MC, S700MC, S960MC, and S1100MC: Different cooling times may affect the balance between pearlite and other phases, influencing the steels' mechanical properties.

Martensite Phase

S700MC, S960MC, and S1100MC: The cooling rate is crucial for martensite formation. Shorter cooling times may result in higher martensite content, contributing to increased hardness and strength.

Austenite Phase

S355MC: Cooling times influence the transformation of austenite, impacting the final microstructure and heat treatment response. S500MC, S700MC, S960MC, and S1100MC: The cooling rate is essential for controlling the amount of retained austenite, which can affect the steel's properties.

Other Microstructural Changes

All Samples: Different cooling times introduce variations in grain size, precipitation, and recrystallization, influencing the overall microstructure. These changes play a role in determining the final mechanical properties [6–10].

Based on the findings from the experimental analysis, distinct microstructural phases were identified for each steel grade under varying Gleeble modeling conditions. For S355MC subjected to a 20 s cooling time and S500MC with 5 s and 10 s cooling times, the base metal exhibited a combination of ferrite and pearlite phases. In the heat-affected zone of S355MC, similar phases were observed, accompanied by inclusions, while in S500MC, a slight increase in pearlite content was noted. Within the Gleeble-modeled region of these samples, consistent phases were identified, complemented by discernible precipitation from chromium.

For S700MC with a 5 s cooling time, the base metal displayed two predominant phases, namely ferrite and pearlite, alongside TiN inclusions. In the heat-affected zone, analogous phases were observed, with the introduction of carbides. The central region of the samples, where Gleeble modeling was applied, revealed the presence of TiN in conjunction with ferrite and carbides.

In the case of S960MC subjected to a 15 s cooling time, the base metal exhibited a notable bainite phase. The heat-affected zone indicated a transformation from bainite to ferrite and pearlite, while within the Gleeble-modeled section, the re-emergence of bainite was observed.

For S1100MC with a 15 s cooling time, the base metal revealed the coexistence of austenite and bainite phases, transitioning predominantly to bainite in the central part of the sample. These findings contribute to a comprehensive understanding of the intricate microstructural variations in different steel grades under diverse Gleeble modeling conditions, providing valuable insights for applications in the automotive industry. See Figs. 7, 8, 9 and 10.

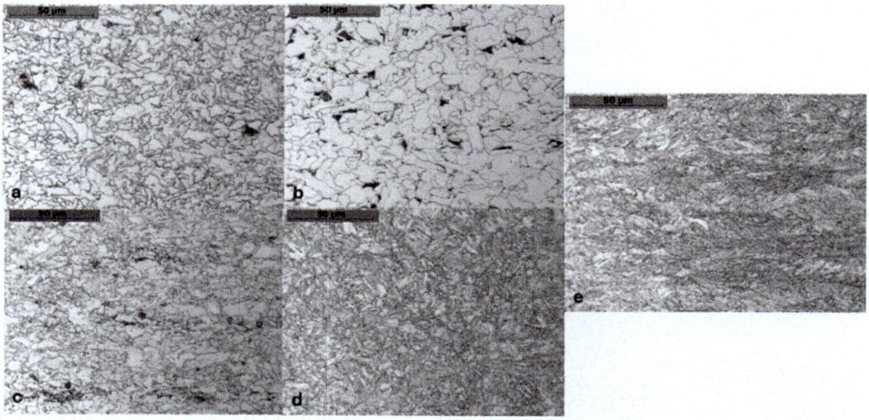

Fig. 7 Base metal for: **a** S355MC/20s, **b** S500MC/10, **c** S700MC/5s, **d** S960MC/15s, **e** S1100MC/15S

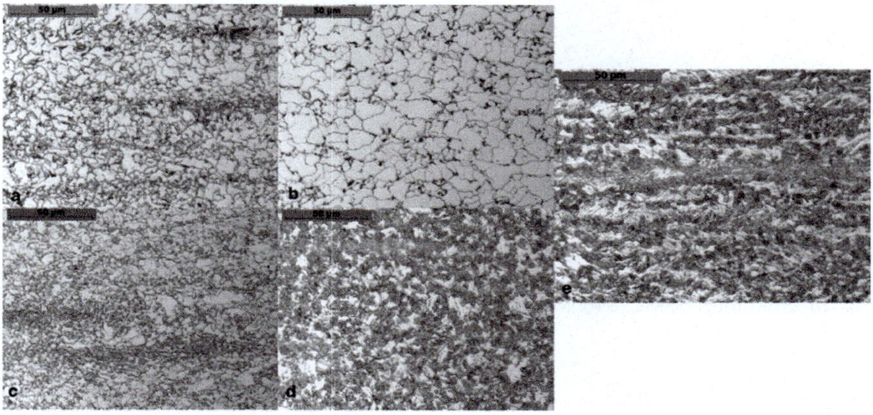

Fig. 8 Heat affected zone for: **a** S355MC/20s, **b** S500MC/10, **c** S700MC/5s, **d** S960MC/15s, **e** S1100MC/15S

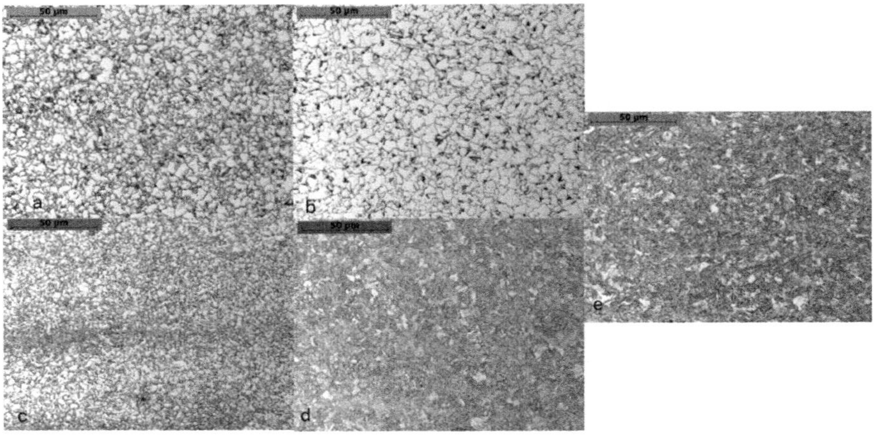

Fig. 9 Middle of the samples for: **a** S355MC/20s, **b** S500MC/10, **c** S700MC/5s, **d** S960MC/15s, **e** S1100MC/15S

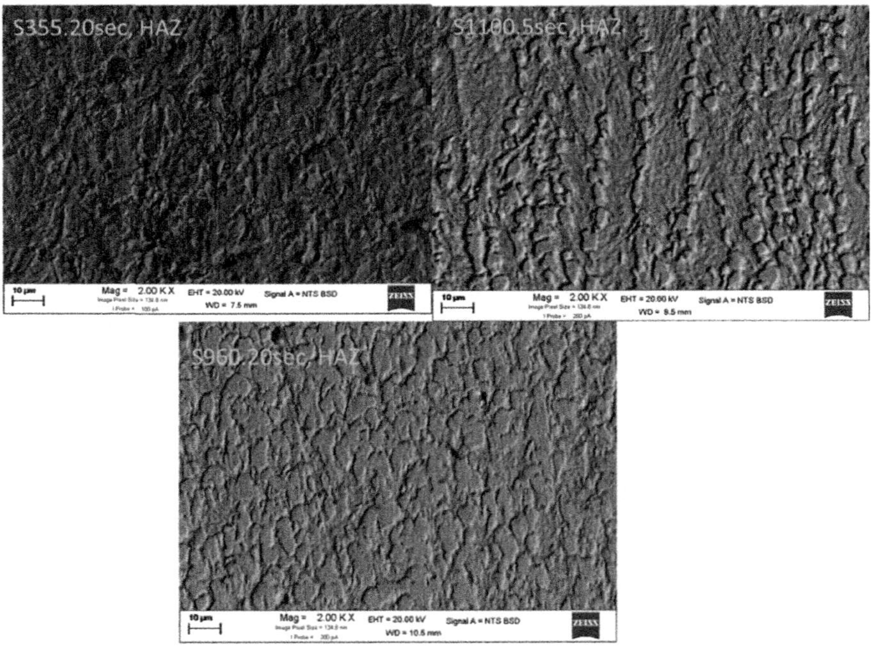

Fig. 10 SEM for S355MC, S960MC and S1100MC

9 Conclusion

The use of high-strength steels is becoming more and more widespread in industry, High-strength steels are susceptible to cold cracking but not in case of S355MC, S500MC and S700MC.

These results will help the engineers/technologists and users of HSS for the improvement of research data and can be implemented in cranes & lifting processes and vehicles industry (trucks etc.)

References

1. Patel KS, Rodriguez AB, Microstructural evolution in high-strength steel alloys under Gleeble modeling. Metall Trans
2. Węglowski MS, Zeman M (2014) Prevention of cold cracking in ultra-high strength steel Weldox 1300. Arch Civil Mech Eng 14:417–424
3. Thomas GA, Speer JG, Matlock DK (2011) Quenched and partitioned microstructures produced via Gleeble simulations of hot-strip mill cooling practices. Metall Mater Trans A 42:3652–3659
4. Lee J (2022) Comparison of mechanical properties between HSS Gleeble simulated specimen and welded specimen
5. Bagul AD, Dubey DK (2021) Effects on hardness by various heat treatments processes on HSS tool bit. Mater Today Proc 43:694–699
6. Jehn HA, Hofmann S, Rückborn VE, Münz WD (1986) Morphology and properties of sputtered (Ti, Al) N layers on high-speed steel substrates as a function of deposition temperature and sputtering atmosphere. J Vac Sci Technol, A: Vac, Surf Films 4(6):2701–2705
7. Tchuindjang JT, Lecomte-Beckers J (2008) Investigation on retained delta ferrite occurrence and its influence on HSS and Semi HSS Strengthening after thermomechanical routes
8. Wang M, Zhang F, Yang Z (2017) Effects of high-temperature deformation and cooling process on the microstructure and mechanical properties of an ultrahigh-strength pearlite steel. Mater Des 114:102–110
9. Caballero FG, Bhadeshia HKDH, Mawella KJA, Jones DG, Brown P (2001) Design of novel high strength bainitic steels: Part 1. Mater Sci Technol 17(5):512–516
10. Lambert-Perlade A, Gourgues AF, Pineau A (2004) Austenite to bainite phase transformation in the heat-affected zone of a high strength low alloy steel. Acta Mater 52(8):2337–2348

Explosive Welding of Metals for Electronic Applications

Lóránd Kugyela, Norbert Daruka, and Tünde Anna Kovács

Abstract In the literature, the available studies about blast loads, usually describe the harmful effects of the explosion. The blast load application in metal technologies is not collected as technology processes. In the engineering practice, explosive processes are not standard technologies. Traditional metal technologies such as welding, forming, and hardening are well studied and introduced in several literatures. This review is made to introduce explosive welding technology for engineering applications. It focused on explosive welding processes knowledge. The goal is to summarize the most essential information about the processes' parameters and applicability. The explosive welding parameter determination is introduced with theoretical and empirical background on the base of the literature.

Keywords Explosion · Cladding · Welding · Joining

1 Introduction

Solid-state metal technologies are well-known and applied metalworking processes. The metal solid-state processes are usually used for volume shaping, hardening and welding. The metal forming is made by hammering, rolling, pressing etc. The cold working processes cause a hardening of the metals. Cold rolling is a suitable process to earn the wanted hardness of the metal sheets (standard temper grades are e.g. full hard, ½ hard, ¼ hard, etc.). Welding is a traditional joining process that establishes a cohesion joint between the metal parts. In industrial practice, several solid-state welding processes are applied. The base property of metal shaping and solid-state welding is the formability of the metals. Metal formability means its ability to be

L. Kugyela
Doctoral School on Safety and Security Sciences, Obuda University, Budapest, Hungary

N. Daruka · T. A. Kovács (✉)
Banki Donat Faculty of Mechanical and Safety Engineering, Obuda University, Budapest, Hungary
e-mail: kovacs.tunde@bgk.uni-obuda.hu

T. A. Kovács et al. (eds.), *The Impact of the Energy Dependency on Critical Infrastructure Protection*, Advanced Sciences and Technologies for Security Applications, https://doi.org/10.1007/978-3-031-78544-3_35

shaped by metal forming without damage (e.g., breaking or cracking) [1]. The formability can also be predictable based on the knowledge of the crystal structure of the metal (face-centred cubic structure metals have good formability), and by the metal's mechanical properties (hardness, yield strength, tensile strength, etc.). These mechanical properties are ductility, malleability and hardness. The ductility is tested by mechanical testing e.g. by tensile test (specific elongation, specific reduction in cross-section), bend test (ability to resist cracking during one continuous bend) and the formability is tested e.g. by the Erichsen cupping test or with deep-drawing test [2].

The traditional forming processes are hot forming and cold forming. The hot forming technologies do over the recrystallization temperature. At this temperature, the metal forming stress is lower than the yielding stress of the metal and during the hot forming the metal continuously recrystallises. The metal working at ambient temperature (up to re-crystallization temperature), with a static low deformation rate over the yield strength results in the hardening of the metal. The cold working increases the dislocation density [3], the yield strength, and the hardness. The crystal structure of the cold-worked metal is changing. In the case of the austenite structure steels, the work hardening can affect phase transformation [3]. The austenite crystal structure (face-centred cubic) deforms and changes to a martensitic tetragonal structure [4]. In the case of the metal cold working processes, it needs to take account of the work-hardening effect [5].

In the joining processes well know several welding processes. It can be divided into two groups the welding processes. One of the welding processes groups is the fusion welding process when the metal joining is made by heat which can melt the samples. It knows several fusion welding processes such as arc welding, plasma transferred arc welding, laser welding, electron beam welding, etc. [6].

The other group of welding processes is solid-state welding, which occurs when the process temperature is less than the melting temperature of metal. Solid-state welding is an excellent opportunity to join similar and dissimilar metals (ultrasonic metal welding, friction stir welding, explosive welding, etc.). During the solid-state welding process, the joint is established by plastic deformation [7, 8]. The plastic deformation as a function of the process temperature establishes different properties of joints [9]. The structure of the welded joint is typical for the type of joint.

The base of the metal hardening is usually made by phase transformation by increasing dislocation density or precipitation hardening treatment [10]. The phase transformation in the case of the steels means austenite–martensite phase transformation by heat treatment or cold working. Precipitation hardening is a unique heat treatment process in the case of metals [10].

For traditional metal forming and solid-state welding in industrial applications use a static load. It can follow during cold rolling or ultrasonic welding processes. The material properties after the solid-state processing depend on the strain rate, temperature and stress state. The strain rate can increase the hardness, decrease the ductility and formability and cause fracture during the solid-state deformation. This is the reason why these processes are made by standard speed. The high-strain-rate process's physical background is different from the static one. The plastic deformation energy can increase by the deformation rate increasing. The forming strength increases as a function of the forming rate. The ratio between the forming rate and the forming strength can be described by the Alder-Phillips equation [11].

Metal technologies are complex processes based on chemical, physical, metallurgical, static strain, and high-strain-rate sciences. Also, it needs to understand the high-strain-rate metal processes that source the properties of the explosion (explosive materials) and its effects [12, 13]. In industrial practice, the rail frog (manganese steel or Hadfield steel) hardening is made by an explosive hardening process. The high strain rate of the explosion caused the austenite phase transformation to martensite. These phenomena affect the surface hardening of the rail frog [14, 15]. Explosive welding is useful for cladding sheets (e.g. carbon steel with titan, aluminium with copper, etc.). To understand the shock wave phenomena in metals needs to know some examples of the useful explosive materials and their properties [16–18]. This review aims to collect the relevant literature and the most important knowledge of blast load application in explosive welding.

2 Explosive Materials and the Affected High-Strain-Rate

It needs to group the explosive materials as a function of the typical properties (explosion rate, density, established gases, established heat, kind of aggregate, etc.) [16]. The definition of explosive materials is the next: after the initiation energy effect happens through a self-supported exothermic chemical reaction. For industrial practice, the reason for the use of explosive materials explosion is a chemical reaction that establishes CO, CO_2 and N_2 gases, (several times H_2O vapour) and heat.

This chemical explosion reaction is self-supporting [17, 18]. The transformation speed of the explosion depends on the chemical composition. As a function of the explosion speed, it can define four groups (Table 1). In industrial practice, the used materials need to be safe, which means the explosive materials can't explode due to heat, beat, flame or other effects. The process starts with the support of a special primary explosive material, a detonator which initiates the explosion of the industrial explosive material.

The ignition of the detonator is possible by flame, electric effect, mechanical effect, or chemical effect, as a function of the kind of detonator [17–20]. The explosive materials used in the mining, demolition of objects, metal industry and other industrial area [21, 22]. The explosion wave shows a typical characteristic shown in Fig. 1a [20]. The kinetics of the detonation established nascent gas pressure shows

Table 1 The kinds of explosions as a function of the explosion speed

Explosion kind	Primary	Secondary	Rapid burning	Burning
The velocity of detonation v_d (km/s)	Min. 1–10	0.1–1	0.0001–0.001	Less than 0.0001

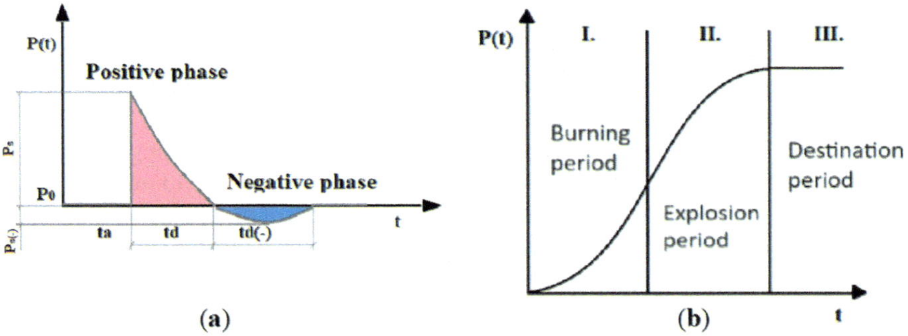

Fig. 1 The explosion characteristic (**a**) [20] and the established nascent gas pressure kinetic (**b**) [18] diagrams

three periods, the burning period, the explosion period, and the destination period shown in Fig. 1b [18].

For manufacturing, it needs to use the energy of the detonation period gas pressure, which can be calculated from Eq. (1) [23].

$$P = v_d^2 \cdot \rho_0 \cdot \frac{\rho - \rho_0}{\rho} \text{ (Pa)} \tag{1}$$

where:

P nascent gas pressure (Pa)
v_d detonation velocity (m/s)
ρ_0 density of the explosive material (g/cm^3)
ρ density of the explosion established gas (g/cm^3).

Some explosive materials and their properties are shown in Table 2.

Table 2 Explosive materials with their properties

Explosive material	Detonation velocity (m/s)	Density (g/cm^3)	Ref.
TNT	6800–6930	1.58	[24, 25]
Elbar	3500–3800	0.75	[26, 27]
PETN (cord)	7000–8400	14 g/m	[28]
Powdery emulsion	2200	0.8	[26, 29]
ANFO (ammonium nitrate + oil)	2500–3500	0.8	[24–26, 25]
Dynamite	3000	1.2	[25]
Ammonium nitrate (pure)	5270	1.725	[25]

3 Methods of Explosive Welding

Explosive welding is a suitable solid-state welding process for similar and dissimilar joining. The base of the welding process is the explosion-established shock wave-affected plastic deformation [26, 31–33]. The high strain rate of the deformation is the key to this process. The deformation rate depends on the used explosive material quantity and quality. Explosive welding is useful for cladding for similar and dissimilar joining. Several dissimilar joints are established with the explosive welding process even though these metals have different melting points and almost impossible to weld them with each other by fusion welding [34].

Two setups are commonly used for this process. The difference between the two configurations is the position of the parent plate and the flyer plate [31–33]. A useful configuration for small-size samples is the oblique configuration in Fig. 2a and for high-size plates can use the parallel configuration shown in Fig. 2b.

The welding joint is made by a jet formation, which is determined by the collision angle. To earn a suitable welded joint, the explosion-generated pressure-initiated stress needs to be sufficient to exceed the dynamic elastic limit of the material. For explosive welding or cladding, the welding parameters used are critical to earning suitable joints. The parameters of explosive welding are explosion speed, the density of the explosive material, and the configuration setup. The usable explosive materials

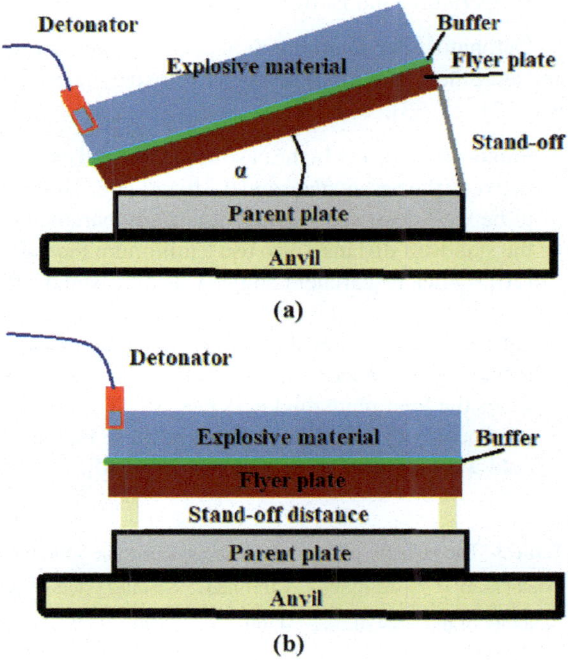

Fig. 2 The configurations for explosive cladding with oblique (**a**) and parallel (**b**) setups

can be of several types, with different densities, aggregates, and explosion speeds. The explosive welding setup segments are the next, according to Fig. 2:

- Parent plate (base plate): This plate is standing on the anvil and it will be joined by the flyer (cover) plate during the welding processes. The surface of the parent plate needs to be polished and cleaned before the welding process [26, 31–34].
- Flyer plate (upper plate): This plate is placed above the parent plate and during the welding process it impacts the parent plate and joins with it. Usually in the explosion welding setups the selection of flyer plate and parent plate is done based on a mass per unit area, whoever the lower density, tensile strength and higher formability are placed as flyer plate [26, 27, 29, 35–38].
- Buffer sheet: It is an optional component of the setup. It can be found on the surface of the flyer plate and during the explosive welding process, this sheet (PVC, rubber, etc.) protects the flyer plate from damage [26, 27, 32, 35, 39].
- Explosive material: The controlled energy source which during the explosion accelerates the flyer plate and joins with the parent plate by the high energy impact. The explosive material can be characterized by its velocity of detonation and density. The most applicable explosive materials for explosive welding explosion velocity in the range of 2000–3000 m/s [23, 26, 31, 38]. The collision velocity (v_c) must be lower than the velocity of sound in the material [26, 38]. The sound velocity of a plane longitudinal wave in some bulk materials is collected in Table 3. The speed of the sound depends on the kind of media and the geometry, temperature, and isotropy of the media.

Detonator: The main function of the detonator is to initiate the explosive material and start the explosion. The used explosive material usually can't explode any other way [22, 26, 33].

- Stand-off distance: In the case of the parallel explosion welding setups the distance between the parent plate and the flyer plate. The size of the stand-off distance needs to be well-designed because it is a key parameter. One of the rules of thumb for the stand-off distance is to use a minimum the half size of the flyer plate thickness [26]. Other researchers suggest using a stand-off distance twice the flyer plate thickness and for the thicker flyer plates (up to 13 mm) the stand-off distance is equal to the flyer plate thickness [32]. Other researchers concluded that to establish a stable wavy bond zone needs to use a stand-off distance between 1.0 and 1.5 times the flyer plate thickness [23, 36]. The stand-off distance is a very important parameter of explosion welding because of the welded joint quality, mostly the cold working hardening phenomenon [30].

Table 3 The velocity of longitudinal waves in some solid media at room temperature [40]

Metal kind	Al rolled	Cu rolled	Steinless steel	Steel (C = 1%)	Ni	Sn rolled
Velocity of the sound (m/s)	6420	4760	5790	5940	6040	3320

- Collision angle: This is a small angle between the flyer plate and the parent plate (Fig. 2a) [26, 31–33]. Several researchers define α generally 5°, but it can be a maximum of 25° [19, 31, 36]; it can find suggestions as a function of the metal kind, e.g. 2.5° for steel/steel, 4° for nickel/steel, and 7° for titanium/steel [38]. At the same time, from purely geometrical considerations, the following relationship can be established between the impact velocity (v_p), the impact angle (α) and the collision point velocity (v_c) (2) [41, 42]:

$$v_p = 2 \cdot v_c \cdot \sin\left(\frac{\alpha}{2}\right) \tag{2}$$

A similar approach to determining the lower boundary of welding has been proposed by Deribas and Zakharenko (3) [42, 43]:

$$\alpha = k_1 \cdot \sqrt{\frac{HV}{\rho \cdot v_c^2}} \tag{3}$$

where HV is the Vickers hardness, the impact angle α is obtained in radians, and ρ is the density of the material. The constant k_1 is in the range of 0.6 to 1.2 and depends on the degree of the surface preparation.

- Collision velocity: The speed of the collision (v_c) is limited to establish a suitable welded joint. In the case of several materials, it needs to be less than the sound velocity in the materials [18, 23, 38].

4 Explosive Welding Parameters for Parallel Setup

For determining the parameters of the parallel explosive welding process it needs to calculate the volume of the explosive material and the setup parameters. During the explosive welding, the flyer plate and the parent plate impact velocity at the point of the collision are v_c. It can determine the pressure impulse (J) acting on a unit area (4) [18, 23].

$$J = \int_0^t P \cdot \mathrm{dt} \tag{4}$$

where:

P is the pressure of the detonation (Pa)
T pressure effect duration (s).

The P pressure depends on the explosive material parameters, and the pressure duration depends on the explosive material volume per unit area. It can see that the pressure duration increases by way of the explosive material volume increasing per unit area [23, 44, 45]. The pressure impulse can be determined from the explosive material volume per unit area and the density of the explosive material ρ_e (g/cm^3), when it needs to use a k (–) explosion ratio. It can calculate the pressure impulse with the next Eq. (5).

$$J = k \cdot T \cdot \rho_e \cdot t \tag{5}$$

M_f (g/cm^2) the flyer plate mass per unit area can be determined with the Eq. (6):

$$M_f = \rho_f \cdot l_f \tag{6}$$

where:

ρ_f density of the flyer plate (g/cm^3)
l_f thickness of the flyer plate (cm).

On the base of the literature, it can be assumed that the collision speed is proportional to the ratio of the mass of the explosive to the cladding plate [18, 23, 44, 45]. Figure 3 shows the ratio between the collision velocity v_c and the Q (ratio of the explosive material) M_e (mass per unit area) and the flyer plate M_f (mass per unit area) [23] (7).

$$Q = \frac{M_e}{M_f}(-) \tag{7}$$

The ratio Q can be between $0.5 < Q < 1.5$. Also needs to determine the stand-off distance l_s, which can be on the base of the literature between $0.5 \cdot l_f < l_s < 1.6 \cdot l_f$, where l_f is the thickness of the flyer plate [23].

The volume of the explosive is calculated from the explosive mass M_e and the density of the explosive material ρ as a function of the Q ratio (8):

$$M_e = Q \cdot M_f \tag{8}$$

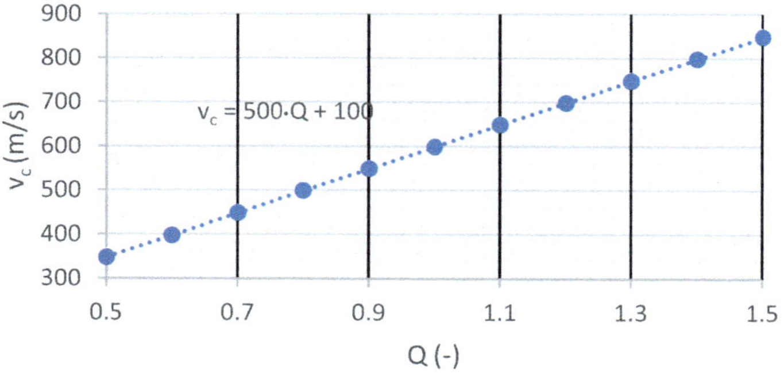

Fig. 3 The collision velocity as a function of the Q ratio [23]

If Q = 1, the thickness l_e (cm) of the explosive material is calculated with the explosive material density ρ_e and with the flyer plate thickness lf and density ρ_f, from the Eq. (9):

$$M_e = M_f \tag{9}$$

Replace the M_f from the Eq. (10) and

$$M_e = \rho_f \cdot l_f \tag{10}$$

The mass of the explosive material (mass per unit area) is also true (11):

$$M_e = \rho_e \cdot l_e \tag{11}$$

The thickness of the explosive material is calculated by Eq. (12):

$$l_e = \frac{\rho_f \cdot l_f}{\rho_e} \tag{12}$$

5 Explosive Welding Parameters for Oblique Setup

The parameter determination is the same as in the case of the parallel setup; only because of the oblique setup does it need to determine the collision angle. It is a small angle α, generally between 2.5° and 25° (Fig. 2a). During the welding process, the flyer plate and the parent plate collide and establish a different jet angle β. Based on the literature, the welding joint occurs if the collision velocity v_c is between 1500 and 7000 m/s and the collision angle is between 2.5° and 31° [25, 26, 38, 46]. The collision velocity is calculated from the Eq. (13) [25, 26, 38, 46].

$$v_c = \frac{\beta}{10} + 5.5 \text{ (m/s)} \tag{13}$$

The thickness of the explosive material is calculated by Eq. (12).

6 The Final Part of the Process

In the case of the explosive welding process, it can use several different metals (with different mechanical and chemical properties) and numerous explosive materials. The process parameters are determined based on the theoretical equations and the empirical laws. After the welding process, it needs to cut the beginning and the ending part of the product to earn uniform geometry. The testing of the welded joint is done by destructive and non-destructive testing. Useful non-destructive tests are visual inspection, ultrasonic testing, and penetration testing.

Some metal combinations made by explosive welding are summarized in Table 4. It can be seen that several metal combinations are available with the explosive welding process. Destructive tests for evaluating the explosive welded joints are the hardness test, tensile test, bending test and macroscopic and microscopic tests. The cross-section of the welded joint is typical, it can detect the explosion-established pressure wave resulting in plastic deformation (Fig. 4).

7 Results and Discussion

Explosive welding is a useful process for similar or dissimilar joining. Several research results are available in this process. Researchers use different explosive materials for experiments, so it is difficult to create a standard technology for explosion welding. In this work, the base rules for the explosive welding technology are introduced. The most important calculation methods for determining the explosive welding setup parameters are introduced. In the case of this process, the parameter determination needs to use some empirical laws based on the experimental practice. The explosives must be handled and stored professionally and safely. Even if the explosive materials are safe the detonator that initiates the explosion is dangerous. The dissimilar joining by way of this process is cost-effective in the case of a unique process or an unusual common geometry.

Table 4 Metal combinations by explosive welding (with references)

	Flyer plate materials										
Parent plate materials	Ta	Mg	Au	Ag	Nb	Ti	Al	Cu	Ni	Mild steel	Stainless steel
Mild steel				[34, 47]		[31] 37 39]	[31, 32]	[31, 38, 39, 30, 25]	[38]	[19]	[31, 33, 34, 48, 25]
Stainless steel						[27, 35]	[49, 36]	[19, 30]	[38]		[38]
Cu	[34, 47, 50–52]				[53, 52]	[34, 51]	[54, 28, 55, 25]	[34, 51]	[53]	[19, 25]	[30]
Al	[34, 47, 50]	[53, 56]				[29, 57, 58]					
Ti							[47]				[27, 35, 58]
Nb								[34]			
Ni			[59]								
Pt						[53]					

Fig. 4 Cross section of an explosive welded copper, stainless steel joint (author)

References

1. Valberg H (2010) Formability and workability. In: Applied metal forming: including FEM analysis. Cambridge University Press, Cambridge, United Kingdom, pp 127–138. https://doi.org/10.1017/CBO9780511801907.011
2. Askeland DR (2010) The science and engineering of materials, 6th edn. Cengage Learning. Inc., Stamford, USA, pp 198–235
3. Hansen N, Jensen DJ (2014) Development of microstructure in FCC metals during cold work. Philos Trans R Soc A 357:1447–1469. https://doi.org/10.1098/rsta.1999.0384
4. Chiang J et al (2011) Effect of microstructure on retained austenite stability and work hardening of TRIP steels. Mater Sci Eng, A 528:4516–4521. https://doi.org/10.1016/j.msea.2011.02.032
5. Dománková M, Kocsisová E, Pinke P, Slatkovsky I (2013) Effect of deformation on sensitisation process in austenitic stainless steel AISI 316. Mater Sci Technol 13:1–10. ISSN 1335-9053
6. MSZ EN ISO 4063:2016
7. Wenya L, Vivek P (2022) Solid State Welding for Fabricating Metallic Parts and Structures. Encycl Mater Met Alloys 4:246–259. https://doi.org/10.1016/B978-0-12-819726-4.00012-0
8. Vural M (2014) Welding processes and technologies. Comprehensive Mater Process 6:3–48. https://doi.org/10.1016/b978-0-08-096532-1.00603-8
9. Gao CY, Zhang LC (2010) A constitutive model for dynamic plasticity of FCC metals. Mater Sci Eng, A 527:3138. https://doi.org/10.1016/j.msea.2010.01.083
10. Dománková M, Bártová K, Slatkovsky I, Pinke P (2016) High nitrogen austenitic stainless steel precipitation during isothermal annealing, Mater Eng - Materiálové inžinierstvo 23:84–89. ISSN 1338-6174 (online version)
11. Rácz P, Göbl N, Horváth D, Mamalis AG (2013) Aspects of electrodynamic forming processes. Mater Sci Forum 767:126–131. https://doi.org/10.4028/www.scientific.net/msf.767.126
12. Figuli L, Bedon C, Zvaková Z, Jangl Š, Kavický V (2017) Dynamic analysis of a blast loaded steel structure. Procedia Eng 199:2463–2469. https://doi.org/10.1016/j.proeng.2017.09.388
13. Figuli L, Papán D (2014) Single degree of freedom analysis of steel beams under blast loading. Appl Mech Mater 617:92–95. https://doi.org/10.4028/www.scientific.net/amm.617.92
14. Mayer G (1992) New directions in research on dynamic deformation of materials. in shock-wave and high-strain-rate phenomena in materials, 1st ed. Marcel Dekker Inc., New York, USA, pp 33–48
15. Kovacs-Coskun T (2016) Explosive surface hardening of austenitic stainless steel. IOP Conf Ser Mater Sci Eng 123:012010. https://doi.org/10.1088/1757-899x/123/1/012010

16. Staudhammer KP, Frantz CE, Hecker SS (1981) Shock waves and high strain rate phenomena in metals. Plenum, New York, USA, pp 91-112
17. Daruka N, Kovács Z (2013) IEDD—improvised explosive device disposal. In: International conference on military technologies—ICMT 2013, pp 383–390
18. Lukács L, Szalay A, Zádor I (2012) Explosive forming and aerospace. Repüléstudományi Közlemények XXIV(2):431–446
19. Bahrani AS, Black TJ, Crossland B (1967) The mechanics of wave formation in explosive welding. Proc R Soc A Math Phys Eng Sci 296(1445):123–136. https://doi.org/10.1098/rspa.1967.0010
20. Mamalis AG, Szalay A, Vaxevanidis NM, Manolakos DE (1998) Fabrication of bimetallic rods by explosive cladding and warm extrusion. J Mater Process Technol 83(1–3):48–53. https://doi.org/10.1016/s0924-0136(98)00042-9
21. Figuli L, Kavicky V, Jangl S, Zvakova Z (2018) Comparison of the efficacy of homemade and industrially made Anfo explosives as an improvised explosive device charge. Communications 20(2):23–27
22. Ivančo M, Erdélyiová R, Figuli L (2019) Simulation of detonation and blast waves propagation. Transp Res Procedia 40:1356–1363. https://doi.org/10.1016/j.trpro.2019.07.188
23. Lukács L, Rácz P (2013) High energy metal working, 1st edn. Nemzeti fejlesztési ügynökség, Budapest, Hungary, pp 115–202
24. Lucia F, Jangl S, Zvakova Z, Kavicky V (2016) Experimental measurements of ANFO explosives. Explosive Ordnance Disposal Centre of Excellence, NATO (2016). www.eodcoe.org (17 July 2023)
25. Ghomi TM (2009) Impact wave process modeling and optimization in high energy rate explosive welding. Mälardalen University Press Licentiate Theses No. 106 ISSN 1651-9256
26. Sherpa BB, Rani R (2021) Explosive welding process to clad materials with dissimilar metallurgical properties. Mater Flow Anal, 1–14. https://doi.org/10.5772/intechopen.94448
27. Kahraman N, Gülenç B, Findik F (2005) Joining of titanium/stainless steel by explosive welding and effect on the interface. J Mater Process Technol 169(2):127–133. https://doi.org/10.1016/j.jmatprotec.2005.06.045
28. Mamalis AG, Szalay A, Vaxevanidis NM, Manolakos DE (1998) Fabrication of bimetallic rods by explosive cladding and warm extrusion. J Mater Process Technol 83:48–53
29. Fang Z, Shi C, Shi H, Sun Z (2019) Influence of explosive ratio on morphological and structural properties of Ti/Al clads. Metals 9(2):119. https://doi.org/10.3390/met9020119
30. Durgutlu A, Okuyucu H, Gulenc B (2008) Investigation of the effect of the stand-off distance on interface characteristics of explosively welded copper and stainless steel. Mater Des 29(7):1480–1484. https://doi.org/10.1016/j.matdes.2007.07.012
31. Crossland B, Bahrani AS (1968) Fundamentals, of explosive welding. Contemp Phys 9(1):71–87. https://doi.org/10.1080/00107516808204394
32. Sherpa BB, Kumar PD, Barta U, Upadhyay A, Agarwal A (2019) Study of the explosive welding process and applications. In: Advances in applied physical and chemical sciences-a sustainable approach, pp 33–39. ISBN 978-93-83083-72-5
33. Buijs NW (2010) Explosive welding of metals in a vacuum environment, pp 1–4. www.stainless-steel-world.net (10 July 2023)
34. Greenberg BA, Ivanoc MA, Kuzmin SV, Lísak VI (2020) Explosive welding. In: Process and structures, 1st edn. Taylor and Francis, Boca Raton, USA, pp 14–190. ISBN 13: 978-0-367-35578-4
35. Findik F (2011) Recent developments in explosive welding. Mater Des 32(3):1081–1093. https://doi.org/10.1016/j.matdes.2010.10.017
36. ASM Handbook 1992 vol 6, Welding brazing and soldering. ASM International, USA, pp 525–537
37. Mousavi AAA, Al-Hassani STS (2005) Numerical and experimental studies of the mechanism of the wavy interface formations in explosive/impact welding. J Mech Phys Solids 53:2501–2528

38. Blazynski TZ (1983) Explosive welding, forming and compaction, 1st edn. Applied Science Publisher, England, Essex, pp 23–268
39. Raghukandan K (2003) Analysis of the explosive cladding of Cu–low carbon steel plates. J Mater Process Technol 139(1–3):573–577. https://doi.org/10.1016/s0924-0136(03)00539-9
40. Lide DR (ed) Handbook of chemistry and physics, 87th edn. Taylor and Francis Group LLC (2007)
41. Wittman RH (1973) The influence of collision parameters of the strength and microstructure of an explosion welded aluminium alloy. In: Proceedings of the 2nd international symposium on use of an explosive energy in manufacturing metallic materials, pp 153–168
42. Bataev IA, Tanaka S, Zhou Q, Lazurenko DV, Jorge Junior AM, Bataev AA, Hokamoto K, Mori A, Chen P (2019) Towards better understanding of explosive welding by combination of numerical simulation and experimental study. Mater Des 169. ISSN 0264-1275. https://doi.org/10.1016/j.matdes.2019.107649
43. Deribas AA, Zakharenko ID (1974) Surface effects with oblique collisions between metallic plates. Combust Explosion Shock Waves 10:358–367
44. Hirsch E (1995) On the inconsistency of the asymmetric-sandwich gurney formula when used to model thin plate propulsion. Propellants, Explos, Pyrotech 20:178–181
45. Catovic A (2020) An overview of Gurney method for estimating the initial velocity of fragments for high explosive munition. Defence Secur Stud 1(1):16–28
46. Abrahamson GR (1961) Permanent periodic surface deformations due to a traveling jet. J Appl Mech 28(4):519. https://doi.org/10.1115/1.3641777
47. Greenberg BA, Ivanov MA, Inozemtsev AV, Patselov AM, Pushkin MS, Vlasova AM (2015) Microheterogeneous structure of local melted zones in the process of explosive welding. Metall and Mater Trans A 46(8):3569–3580. https://doi.org/10.1007/s11661-015-2968-3
48. Kacar R, Acarer M (2004) An investigation on the explosive cladding of 316L stainless steel-din-P355GH steel. J Mater Process Technol 152(1):91–96. https://doi.org/10.1016/j.jmatprotec.2004.03.012
49. Carvalho GHFSL, Galvão I, Mendes R, M. Leal R, Loureiro A (2020) Aluminum-to-steel cladding by explosive welding. Metals 10(8):1062. https://doi.org/10.3390/met10081062
50. Greenberg BA, Ivanov MA, Inozemtsev AV, Kuz'min SV, Lysak VI, Pushkin MS (2015) Evolution of interface relief during explosive welding: transitions from splashes to waves. Bull Russ Acad Sci Phys 79(9):1118–1121. https://doi.org/10.3103/s1062873815090075
51. Pushkin MS, Greenberg BA, Ivanov MA, Inozemtsev AV, Patselov AM, Besshaposhnikov YP (2020) Microstructure of joints Cu–Ta, Cu–Ti, Cu–Cu, produced by means of explosive welding: fractal description of interface relief. Compos Interf, 1–14. https://doi.org/10.1080/09276440.2020.173688
52. Parchuri PK, Kotegawa S, Yamamoto H, Ito K, Mori A, Hokamoto K (2019) Benefits of intermediate-layer formation at the interface of Nb/Cu and Ta/Cu explosive clads. Mater Des 166:107610. ISSN 0264-1275. https://doi.org/10.1016/j.matdes.2019.107610
53. Chadwick MD, Jackson PW (1983) Explosive welding in planar geometries. In: Explosive welding, forming and compaction, pp 219–287. https://doi.org/10.1007/978-94-011-9751-9_7
54. Lokaj J, Sahul M, Sahul M, Caplovic L (2022) Influence of heat treatment on the properties of explosively welded Cu-Al joints. High Energy Mater 14:4–13. https://doi.org/10.22211/matwys/0228
55. Kaya Y (2018) Investigation of copper-aluminium composite materials produced by explosive welding. Metals 8(10):780. https://doi.org/10.3390/met8100780
56. Mroz S, Stradomski G, Dyja H et al (2015) Using the explosive cladding method for production of Mg-Al bimetallic bars. Archiv Civ Mech Eng 15:317–323. https://doi.org/10.1016/j.acme.2014.12.003
57. Li Y, Liu C, Yu H, Zhao F, Wu Z (2017) Numerical simulation of Ti/Al bimetal composite fabricated by explosive welding. Metals 7(10):407. https://doi.org/10.3390/met7100407

58. Gullino A, Matteis P, D'Aiuto F (2019) Review of aluminum-to-steel welding technologies for car-body applications. Metals 9(3):315. https://doi.org/10.3390/met9030315
59. Bechtold G, Michael I, Prümmer R (1977) Explosive cladding with gold. Gold Bull 10:34–37. https://doi.org/10.1007/BF03215424

The Relationship Between Situational Crime Prevention and Road Safety Defending Against Environmental Threats

József Krenner

Abstract The aim of the research is to examine the relationship between situational crime prevention and road safety, with a specific focus on environmental factors and pedestrian safety. The research endeavors to develop a model that aids in identifying risk factors and their underlying causes within the traffic environment. The model analyzes the built environment, accident statistics, and legal regulations, with particular emphasis on pedestrians. Current police methods, campaigns, and responses to pedestrian safety issues are also analyzed. According to the WHO report, millions of individuals lose their lives in road accidents annually, predominantly pedestrians, cyclists, and motorcyclists. Environmental criminology and situational crime prevention offer approaches that analyze the interplay between human behavior and environmental factors, thereby enhancing road safety. The situational prevention method can be effectively applied to enhance road safety, considering the combined influence of various factors in the occurrence of traffic accidents.

Keywords Situational crime prevention · Pedestrian road safety · Environmental criminology

1 Introduction

According to the WHO Global Status Report 2022 [1], 1.35 million people are killed in road traffic crashes every year, and the number of seriously injured is estimated to reach 50 million. It is the leading cause of death in the 5–29 age group and ranks eighth in the overall population. More than half of the victims are pedestrians, cyclists or motorcyclists.

Pedestrians are a particularly vulnerable group as they are the most defenceless in the event of a collision. The above-average vulnerability of pedestrians is due to

J. Krenner (✉)
Doctoral School of State and Law, University of Győr, Győr, Hungary
e-mail: jozsef.krenner@gmail.com

T. A. Kovács et al. (eds.), *The Impact of the Energy Dependency on Critical Infrastructure Protection*, Advanced Sciences and Technologies for Security Applications, https://doi.org/10.1007/978-3-031-78544-3_36

the fact that their bodies are in contact with the vehicle in virtually all cases of a road accident, and the subsequent fall onto the road can have serious consequences [2]. Of the safety issues in pedestrian transport, the leading risk factors are almost invariably related to environmental and situational elements. Drawing on the methodology of criminology, this research explores how situational crime prevention can provide solutions to improve safety.

2 The Conceptual Basis of Situational Crime Prevention

The basic concept of situational crime prevention is the same as that of accident prevention. Most importantly, human behaviour should be able to recognise risk situations in time and be defensive [3]. It is statistically proven that the people who cause accidents are not basically "criminals", they do not participate in road traffic with the intention of harming others by breaking the rules [4]. However, the consequence of non-compliance with traffic rules is that accident situations arise in which the tripartite traffic system (human-vehicle-environment) must be examined in a complex way to determine what led to the dangerous situation:

- human behaviour (non-compliance with rules),
- failure to comply with a given rule,
- or a characteristic of the environment that creates a hazardous situation.

Situational crime prevention is nothing else but a set of efforts to reduce the number of occasions that lead to crime [5]. From this perspective, it is necessary to investigate what circumstances in transport lead to dangerous situations that lead to crime, in the case of this research, traffic crime. In this context, from the perspective of the person who caused the situation and the person who is the victim of the situation, how the situation could have been prevented, taking into account the environmental factors.

From the very beginning, the science of criminology has examined the role of the victim in the occurrence of crime. This research has made it possible to make crime prevention an integral part of the process of telling potential victims what they should and should not do to avoid becoming victims. This can be seen as much in the context of transport, accidents in general, as in the proposals and measures to prevent crimes against property or life [3].

Nor should it be overlooked that the commission of crimes (in this case, non-compliance with traffic rules) is not the insidious villainy of despicable villains, but a behavioural error of individuals like ordinary people [5], who in the course of their routine activities exploit certain opportunities and seek to take advantage of non-compliance with certain rules. However, it is utopian to assume that all drivers will obey all traffic rules in all cases, leaving only better protection of persons and property, which is what science believes can be achieved by applying so-called situational crime prevention [6].

3 The Relationship Between Situational Crime Prevention and Road Safety

The complex system of transport involves many risks, is part of everyday life and has a very diverse set of stakeholders. As a result, some groups are more vulnerable because they are defenceless against other participants. These include pedestrians, cyclists and motorcyclists. Pedestrians are a particularly vulnerable group, as they are the most vulnerable to other road users and have the most varied knowledge of traffic rules (as anyone can be on foot).

The dynamic nature of traffic creates situations which can result in an unexpected event, an accident or even a proximity offence. The situational crime prevention approach is based on the recognition that crime does not occur in equal proportions, but happens more frequently in certain places, at certain times and in certain circumstances. This method can be used to make traffic safer.

The hypothesis of the research is that, using the situational crime prevention method, a model can be constructed to investigate whether a given traffic-relevant environment (road crossing, pedestrian crossing, roundabout, road section inside and outside a residential area) harbours a risk factor that can be reduced or even eliminated by changing the environment. The model would take into account the built environment and accident statistics in a complex way, as well as the legal framework, looking at the legal standards for road users and traffic regulation, with a particular focus on pedestrians.

4 Possible Responses from Criminology and/or Situational Crime Prevention to Enhance Pedestrian Safety

Possible responses from environmental criminology, which will be analysed in detail in this research, include:

- The relationship between environmental criminology and road safety focuses on the effects of human activities that affect the environment and road safety;
- To improve road safety, it is important to analyse human behaviour and environmental factors such as infrastructure, cars and transport policies;
- Human behaviour plays a major role in road safety. An environmental criminology approach can help to understand behaviours such as speeding, drink-driving or inattention that increase the risk of traffic accidents. Such behavioural patterns require consideration of human factors aspects to solve road safety problems;
- Possible responses to situational crime prevention;
- Situational crime prevention can be effectively applied to improve road safety, as traffic accidents are often caused by a combination of factors, such as weather conditions, vehicle condition, driver behaviour, deficiencies in transport infrastructure, etc.

5 The Core Values of Research

The research hypothesis is that there is a link between road safety and situational crime prevention. That is, in making transport safer, especially for pedestrians, situational crime prevention can be a powerful tool to help protect against environmental threats. Furthermore, traffic accidents and crimes can be linked to situational circumstances in addition to, or even separately from, the human factor. At international level, it is recognised that a complex transport system must be viewed in a complex way and that all factors—human, environmental or local—which may influence road safety must be taken into account at the planning stage. Finland is a world leader in this field, with a very low fatality rate thanks to Helsinki's transport concept.

6 The Finnish Model

Finnish traffic is extremely safe, with only 219 people killed on Finnish roads in 2021, or four deaths per 100,000 inhabitants. The number of fatalities fell by 50% between 2001 and 2019, when Helsinki made international news by not having a single road accident resulting in a pedestrian or cyclist fatality for a full year (2019) (two in 2021, compared to 22 in 1990).

Like its neighbours Norway and Sweden, which are the birthplaces of the Vision Zero road safety movement, Finland's roads are significantly safer than they have been for several decades [7].

One of the most significant improvements has been the lowering of speed limits: from 2019, the maximum speed limit in Helsinki basically be 30 km/h, 40 km/h in suburban areas and 50 km/h on designated main roads only. It is worth noting that when the speed limit is lowered from 40 to 30 km/h, the risk of fatality in a collision between a vehicle and a pedestrian is halved and reduced to just under 5% [8].

Thanks to the above, after 2019, when no pedestrian or cyclist was killed in a road traffic accident in Helsinki [9], the number of fatalities remains low, with only eighteen fatal road accidents between 2020 and 2022 [10], which are convincing proof that the methods used by the Finns are effective and that the Vision Zero project is working.

It can be said that the Finnish model essentially adapts traffic rules to the conditions and the environment. Helsinki has built up a substantial public transport infrastructure to reduce car traffic [11], is open to new innovative solutions (e.g. electric scooters), but regulates their use as soon as they are introduced, making the limits clear (where and under what conditions they can be used).

7 Vision Zero

Vision Zero [12] is a global movement to improve road safety and ultimately to ensure that no one is killed or seriously injured just for being on the road. It covers all road users: drivers, cyclists, pedestrians.

The Vision Zero Network [13] is a non-profit campaign that brings together businesses, communities and governments at all levels to advocate for better rules/policies and road conditions to change the misconception that fatalities are inevitable. The Vision Zero campaign believes that all people have the right to travel safely in their communities, and that system designers and policy makers have a shared responsibility to ensure safe travel conditions. The movement was launched by the Swedish Parliament in 1997, and since then all transport investments in the country have been implemented according to Vision Zero standards. The target set at that time of zero deaths on Swedish roads by 2020 was not achieved, but there has been a 74% reduction in fatalities (1990: 772, 2020: 204). The method has been gradually adopted, first by neighbouring countries (Finland, Sweden), then by the European Union, the United States, Canada and India. In addition, some countries have adopted the ideas but have launched their own campaign under a different name (e.g. Road to Zero in New Zealand).

8 The Project

For too long, humanity has considered fatal and serious injury road accidents to be an unavoidable side-effect of modern life. Although they are called 'accidents', the reality is that these tragedies can be prevented by a proactive, preventative approach that puts road safety first as a public health issue. The loss of life due to crashes comes at a tragic cost that, in addition to personal loss, also deeply affects the community, including: personal economic costs and emotional trauma for those affected; and significant expenditure of taxpayer dollars for emergency response and long-term health care costs.

9 A New Vision for Safety

Vision Zero is a significant departure from the status quo in two main ways. Vision Zero recognises that people sometimes make mistakes, so the road system and associated policies must be designed to ensure that these inevitable mistakes do not lead to serious injuries or deaths. This means that system designers and policy makers are expected to improve the environment, policies (such as speed control) and other related systems to reduce the severity of accidents. Vision Zero is a multi-disciplinary approach that brings together diverse and necessary stakeholders to address this complex problem. In the past, there has been no meaningful cross-disciplinary collaboration between local transportation planners and engineers, policy makers, and public health professionals. Vision Zero recognises that many factors contribute to safe mobility—including roadway design, speed, road user behaviour, technology and policies—and sets clear targets to reduce fatalities and serious injuries to zero. Vision Zero, according to its creators, is not a slogan or a marketing ploy, not even a programme. Vision Zero is a fundamentally different approach to road safety.

Communities that want to succeed with this project must accept that the way they have managed transport is not enough, and that radical change is needed, as follows.

- Leadership, collaboration, and accountability must be established and maintained, especially among a diverse group of stakeholders that includes proximity professionals, policy makers, public health officials, police, and community members.
- Collect, analyze and use data to understand trends and the potential disproportionate impact of traffic fatalities on certain populations.
- Focus on equity and community engagement.
- Setting the maximum speed limit to a safe level.
- Setting a timetable to reduce road deaths and serious injuries to zero, leading to prioritisation and accountability, and ensuring transparency of progress and challenges.

10 Summary

The theoretical foundations of traffic safety research are built around environmental criminology, routine activity and crime theory [14], situational crime prevention, community involvement and crime prevention strategies. These theories and approaches help us to understand how to improve road safety, particularly in the area of pedestrian protection. This research aims to combine these theoretical foundations in order to develop effective strategies for improving road safety. The research will take into account positive experiences from international practice, in this case Finland, and the alternative approach offered by the Vision Zero project.

References

1. Road traffic mortality. https://www.who.int/data/gho/data/themes/topics/topic-details/GHO/road-traffic-mortality. Last accessed 15 Jan 2024
2. ORFK-OBB. https://kreszvaltozas.hu/baleset_megelozes/gyalogos-balesetek-alakulasa/. Last accessed 16 Jan 2024
3. Irk F (2015) Viktimológia, közösségi és szituációs bűnmegelőzés. Belügyi Szemle 63(4):5–20
4. Lange R (1970) Das Rätsel Kriminalität – Was wissen wir vom Verbrechen? Metzner, Frankfurt am Main / Berlin

5. Gönczöl K, Kerezsi K, Korinek L, Lévay M (2016) Kriminológia – Szakkriminológia. Wolters Kluwer Kft., Budapest
6. Viski L (1973) Kriminálszociológia és közlekedési bűncselekmények. Jogtudományi Közlöny 28(7–8):365–374
7. Statistics Finland. https://stat.fi/til/ton/2021/12/ton_2021_12_2022-01-20_tie_001_en.html. Last accessed 16 Jan 2024
8. The hit rate graph. https://extrat.liikenneturva.fi/tormaysnopeus/en/gfx/deathprobability_en.svg. Last accessed 16 Jan 2024
9. Helsinki has no pedestrian fatalities in 2019. https://www.eltis.org/in-brief/news/helsinki-has-no-pedestrian-fatalities-2019. Last accessed 16 Jan 2024
10. Accident road map Finland. https://mobilityanalytics.ramboll.com/onn/poliisi/?force=1. Last accessed 16 Jan 2024
11. Helsinki traffic statistics. https://kaupunkitieto.hel.fi/fi/liikenne. Last accessed 16 Jan 2024
12. What is Vision Zero? https://visionzeronetwork.org/about/what-is-vision-zero/. Last accessed 16 Jan 2024
13. What is the Vision Zero Network? https://visionzeronetwork.org/about/vision-zero-network/. Last accessed 16 Jan 2024
14. Cohen L, Felson M (1979) Social change and crime rate trends: a routine activity approach. Am Sociol Rev 44(4):588–608

Mind the Gap: Introducing the "Generation Gap's Problem" in Cybersecurity

Szandra Anna Laczi and Valéria Póser

Abstract Following on from my earlier research, this article aims to raise awareness of the problem of digital parenthood and cybersecurity. It presents the problems caused by the rapid and dynamic digital growth and tries to identify the problem of digital natives/digital immigrants in relation to cybersecurity, and then describes the digital and generation gap between age groups and its implications for cybersecurity.

Keywords Digital patenting · Digital gap · Cybersecurity

1 Introduction

The present era is witnessing an unprecedented increase in digital development, bringing profound changes to the dynamics of everyday life and child-rearing. The pervasive integration of digital tools into different aspects of human life has brought about a paradigm shift, changing the way individuals go about their daily lives, with a huge impact on children's lives and health [1]. In the midst of these transformative developments, a particularly important issue arises: the generation gap. Due to the wide reach and rapid development of digital technologies, there is a disparity between the technological literacy and adoption of different age groups, resulting in a perceived generation gap.

The implications of the digital generation gap extend beyond mere technological proficiency, permeating fundamental aspects of interpersonal relationships, communication, and cultural understanding. The older generation often finds itself grappling with the swift pace of technological advancements, leading to a palpable disconnect with the digital fluency effortlessly exhibited by younger counterparts. Consequently, this intergenerational discord may hinder effective communication and mutual comprehension between parents and their digitally native children.

S. A. Laczi (✉) · V. Póser
Doctoral School of Applied Informatics and Applied Mathematics, Obuda University, Budapest, Hungary
e-mail: laczi.szandra@uni-obuda.hu

T. A. Kovács et al. (eds.), *The Impact of the Energy Dependency on Critical Infrastructure Protection*, Advanced Sciences and Technologies for Security Applications, https://doi.org/10.1007/978-3-031-78544-3_37

Moreover, the generation gap in the digital realm extends to the realm of parenting, with challenges arising in navigating the nuances of digital upbringing. Parents are confronted with the task of reconciling traditional parenting values with the novel challenges posed by the digital landscape, such as cyber threats, online content exposure, and the potential impact of prolonged screen time on child development [1, 2].

As the chasm widens between the digital experiences of different age cohorts, there is an imperative for society to address this burgeoning issue systematically. The following figure for 2022 shows (although it includes data for Poland) that almost all children aged 9–10 already have their own smartphone.

It is well observed in Fig. 1 that children start using digital devices (tablets, smartphones, computers) at an early age. The graph shows that 87.2% of children start using mobile phones before the age of 10 [3]. It should be the responsibility of the parent to control them and ensure safety at all times. To achieve this, it is necessary to map and understand the generation gap. Strategies must be devised to bridge the digital generation gap, fostering a harmonious coexistence between the technologically seasoned younger generation and their more digitally reserved counterparts. Only through a nuanced understanding of the multifaceted dimensions of this issue can society hope to navigate the intricacies of digital growth and its profound implications on child-raising and interpersonal relationships.

In this paper, we seek to answer the following questions:

(1) *What is the generation and digital gap and how does it relate to cybersecurity?*

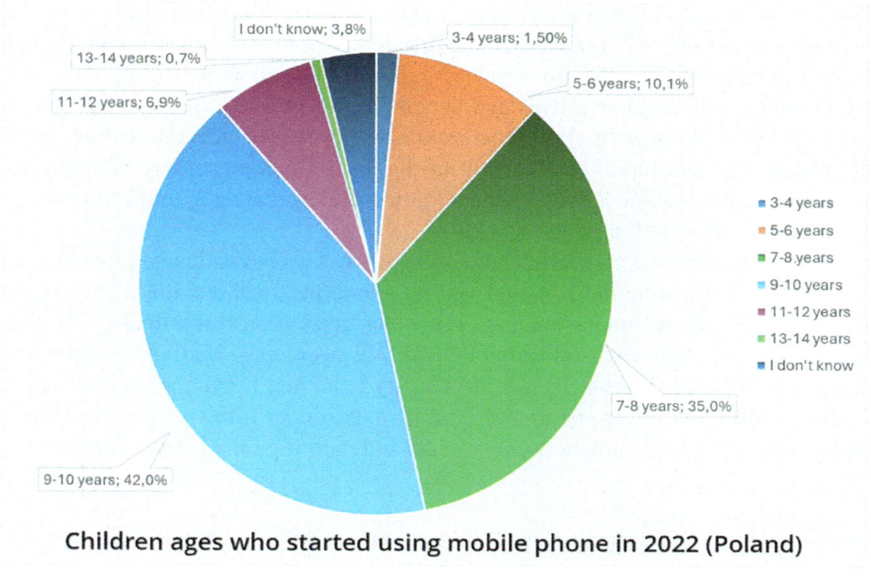

Fig. 1 Children started using mobile phone in Poland in 2022 [3]

(2) *What are the solution (if any) to narrow the gap?*

2 The Generation Gap

A. *Who are digital immigrants and digital natives?*

First, we need to define who are digital natives and who are digital immigrants. To do this, we will use Marc Prenksy's 2001 study, which introduces these definitions into the public consciousness [4]. We realize that this article may seem old, but its relevance has not diminished to this day, so it is perfectly suitable for setting up definitions. To understand the digital divide and its importance for cybersecurity, we need to understand the problem and how it relates to digital education, digital parenting and security awareness. For the sake of simplicity, we will quote Prensky's definitions verbatim, followed by a brief explanation.

> But the most useful designation I have found for them is Digital Natives. Our students today are all "native speakers" of the digital language of computers, video games and the Internet. [4]

What does this mean? It means that the generation growing up is a generation that learns the digital world as their mother tongue, grows up with digital achievements, and has a much higher adaptability. There is also an enormous body of literature on the challenges of the digital, as well as the fact that the well-being of modern people is highly dependent on the threats of the digital world around them [5]. This "mother tongue" often creates problematic situations in education, as members of generations Z (1995–2010) and Alpha (2011–2025) are already part of the digital age [6]. The highlight of the problem lies in the growing disparity between traditional educational approaches and the evolving digital experiences and preferences of Generations Z and Alpha. The metaphorical use of "mother tongue" emphasizes that these generations are inherently fluent in the language of the digital age due to their exposure and immersion in digital advancements from a young age. This proficiency and comfort with the digital realm create a mismatch with conventional educational methods that may not align with the technological context in which these individuals operate. As a result, there is a risk of educational challenges, hindering the optimal development and learning experiences of individuals within Generations Z and Alpha.

> Those of us who were not born into the digital world but have, at some later point in our lives, become fascinated by and adopted many or most aspects of the new technology are, and always will be compared to them, Digital Immigrants. [4]

Digital immigrants and digital natives represent two distinct groups in the contemporary digital landscape, particularly within educational contexts. Digital immigrants, often older individuals, did not grow up with ubiquitous digital technology. They have had to adapt to new forms of communication, information processing, and learning methodologies that are inherent in the digital age. This adaptation can be challenging, as it involves not only acquiring new skills but also a shift in mindset and approach. Prensky's work is often cited in academic discussions about the impact of digital technology on learning and communication styles. He argues that digital immigrants, having not grown up with digital technology, often retain a "pre-digital" mindset. This affects how they approach technology, often perceiving it as an add-on rather than an integral part of life and learning. In contrast, digital natives, who have grown up immersed in digital technology, see it as an integral part of their lives, influencing their approaches to communication, learning, and information processing [4].

B. *What is digital gap?*

Before moving on, it is important to clarify the difference between two further concepts. One is the "digital divide" and the other is the "generation gap". If we had to come up with a definition of what a digital divide really is, we would define it as the inequality between individuals, communities, geographical regions, depending on the information and communication technologies (ICTs) and access to them. This gap is influenced by a number of factors, such as [7–9]:

- economic differences
- geographical differences
- social and cultural differences
- educational differentiation.

The digital divide has many elements, a theme that is also addressed in Aissaoui's 2021 literature review [7]. The author identifies three levels of digital divide based on the studies reviewed, which are shown in Fig. 2.

Building on this, and taking into account other studies, we have tried to identify the key elements of the digital gap (or digital divide):

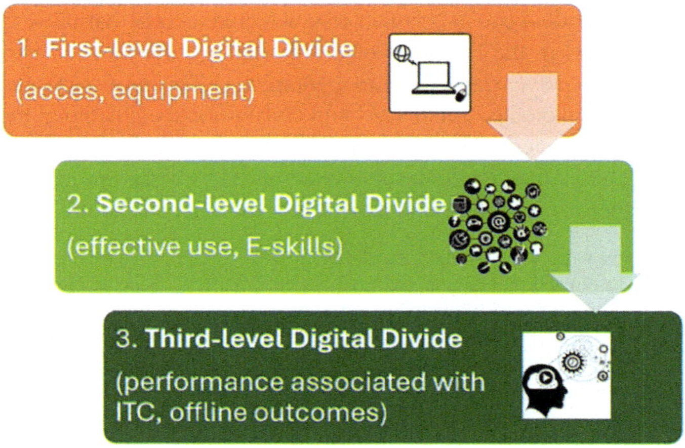

Fig. 2 Digital divide levels [7]

1. **Access to technology**: by this we mean access to good quality IT and digital tools, as well as access to the internet. In many cases, access to these is limited or non-existent in more backward regions. This is clearly shown in the figure above as the first level of digital divide.
2. **Digital literacy (i.e. usage)**: if the first level can be achieved, i.e. the necessary tools and technologies are available, then the availability of the skills to use them is also crucial. This includes the ability to navigate the internet and understand digital content (more on how this affects the understanding of digital threats later). This is the second level in the diagram.
3. **Content availability and language differences**: we would put this point in the second level, as a lot of content is only available in a given language, thus limiting understanding. Language barriers can further widen the digital divide.
4. **Economic factors**: we would assign this factor to level one, as expensive IT equipment or internet access may not be easily accessible to those on lower incomes, which also widens the gap.
5. **Socio-cultural factors**: the use of technology varies by region. This is further exacerbated by the presence of traditional gender roles and social differentiation (whether geographical or sociological).
6. **Mobilising digital resources**: this is the third level in the diagram, suggesting that in addition to access to and use of ICT, it is important to consider the level of skills users have, how they can make use of the technology available, how individuals and the economy can also use the opportunities available to grow. This category is a little more abstract, but an important part of the digital divide.

The above shows how significant and complex the digital divide is. In the digital age, it is extremely important to study and understand it because the problems caused by the digital divide affect our daily lives. Bridging the digital divide is a shared social responsibility that can be improved by providing access and education. In order to offer a new perspective, we have plotted the distribution of the levels in a separate figure, with the aim of providing a new perspective for understanding the above (Fig. 3).

C. *What is generation gap?*

We will also look at the problem of the generation gap, following the example of the previous ones. First, we will try to define what we call the generation gap as the differences in attitudes, values and perceptions between digital immigrants and digital natives. Why is there a generation gap? Is it simply caused by age? The differences are mainly due to the different experiences, social changes, technological advances and cultural changes that each generation experiences over the years. Why is it important to understand this? To understand young people better and to help young people understand themselves better, it is necessary to look closely at the generation gap [10].

Key elements of the gap are:

1. **Digital/technological literacy**: As we mentioned earlier, the young generations—especially Z and Alpha—are digital natives, they have grown up on digital

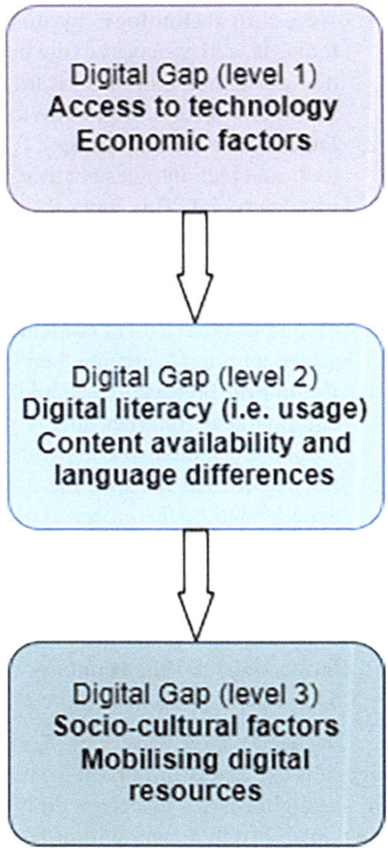

Fig. 3 Digital gap levels (*source* author)

technology, in Prensky's terminology we used earlier: it is their mother tongue. For them, there was no need to adapt to the digital space because they were born into it. Their adaptation to the constantly evolving digital space is much more flexible than that of digital immigrants.

2. **Communication styles** [11]: a recent study by Rathi and Kumar, 2023, highlights (although it mainly looks at different generations from a work perspective) the differences in communication between generations. The study also reveals that different generations prefer different modes of communication, with younger generations preferring written communication (text message, email, social media) and older generations preferring face-to-face interactions [12, 13]. A visualisation of this can be seen in the—very telling—graph (Fig. 4).

3. **Cultural and societal values:** every individual is influenced by different social, historical and cultural influences during his/her personal development, which fundamentally affect his/her values, attitudes towards different issues or even digital development. These values may differ from one individual to another,

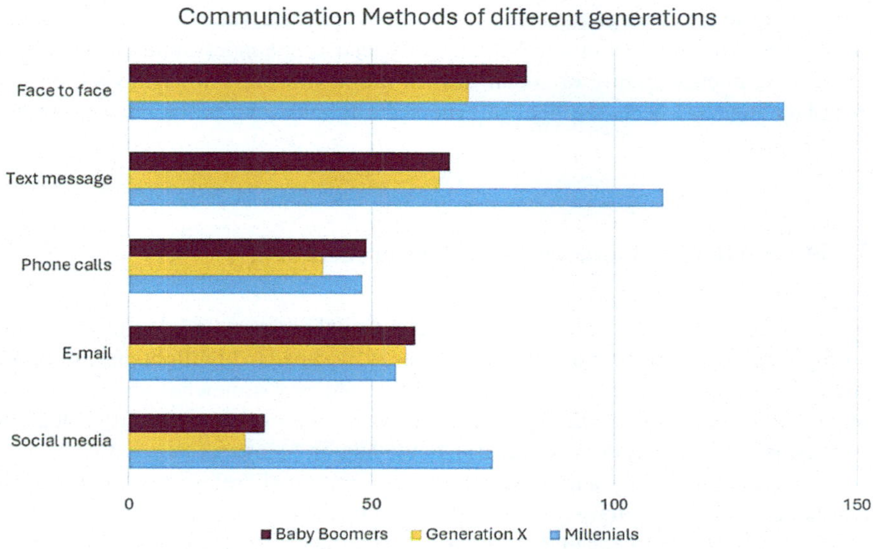

Fig. 4 Communication channel preferences [12]

from one generation to another, which can lead to a widening of the generation gap.

4. **Work attitudes and values:** as we have already referred to a very recent study by Rathi and Kumar [11], which looks at this very issue, we felt it was important to include this as an elementary one in the generation gap. Preferences for work-life balance among digital immigrants and digital natives are different, and even significant. Older generations value job stability, but younger generations are easily switched, seeking personal fulfilment and a better work-life balance. It is easy to see that this problem can also cause conflict between generations, widening the gap.

D. *How digital gap and generation gap connected and why is this a real problem?*

The digital divide and the generation gap are closely linked. We would highlight technological skills and adaptation in particular. From the foregoing, it is clear that the digital divide is significantly influenced by the generation gap, as the older generation (digital natives) adapt to digital tools and technologies in a variable way. Those who did not grow up in the presence of digital technology face new challenges in adapting to new digital platforms, which the younger generation (digital natives) can master more quickly and with greater adaptability. It is important to underline that it is not that digital natives know everything, but that they adapt more flexibly and faster. This lack of flexibility, which often characterises the older generation, can even lead to digital exclusion [14].

At their intersection are communication barriers, learning and working styles, and the personal and professional use of technology. They interact continuously and have a significant impact on many aspects of everyday life, such as work, education, family and social interactions. They can easily lead to misunderstandings which can result in weakened family or social ties. In the next section we will look at how this phenomenon affects the field of cybersecurity and what problems arise in this specific area.

3 How All This Connected to Cybersecurity?

To shed more light on the relationship between digital natives, digital immigrants and the digital and generation gap with cybersecurity, we need to highlight one important factor. This factor lies in awareness, which to date is a somewhat underestimated area of cybersecurity, even though one of the basic tenets of cybersecurity is that the weakest point of any system is necessarily the human being.

Let me go into this for a moment because it is extremely relevant to my research. This concept (which we have highlighted above in italics for emphasis) stems from the observation and experience that technical security measures can be as well organised and robust as they are when compromised by human error, inattention, gullibility or other behaviours. This can, does and will happen at many levels of cyber security. So how can we protect against it? By developing, shaping and educating awareness.

A. *Importance of awareness*

Digital awareness is essential, as it involves assessing and understanding the benefits and risks associated with digital tools and technologies. Digital awareness implies that the user is aware of the potential dangers of the digital space (phishing, fraud, cyberbullying). Digital awareness enables individuals and organisations to navigate more responsibly and safely in the digital space and world. Why is it so important to have the right level of security awareness? It's to ensure that we can properly exploit the benefits of digital technologies while minimising the risks.

Let's connect all this to what we have done before. Digital awareness is a significant cross-generational divide. Older generations, who did not grow up with the internet and digital technologies, lack basic knowledge and skills, while younger generations do. Nevertheless, several studies have found that older generations are more vulnerable to cyber-families and attacks [15]. Surprisingly, however, Sarno et al. found that older people are more suspicious, for example, they mark suspicious content as spam more often, and thus performed better on surveys than their younger counterparts [16]. Overall, therefore, it can be concluded that although the knowledge of older age groups may be lower, young people are more prone to underestimate potential risks and their consequences. An excellent example of this is the phenomenon of oversharing (excessive sharing of information on different social media platforms), which has also been investigated in several studies [17].

How can this kind of awareness be acquired? Through continuous education and awareness-raising activities. We have basically divided this into three levels:

1. **Basic security awareness education for all ages**: including best practices, basic digital competencies
2. **Promoting intergenerational dialogue**: We attach particular importance to this point, as in many cases it is the lack of communication that causes human vulnerability. In the next phases of my research, we will focus on this, looking at the responsibility of parents to help their children better understand the consequences of their online activities, including, of course, raising awareness among the older generation.
3. **Institutional education**: the state, institutions and organisations should work together to support the development of safety awareness.

My diagram shows how the areas discussed above relate to each other (Fig. 5).

B. *Connection to parenting*

While it may have seemed to suggest that the younger generation is more digitally literate, we have argued that they have only gained an advantage in their use and understanding of technology [16, 17]. This suggests that the link between parental engagement and the challenge of the generation and digital divide is significant. It requires parents to be aware of the risks and the severity of their potential consequences (e.g. cyberbullying can lead to suicide [18]). Parents need to educate themselves to recognise the online (cybersecurity) threats to their children and to be aware of possible solutions.

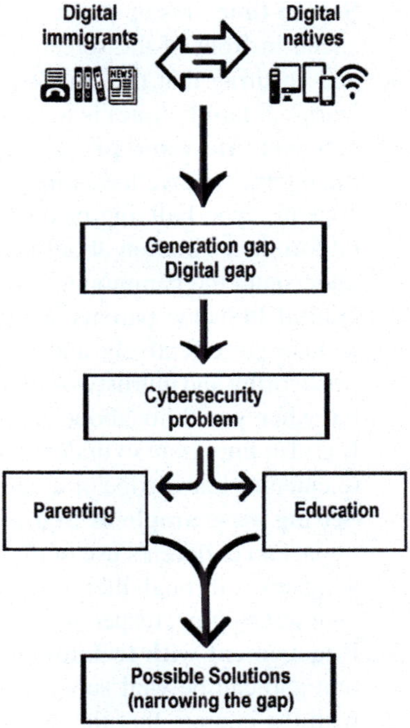

Fig. 5 Connection illustration (*source* author)

We believe that one of the most important pillars of parenting is communication. Strong guidance and awareness, proper education could go a long way in keeping the growing generations safe online. In the next phases of my research, we will assess parents' cybersecurity knowledge, creating metrics to quantify the level of knowledge parents have. Effective parenting involves bridging the generation gap, fostering open communication and setting healthy, appropriate boundaries. This dual focus can help create a safer environment for children and equip them with the skills they need to navigate the digital world responsibly.

C. *Defining digital parenting*

To separate traditional parenthood from digital parenthood (which are, of course, closely related), we need to identify the roles in which they differ. In traditional parenting we mean the care, guidance and support needed to help children become independent adults. This covers a wide range of activities, from meeting physical needs (food, home), to passing on social skills and values, to providing emotional support.

Digital parenting, on the other hand, can perhaps be defined as a more modern aspect of traditional parenting, which involves managing and managing the use of digital devices and technologies (smartphones, tablets, computers and other devices) and the associated risks, but it also involves great challenges [19].

Below, we have collected some of the issues that a parent will encounter, manage and be aware of in digital education:

1. **Screen time**: it is up to the parent to decide how much screen time the child can spend in front of the screen. It is an important element in avoiding addictions and ensuring that the child's healthy development (physical play, social skills, learning, family time) is not compromised.
2. **Internet safety and privacy**: educating children about the risks associated with the internet (as we have already touched on), such as exposure to inappropriate content, cyberbullying or oversharing. It is important for parents to be able to talk to their children about privacy and staying safe online, and to make them understand the paramount importance of this.
3. **Digital literacy**: parents should help their children to understand how to use technology effectively and responsibly, including how to distinguish between trustworthy and untrustworthy sources of information [20]. This may seem trivial, but when you think about it, it is a difficult and complex task, even for an adult. It is also important to understand the digital footprint and how to use technology to enhance knowledge or creativity.
4. **Setting an example of digital behaviour** [21]: parental care has a significant impact on children's use of digital technology, so parents need to set an example themselves through their own use of technology, including by allowing for their own screen time, respecting privacy and engaging in positive online interactions.
5. **Engagement with technology**: mutual understanding and bridging the generation gap can be well supported by a parent's ability to understand and engage with the content that the child likes online. Showing sufficient engagement in

learning about them: whether it's playing games together, watching videos or learning about the platforms and services the child uses. This kind of behaviour can support early detection and prevention of risks.

4 Briding the Gap: Theories to Solve the Problem

This raises the question of what opportunities we have to bridge the gap in cybersecurity that corresponds to the different digital literacies of different age groups. Some possible solutions are outlined below:

- **Education**: as we mentioned earlier under the role of parents, education and training is perhaps the most important. We would stress that education not only of children but also of parents is a cardinal point for awareness raising. Education must therefore also take place on an individual and school level, and it would therefore be desirable to integrate cybersecurity into education as soon as possible.
- **User-friendly security tools**: there is a slow but steady trend for large companies (Apple [22], Microsoft [23]) to integrate cybersecurity and child protection into digital devices. This forward-looking trend shows how important this topic is. Efforts should be made to promote the use of cybersecurity, with easy-to-use, widely available software and support that requires minimal user intervention.
- **Awareness campaigns**: awareness of cybersecurity threats and best practices can be increased through campaigns targeting different demographic groups in different formats. Efforts should be made to make campaigns as easy to understand as possible for the target age groups.
- **Mentoring programmes**: mentoring programmes can be effective, whereby cybersecurity experts guide less experienced individuals (be they children, parents or older age groups).
- **Accessible resources**: providing online tutorials, guides and frequently asked questions (FAQs) to suit different learning styles and preferences, in a way that makes cybersecurity and its potential applications easily accessible to all.
- **Inclusive policymaking**: to assess the needs of different generations and involve their representatives in the cybersecurity policymaking process, to increase understanding and effectiveness, and to transfer knowledge faster and more smoothly.
- **Fostering intergenerational dialogue**: promoting discussion forums, dialogue where people from different generations can share their experiences and concerns about cybersecurity. This can promote understanding and common problem-solving.
- **Regular updates**: cybersecurity is a constantly evolving, dynamic field. Regular knowledge updates are essential to keep up to date with threats and risks.
- **Tailored user experiences**: the potential of using artificial intelligence (my research area intends to focus on this area in the future) to raise awareness of cybersecurity, and enhancing digital experiences to match the level of education would make cybersecurity easier to manage.

- **Support and helplines**: setting up dedicated support points where people (whether children, young people, adults or elderly) can get help with their cybersecurity issues and mental health (e.g. in case of cyberbullying).

5 Conclusion

Overall, the generation and digital divide interacts with digital parenthood and cybersecurity. This phenomenon deserves in-depth investigation. This divide encompasses the inequalities in digital literacy and access to technological resources between different generations, which significantly affects the way digital parenting is understood and understood.

First, we introduced the terminology of digital immigrants and digital natives, followed by a definition of the generation gap and the digital divide. We found that the generation gap refers to the differences in experiences, attitudes and competences with digital technologies between older and younger generations. Older generations may not have grown up in the ubiquitous presence of digital technology and thus potentially lack innate knowledge of these tools. This disparity can lead to a gap in understanding and engagement with the digital world, a critical aspect of digital parenting where the role is to guide and protect younger generations in the digital world.

We then focused on their relationship to cybersecurity, where they become very prominent (answering our first question). It can be said that the generation gap fundamentally affects the way we assess and understand cybersecurity risks, and we should strive to narrow this gap to ensure children are safe online. The digital divide further complicates this dynamic. Unequal access to the internet and modern digital tools, socio-economic status, geographical location and educational opportunities are barriers to effective cybersecurity risk reduction. All these factors together play a crucial role in digital parenting and in creating the right cybersecurity environment. It influences not only the ability of parents to effectively manage and protect their children in the digital space, but also the overall approach and strategies used to address cybersecurity challenges. Understanding and bridging these gaps is essential to increasing digital literacy and promoting a safer digital environment for future generations. In our article we have highlighted a number of possible solutions to narrow the gaps (thus answering our second question). These possible solutions need further fine-tuning, further elaboration and are important for societal well-being and security.

Our long-term goal is to better map these gaps, to use our research to highlight parents' cybersecurity preparedness, and to develop further recommendations and practices to narrow the gaps.

References

1. Dhanya VJ (2020) Impact of digital devices on children. Eur J Mol Clin Med 7(11). ISSN 2515-8260
2. Council on Communications and Media (2016) Media use in school-aged children and adolescents. Pediatrics 138(5):e20162592. https://doi.org/10.1542/peds.2016-2592
3. Share of children who started using a mobile phone in Poland in 2022, by age group (www.statista.com). Last viewed: 12 Jan 2024, 12:38. https://www.statista.com/statistics/1026972/poland-mobile-usage-among-children-by-age/
4. Prensky M (2001) Digital natives, digital immigrants Part 1. On the Horizon 9(5):1–6
5. Belinskaya EP, Shaekhov ZD (2023) Psychological Well-Being and Adaptation to the Risks of Digital World at a Young Age
6. Tafonao T, Saputra S, Suryaningwidi R (2020) Learning media and technology: generation Z and Alpha. Indonesian J. Instruct. Media Model **2**(2):89. https://doi.org/10.32585/ijimm.v2i2.954. LicenseCC BY-SA 4.0 2020
7. Aissaoui N (2021) The digital divide: a literaturereview and some directions forfuture research in light of COVID-19. Glob. Knowl. Memory Commun. https://doi.org/10.1108/GKMC-06-2020-0075
8. DiMaggio PJ, Hargittai E, Celeste C, Shafer SJ (2004) From unequal access to differentiateduse: a literature review and agenda for research on digital inequality. In: Neckerman K (ed) Social inequality. Russell Sage Foundation, New York, NY, pp 355–400
9. DiMaggio P, Hargittai E, Celeste C, Shafer S (2004) From unequal access to differentiated use: a literature review and agenda for research on digital inequality. In: Neckerman K (ed) Social inequality. Russell Sage Foundation, New York, NY, pp 355–400
10. Lawati S (2019) Understanding the psychology of youths: generation gap. Int J Psychol Counsel 11(6):46–58. https://doi.org/10.5897/IJPC2019.0568
11. Rathi S, Kumar P (2023) Differences in work value, communication style, and leadership style among generational cohorts at the workplace: an overview. In: Role of human resources for inclusive leadership, workplace diversity, and equity in organizations lab: Soni Rathi's Lab. https://doi.org/10.4018/978-1-6684-6602-5.ch010
12. Downs H (2019) Bridging the gap: how the generations communicate. Concordia J Commun Res 6, Article 6. https://doi.org/10.54416/SEZY7453
13. Leung L (2013) Generational differences in content generation in social media: the roles of the gratifications sought and of narcissism. Comput Hum Behav 29:997–1006
14. Tomczyk L, Mascia ML, Gierszewski D, Walker C (2023) Barriers to digital inclusion among older people: a intergenerational reflection on the need to develop digital competences for the group with the highest level of digital exclusion. Innoeduca. Int J Technol Educ Innov 9(1):5–26. ISSN 2444-2925. https://doi.org/10.24310/innoeduca.2023.v9i1.16433
15. Shang Y, Wu Z, Du X, Jiang Y, Ma B, Chi M (2022) The psychology of the internet fraud victimization of older adults: a systematic review. Front Psychol 13. https://doi.org/10.3389/fpsyg.2022.912242
16. Sarno DM, Lewis JE, Bohil CJ, Neider MB (2019) Which phish is on the hook? Phishing vulnerability for older versus younger adults. Hum Factors J Hum Factors Ergon Soc 62(5). https://doi.org/10.1177/0018720819855570
17. Shabahang R, Shim H, Aruguete MS, Zsila Á (2022) Oversharing on social media: anxiety, attention-seeking, and social media addiction predict the breadth and depth of sharing. Psychol Rep, OnlineFirst 2022. https://doi.org/10.1177/00332941221122861
18. Schonfeld A, McNiel D, Toyoshima T, Binder R (2023) Cyberbullying and adolescent suicide. J Am Acad Psychiat Law Online, JAAPL.220078-22. https://doi.org/10.29158/JAAPL.220078-22
19. Zhao P, Bazarova NN, Valle N (2023) Digital parenting divides: the role of parental capital and digital parenting readiness in parental digital mediation. J Comput-Mediated Commun 28(5). https://doi.org/10.1093/jcmc/zmad032

20. Topal R, Shargh F (2023) Teaching students how to find and identify reliable online sources: a series of exercises. J Polit Sci Educ 19(27):1–10. https://doi.org/10.1080/15512169.2022.2163899
21. Wahyuningrum E, Suryanto S, Suminar DR (2020) Parenting in digital era: a systematic literature review. J Educ Health Commun Psychol 9(3):226–258. https://doi.org/10.12928/jehcp.v9i3.16984
22. Apple Child Safety features. Last viewed: 14 Jan 2024 18:29. https://support.apple.com/en-us/HT201304
23. Microsoft Family Safety. Last viewed: 14 Jan 2024 18:30. https://www.microsoft.com/en-us/microsoft-365/family-safety

Health and Safety Considerations Before the Modern Times

Kornélia Lazányi

Abstract Current article explores the historical evolution of health and safety considerations from ancient times to the medieval ages, with a particular focus on safety within the realm of health practices. It delves into how early thinkers, such as Zarathustra, Hippocrates, and Vitruvius, laid foundational principles that continue to influence modern safety practices alongside health standards. Transitioning to the medieval era, amidst burgeoning urbanization and limited medical understanding, the discourse on health intertwines with safety principles, emphasizing the importance of environmental factors, lifestyle choices, and overall well-being. Figures like Avicenna, Hildegard of Bingen, Maimonides, Roger Bacon, Guy de Chauliac, and Bartolomeo Platina subtly interweave health considerations with safety consciousness, recognizing the intrinsic link between them. Despite the emergence of formalized safety science later, their contributions underscore the importance of safety in promoting individual and community health. Reflecting on this historical journey underscores the timeless relevance of holistic approaches to safeguarding health, emphasizing the interconnectedness of safety measures with broader health practices. Engaging with historical wisdom offers profound insights into ensuring the safety of communities and workplaces, inspiring continued commitment to innovation, collaboration, and vigilance in protecting human well-being throughout history and into the future.

Keywords Health · Safety · Ancient times · Medieval ages

1 Introduction

This exploration into the perspectives on health and safety presents a captivating journey through the corridors of time and the vast geographical expanses of human knowledge. As we delve into the insights of figures such as Hippocrates, Vitruvius,

K. Lazányi (✉)
John von Neumann Faculty of Informatics, Obuda University, Budapest, Hungary
e-mail: lazanyi.kornelia@nik.uni-obuda.hu

T. A. Kovács et al. (eds.), *The Impact of the Energy Dependency on Critical Infrastructure Protection*, Advanced Sciences and Technologies for Security Applications, https://doi.org/10.1007/978-3-031-78544-3_38

Avicenna, Hildegard of Bingen, Maimonides, Roger Bacon, Guy de Chauliac, and Bartolomeo Platina, the article embarks on a fascinating voyage that spans across different cultures and epochs.

Geographically dispersed across Persia, Europe, and the Middle East, these thinkers have left an indelible mark on the understanding of health and safety, contributing to a tapestry of knowledge that transcends borders. In terms of time, the below mentioned authors and sources offering glimpses into the medieval times, into a world where every day norms were vastly different from our contemporary understanding. The practices described by these authors provide insights into a time when empirical observation, holistic well-being, and meticulous surgical procedures coexisted with what we might consider unconventional methods today.

The everyday 'normal' in medieval society was shaped by a confluence of cultural, religious, and philosophical perspectives, often diverging from our modern sensibilities. The nuanced approaches to health and safety outlined by these thinkers reflect a rich tapestry of historical practices, each contributing to a mosaic that challenges our preconceptions and invites us to appreciate the complexity of the understanding of human safety.

As we journey through these distinct perspectives, it becomes evident that the pursuit of well-being was a common thread uniting diverse civilizations. While the specifics of daily life may have varied, the shared emphasis on maintaining balance, interconnectedness, and preventive measures serves as a testament to the enduring quest for a healthier, safer existence. This essay invites readers to immerse themselves in the intellectual landscapes of selected medieval thinkers recognizing the profound impact of their contributions on the evolving concepts of health and safety. Through their lenses, we gain not only historical insights but also a deeper appreciation for the kaleidoscopic nature of human knowledge and the perennial pursuit of well-being across time and space.

2 Health and Safety Perspectives of Ancient Times

The formalized study of organizational safety, known as safety science, is a distinctly modern discipline that gained prominence in the twentieth century, particularly during the Industrial Revolution and beyond. The term "safety science" encapsulates the systematic and interdisciplinary examination of safety within organizational contexts. However, the concept of health and safety has always been a cornerstone of human civilization, evolving through the ages from ancient directives to modern regulations [1]. This essay examines the historical progression of health and safety considerations, illustrating how ancient wisdom and practices have informed and shaped contemporary approaches to ensuring the well-being of communities and workplaces. From the spiritual cleanliness codes in Zarathustra's Videvdad to Hippocrates' early medical ethics and the architectural innovations of Vitruvius, each era contributed layers to the foundation of today's health and safety standards. By tracing the lineage of these concepts back to their origins, we uncover a continuous thread of human endeavor aimed at mitigating risks and enhancing life quality, underscoring the timeless relevance of health and safety in human progress.

The Videvdad, attributed to Zarathustra (Zoroaster), while being renowned for its spiritual and moral teachings, with its emphasis on purity and cleanliness, influences even current times' principles of health and safety [2]. Through guidelines for personal hygiene and cleanliness in surroundings, the text indirectly addresses the importance of preventing disease transmission and promoting community well-being [3]. Furthermore, while not explicit, the Videvdad contains directives related to sanitary practices. These include guidelines for waste disposal and food preparation, essential for maintaining health standards in ancient societies. Additionally, the Videvdad advocates for environmental stewardship, urging the protection of water sources. Such practices reflect a holistic approach to health and safety, considering both individual and communal well-being [4].

Ancient thinkers such as Hippocrates, revered as the "Father of Medicine" have also greatly contributed to the understanding of the interconnectedness of lifestyle, environment, and well-being [5]. Alcmaeon of Croton, who lived around 470 BC, became the earliest Greek physician to propose that the health of individuals could be affected by the quality of water [6]. The Hippocratic treatise "Airs, Waters, Places," composed around 400 BC, also extensively discusses various aspects including the origins, characteristics, and impacts on health associated with different sources of water. However, Hippocrates also placed emphasis on balance and moderation in diet and activity underscoring the importance of preventive measures for maintaining health. Moreover, Hippocrates' observations on environmental factors influencing health hinted at early considerations of workplace well-being, reflecting a holistic understanding of human flourishing.

In line with this, alongside medical insights, ancient civilizations demonstrated an awareness of health and safety considerations in architectural design [7]. Ensuring the provision of safe water has been a timeless imperative for built environments, deeply intertwined with the promotion of health and safety. From ancient times, communities expended considerable resources and labor to secure clean water sources essential for sustaining life [8]. Archaeologists have uncovered remnants of water pipes, toilets, and sewage systems in the remains of past civilizations across the globe, spanning from India to Rome and from Greece to Egypt to South America [9]. Some of these artifacts date back over 4000 years, illustrating the longstanding efforts to address sanitation and water management. However, along with its vital role, water also presented significant challenges to public health and safety [10]. Waterborne diseases and the long-term ramifications of improper waste disposal underscored the critical importance of managing water resources effectively to safeguard the well-being of populations.

Vitruvius, a Roman architect and engineer, articulated a complex set of principles of structural integrity and environmental suitability in his writings [11]. His emphasis on natural light, ventilation, and sanitation foreshadowed modern principles of ergonomic design and sustainable architecture. Despite the limitations of ancient building techniques, Vitruvius' emphasis on the relationship between design and human well-being laid a foundation for subsequent developments in architectural safety [12]. However, the realization of health-related safety hazards is not only limited to doctors and architects. Mold infestation was a prevalent issue in ancient structures, vividly described in the Old Testament as a "plague" appearing as greenish or reddish growth on walls (Leviticus 14:33–45).

Furthermore, the realities of ancient work environments also reveal pragmatic responses to occupational hazards. From labor-intensive industries to mining and manufacturing, workers faced various risks without the benefit of formalized safety regulations [13]. The foundations of health and safety were laid as early physicians observed the impact of working and living conditions on the well-being of individuals, particularly among slaves and serfs. Ancient Egyptian depictions and literature vividly illustrate the harsh realities faced by various professions. (Blacksmiths worked near furnaces without protection, masons toiled without adequate clothing, weavers struggled to breathe in cramped spaces, laundry workers faced dangers near water bodies, and dye workers experienced fatigue and discomfort from their tasks [13]. These observations underscored the importance of addressing social and environmental factors in promoting health and safety in the workplace [14]. However, guilds and trade associations provided informal training programs and support networks aimed at mitigating risks and imparting safety knowledge [15]. Some countries such as Babylonia had central regulations regarding medical practices and health related hazards, while other countries, such as Greece have created a wide-spread network of trained professionals guided by the Hippocratic Oath [16]. Although these efforts may appear rudimentary by modern standards, they reflect a collective wisdom that laid the groundwork for subsequent advancements in workplace safety standards and labor rights movements [17].

In evaluating these ancient perspectives on health and safety, it's essential to recognize both their contributions and limitations. While figures like Hippocrates and Vitruvius provided valuable insights into preventive measures and architectural design, ancient societies grappled with disparities in access to safety protections and societal attitudes towards certain occupations. By examining historical precedents through a critical lens, contemporary practitioners can glean valuable insights into the evolving nature of health and safety considerations across civilizations.

In conclusion, the study of health and safety has deep roots in ancient times, where early thinkers, architects, and workers grappled with the complexities of maintaining well-being in diverse contexts. By exploring these ancient perspectives, we gain a deeper appreciation for the enduring relevance of holistic approaches to health and safety that transcend temporal and cultural boundaries.

3 Medieval Health and Safety Considerations

Comparing the health-related challenges and safety perspectives of the medieval ages with those of ancient times reveals both similarities and differences in approaches to addressing health and safety concerns. Ancient sources emphasize the importance of purity, cleanliness, and environmental stewardship in promoting health and safety. They recognize the interconnectedness of lifestyle, environment, and well-being, advocating for preventive measures and community well-being.

In terms of sanitation practices, both ancient and medieval societies demonstrated awareness of the importance of waste disposal, food preparation, and environmental protection. While ancient civilizations lacked formalized safety regulations, they have developed pragmatic responses to occupational hazards, as observed in the realities of ancient work environments. However, the level of organization and regulation varied across regions and time periods. Medieval societies have developed strategies to mitigate risks and provide safety training for workers [18]. Interestingly, the role of physicians, government officials, and communities in addressing public health issues such as epidemics, sanitation, and quarantine measures were almost equally important, calling attention to the holistic nature and health related safety and its potential solutions [19]. In line with this Cohn's [20] work underscored the collaborative efforts required to safeguard the health and safety of medieval populations amidst the threat of infectious diseases and environmental hazards.

Urbanization and sanitation were significant challenges in both periods. Both ancient civilizations and medieval societies faced significant challenges from infectious diseases [19]. In both periods, crowded living conditions, inadequate sanitation, and limited medical knowledge contributed to the spread of diseases such as cholera, typhoid fever, and tuberculosis, but medieval cities faced greater issues due to overcrowding and inadequate infrastructure. While ancient cities had some sanitation systems such as public baths and sewage drains, medieval cities lacked comprehensive sanitation measures, leading to higher rates of disease transmission and environmental pollution [21].

In ancient civilizations, temples and religious centers served as centers of healing and medical care, providing treatments and rituals to restore health. In medieval Europe, monasteries and convents played a similar role in providing healthcare and social support, but religious institutions also promoted public health measures and hygiene practices as moral imperatives, reflecting the influence of religious beliefs on health practices during this era. In line with this, it is important to note that the role of social and religious institutions differed between ancient and medieval societies in addressing health and safety concerns. While both ancient and medieval societies had limited medical knowledge, there were differences in medical practices and approaches to healthcare. Ancient civilizations such as the Greeks and Egyptians made significant advancements in medical theory and practice, including surgical techniques and pharmacology. In contrast, medieval medicine was heavily influenced by religious beliefs and superstitions, with treatments often based on herbal remedies, bloodletting, and prayer [22].

A good example of this is the Tacuinum Sanitatis, which can be considered the heart of medieval health practice—a medieval health handbook offering insights into health maintenance in Europe during the Middle Ages [23]. This manuscript provided practical advice for everyday health, medieval medical treatises and guidance on maintaining well-being through diet, exercise, and hygiene practices, reflecting a pragmatic approach to health promotion. However, it also providing an undisputable proof of the influence of medieval beliefs and superstitions on medical practices, underscoring the intersection of science and spirituality in medieval health discourse [24]. While the book was endeavoring to provide the theoretical underpinnings of medical practices recommended and prescribing the role of physicians, surgeons, and apothecaries in treating illness and injury, from a modern point of view it is efficiently highlighting the limitations of medieval medicine, including the prevalence of ineffective treatments and the persistence of supernatural explanations for disease [25].

In contrast, the Islamic world during the medieval period witnessed significant advancements in health and medicine [26], Islamic scholars and physicians made substantial contributions to medical knowledge, including advancements in surgery, pharmacology, and public health. What is more, the integration of diverse medical traditions in the Islamic world fostered innovative approaches to healthcare that later even influenced medical practices in medieval Europe [27].

In summary, although the health-related approaches in the medieval ages share some commonalities with those of ancient times, they also exhibit unique characteristics shaped by the socioeconomic, cultural, and religious contexts of the medieval period. While the medieval period saw the continuation and preservation of medical knowledge from ancient civilizations such as Greece, Rome, and the Islamic world it is undeniable that the medieval period was marked by significant challenges and shortcomings in healthcare and safety practices. The medieval ages were characterized by widespread disease, unsanitary living conditions, and limited medical understanding. Therefore, while it may be overly simplistic to characterize the medieval ages solely as a "dark era" in regards to health and safety, it is important to acknowledge the significant challenges and shortcomings that existed alongside the limited progress and innovations made during this period.

4 People Determining the Mainstream of Health and Safety in the Medieval Ages

In this realm of medieval health and safety considerations, prominent figures such as Avicenna, Hildegard of Bingen, Maimonides, Roger Bacon, Guy de Chauliac, and Bartolomeo Platina made significant contributions, each offering a distinctive perspective that collectively shaped the discourse on health and indirectly influenced early safety considerations.

Avicenna (Ibn Sina) (980–1037), a Persian polymath of the late medieval period, played a pivotal role in shaping health discourse. His renowned work, "The Canon of Medicine," not only synthesized medical knowledge from diverse cultures but also laid the foundation for a holistic perspective on health that inherently encompassed safety principles [28]. Avicenna's meticulous approach to medical practice, incorporating an understanding of environmental factors, cleanliness, and preventive measures, hinted at safety considerations. Though not explicitly framed within the modern concept of safety, his emphasis on maintaining a balanced and harmonious environment for health implicitly introduced elements of safety consciousness [29]. Avicenna's holistic approach underscored the interconnectedness of health and well-being, indirectly influencing early considerations of safety within medical practices and environmental factors.

Hildegard of Bingen (1098–1179), a multifaceted medieval mystic, healer, and abbess, contributed to the discourse on health. Her extensive writings, covering herbal medicine, natural remedies, and holistic well-being, emphasized a comprehensive approach to health [30, 31]. While her primary focus was on healing practices and spiritual matters, Hildegard's profound understanding of the interconnectedness between nature, health, and the human body hinted at principles extending beyond individual health. Although not directly associated with safety science, Hildegard's emphasis on natural remedies and holistic well-being indirectly addressed the importance of maintaining a harmonious environment for overall health [32]. This emphasis laid the groundwork for future discussions on the significance of broader environmental and lifestyle factors in ensuring health and safety [33, 34].

Maimonides (1138–1204), a Jewish philosopher and physician, significantly influenced medieval medicine with his comprehensive work, "The Book of Healing" [35]. His impact resonated across both Jewish and Islamic medical traditions, sharing a cultural bridge with the broader medical discourse of the time. Although Maimonides' writings did not specifically concentrate on health hazards, they provided invaluable insights into prevailing medical practices. In parallel with Avicenna and Hildegard, Maimonides emphasized the interconnectedness of the body and the importance of preventive measures in maintaining overall health [36]. These shared perspectives, articulated in distinct cultural and religious contexts, collectively contributed to the broader medieval understanding of holistic well-being. While not explicitly delving into the modern concept of safety, the emphasis on preventive measures by these influential figures hinted at considerations integral to evolving concepts of health and safety in subsequent centuries [37].

Roger Bacon (c. 1219–1292), an English philosopher and Franciscan friar, played a pivotal role in shaping medieval scientific thought. His empirical approach to understanding the natural world laid the foundation for advancements in various fields, including medicine [38]. In harmony with Avicenna, Bacon's empirical methodologies subtly intertwined with the evolving concepts of health and safety. The emphasis on systematic observation and experimentation, shared with Avicenna's holistic approach, suggested a recognition of the interplay between natural elements and well-being. While not explicitly venturing into safety science, Bacon, in collaboration with other medieval thinkers, contributed to a paradigm shift that would later underpin modern scientific approaches to health and safety [39]. The shared emphasis on systematic and evidence-based methodologies formed a connecting thread among these influential figures, contributing collectively to the broader understanding of health during the medieval era.

Guy de Chauliac (c. 1300–1368), a renowned medieval surgeon and physician, left an enduring mark on the history of medieval medicine through his comprehensive work, "Chirurgia Magna" [40] This echoes the collaborative spirit seen in the works of his contemporaries such as Avicenna and Hildegard of Bingen. While Guy de Chauliac's primary focus was on advancing surgical techniques and medical procedures, his contributions subtly intertwined with evolving notions of health and safety during the medieval era. In alignment with the holistic perspectives of Avicenna and Hildegard, Chauliac's meticulous descriptions of medical practices in "Chirurgia Magna" underscored the importance of precision, hygiene, and a structured approach to medical interventions [41]. These shared principles suggested an emerging recognition of the interconnectedness between procedural accuracy and patient well-being, forming a collective narrative within medieval medical thought [42].

Bartolomeo Platina (1421–1481), an Italian humanist and papal librarian, contributed to culinary and health treatises. His work, "De honesta voluptate et valetudine" (On Right Pleasure and Good Health), delved into the intersection of food, health, and well-being. While not directly linked to safety science, Platina's treatise addressed the importance of balanced nutrition, culinary hygiene, and the impact of lifestyle choices on health [38]. In exploring the connections between diet and well-being, Platina unintentionally contributed to the broader understanding of factors influencing human health and safety [43].

In summary, these medieval figures, through their diverse contributions, collectively shaped the discourse on health and safety, offering insights into interconnected considerations that would later inform evolving concepts in subsequent centuries. Despite these contributions, it is crucial to recognize the limitations of attributing modern safety science concepts to figures from the medieval ages. The medieval worldview and understanding of medicine and well-being differed significantly from contemporary perspectives. Safety, as a systematic and interdisciplinary field, emerged more prominently in response to the challenges posed by industrialization and technological advancements during the modern era. The foundational principles laid down by these medieval figures, albeit within the context of their time, demonstrated a nascent awareness of the importance of health, hygiene, and holistic well-being—themes that would later become integral to the development of safety science.

5 Summary and Conclusions

The historical journey of health and safety considerations is a testament to the enduring human endeavor to understand and mitigate the risks posed by the natural and built environment, as well as by the very nature of human work and lifestyle. It highlights the interconnectedness of health and safety with broader social, environmental, and economic factors, necessitating a holistic and integrated approach. The exploration of historical sources has unveiled a tapestry of insights, interwoven with diverse perspectives and historical contexts. Across the epochs and geographies explored, the common thread of prioritizing well-being emerges, intricately intertwined with evolving notions of safety. This historical perspective not only enriches our understanding of the evolution of health and safety practices but also inspires a continued commitment to innovation, collaboration, and vigilance in protecting human well-being in an ever-changing world.

Current article endeavors to highlight how the medical aspects of safety were enriched in the ancient times and how the medieval ages and its religious and philosophical streams have impacted the practices and approaches of the medical practitioners and other people (municipalities, religious groups, such as monasteries and other societies) taking part of the organization and delivery of medical treatment and prevention. Accordingly, the article begins with an exploration of ancient sources, where figures like Zarathustra, Hippocrates, and Vitruvius laid foundational principles that continue to resonate in modern health and safety practices. Their emphasis on purity, cleanliness, and environmental stewardship underscored the holistic approach to safeguarding individual and communal well-being. These ancient thinkers, albeit without formalized safety regulations, recognized the intrinsic connection between health and the surrounding environment, paving the way for subsequent discussions on safety consciousness.

Transitioning to the medieval ages, the paper explores the complex tapestry of health-related challenges and safety perspectives. While similarities with ancient times exist, such as the recognition of sanitation practices and preventive measures, medieval societies grappled with unique challenges amidst the backdrop of burgeoning urbanization and limited medical understanding. Despite the prevailing limitations, influential figures like Avicenna, Hildegard of Bingen, Maimonides, Roger Bacon, Guy de Chauliac, and Bartolomeo Platina contributed nuanced perspectives that subtly intertwined health considerations with safety principles. Their works underscored the interconnectedness between environmental factors, lifestyle choices, and overall well-being, hinting at a nascent awareness of safety consciousness within medieval discourse.

Of particular significance is the role of individuals in determining the mainstream of health and safety in the medieval ages. Avicenna's holistic approach, Hildegard's emphasis on natural remedies, Maimonides' preventive measures, Bacon's empirical methodologies, Chauliac's meticulous descriptions, and Platina's exploration of diet-health connections collectively shaped the discourse on health and indirectly influenced early safety considerations. While safety science as a formalized discipline emerged later, the foundational principles laid down by these medieval figures underscored the intrinsic connection between health and safety, hinting at a holistic understanding that transcended temporal boundaries.

Drawing upon these insights, it becomes evident that the pursuit of health and safety has been an enduring endeavor throughout human history, marked by both continuity and change. As we reflect on this journey, it is imperative to acknowledge the complexities and nuances inherent in historical perspectives on health and safety. While the past may offer valuable lessons and insights, it also reminds us of the evolving nature of knowledge and the imperative of adapting to changing circumstances. In line with this, this exploration underscores the timeless relevance of holistic approaches to health and safety, highlighting the interconnectedness of individual well-being, societal norms, and environmental factors across time and space. By engaging with the wisdom of the past, we gain a deeper appreciation for the multifaceted nature of human endeavors aimed at safeguarding the health and safety of communities and workplaces alike.

References

1. Hale A, Hale M (2003) The origins of total safety management: a review of literature. Saf Sci 41(2–3):155–165
2. Sigerist HE (1987) A history of medicine, vol 2. Oxford University Press, Oxford
3. Herzfeld E (1947) Zoroaster and his World, vol 1. Princeton University Press, Princeton
4. Elgood C (2010) A medical history of Persia and the Eastern Caliphate: from the earliest times until the year AD 1932. Cambridge University Press, Cambridge
5. Smith JA (2000) Health and medicine in Ancient Greece: from Hippocrates to Galen. Oxford University Press, Oxford
6. Vuorinen HS, Juuti PS, Katko TS (2007) History of water and health from ancient civilizations to modern times. Water Sci Technol Water Supply 7(1):49–57
7. Dannenberg AL (2011) Making healthy places: designing and building for health. Island Press-Center for Resource Economics, Washington, DC
8. Frumkin H, Wendel AM, Abrams RF, Malizia E (2011) An introduction to healthy places. In: Making healthy places: designing and building for health, well-being, and sustainability, pp 3–30. Island Press-Center for Resource Economics, Washington, DC
9. Rosen G (1993) A history of public health, Exp. edn. Johns Hopkins University Press, Baltimore. Originally published 1958
10. Backer LC (2015) Community design for water quantity and quality. In: Making healthy places: designing and building for health, well-being, and sustainability, pp 91–105
11. Brown MC (2015) Vitruvius: ten books on architecture. Cambridge University Press, Cambridge
12. Kleindorfer PR, Kunreuther H (eds) (1999) Health and safety in the workplace: historical trends and future directions. Cambridge University Press, Cambridge
13. Abrams HK (2001) A short history of occupational health. J Public Health Policy 22(1):34–80
14. Parkins MD (2001) Pharmacological practices of ancient Egypt. In: Whitelaw WA (ed) History of medicine days. The University of Calgary, Faculty of Medicine, pp 5–12
15. Ahmad I, Sattar A, Nawaz A (2016) Occupational health and safety in industries in developing world. Gomal J Med Sci 14(4):223–228
16. Sigerist HE (1934) Medical societies, past and present. Yale J Biol Med 6(3):351
17. Toole TM, Schneider GP (1994) Occupational safety and health in the age of high technology: forging new frontiers. Greenwood Publishing Group, Westport
18. Nanda S (2016) Health care in Medieval times. Sci Horizon 13
19. Robb J, Cessford C, Dittmar J, Inskip SA, Mitchell PD (2021) The greatest health problem of the Middle Ages? Estimating the burden of disease in Medieval England. Int J Paleopathol 34:101–112
20. Cohn Jr S (2017) Plague violence and abandonment from the Black Death to the early modern period. Annales de démographie historique 2:39–61
21. Ciecieznski NJ (2013) The stench of disease: public health and the environment in Late-Medieval English towns and cities. Health Cult Soc 4(1):91–104
22. Wagner C, De Gezelle J, Komarnytsky S (2020) Celtic provenance in traditional herbal medicine of Medieval Wales and classical antiquity. Front Pharmacol 11:105
23. Mendelsohn LD (2013) The Tacuinum Sanitatis: a medieval health manual. CUNY Academic Works, New York
24. Demaitre L (2013) Medieval medicine. Bloomsbury Publishing, London
25. Geltner G, Coomans J (2023) The healthscaping approach: toward a global history of early public health. Hist Methods J Quant Interdiscip Hist 56(1):18–33

26. Pormann PE, Savage-Smith E (2007) Medieval Islamic medicine. J Br Islamic Med Assoc 1(1):43–54
27. Page S (2023) Entangled magic in the Medieval Latin West. Entangled Religions 14(3)
28. Geller MJ (2024) Commentaries from Babylon to Byzantium: Gleanings from the MPWIG. Max Planck Institute of Wissenschaftsgeschichte, Berlin
29. Lakhtakia R (2014) A trio of exemplars of medieval Islamic medicine: Al-Razi, Avicenna and Ibn Al-Nafis. Sultan Qaboos Univ Med J 14(4):455–459
30. Throop P (1998) Hildegard of Bingen's Physica: the complete English translation of her classic work on health and healing. Healing Arts Press, Rochester, VT
31. Butcher CA (2007) Hildegard of Bingen: a spiritual reader. Paraclete Press, Brewster, MA
32. Hay KA (2008) Hildegard's medicine: a systematic science of Medieval Europe. In: Proceedings of the 17th annual history of medicine days
33. Weed S (1989) Wise woman herbal: healing wise. Ash Tree Publishing, Woodstock, NY
34. Wood M (2009) The earthwise herbal: a complete guide to new world medicinal plants. North Atlantic Books, Berkeley
35. Rosner F (2002) The life of Moses Maimonides, a prominent medieval physician. Einstein Quart J Biol Med 19:125–128
36. Segal I, Blazer S (2020) The Maimonides model for a regimen of health: a comparison with the contemporary scenario. Rambam Maimonides Med J 11(4)
37. McVaugh MR (2012) The medieval appropriation of Maimonides. IMF-CSIC Working Papers. Available at: https://api.semanticscholar.org/CorpusID:161769741
38. Wear A (2000) Knowledge and practice in English medicine, 1550–1680. Cambridge University Press, Cambridge
39. Power A (2006) A mirror for every age: The reputation of Roger Bacon. Engl Hist Rev 121(492):657–692
40. Pilcher JE (1895) Guy de Chauliac and Henri de Mondeville—a surgical retrospect. Ann Surg 21(1):84
41. Siraisi NG (1990) Medieval and early renaissance medicine: an introduction to knowledge and practice. University of Chicago Press, Chicago
42. Cunningham A (1997) The anatomy of the nervous system from the eyes to the soul. Brill, Leiden
43. Ceron A (2015) A humanist in the kitchen. Platina's De honesta voluptate et valetudine. Doctor Virtualis 13

Introduction to the Institutional ArtWork Protection and Public Collection Storage

László Lőrincz

Abstract Institutional ArtWork Protection plays a critical role in the protection of public collections, both for temporary and permanent exhibitions. Therefore, Institutional ArtWork Protection requires a multidisciplinary approach, which includes not only the protection of the property and the inventory of artefacts, but also the complete object protection of the institution, as well as professional cooperation between the necessary disciplines, including continuous training. In the context of Artwork Protection in public collections institutions, storage is a critical factor for the security of artworks. According to the current professional principles, ArtWork Protection is an integral part of the exhibition and storage, which involves the storage of artefacts in a way that minimises the risk by taking into account a number of risk factors.

Keywords ArtWork Protection · Institutional ArtWork Protection · Public collection storage

1 Introduction

The background for the choice of the topic of my doctoral thesis is, on the one hand, that since 2015, as the head of the Security Department of the Museum of Fine Arts—Hungarian National Gallery, we have had the opportunity to get acquainted with the theoretical concept of Institutional ArtWork Protection and its practical complexity. And from multiple perspectives, such as artists, curators, art directors and conservators in charge of collections, and professionals responsible for the protection of artworks. After all, in a temporary exhibition, not only the work of art, but almost always the exhibition space itself, functions as part of the final installation, as an artistic creation. For this reason, artists, curators and restorers take a very different

L. Lőrincz (✉)
Doctoral School on Safety and Security Sciences, Obuda University, Budapest, Hungary
e-mail: lorincz.laszlo@uni-obuda.hu

T. A. Kovács et al. (eds.), *The Impact of the Energy Dependency on Critical Infrastructure Protection*, Advanced Sciences and Technologies for Security Applications, https://doi.org/10.1007/978-3-031-78544-3_39

approach to the protection of works of art, while for them too, the protection of the work (s) is the most important thing, regardless of everything else [1].

On the other hand, in 2017, the City Park Ltd. In 2017, I audited the security concepts of all the planned (New Hungarian National Gallery) and completed (National Museum Restoration and Storage Centre, Hungarian House of Music, Olof Palme House) investments of the Liget Budapest Project.

Finally, at the beginning, I found regulatory and procedural problems and shortcomings in relation to the protection of artefacts, mainly at the domestic level, which are of concern to experts at the international level, such as the International Council of Museums (ICOM) International Committee for Museum Security (ICMS) and the Security of Major Museums Europe (SOMME).

2 Institutional ArtWork Protection

The topic of my basic research is *Institutional ArtWork Protection in the context of temporary exhibitions*, within the framework of which I aim to study and compare Hungarian and international institutions, which provides an opportunity to learn about the complete Institutional ArtWork Protection, both in terms of asset protection and conservation protocols [1].

2.1 Institutional

When it comes to the topic of my basic research, the first question that regularly comes up is why I chose the institutional sector and not the private sector as my research topic.

The reason for my choice of topic is very simple, since artworks, as in all countries of the world, are predominantly state-owned or in state custody, i.e. in museum deposit [2], while privately owned artworks are the treasures of art collectors and are either safely hidden from the outside world or are also deposited in a museum.

Artworks may also be placed in public collections for the following reasons, among others:

Storage and exhibition: where special conditions are created for the proper preservation and storage of artefacts, such as the right temperature, humidity and lighting to minimise damage or premature ageing. On the other hand, they provide the necessary fire protection in addition to adequate protection.

Research and study: public collections institutions allow researchers, historians, artists and other experts to study artefacts. They also help researchers, historians and historians of art and science.

Education and training: public collections institutions use artefacts for educational purposes. School groups, students, teachers can visit these places and use the artefacts to teach lessons or to implement other forms of education, such as creative or creative classes.

Reaching and building community: public collection institutions make artworks accessible to the community, which can connect with and share messages with visitors through community events, cultural events and exhibitions. After all, through exhibitions, programmes and events, they give people the opportunity to meet and study different works of art directly.

Preservation of cultural values: public collection institutions contribute to the preservation of the artistic, historical and cultural heritage of a given culture or period.

These reasons show that institutions are not only physical repositories, but also centres for the preservation, study, education and community participation of cultural heritage, which in almost all cases requires specific and unique solutions for the ArtWork Protection.

2.2 ArtWork Protection

In the institutional system of public collections, the concept of Institutional ArtWork Protection covers two broad areas: the protection of the property of artefacts and the conservation of the artefacts.

Property Protection of Works of Art

The protection of works of art requires special attention, as the value of works of art is not only material, but also intangible. The ArtWork Protection includes physical and security measures to minimise damage, theft or other adverse effects such as vandalism. These complex conservation protocols include, but are not limited to: display, storage, warehousing, packaging and transport.

In addition to mechanical protection (bars, special doors and locks), security alarm and camera systems will be installed in public collection institutions, which, when combined, can provide a higher level of protection. An important role is played by monitoring and regulating access to, and occupation of, areas containing artefacts, for example by means of access control systems. A particularly important role is played by security protocols that ensure that only authorised persons have access to certain areas (warehouse, storage, packing), so that only authorised persons have access to the artefacts.

The continuous preventive and safety-conscious training of security staff should be highlighted, as well as the knowledge of a pre-prepared emergency plan and its annual exercise.

In addition to foresighted, artefact-specific planning and continuous monitoring of the exhibition, it is also crucial to ensure the necessary storage security. The primary objective of warehouse security is to protect valuable objects, artefacts or documents from theft, damage or other threats, but it also means protecting the physical and mental safety of the storage areas, objects and the people working in them. Therefore, security measures should be applied in combination, depending on the value and nature of the individual artefacts.

Conservation of Works of Art

The conservation of artefacts is the sum of measures taken by public collection institutions to preserve the value of artefacts and their existing conditions in the long term. Conservation covers a wide range of measures, including environmental conditions of works of art during storage, restoration and conservation, on which there is already a large body of research, publications and legislation.

I would like to highlight just a few key aspects of the conservation of artefacts. Maintaining the right temperature and humidity is essential for the protection of artefacts, but we must not forget about harmful UV radiation and excessive light, which can also damage artefacts. Therefore, regular inspections and surveys will allow us to prevent possible damage or changes to the artefacts. The restoration and conservation of these changes, damages or potential damages should be carried out by qualified personnel, following the necessary safety protocols.

From the above, it is clear that the conservation of artefacts is a complex process that requires an integrated approach, as the measures to be taken depend on the nature of the artefact and the expertise and tools required to restore it.

2.3 Temporary Exhibitions

A temporary exhibition is an exhibition of art or an exhibition focusing on a particular theme that lasts only for a limited period of time and is usually temporary (not usually repeated). Temporary exhibitions can be held in museums, art galleries, libraries or other public collections or cultural institutions.

Temporary exhibitions are an opportunity for public collection institutions to regularly offer visitors fresh content and to highlight current artistic or cultural themes. In doing so, they contribute to cultural diversity and maintain visitor interest in the public collection institution.

Therefore, planning for the protection of a temporary or even a new permanent exhibition is not just a matter of defining a standard asset protection plan or security protocol, but much more than that, because it involves preparing for very different risk factors, hazards and environmental conditions. The frequency of possible risk events and the extent of possible damage vary during the organisation of temporary exhibitions and from the start to the end of the project, so the structure of the protection system put in place must be flexible and its elements must be defined in proportion to the threat.

3 Public Collection Storage

Public storage involves the storage, care and preservation of valuable artefacts. When stored in warehouses or depots, care must be taken to ensure appropriate environmental conditions for public collections, adequate security measures and the prevention of possible damage.

Public repository management should also include stock management, organisation and documentation. Institutions should monitor the condition of materials and, where necessary, carry out conservation or restoration interventions to preserve materials.

3.1 Definition of Public Collection Storage

Warehouse security is a concept that refers to the physical and mental protection of storage locations, objects and the people who work in them [4–6] The primary purpose of warehouse security is to protect valuable objects, artefacts or documents from theft, damage or other threats. Warehouse security includes a number of important elements, including:

Physical security: the physical security of the warehouse means the protection of the building, the fence, doors and windows. Measures to restrict access and movement within the warehouse.

Security protection: alarms, signalling systems, security cameras (CCTV), access control systems and robots (humanoid, robot dog, operator) and the use of a security control centre and monitor room.

Fire safety: fire safety is an important part of warehouse security. This includes the installation of fire extinguishers, the use of fire alarm systems, firefighter training and rapid response in the event of a fire. A particularly important task is the choice of the appropriate extinguishing technology, which can be addressed in a separate case-specific study.

Anti-theft protection: physical protection measures such as locks, doors and alarm systems may be used as part of the warehouse's protection against theft. In addition, internal control procedures, asset tracking and regular inventory will also help to protect against theft. A layered security policy should be developed covering basic protection and specific additional artefact-specific protection.

Monitoring environmental conditions: it is important to maintain optimal environmental conditions for artefacts or "sensitive" objects. This includes temperature and humidity control, protection against airborne dust, UV radiation and other harmful effects. Current energy policy challenges mean that sustainability and renewable energies must also be taken into account.

Data protection: protecting data and documents stored in the warehouse is also key. Electronic systems, backups and access control options help to keep data secure. A number of standards and recommendations from the business sector can be taken up, for example ISO 27001 [6].

Preventive personal protection and safety awareness training: educating and training employees on safety procedures and how to deal with emergency situations is vital. The provision of personal protective equipment for warehouse workers also contributes to physical safety.

Cybersecurity: cybersecurity in digital warehouses is also of paramount importance. Protecting against stored data and its accessibility is critical to warehouse security.

Warehouse security includes all measures that prevent hazards and risks, effectively protect assets and the safety of people working in the warehouse. Specialist security experts and designers are usually involved in the design and maintenance of security measures.

3.2 Risks in Public Collection Storage

There are many risks involved in the safe storage of public collections. These risks can be grouped according to their characteristics. A secure repository requires a risk analysis covering each of these groups and risks. Risks should be assessed both generally and individually.

In the case of a general survey, the analyses are those for the warehouse, whereas in the case of an individual survey, the individual artefact and the warehouse are analysed together and separately. The former is to assist in the planning of the safety of the warehouse and the latter is to assist in the procedure for a particular exhibition or entry into storage. Security risks in public storage:

Physical hazards: theft, damage and vandalism, fire and explosion, water damage, equipment failure, other natural disasters.

Environmental factors: temperature and humidity, light and UV radiation, airborne dust, acid/alkali.

Pest: rodent, mosquito, fungus/bacteria.

Access problems: unauthorised access, improper handling.

Digital threats: cybersecurity threats, data erasure and loss.

Budget and resource constraints: institutions with low budgets, lack of equipment and technology.

4 Conclusions

The public collection as a collection term includes libraries, archives, museum institutions, image and sound archives owned or founded by the state, local government, national minority governments, public corporations and public foundations [7].

In Hungary, there are a total of 1710 public collection institutions, so there is no question that the number of public collection institutions and the number of institutional artefacts they hold is significant in Hungary as well.

In Hungary, public museum repositories are mostly state-owned, with a small number of municipal, religious or private repositories, and the real estate holdings of rural museums are owned by municipalities.

Public collections storage is a function of cultural institutions that fall into the category of public collections, such as libraries, museums or archives. The purpose of public collections is to collect, preserve, document and make available to the community materials of cultural, artistic or scientific value, and for this purpose the storage of artefacts is essential.

In a public collection institution, storage is a critical factor in the ArtWork Protection to ensure their safety. In addition to the general considerations, the site and the artifact must be taken into account when designing appropriate protection. According to current professional principles, storage is an integral part of the protection of artefacts, in addition to exhibition, whereby the storage of artefacts is reduced to the expected/accepted level of risk by taking into account a number of risk factors.

These public institutions hold rich collections covering many aspects of culture, art, history and national values. Public collections can take many forms, including museums, libraries, archives, galleries and other similar institutions that preserve valuable artefacts and documents for future generations, and it is therefore necessary to address Institutional ArtWork Protection.

References

1. Lőrincz L, Berek T (2023) The role of closed-circuit television in institutional artwork protection as a subsystem of smart city. Interdiscip Descr Complex Syst 21(4):394–395
2. 19/2015. (IV.2.) EMMI regulation on the museum deposit. Available from 02 May 2015
3. László Z (2008) What rules should be taken into account for the storage of goods in warehouses. National Institute for Vocational and Adult Education
4. Hofstra N, Petkova B, Dullaert W, Reniers G, de Leeuw és S (2018) Assessing and facilitating warehouse safety. Saf Sci 105:134–148. https://doi.org/10.1016/j.ssci.2018.02.010
5. Halawa F, Dauod H, Lee IG, Li Y, Yoon SW, Chung SH (2020) Introduction of a real time location system to enhance the warehouse safety and operational efficiency. Int J Prod Econ 224:107541. https://doi.org/10.1016/j.ijpe.2019.107541
6. ISO standard 27001 (2022) Information security, cybersecurity and privacy protection Information security management systems
7. 1997. évi CXL. Act on Museums, Public Libraries and Public Culture. The (o.) point of Annex 1

Measurability Aspects of the Effectiveness of Cybersecurity Awareness Programs

Sandor Magyar and Imre Dobak

Abstract Users are the easiest targets for attacks from cyberspace. Preventive actions play a significant role in protection, one form of prevention is Cybersecurity Awareness. To get feedback on the effectiveness of awareness, measurements need to be taken. Measuring the results and effectiveness of security awareness programs is more difficult than follow-up surveys after training and education. The impact of awareness-raising on cognitive abilities must be measured, i.e. the extent to which the correct response occurs after threats are perceived and recognised. This publication presents several methods and solutions for measuring information security awareness programs, as well as evaluation methods to assess their effectiveness.

Keywords Awareness · Cybersecurity · Education · Information security

1 Introduction

One of the most effective ways to protect against threats from cyberspace is prevention, including awareness. As 74% of all breaches include the human element, [1] this area is particularly important, especially given the challenges of the spread of emerging and disruptive technologies.

An important element in a complex approach to cybersecurity is to increase human awareness of cybersecurity in almost all aspects of life. Security-conscious behavior is essentially an internal motivation to increase security, understand the cybersecurity context, recognize and avoid risks, and protect oneself.

Its importance is indisputable nowadays, as the labor market is facing a significant shortage of cybersecurity professionals, thus raising the value of general awareness

S. Magyar (✉) · I. Dobak
Institute of National Security, Ludovika University of Public Service, Ludovika Tér 2, 1083 Budapest, Hungary
e-mail: magyar.sandor@uni-nke.hu

I. Dobak
e-mail: dobak.imre@uni-nke.hu

T. A. Kovács et al. (eds.), *The Impact of the Energy Dependency on Critical Infrastructure Protection*, Advanced Sciences and Technologies for Security Applications, https://doi.org/10.1007/978-3-031-78544-3_40

in society and targeted training and preparation of those who already have IT and related skills [2].

Certainly, the learning pyramid also shows that there are different rates of knowledge retention for different forms of training in each of the traditional (passive) teaching methods and teaming (active) teaching areas [3].

The widely researched issue is essentially one of the different forms of training and education, ranging from traditional forms of training to specific awareness programs on cyber security threats. Regardless of their internal content, their importance is not only to increase awareness of the specific threats and ways of avoiding them but also to shape attitudes. This can be seen as the basic principle of these programs, on the basis of which it is possible to transfer knowledge of different depths and to make it known to the participants.

However, it may be called into question to what extent these programs achieve their objectives, and to what extent the different forms of knowledge transfer help in the acquisition of knowledge. How can the necessary knowledge be transferred and fixed in the most effective way possible in a constantly changing and evolving cybersecurity environment, what are the most effective education, training, and awareness methods?

The results of all this can be assessed by measuring the results of security awareness, which is also a complex issue that goes beyond the testing of knowledge. On the security awareness side, however, measurement may not be so straightforward because the difference between knowledge and awareness may have specific characteristics. In the case of knowledge, we can talk about a simple test that can be measured well, while awareness, rather, can be validated more effectively by simulations. Recently, several valuable research results and scientific publications have been published in this field [4].

2 The Arenas of Cyber Awareness

Traditional forms of education are seen as a basic element of training and education, but in our digital world (especially after the COVID pandemic) they have changed significantly. The environment for education and training has changed and, in addition to face-to-face forms of education, online and other IT-based training solutions, which were already available in the field of cyber security, have become more prevalent [5]. The effectiveness of e-learning training, in particular AI-enhanced training, is also showing an increasing trend in identifying weaknesses in machine learning. These identified weaknesses can also be adequately addressed by online learning materials. The significance of this in the field is not only a change in the form of education, as the threats and challenges themselves come from this medium, thus offering many opportunities for participants to acquire new knowledge through up-to-date, relevant examples using modern methodology. Without going into their forms and various methods, the measurement of the effectiveness of awareness programs and their methods are important for us in this study.

In the case of a company, security awareness can be focused on increasing the security culture at the organizational level, which is the sum of security-conscious attitudes and behaviors at the individual level and can refer to the overall "resilience" of the whole organization.

It is understood that all forms of education and training have advantages and disadvantages, which may be the basis for their application to a particular target group, and that the limited resources and time available for training are often mentioned. However, these should not be a reason for cyber security awareness programs to be interesting, interactive, and effective. In cybersecurity education that truly forms the mindset of the individual, it is important to enhance critical thinking, and interactivity, and develop and strengthen related logical skills [6]. It is no coincidence that practical, situational programs are increasingly becoming the focus of awareness training, which can even be complemented by cybersecurity board games. However, advances in technology support a concrete range of possibilities, so chatbots already appearing on everyday IT endpoints are already enhancing awareness-raising opportunities [7].

3 Objectives and Possibilities for Measuring the Results and Effectiveness of Security Awareness Programs

Regardless of the internal content of cybersecurity education, training, and other awareness structures, an important aspect at the end of the process is to measure effectiveness. The difficulty arises from the fact that, while traditional training often measures the level of learning of the educational material, security awareness training aims not only to transfer knowledge but also to "shape attitudes". On the one hand, they are used to assess the effectiveness of the chosen methodology and, on the other hand, they provide feedback on the participants' knowledge. An important aspect is that these measurement results may indicate the need for possible modification of the program structure, change of the internal content, or other training methodologies. They can also help to replace ineffective solutions and make necessary improvements to internal content in a rapidly evolving area of cybersecurity challenges that require continued attention even for programs that are deemed effective.

However, measuring effectiveness is difficult, especially when there are also questions about the "durability" of a security-conscious approach. Simply stated, just because someone passes an exam in a particular area of cybersecurity knowledge does not necessarily mean that their cybersecurity awareness behavior will be permanent. All of this indicates that progress requires more measurements, and baseline data (benchmarks) against which to measure either the acquisition of knowledge of a security awareness program or the 'durability' of the maintenance of conscious behavior.

In a large corporate organization, there are several ways of measuring the initial level of security awareness, ranging from traditional methods of broad-based surveys (tests, questionnaires, exercises, interviews) to surveys of a selected sample (representative sampling), and for example, employee responses to security incidents created for testing purposes. Regardless of the type and form in which the measurements are carried out, the aim is to establish the level of safety awareness at an organizational level, which will serve as a reference at the end of any further education, training, or awareness-raising process.

For measurements, it is necessary to compare the results achieved against a reference point, which identifies the level of development after training and education. In the case of traditional training, examination-type measurements can be effective, but in the case of awareness, which targets the cognitive domain, social engineering-type audits are more likely to identify the results of the knowledge acquired. Effectiveness can be optimally determined by the same type of input and output measurements. If the initial measurement is done with tests, then the tests are the most objective way to compare the progress of the participants in the program. Targeted programs, such as the detection of phishing attempts, can be identified measurably by audit-type reference measurement.

The international literature comprehensively addresses the complex issue of cybersecurity awareness programs, their design, and aspects related to awareness [8]. Methods and solutions for measuring the stated objectives of programs can be considered a specific subtopic of these, and we will formulate the following aspects concerning this:

- Traditional forms of measurement include tests and various final tasks, which can be presented as a concluding part of the awareness programs. These include targeted interviews, to get the views of participants, which can also provide valuable feedback.
- The measurement aspects can be multifaceted so that in addition to existing security knowledge, they can include knowledge of security aspects of device use, the degree of motivation, or even other aspects of awareness, the familiarity of different sources, and behavior [9]. Another study [10], in the field of cyber security, assesses awareness through the measurement of existing "knowledge", "self-observation of skills", "actual abilities and behavior", "attitude".
- The complexity of surveys with quantitative and qualitative elements is indicated by the fact that awareness is related to the IT knowledge of the participants, their IT tools, and their Internet usage habits [11]. As part of the survey, the age or even the educational level of the participants is an important element, assuming the general finding that older and more educated people have more experience in certain areas, which is however clear due to the specificities of cyberspace. In cyberspace, there are of course several other areas of influence that may contribute to awareness, among which we would highlight the cybersecurity incidents suffered. Unfortunately, experiencing cyber security incidents is a highly effective 'awareness-raising' exercise for individuals and organizations, from which lessons learned can serve as an example for others.
- Employees response to practical tasks, even to unsolicited fictional security incidents created for testing purposes, which can measure the presence/or absence

of a higher level of conscious thinking. The significance of its application can be seen in its unexpected nature, in its similarity to real security incidents, and can allow the level of security awareness of a given corporate community to be tested, even periodically (randomly).

- Exercise can also be an effective form of measurement. This solution is a proven method in the military and cyber defense fields, where participants can learn the importance of awareness and conscious behavior through scenario-based exercises. In this context, several areas such as crisis communication, media news management, etc. can be tested, which also provides feedback on the level of awareness. With the possibility of a comprehensive approach to understand systemic effects, awareness-raising on other important areas can also be used, where the development of cooperation between different actors is established. An advantage can be that, in addition to the theoretical educational elements, practical examples are more durable in the longer term and often the extent of theoretical knowledge acquisition can, in our opinion, be better measured through concrete examples. Fictional events, real-life situations, and examples can all be part of these practical forms.

The use of chatbots can be an effective tool to raise awareness. Artificial intelligence already contributes effectively to awareness. However, their use can provide additional feedback in the area of statistical data analysis of specific actions and proposals. The trend lines provided by chatbots can provide useful information in the qualitative domain, based on their numerosity or risk results. They can help identify areas of deficiency that need to be improved during security awareness. In our opinion, several aspects of information security awareness need to be examined. The following factors could form elements of a complex evaluation method to assess the effectiveness of awareness programs:

- Self-assessment: participants self-awareness of cybersecurity issues, which may not be an adequate measure on its own, but can be linked to the recognition of motivation and conscious thinking when measured over several periods.
- Motivation: the internal need for conscious behavior and conduct in the field of cybersecurity. This may include individual recognition of the importance of challenges and risks and the emergence of a need for development (e.g. training). In cyberspace, conscious behavior reflects a need for security, open to security-conscious thinking, and opportunities to enhance it (e.g. training, education).
- Competences: different individual and professional skills to protect against cybersecurity threats in the daily operations in cyberspace. These are related to the complex set of cybersecurity knowledge and the need to consciously monitor the constantly evolving challenges.
- Knowledge: in our dynamically evolving world, digital competencies and different levels of internet and device usage are aspects of this. Acquiring knowledge of the threats, forms of attack, and defensive solutions related to cybersecurity in varying degrees of depth. Its internal content is multi-level, requiring continuous development and follow-up. Under this aspect, education can help to identify and defend against challenges in a more conscious and targeted way.

- Existing skills: these include generational differences in the use of particular solutions and applications in cyberspace, which can mean different knowledge and skills in particular areas.
- Overload, task completion under time pressure: Stress at work as well as time factors can also affect awareness, as the pressure to comply can lead to non-conscious clicks, but measuring this is extremely difficult without simulating real environments and situations. The probability of errors occurring increases with overload. Measuring this can also be a tool, especially for managerial awareness.

Another specificity of awareness is that different themes and possibly different methods are effective for different levels and areas. Thus, tailored programs and measurement methods may be needed for the senior management level, the user level, and the IT operator/developer level. These may also influence the measurement options and methods. Measurement at the management level is one of the most challenging tasks, especially in the area of communicating results. Confronting managers with weaknesses and deficiencies works effectively for managers who are committed to security.

Measurement can also be influenced by measuring the level of awareness of the individual or the group. At the individual level, it is a measure of one's own individual areas of awareness. However, when measuring a group, it is also necessary to take into account the influencing factors that affect each other. The role played in the group, such as the 'influencer effect', the leadership role, and the opinion of the person who is perceived to have more experience, can all influence the safety awareness of the group. The effectiveness of measurement can be effectively achieved by treating the group as a group, the feedback from which, at both group and individual level, can usefully be presented.

Objective measurement can be when the target audience is not aware of either the initial or the final measurement being taken. This can be done, for example, by sending a phishing letter to users after an internal authorization has been conducted, and then analyzing the execution of the phishing again after the awareness program has been sent to users.

4 Summary

In order to avoid having to identify the effectiveness of security awareness in the future based on the incidents suffered, it is necessary to provide feedback on the effectiveness of security awareness and to carry out appropriate measurements. Without measurements, it is not possible to identify the weaknesses of individuals or even groups or the themes for further awareness programs.

Of course, in the case of awareness training, the knowledge given back in the classical examination may not yet lead to security-conscious behavior in given situations, so measurement also provides a kind of orientation in the area of effectiveness, in the identification of gaps. It is necessary to focus on the achievement in the cognitive domain during awareness, which focuses on understanding the effects of negative processes after suffering damage rather than fear of punishment after an incident.

There are several ways to perform measurements, which can be both passive and active. Of course, active methods, such as in education, can lead to greater accuracy. The range of these methods should ensure that the effectiveness of awareness, and hence the identification of threats and the effectiveness of responses to them, is appropriate. Questionnaires and written tests typically rely on knowledge, which cannot be perfectly matched to the response in a given situation when measuring awareness.

There are several possibilities for measurement, but the method used to generate the reference data is also an effective way of back-testing. The initial knowledge measured in questionnaires and tests is best compared by conducting similar tests after the awareness program has been conducted.

References

1. Enterprise V (2023) Verizon data breach investigations report
2. Tăbuşcă A, Garais G, Enăceanu A (2022) Cybersecurity education—the new literacy. J Inf Syst Oper Manag 16(2):263–283
3. Maine B. The learning pyramid by the national training laboratories, USA
4. Marnewick A, von Solms S (2022) The development of cybersecurity awareness measurement model in the water sector. In: International conference on cyber warfare and security, vol 17, pp 211–218. https://doi.org/10.34190/iccws.17.1.43
5. Yurdakal İ (2023) Digital awareness in the digital age: awareness scale regarding digital transformation in education (Das): validity and reliability study 3:138–157
6. Sareen A, Jasaiwal S (2021) Need for cyber security education in modern times. Int J Multidiscip Trends 3(2):188–191. https://doi.org/10.22271/multi.2021.v3.i2c.179
7. Abdelhamid S, Mallari T, Aly M (2023) Cybersecurity awareness, education, and workplace training using socially enabled intelligent chatbots. In: Guralnick D, Auer ME, Poce A (eds) Creative approaches to technology-enhanced learning for the workplace and higher education. TLIC 2023. Lecture notes in networks and systems, vol 767. Springer, Cham. https://doi.org/10.1007/978-3-031-41637-8_1
8. Huraj L, Lengyelfalusy T, Hurajová A, Lajčin D (2023) Measuring cyber security awareness: a comparison between computer science and media science students, TEM J 12(2):623–633. ISSN 2217-8309. https://doi.org/10.18421/TEM122-05
9. Zwilling M, Lesjak D, Natek S, Phusavat K, Anussornnitisarn P (2019) How to deal with the awareness of cyber hazards and security in (higher) education?, thriving on future education, industry, business and society. In: Proceedings of the MakeLearn and TIIM international conference, ToKnowPress
10. Pooja PR, Shashidhar R (2022) Evaluation of student's awareness towards cyber security. Phronimos: Kiams J 2(4):33. ISSN: 2583-3839
11. Karaca İ, Söner Ö (2023) An evaluatıon of students cybersecurıty awareness in the marıtıme industry. Int J 3D Printing Tech Dig Ind 7(1):78–89

Safety Technological Testing of the Compound Bow Structure

László József Mónus

Abstract In this work, I present the change in the structure of compound bows. I studied the design of bows in detail. After examining the material and production technology of the bow arms, the modeling of the bows is also discussed. After researching the literature, I used them to examine the sports equipment. In order to learn about the important parameters of bow limbs, we carried out material tests. We tested three major brands of compound bows. We tested by making lightened test specimens from the bow limbs. Bending tests and tensile tests clearly showed the properties of the bow arms. The bending test and tensile test results of the three major brands were detailed and compared, and they were presented with common diagrams. Thanks to the microscopic images, the bow manufacturing technology was clearly defined. After summarizing the results, the first conclusion was formulated and the most important plans for the near future were also discussed.

Keywords Compound bow · Safety technology · Carbon fiber · Fiber glass · Tensile test · Bending test

L. J. Mónus (✉)
Doctoral School on Materials Sciences and Technologies, Obuda University, Budapes, Hungary
e-mail: monuslaszlojozsef@gmail.com

T. A. Kovács et al. (eds.), *The Impact of the Energy Dependency on Critical Infrastructure Protection*, Advanced Sciences and Technologies for Security Applications, https://doi.org/10.1007/978-3-031-78544-3_41

1 Evolution of the Structure of Compound Bows

1.1 Compound Bows in the 1960s and 70s

In the 1960s, the first versions of compound bows were made. These bows are made of wood, plastic or aluminum. Due to their name, they contained cams at the end of the bow arms, the cams had several functions. Compared to today's cams the cams of the 1960s bows were much smaller and provided a 1, 5—3 times smaller gear ratio. The cams also helped the archer to maintain the draw length when the bow is drawn, and the shooter can transfer the greatest energy to the arrow through the bow. It also extends the archer's time during release and helps keep the arrow at the maximum draw length until the moment of release (Fig. 1).

The bow arms(limbs) were made of wood laminated with fiberglass according to the technology of the time. Their disadvantage was that their structure was weaker, they were easily damaged, and they were extremely sensitive to weather conditions [2–4] (Fig. 2).

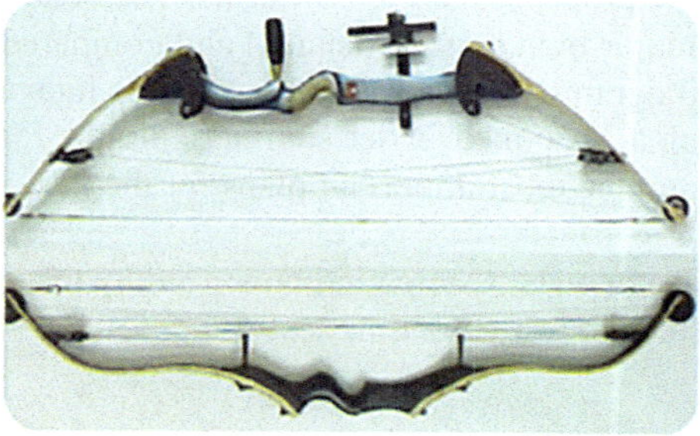

Fig. 1 Compound bows in the 1960s [1]

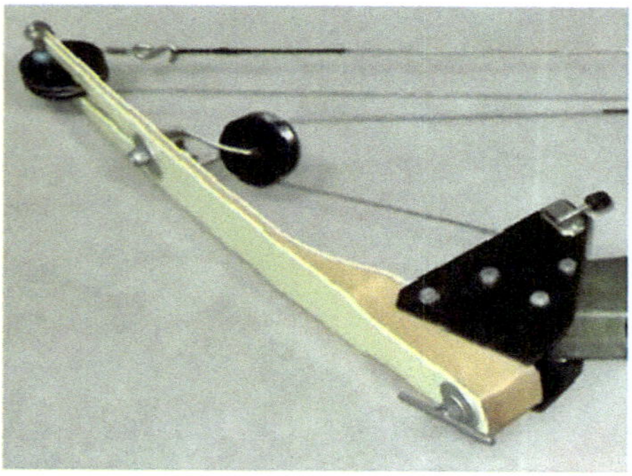

Fig. 2 The CAM and bow arm (limb) of the 1960s structure [1]

1.2 Compound Bows in the 1980s and 90s

In the 1980s and 1990s, increasing the performance of bows became a priority. They would like their bows to become faster, stronger, more durable, more resistant to environmental conditions and capable of covering greater distances. The design of the bows, apart from the handle, the bow no longer contained wood (Fig. 3).

The need to reduce the weight of the structure, to increase its stability, to absorb and dampen rising tensions and vibrations has increased. In the 1980s, newer designs exploded into bow production. Ideas about the ideal bow have changed, the spread of CNC technology and carbon fiber-reinforced fabrics has opened the way for designers, and better and better materials have resulted in high-performance structures. The biggest, striking differences compared to contemporary designs are the length of the coiled bows, the size of the coils, the size of the supporting tension threads, and the natural material of the bow string. These bows were much longer and required a longer draw length than today's. The reason for this is that the construction was not tight enough, so they tried to compensate for the shortcomings with the help of longer bow arms and greater bow strength (weight). This made the bow much more vulnerable, it was at this time that the passion for technical competition really came into play in the sport, so this period brought the most material problems, bow arm, and auger breaks string and tension string breaks, as well as athlete injuries. They often used almost farcical grip shapes and stretched bow shapes in the hope of serious success. [2–4]

Fig. 3 Compound bows in the 1980s and 90s [2]

1.3 Today's Compound Bows

The most modern recurve bows are built in the following way. The cams have become larger, eccentric, and provide the shooter with a 6–9 × gear ratio. Its operating principle is the following: the cam has to be tipped over its axis but after that it is easy to hold (Fig. 4).

The greater cam tipping makes it easier to hold the bow. In competitions, the pulling force of the bows was limited, the constructions became shorter and "tighter" for more dynamic solutions. The bow string made of synthetic materials like dyneema and the stiffening strings of the bow arms helped to provide stability against the stresses that occurred. Fiberglass, glass fabric and carbon fiber, carbon fabric have become the alpha and omega of the new design bows. development has made shooting more balanced and faster, and bows have a longer range [2–4] (Fig. 5).

2 Initial Tests and Emerging Problems

The biggest threat to archers and the structure of bows is the empty solution the so-called "dry fire". The empty release occurs when the arrow slips out of the bow string for some reason, in this case the load of the shot does not go through the arrow body but attacks the structure of the bow. In this case, both the compound bow and the athlete who uses it can get serious injuries. The failure factor is further increased by the fashionable firing mechanism of the "thumb release machine", which provides a perfect solution at the press of a single button. Bow manufacturers often promise that their bows can withstand more than 100 dry fires, but in reality this cannot be said at all. Either the physical integrity of the shooter is affected, or the structural defect can cause damage of thousands of euros.

Fig. 4 The modern eccentric compound bow cam

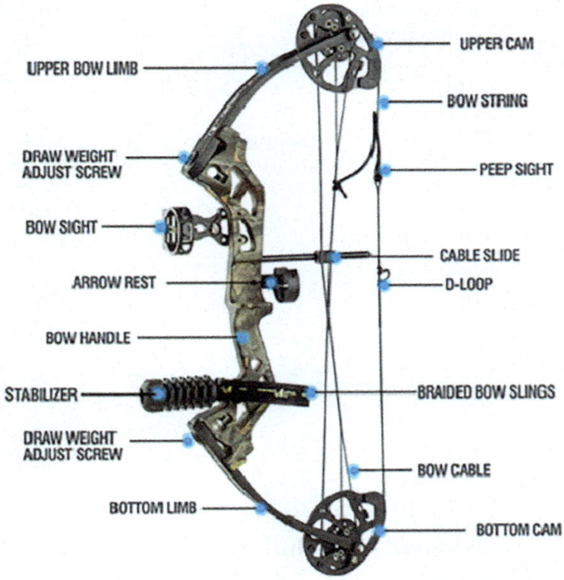

Fig. 5 Structure of the modern compound bow [5]

In the past year, we tested several compound bows with empty solution and we got unexpected results during our tests. It turned out that, regardless of the bow manufacturer and construction, damage of a completely similar nature occurred in the bow arms, as a result of the empty solution. Each time the damage shown in Fig. 7 occurred in the reinforced carbon fiber bow arm [6] (Fig. 6).

Fig. 6 Damage to the arm of the compound bow after empty release [6]

Fig. 7 Top view of the cleavage in the microscope image

We examined three different types of compound bows, with some types the problem shown in Fig. 7. Occurred as a result of the serial loading, and with the other types the cleavage occurred as a result of the empty solutions. Tested the major brands, in all cases the bow limbs split lengthwise at the end closer to the cam, but before the cam, regardless of the shape of the bow arm. The cleavage even attacked the inner layer of the composite bow arm. The cleavage creates an irreparable condition in the bow arm, which must be replaced afterwards, and on the other hand, the broken pieces can be dangerous for the physical integrity of the shooter and the competitors standing next to the shooter. The damaged bow arms were subjected to microscopic examination, we wanted to examine the microstructure of the fractures. Figures 8 and 9 show the top view of the longitudinal splitting of the bow arm with a microscope.

Figure 9. shows a side view of the cleavage with a microscope. It can be seen that the upper glass fiber layer has been completely replaced. In this case, the cleavage was approximately 26 mm long, but in several cases it exceeded 40–50 mm in length.

3 Composite Material of the Bow Limbs

3.1 Laminated Bow Limbs

The bow arm can be made with three types of technology. It can be made by lamination, which is now the least effective technology of the three. The modern bows produced in this way are very sensitive, they are less resistant to the stresses that occur, and they are extremely sensitive to weather conditions. In addition, the available performance is also more limited. [6, 7]

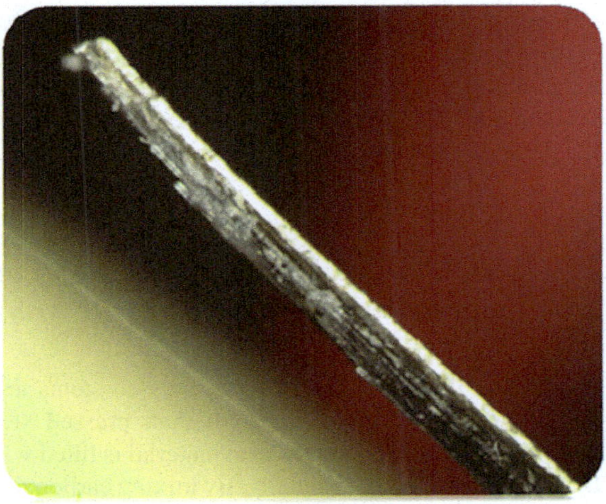

Fig. 8 Microscope image of the split from the side

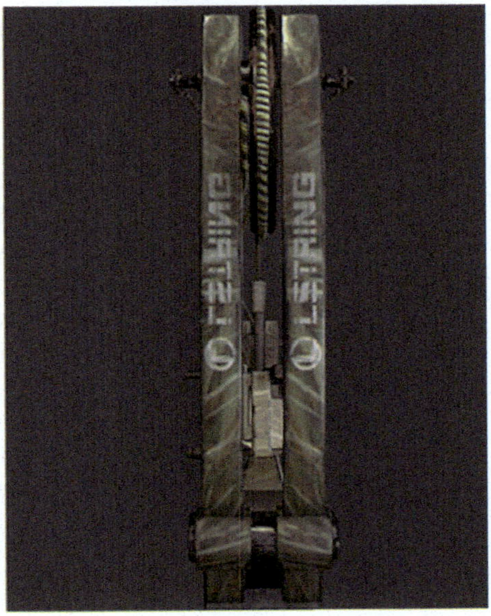

Fig. 9 The arms of the bow from the top view

3.2 Solid Fibre-Reinforced Fiberglass/Carbon Fiber-Based Extruded Composite Arms

This type starts with chopped glass fibers and carbon fibers suspended in molten epoxy resin, this hot paste is forced under high pressure through a rectangular extrusion die to form a long rod. Let this bar cool. This bar is mostly the right width and thickness. Requires minimal post-machining and contouring to provide specific bow arm bends. Major manufacturers use this technology. Since the fibers lie linearly in one direction while they are pushed through the extrusion die, their torsional stability is low and not quite adequate. As a result, it is particularly sensitive to the spread of cracks, cracks, abrasions, scratches, even in the case of a sports equipment used in the field, can develop and spread in seconds [6–10].

3.3 Solid Fiber-Reinforced Fiberglass/Carbon Fiber-Based Compression Molded Composite Arms

Comparing the two latter designs, pressed stems are definitely more durable and provide greater torsional strength than pressed stems. During the compression molding technology, the bow arm material is filled with fine woven fiberglass fabric. Both vertically and horizontally. By mixing carbon fiber through an epoxy bath, they are separated into several points horizontally like a spindle. It is then wound between two rods, the rods have a horizontal drive shaft, the two rods rotate, so the fabric can be wound back between the rods at several points. At the same time, the bars oscillate forward and backward, so the fabric lies on the bars at certain points with a width of 30–50 mm. The machine operator cuts these fabric strips, takes the transparent mass between two bars and lays it in the press tool. The press is then closed and after a specified time, at a specified pressure and temperature, the epoxy sets. The semi-finished bows are then removed from the press and allowed to cool to room temperature. The bow limbs taken out of the press tool only need to be shaped to a small extent, they are almost ready to shape after being taken out. The fibers of the bow arms made in this way were oriented very favorably due to the fabric tape. For this reason, their torsional stability is higher than the ends of the bow arms and the load of the tilting worm gears is better tolerated. At the same time, they are also less sensitive to weather conditions [6–10] (Fig. 9).

In the case of the latter two technologies, we can easily check where the fracture or crack originates. The surface to be tested is cut with high precision from the bow arm, and then the epoxy resin is burned out of it. After that, the location and the direction of the cleavage are clearly visible.

4 Modeling the Compound Bow

In order to further develop the composite material of the bow arm, as well as to protect the construction from structural damage and to prevent injuries to athletes, we need to know the exact material properties of the bow arm and the parameters that characterize them. To do this, first I tried to model the structure. In order to determine the stresses arising in the bow and affecting the bow arm, we need to know its most important material characteristics. In the coming period, I would like to define and

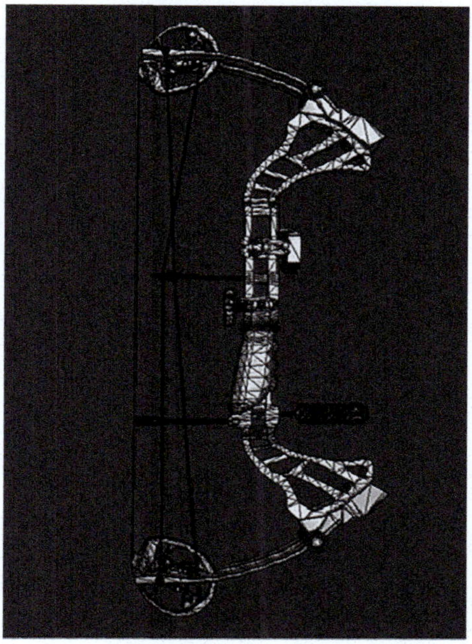

Fig. 10 The model of the compound bow

model the most important parts of the bow as accurately as possible according to the results obtained from the material tests. In this way, we can get to know the factors that can be changed to create a bow arm that is more ideal than before. This can change the design of compound bows in terms of safety and performance as well. Figure 11 shows the model of the recurve bow, the microscopic images of the bow arm of which we have already seen, and we performed material tests on them as described below. With the appropriate material characteristics, by simulating the construction with a finite element, I would get a clear picture of the magnitude and type of stresses on the structure. We would find out which parts should be changed and what kind of work and costs this would entail in terms of structural design, material changes, and material development (Fig. 10).

5 Material Tests

5.1 Three-Point Bending Test

In order to determine the material characteristics of the bow arm, we need to perform material tests. To this end, I subjected the bow arms to a three-point bending test. I wanted to test the products of several bow manufacturers for resistance to bending

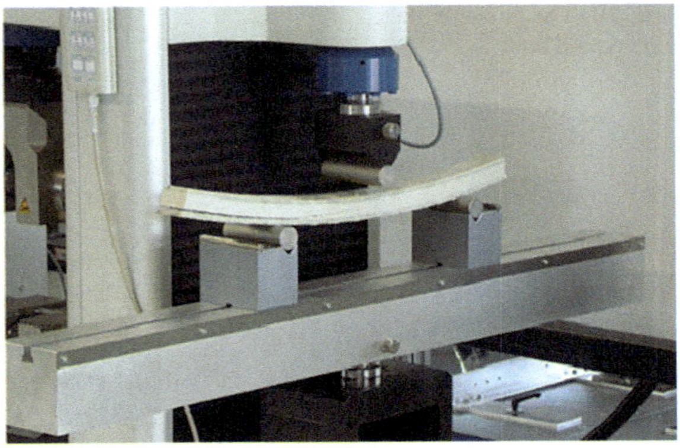

Fig. 11 Bending stand of a three-point bending test [11]

stress. The tests were carried out in the test laboratory of the Kermi ÉMI-TÜV in Szentendre (Fig. 11).

We performed the tests on the arms of top-class bows. Three-point bending was performed on PSE Xforce, Vulture and Hori-Zone bow arms, during the measurements we recorded the bending of the arms depending on the bending force acting on the bow arms, knowing the support distance (Fig. 12 and 13).

After recording the measured values, we performed the test on 5 bow arms and 5 specimens. Similarly, according to the diagrams shown in Figs. 13 and 14, the diagram of the Vulture bow arm was also developed.

The following test values were obtained:

On the Hori-Zone 1 workpiece, the specimen broke at a maximum bending force of $\mathbf{F_{max\,1} = 5480\ N}$, at which point the bow arm produced a deflection of $\mathbf{dL_1 = 10.6\ mm}$.

On the Hori-Zone 2 workpiece, the specimen broke at the maximum bending force of $\mathbf{F_{max\,2} = 5770\ N}$, at which point the bow arm produced a deflection of $\mathbf{dL_2 = 9.6\ mm}$.

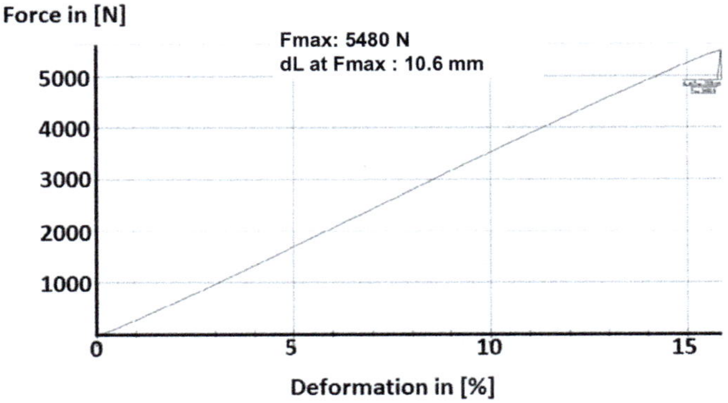

Fig. 12 Three-point bending test results of the Hori-Zone bow limb

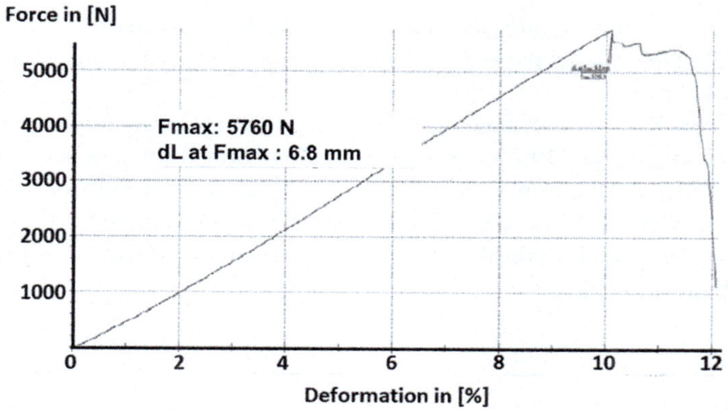

Fig. 13 Three-point bending test results of the PSE Xforce bow limb

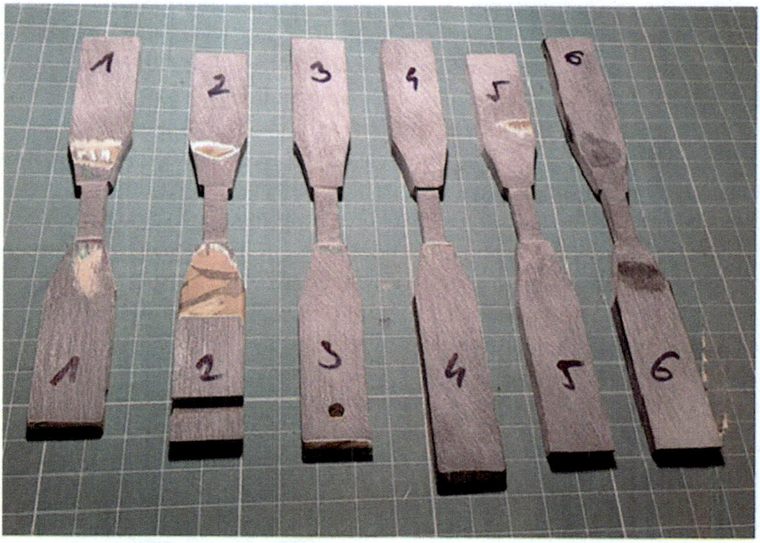

Fig. 14 The lightened test speciments

In the case of PSE Xforce, the specimen broke at the maximum bending force of $\mathbf{F_{max\,3} = 5760\ N}$ on workpiece 3, at which point the bow arm produced a deflection of $\mathbf{dL_3 = 6.8\ mm}$.

In the case of PSE Xforce, the specimen broke at the maximum bending force of $\mathbf{F_{max\,4} = 5790\ N}$ on workpiece 4, at which point the bow arm produced a deflection of $\mathbf{dL_4 = 6.9\ mm}$.

In the case of the Vulture, the specimen broke at a maximum bending force of $\mathbf{F_{max\,5} = 5760\ N}$ on workpiece 5, at which point the bow arm produced a deflection of $\mathbf{dL_5 = 6.8\ mm.}$

So the PSE Xforce, withstand the highest maximum bending force at the moment of its break ($\mathbf{F_{max\,4} = 5790\ N}$) and the Hori-Zone bow arms produced the highest deflection ($\mathbf{dL_1 = 10.6\ mm.}$) during the tests. Based on the results of our measurements, we determined with the measuring staff of Kermi that these test specimens were particularly weak against bending stress, even among the composite materials, none of them stood out.

5.2 *Tensile Test*

After the tests, we also wanted to perform a tensile test on the pieces, I returned to the Kermi testing center with new bow arms to examine their tensile strength, their elongation during uniaxial tensile stress, and other important parameters. In the first week, we measured the bow arms with a Zwick Z010 digital tensioning machine. During the tests, however, we did not reach any results, the bow arms did not break in the tearing machine, a week later we also tried on a machine with a load capacity of 100 kN, which also did not lead to results. The pieces turned out to be too strong, their tensile strength is enormous. At that time, there were no more digital devices left in the laboratory, as the larger 850 kN tensile machine with hydraulic grip is suitable for manual measurement. The bow limbs didn't break it just de-laminated at the grips. In order to obtain tangible data, I made test specimens from the bow arms, simplified their structure in the following way, in order to obtain high-precision tensile diagrams and test results on lower-powered digital tensile machines [12, 13] (Fig. 15).

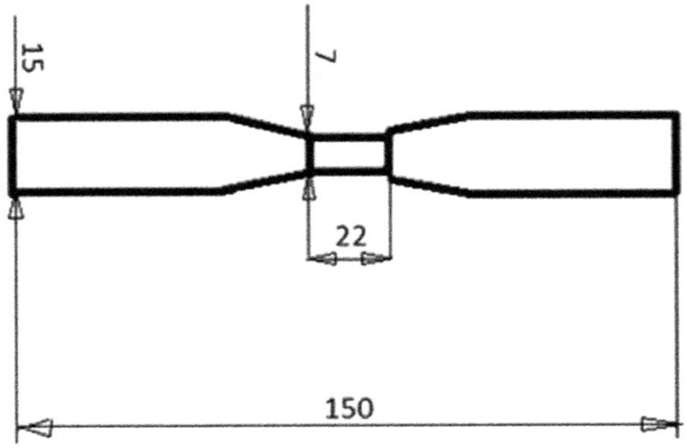

Fig. 15 Main dimensions of the lightened specimens (thickness 1, 5 mm)

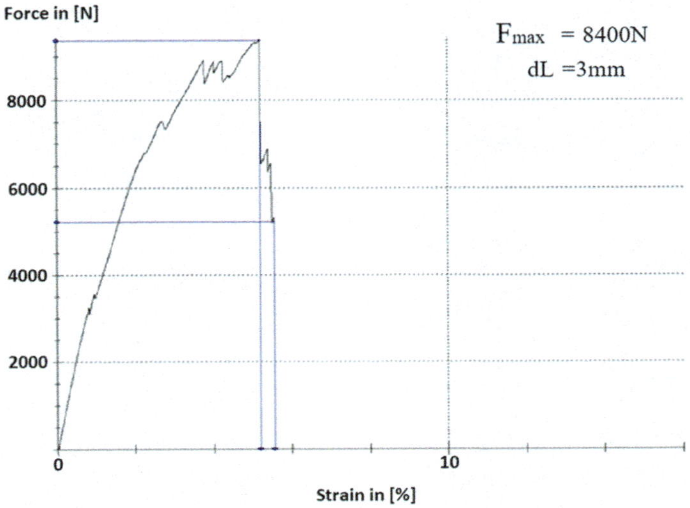

Fig. 16 The tensile diagram of specimen 1

5.3 *The Results of the Tensile Tests*

PSE Xforce Bow Arm's force-strain diagram (1) (Fig. 16).

During the tensile test of the PSE Xforce specimen, we measured the maximum tensile force $F_{max\ 1} = 8400$ N, at which point the equipment showed $dL_1 = 3$ mm strain. The starting cross section was $S_0 = 10.5 mm^2$,

The ultimate strenght (R_1max) was:

$$R_1\ max = \frac{F_{max\ 1}}{S_0} = \frac{8400\ \mathbf{N}}{10.5\ mm^2} = 800\ MPa$$

Hori-Zone Bow Arm's force-strain diagram (2) (Fig. 17).

During the tensile test of the Hori-Zone specimen, we measured the maximum tensile force $F_{max\ 2} = 7560$ N, at which point the equipment showed $dL_2 = 2.8$ mm strain. The starting cross section was $S_0 = 10.5\ mm^2$,

The ultimate strenght (R_2max) was:

$$R_2\ max = \frac{F_{max\ 2}}{S_0} = \frac{7560\ \mathbf{N}}{10.5\ mm^2} = 720\ MPa$$

We tested the Vulture bow limbs as well, we arranged the tensile diagrams in one large common diagram for easier comparison. During the tensile test of the Vulture specimen, we measured the maximum tensile force $F_{max\ 3} = 9360$ N, at which point the equipment showed $dL_2 = 4.1$ mm strain. The starting cross section was $S_0 = 10.5\ mm^2$ (Fig. 18).

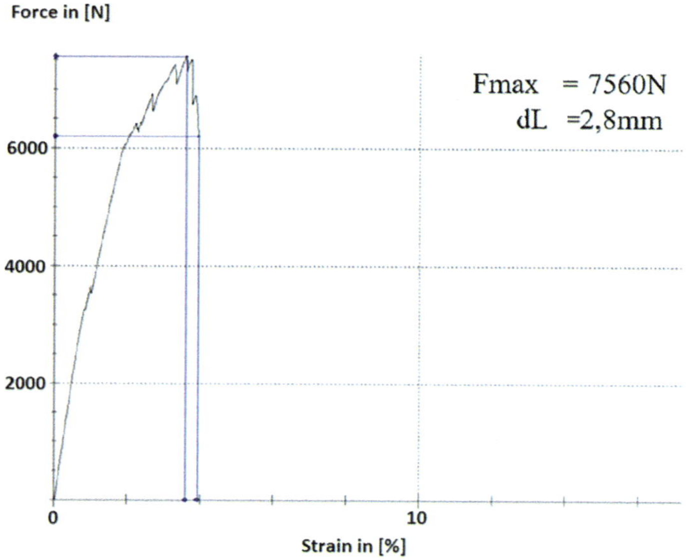

Fig. 17 The tensile diagram of specimen 2

$$R_3 \max = \frac{F_{\max 3}}{S_0} = \frac{9360\ \mathbf{N}}{10.5\ \text{mm}^2} = 891.42\ \text{MPa}$$

Nr	Specimen identifier	F_{max} N	dL at F_{max} mm	F_{Break} N	dL at break mm
1		8400	3,0	3600	3,9
2		7560	2,8	6200	3,1
3		9360	4,1	5220	4,4

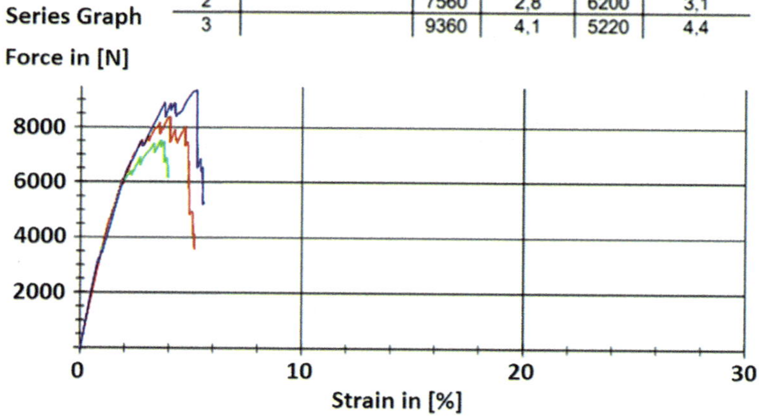

Fig. 18 The specimens on a common comparative tensile diagram

Fig. 19 Microscope image of the PSE Xforce bow limb

From the comparison, it can be seen that there are major differences between the products of two manufacturers, the tensile tests were performed on a Zwick250 digital tensile machine, in accordance with the ISO 4674-1 standard, the test temperature in the measuring laboratory was 20 °C. Considering the composite material, we worked with a tensile speed of 5 mm/min, the Pre-Load was 1 N in all cases.

We visually inspected the fragments of the torn specimens in the laboratory and recorded the following images (Figs. 19 and 20):

From the pictures of the fragments, it can be recognized that among the bow arm production technologies described earlier in this work, the examined bow limbs were made according to the extruded technology. This technology is much more sensitive to torsional forces, which is why this problem occurred so regularly in the case of these sports equipment. The fibers embedded in carbon and fiberglass epoxy arranged in the longitudinal direction are clearly visible, however according to my current knowledge, this is exactly the weakness of this technology.

6 Conclusion and Near Future Plans

The results measured during tensile tests are still very recent data. In addition to these, I have already sent several new workpieces to the measuring laboratory, on which we will perform shear and fatigue tests in the near future, and I got the idea of a fatigue test that can be applied specifically to composite materials from the experts at Kermi. As far as material development and material tests are concerned, I would

Fig. 20 Microscope image of the Vulture bow limb

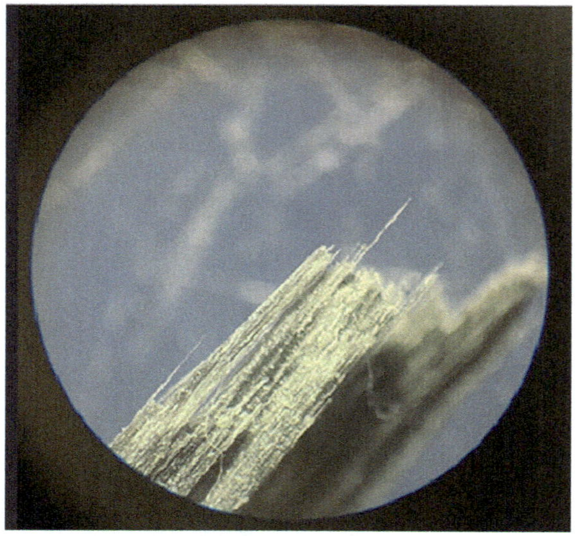

also like to examine the compression molded material structures made with the previously presented "Compression Molded" technique. I think, according to my current knowledge, that the structure of the bow arms will achieve better material characteristics and be more resistant if built according to this technology. Furthermore, I would like to investigate the elongation of the extreme local threads, the sensitive behavior of the pieces to various finely shaped processing. In my next works, I plan to try the installation and testing of new materials, for example Kevlar fibers, as well as "Dyneema" fabric/fibers, which, as far as I know, are already used in the structure of recurve bows, namely in the fibers that build the strings. [12, 13]

References

1. Source Image, ArcheryTalk.com. https://www.archerytalk.com/threads/vintage-compound-bow-collection.1621201/page-4
2. Jan H Sachers, History of Compound bows. https://www.bow-international.com/features/the-history-of-compound-bows/
3. Mitch Vaughan, Modern Archery Compound Bow. https://www.scribd.com/document/129528700/Modern-Archery-Compound-Bow
4. Mihai Robert, SIL Hybrid Bow. https://www.scribd.com/document/487272204/SIL-hybrid-bow
5. Source Image, Biztonságtechnika az íjászatban. http://oieminta.ingyenweb.hu/keret.cgi?/jobbalsoalapok.htm
6. Peter Campbell, What to do if You Accidentally Dry Fire a Bow. https://archershub.com/what-to-do-if-you-accidentally-dry-fire-a-bow/ (last access 15.12.2024.)
7. Compound Bow Limbs, Laminated or Solid Glass/Graphite? https://engin1000.pbworks.com/f/Bow+Limb+Manufacturing.pdf (last access 15.12.2024.)

8. Marc Alvin Lim, Carbon fiber reinforcement 2008. https://www.scribd.com/document/58160353/Carbon-Fiber-Reinforcement
9. Tehreem Imran, Fiber Glass 2008. https://www.scribd.com/document/552625300/Fiber-Glass
10. Quazi TH, Shubhra Glass Fiber NR Composite. https://www.scribd.com/document/512523382/Glass-Fiber-Nr-Composite
11. Source Image, Grimas Homepage. https://grimas.hu/termek/harom-es-negypontos-hajlitoallvanyok/
12. Callister Jr WD, Rethwisch DG (2000) Material science and engineering
13. Ashby MF, Jones DRH (1980) Engineering materials, vol 1

Cybersecurity Challenges in the Age of Drones: Navigating the Integration of Unmanned Aerial Vehicles with Cyberspace

Zoltán Nyikes and László Tóth

Abstract This article explores the security challenges related to drones in cybersecurity. It provides a historical overview of drone development and its diverse applications. The narrative underscores the increasing integration of drones with cyberspace. Protection against unauthorised access and potential cyber threats becomes a central theme. The article emphasises securing drone positioning systems, communication channels, and data storage. Proposed solutions include employing multiple Global Navigation Satellite Systems (GNSS) and implementing cryptographic measures. The Identification Friend or Foe (IFF) system is examined as a means to enhance drone safety. TEMPEST solutions and a self-destruct system are recommended to minimize risks. Ongoing advancements in drone cybersecurity are highlighted, including encryption, integrated defense systems, education, monitoring, regulations, simulations, and collaboration. The article concludes by presenting future trends and areas for improvement, stressing the importance of comprehensive security strategies for the responsible integration of drones into modern society.

Keywords Drones · Cyber security · Critical infrastructure

Z. Nyikes (✉)
Milton Friedman University, Budapest, Hungary
e-mail: nyikes.zoltan@uni-milton.hu

L. Tóth
Banki Donat Faculty of Mechanical and Safety Engineering, Obuda University, Budapest, Hungary

T. A. Kovács et al. (eds.), *The Impact of the Energy Dependency on Critical Infrastructure Protection*, Advanced Sciences and Technologies for Security Applications, https://doi.org/10.1007/978-3-031-78544-3_42

1 Drones and Cyberspace

1.1 Drones

Drones (unmanned aerial vehicles—UAVs) are becoming increasingly popular in various industries, including military, civil, and commercial applications. The history of drones dates back to the 1900s when Nikola Tesla first experimented with wireless remote control. During the Second World War, drones were first used for military purposes, notably exemplified by the V-1 flying bomb. The AQM-34 Ryan Firebee, developed in the United States in the 1960s, was one of the early drones for military use. In recent decades, technological advances and cost reductions have facilitated the widespread commercial use of drones. In agriculture, drones are frequently employed to monitor fields and manage crops [1]. The use of drones in infrastructure monitoring, research and rescue operations is also growing significantly. In the entertainment industry, drones are used for filming, aerial photography, and various events. Legal regulations for drone flights vary by country and govern aspects such as altitude and flight zones. Beyond industrial and commercial applications, drones are also gaining popularity among hobbyists for personal use. The development and proliferation of drones pose new challenges for air traffic control, cyber defence, and the private sector [2].

1.2 Cyberspace

The history of cyberspace dates back to the early 1960s when the US government began developing a network called ARPANET, the precursor to the internet. In the 1980s, cyberspace spread from military and scientific infrastructure to the civilian sector. In the 2000s, cyberspace took on a new dimension with the widespread deployment of Internet technologies and services. The protection of cyberspace has become increasingly important in parallel with the growing number of cybercrime, espionage, and cyber-attacks. Cybersecurity strategies have evolved from the early days of firewalls and anti-virus programs to systems capable of handling complex threats. States and companies are investing significant resources in cyber security measures to strengthen their defences. Cyberspace is constantly changing and evolving, driven by the introduction of new technologies and the interconnectivity of devices. Cyber warfare and cyber espionage are phenomena that take place in cyberspace and affect international relations. Ethical hackers and cybersecurity professionals play a crucial role in safeguarding cyberspace by identifying and addressing potential vulnerabilities. In the future, the introduction of quantum cyber technologies and the development of artificial intelligence may create further challenges and opportunities in cyberspace protection [3].

1.3 The Link Between Cyberspace and Drones

Protecting drone cybersecurity is essential, as many drones can be remotely controlled online. Their vulnerability to cyber threats makes drones easily vulnerable to various attacks. Connectivity to cyberspace makes drones vulnerable to unauthorized access, which can compromise data or functionality. Cybercriminals could potentially take control of drones, leading to dangerous consequences. Unauthorised access creates the potential for theft or manipulation of data recorded by drones. Cyber-attacks can disrupt the navigation of drones, leading to accidents or disasters. Cyber threats to drones monitoring critical infrastructure or sensitive areas can compromise not only the safety of the drones themselves but also that of the targets they are observing [4]. When used in military applications, insufficient cybersecurity can lead to significant national security risks. Attacks by cybercriminals can easily render drones uncontrollable, potentially leading to deliberate damage. Cyber threats can damage the drones' communication channels, preventing safe control. In the event of cyber espionage, sensitive information collected by drones could fall into unauthorised hands. Protecting the cybersecurity of unauthorised drones can prevent the damage they can cause to other drones or infrastructure [5]. Making drones resistant to cyber-attacks is necessary to ensure the data integrity from drones. Cyber security protection helps prevent intentional interference or manipulation of drones. The lack of cybersecurity of sensors and cameras connected to drones can lead to privacy concerns. Because of their vulnerability to cyber threats, drone protection increases social and economic stability. The connectivity of drones to cyberspace enables their use for disaster response purposes, and without protection, they can be vulnerable. Inadequate cybersecurity could jeopardize the integration of drones into the aviation system. Protection against cyber-attacks can enhance the reliability of remote control and navigation systems used by drones. Protecting the cyber security of drones contributes to the responsible and safe use of technology in modern society. It can therefore be concluded that drones are inseparable from cyberspace and that they are a manifestation of it [6].

2 Cyber Defence Capabilities of Drones

2.1 Protected Geolocation System

The safe operation of drone positioning systems is critical for their widespread applications and integration into everyday life. Without sensitive and accurate positioning, drones are not only compromised in their effectiveness but also become a potential source of danger. In the following sections, we will discuss the importance of security in drone positioning systems, the risks associated with its absence, and potential solutions to enhance the reliability of these systems. Making drones' positioning systems safer is necessary primarily for navigation and precise targeting. Drones need to accurately track their spatial position and direction, mainly when operating over their flight areas or critical infrastructure [7]. The lack of reliability of the positioning system can have serious consequences, such as collisions with other aircraft, violations of flight restrictions or even catastrophic accidents. In cyber security, drone positioning systems can also be vulnerable. Unauthorised access, data manipulation or even remote control can compromise the security of drones and their environment. Attempts by cybercriminals or malicious actors to manipulate the positioning systems of drones could deliberately interfere with navigation, which could compromising the safety of drones and land objects. One possible solution is redundancy and reinforcement of GNSS (Global Navigation Satellite System) systems. The combined use of several GNSS systems, such as GPS, GLONASS, Galileo or BeiDou, can improve the accuracy and resilience of positioning systems to cyber threats. In addition, the incorporation of ground-based navigation reference points and sensors can also help to increase redundancy and reliability.

The other solution is introducing cryptographic and cybersecurity measures in drone positioning systems. The use of enhanced authentication and encryption protocols in communication between drones and location systems, along with protecting the integrity and authenticity of incoming data, is crucial. These measures can help make such systems more resilient to external cyber threats. Cryptographic solutions will be explored in detail in a separate section. Therefore, the safe operation of drone positioning systems not only ensures more efficient and secure drone operations but also protects the environment and the community. Continuous innovation and the involvement of cybersecurity experts are crucial to developing a cyber defence strategy for drones and minimising threats.

2.2 Protected Communication Channel

The safe operation of a drone's communication system is crucial for the efficient and safe control of drones and safe communication with the environment and other aircraft. The lack of a secure communication system poses serious risks. The following sections explain why security is important, the potential risks, and solutions to improve drone communication systems. The absence or vulnerability of drone communication systems raises several problems. Unauthorised access, communication interference or malicious interference can have serious consequences. One potential risk is the loss of remote control, which could cause drones to become uncontrollable or to deviate from their programmed routes [8].

Cybersecurity risks are also significantly increased due to the weaknesses or vulnerabilities of drones' communication systems. Cybercriminals or malicious hackers could launch targeted attacks against communication channels and thus disrupt the regular operation of drones [9]. Data theft and the leakage of sensitive information collected by drones can have serious consequences, including compromising privacy and security. One solution is cryptographic protection of the drone communication system and strong authentication. The use of encrypted communication channels can help prevent unauthorised access and direct interference in data exchanges between drones and control centres. Robust authentication procedures ensure that only authorised users can control the drones [10].

Encrypted communication

The security of the drone communication system can be further enhanced through the use of encrypted communication solutions, as mentioned earlier. These technologies allow data to be sent and received in such a way that it can only be understood by the designated recipients and cannot be observed or understood by third parties in an accessible form. The first step to setting up a secure communication system is to use cryptography. Instead of being sent as plain text, the data is encrypted using a mathematically complex algorithm that only parties with the correct secret key can decipher. This prevents unauthorized access to the data and ensures that its content remains understandable only to authorized parties [11].

End-to-end cryptography can be particularly practical for drones. In this case, the communication between the source and target is fully encrypted and only decrypted at the endpoints. Since communication between the drone and the controller is secure, potential attackers or unauthorized individuals cannot access the encrypted data, even if it is intercepted. Another crucial aspect is secure key management. Keys are integral to the cryptographic process, and their secure storage and exchange are essential for the system's reliability. Using key management protocols and cryptographic procedures can help prevent unauthorised access or compromise of keys [12].

The third measure is implementing regular updates to cryptographic communication solutions. As cryptographic systems continuously evolve, updates are essential to address new security challenges. Timely updates to software and firmware ensure that the cryptography used always meets the latest security standards.

Frequency-hopping system

The drones' proprietary communication system can be enhanced by frequency hopping, increasing its security against external interference and unauthorised access. Frequency hopping is a technique whereby the communication frequency changes over time and, the two communicating units synchronously track this change. This method can effectively reduce cyber security threats and other types of interference. Below, we will illustrate how frequency hopping can be applied to drone communication systems [13].

Dynamic Frequency Hopping. The system dynamically changes the frequency through which the drones communicate. This change is rapid and regular, and both ends of the communication system must follow it. Dynamic frequency hopping makes it difficult for external intruders to continuously monitor or jam the signal.

Pseudo-random Frequency Hopping. The system uses a pseudo-random algorithm for frequency hop. This method ensures that frequency hops appear random, making it difficult for unauthorized parties to design deliberate access or interference.

Synchronized Frequency Hopping. Synchronized frequency hopping between drones allows all communicating units to use a standard frequency at a given time and then change the frequency together. This increases the efficiency and coordination of the system.

Implementing communication protocols. Secure communication protocols, such as cryptographic techniques and authentication, should be used with the hopping system to prevent unauthorised access and manipulation of data. Using redundant communication channels can also contribute to the safe operation of drones. Simultaneous using multiple communication systems, such as Wi-Fi, radio frequency and mobile networks, increases the system's resilience to possible interference or attacks. Continuous updates and increased software security are also essential to protect the communication system of drones. Regular updates and updated software that closes security holes can minimise potential vulnerabilities that cybercriminals can exploit.

2.3 IFF—Identification, Friend or Foe

The IFF (Identification Friend or Foe) system is an effective tool for identifying drones and enhancing their safe operation. It differentiates between friendly and hostile entities in airspace and military areas, aiding organizations and systems in operating more safely and efficiently. The following section details how the IFF system can be utilized on drones to improve safety [14].

Set Identification Tags. Each drone is equipped with a unique identification tag that can be detected by the IFF system. This identification can be a code or a mark carried by the drone.

IFF Radio communication. Drones equipped with an IFF system can communicate via radio with ground stations and other aircraft. The IFF system enables drones to communicate and signal clearly with other systems in their environment. The IFF system can be enhanced with frequency hopping techniques to help prevent interference and unauthorized access. Cryptographic protection can also be integrated into the IFF system. This ensures that communications are safeguarded by encryption, making the data interpretable only by authorized systems. These measures prevent unauthorized access and safeguard identification data.

Integration of the IFF System with Air Traffic Management Systems. The IFF system for drones can be integrated with air traffic management systems. This integration will enable air traffic controllers to identify and track drones in the airspace, thereby minimizing potential conflicts and enhancing flight safety.

Adherence to Standards and Protocols. When implementing an IFF system, adhering to industry standards and protocols is crucial to ensure compatibility with other aircraft and ground systems.

2.4 *Tempest*

Applying TEMPEST (it's not an acronym, it's the name for protection against compromising radiation) solutions can enhance the safety of drone operations, particularly by protecting against compromising emissions. TEMPEST is a system of standards that includes protection measures against electromagnetic emissions. The following describes how these measures can be applied to drones [15].

Electromagnetic Noise Filtering. Integrate electromagnetic noise filters into the electronic components of drones to reduce interference and enhance system performance. These filters can help reduce electromagnetic noise, making it difficult for outside observers to intercept drone activity or communications.

Physical Isolation. Sensitive electronic components should be physically isolated from other systems to minimize electromagnetic interference and prevent emissions from being detected by unauthorized individuals.

RFI (Radio Frequency Interference) Detection. Deploy sensors and systems to counter radio frequency interference (RFI) that can instantly detect and alert you to external electromagnetic interference. Timely detection enables a rapid response and appropriate action. Additionally, employing TEMPEST solutions to manage electromagnetic emissions can enhance drone safety by minimizing cybersecurity risks and reducing external interference.

2.5 *Self-Destruct System*

To enhance drone safety, a built-in self-destruct system can be valuable, particularly for protecting data and sensitive information. The following section will explain how such a system can be implemented to provide active protection in case of drone loss or unauthorized access [16].

Data Encryption. The self-destruct system ensures that all stored data is encrypted. Sensitive information, such as maps and other data, is stored on the drone only in encrypted form.

Remote Access Unauthorized. The self-destruct system includes a mechanism that enables the owner to remotely activate self-destruct or delete stored data if the drone is lost or falls into unauthorized hands.

Exit Code or Signature. A unique exit code or signature, known only to the authorized owner, can be assigned to the self-destruction system. This code or signature allows the owner to remotely initiate self-destruction or delete data.

Geolocation Tracking and Remote Control. A geolocation tracker can be integrated into the self-destruct system, enabling the owner to monitor the drone's exact location. Additionally, remote control capabilities allow the owner to restrict the drone's movement or initiate self-destruction if it falls into unauthorized hands.

Refining the Self-Destruct Process. The self-destruct system can be enhanced with refinements to increase efficiency, such as causing physical damage, erasing memory, or completely shutting down the drone's power.

Remote Notification System. When the self-destruct system is activated, the owner should be notified immediately. This can be done through a push notification on a mobile app or an email alert, allowing the owner to respond quickly.

Self-Destruct Conditionality. A self-destruct system can include conditional triggers for activation. For example, it may activate if the drone deviates significantly from its programmed route or if it does not receive confirmation from the owner within a specified time frame.

Documented Activity and History. The self-destruct system can log all activities and events, including the reason for activation, the exit code or signature, and any remote access attempts. This documentation can also serve legal or security purposes. Integrating a self-destruct system helps prevent unauthorized access and protects sensitive data on drones, thereby enhancing security in compromised situations. However, it is crucial that such systems are carefully designed and thoroughly tested to minimize the risk of failure or misuse.

2.6 Protected Data Recording and Storage System

Enhancing drone security can be achieved through the use of a secure data recording and storage system. This system contains the drone's flight data, communications, control, and data recorded by sensors. The following will show how such a system can be implemented to maximise safety [17].

Hardware Cryptographic Modules. Integrate hardware cryptographic modules into your data capture and storage system. These modules provide a high level of encryption and prevent unauthorized access, even in cases of physical tampering.

Multi-layer encryption. Use multi-layer encryption methods to protect your data. Flight data, communications, control, and sensor data should all be encrypted separately and only decrypted with authorized keys.

Biometric Authentication. Implement biometric authentication methods, such as fingerprint or facial recognition, to ensure that only the rightful owner or authorized individuals can access the recorder and storage.

Authorisation management. Implement robust authorization management in the recording and storage system. Ensure that only users or devices explicitly authorized have access to the data.

Physical Security. Place the data recording and storage system in a physically secure location within the drone. For example, position the device that stores flight data and sensitive information in an area that is difficult to access and protected against unauthorized tampering.

Remote Data Access Control. If necessary, implement remote data access control mechanisms. This enables the owner to remotely manage and access the data stored on the drone in exceptional circumstances or if the drone is lost.

Logging and Alerts. Record all data access and manipulation attempts, and alert the owner if unauthorized access or tampering with the recording system is detected.

Automatic Lockdown in case of an attempt. If the recording system detects an unauthorized access attempt, it will automatically lock and notify the owner or authorized organization of the security incident.

Restore Points. Regularly create restore points to enable data recovery in the event of a system failure or unauthorized access.

Regular Updates and Security Audits. Security measures should be updated regularly, and regular security audits should be performed on the recording and storage system. In addition, be involved in following industry security recommendations and updates.

3 Current Developments and Research Trends

The future of drone cybersecurity is closely linked to broader trends in cybersecurity and information security. Current developments and research directions focus on the following areas to further improve the cyber protection of drones.

Develop encryption technologies that consider the specific characteristics of the airborne environment to provide a higher level of encryption of the data collected to prevent unauthorised access and interception.

Integrated Defence Systems. Develop integrated protection systems designed to detect and safeguard drones against various cyber-attacks, such as remote malware and network intrusions [18].

Cyber Defence Education and Awareness. Proper cybersecurity education and awareness for drone operators and users are crucial. It is essential that they understand basic cyber-attacks and know how to defend against them.

Continuous Monitoring and Detection. Special attention should be given to continuous monitoring and anomaly detection within drone networks and systems. Utilizing artificial intelligence and machine learning technologies can facilitate automated threat analysis and enable immediate responses.

Regulations and Compliance. Governments and regulators must continuously update regulations and policies related to drone cybersecurity. Companies and organizations must keep abreast of these changes and comply with regulations.

Simulations and Exercises. Conduct practical simulations and tests on drone cybersecurity systems to prepare them for real-world attacks and incidents.

Cooperation and Information Exchange. Industry stakeholders, research institutes, and government agencies must collaborate and share information on best practices and current threats related to drone cybersecurity. Based on the previous points, it is recommended to develop comprehensive security strategies and support ongoing research and development in cybersecurity to enhance drone defense. Regular reviews and swift responses to evolving threats are crucial for maintaining drone security. While the future of drone cybersecurity will present numerous challenges, it also offers significant opportunities for improvement and research. Some possible directions for improvement and suggestions for further enhancing drone cyber security are listed below.

AI and Machine Learning Based Detection. Employ AI and machine learning algorithms to automatically detect cyber-attacks and threats [19]. Such systems can learn from business as usual and detect unusual or suspicious activity [20].

RFI (Radio-Frequency Interference) Detection. Develop effective RFI detection and protection systems to detect and prevent electromagnetic interference and jamming attempts against drones. Deploy drone-specific Intrusion Prevention Systems that actively detect and block interference from malicious or unauthorized drones.

Blockchain-based Security. Employ blockchain technology to securely store and authenticate data collected by drones. This approach enables easy verification of data integrity and authenticity.

Quantum Security. Develop and deploy quantum security solutions alongside traditional cryptographic methods, as the future growth of quantum computing capacity may pose risks.

Drone Network Security. Develop network security solutions to secure the entire drone fleet. A centralized system allows for more effective monitoring and management of network events.

Data Protection and Anonymization. Research to anonymize the data collected by drones and protect personal information while still allowing you to process the information you need.

Automated Security Updates. Develop automated update mechanisms that quickly and efficiently apply security patches to drone systems and software.

4 Summary

The article discusses the security challenges associated with drones in the context of cyber defence. It examines the historical development of drones, their expanding use across various sectors, and the legal frameworks that regulate their operations. The integration of drones with cyberspace introduces cybersecurity concerns, highlighting the need to protect against unauthorized access and potential cyber threats.

The focus then shifts to safeguarding drone positioning systems, communication channels, and data recording and storage systems. The vulnerabilities in these areas could lead to accidents, unauthorized control, or data manipulation. The text proposes solutions such as using multiple Global Navigation Satellite Systems (GNSS), implementing cryptographic measures, and employing secure communication protocols.

Furthermore, the discussion extends to the Identification Friend or Foe (IFF) system, which can enhance the safety of drone operations by helping identify and track drones in airspace. Implementing TEMPEST solutions and a self-destruct system is recommended to minimize electromagnetic emissions and protect sensitive data in the event of loss or unauthorized access.

The article highlights the need for continuous developments in drone cybersecurity, emphasizing encryption technologies, integrated defence systems, education, monitoring, regulations, simulations, and cooperation among stakeholder cooperation. It concludes by presenting future trends and potential directions for improvement, including the use of AI and machine learning, blockchain-based security, quantum security, network security, data protection, anonymization, and automated security updates. Overall, the text underscores the need for comprehensive security strategies to ensure the responsible and safe integration of drones into modern society.

References

1. Carou D (2021) The aerospace sector. In: Aerospace and digitalization. Springer briefs in applied sciences and technology. Springer, Cham. https://doi.org/10.1007/978-3-030-67849-4_2
2. Custers B (2016) Drones here, there and everywhere introduction and overview. In: Custers B (eds) The future of drone use. Information technology and law series, vol 27. T.M.C. Asser Press, The Hague. https://doi.org/10.1007/978-94-6265-132-6_1
3. Krämer NC, Neubaum G, Eimler SC (2017) A brief history of (social) cyberspace. In: Holyst J (eds) Cyberemotions, understanding complex systems. Springer, Cham. https://doi.org/10.1007/978-3-319-43639-5_2
4. Omolara AE, Alawida M, Abiodun OI (2023) Drone cybersecurity issues, solutions, trend insights and future perspectives: a survey. Neural Comput Applic 35:23063–23101. https://doi.org/10.1007/s00521-023-08857-7
5. Poursoltan M, Pinède N, Vallespir B, Traore MK (2022) A new modeling framework for cyber-physical and human systems. In: 2022 Annual modeling and simulation conference (ANNSIM), pp 90-101, San Diego, CA, USA. https://doi.org/10.23919/ANNSIM55834.2022.9859402.

6. Sanger DE (2017) Cyber, drones, and secrecy. Underst Cyber Confl 14:61–78
7. Barreiro JM, Lara JA, Manrique D, Smith P (2023) Towards the portability of knowledge in reinforcement learning-based systems for automatic drone navigation. PeerJ Comput Sci 9:e1402. https://doi.org/10.7717/peerj-cs.1402
8. Benjamin G (2022) Drone culture: perspectives on autonomy and anonymity. AI Soc 37(2):635–645
9. Sabuwala, NA, Daruwala, RD (2023) Drones: architecture, vulnerabilities, attacks and countermeasures. In: Sharma H, Saha AK, Prasad M (eds) In: Proceedings of international conference on intelligent vision and computing (ICIVC 2022). ICIVC 2022. Proceedings in adaptation, learning and optimization, vol 17. Springer, Cham. https://doi.org/10.1007/978-3-031-31164-2_18
10. Derhab A et al (2023) Internet of drones security: taxonomies, open issues, and future directions. Vehicular Communications 39:100552
11. Haque MS, Chowdhury MU (2018) A new cyber security framework towards secure data communication for unmanned aerial vehicle (UAV). In: Security and privacy in communication networks: Securecomm 2017 international workshops, ATCS and SePrIoT, Proceedings, vol 13. Springer International Publishing Niagara Falls, ON, Canada, pp 113–122, 22–25 Oct
12. Pyzynski M, Balcerzak T (2021) Cybersecurity of the unmanned aircraft system (UAS). J Intell Robot Syst 102:35. https://doi.org/10.1007/s10846-021-01399-x
13. Luo H et al. ESCM: an efficient and secure communication mechanism for UAV networks. IEEE Trans Netw Serv Manage. https://doi.org/10.1109/TNSM.2024.3357824
14. Carou D (2021) The aerospace sector. In: Aerospace and digitalization. Springer briefs in applied sciences and technology. Springer, Cham. https://doi.org/10.1007/978-3-030-67849-4_2
15. Crippa P (2022) Cyber security and drones. Handb Secur Sci 619–633
16. Radovanovic M, Petrovski A, Jokic Z, Aleksic A (2022) Use of unmanned aerial vehicles integrated into the C4IRS system in modern combat operations. In: 8th International scientific professional conference security and crisis management–theory and practice (Secman)-safety for the future, Sremska Kamenica, pp 88–98
17. Siddappaji B, Akhilesh KB (2020) Role of cyber security in drone technology. In: Akhilesh K, Möller D (eds) Smart technologies. Springer, Singapore. https://doi.org/10.1007/978-981-13-7139-4_13
18. Alizadeh J, Noorani Asl SH (2024) Practical key recovery attack on authentication and key agreement protocol SecAuthUAV in a network of drones and provide an improved protocol. 12(2):51–59. URL: http://monadi.isc.org.ir/article-1-253-en.html
19. Gonçalves PC (2020) Cyberspace and artificial intelligence: the new face of cyber-enhanced hybrid threats. IntechOpen. https://doi.org/10.5772/intechopen.88648
20. Hwang YS et al (2023) How does ChatGPT evaluate the value of spatial information in the 4th industrial revolution? Spat Inf Res. https://doi.org/10.1007/s41324-023-00567-5

Model of the Thermal Effect of a Spill Fire on an Oil Product-Containing Tank

Volodymyr Oliinik and **Oleksii Basmanov**

Abstract Chemical industry companies show a tendancy to consolidation and it results in a continouous increase in the concentration of hazardous chemicals in a relatively small space. In spite of safety measures taken by different branches of industry, storage and transportation agencies, major accidents still occur. A significant percentage of such accidents can be attributed to the spill of flammable liquid and its ignition. The heat flow from the fire to the neighboring technological facilities can result in the cascading spread of the fire to these facilities due to the heating of the metal structures that contain combustible liquids attaining self-ignition temperatures therein. This paper deals with a construction of a model of the thermal effect of an oil spill fire on a neighboring oil product-containing tank. This model is based on a differential equation of a parabolic type that describes the distribution of heat along the wall and roof of the tank. This model takes into account radiative and convective heat exchange with the fire, the environment, the liquid in the tank and the vapor-air mixture in its gas space. The finite difference method was used to derive the temperature distribution. The obtained results can be used to assess the consequences of the thermal impact of the fire and to take a decision as for the need for tank cooling.

Keywords Oil spill · Oil tank · Spill fire

1 Background

A significant number of emergency situations that arise during the transportation, processing and storage of oil and oil products result from an emergency liquid spill [1]. In this case, the appearance of an ignition source and the ignition of liquid vapors above the spill pose the main danger. The thermal effect of the fire on nearby technological objects and in particular, on containers with flammable liquids presents the difficulty for localization and liquidation of such fires. Steel walls heated to the

V. Oliinik (✉) · O. Basmanov
National University of Civil Defence of Ukraine, Kharkiv, Ukraine
e-mail: oleinik@nuczu.edu.ua

T. A. Kovács et al. (eds.), *The Impact of the Energy Dependency on Critical Infrastructure Protection*, Advanced Sciences and Technologies for Security Applications, https://doi.org/10.1007/978-3-031-78544-3_43

self-ignition temperature of liquid vapors can lead to an explosion of the vapor-air mixture in the tank gas space. In addition, the thermal effect of fire can result in the loss of strength by steel structures [2] and depressurization of flanged joints. Another consequence of accidents related to the spillage and burning of oil products is the pollution of underground and river waters due to the seepage of liquid deep into the soil [3] and the release of hazardous substances into the atmosphere [4]. Spreading over considerable distances, these significantly affect the state of the air [5]. There is also a possibility of fire spreading to neighboring natural landscapes [6]. One of the important causes of major accidents is the "domino effect" [7]. According to statistical data [8], collected for the accidents with 50 year-old history of studies, the "domino effect" is mainly characteristic of liquefied gas, oil, and oil products. Accidents frequently occur at technological plants (38.5%), warehouses (33%) and during transportation (10.6%). Flammable liquid fires are one of the dominant development scenarios (42%) for industrial accidents. In addition, as for the emergency situations related to fires, about 60% of them are fires involving flammable liquids [9].

In [10], the thermal effect of a fire in an oil product-containing tank on the neighboring tank is studied using the Fire Dynamics Simulator (FDS). The temperature distribution in the updrafts above the combustion center and the influence of wind speed have been analyzed. However, this scientific paper gives no consideration to the spilled liquid fire. The thermal effect of a fire in an oil product-containing tank on a similar neighboring tank was studied in [11]. Using the thermal conductivity equation, the temperature distribution inside the tank wall was determined. The model takes into account the radiation heat exchange with the fire torch and the heated walls of the burning tank. Convection heat exchange is neglected in this research. In [12], a model of the thermal effect of a flame on a nearby oil product-containing tank has been proposed. At the same time, the lower part of the flame has a high temperature, and the upper part has a low temperature. The drawback of the model is that it fails to take into account different temperatures on the outer and inner surfaces of the tank. In [13], consideration is given to the effect of flame pulsations on the heat flow from a fire. The strength of the tank structures heated under the influence of fire has been studied. However, no consideration is given to the consequences of heating steel structures to the self-ignition temperature of the vapors of petroleum products. In [14], the authors analyse the effect of a fire in a nearby tank and a fire- in-collapse on a oil product-containing tank, but the effect of wind on the heat flow from the fire was not studied.

In [15], consideration is given to the specific features of combustion of the spills of flammable liquids both on the water surface and on solid surface. However, no consideration is given to the thermal effect of the fire on neighboring objects. In [16], a numerical simulation was carried out for the simultaneous burning of several adjacent pools, i.e. pool fire. In [17], a model was constructed for the liquid burning and extinguishing with the liquid spreading on a horizontal surface and having the shape of a circle. However, such a situation is idealized: even a slight slope of the surface leads to the stretching of the spill area along the slope direction [18]. In [18], the liquid seepage deep into the soil is also taken into account, and in [19] the procedure for experimental determination of seepage parameters has been described. And again, no consideration is given to the parameters of the flame over the burning and spreading liquid. Paper [20] describes the model of the convection component of the heat flow from a spill fire of arbitrary shape, but the radiation component of the heat flow, which is predominant, is not considered in this research.

Thus, the cascading spread of fire caused by spillage and ignition of flammable liquid is an urgent problem. One of the methods of counteracting such an accident development is the protection of nearby technological facilities by cooling them with water. This, in turn, requires an assessment of the thermal effect of fire on them.

The purpose of this research is to construct a model for the thermal effect resultant from a spilled flammable liquid fire on a vertical oil product-containing steel tank.

2 Mathematical Model

2.1 Tank Surface Temperature Distribution

One of the specific features of the heating of the tank wall is that its lower part is in contact with the liquid in the tank (wetted part of the wall, Fig. 1), and the upper part is in contact with the vapor-air mixture in the gas space of the tank (dry part of the wall, Fig. 1).

To construct a model of the thermal effect of a spill fire on a vertical oil product-containing steel tank, we will proceed from the following assumptions:

1. The outer surface of the tank wall participates in radiant heat exchange with the torch surface and the surrounding environment, as well as in convective heat exchange with the air environment.

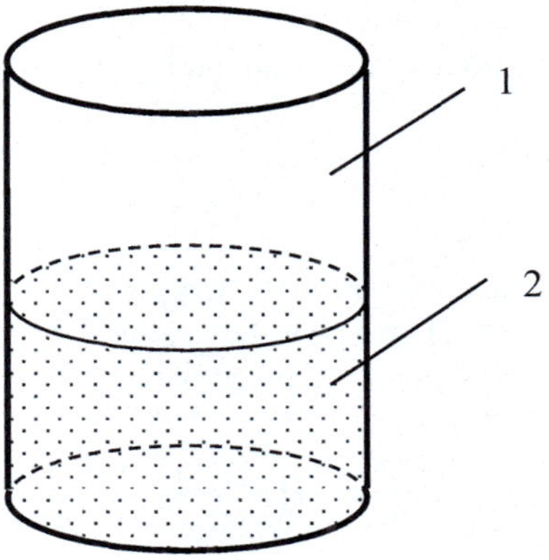

Fig. 1 Dry (1) and wet (2) parts of the wall for the liquid-filled tank

2. The inner surface of the dry part of the wall participates in radiant heat exchange with the surface of the liquid poured into the tank and other points on the inner surface of the wall, as well as in convection heat exchange with the steam–air mixture in the gas space of the tank.
3. The inner surface of the wetted part of the wall participates in convection heat exchange with the liquid poured into the tank.
4. The temperature on the outer and inner surfaces of the wall is the same.
5. Heat spreads along the tank wall due to the thermal conductivity of the wall material (steel).

Assumption 4 implies that for the tank wall we have

$$\frac{\partial T}{\partial r} = 0, \tag{1}$$

where T is the temperature of the tank wall.

A similar situation occurs for the tank roof: considering the temperature to be the same throughout the entire thickness of the roof, we obtain

$$\frac{\partial T}{\partial z} = 0. \tag{2}$$

The equation of heat conduction in a cylinder is expressed as

$$\frac{\partial T}{\partial t} = a\left(\frac{1}{r}\frac{\partial}{\partial r}\left(r\frac{\partial T}{\partial r}\right) + \frac{\partial^2 T}{\partial z^2} + \frac{1}{r^2}\frac{\partial^2 T}{\partial \varphi^2}\right), \tag{3}$$

where T(r, φ, z) is the temperature (Fig. 2); a is thermal diffusivity:

$$a = \frac{\lambda_c}{c_c \rho_c};$$

λ_c, c_c, ρ_c are the thermal conductivity, the heat capacity and the density of steel, respectively.

For the tank wall, taking into account (1), expression (3) turns into

$$\frac{\partial T}{\partial t} = a\left(\frac{\partial^2 T}{\partial z^2} + \frac{1}{R^2}\frac{\partial^2 T}{\partial \varphi^2}\right); \quad 0 < z < H; \quad 0 < \varphi < 2\pi,$$

where T(φ, z) is the temperature of the tank wall at the point with coordinates (φ, z); R, H are the radius and height of the tank, respectively. Taking into account the radiation and convection heat transfer of the wall, we get

$$\frac{\partial T}{\partial t} = a\left(\frac{\partial^2 T}{\partial z^2} + \frac{1}{R^2}\frac{\partial^2 T}{\partial \varphi^2}\right) + \frac{q}{c_c \rho_c \delta}; \quad 0 < z < H; \quad 0 < \varphi < 2\pi, \tag{4}$$

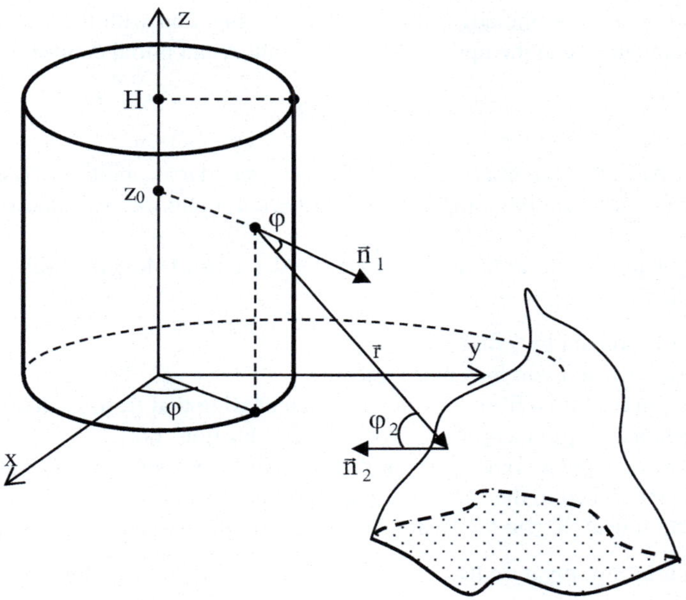

Fig. 2 Selecting the coordinate system when modeling the heating of the tank wall and roof

where δ is the thickness of the tank wall. Substituting (2) into (3) we get the temperature distribution equation for the tank roof.

$$\frac{\partial T}{\partial t} = a\left(\frac{1}{r}\frac{\partial}{\partial r}\left(r\frac{\partial T}{\partial r}\right) + \frac{1}{r^2}\frac{\partial^2 T}{\partial \varphi^2}\right); \quad 0 < r < R; \;\; 0 < \varphi < 2\pi.$$

Taking into account the radiative and convective heat transfer of the roof, the last equation turns into

$$\frac{\partial T}{\partial t} = a\left(\frac{\partial^2 T}{\partial r^2} + \frac{1}{r}\frac{\partial T}{\partial r} + \frac{1}{r^2}\frac{\partial^2 T}{\partial \varphi^2}\right) + \frac{q}{c_c \rho_c \delta}; \quad 0 < r < R; \;\; 0 < \varphi < 2\pi \qquad (5)$$

The lower edge of the wall is assumed to be thermally insulated. Then the boundary condition for the tank wall will be expressed as

$$\left.\frac{\partial T}{\partial z}\right|_{z=0} = 0. \qquad (6)$$

The temperature at the edge of the roof at the junction with the wall coincides with the temperature at the upper edge of the wall. As an initial condition we take

$$T|_{t=0} = T_0. \tag{7}$$

Hence, partial differential Eqs. (4), (5) with boundary condition (6) and initial condition (7) describe the temperature distribution along the wall and roof of the tank.

The heat flow at an arbitrary point on the dry part of the tank wall or its roof consists of

- heat flow radiated by a fire;
- heat flow radiated into the environment;
- convection heat flow from an ambient air (including that from heated air masses and combustion products rising above the combustion spot);
- heat flow radiated by other parts of the inner surface of the tank wall and roof;
- heat flow radiated to the surface of the liquid in the tank;
- convection heat flow to the steam–air mixture in the gas space of the tank.

For an arbitrary point on the wetted wall of the tank, the heat flow consists of

- heat flow radiated by a fire;
- heat flow radiated into the environment;
- convection heat flow from an ambient air (including that from heated air masses and combustion products rising above the combustion spot);
- convection heat flow to the liquid poured into the tank.

Hence, the heat flux density q at an arbitrary point on the dry part of the wall or roof of the tank is expressed as

$$q = q_1 + q_2 + q_3 + q_4 + q_5 + q_6 + q_7, \tag{8}$$

where q_1 is the heat flux density radiated by the fire; q_2 is the density of heat flow radiated into the environment; q_3 is the convection heat exchange with ambient air and combustion products; q_4 is the heat flux density radiated by other points of the inner surface of the wall through the gas space of the tank; q_5 is the heat flux density radiated by the inner surface of the tank roof; q_6 is the heat flux density radiated by the liquid surface; q_7 is the convection heat exchange with a steam–air mixture in the gas space.

For a wetted wall, the heat flux density is described by the expression

$$q = q_1 + q_2 + q_3 + q_8,$$

where q_8 is the heat flow density due to the convection heat exchange with the liquid in the tank.

2.2 Estimation of Heat Flux by Radiation

The heat flux density from a fire is described by the Stefan-Boltzmann law

$$q_1 = c_0 \varepsilon_f \varepsilon_w \left(\left(\frac{T_f}{100} \right)^4 - \left(\frac{T}{100} \right)^4 \right) \psi_f, \tag{9}$$

where $c_0 = 5.67$ W/(m^2 K^4) is the constant; ε_f, ε_w are the degrees of blackness of the radiating surface of the torch and the wall of the tank, respectively; T_f is the torch surface temperature; T is the temperature of the surface of the tank wall at a given point; ψ_f is the mutual irradiation coefficient between the torch and the appropriate point on the tank surface

$$\psi_f = \frac{1}{\pi} \iint\limits_S \frac{\cos\varphi_1 \cos\varphi_2}{r^2} dS, \tag{10}$$

where φ_1, φ_2 are the angles between the radius vector r connecting the points on the torch surface and on the tank wall, and the normal vectors n_1, n_2 to the surfaces at these points (Fig. 2). At the same time, the integration is performed only over that part of the torch surface for which the cosine values in the numerator (10) are positive.

Taking into account that

$$\cos\varphi_1 = \frac{(\vec{r}, \vec{n}_1)}{\left|\vec{r}\right| \cdot |\vec{n}_1|}; \quad \cos\varphi_2 = \frac{(\vec{r}, \vec{n}_2)}{\left|\vec{r}\right| \cdot |\vec{n}_2|},$$

where (r, n) is the scalar product of vectors, dependence (10) can be written as

$$\psi_f = \frac{1}{\pi} \iint\limits_S \frac{(\vec{r}, \vec{n}_1)(\vec{r}, \vec{n}_2)}{|\vec{n}_1| \cdot |\vec{n}_2| r^4} dS. \tag{11}$$

If the radiating surface is preset in the parametric form, then the surface element can be written as

$$dS = \sqrt{A^2 + B^2 + C^2} du dv,$$

where A, B, C are derived from the formulas

$$A = \frac{\partial y}{\partial u}\frac{\partial z}{\partial v} - \frac{\partial z}{\partial u}\frac{\partial y}{\partial v}; \quad B = \frac{\partial z}{\partial u}\frac{\partial x}{\partial v} - \frac{\partial x}{\partial u}\frac{\partial z}{\partial v}; \quad C = \frac{\partial x}{\partial u}\frac{\partial y}{\partial v} - \frac{\partial y}{\partial u}\frac{\partial x}{\partial v}.$$

Then the formula (11) is expressed as

$$\psi_f = \frac{1}{\pi} \iint_{\Omega} \frac{\left[A(x_1 - x_2) + B(y_1 - y_2) + C(z_1 - z_2)\right]}{\left[(x_2 - x_1)^2 + (y_2 - y_1)^2 + (z_2 - z_1)^2\right]^2} \times \left[e_x(x_2 - x_1) + e_y(y_2 - y_1) + e_z(z_2 - z_1)\right] dudv. \quad (12)$$

where (x_1, y_1, z_1) is a point on the heated surface; (e_x, e_y, e_z) is the unit normal vector to the surface at this point; (x_2, y_2, z_2) is a point on the flame surface. At the same time, the integral in (12) is calculated only for those points of the spill area Ω, in which both factors in square brackets in the numerator exceed zero.

Let the center of the tank base be located at the point $(x_0, y_0, 0)$. Then the equation of the surface of its wall in parametric form is expressed as

$$\begin{cases} x_1 = x_0 + R\cos\varphi, \\ y_1 = y_0 + R\sin\varphi, \\ 0 \le z_1 \le H. \end{cases} \quad (13)$$

In this case, the normal vector to the surface is written as

$$\begin{cases} e_x = \cos\varphi, \\ e_y = \sin\varphi, \\ e_z = 0. \end{cases} \quad (14)$$

Taking (14) into account, expression (12) turns into

$$\psi_f = \frac{1}{\pi} \iint_{\Omega} \frac{\left[A(x_1 - x_2) + B(y_1 - y_2) + C(z_1 - z_2)\right]}{\left[(x_2 - x_1)^2 + (y_2 - y_1)^2 + (z_2 - z_1)^2\right]^2} \times \left[\cos\varphi(x_2 - x_1) + \sin\varphi(y_2 - y_1)\right] dudv.$$

Similarly, the equation for the tank roof surface in parametric form is expressed as.

$$\begin{cases} x_1 = x_0 + r\cos\varphi, \\ y_1 = y_0 + r\sin\varphi, \\ z_1 = H, \end{cases}$$

And the normal surface vector.

$$\begin{cases} e_x = 0, \\ e_y = 0, \\ e_z = 1. \end{cases}$$

Then, for the tank roof the radiation coefficient will be expressed as.

$$\psi_f = \frac{1}{\pi}\iint\limits_{\Omega} \frac{[A(x_1 - x_2) + B(y_1 - y_2) + C(H_1 - z_2)]}{[(x_2 - x_1)^2 + (y_2 - y_1)^2 + (z_2 - H_1)^2]^2}(z_2 - H_1)dudv.$$

The density of the flux entering the environment is described by the expression

$$q_2 = c_0\varepsilon_w\left(\left(\frac{T_0}{100}\right)^4 - \left(\frac{T}{100}\right)^4\right)(1 - \psi_f), \tag{15}$$

where T_0 is the ambient temperature.

The heat flux density from other points on the inner surface of the tank wall is described by the Stefan-Boltzmann law

$$q_4 = \frac{c_0\varepsilon_w^2}{\pi}\iint\limits_{S_w} \frac{\cos\varphi_1\cos\varphi_2}{r^2}\left(\left(\frac{T_w}{100}\right)^4 - \left(\frac{T}{100}\right)^4\right)dS_w, \tag{16}$$

where T_w is the temperature at other points of the tank wall; the integral is calculated over the entire inner surface of the dry part of the wall S_w. In (16), in contrast to (9), (10), the temperatures are under the integral sign, since the wall temperature takes different values at different points. Note that the density of the heat flow to a point on the inner surface of the roof is also given by expression (16).

Similarly, the density of the heat flux from the tank roof is described by the expression

$$q_5 = \frac{c_0\varepsilon_w^2}{\pi}\iint\limits_{S_R} \frac{\cos\varphi_1\cos\varphi_2}{r^2}\left(\left(\frac{T_R}{100}\right)^4 - \left(\frac{T}{100}\right)^4\right)dS_R, \tag{17}$$

where T_R is the temperature at a given point on the tank roof S_R. Note that there is no radiation heat exchange between the points on the tank roof, i.e. $q_5 = 0$ for them.

Density of the heat flux from the liquid surface

$$q_6 = c_0\varepsilon_f\varepsilon_\ell\left(\left(\frac{T_\ell}{100}\right)^4 - \left(\frac{T}{100}\right)^4\right)\psi_\ell, \tag{18}$$

$$\psi_\ell = \frac{1}{\pi}\iint\limits_{S_\ell} \frac{\cos\varphi_1\cos\varphi_2}{r^2}dS_\ell, \tag{19}$$

where T_ℓ is the temperature of the liquid in the tank; ψ_ℓ is the mutual radiation coefficient between a point on the inner surface of the tank wall or roof and the liquid surface; S_ℓ is the liquid surface. The temperature of the liquid is considered constant throughout its volume because of its irregularities due to mixing by convection fluxes.

2.3 Convection Heat Flow

The heat flow density due to convective heat transfer with the ambient air is determined by Newton's law

$$q_3 = \alpha_3(T_a - T), \tag{20}$$

where T_f is the temperature of air masses. As a matter of fact, these can be heated combustion products; α_3 is the convective heat transfer coefficient with ambient air masses.

Heat flow density due to the convective heat transfer with the steam–air mixture in the gas space of the tank is expressed as

$$q_7 = \alpha_7(T_m - T), \tag{21}$$

where α_7 is the convective heat transfer coefficient of the inner surface of the wall with the steam–air mixture; T_m is the temperature of the steam–air mixture. The temperature of the mixture is assumed to be the same throughout the entire volume of the gas space, since its nonuniform heating is leveled out by arising convection movements.

The heat flux density due to the convection heat exchange of the inner surface of the wall and the liquid in the tank has the form

$$q_8 = \alpha_8(T_\ell - T), \tag{22}$$

where α_8 is the convective heat transfer coefficient

3 Numerical Simulation

To solve the tank wall heating Eq. (4) with boundary conditions (6), we apply the finite difference method. For this purpose, let's cover the tank wall with a grid with nodes (φ_i, z_i) with a step $\Delta\varphi$ along the angle φ and a step Δz along the z axis:

$$\varphi_i = i\Delta\varphi; \quad i = \overline{0..n_1 - 1}; \quad \Delta\varphi = \frac{2\pi}{n_1}. \tag{23}$$

$$z_j = j\Delta z; \quad j = \overline{0..n_2}; \quad \Delta z = \frac{H}{n_2}. \tag{24}$$

We will cover the tank roof with a grid with nodes (r_k, φ_i) with a step Δr along the r axis, filling the step $\Delta\varphi$ in the same way as for the wall.

$$r_k = k\Delta r; \quad k = \overline{0..n_3}; \quad \Delta r = \frac{R}{n_3}. \tag{25}$$

The number of nodes n_1 along the φ angle, n_2 along the z axis, n_3 along the r axis is chosen in such a way that

$$\Delta z \approx R\Delta\varphi \approx \Delta r. \tag{26}$$

The fulfillment of condition (26) implies that the points (φ_i, z_i) form a grid on the tank wall, for which the distance between the nodes in the horizontal direction is approximately equal to the distance between the nodes in the vertical direction. To fulfill condition (26), it is sufficient to preset the step Δx and choose n_1, n_2, n_3 according to the formulas.

$$n_1 = \frac{2\pi R}{\Delta x}; \quad n_2 = \frac{H}{\Delta x}; \quad n_3 = \frac{r}{\Delta x}, \tag{27}$$

where n_1, n_2, n_3 are rounded to whole values.

Let's choose a step along the time axis Δt. Then, at grid nodes (φ_i, z_j), the time derivative can be approximated by an expression

$$\frac{\partial T}{\partial t} \approx \frac{T_{ij\,m+1} - T_{ijm}}{\Delta t}, \tag{28}$$

where $T_{ijm} = T(\varphi_i, z_j, t_m)$. To approximate other derivatives by spatial coordinates in the equation for the tank wall (4), we use the expressions

$$\frac{\partial^2}{\partial\varphi^2} T(\varphi_i, z_j, t_m) \approx \frac{1}{\Delta\varphi^2}\left(T_{i+1\,jm} - 2T_{ijm} + T_{i-1\,jm}\right); \tag{29}$$

$$\frac{\partial^2}{\partial z^2} T(\varphi_i, z_j, t_m) \approx \frac{1}{\Delta z^2}\left(T_{i\,j+1\,m} - 2T_{ijm} + T_{i\,j-1\,m}\right). \tag{30}$$

Then the differential Eq. (4) can be written in finite differences

$$\frac{T_{ij\,m+1} - T_{ijm}}{\Delta t} = a\left(\frac{T_{i\,j+1\,m} - 2T_{ijm} + T_{i\,j-1\,m}}{\Delta z^2} + \frac{T_{i+1\,jm} - 2T_{ijm} + T_{i-1\,jm}}{R^2\Delta\varphi^2}\right) + \frac{q^{ijm}}{c_c\rho_c\delta}, \tag{31}$$

where q_{ijm} is the heat flux density falling on the point (φ_i, z_j) on the tank wall at time t_m.

To approximate the derivatives $\partial T/\partial r$ and $\partial^2 T/\partial r^2$ we use the expression

$$\frac{\partial}{\partial r}T(r_k, \varphi_i, t_m) \approx \frac{T_{k+1\,i\,m} - T_{k-1\,i\,m}}{2\Delta r}; \tag{32}$$

$$\frac{\partial^2}{\partial r^2}T(r_k, \varphi_i, t_m) \approx \frac{T_{k+1\,i\,m} - 2T_{kim} + T_{k-1\,i\,m}}{\Delta r^2}. \tag{33}$$

Then Eq. (5) in finite differences will take the form

$$\frac{T_{ki\,m+1} - T_{kim}}{\Delta t} = \frac{a}{\Delta r^2}\left[T_{k+1\,i\,m} - 2T_{kim} + T_{k-1\,i\,m} + \frac{T_{k+1\,i\,m} - T_{k-1\,i\,m}}{2k} + \frac{T_{k\,i+1\,m} - 2T_{kim} + T_{k\,i-1\,m}}{k^2\Delta\varphi^2}\right] + \frac{q^{kim}}{c_c\rho_c\delta}, \tag{34}$$

where q_{kim} is the heat flux density falling on the point (r_k, φ_i) on the tank roof at time t_m.

4 Discussion

When constructing a heating model for the steel structures of the tank, its surface is conventionally subdivided into the three parts with different conditions:

- the wetted part of the wall that is in contact with the oil product.
- the dry part of the wall that is in contact with the steam–air mixture.
- the tank roof.

When burning a spilled combustible liquid, the highest heat flow density falls on the lower part of the tank. But this part of the wall is cooled due to contact with the liquid inside the tank (Fig. 1) and therefore its temperature does not attain a dangerous value. The dry part of the wall has a slightly lower heat flow density, but the cooling effect from the steam–air mixture is 1or 2 orders of magnitude less compared to the cooling effect of the liquid. The tank floor can be heated by a fire in an adjacent tank, but when a spill burns on the ground, the flame height is usually lower than the tank height. That is why the heating of the dry wall poses the greatest danger. In particular, it follows that the lower the level of the oil product in the tank, the greater the danger it is exposed to.

The model of the temperature distribution on the steel surface of the tank consists of differential equations of thermal conductivity (4) and (5) that describe the heat distribution along the wall and roof, respectively, and additionally take into account heat exchange with the environment. The assumption of the same temperature value on the outer and inner surfaces of the wall and roof of the tank is based on a number of studies carried out for the temperature distribution along the depth of the tank wall. In particular, it is shown in [21] that the temperature difference on the inner and outer surfaces of a steel wall with a thickness of 8 mm does not exceed 1.5°, while the temperature of the wall exceeds 250 °C. The obtained results can be explained by a small thickness of the wall and high thermal conductivity.

The constructed model takes into account the radiant and convective heat exchange of the wall and roof of the tank with the fire, as well as the radiant and convective heat exchange with the environment, the liquid in the tank and the vapor-air mixture in its gas space.

For the practical use of the obtained model, it is necessary to solve the differential equations (4)–(5) under the appropriate boundary and initial conditions. For this purpose we use the finite difference method. A grid was constructed on the wall and roof of the tank according to formulas (23)–(25). Approximation of the derivative with respect to time (28) and the derivatives with respect to the variables r, z, φ using expressions (29), (30), (32), (33) allows us to proceed to finite difference Eqs. (31), (34).Hence, the sequential application of formulas (31), (34) allows us to obtain the distribution of temperatures along the wall and roof of the tank at successive moments of time.

The obtained results can be used to determine the consequences of the thermal effect of a fire on the vertical steel tank filled with a flammable liquid.

5 Conclusions

A heating model was constructed for the vertical oil product-containing steel tank exposed to the thermal effect of a spill fire. The model is based on the differential equation of thermal conductivity that describes the temperature distribution along the wall and roof of the tank. The model takes into account radiation and convection heat exchange with the fire, the environment, the liquid in the tank and the vapor-air mixture in its gas space. For practical use of the model, the finite difference method is applied. It allows us to find the temperature distribution at any time and determine the need for the tank protection.

References

1. Raja S, Tauseef SM, Abbasi T (2018) Risk of fuel spills and the transient models of spill area forecasting. J Fail Anal Prev 18:445–455
2. Vasilchenko A, Otrosh Yu, Adamenko N, Doronin E, Kovalov A (2018) Feature of fire resistance calculation of steel structures with intumescent coating. In: MATEC Web of Conferences 230:02036
3. Loboichenko V, Strelec V (2018) The natural waters and aqueous solutions express-identification as element of determination of possible emergency situation. Water Energy Int 61(9):43–50

4. Kustov MV, Kalugin VD, Tutunik VV, Tarakhno EV (2019) Physicochemical principles of the technology of modified pyrotechnic compositions to reduce the chemical pollution of the atmosphere. Voprosy khimii i khimicheskoi tekhnologii 1:92–99
5. Popov O, Iatsyshyn A, Kovach V, Artemchuk V, Kameneva I, Taraduda D, Sobyna V, Sokolov D, Dement M, Yatsyshyn T (2020) Risk assessment for the population of Kyiv, Ukraine as a result of atmospheric air pollution. J Health Pollut 10:25
6. Mygalenko K, Nuyanzin V, Zemlianskyi A, Dominik A, Pozdieiev S (2018) Development of the technique for restricting the propagation of fire in natural peat ecosystems. East-Eur J Enterp Technol 1(10):31–37
7. Huang K, Chen G, Khan F, Yang Y (2021) Dynamic analysis for fire-induced domino effects in chemical process industries. Process Saf Environ Prot 148:686–697
8. Hemmatian B, Abdolhamidzadeh B, Darbra R, Casal J (2014) The significance of domino effect in chemical accidents. J Loss Prev Process Ind 29:30–38
9. Fabiano B, Caviglione C, Reverberi AP, Palazzi E (2016) Multicomponent hydrocarbon pool fire: analytical modelling and field application. Chem Eng Trans 48:187–192
10. Elhelw M, El-Shobaky A, Attia A, El-Maghlany WM (2021) Advanced dynamic modeling study of fire and smoke of crude oil storage tanks. Process Saf Environ Prot 146:670–685
11. Semerak M, Pozdeev S, Yakovchuk R, Nekora O, Sviatkevich O (2018) Mathematical modeling of thermal fire effect on tanks with oil products. MATEC Web of Conferences 247(00040)
12. Espinosa SN, Jaca RC, Godoy LA (2019) Thermal effects of fire on a nearby fuel storage tank. J Loss Prev Process Ind 62(103990)
13. Li Y, Jiang J, Zhang Q, Yu Y, Wang Z, Liu H, Shu C-M (2019) Static and dynamic flame model effects on thermal buckling: fixed-roof tanks adjacent to an ethanol pool-fire. Process Saf Environ Prot 127:23–35
14. Ahmadi O, Mortazavi SB, Pasdarshahri H, Mohabadi HA (2019) Consequence analysis of large-scale pool fire in oil storage terminal based on computational fluid dynamic (CFD). Process Saf Environ Prot 123:379–389
15. Liu J, Li D, Wang Z, Chai X (2021) A state-of-the-art research progress and prospect of liquid fuel spill fires. Case Stud Therm Eng 28:101421
16. Nemalipuri P, Singh V, Das HC, Pradhan MK, Vitankar V (2023) Consequence analysis of heptane multiple pool fire in a dike. Proc Inst Mech Eng, Part C: J Mech Eng-Ing Sci
17. Abramov Y, Basmanov O, Krivtsova V, Salamov J (2019) Modeling of spilling and extinguishing of burning fuel on horizontal surface. Naukovyi Visnyk NHU 4:86–90
18. Oliinik V, Basmanov O (2023) Model of spreading and burning the liquid on the soil. Probl Emerg Situat 1(37):18–30
19. Abramov Y, Basmanov O, Khmyrov I, Oliinik V (2022) Justifying the experimental method for determining the parameters of liquid infiltration in bulk material. East-Eur J Enterp Technol 4/10(118):24–29
20. Abramov Y, Basmanov O, Oliinik V, Khmyrov I, Khmyrova A (2022) Modeling the convective component of the heat flow from a spill fire at railway accidence. EUREKA: Physics and Engineering, 6:128–138
21. Basmanov O, Maksymenko M (2022) Modeling the thermal effect of fire to the adjacent tank in the presence of wind. Probl Emerg Situat 1(35):239–253

Advanced Object Recognition Using Drones

Arnold Őszi

Abstract We can use traditional high-resolution color cameras mounted to the drone to recognize objects. This article shows what other breakthrough options we have and what benefits they bring. We can create bird eye view images, and also, we can measure distances accurately from the air. The article covers the sensors used for Field mapping, orthophoto, 3D modelling. The article introduces LIDAR technology, what can be used to create a very accurate 3D map of the surface. Ground-penetrating radar can be used to see what is under the surface. It can detect buried objects and underground structures from the air. The article also covers thermal imaging. It has lots of possibilities when we use it in a drone. It can detect objects way better than regular color cameras. It can not only detect leakage, but it can see what happened in the past. It can see not in the dark, but in smoke and fog as well. Drones can be equipped with multi-spectral and hyper-spectral cameras, so the object recognition can be way better and much more accurate. Also we can use drones for calculating size, surface and volume. We can also measure the structure of walls by using contact devices attached to the drone. The article also introduce some special usage of the drones.

Keywords Drones · Sensors · UAVs · Imaging · Lidar · Radar · Machine vision · Machine sensing

1 Introduction

Drones, also known as unmanned aerial vehicles (UAVs), are equipped with various sensors to achieve a wide range of goals.

Nowadays we have a lot of opportunities to do specific measurements. These support engineers for more accurate planning. Using drones, we can do several measurements even in dangerous or hardly accessible environment. In most cases it

A. Őszi (✉)
Obuda University, Budapest, Hungary
e-mail: oszi.arnold@bgk.uni-obuda.hu

T. A. Kovács et al. (eds.), *The Impact of the Energy Dependency on Critical Infrastructure Protection*, Advanced Sciences and Technologies for Security Applications, https://doi.org/10.1007/978-3-031-78544-3_44

is expensive to do these tasks, but since the drone technology available, we can do it faster and for less money. Doing these measurements requires specific hardware. In this paper I will cover these, and I will give detailed description for each, highlighting how these works and what these can be used.

These sensors are critical for navigation, control, and doing special measurements. In this article I cover the sensor types what are commonly used today in the UAV industry. These sensors can be divided in two categories. The fist category is the main sensors. These collect data from the environment for further process like 3D mapping, measuring, surface analysis and so on. The second category of the sensors is used for navigation and maneuvering the drone itself. This data can be used to object avoidance, collision avoidance or even autonomous flight. The article focuses only on the sensors, its functions, types and their operation. The paper not cover the data analysis and processing.

2 Main Measurement Sensors

This section introduces the sensors, which are used for main data collection. Its measurement is the aim of the flight. The sensors must be lightweight and compact enough to not interfere with the drone's flight capabilities. These sensors are mounted on drones, which are programmed to fly through the area of interest. After or even during the flight these data is processed and give us the measurement information we need. There are a lots of sensor types can be used on drones. This section introduces these ones and gives a description about their function and main usage.

2.1 RGB Camera

It can be used to capture airborn images. This is the most known and most used sensor on drones. The size, weight, resolution and the price can vary in a wild range. The live image can be streamed to a ground station, so the pilot can do corrections for better image composition. The data can be used for visual inspections, to create maps, 3D surface reconstruction or detecting a fire just to name a few. For example it is possible to create a 3D model of a building or a vehicle [1].

2.2 Multispectral Camera

It uses not only the red, green and blue channels of the visible electromagnetic spectrum, but also one or more infra-red frequency ranges as well. Using multispectral imaging, the identification of the objects are more accurate. Each material have a different reflection and absorption in the spectrum scale, knowing these it will be more successful to identify them. For example, in agriculture it is possible to determine the health or the infections of the agricultural field [2, 3].

2.3 Hyperspectral Camera

It captures images using high number of narrow bands of the spectrum. Compared to the normal RGB cameras, in hyperspectral ones there are 100 or 1000 bands available instead of three. Using this extra information, the recognition of materials is way more accurate compared to the RGB cameras. Using this technology, it is possible to recognize given materials and the amount of them.

It mostly used in agriculture, environmental monitoring, locate persons, water quality monitoring and mineral exploration [4–6].

2.4 Thermal Camera

Detect infrared radiation to capture thermal images. Each material has a temperature and it radiates infra energy into the environment. Thermal camera can see this radiation and the temperature can be seen on the image. It has no color information, but the shapes of the object also can be seen. If a person warms up a seat for example, using thermal camera it can see it so we can see the past events as well. This camera can see in dark, fog and smoke as well.

Applied in search and rescue, building inspections, and monitoring wildlife [7].

2.5 Laser Range Finders

Using these devices, it is possible to measure the distance and the speed of the moving objects on the surface, like the speed of the ground vehicles. It is an active device, it emits laser light, then measure the reflection and calculates the distance of the object. This sensor usually does not work in rain, fog or smoke. The data of laser range finders most frequently used as the basic information for 3D mapping of the surface or 3D object.

2.6 LiDAR Sensor

Light Detection and Ranging (LiDAR) sensor technology continuously scan the environment and measure the distance of each point in the environment from the sensor. Using this information and the location information of the drone, 3D maps can be created. This 3D map has a very high resolution compared to other 3D mapping techniques. Also, it is possible to scan a forested area and the vegetation can be turned off in order to see the soil surface under the vegetation area.

LiDAR sensors enable drones to create high-precision 3D maps. It is also possible to measure distances in between objects or measure length of objects like roads, trees or buildings [8, 9].

2.7 Radar

It transmits electromagnetic waves into the environment. These waves reflect from objects. It referred to as echoes. This echo is received by the antenna, then it is processed and analyzed. The reception time of the radio waves is used to calculate the distance of the object from the radar. The change in frequency of the returned signal, due to the Doppler effect, can be used to determine the speed of the object. Usually radar systems uses the same antenna for sending and receiving the radio wave energy.

Using Radar 3D mapping is also possible, but with a lower resolution compared to the Lidar technology [10].

2.8 Ground-Penetrating Radar

A Ground-penetrating Radar (GPR) system sends high-frequency (typically polarized) radio waves into the ground from a transmitter. The frequency of these waves usually ranges from 10 MHz to 2.6 GHz. These electromagnetic waves reflects back and the antenna receive the signal. GPR uses radar pulses to image the subsurface. Using GPR it is possible to detect underground caves, metals, building structures, basements, water, bedrock and all kinds of buried objects.

Ground Penetrating Radar equipped on drones is a relatively new solution, that combines the capabilities of GPR with the mobility and accessibility of drones. In this way it is possible to monitor areas where a person cannot walk like bushes snow or water [11, 12].

2.9 Environmental and Gas Sensors

Drone-mounted gas sensors analyzes the chemical composition of air materials and detect specific gases in the air. They are particularly useful in areas that are difficult, dangerous, or otherwise impossible to reach with traditional gas sensing methods. Using drones is the most cost effective way to measure the gas ingredients in higher altitudes. This could be used over industrial sites, landfill areas, agricultural fields, or disaster zones [13–15].

2.10 *Capture Audio Data. Acoustic Sensing*

Acoustic sensors enable drones to monitor and analyze sound patterns in natural environments, supporting ecological studies and biodiversity assessments. It is important to install the microphone away from the rotors to make the sound clear of propeller sound. Can be used to locate people, measure sound pollution in cities or industrial sites. Also, mechanical malfunctions can be found in industrial sites, if they have a specific sound.

2.11 *Radio Signals*

Detecting radio signals using drones. In practical applications, drones equipped with RF sensors, such as antennas, can be used for measuring signals for various purposes like inspecting antenna performance or detecting emissions from other sources. Such drones can measure signal levels, analyze the vertical plane pattern of target antennas, and collect data on signal reflection effects. RF sensors are most effective in open fields with little interference. In urban settings with buildings and metal structures, RF signals can bounce and provide incorrect information, requiring the installation of multiple sensors for adequate coverage.

2.12 *Water Quality Sensors*

Measure parameters such as pH, turbidity, or dissolved oxygen. It is very useful solution to do these measurements with UAV-s, because in most cases it is not possible to reach the water surface for humans. Using drones, it is more effective to monitor the water quality. It can measure the temperature, Ph, conductivity, and other parameters of the water. The method is used in environmental monitoring, agriculture and industry [16, 17].

2.13 *Sample Collection Possibilities*

Drones are able to collect samples for later analysis. It can collect water, soil or air samples. It closes the collected sample into a kind of test tube and it transports to the landing spot. The drone has to move a kind of robotic arm, what does the sample collection. Using drones it can be effective, safe and fast to collect the chosen sample [18].

2.14 Contact Measurements

Drones can do contact measurements as well. For doing this, the sensor has to physically touch the investigated surface. In a lot of cases, it is hardly possible to do these measurements personally. Using drones, the whole measurement process will be safer and more cost-effective [19].

Contact ultrasonic measurement

Contact ultrasonic measurements are mainly used for detecting structural defects in buildings or bridges. It generates sound waves, what is reflects from the structure. Using contact ultrasonic measurements cracks and other internal damages can be detected in the material.

Contact Hardness measurement

It measures the hardness of a surface. Usually metal structures are measured, but any other material can be tested like wood or plastic. The most frequently used methods are Brinell, Vickers and Rockwell. During these tests the connection in between the drone and the measures surface must be rigid.

Contact Layer thickness measurement

Using this measurement, the thickness of coatings or other layers can be measured. There are several techniques available for thickness measurement like ultrasonic, x-ray fluorescence, magnetic, eddy current or beta backscatter just to name a few [20].

3 Drone Position Sensors for Navigation and Maneuvering

Improved flight control software enables precise control and stabilization, even in challenging conditions. This contributes to the safety and reliability of drone operations.

It's important to note that the specific sensors required may vary depending on the drone's intended use and the characteristics of the indoor or outdoor environment. Additionally, advancements in technology may lead to the development of new sensor types or improvements in existing ones for drone applications.

Advanced GPS systems, obstacle detection sensors, and sophisticated algorithms helps drones to navigate autonomously. This autonomy is crucial for tasks such as mapping, surveillance, and delivery services [21].

Indoor drone flights pose specific challenges compared to outdoor flights, primarily due to the absence of GPS signals and the confined space. Indoor Positioning System (IPS) technologies, such as Wi-Fi-based positioning or visual-based systems, can provide additional localization information when GPS signals are unavailable.

To navigate and gather information indoors effectively, drones typically require specialized sensors. This paper introduces the sensors what are commonly used for indoor drone flights.

3.1 Inertial Measurement Unit

The Inertial Measurement Unit (IMU) usually consist accelerometers and gyroscopes and sometimes magnetometers. The IMU provides information about the drone's acceleration, orientation, and velocity. An IMU typically measure the drone's acceleration and rotation. This information is crucial for stabilizing the drone and maintaining control.

Magnetometers measure changes in strength and direction of the magnetic field, aiding in orientation and navigation. They can help drones maintain a stable heading in indoor and outdoor environments [22].

3.2 Barometer

Barometers measure atmospheric pressure to estimate altitude. These sensors enhance altitude control and flight stability. Barometric altimeters measure atmospheric pressure changes to estimate changes in altitude. This sensor helps drones maintain a stable vertical position when flying indoors and outdoors [23, 24].

3.3 Camera

RGB cameras also can be used as an optical flow sensor. Optical flow sensors use cameras to detect the movement of ground and estimate the drone's position relative to those points. These are particularly useful for maintaining altitude and position when GPS signals are unavailable [25, 26].

3.4 Lidar Technology

Using LiDAR, the objects in the environment of the drone can be seen. The size and the distance of the objects are accurately measured during the drone flight.

LiDAR-equipped drones can navigate through complex environments by detecting and avoiding obstacles, ensuring safe and efficient operations. It can greatly increase its obstacle avoidance abilities [27, 28].

3.5 Ultrasonic or Laser Range Finders

Ultrasonic or laser sensors help drones measure distances to obstacles or surfaces in their vicinity. They are crucial for obstacle avoidance and maintaining a safe distance from walls and other objects. Ultrasonic sensor will not work in loud areas and long distances, while laser devices will not work in smoke, fog and rain.

3.6 Radio Signals

Drones can be equipped with radio frequency (RF) sensors to detect other drones by passively listening to the radio frequency spectrums in which they communicate with their controllers. Common communication bands for drones include the 2.4 and 5.8 GHz bands, along with others like 1.2 and 1.3 GHz. RF sensors can identify signals generated by a drone and its controller by referencing a database of communication protocols. This method can provide information such as drone and pilot location [29].

GPS navigation is a special kind of radio navigation. It receives signals from global navigation satellites to determine the drone's location. It is essential for navigation, mapping, and surveying applications. The GPS sometimes is not enough so Glonass, Galileo, BeiDou, and other satellites are also can be used for better accuracy.

3.7 Radar Systems

Use radio waves to detect objects and determine their distance and speed. Radar sensors enable drones to operate in adverse weather conditions and low visibility.

Applied in collision avoidance systems and long-range detection. It makes the drone suitable for applications such as surveillance and security. Radar systems assist in terrain following, allowing drones to maintain a consistent altitude above the ground, which is essential for applications like agricultural spraying.

3.8 Acoustic Radar

It is possible to the techniques what bats use for navigation. The drone generates a sound then it reflects beck from objects and the environment can be mapped. It has lower resolution compared to lidar systems [30].

The previously mentioned ultrasonic sensors are very similar devices. They measure distance by emitting ultrasonic waves. It has low price, but low accuracy. Used for altitude control, obstacle avoidance, and indoor navigation [31].

4 Conclusion and Further Research Works

This article focused only on the sensors. Now the reader can see the wide variety of sensors what can be used on drones. Usually these sensors used together, so the measurements can give a lot more data. Sensor fusion is another technique to improve the measurements. It was not part of this paper, but using it the measurements will be a lot more effective.

Drone sensor technologies have ushered in a new era of possibilities, expanding the scope of applications across various domains. Continued research and innovation in sensor development, coupled with advancements in data processing, will further propel the capabilities of drones, unlocking new horizons for aerial data collection.

References

1. Daniyar K, Olga S, György Gy (2023) Computer vision-based fire detection using enhanced chromatic segmentation and optical flow model. Acta Polytechnica Hungarica 20(6). https://acta.uni-obuda.hu/Kaliyev_Shvets_Gyorok_135.pdf
2. Furukawa F, Laneng LA, Ando H, Yoshimura N, Kaneko M, Morimoto J (2021) Comparison of RGB and multispectral unmanned aerial vehicle for monitoring vegetation coverage changes on a landslide area. Drones 5:97. https://doi.org/10.3390/drones5030097
3. Zhao X, Li Y, Chen Y, Qiao X, Qian W (2023) Water chlorophyll a estimation using uav-based multispectral data and machine learning. Drones 7:2. https://doi.org/10.3390/drones7010002
4. Olivetti D, Cicerelli R, Martinez J-M, Almeida T, Casari R, Borges H, Roig H (2023) Comparing unmanned aerial multispectral and hyperspectral imagery for harmful algal bloom monitoring in artificial ponds used for fish farming. Drones 7:410. https://doi.org/10.3390/drones7070410
5. Fawcett D, Panigada C, Tagliabue G, Boschetti M, Celesti M, Evdokimov A, Biriukova K, Colombo R, Miglietta F, Rascher U et al (2020) Multi-scale evaluation of drone-based multispectral surface reflectance and vegetation indices in operational conditions. Remote Sens 12:514. https://doi.org/10.3390/rs12030514
6. Goyens C, Lavigne H, Dille A, Vervaeren H (2022) Using hyperspectral remote sensing to monitor water quality in drinking water reservoirs. Remote Sens 14:5607. https://doi.org/10.3390/rs14215607
7. Vlaminck M, Diels L, Philips W, Maes W, Heim R, Wit BD, Luong H (2023) A multisensor uav payload and processing pipeline for generating multispectral point clouds. Remote Sens 15:1524. https://doi.org/10.3390/rs15061524

8. Wallace L, Lucieer A, Watson C, Turner D (2012) Development of a UAV-LiDAR system with application to forest inventory. Remote Sens 4:1519–1543. https://doi.org/10.3390/rs4061519
9. Lee J, Jo H, Oh J (2023) Application of drone lidar survey for evaluation of a long-term consolidation settlement of large land reclamation. Appl Sci 13:8277. https://doi.org/10.3390/app13148277
10. Lee CU, Noh G, Ahn B, Yu J-W, Lee HL (2019) Tilted-beam switched array antenna for uav mounted radar applications with 360° coverage. Electronics 8:1240. https://doi.org/10.3390/electronics8111240
11. Vergnano A, Franco D, Godio A (2022) Drone-borne ground-penetrating radar for snow cover mapping. Remote Sens 14:1763. https://doi.org/10.3390/rs14071763
12. Lijcklama à Nijeholt L, Kronshorst TY, Teeffelen KV, van Manen B, Emaus R, Knotter J, Mersha A (2023) Utilizing drone-based ground-penetrating radar for crime investigations in localizing and identifying clandestine graves. Sensors 23:7119. https://doi.org/10.3390/s23167119
13. Park J, Jumu F, Power J, Richard M, Elsahli Y, Jarkas MA, Ruan A, Luican-Mayer A, Ménard J-M (2022) Drone-mountable gas sensing platform using graphene chemiresistors for remote in-field monitoring. Sensors 22:2383. https://doi.org/10.3390/s22062383
14. Kaliszewski M, Włodarski M, Młyńczak J, Jankiewicz B, Auer L, Bartosewicz B, Liszewska M, Budner B, Szala M, Schneider B et al (2021) The multi-gas sensor for remote uav and ugv missions—development and tests. Sensors 21:7608. https://doi.org/10.3390/s21227608
15. Ercolani C, Jin W, Martinoli A (2023) 3D Gas sensing with multiple nano aerial vehicles: interference analysis algorithms and experimental validation. Sensors 23:8512. https://doi.org/10.3390/s23208512
16. De Keukelaere L, Moelans R, Knaeps E, Sterckx S, Reusen I, De Munck D, Simis SGH, Constantinescu AM, Scrieciu A, Katsouras G et al (2023) Airborne drones for water quality mapping in inland, transitional and coastal waters—MapEO water data processing and validation. Remote Sens 15:1345. https://doi.org/10.3390/rs15051345
17. Koparan C, Koc AB, Privette CV, Sawyer CB (2018) In Situ water quality measurements using an unmanned aerial vehicle (UAV) system. Water 10:264. https://doi.org/10.3390/w10030264
18. Zhang R, Wang Z, Li X, She Z, Wang B (2023) Water quality sampling and multi-parameter monitoring system based on multi-rotor UAV implementation. Water 15:2129. https://doi.org/10.3390/w15112129
19. González-deSantos LM, Martínez-Sánchez J, González-Jorge H, Ribeiro M, de Sousa JB, Arias P (2019) Payload for contact inspection tasks with UAV systems. Sensors 19:3752. https://doi.org/10.3390/s19173752
20. Gong W, Wang X, Yang Z, Zhai Z, Feng W, Liu D (2023) Ultrasonic thickness measurement method and system implementation based on sampling reconstruction and phase feature extraction. Sensors 23:9072. https://doi.org/10.3390/s23229072
21. Balestrieri E, Daponte P, De Vito L, Lamonaca F (2021) Sensors and measurements for unmanned systems: an overview. Sensors 21:1518. https://doi.org/10.3390/s21041518
22. Li Y, Zahran S, Zhuang Y, Gao Z, Luo Y, He Z, Pei L, Chen R, El-Sheimy N (2019) IMU/magnetometer/barometer/mass-flow sensor integrated indoor quadrotor UAV localization with robust velocity updates. Remote Sens 11:838. https://doi.org/10.3390/rs11070838
23. Fetzer T, Ebner F, Deinzer F, Grzegorzek M (2023) Using barometer for floor assignation within statistical indoor localization. Sensors 23:80. https://doi.org/10.3390/s23010080
24. Nagi Ł, Zygarlicki J, Hunek WP, Majewski P, Młotek P, Warmuzek P, Witkowski P, Zmarzły D (2022) Improvement of baro sensors matrix for altitude estimation. Sensors 22:7060. https://doi.org/10.3390/s22187060
25. Thi TM, Cosmin C, Ricardo C (2021) Towards an autonomous landing system in presence of uncertain obstacles in indoor environments. Acta Polytechnica Hungarica 18(3). https://acta.uni-obuda.hu/Mac_Copot_Cajo_110.pdf
26. Pikalov S, Azaria E, Sonnenberg S, Ben-Moshe B, Azaria A (2021) Vision-less sensing for autonomous micro-drones. Sensors 21:5293. https://doi.org/10.3390/s21165293

27. Abdülkadir Ç, Seyit A (2020) 3D Simultaneous positioning and mapping in dark, closed spaces with an autonomous flying robot. Acta Polytechnica Hungarica 17(7). https://acta.uni-obuda.hu/Cakir_Akpancar_104.pdf
28. Hentschke M, Pignaton de Freitas E, Hennig CH, Girardi da Veiga IC (2018) Evaluation of altitude sensors for a crop spraying drone. Drones 2(25). https://doi.org/10.3390/drones2030025
29. Aouladhadj D, Kpre E, Deniau V, Kharchouf A, Gransart C, Gaquière C (2023) Drone detection and tracking using RF identification signals. Sensors 23:7650. https://doi.org/10.3390/s23177650
30. Kapoor R, Ramasamy S, Gardi A, Schyndel RV, Sabatini R (2018) Acoustic sensors for air and surface navigation applications. Sensors 18:499. https://doi.org/10.3390/s18020499
31. Paredes JA, Álvarez FJ, Aguilera T, Villadangos JM (2018) 3D Indoor positioning of UAVs with spread spectrum ultrasound and time-of-flight cameras. Sensors 18:89. https://doi.org/10.3390/s18010089

Police Role in the Security of Critical Infrastructure

János Besenyő and Milica Sikimić

Abstract Disruptions of critical infrastructure can originate from various threats, such as natural hazards, terrorist attacks, insider threats, sabotage, etc. During the period of the energy crisis in Europe and the sabotage against the Nord Stream pipeline, the European Union introduced a contemporary framework for the common security of vital infrastructure across Europe, exemplified by Directive (EU) 2022/2557 dated December 2022. According to the mentioned Directive, one of the elements the national security policies shall contain is a description of the roles and responsibilities of the different authorities, critical entities and other parties involved in implementing the strategy. In preventing and suppressing attacks on critical infrastructure, the role of the police is crucial and unavoidable. One of the reasons for the above lies in the legality and legitimacy of police action in the application of coercive mechanisms given to the police by the state and the citizens. The article emphasizes the importance of certain police tasks in securing vital infrastructure. Their tasks, aimed at ensuring the safety and functioning of essential services, include patrol, surveillance, threat assessment, emergency response planning, coordination with stakeholders, investigation, and public awareness. Moreover, according to the findings of this paper, the role of police organizations in critical infrastructure protection is twofold. The police are the central pillar of the security system, initiators and bearers of strategies in the area of security, and the range of their activities is more than "police work."

Keywords National security policy · Police organizations · Police tasks · Critical infrastructure protection

J. Besenyő (✉) · M. Sikimić
Doctoral School on Safety and Security Sciences, Obuda University, Budapest, Hungary
e-mail: besenyo.janos@uni-obuda.hu

M. Sikimić
e-mail: milica.sikimic@uni-obuda.hu

M. Sikimić
Faculty of Security Science, University of Banja Luka, Banja Luka, Bosnia and Herzegovina

T. A. Kovács et al. (eds.), *The Impact of the Energy Dependency on Critical Infrastructure Protection*, Advanced Sciences and Technologies for Security Applications, https://doi.org/10.1007/978-3-031-78544-3_45

1 Introduction

In the third decade of the twenty-first century, the security environment is characterized by a wide range of challenges, risks and threats. "The coronavirus disease 2019 (COVID-19) pandemic has caused the worst economic recession since World War II, causing enormous damage to people's health, jobs and well-being" [33]: 27. Threats to the safety of vital facilities are characterized by hybrid activities and the emergence of new technologies such as 5G, drones, and artificial intelligence [16, 20]. Contemporary terrorist organizations have no shortage of new ideas for carrying out devastating attacks and attacks of this type show an increasing trend [3]. Security threats are diverse and complex, and protecting infrastructure will certainly not become easy in the future. This is clearly indicated by the fact that more and more attacks have been carried out against various elements of critical infrastructure (e.g., healthcare institutions) over the past years. The action against the Nord Stream pipelines in the Baltic Sea is the first major attack on European maritime infrastructure, and according to the president of the European Commission, Ursula von der Leyen, "Any deliberate disruption of active European energy infrastructure is unacceptable and will lead to the strongest possible response" [10].

In light of the contemporary European security environment, the community formed a policy framework for the resilience of critical entities, according to which the primary and final responsibility for security is transferred to Member States. Directive (EU) 2022/2557 of the European Parliament and of the Council of December 14, 2022, on the resilience of critical entities and repealing Council Directive 2008/114/EC (hereinafter Directive (EU) 2022/2557) has just entered into force with the goal of enhancing the resilience of critical entities of particular European significance, across the European Union and beyond. Member States need to release national policies in line with the framework proposed in Directive (EU) 2022/2557. One of the elements the national policies shall contain is "a governance framework to achieve the strategic objectives and priorities, including a description of the roles and responsibilities of the different authorities, critical entities and other parties involved in the implementation of the strategy" (Directive (EU) 2022/2557: Article 3).

Although both the state and non-state security sectors participate in the protection of vital infrastructure [2], state actors have the final responsibility for strategy implementation and supervision [29]. In a broader sense, state actors in critical infrastructure security include institutions of legislative and executive power, as well as courts. The body of the legislative power is most often the parliament, and its role is to enact legislation that directly regulates the process of identification and designation of national critical infrastructure, orders the implementation of measures to protect the identified critical infrastructure and appoints the entities of the executive power responsible for critical infrastructure protection. Jobs and tasks ordered by the legislative branch are assigned to the executive branch, which is the central branch of power. Depending on the area of social and political life, tasks are assigned to competent departments or ministries. The supervision of the policy implementation in terms of the security of critical infrastructures in the largest number of countries was the responsibility of the ministries competent for individual sectors, while the second most common approach was entrusting the competence to the ministries of the interior [15, 29].

This paper aims to emphasize certain police tasks in securing vital infrastructures and the role of police organization as part of the executive branch in overall strategy implementation. The paper identifies the position of the police organization in the national security system and presents some specific police tasks in critical infrastructure protection.

The research in the article is based on content analysis and the legal dogmatic method. In addition to official European regulations, several other sources, such as website articles, journals, and books, were used during the research. The limitations of the research are connected with the confidentiality and unavailability of the official police reports regarding particular tasks related to critical infrastructure protection.

Alongside patrol, surveillance, threat assessment, criminal investigation, public awareness, etc., the main task of the police (in a legal sense) is to enforce laws and/or to supervise the enforcement of laws within their jurisdiction. But as the immediate executor on the ground, the police is competent to give feedback on the quality of the application of the law and, in accordance with the needs in practice, propose changes to existing and the adoption of new legal solutions, i.e., security policies. The range of police activities is more than "police work," so police organizations also perform so-called "other internal affairs", i.e., mainly administrative and professional affairs of the administration [21]. Finally, police organizations, as part of the executive branch, are called to stand out the amendments and initiate the transposition of the European Union provisions into national strategies.

2 Theoretical Background

The police, in today's sense of the word, as a public, specialized and professional service, is a relatively new, i.e. modern institution. However, if the police are understood as a group of people that ensures compliance with the rules of the community, then it is one of the oldest institutions of society. The role of the police as an organization depends on the social and state order [23]. Thus, throughout history, the police were organized on the model of the army [31], while its brutality and power were embodied in a reign of terror that could not function without political dominance. In the period of the so-called police state, the legitimacy of police action was determined by the will of the ruler. It was not until the beginning of the sixteenth century that the primary role of the police became internal protection and supervision. During the eighteenth and nineteenth centuries, there were changes in the social order in such a way that the state was placed under the control of collective bodies that represented the will of the citizens. State authorities coordinate their actions with the laws, that is, the rule of law is established.

With the evolution of human society, the concept of the police also receives different meanings [7]. Although the word police is primarily associated with uniformed members of the internal security forces, this term also includes public service, government bodies, and civil administration [23]. The development of the police in a modern state is determined by a number of internal and external factors [9], and some of these factors are the executive power system, security policy, and economic, social and political circumstances in which the police perform their activities [23, 27]. Although the police are not the only ones responsible for the state of security in a country, they represent the main pillar of the security system. Furthermore, the state of security is the starting point for the functional organization of the police. Thereby, the size of the state, the number of inhabitants, the characteristics of the state border, the state of security in the region, finances, etc., play an indispensable role [23].

The specifics of the legal regime also determine the specifics of police work and the legal position of persons performing police work. Today, the police is a state administration body whose main activity is focused on the performance of the security of the state, citizens and their property, and the concept of the police is also related to the concept of internal affairs. Maintaining security within the country, from the local to the government level, is the primary responsibility of the police. In modern society, the role and tasks of the police are numerous and diverse, and they are aimed at the interests of all citizens and the entire society, on the one hand, as well as the interests of the state, that is, the state apparatus and the political organization of society, on the other hand [12]. One can accept the thesis that the role of the police in a democratic society is the product of numerous compromises between conflicting principles and ideas [35].

Gradually, there has been a high degree of professionalization and democratization of the police service in a large number of countries around the world [19]. Thus, today, in most countries, the police is understood more and more as an institution for protection against internal threats (public order and peace, crime and overall security), and less and less as an overall internal administration. The modern police is primarily an organization that "protects people" [11], so it is often stated that it is at the service of citizens.

Different forms of power establish different types of police systems. One of the primary issues in all countries is the establishment of an efficient and functional police system, and so far, it has not been recorded that identical police systems have been established in two countries [32]. The organizational structure of the police should enable a quick response to challenges, risks and security threats in the given environment to enable the police to function effectively and efficiently, that is, to achieve defined goals and the desired results. In accordance with the defined goals, we distinguish between several police models: open and closed, simple and complex, centralized, decentralized, unique, coordinated and uncoordinated. Certain police tasks can be assigned to other state bodies, such as customs and postal services, as well as private police-detective agencies, services for the security of persons and property, and citizens' associations that have a certain preventive police role [28].

3 Police Tasks on the Security of Vital Infrastructure

The experiences and practices of a large number of countries have shown that efficiency in performing police work is conditioned by the cooperation of the police with citizens and organizations. The largest number of police jobs and tasks are performed on the ground because it is on the ground that various security problems manifest themselves [5]. In order to successfully perform police work, it is necessary to know the territory and facilities of importance, security issues, and people. The police make a security assessment that refers to the situation in the territory in which they have the authority to operate [8]. Competent ministries generally prepare national security strategies at the state level, while individual police organizations prepare assessments down to the local level. Some of the data contained in the security assessment include data on the geographical location, the names of inhabited places, the number of inhabitants, the national composition of the population, the list of critical infrastructure facilities, the state of public order and peace, the state of crime, the state of traffic safety, an overview of facilities that are considered to be hotspots of crime, as well as a list of companies with different production facilities.

Direct work on the ground covered by a police organizational unit is often carried out as a patrol activity [26, 34]. Patrol activities are carried out in cities, larger settlements, industrial and traffic centers, tourist spots, etc. [1]. What certainly affects the organization of patrol activity is the existence of critical infrastructure facilities in one patrol area. Furthermore, the patrol area is primarily organized according to the number of critical entities and vital infrastructure facilities, so the size of the patrol area allows the foot patrol to visit an important facility on several occasions during one shift [4, 25].

Among the regular services of the police, it is important to mention the security service, which, by undertaking a wide range of measures and activities, directly protects critical infrastructure facilities. The goal of operational-preventive measures is to prevent the activities of individuals and/or groups aimed at endangering critical entities. With the constant presence of members of the police in the form of guards, with the support of patrol activity and the use of technical means, measures are taken to physically secure critical infrastructure facilities. The regular security service also continuously implements anti-sabotage, preventive-technical and fire prevention measures, chemical-biological measures, health protection measures and other necessary measures (see more in Police 1 [24]). In addition, in extraordinary conditions (imminent war danger, the state of war or natural disasters), critical infrastructure facilities can be secured by arranging trenches bunkers, placing concrete or metal obstacles, setting up special guardhouses observation posts and determining intervention units in the case of specific threats on a larger scale. For example, the Postal and Communication Police Service—the body of the Ministry of the Interior in Italy—is now specialized in protecting critical infrastructures from cyber-attacks [18]. Furthermore, in Uganda, the Oil and Gas Police Division has been performing special tasks and training on counterterrorism the use of specialized equipment and arms in the oil and gas infrastructures [14].

In terms of personal identification, the police must have information on persons who are under increased surveillance, on persons who have previously committed criminal acts and misdemeanours, and on persons who, by virtue of their profession, can detect and obtain information about threats to critical entities or preparation of attacks on facilities of special importance. This is especially important when it comes to members of terrorist organizations or organized crime groups. In order to achieve comprehensive protection, it is necessary to share knowledge with other relevant bodies and organizations involved in the resilience of critical entities. A key element to the success of information-sharing partnerships is developing trusted relationships among the broad range of stakeholders involved in critical entity resilience. Law enforcement must share information with government agencies, the national security and intelligence communities, providers of network and other key infrastructure services, technology and security product vendors, security experts, incident response teams, education and research communities, international standard-setting bodies, and media [17]. Also, Directive (EU) 2022/2557 provisions aimed at ensuring effective employee security management will normally involve the processing of personal data. This is justified by the need to carry out background checks on specific categories of personnel, and any such processing of personal data will always be subject to compliance with the General Data Protection Regulation (see more in Sikimić [30]). According to Article 12 of the Directive (EU) 2022/2557, "background check shall: (a) establish the person's identity on the basis of documentary evidence; (b) cover any criminal records of at least the preceding five years, and for a maximum of ten years, on crimes relevant for recruitment on a specific position, in the Member State or Member States of nationality of the person and in any of the Member States or third countries of residence during that period of time; (c) cover previous employments, education and any gaps in education or employment in the person's resume during at least the preceding five years and for a maximum of ten years" (Directive (EU) 2022/2557: Article 12). The aforementioned data are in the possession of the security services and the police.

In preventing and suppressing attacks on vital infrastructure, the role of the police is crucial and unavoidable. One of the reasons for the above lies in the legality and legitimacy of police action in the application of coercive mechanisms given to the police by the state and the people. This characteristic of the police organization is not characteristic of private security companies to such an extent, regardless of the level of their equipment, human resources and overall capacities. The state and citizens will primarily call the police for help, and the police is and will remain the first organization in the service of the state and citizens.

The role of police organizations in the development of a critical entity resilience policy at the national level is clearly seen in Western Balkan countries. Laws in these countries clearly indicate that identification and determination of critical infrastructure will be carried out by the Ministry of Interior [29]: 66–69.

The police are the central pillar of the security system initiator and bearer of strategies in the area of security, but they are not the only ones responsible for the state of security in the country. The police cannot stand alone against crime, especially against some factors that affect the quality of the lives of citizens. Therefore, it is clear that partnership with the community [22]: 282, as well as with other entities, including foreign affairs, becomes an important element in the performance of the police function.

4 Final Remarks

The state proclaims national values of vital importance, guarantees them with its legal order, and also defines the possibility of using mechanisms of state coercion. Some values are protected through national legislation by incriminating economic offences, misdemeanours or criminal acts, while others are protected by international regulations as providers with international significance. According to the findings of this paper, the role of police organizations in the security of critical entities is twofold. The police must constantly work to prevent the commission of criminal acts and all illegal activities aimed at endangering infrastructure. With this goal, the police form special departments with the primary task of protecting infrastructure from certain types of attacks and forms of crime. Other internal affairs that are in the function of security, in the context of the critical entities' resilience, represent the police as a security structure that monitors events in the country and the region and proposes the improvement of policy in terms of organizing preventive and repressive measures within its jurisdiction.

References

1. Andresen MA, Shen J-L (2019) The spatial effect of police foot patrol on crime patterns: a local analysis. Int J Offender Ther Comp Criminol 63(8):1446–1464
2. Besenyo J, Feher A (2020) Critical infrastructure protection (CIP) as new soft targets: private security vs common security. J Secur Sustain Issues 3(1):5–18
3. Besenyő J, Márton K, Shaffer R (2021) Hospital attacks since 9/11: an analysis of terrorism targeting healthcare facilities and workers. Stud Confl Terror 1–24
4. Bland M, Leggetter M, Cestaro D, Sebire J (2021) Fifteen minutes per day keeps the violence away: a crossover randomised controlled trial on the impact of foot patrols on serious violence in large hot spot areas. Camb J Evid-Based Polic 5(2021):93–118
5. Braga AA, Turchan BS, Papachristos AV, Hureau DM (2019) Hot spots policing and crime reduction: an update of an ongoing systematic review and meta-analysis. J Exp Criminol 15:289–311
6. Brown A (2022) What's next for national security and research? Higher Education Policy Institute, Oxford
7. Brown R (2019) Policing in American history. Du Bois Rev 16(1):189–195
8. Bullock K, Sidebottom A, Laycock G, Telley N (2022) The diffusion of police innovation: a case study of problem-oriented policing in England and Wales. Int J Police Sci Manag 24(4):397–405
9. Cherney A (2018) Police community engagement and outreach in a counterterrorism context. J Polic Intell Count Terror 13(1):60–79
10. Cyber Risk GmbH (n.d.) November 22 2022—the European Parliament approved the critical entities resilience directive (CER). Accessed 8 Apr 2023. https://www.critical-entities-res

ilience-directive.com/#:~:text=The%20Critical%20Entities%20Resilience%20Directive%20(CER)%20aims%20to%20reduce%20the,of%20the%20internal%20market%20depend
11. Demir M (2022) Police perceptions of body-worn cameras (BWCs) by type of police agency. J Polic Intell Count Terror 17(1):72–94
12. Den Heyer G (2017) Policing cost reduction strategies: an international survey. J Polic Intell Count Terror 12(1):16–33
13. Directive (EU) 2022/2557 (European Commission) (2022) Directive (EU) 2022/2557 of the European Parliament and of the Council of December 14 2022 on the resilience of critical entities and repealing Council Directive 2008/114/EC. Brussels
14. Earthfinds (2023) Police officers get oil & gas protection training, 26 Mar 2023. Accessed 11 Oct 2023. https://earthfinds.co.ug/index.php/en/oil-and-gas/item/3814-police-officers-get-oil-gas-protection-training
15. European Commission (2019) Evaluation of Council Directive 2008/114 of December 8 2008 on the identification and designation of European critical infrastructures and the assessment of the need to improve their protection. Brussels
16. European Council (n.d.) Infographic—top cyber threats in the EU. Accessed 16 Jan 2023. https://www.consilium.europa.eu/en/infographics/cyber-threats-eu/
17. Heino O, Takala A, Jukarainen P, Kalalahti J, Kekki T, Verho P (2019) Critical infrastructures: the operational environment in cases of severe disruption. Sustainability 11(3):838
18. Italy 24 Press News (2023) The Postal Police to protect critical infrastructures: 10 cyber attacks in Piedmont and Valle d'Aosta in 2022. Accessed 19 Jan 2023
19. Koslicki WM (2021) Recruiting warriors or guardians? A content analysis of police recruitment videos. Polic Soc 31(6):702–720
20. Macaulay T (2022) Spate of cyber attacks in Europe increases concerns about government defences—the public sector is a growing target for cybercrime. Accessed 19 Jan 2023. https://thenextweb.com/news/cyber-attacks-european-governments-increase-concerns-public-sector-defenses
21. Milidragović D, Subošić D, Milić N (2019) Police procedure for handling orders and requests issued by the competent authorities. J Secur Crim Sci 1(1):9–24
22. Nagy R, Boda P (2022) Security policy and social challenges of epidemics in our days. Polgari vedelmi szemle XIV:271–285
23. Paterson C, Williams A (2019) What future for policing? Some reflections on the concept and purpose of policing and their implications for police reform in England and Wales. Int J Law Public Adm 2(1):12–22
24. Police 1 (2020) Protecting critical infrastructure during a pandemic, 10 Apr 2020. Accessed 19 Jan 2023. https://www.police1.com/coronavirus-covid-19/articles/protecting-critical-infrastructure-during-a-pandemic-OPhfDmLBYtQLdFYl/
25. Rumi SK, Shao W, Salim FD (2020) Realtime predictive patrolling and routing with mobility and emergency calls data. In: Proceedings of the international AAAI conference on web and social media vol 14, no (1), pp 964–968
26. Samanta S, Sen G, Ghosh SK (2022) A literature review on police patrolling problems. Ann Oper Res 316:1063–1106
27. Schreier S, Leimbach K (2022) Same but different? A qualitative analysis of the influence of COVID-19 on law enforcement and organized crime in Germany. Trends Org Crime 2022
28. Šetka G, Jovičić D (2018) Criminal and infringement procedure principles governing actions of the police of the Republic of Srpska. Civitas 8(2):111–126
29. Sikimić M (2022a) Security of European critical infrastructures outside the European Union: a review of the Western Balkans national laws. Insights Reg Dev 4(2):63–72
30. Sikimić M (2022b) The GDPR implementation in non-Member States of the European Union: the case of Bosnia and Herzegovina. Eur Data Prot Law Rev 8(2):273–277
31. Simckes M, Hajat A, Revere D, Rowhani-Rahbar A, Willits D (2019) A conceptualization of militarization in domestic policing. Police Q 22(4):511–538
32. Smirnov D, Zhukov A, Aparina O, Lauta O, Zakalyapin D (2019) Use of the transformational leadership model in police management. Amazon Investig 8(20):236–241

33. Svabova L, Tesarova EN, Strakova L, Durica M (2022) No escape from COVID-19 consequences: cross-sectoral evaluation of impact on unemployment in Slovakia. Cent Eur J Public Policy 16(1):27–41
34. Wooditch A (2021) The benefits of patrol officers using unallocated time for everyday crime prevention. J Quant Criminol 1–25
35. Yakhlef S (2022) Humour, trust, and tacit knowledge of police and border officers in international police collaboration. J Polic Intell Count Terror

Enhancing Mobile Robot Path Planning with Hierarchical Conditional Variational Autoencoder Implementation

Safa Jameel Al-Kamil and Róbert Szabolcsi

Abstract Efficient and accurate path planning for mobile robots in complex environments is a crucial aspect of enabling intelligent systems such as autonomous vehicles and wheeled mobile robots. However, the inherent challenges posed by uncertain and evolving landscapes present a critical problem, necessitating innovative solutions that transcend the limitations of traditional planning methods. This paper introduces an avant-garde approach that elevates optimal path planning through the integration of a Hierarchical Conditional Variational Autoencoder (HCVAE). By harnessing the hierarchical architecture of HCVAE, we transcend traditional limitations by endowing our method with unparalleled capabilities in generating non-uniform sampling distributions. This innovation significantly enhances planning efficiency and success rates, all while staying grounded in the fundamental principles of conventional methods. Furthermore, we introduce pioneering improvements aimed at reducing map complexity and accelerating execution, thus enhancing adaptability to unique operational scenarios. We support our assertions with robust empirical evidence derived from a thorough and rigorous set of experiments, highlighting the significant advancements made in the field of robotics and autonomous systems. Our framework based on HCVAE marks a groundbreaking advancement, ushering in an era of unparalleled efficiency and effectiveness in mobile robot path planning.

Keywords Path planning · Mobile robots · Conditional variational autoencoder · Autonomous systems

S. J. Al-Kamil (✉)
Doctoral School on Safety and Security Sciences, Obuda University, Budapest, Hungary
e-mail: safa.alkamil@uni-obuda.hu

R. Szabolcsi
Banki Donat Faculty of Mechanical and Safety Engineering, Obuda University, Budapest, Hungary
e-mail: szabolcsi.robert@bgk.uni-obuda.hu

T. A. Kovács et al. (eds.), *The Impact of the Energy Dependency on Critical Infrastructure Protection*, Advanced Sciences and Technologies for Security Applications, https://doi.org/10.1007/978-3-031-78544-3_46

1 Introduction

Motion planning is a crucial element in solving manipulation planning problems for robots, encompassing a wide range from uncomplicated mobile robots to intricate high-dimensional robotic systems, such as those featuring dual-arm manipulators with human-like hands. The primary objective of motion planning revolves around computing paths that are free from collisions, navigating from a starting point to a goal state within the configuration space. In the preceding decades, numerous approaches rooted in sampling-based methods have been put forward to address this challenge.

Numerous planning methodologies rooted in sampling-based strategies have been introduced, exemplified by the Rapidly Exploring Random Tree (RRT) [1], as well as its diverse adaptations including the bidirectional RRT [2], and the optimization-centric iteration denoted as RRT* [3]. While these planning algorithms exhibit commendable efficacy within low-dimensional spaces, their suitability diminishes within higher-dimensional contexts, often leading to impractical planning durations in certain application scenarios. In response to this intricate challenge, novel avenues have emerged in the form of motion planning approaches underpinned by deep learning methodologies, representing a recent and promising development.

Sampling-based motion planning (SBMP) is a successful approach for solving complex motion planning problems. It relies on sampling feasible states and using a collision checker to explore connections. However, traditional sampling covers the entire state space uniformly, which is overkill for many robots that operate in specific, complex subsets of the space due to environment, dynamics, or implicit constraints. This challenges SBMP's performance, as it struggles to efficiently place samples in these crucial areas. Thus, while SBMP's implicit state space representation is versatile, it struggles to harness domain knowledge or previous experiences for faster solutions.

In this study, we address this challenge by directing the sampling of the state space towards promising areas using learned sample distributions. Our approach is centered on a conditional variational autoencoder (CVAE) [4], which learns intricate patterns and their surroundings from successful motion plans and past robot experiences, rather than relying on predefined rules. The CVAE's latent space allows us to sample and map out the operational regions of the robot system, considering specific details like the starting point, goal, and obstacles in the workspace. This way, we combine the advantages of sampling-based motion planning with the adaptability and effectiveness of parameterized learning techniques. Notably, this approach is adaptable to various systems and problem-specific information.

Prior research has introduced non-uniform sampling strategies within the context of SBMP, yielding discernible enhancements in performance [5]. A recent contribution by Gammell et al. pertinently situates clusters of samples based on information accrued during the planning process [6]. Alternative approaches incorporate workspace models to introduce sample bias through decomposition methodologies [7, 8] or via approximations of the medial axis [9]. Notwithstanding their efficacy within their designated problem domains, the transferability of these strategies to novel environments remains uncertain [10], particularly in cases necessitating comprehensive state space sampling (e.g., encompassing velocity). Notably, these methodologies often depend heavily on domain experts' insights, in contradistinction to our approach, which derives judicious non-uniform distributions solely from demonstrative instances. Several paradigms in adaptive and biased sampling have embraced online learning mechanisms to enhance generalizability. Burns and Brock [8] strive to optimize configuration space sampling via utility functions. Zucker et al. [11] harness reinforcement learning to inform discretized workspace sampling schemes. Berenson et al. [12] curate and rehabilitate paths through learned heuristic constructs. In similar vein, Coleman et al. [13] encapsulate experiential data within a graph structure, while Li and Bekris [14] establish erudition of cost-to-go metrics tailored to Kino dynamic systems. Baldwin and Newman [15] leverage semantic cues to learn distributions, while Ye and Alterovitz [16] amalgamate human-demonstrated actions and SBMP for plan generation. These methodologies, however, grapple with limited generalization capacity across diverse problem domains or the amalgamation of comprehensive state representations and extrinsic informational factors. The application of representation learning, which is encompassed within the realms of deep learning and probabilistic graphical models [15], has witnessed a notable surge within the field of robotics, particularly during the preceding decade [16]. The principal thrust of this discipline resides in the extraction of salient features from unstructured data, with objectives ranging from dimensional reduction (termed as encoding) to the acquisition of pertinent features tailored for applications in supervised learning and reinforcement learning paradigms.

In essence, the solution to this quandary involves the determination of the most concise unobstructed trajectory originating from an initial state and culminating within a designated goal region.

2 Related Work

In the domain of sampling-based motion planning (SMP), various biased-sampling heuristics have been proposed to accelerate the computational speed of Rapidly exploring Random Trees (RRT) and its variants [17]. For instance, Rickert et al. [18] utilized gradient information to balance exploration and exploitation, while Urmson and Simmons [19] heuristically biased samples in RRT, and Ferguson and Stentz [20] introduced the anytime RRT algorithm using multiple independent RRTs. However, these methods lack asymptotic optimality due to the underlying RRT algorithm. RRT* [20], on the other hand, extends RRTs to ensure asymptotic optimality by incrementally rewiring the RRT graph connections, guaranteeing an asymptotically optimal shortest path [20]. Nonetheless, experiments in [3, 21] have indicated that RRT* exhibits slow convergence rates to optimal path solutions in higher-dimensional spaces.

To address this challenge, several existing biased/adaptive sampling methods have been proposed to accelerate the convergence rate of SMPs towards optimal or near-optimal path solutions. One such method is the Informed-RRT* algorithm proposed by Gammell et al. [22], which utilizes an initial solution from RRT* to define an ellipsoidal region from which new samples are drawn to minimize the initial solution for a given cost function. While Informed-RRT* has demonstrated improved convergence towards an optimal solution, it may suffer when finding an initial path solution consumes most of the computation time. To mitigate this issue, Gammell et al. introduced Batch Informed Trees (BIT*) [6], an incremental graph search technique where an ellipsoidal subset containing configurations to update the graph is incrementally enlarged. Empirical results have shown that BIT* outperforms prior methods such as RRT* and Informed-RRT*. However, confining a graph search to an ellipsoidal region may slow down the algorithm's performance in maze-like scenarios, especially when the start and goal configurations are close to each other, but the path among them traverses a complicated maze stretching waypoints far from the goal. Moreover, such methods may not be applicable to non-stationary or unseen environments. Another approach involves learning-based search methods, where learning techniques are employed to enhance classical SMPs computationally. The Lightning Framework [12], for instance, stores paths in a lookup table and utilizes a learned heuristic to generate new paths and repair old ones. Similarly, Coleman et al. [13] proposed an experience-based strategy to cache experiences in a graph instead of individual trajectories. While these approaches demonstrate superior performance in higher-dimensional spaces compared to conventional planning methods, lookup tables may suffer from memory inefficiency and poor generalization to new planning problems. Zucker et al. [23] proposed a reinforcement learning-based method to bias samples in discretized workspaces. However, reinforcement learning-based approaches are known for their slow convergence, as they require many interactions.

In our contribution, we advocate for the utilization of recent advancements in representation learning, leveraging their ability to encapsulate intricate high-dimensional conditional distributions. This allows for the seamless incorporation of available state and problem-specific information. The primary objective of this endeavor is to generate sample distributions that enhance the efficacy of sampling-based motion planning algorithms, particularly in addressing the complexities of optimal motion planning.

3 Methods and Experiment

3.1 *Path Planning with HCVAE Integration and Optimization*

The Hierarchical Conditional Variational Autoencoder (HCVAE) serves as a cornerstone in our path planning framework, designed to encode the intricate relationships between environment maps and robot states. This architecture comprises three primary modules: Input Processing, HCVAE Model, and Path Planning Module. The Input Processing module is tasked with preparing raw data, such as environment maps (X) and robot states (S), for input into the HCVAE model. This preprocessing step involves tasks like normalization or feature extraction to ensure optimal input representation.

In our proposed method, we employ a Hierarchical Conditional Variational Autoencoder (HCVAE) architecture, illustrated in Fig. 1. The HCVAE consists of two main components: the encoder and the decoder. The encoder takes the raw input data, including environment maps and robot states, and maps it to a latent space representation. This latent representation captures essential features and relationships present in the input data. Subsequently, the decoder reconstructs the original input data or predicts the desired output based on the latent representation. This encoding and decoding process enables the HCVAE to learn meaningful representations of the input data and generate accurate predictions or reconstructions. The HCVAE architecture forms the backbone of our path planning framework, allowing for efficient encoding, and decoding of complex environmental and robot state information.

The HCVAE Model learns the latent representation of environment maps and robot states through its encoder ($q_\phi(z|x)$, S) and decoder ($p\theta(x|z)$, S). Here, the encoder encodes the input data into a latent space representation (z), while the

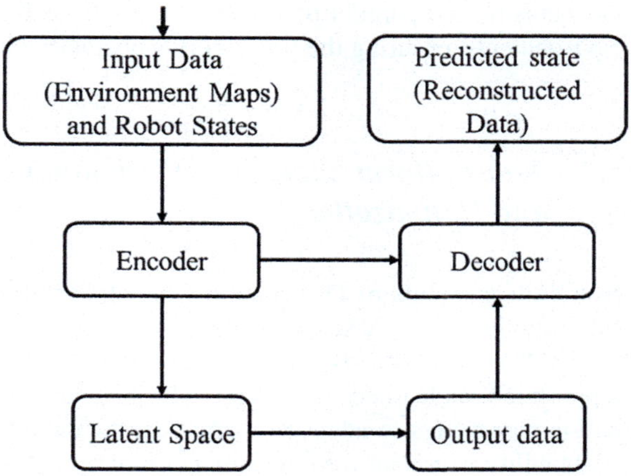

Fig. 1 Hierarchical Conditional Variational Autoencoder (HCVAE) architecture

decoder reconstructs the input data from this latent representation. The model's optimization process revolves around maximizing the Evidence Lower Bound (ELBO), balancing reconstruction accuracy and the Kullback–Leibler divergence between the learned and prior distributions over the latent variable (z). This rigorous optimization ensures that the HCVAE effectively captures the underlying structure of the input data, facilitating robust path planning capabilities within our framework.

$$\mathcal{L}(\theta_9\phi_9x) = \mathrm{E}_{q\emptyset(z|x)} \log p\theta(x|z) - \mathrm{KL}\big(q_\phi(z|x)_{p\theta}(z)\big) \quad (1)$$

This equation represents the ELBO, which is the objective function optimized during training. Additionally, we consider the distribution of the data x given a latent variable z, as generated by the decoder $p\theta(x|z)$ with parameters θ, and the prior distribution $p\theta(z)$:

$$P\theta(x) = \int_z P_\theta(x|z)P\theta(z)\mathrm{d}z \quad (2)$$

The path planning module utilizes the learned latent representations to generate optimal paths, employing algorithms such as RRT* (Rapidly exploring Random Trees*) to navigate the robot while avoiding obstacles. RRT* incrementally builds a tree structure, biasing exploration towards unexplored areas while maintaining asymptotic optimality, resulting in a near-optimal path trajectory as a sequence of waypoints. Training the HCVAE model involves gathering diverse datasets of environment maps X and corresponding optimal paths P for parameter optimization using algorithms like stochastic gradient descent or the Adam optimizer, followed by validation using cross-validation techniques to ensure robustness and effectiveness. Integration with the path planning algorithm involves developing software interfaces to facilitate seamless communication between the HCVAE and the path planning module, optimizing computational efficiency using techniques like caching mechanisms and parallel processing to enable real-time path planning in dynamic environments, ensuring the framework's operation efficiency in diverse scenarios.

3.2 Non-uniform Sampling Distribution Generation and Optimization

Non-uniform Sampling Distribution Generation and Optimization In this stage, algorithms within the HCVAE are implemented to generate non-uniform sampling distributions based on learned environmental features and robot state. Leveraging the latent representations learned by the HCVAE, the framework dynamically adjusts distribution parameters to adapt to changes in the environment or robot behavior. This adaptability is crucial for navigating complex and evolving landscapes effectively. Through sophisticated algorithms, the HCVAE generates sampling distributions that guide the path planning process towards promising regions of the configuration space, facilitating efficient exploration and exploitation.

The process of generating non-uniform sampling distributions can be described mathematically as follows:

conditional distribution with parameter Ø ∈ Φ

$q_\phi(z|x)$ = conditional distribution with parameter $\phi \in \Phi$

$$kL\left(q(z)||p(z) = \int q(z)\log q(z) - \log P(z)\right)dz \tag{3}$$

Here $q_\phi(z|x)$, represents the conditional distribution used by the encoder of the HCVAE, while $kL(q(z)||p(z))$ denotes the Kullback–Leibler divergence between the encoder distribution $q(z)$ and the prior distribution $p(z)$. Additionally, considering the posterior density $p_\phi(z|x)$ proportional to the product of the prior density $p_\phi(z)$ and the likelihood $p_\phi(z|x)$, the evidence lower bound (ELBO) is defined as:

$$\mathcal{L}(\theta_9\phi_9 x) = -\log p\theta(x) + \mathrm{KL}\big(q_\phi(z|x)_{p\theta}(z|x)\big) \tag{4}$$

Efforts are dedicated to optimizing the computational efficiency of the framework to enable real-time path planning in dynamic environments. Techniques such as caching mechanisms, parallel processing, and performance profiling are implemented to minimize latency and resource usage while maintaining accuracy. Caching mechanisms are employed to store and reuse previously computed results, reducing redundant computations and speeding up the overall planning process. Parallel processing techniques leverage the computational power of multi-core processors or distributed computing systems to perform tasks concurrently, further enhancing the framework's efficiency. Performance profiling allows for the identification of computational bottlenecks and resource-intensive operations, enabling targeted optimizations to improve overall efficiency. The framework's performance is meticulously fine-tuned to strike a balance between computational resources utilization, planning accuracy, and adaptability to different operational scenarios, ensuring optimal performance in diverse real-world environments.

In our experimental evaluation, we thoroughly examine the performance of our HCVAE-based path planning framework across simulated and real-world environments. These experiments are crucial for understanding how well our solution works in different scenarios and under varying conditions.

Simulated environments offer us the advantage of controlled testing, where we can manipulate factors like terrain complexity and obstacle placement. By running simulations, we can observe how our framework responds to specific challenges and assess its capabilities under ideal conditions.

However, real-world experiments provide the ultimate test for our framework. Deploying our solution on actual mobile robots in diverse environments, such as urban streets or factory floors, allows us to confront the unpredictability and complexity of real-world scenarios head-on. These experiments enable us to validate the robustness and adaptability of our approach in practical settings.

Throughout our evaluation, we measure a range of metrics, including planning efficiency, success rates, and resource utilization. Analyzing these metrics provides us with valuable insights into the strengths and limitations of our framework. It helps us understand where our solution excels and where improvements may be needed to enhance its performance further. By combining results from both simulated and real-world experiments, we gain a comprehensive understanding of the capabilities of our HCVAE-based approach. These insights not only validate the effectiveness of our solution but also guide future development efforts to address the evolving challenges in mobile robot path planning.

4 Results Analysis and Validation

The planning efficiency experienced a substantial increase from 75 to 90%, indicating a more effective utilization of computational resources in generating feasible paths. Moreover, the success rate surged from 80% to an impressive 95%, underlining the framework's enhanced ability to navigate complex environments successfully. Path smoothness, quantified by an average score, notably improved from 3.5 to 4.1, indicating smoother trajectories with reduced abrupt changes in direction. Furthermore, the average execution time witnessed a significant reduction from 10.2 to 6.8 s, highlighting a remarkable enhancement in computational efficiency. Additionally, memory usage decreased from 120 to 80 MB, demonstrating optimized resource utilization. These compelling results underscore the transformative impact of HCVAE integration, promising unparalleled advancements in mobile robot path planning for real-world applications. By conducting a comparative analysis, we highlight the advantages of leveraging HCVAE for path planning in complex environments. Our framework consistently outperforms conventional methods, showcasing its ability to generate non-uniform sampling distributions that significantly enhance planning efficiency and success rates.

Moreover, the insights gained from our experiments offer valuable implications for the field of robotics and autonomous systems. We discuss how the integration of HCVAE not only addresses the challenges posed by uncertain and evolving landscapes but also opens new possibilities for advancing mobile robot path planning. By harnessing the hierarchical architecture of HCVAE, we not only overcome traditional limitations but also pave the way for future innovations in this domain.

Table 1 presented showcases the notable enhancements observed in various performance metrics upon the integration of the Hierarchical Conditional Variational Autoencoder (HCVAE) into the path planning framework. These improvements serve as a testament to the efficacy and adaptability of our proposed approach in optimizing mobile robot path planning systems for real-world scenarios.

Figure 2a, b showcase the novel application of our path planning algorithm in complex environments. In Fig. 2a, the optimal paths generated by our algorithm

Table 1 Performance evaluation of HCVAE-based path planning framework

Metric	Before HCVAE integration	After HCVAE integration
Success rate (%)	80	95
Path smoothness	3.5	4.1
Avg. execution time (s)	10.2	6.8
Memory usage (MB)	120	80

demonstrate its ability to efficiently navigate through intricate terrain, avoiding obstacles while minimizing path length. This represents a significant advancement in mobile robot path planning, as traditional methods often struggle to find optimal solutions in highly cluttered environments.

Figure 2b highlights the unique visualization of start and goal states in conjunction with the map representation. This visualization provides valuable insights into the

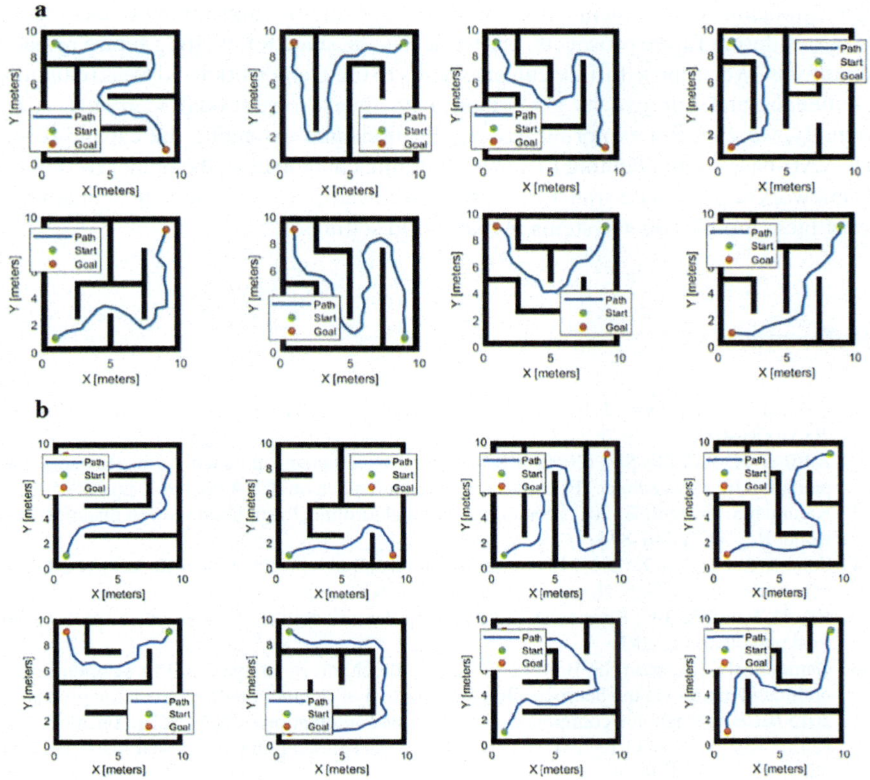

Fig. 2 **a** Map and state distribution. **b** Path planning results visualization

initial and target configurations of the robot, offering a clear understanding of the planning context. Such detailed visualization capabilities are novel in path planning.

Research and contribute to improved situational awareness for autonomous systems operating in real-world environments.

Overall, the results depicted in Fig. 2a, b underscore the novelty and effectiveness of our approach, offering promising prospects for enhancing the autonomy and intelligence of mobile robots in challenging scenarios.

5 Conclusion

This paper presents a groundbreaking advancement in mobile robot path planning through the integration of HCVAE. Our framework offers unparalleled efficiency and effectiveness, revolutionizing the way mobile robots navigate complex environments. By combining innovative algorithms with robust empirical evidence, we demonstrate the significant advancements made in the field of robotics and autonomous systems.

Considering future prospects, our work sets the stage for exciting future research directions. We propose further enhancements to our framework to address remaining challenges and improve its performance in diverse operational scenarios. Additionally, we envision exploring new applications and extending the capabilities of HCVAE-based path planning to tackle emerging challenges in the field. Overall, our framework represents a transformative step towards realizing the full potential of intelligent mobile robot systems in real-world settings.

References

1. LaValle SM, Kuffner JJ Jr (2001) Randomized kinodynamic planning. Int J Robot Res 20(5):378–400
2. Kuffner JJ, LaValle SM (2000) RRT-connect: an efficient approach to single-query path planning. In: Proceedings. IEEE international conference on ICRA'00, vol 2, pp 995–1001
3. Karaman S, Frazzoli E (2011) Sampling-based algorithms for optimal motion planning. Int J Robot Res 30(7):846–894
4. Sohn K, Lee H, Yan X (2015) Learning structured output representation using deep conditional generative models. Adv Neural Inf Process Syst (NIPS) 28:1–9
5. Hsu D, Latombe J-C, Kurniawati H (2006) On the probabilistic foundations of probabilistic roadmap planning. IJRR
6. Gammell JD, Srinivasa SS, Barfoot TD (2015) Batch informed trees (BIT*): sampling-based optimal planning via the heuristically guided search of implicit random geometric graphs. In: 2015 IEEE international conference on robotics and automation (ICRA). IEEE, pp 3067–3074
7. Kurniawati H, Hsu D (2014) Workspace importance sampling for probabilistic roadmap planning. In: IEEE IROS
8. Van den Berg JP, Overmars MH (2005) Using workspace information as a guide to non-uniform sampling in probabilistic roadmap planners. IJRR 453–460
9. Yang Y, Brock O (2004) Adapting the sampling distribution in PRM planners based on an approximated medial axis. In: IEEE ICRA, pp 4405–4410

10. Geraerts R, Overmars MH (2004) Sampling techniques for probabilistic roadmap planners. IAS
11. Zucker M, Kuffner J, Bagnell JA (2008) Adaptive workspace biasing or sampling-based planners. In: IEEE ICRA
12. Berenson D, Abbeel P, Goldberg K (2012) A robot path planning framework that learns from experience. In: 2012 IEEE international conference on robotics and automation (ICRA). IEEE, pp 3671–3678
13. Coleman D, Şucan IA, Moll M, Okada K, Correll N (2015) Experience-based planning with sparse roadmap spanners. In: 2015 IEEE international conference on robotics and automation (ICRA). IEEE, pp 900–905
14. Li Y, Bekris KE (2011) Learning approximate cost-to-go metrics to improve sampling-based motion planning. In: IEEE ICRA
15. Baldwin I, Newman P (2010) Non-parametric learning for natural plan generation. In: IEEE IROS, pp 4311–4317
16. Ye G, Alterovitz R (2015) Demonstration-guided motion planning. In: ISRR, pp 1–16
17. LaValle SM (1998) Rapidly exploring random trees: a new tool for path planning
18. Rickert M, Brock O, Knoll A (2008) Balancing exploration and exploitation in motion planning. In: 2008 IEEE international conference on robotics and automation. ICRA 2008. IEEE, pp 2812–2817
19. Urmson C, Simmons R (2003) Approaches for heuristically biasing RRT growth. In: Proceedings 2003 IEEE/RSJ international conference on intelligent robots and systems, 2003 (IROS 2003), vol 2. IEEE, pp 1178–1183
20. Ferguson D, Stentz A (2006) Anytime RRTs. In: 2006 IEEE/RSJ international conference on intelligent robots and systems. IEEE, pp 5369–5375
21. Qureshi AH, Ayaz Y (2016) Potential functions-based sampling heuristic for optimal path planning. Auton Robot 40(6):1079–1093
22. Gammell JD, Srinivasa SS, Barfoot TD (2014) Informed RRT*: optimal sampling-based path planning focused via direct sampling of an admissible ellipsoidal heuristic. In: 2014 IEEE/RSJ international conference on intelligent robots and systems (IROS 2014). IEEE, pp 2997–3004
23. Zucker M, Kuffner J, Bagnell JA (2008) Adaptive workspace biasing for sampling-based planners. In: 2008 IEEE international conference on robotics and automation. ICRA 2008. IEEE

EEG-Based Biometric Identification Using Beta-Brainwaves with a New Methodology

Réka Veronika Sallay and Arnold Őszi

Abstract In this paper, we introduce an Electroencephalography (EEG) based biometric identification method. Our research focuses on new methods, what the previous researchers did not cover. The new method we introduce is more accurate and reliable than the previous ones. In today's world, information security and impersonation are a common and exceptionally important problem. Everyday, we come face to face with multiple forms of impersonation, like a hijacked social media account or in more severe cases, identity theft. Biometric identification is becoming an increasingly important field, with new methods racing against each other. EEG is a method to record the electrogram of the electrical activity of the brain. The biological signal detected by the EEG has proven to be a relatively low error risk way of biometric identification. The previous researches focused on identification based on general brain activity or the Alpha-brainwaves. In this research paper we focused on identification based on Beta-brainwaves, as well as a new signal processing method using Discrete Fourier-Transformation (DFT).

Keywords EEG biometric identification · Brainwaves · Beta-brainwaves · Discrete Fourier-transformation

1 Motivation

In recent years, personal authentication has become the number one pressing issue. We face the all-time threat of identity theft and impersonation. With such an urgent issue, biometric identification methods have skyrocketed. Particularly in the fields of fingerprint, voice, and face recognition. However, the technology based in these areas is not without faults and difficulties.

A relatively new research focus is on brainwaves in biometric identification methods. Traditional brainwave authentication methods often rely on complex

R. V. Sallay · A. Őszi (✉)
Obuda University, Bécsi Street 96/B, 1034 Budapest, Hungary
e-mail: oszi.arnold@bgk.uni-obuda.hu

T. A. Kovács et al. (eds.), *The Impact of the Energy Dependency on Critical Infrastructure Protection*, Advanced Sciences and Technologies for Security Applications, https://doi.org/10.1007/978-3-031-78544-3_47

processes (for example the use of AI technologies) and expensive systems (using expensive models of medical grade EEGs). These existing methods are also based on alpha-, or a mix of alpha- and other-brainwaves [1–6].

Our novel approach offers a more cost-effective, quicker, and reproducible method, achieved through innovative signal detection and processing techniques with high accuracy.

This research is significant because of the potential impact it can have on our everyday lives. By the simplification of the implementation, the new method gives way to multiple new exciting possibilities, such as, but not limited to, secure access control in corporate environments and continuous authentication in consumer electronics. Moreover, the cost-effectiveness of this solution is a gateway to great technological advancement for various socioeconomic backgrounds, creating a more diverse user base of modern security techniques.

2 Background

To understand the problem fully, we should first familiarize ourselves with a few biological and technical definitions. We will introduce some necessary background information.

2.1 Biological Background

Brainwaves or neural oscillations are repetitive electrical patterns caused by the synchronized firing of neurons in the brain. There are different oscillation activities in individual neurons and neural ensembles.

Oscillatory behavior in individual neurons: Neurons can exhibit oscillatory behavior either in their membrane potential (the difference in electric charge between the inside and outside of the neuron) or in the rhythmic patterns of action potentials (brief electrical impulses) they generate. These oscillations can lead to rhythmic activation of neighboring neurons through synaptic connections [7].

Oscillatory behavior in neural ensembles: (Refer to groups of neurons that work together to perform specific functions.) When large numbers of neurons within these ensembles synchronize their activity, it can give rise to macroscopic oscillations. These macroscopic oscillations can be detected and recorded using techniques such as electroencephalography (EEG), which measures electrical activity in the brain [7].

The interaction between neurons can produce oscillations at frequencies different from the firing frequency of individual neurons. This means that while individual neurons may fire at a certain rate, the collective activity of interconnected neurons can give rise to oscillations at different frequencies [7, 8].

Types of brainwaves: There are also different types of brainwaves since the synchronized firing of neurons results in measurable wave patterns categorized by frequency. These categories are Alpha (8–12 Hz), Beta (12–30 Hz), Gamma (30–100 Hz), Delta (0.5–4 Hz) and Theta (4–8 Hz) brainwaves [7–11].

Most of the already published research focuses on Alpha brainwaves. A calm and aware state of awareness is linked to alpha waves. When the mind is at ease, as it is during meditation or gentle relaxation, they are common. When someone's eyes are closed, but they are not asleep, alpha waves can also be seen [7–9].

Our novel method makes Beta brainwaves the central and only brainwave. Concentrating and analytical, active thoughts are connected to beta waves. When awake, they are in charge, particularly when making decisions, solving problems, or concentrating on mental activities [7–9]. We chose this brainwave because of these associations and because of the possibility of making a better device for everyday authentication purposes, and last but not least this type of brain wave is not explored in modern research associated with identification.

2.2 *Technical Background*

Techniques for measuring brainwaves entail recording and examining the electrical activity produced by brain neurons. Electrodes are applied to the scalp using the widely known EEG technique to detect and record electrical signals, providing real-time information about brainwave patterns. Advanced devices called EEG machines are made to precisely record, amplify, and capture electrical activity in the brain. These devices use an electrode array that obeys the standards, such as the 10–20 system to identify small voltage variations caused by synchronized neuronal firing [1, 2, 5, 8].

3 The Measurement Device

In contrast to the already existing researches that use a complicated and expensive EEG system for their measurement, we used a novel approach in the sense that we built our own simplified EEG that is specifically made to only measure the beta brainwave range.

The device is a simple data acquisition device. It was based on the designs detailed in [12, 13]. Our design can be broken down into six parts (Fig. 2).

The 1st stage is the input stage. It has only 3 electrodes out of which only two receive and forward data to the device, and the third one connects to the ground.

In the 2nd stage, is the 60 Hz notch filter. Its task is to remove as much interference as possible from our circuit before adding additional gain.

In the 3rd stage, we used a 12 Hz (cut-off frequency) multiple feedback 2nd order high-pass filter. Its goal is to get rid of signals that are below 12 Hz, which we consider noise, and also to eliminate interference from the galvanic skin response throughout the head and enable clearer monitoring of desired brain wave activity, which are beta waves.

In the 4th stage, which we call the 30 Hz multiple feedback 2nd order low-pass filter. It is used to remove anything above 30 Hz because beta wave information stops at that frequency (it works very similarly to the high-pass filter).

In the 5th stage, we calculate the gain, and the potentiometer acts as a variable resistor, adjusting the circuit's gain linearly. Adjusting the potentiometer during readings ensures voltage stability and minimizes digital reading errors without demanding maximum amplification.

In the 6th stage, stage six, it is simply just another 60 Hz notch filter. The data will still have a significant amount of noise in it at this point, despite having gone through all the earlier filtering stages. We will run it through a second notch filter, the same one as the previous one, with a center frequency of 60 Hz, to correct this. There will be some noise in the final data, but we will reduce that in a 2-stage noise reduction method in the software.

It runs on 9 V batteries, to reduce the internal noise in the device. It is connected to the computer with an audio jack cable (Figs. 1 and 2).

Fig. 1 Real life image of the completed simplified EEG based beta brainwave measurement device

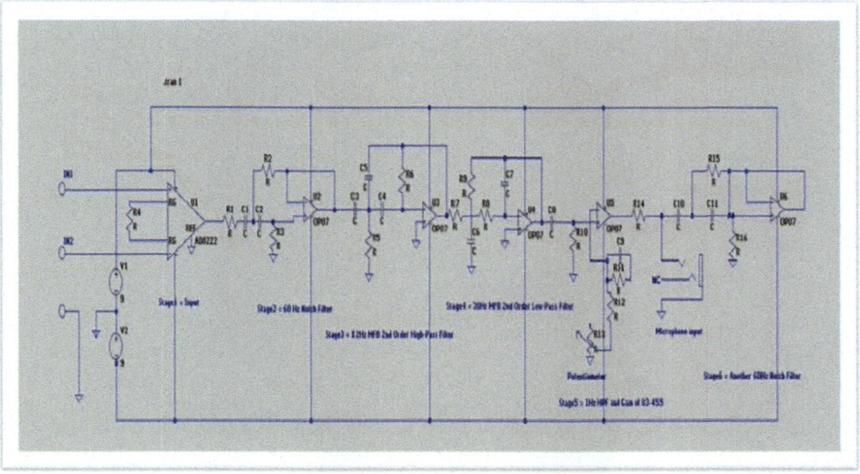

Fig. 2 Schematic drawing of the measurement device

4 Measurement Process

The measurement process was quite a simple procedure. For our purpose we chose individuals that are between the ages of 26–50 years, this decision was made based on the fact that the human brain, especially the frontal cortex, finishes development around the age of 25 [14]. None of the subjects had any diagnosed history or severe cases of mental illness that would influence the brainwaves and complicate the measurement [11, 15].

We asked the subjects to sit comfortably in a chair, but to imagine a ball and to turn it over in their mind. The subjects were silent and motionless, relaxed but concentrating and awake. Before the measurement began, we did a few tests to find out the baseline noise level in the room. We placed the electrodes on the subject's scalp in the general positions seen in Fig. 3. We chose those areas especially since frontal beta waves are easily measurable there [8, 11, 16].

We highlighted a few possible electrode placements with green squares and the most used ones with red circles (Fig. 3). The most chosen areas were in the orbital and frontal regions, for the ground electrode we chose the mastoid area. In a few subjects, we also made measurements on the occipital region. However, we have chosen not to measure there due to the restricted accessibility (usually that area is mostly covered by hair). For each subject, we did 3 measurements, each lasting around 0.4 s. During this time, we have roughly gotten one hundred thousand points per measurement.

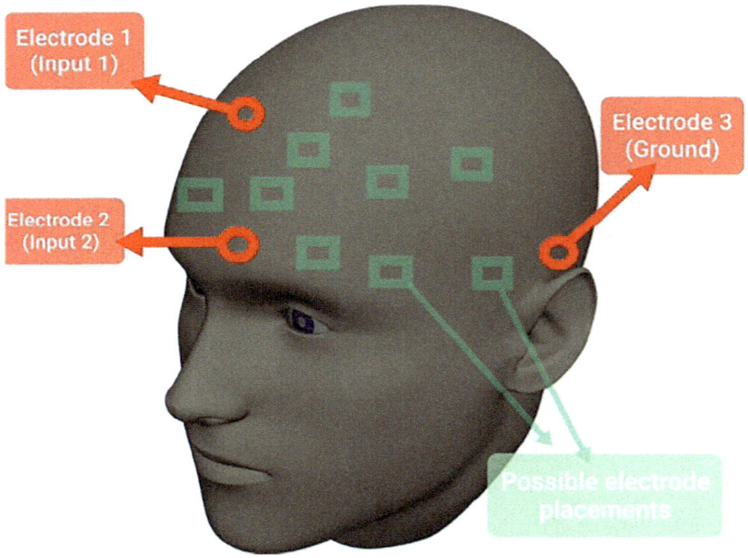

Fig. 3 EEG placement used in measurements

5 Novel Identification Method

During the measurements which we controlled from the computer connected to the measurement device, we used a program called Wolfram Mathematica.

Wolfram Mathematica's built-in functions made it possible to process the raw data we acquired. Firstly, we subtracted the reference measurement (baseline noise level measurement) from the official measurement we conducted on the subjects. After we had the raw signal with reduced noise (Fig. 4), we Discrete Fourier Transformed the measured signals one-by-one. We can see the signal after DFT in Fig. 5.

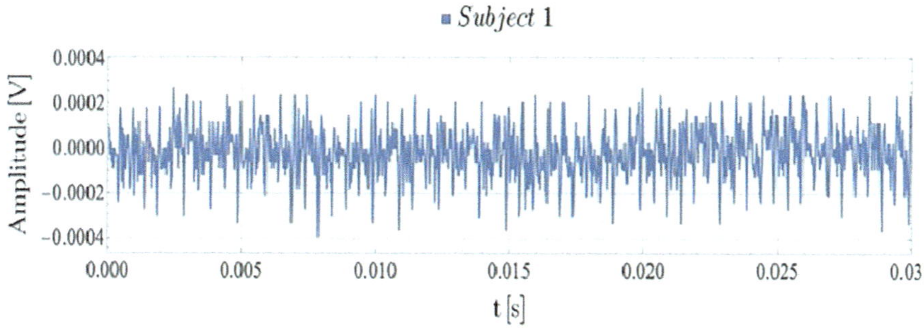

Fig. 4 Raw signal after subtraction of reference signal

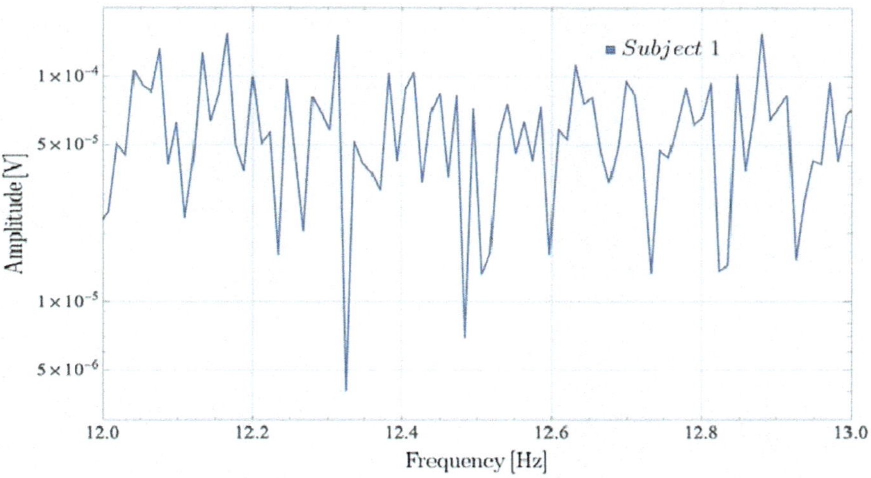

Fig. 5 Zoomed in signal after DFT

After DFT we have averaged the signals of a subject together (see in Fig. 6). Following this, we can compare 2 subjects' averaged signals (see in Fig. 7).

For the identification, we used an intelligent algorithm that searched for characteristic points that were specific to the subjects. By specific, we mean the specific frequency components have the same amplitude in each measurement within a reasonable error limit. Because of the huge amount of data, a faster approach was needed than Wolfram Mathematica to compare and identify the characteristic frequencies. For these calculations, we wrote a fast C++ program. It is important to note that these frequencies are unique for each person. This means for identification you need to know the subject you want to identify and the frequencies characteristic to

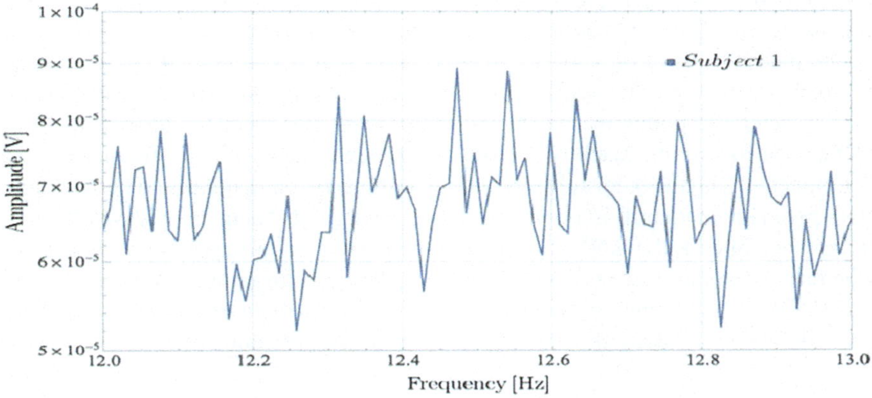

Fig. 6 DFT-d signal after averaging (zoomed in)

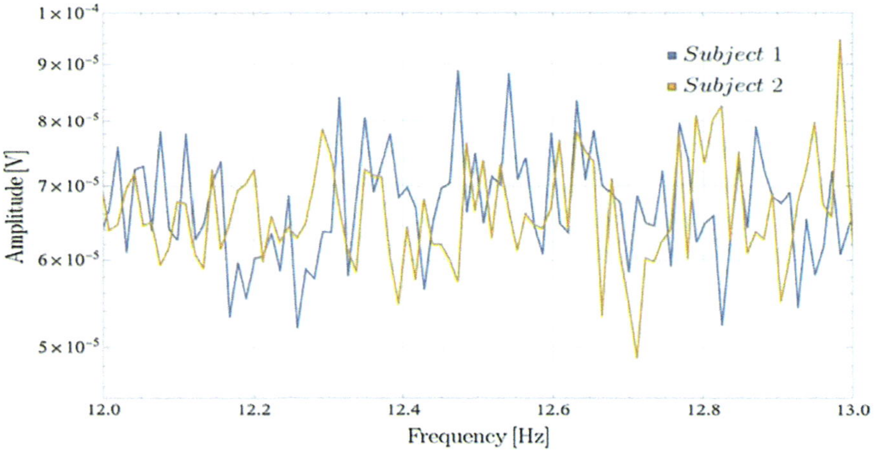

Fig. 7 Comparison of 2 subjects' averaged signals (zoomed in)

the specific sample. Because these frequencies can be quite few (in our measurements there were < 30 characteristic frequencies per subject within error) the identification process can be much faster than similar highly secure identification techniques.

6 Results

During our measurements, we studied 10 individuals. On each individual, we measured multiple times and used the process described in the previous section. We used an error margin of 1%, meaning that if in two distinct measurements, two amplitudes on the same frequency were equal within that range, it was considered a match. After the data evaluation and the characteristic point finding, we found that using the 1% error margin, a maximum of 30 characteristic points were found in the data measured on each individual. After that, to identify them, we needed to compare the data of the different individuals. We found that in the case where 30 characteristic points were detected, only 2 matches were found with the other subjects at maximum. Each of the others had fewer characteristic frequencies and less matches with the other subjects overall, so this gives us a minimum percentage of 94% precision in identifying a person based on purely just beta brain waves. The true acceptance rate (TAR) is 94%, and while there wasn't a false positive identification (the false acceptance rate (FAR) being 0% so far), we can estimate a 6% equal error rate (EER) based on measurements. This exceeded all of our expectations because it proved to be better than most other high-security identification techniques.

7 Summary

During our research the goal was to find a new identification method that is faster, easier, and can be done without expensive equipment, using a simplified version of an EEG. Our approach was to use only beta-brainwaves, which has not been done before for identification purposes since all other published research is mostly based on alpha and a combination of alpha and other brainwaves. We chose the beta-brainwaves (12–30 Hz) because it is associated with awareness, wakefulness, cognitive and motor functions, and also because they haven't been studied yet, even though its promising properties.

For the measurements, we have decided to use a simplified EEG device that is built specifically to only measure beta-brainwaves. We chose this option to reduce the cost, and to make it quicker and more reproducible. This way, the device is available in more diverse socio-economic backgrounds, thus making a bigger impact on safety globally.

The device can be divided into six main stages: 1st stage—Signal input, 2nd stage to − 60 Hz Notch filter, 3rd stage—12 Hz MFB second-order high-pass filter, 4th stage to − 30 Hz MFB second-order low-pass filter, 5th stage—1 Hz HPF and gain calculation, 6th stage—another 60 Hz notch filter. The device runs on 9 V batteries to reduce internal noise and connects to a computer using an audio jack cable. For the measurements, we used Wolfram Mathematica to measure using the microphone input of a computer and using the software, we subtracted the background noise from the measurements, then we took the DFT of all the measurements. Following that, we averaged the DFT-d signals. After that, using a special fast algorithm and the C++ language, we found the characteristic frequencies for the subjects, and then we compared these frequencies with the other subjects. The greatest achievement of our research is that, using this method, a lot of characteristic points can be found for each individual and using these points 94% of accuracy can be achieved.

This exceeded our original expectations and clearly shows that our device and the novel method is a reliable way of identification.

We showed that biometric identification is possible based only on beta-brainwaves using a simplified EEG device. This is important because it makes a usually very expensive and time-consuming identification method generally more easily accessible and cost-effective, which is invaluable in our world today.

References

1. Miyamoto C, Baba S, Nakanishi I (2009) Biometric person authentication using new spectral features of electroencephalogram (EEG). In: 2008 international symposium on intelligent signal processing and communication systems, ISPACS 2008, pp 1–4. https://doi.org/10.1109/ISPACS.2009.4806762
2. Nakanishi I, Maruoka T (2020) Biometrics using electroencephalograms stimulated by personal ultrasound and multidimensional nonlinear features. Electronics 9:24

3. Singhal GK, RamKumar P (2007) Person identification using evoked potentials and peak matching. In: Proceedings of 2007 biometric symposium
4. Ravi VR, Palaniappan R (2005) Recognition individuals using their brain patterns. In: Proceedings of the 3rd international conference on information technology and applications
5. Marcel S, Millán Jdel R (2007) Person authentication using brainwaves (EEG) and maximum a posteriori model adaptation. IEEE Trans Pattern Anal Mach Intell 29(4):743–752
6. Poulos M, Rangoussi M, Chrissikopoulos V, Evangelou A (n.d.) Parametric person identification from the EEG using computational geometry. In: Proceedings of ICECS 1999 6th IEEE international conference on electronics, circuits and systems (Cat. No. 99EX357). ICECS'99 (1999)
7. Whittington MA, Traub RD, Adams NE (2019) A future for neuronal oscillation research. Brain Neurosci Adv 2:2398212818794827
8. Arias-Cabarcos P, Fallahi M, Habrich T, Schulze K, Becker C, Strufe T (2023) Performance and usability evaluation of brainwave authentication techniques with consumer devices. ACM Trans Priv Secur 26(3). Article 26
9. Abhang PA, Gawali BW, Mehrotra SC (2016) Technical aspects of brain rhythms and speech parameters. In: Introduction to EEG- and speech-based emotion recognition, pp 51–79
10. Hosťovecký M, Babušiak B (2017) Brain activity: beta wave analysis of 2D and 3D serious games using EEG. J Appl Math Stat Inform 13(2):39–53
11. Kropotov JD (2009) Beta rhythms, chap 3. In: Quantitative EEG, event-related potentials and neurotherapy. Academic Press, pp 59–76
12. AUTODESK instructables, DIY EEG (and ECG) circuit. Accessed 2024/02/11. URL: https://www.instructables.com/DIY-EEG-and-ECG-Circuit/
13. MIT Fab Lab (2015) Final project: DIY EEG: EEG: cheap and small, fall 2015. Accessed 2024/02/11. https://fab.cba.mit.edu/classes/863.15/section.Harvard/people/Ricker/htm/Final_Project.html
14. Arain M, Haque M, Johal L, Mathur P, Nel W, Rais A, Sandhu R, Sharma S (2013) Maturation of the adolescent brain. Neuropsychiatr Dis Treat 9:449–461
15. Kropotov JD (2016) Beta and gamma rhythms. In: Functional neuromarkers for psychiatry, pp 107–119
16. Riera A, Soria-Frisch A, Caparrini M et al (2007) Unobtrusive biometric system based on electroencephalogram analysis. EURASIP J Adv Signal Process 2008:143728

Examination of UV Radiation During Inert Gas Arc Welding

Márton Schramkó, Péter Pinke, Dávid Dunavölgyi, and Tünde Anna Kovács

Abstract In the arc welding process, precise consideration of various parameters is essential to ensure the formation of optimal joints. Accordingly, welding parameters play a critical role throughout the procedure. Concurrently, heightened attention has been directed towards mitigating the adverse effects inherent to welding processes, spurred by advancements in our understanding of their impacts. Given its widespread industrial application, our research focuses primarily on arc welding, with a specific emphasis on measuring ultraviolet (UV) radiation emitted during the process. Among welding methodologies, gas-shielded arc welding predominates and warrants scrutiny regarding UV radiation emission, particularly as previous studies have indicated lower UV intensity compared to other welding techniques. Our principal objective is to devise an appropriate methodology for quantifying UV radiation emitted during arc welding. Notably, arc welding is extensively utilized in mechanized and automated industrial settings, underscoring the necessity of delineating safety parameters concerning harmful radiation exposure during industrial applications. However, achieving this necessitates a comprehensive understanding of UV radiation emitted by various gases in a plasma state, for which limited empirical data are available in existing literature. Given the predominant focus on specific gas compositions in extant research, we recognize the significance of developing a measurement protocol capable of assessing and contrasting UV radiation emissions across different gases. Our discourse elucidates the significance of this endeavor and outlines the strategic blueprint for our experimental approach.

Keywords Welding gases · Ultraviolet radiation · Arc welding · UV

M. Schramkó (✉)
Doctoral School on Materials Sciences and Technologies, Obuda University, Budapest, Hungary
e-mail: schramko.marton@bgk.uni-obuda.hu

P. Pinke · D. Dunavölgyi · T. A. Kovács
Banki Donat Faculty of Mechanical and Safety Engineering, Obuda University, Budapest, Hungary
e-mail: kovacs.tunde@bgk.uni-obuda.hu

T. A. Kovács et al. (eds.), *The Impact of the Energy Dependency on Critical Infrastructure Protection*, Advanced Sciences and Technologies for Security Applications, https://doi.org/10.1007/978-3-031-78544-3_48

1 Definition and Effects of UV Radiation

1.1 Ultraviolet Radiation

Ultraviolet (UV) radiation, positioned between visible light and X-rays on the electromagnetic spectrum, encompasses wavelengths spanning from 10 to 400 nm. Not perceptible to the human eye owing to its short wavelength and high frequency, UV radiation derives its designation from its position beyond the violet end of the visible light spectrum. As part of the electromagnetic spectrum, UV radiation is categorized into three distinct regions based on wavelength: UV-A (315–400 nm), UV-B (280–315 nm), and UV-C (100–280 nm).

Furthermore, the effects of UV radiation on biological and material systems vary significantly depending on the specific wavelength range. UV-A radiation, for instance, is primarily associated with skin aging and contributes to the formation of wrinkles, while UV-B radiation is a primary cause of sunburn and skin cancer. UV-C radiation, with the shortest wavelengths, possesses germicidal properties and is utilized for disinfection purposes in various applications.

Given the diverse effects and potential applications of UV radiation, comprehensive knowledge of its characteristics and behaviors across different wavelength ranges is imperative. Through meticulous investigation and experimentation, we aim to deepen our understanding of UV radiation and its implications for various scientific and technological endeavors. In addition, the scale can be divided into 3 parts:

Far Ultraviolet radiation FUV (100–200 nm)
Medium Ultraviolet radiation MUV (200–300 nm)
Near Ultraviolet radiation NUV (300–400 nm).

These wavelength intervals are employed within a categorization system, more commonly recognized in everyday discourse, which organizes UV radiation in a manner slightly divergent from scientific convention, focusing on its adverse impact.

UV-A radiation, spanning wavelengths from 320 to 400 nm, constitutes the longest wavelength and is considered the least harmful among ultraviolet light spectra. Often referred to colloquially as "black light," UV-A radiation finds prevalent application in inducing fluorescence, eliciting colored luminescence in various artistic contexts. Certain insects and avian species possess the visual capacity to perceive UV-A radiation, while in exceptional cases, select humans, particularly individuals afflicted with Aphakia, may also exhibit sensitivity to this form of ultraviolet radiation.

UV-B radiation, spanning wavelengths from 290 to 320 nm, poses risks of skin burns, heightened susceptibility to skin cancer, and cellular damage upon prolonged exposure. Notably, approximately 95% of UV-B radiation is absorbed by the ozone layer within Earth's atmosphere, mitigating its adverse effects and thereby conferring a degree of protection in daily life.

UV-C radiation, spanning wavelengths from 100 to 290 nm, is characterized by its extreme harmfulness, being nearly entirely absorbed by the Earth's atmosphere on a daily basis. Widely recognized for its germicidal properties, UV-C radiation

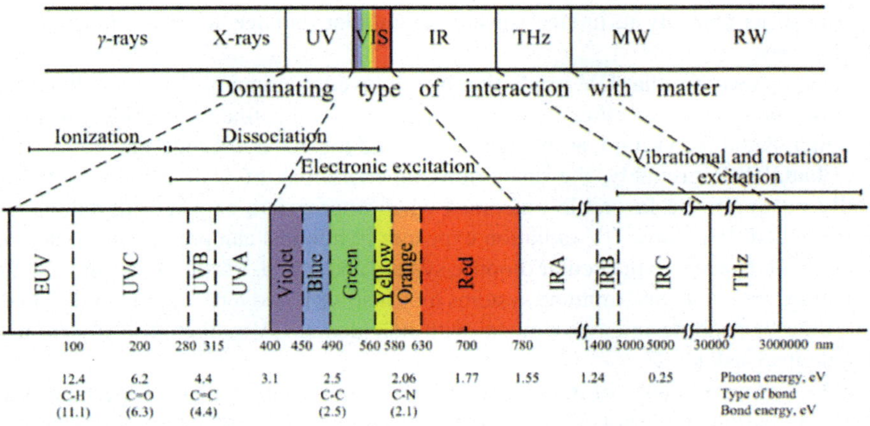

Fig. 1 Spectrum of electromagnetic radiation [1]

is employed as a disinfectant in various applications, including food, air, and water treatment, where it effectively eradicates microorganisms by disrupting the nucleic acids within their cells. However, during welding operations, UV-C radiation poses a significant hazard as the welding environment lacks sufficient atmospheric air to fully attenuate its intensity, exposing welders to potential health risks associated with direct UV-C exposure (Fig. 1).

1.2 Health Problems Caused by UV Radiation

Prolonged exposure to UV-A and UV-B radiation, in the absence of adequate protection measures, poses considerable health risks. For instance, prolonged sun exposure results in the formation of a "tan," whereby melanin accumulates in the skin to absorb UV rays and dissipate their energy as heat. Employing sunscreen is imperative as a preventive measure against UV radiation, as it creates a barrier capable of absorbing UV-A and UV-B wavelengths before they penetrate the skin. Failure to safeguard against prolonged sun exposure may lead to sunburn, characterized by erythema of the epidermis, and over time, heightened susceptibility to skin cancer and other deleterious cellular alterations becomes markedly elevated.

The effects in detail:

Before unequivocally categorizing UV radiation as the most deleterious form of radiation, it is pertinent to acknowledge its beneficial attributes, albeit contingent upon appropriate dosage. Primarily, UV radiation is essential for the synthesis of vitamin D within the body; however, this requisite threshold can typically be attained with brief exposures of 10–15 min to sunlight per week. Externally, UV radiation manifests an antibacterial efficacy, particularly conducive for object sterilization, owing to its capacity to induce substantial cellular damage when administered in high doses.

It is noteworthy that UV irradiation is employed therapeutically to facilitate the healing process in specific medical conditions. Such ailments include psoriasis, eczema, jaundice, vitiligo, atopic dermatitis, and localized scleroderma [1, 2].

ChatGPT Regrettably, the deleterious repercussions of UV radiation substantially outweigh the favorable outcomes when appropriate exposure durations are disregarded. Excessive UV radiation exposure in humans can precipitate acute and chronic detriments to the ocular dioptric system and retina. Furthermore, the integumentary, circadian, and immune systems may also suffer adverse effects. Individuals working or residing in environments with heightened UV radiation levels are at a heightened risk [1, 2].

Distinct wavelengths of light exert varying effects on the human cornea and skin. UV-A radiation, while not eliciting an immediate response, can induce photokeratitis and erythema, particularly in individuals with lighter skin tones, as wavelengths extend from the onset of the UV-B spectrum at 315 nm, peaking rapidly at 300 nm. The skin and eyes exhibit maximal sensitivity to UV radiation within the range of 265–275 nm, situated within the lower UV-C spectrum. Fortunately, atmospheric filtration effectively screens this range; however, occupations involving localized UV radiation generation, such as welding, necessitate the utilization of protective equipment or creams to mitigate associated risks [1, 2].

Excessive exposure to UV-B radiation not only induces sunburn but also contributes to the development of certain types of skin cancer. However, the extent of erythema and ocular irritation, primarily attributed to UV-A radiation, does not reliably predict the long-term effects of UV exposure, despite reflecting direct DNA damage. All categories of UV radiation are implicated in skin damage due to the depletion of vitamin A reserves within the skin. The risk of cancer escalates significantly with exposure to UV-B and UV-C radiation, particularly augmenting the likelihood of melanoma onset. Malignant melanoma, the most lethal form of skin cancer, primarily arises from DNA damage independent of UV-A radiation. UV-C radiation, possessing the highest energy levels, is the most perilous type of ultraviolet radiation, inducing mutagenic and carcinogenic effects. Previous beliefs regarding the innocuousness of UV-A radiation have been dispelled, as it is now acknowledged to contribute to skin cancer via DNA impairment, facilitated by the generation of free radicals such as reactive oxygen species. Although UV-A radiation per se does not directly inflict harm on the body, it generates highly reactive chemical intermediates, including hydroxyl and oxygen radicals, which, in turn, can inflict DNA damage. UV-A radiation exerts an immunosuppressive impact on the entire body, primarily accounting for the immunosuppressive effects attributed to sunlight [3].

Unfortunately, the ocular structures are particularly vulnerable to the deleterious effects of UV radiation. The eye exhibits maximal sensitivity to UV radiation within the lower UV-C band, specifically within the wavelength range of 265–275 nm. Although solar radiation predominantly lacks radiation within this wavelength, it is prevalent in welder's arc lights and other artificial light sources. Exposure to such sources can induce conditions like "welding flash" or "arc eye" (photokeratitis) and precipitate the development of cataracts, pterygium, and pinguecula. UV-B radiation in sunlight, ranging between 310 and 280 nm, also contributes to photokeratitis ("snow blindness") to a lesser degree and can inflict damage upon the cornea, lens, and retina [13].

2 UV Radiation Intensity Comparison in Shielded Gas and Coated Electrode Arc Welding

Figure 2 illustrates two welding processes often compared in UV radiation intensity. On the right, shielded gas arc welding is depicted, while coated electrode arc welding is shown on the left. Notably, shielded gas arc welding emits more intense radiation in the lower half of the light spectrum (up to 200–400 nm), whereas coated electrode arc welding emits predominantly above 400 nm in the visible light range, with minimal emission in the UV range [4].

Shielding gas arc welding is also known as MIG (neutral) or MAG (active). The procedure number is 13.

Throughout the welding process, the machine generates high voltage to initiate a high-temperature electric arc. This arc heats the surface of the raw material, causing it to melt and form a molten pool known as the welding bath. A wire, heated by the arc, melts and is continuously fed into the weld at a controlled speed to maintain welding continuity. The molten filler material integrates into the welding bath, filling gaps between materials to form a joint. Impurities present during the process can adversely affect the quality of the weld, making it sensitive to gas selection. Unlike coated electrode welding, this process does not require rod replacement, making it suitable for long welds and easily automatable. Its relatively quick learning curve and widespread industrial usage further contribute to its popularity. The process's adaptability to various materials and environments enhances its versatility, making it a preferred choice in many industrial applications [6, 11].

2.1 Short Summary of TIG Welding

In arc welding, the thermal energy generated by the electric arc current facilitates the fusion of the base materials and the filler material. If the shielding gas is employed to safeguard the arc, base, and filler materials from atmospheric contamination, thus characterizing the process as arc welding with shielding gas. When the arc forms between the base material and a non-consumable tungsten electrode, the technique is specifically termed tungsten electrode shielded gas arc welding. In tungsten electrode shielded gas welding, the electric arc is established between the base material and the tungsten electrode. Tungsten, renowned for its elevated melting point, stands out as the prime metal for electrode applications. Contemporary practices often incorporate

Fig. 2 Gas metal arc welding and shielded metal arc welding [5]

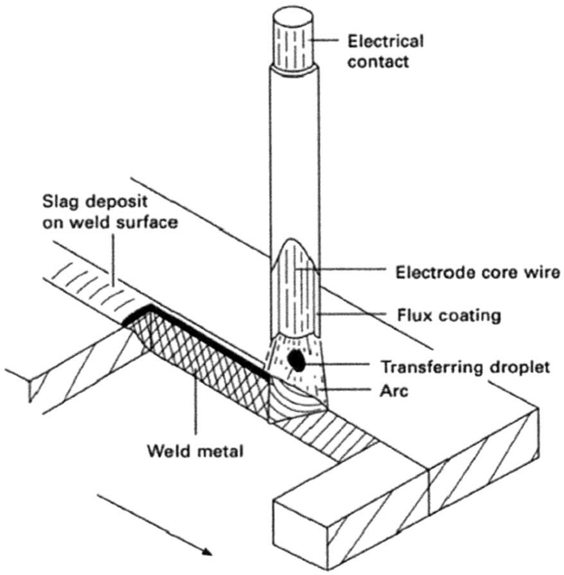

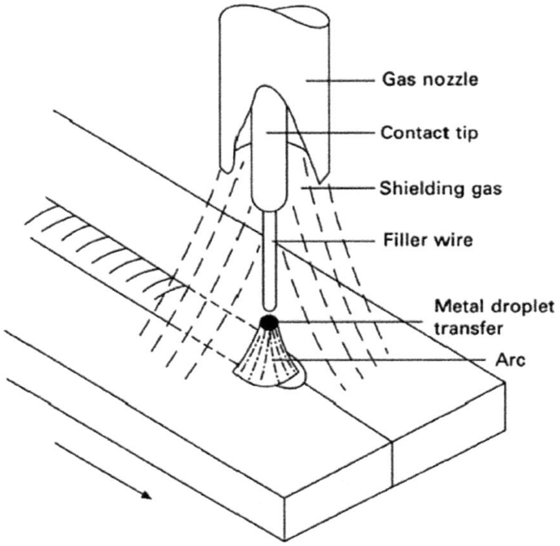

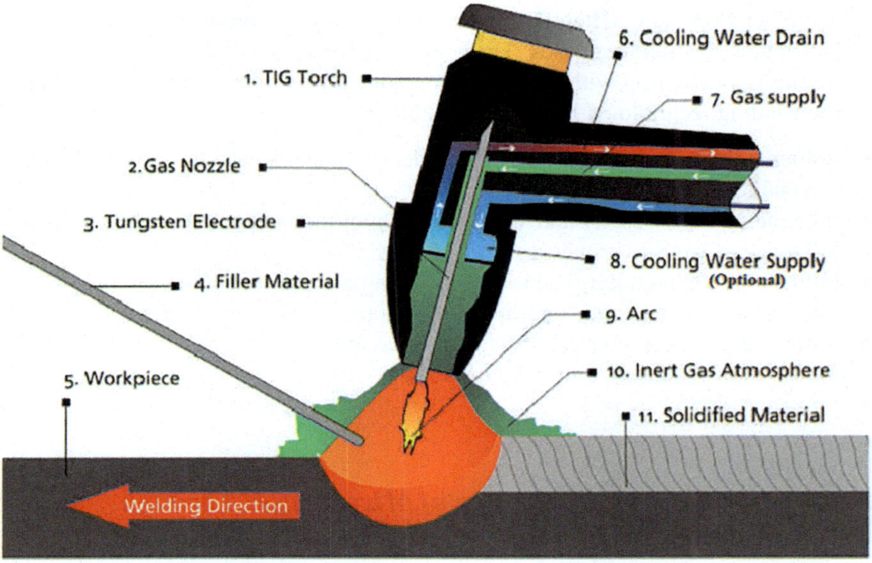

Fig. 3 Gas tungsten arc welding TIG (GTAW) [8]

tungsten alloys with diverse oxides. These tungsten electrodes are designated based on their alloy composition, with their ends color-coded for easy identification. Tungsten electrode shielded gas welding offers versatility in current application, accommodating both continuous and pulsating currents. Under direct current, the arc's power remains consistent, with cooling dynamics primarily influenced by welding speed, given uniform current strength and plate thickness. Employing pulsating current introduces a variable influenced by frequency, impacting arc temperature. At frequencies below 7 Hz, a temperature disparity exists between the scar bath during peak current and base current phases. Above this threshold, temperature equilibrium prevails. Pulsed arc welding categorizes into slow, medium and fast or rapid variations. This modulated approach allows for deeper fusion with peak current and smaller scar pool formation with base current. By discerning these effects, optimal pulsed arc welding parameters can be established. Additionally, pulsed arc welding entails reduced heat input, minimizing deformation and residual stress. The TIG welding process can be seen in Fig. 3 [7, 13].

3 The Role of Shielding Gas

As previously stated, the production of UV radiation tends to be more pronounced in shielding gas processes compared to other arc welding methods. Our observations, derived from measurements, indicate that the radiation level is notably affected by the choice of shielding gas. Therefore, we provide a concise overview of the functions and attributes of shielding gases. These gases serve the dual purpose of preventing atmospheric interaction within the welding zone and exerting significant influence on the welding technology. Key effects include:

- Influencing the characteristics of the electric arc (electricity, flow technology, thermal technology)
- Influencing the viscosity and surface tension of the melt and detached droplets
- Influence of wetting properties
- Influencing the metallurgical reactions taking place in the pool
- Influence of fusion conditions and shape factors of the weld bed
- Influencing the geometry and shape of the seam
- Influence of radiations and radiation losses
- Influence of emissions.

3.1 An Explanation of the Use of Mixed Gases Through Hydrogen and Helium

Hydrogen stands out among the constituents of shielding gases due to its superior thermal conductivity. Within the arc's environment, hydrogen molecules undergo dissociation through an endothermic reaction, elevating them to a higher energy state. Subsequently, at the periphery or beyond the arc's atmosphere, these hydrogen atoms recombine within the cooler shielding gas envelope, releasing the absorbed energy in an exothermic reaction. These processes contribute to heightened heat input into the weld seam. However, owing to the susceptibility of hydrogen to cold cracking, its usage is typically restricted to materials insensitive to hydrogen, such as austenitic stainless steels and nickel-based alloys. Helium, characterized by its inert nature, possesses the second-highest thermal conductivity among gases and can be universally applied across all materials.

Incorporating hydrogen and/or helium into the shielding gas consistently enhances the productivity or quality of welding. Hydrogen's impact surpasses that of helium, as evidenced by achieving effects with 1–2% hydrogen equivalent to 20–30% helium content. This underscores the significant influence of mixture composition on the efficacy of shielding gas incorporation [9].

4 Example for UV Radiation Intensity Measurement Method

UV-C radiation, the most harmful type emitted during welding, is typically absorbed by Earth's atmosphere, minimizing direct exposure. Despite its significant production intensity, research often overlooks it due to its limited ability to penetrate [4].

Existing literature contains studies exploring the impact and scope of welding gases on UV radiation. The prevalent measurement technique, illustrated in Fig. 4, involves assessing the welding seam along a designated direction, with adjustable distances for the measuring instrument. This method is ideal for augmenting an established welding procedure by determining its safety parameters or assessing the characteristic UV radiation propagation curve. Nonetheless, it encounters various complicating factors hindering precise measurements. The continuous movement of the machine during welding induces fluctuations in distance, impeding accuracy. This problem comes from the welding process type. Consequently, this measurement setup proves suboptimal for laboratory investigations [12].

From our point of view the problem with this method is that when you measure it your result will be distorted by smoke and it gives only an answer for the practical radiation. On the other hand it not gave us an easy to measure environment for the specific gases that we wanted to measure. For making a correct model for the UV radiation we need a controlled measuring setup that can help us measure the power output in the case of electric arcs formed in different gas mixtures.

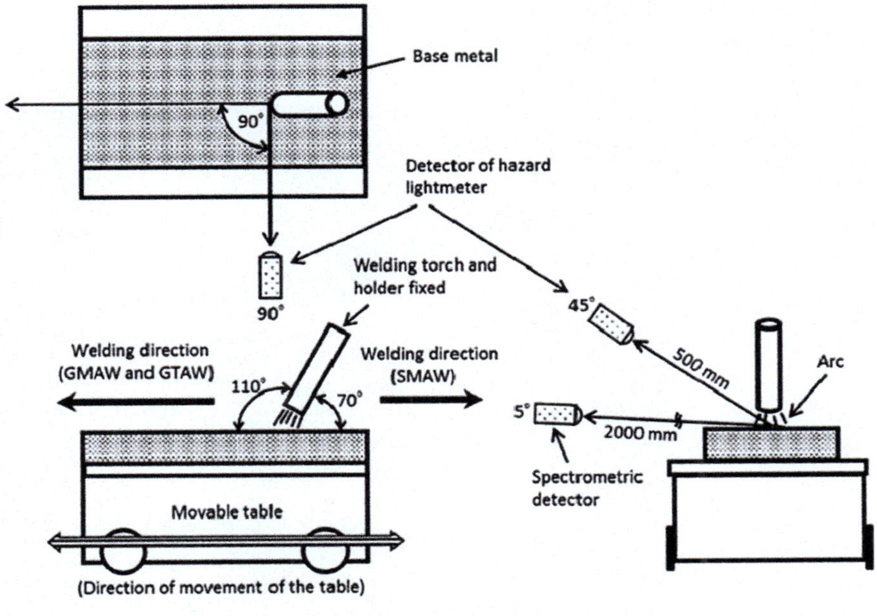

Fig. 4 Measurement method and schematic diagram of its arrangement [10]

5 Our New UV Radiation Intensity Measurement Method

In our idea we wanted to use the TIG process to make an easy to use setup. The idea is to make a chamber where we can measure the flow of the gas what we are using. In this way we can get a clear picture to approximately how much shielding gas in contact with the arc. The setup is made from a welding electrode a gas deflector and 2 wolfram electrode one used for the positive side and one for the negative. The main place where we can make the arc can be seen in Fig. 5. This part can be changed to have a fast deflector of different type. Until now we tested 2 type of deflector and the result is that if we used the commonly used ceramic deflector the electrode still can contact with the air as you can see in Fig. 6. There is oxidation on the ceramic deflector while where Pyrex gas deflector what cover the hole test filed made a better defends against unwanted gases. In our measurement the Pyrex "glass" did not absorb UV radiation so it can be used for future tests.

As you can see in Fig. 7, the test setup is made of the arc generating point and a measuring point in 200 mm from there in this setup we can be sure to measure every gas with the same point and with the clearest possible way. Because in the welding process it is easy to change the ampere of the welding it is very easy to test a variety of different parameter with different gases.

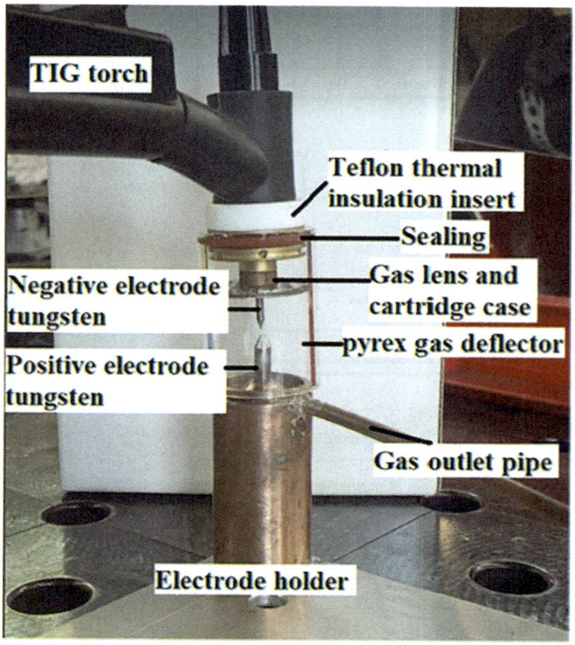

Fig. 5 The test setup (TIG Pyrex deflector cartridge housing with gas deflector)

Fig. 6 Right side is the Pyrex shielding and in the left side is the standard ceramic deflector

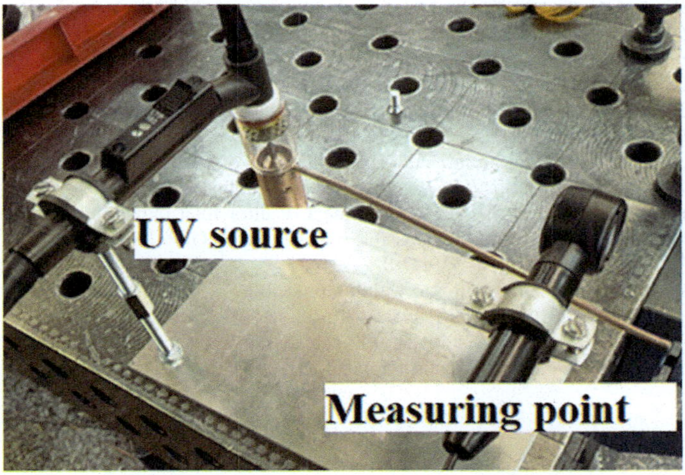

Fig. 7 The new UV measuring setup

6 Summary

We posit that UV radiation measurement during welding holds significance in industrial contexts, particularly for welding Big structures or huge components and maintaining welding robot facilities to define protective boundaries. Our objective was to devise a viable procedure enabling straightforward assessment of UV emission levels from different shielding gases, thus aiding the formulation of welding safety protocols in the future. Extensive measurements are still necessary to yield a comprehensive dataset, essential for delineating the role of UV radiation effectively. This understanding could streamline the implementation of suitable safety protocols within the industry, contributing to enhanced workplace safety and operational efficiency.

References

1. Tuchin VV (2015) Tissue optics and photonics: biological tissue structures. J Biomed Photonics Eng. https://doi.org/10.18287/jbpe-2015-1-1-3
2. Juzeniene A, Moan J (2012) Beneficial effects of UV radiation other than via vitamin D production. Dermato-Endocrinology 4(2):109–117. https://doi.org/10.4161/derm.20013
3. Svobodová AR, Galandáková A, Šianská J, Doležal D, Lichnovská R, Ulrichová J, Vostálová J (2012) DNA damage after acute exposure of mice skin to physiological doses of UVB and UVA light. Arch Dermatol Res 304(5):407–412. https://doi.org/10.1007/s00403-012-1212-x
4. Tenkate TD (1998) Optical radiation hazards of welding arcs. Rev Environ Health 13(3). https://doi.org/10.1515/reveh.1998.13.3.131
5. Norrish J (2006) An introduction to welding processes. In: Advanced welding processes, pp 1–15. https://doi.org/10.1533/9781845691707.1
6. Norrish J (2006) An introduction to welding processes. In: Advanced welding processes, pp 1–15. https://doi.org/10.1533/9781845691707.74
7. Bitay E (2021) Hegesztési alapismeretek. Societatea Muzeului Ardelean. Transylvanian Museum Society, Kolozsvár. ISBN: 9786067392067. https://doi.org/10.36242/mtf-16
8. https://rime.de/en/welding/tig-welding/
9. https://www.lindegas.hu/hu/images/Gyak-tud_Hegesztesi%20vedogazok_2017_tcm71-657896.pdf
10. Takahashi J, Nakashima H, Fujii N, Okuno T (2019) Comprehensive analysis of hazard of ultraviolet radiation emitted during arc welding of cast iron. J Occup Health. https://doi.org/10.1002/1348-9585.12091
11. MacKie RM (2006) Long-term health risk to the skin of ultraviolet radiation. Prog Biophys Mol Biol 92(1):92–96. https://doi.org/10.1016/j.pbiomolbio.2006.02.008
12. Otokpa OE, Usman YB (2013) An assessment of ultraviolet radiation components of light emitted from electric arc and their possible exposure risks. Glob J Pure Appl Sci 19:145–149. https://doi.org/10.43146/gjpas.v19i1.18
13. Norrish J (2006) An introduction to welding processes. In: Advanced welding processes, pp 1–15. https://doi.org/10.1533/9781845691707.58
14. Mihály K (2023) Védőgázok hatása a huzal leolvadására és az anyagátvitelre. Link: https://docplayer.hu/9041583-Vedogazok-hatasa-a-huzal-leolvadasara-es-az-anyagatvitelre.html

Supply Chain Management Using Automatic Guided Vehicle Systems in Smart Factory Logistics

János Simon and László Gogolák

Abstract The current paper presents theoretical and experimental evidence to justify the need of Automatic Guided Vehicle (AGV) systems in supply chain management within smart factories. AGV systems can be integrated with other smart factory technologies, such as manufacturing execution systems (MES) and warehouse management systems (WMS), to optimize the flow of materials throughout the supply chain. For example, AGVs can be used to transport materials to a specific location in the factory based on real-time demand information, or to automatically re-supply materials to a production line as they are consumed. AGVs can also be used to improve the efficiency of the supply chain by reducing the need for human labor and minimizing the need for manual material handling. They can also improve the accuracy of inventory management by providing real-time data on the movement of materials within the factory. In addition, AGVs can be used to improve the safety and ergonomics of the supply chain by reducing the need for workers to manually move heavy loads or operate dangerous equipment. Overall, AGV systems can play an important role in supply chain management within smart factories by automating the movement of materials and goods, improving efficiency, accuracy, and safety and helping factories to be more flexible and responsive to changing market conditions.

Keywords AGV · ERP · Industry 4.0 · MES

J. Simon (✉) · L. Gogolák
Subotica Tech—College of Applied Sciences, Marka Oreskovica 16, 24000 Subotica, Serbia
e-mail: simon@mk.u-szeged.hu

Department of Mechatronics and Automation, Faculty of Engineering, University of Szeged, Mars tér 7, 6724 Szeged, Hungary

T. A. Kovács et al. (eds.), *The Impact of the Energy Dependency on Critical Infrastructure Protection*, Advanced Sciences and Technologies for Security Applications, https://doi.org/10.1007/978-3-031-78544-3_49

1 Introduction

Automatic Guided Vehicle (AGV) systems can play an important role in supply chain management within smart factories by automating the movement of materials and goods. AGVs can be used to transport materials between different stages of production, as well as between different areas of a facility, such as storage areas, assembly lines, and shipping docks. MES and WMS are software systems that are used to manage and optimize the operations of manufacturing and logistics facilities. MES is a software system that is used to manage and control the operations of a manufacturing facility. It is used to track and control the production process, from raw materials to finished goods [1]. MES systems can be used to manage the scheduling, execution, and monitoring of production operations, as well as to provide real-time information on the status of production, inventory, and equipment. This information can be used to optimize production processes and improve efficiency. WMS is a software system that is used to manage and control the operations of a warehouse or distribution center. It is used to track and control the movement of goods within a facility, from receiving and storing, to picking, packing and shipping. WMS systems can be used to manage the scheduling, execution, and monitoring of logistics operations, as well as to provide real-time information on the status of inventory, orders, and shipments [2]. This information can be used to optimize logistics processes and improve efficiency. Both systems can be integrated with other systems such as ERP and SCADA to provide a comprehensive view of the production and logistics operations and to improve the overall performance of the factory. MES and WMS systems can be used together to provide a comprehensive view of the production and logistics operations, allowing factories to optimize their operations and respond quickly to changing market conditions. The smart factory development has five stage design and development process, moving from Initiation and Concept Generation, through to Detailed Design, Validation and Lessons Learnt or wisdom as depicted on Fig. 1.

The initiation and concept generation stage of a project is the first step in the design process. It involves defining the problem or opportunity, and identifying the objectives and requirements for the project. This stage typically includes activities such as market research, brainstorming, and feasibility analysis. The concept generation stage involves the development of multiple potential solutions to the problem or opportunity identified in the initiation stage [3]. These concepts are evaluated based on their feasibility, cost, and alignment with the project objectives and requirements. The detailed design stage involves the further development and refinement of the selected concept, including the creation of detailed drawings, specifications, and calculations. This stage includes activities such as design reviews and the development of prototypes. Validation is the process of testing and evaluating the design

Fig. 1 Design process stages

to ensure that it meets the project objectives and requirements. This stage includes activities such as testing of prototypes, simulations, and pilot projects. The validation process is an ongoing process that continues throughout the design process, with the final validation done before the product is released. Overall, these stages are essential to ensure that the product or project is feasible, efficient, and meets the needs and expectations of the customer or end-user. These stages are iterative and are often revisited as the design process progresses and new information is gathered in working environment [4].

2 Research

This research has the following phases: literature review, data collection, identification of processes being addressed, understanding of existing and new processes, and analyzing and interpreting the results. The goal of this pilot project is to develop a sorting automation system that sorts products based on product weight. The requirements of Industry 4.0 was taken into account and the automation system prepared accordingly. The goal is to create an automated process where reading and human error can be eliminated and each product is weighed. This process is made using PLC and programming. Striving to select the right sensors and actuators for the task, which will hold their own under any circumstances. The operating principle of the tape scale was examined on several lines. The automation process can be monitored, started and stopped using the web interface. The purpose of this is that if a complete automation system is built, it will be enough to monitor from a computer and to be able to set the value to be measured before each process. The Factory I/O was used to create the 3D model, and the Tia portal, which is the programming interface of the SIMATIC PLC, was used to write the program.

2.1 Literature Overview

Chen et al. develops a conceptual framework for a new paradigm called smart factory based on CPS by applying virtual-real mapping and fusion, digital twin, big data-driven, virtualisation, and edge-to-cloud service technology to the manufacturing system [5]. Arifin et al. undertakes the development of an analytical model using a regression technique to estimate the number of vehicles required by a facility [6]. Chavez et al. examines the reported degree of digitalization in implemented solutions in SMEs when handling deviations and analyze the integration of such solutions in their digital transformation process [7]. A huge number of studies address the importance of managing information flows to generate logistics performance improvements, SCM performance, and organizational performance. Elia et al. analyses potential contributions of smart technologies in supporting an efficient prevention process with a focus on near miss detection based on a literature review analysis in the industrial safety [8]. Gnoni et al. proposes to safety managers criteria and suggestions for choosing appropriate CPS for seizing the great opportunities for safety improvement [9]. Grau et al. gives an overview of the evolution of robotics from its beginnings to recent trends like collaborative robotics, autonomous robots, and human–robot interaction [10]. Hofmann et al. discusses the opportunities of Industry 4.0 in the context of logistics management, since implications are expected in this field. The authors pursue the goal of shedding light on the young and mostly undiscovered topic of Industry 4.0 in the context of logistics management, thus following a conceptual research approach [11]. José shows the relevance of the topics Management, Social Issues, and Sustainability, which is not then reflected in publications of articles on Industry 4.0 in journals in these areas [12].

2.2 Research Background

Smart manufacturing engineering is the application of advanced technologies and methodologies to improve the efficiency, productivity, and flexibility of manufacturing operations. It is a key component of Industry 4.0, also known as the fourth industrial revolution [13]. Smart manufacturing engineering involves the integration of technologies such as the Internet of Things (IoT), artificial intelligence (AI), big data analytics, and cloud computing to optimize the performance of manufacturing systems.

Some examples of smart manufacturing engineering applications include:

- Predictive Maintenance: Using IoT sensors and AI algorithms to predict when equipment is likely to fail, allowing for proactive maintenance to be performed before a breakdown occurs.
- Real-time Monitoring: Using IoT sensors and big data analytics to monitor the performance of manufacturing systems in real-time, allowing for early identification of problems and more efficient use of resources.
- Automated Quality Control: Using machine vision and AI algorithms to automatically inspect and test products, reducing the need for human inspection and increasing the accuracy of quality control.
- Flexible Manufacturing: Using cloud computing and IoT to enable manufacturing systems to be quickly reconfigured to meet changing market conditions and production requirements.
- Cybersecurity: Ensuring that smart manufacturing systems are secure from cyber threats by implementing security measures such as firewalls, intrusion detection systems, and encryption.

Smart manufacturing engineering is an interdisciplinary field, involving a combination of engineering, computer science, information technology, and operations management. It requires a deep understanding of manufacturing processes, as well as expertise in advanced technologies such as IoT, AI, and big data analytics [14].

Predictive maintenance (PdM) is a maintenance strategy that utilizes data from sensors, IoT devices and machine learning algorithms to predict when equipment is likely to fail, allowing for proactive maintenance to be performed before a breakdown occurs. This approach is designed to optimize equipment performance and minimize downtime, while reducing maintenance costs. Predictive maintenance is based on the collection of data from various sources such as vibration sensors, temperature sensors, and oil analysis. The data is analyzed using machine learning algorithms to identify patterns and anomalies that may indicate equipment failure. Based on this analysis, maintenance tasks can be scheduled in advance, before the equipment fails [15]. There are several techniques that are used in predictive maintenance, including:

- Vibration Analysis: Monitors the vibration of equipment to detect early signs of failure such as bearing wear or misalignment.
- Infrared Analysis: Uses infrared cameras to detect temperature changes in equipment, which can indicate overheating or other problems.
- Oil Analysis: Monitors the condition of lubricating oils to detect signs of wear or contamination.
- Ultrasonic Analysis: Uses ultrasonic sensors to detect leaks or other problems in equipment.
- Machine Learning: Machine learning algorithms are used to analyze sensor data and identify patterns and anomalies that may indicate equipment failure.

Predictive maintenance can provide many benefits such as:

- Reducing downtime,
- Increasing equipment availability and productivity,
- Improving safety,
- Reducing maintenance costs,
- Improving equipment performance and efficiency.

However, implementing predictive maintenance can be complex and requires a significant investment in technology and expertise. It also requires a cultural change in the maintenance process, from reactive to proactive.

2.3 Overall Architecture of Sustainable Digital Factory

The overall architecture of a sustainable digital factory includes several key components, including:

- IoT and Cyber-Physical Systems (CPS): IoT devices and CPS are used to collect and transmit data from manufacturing systems, providing real-time insights into the performance of equipment and operations [16]. This data can be used to optimize production processes, reduce energy consumption and improve overall efficiency.

- Cloud Computing and Big Data Analytics: Cloud computing and big data analytics are used to process and store the large amounts of data generated by a digital factory. This data can be analyzed to identify trends, optimize production processes and improve decision-making.
- Artificial Intelligence and Machine Learning: AI and machine learning are used to automate tasks, improve forecasting and prediction, and optimize production processes. This can help to improve overall efficiency and reduce costs.
- Cybersecurity: With the increased connectivity and data exchange, cybersecurity becomes a major concern in sustainable digital factories. It is necessary to implement security measures such as firewalls, intrusion detection systems, and encryption to protect systems and data from cyber threats.
- Human–Machine Interfaces (HMI): HMI interfaces are used to provide operators with real-time information on the status of production systems and to allow them to monitor and control the systems.
- Manufacturing Execution Systems (MES) and Enterprise Resource Planning (ERP) Systems: These systems are used to manage and control the production process.

The overall architecture of sustainable digital factory diagram can be divided into four main layers: equipment layer, interaction layer, MES layer, and enterprise layer as depicted on Fig. 2.

The enterprise layer of sustainable digital factory includes ERP, OA (office automation), CRM (customer relationship management), SCM (supply chain management), PDM (product data management), visual Kanban, and war situation center. CRM user orders are transferred to ERP to make production plans. At the same

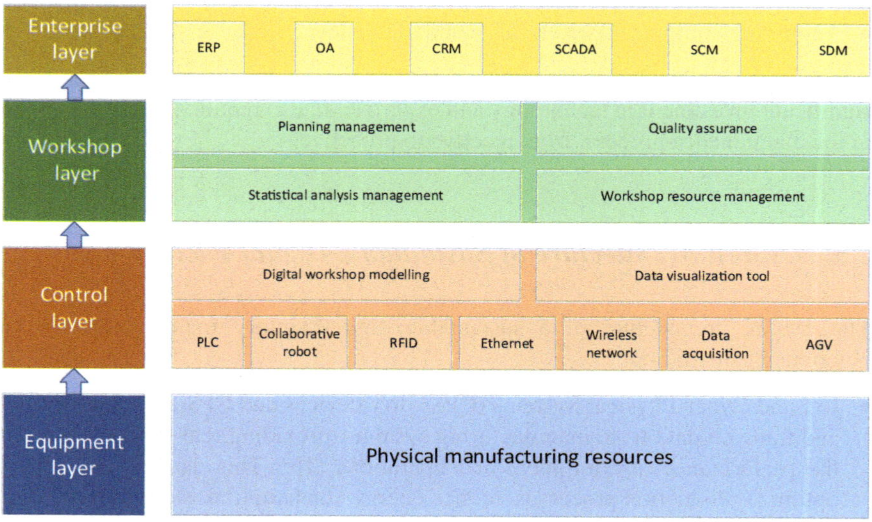

Fig. 2 Overall architecture of smart factory

time, through the OA system, SCM, material order management, R&D (research and development), PDM, and other auxiliary support, the production plan is transferred to the MES level of the factory.

The workshop resource management module includes personnel management, equipment management, tooling management, and data collection.

The control layer is the interaction with basic equipment, mainly including the connection and integration of communication equipment such as DNC (distributed numerical control), switch, wireless network, RFID (radio frequency identification), and configuration of AGV system.

The equipment layer mainly includes all kinds of equipment in the workshop, such as logistics equipment, processing equipment, assembly equipment, welding equipment, storage equipment, etc. This layer includes automation equipment and other data acquisition equipment, mainly including physical manufacturing resources such as collaborative industrial robots, intelligent machine tools, acquisition cards, scanning equipment, AGVs, and sensors.

3 Results

Factory I/O is a software that allows users to create and simulate a virtual manufacturing environment. It can be used to design and test a smart factory concept by simulating the behavior of industrial equipment, such as PLCs, robots, and conveyors can be seen on Fig. 4. The smart factory design concept using Factory I/O software typically includes the following components:

- Virtual PLCs: Factory I/O can be used to simulate the behavior of Programmable Logic Controllers (PLCs), which are used to control and automate industrial equipment.
- Virtual Robots: Factory I/O can be used to simulate the behavior of industrial robots, such as their movements, trajectories, and interactions with other equipment.
- Virtual Conveyors: Factory I/O can be used to simulate the behavior of conveyors, such as their speed, direction, and interactions with other equipment.
- Virtual Sensors: Factory I/O can be used to simulate the behavior of sensors, such as temperature, pressure, and vibration sensors, and their data can be used to monitor and control the manufacturing process.
- Human–Machine Interface (HMI): Factory I/O can be used to create a virtual HMI, allowing users to monitor and control the virtual factory environment.
- Data Analytics: Factory I/O can be used to collect data from the virtual factory.

ERP systems are designed to manage and optimize the overall performance of an organization, by integrating and automating various business processes such as accounting, inventory management, customer relationship management, and production planning. WMS systems, on the other hand, are designed to manage and optimize the operations of a warehouse or distribution center, by automating and optimizing processes such as receiving, storing, picking, packing, and shipping. The integration of ERP and WMS systems allows for a more comprehensive and efficient management of the supply chain, by providing a seamless flow of information between different systems and departments, and allowing for a more accurate and real-time view of inventory levels, production schedules, and customer orders. By integrating inventory data between ERP and WMS systems, organizations can have a more accurate view of inventory levels and can make more informed decisions regarding inventory replenishment, and organizations can more easily track the status of production orders, allowing them to make more informed decisions about production scheduling.

Data centers play a critical role in smart factories by providing the necessary computing and storage resources to support the collection, analysis, and distribution of data. Smart factories rely on a large amount of data from various sources such as sensors, machines, and systems, which must be processed, stored, and analyzed in real-time (Fig. 3). Data centers provide the necessary infrastructure to support these operations. They also play a critical role in enabling the integration of advanced technologies such as IoT, AI, and big data analytics to improve the performance and efficiency of manufacturing operations (Fig. 4).

TIA Portal is a software development environment from Siemens that is commonly used for programming and configuring industrial automation systems such as programmable logic controllers (PLCs), human–machine interfaces (HMIs), and drives. TIA Portal allows users to program, test, and debug automation systems in a virtual environment, which can improve efficiency and reduce development time. TIA Portal and Factory I/O are both widely used in the industrial automation industry, and are considered to be de facto standards for industrial simulation. Both software are used to design and simulate industrial control systems, and both have extensive libraries of industrial components and devices that can be used to create virtual systems. TIA Portal is mainly used to program and configure real PLCs, while Factory I/O is mainly used for simulating industrial control systems. Both software can work together, TIA Portal can be used to program the PLCs and Factory I/O can be used to simulate the entire system, including the PLC. This allows users to test and validate their control system in a virtual environment before it is deployed in the real world.

3.1 Real World Scenario

The real word scenario is conducted in the robotics laboratory at the University of Szeged. The task of the AGV unit is to serve the collaborative robot with the material on the conveyor belt. The project includes four Magni-type AGV units from the Ubiquity Robotics company that are adapted to the needs of the environment (Fig. 5).

The Magni robot by Ubiquity Robotics is a mobile robot platform designed for research and development, as well as for use in industrial and educational settings. The robot is 417.40 × 439.09 × 265 mm in size and weighs around 13 kg. It is

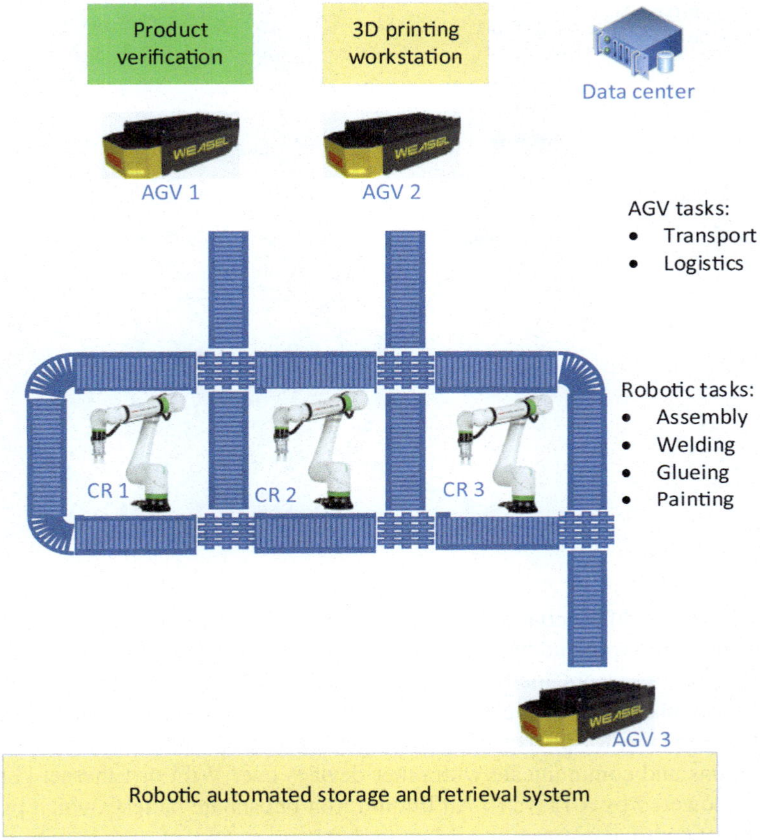

Fig. 3 Smart factory design concept logistics

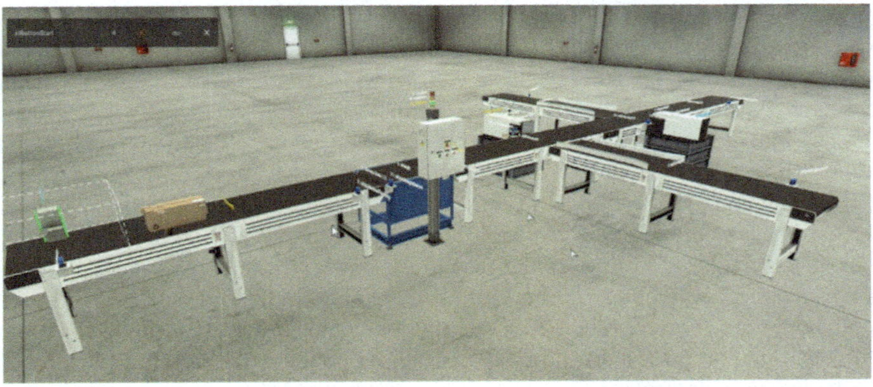

Fig. 4 Smart factory design example

Fig. 5 Magni-type AGV units from the Ubiquity Robotics

equipped with a differential drive system, which allows it to move in any direction and rotate on the spot and has a top speed of about 1 m/s. Magni robot is equipped with a variety of sensors, including a LIDAR sensor for localization and mapping, an IMU for orientation, and several cameras for object detection and visual inspection. It is powered by a Raspberry Pi 4, which can be used to run a variety of software applications and communicate with other devices over WiFi or Ethernet [17]. The robot is powered by a 12 V, 10 Ah lithium iron phosphate battery, which provides up to 4 h of continuous operation. Magni robot has a variety of connectivity options, including WiFi, Ethernet, USB, and serial interfaces, also comes with a variety of software tools, including a web-based interface for controlling the robot and a software development kit (SDK) for writing custom applications [18, 19]. The robot can carry a payload up to 100 kg. The Magni robot is designed to be flexible and customizable, which makes it well suited for a wide range of applications, including research, education, and industrial automation.

The Magni robot by Ubiquity Robotics can be used in collaboration with a UR5 robot arm, which is a popular industrial robot arm manufactured by Universal Robots (Fig. 6). When using the Magni robot in collaboration with a UR5, the Magni can be used as a mobile base for the UR5, allowing it to move around the factory floor, and transport the UR5 to different locations. The UR5 can then be used for tasks such as assembly, pick and place, and welding. One example of a use case for this collaboration is in a warehouse or distribution center, where the Magni robot can move around and transport the UR5 to different locations to pick and place items from shelves and storage areas. The UR5 can also be used for tasks such as packaging, sorting and palletizing. Another example is in a manufacturing plant, where the Magni robot can move around and transport the UR5 to different stations, where it can be

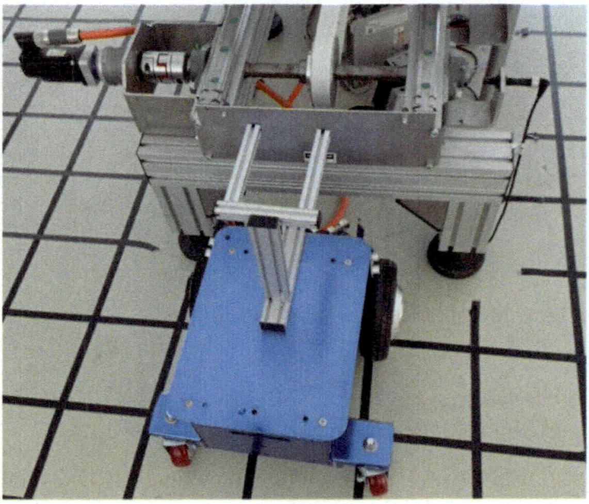

Fig. 6 Magni-type AGV aligns to the conveyor belt

used for tasks such as assembly, welding, and painting. The collaboration between the Magni robot and UR5 provides a flexible and efficient solution for automating a wide range of tasks in different industries. The combination of the Magni's mobility and the UR5's precision and dexterity can help to increase productivity and accuracy, while also reducing labor costs.

4 Discussion

Based on these case studies, a common framework is proposed, defining the rules or main steps, and possibly equally suitable for addressing the incorporation of AGVs in other smart factories.

KPIs (Key Performance Indicators) are metrics that are used to measure the performance of an AGV (Automatic Guided Vehicle) implementation. These metrics can be used to evaluate the success of the implementation, identify areas for improvement, and make decisions about future investments in the technology. Some examples of KPIs for AGV implementation are given in Fig. 7.

By monitoring these KPIs, organizations can identify areas where the AGV implementation is performing well and areas where improvements can be made. This information can be used to make decisions about future investments in the technology and to continuously improve the performance of the AGV system. As the pilot project has proved, the return on investment of replacing man-driven logistic vehicles by AGVs is quite short, in less than five years.

KPIs (Key Performance Indicators) are metrics that can be used to evaluate the performance of a smart factory. These metrics can provide insights into the efficiency, productivity, and overall performance of the factory, and can be used to identify

Key Performance Indicators (KPIs)			
Delivery on Time	**Delivery in Full**	**Cost of Delivery**	**Qualitative KPIs**
• Delay • Oversupply – earlier delivery	• Incomplete deliveries • Lost deliveries • Damaged items	• Operating costs • Maintenance costs • Costs of accidents and repairs • Safety & ergonomics • costs	• Ergonomic aspects • Motivation • Standardization • Visualization

Fig. 7 Model of KPIs for AGV implementation

areas for improvement. Some examples of KPIs that can be used to evaluate the performance of a smart factory include:

- OEE (Overall Equipment Effectiveness): This metric measures the overall performance of the factory's equipment and systems, taking into account factors such as availability, performance, and quality.
- MTBF (Mean Time Between Failures): This metric measures the average time between equipment failures, and can be used to identify equipment that is prone to breakdowns or maintenance issues.
- TEEP (Total Effective Equipment Performance): This metric measures the overall efficiency of the equipment and systems in the factory, taking into account factors such as equipment uptime, production speed, and quality.
- Downtime: This metric measures the amount of time that the factory is not in production due to equipment failures, maintenance, or other issues.
- WIP (Work in Progress): This metric measures the amount of work that is currently in progress in the factory, and can be used to identify bottlenecks in the production process.
- Yield: This metric measures the percentage of products that are produced without defects, and can be used to identify quality control issues in the factory.
- Energy Consumption: This metric measures the energy consumed by the factory, and can be used to identify opportunities for energy savings and to evaluate the environmental impact of the factory.
- Lead Time: This metric measures the time it takes for an order to be fulfilled, and can be used to identify opportunities for improving the speed and efficiency of the production process.

By monitoring and analyzing these KPIs, organizations can gain a better understanding of the performance of their smart factory and identify areas where improvements can be made. This can help to optimize production processes, improve efficiency and reduce costs, and increase the overall competitiveness of the factory.

5 Conclusion

This article presented the possibility of using smart factory simulation software to improve the smart factory design concept and logistics efficiency. It can be concluded that research results have shown that the determinants of logistics performance determine the levels of supply chain integration, as an indicator of SCM performance, which directly involves organizational performance via marketing and financial performance. AGVs can be integrated into a smart factory environment to support the operations of collaborative robots. AGVs can be used to transport materials, tools, and equipment to the collaborative robots, allowing them to perform their tasks more efficiently and effectively. In a smart factory environment, AGVs can be equipped with sensors such as cameras and LIDAR, to navigate and observe the collaborative robots and the surrounding environment. These sensors can be used to detect and avoid obstacles, as well as to locate and track the collaborative robots. AGVs can communicate with the collaborative robots through various communication protocols such as OPC-UA, ROS and MQTT, to coordinate their movements and tasks. AGVs can also receive commands from the collaborative robots, for example, the AGV can be directed to pick up a specific part or tool and bring it to the collaborative robot. The AGVs can also monitor the status and performance of the collaborative robots and send the data to the central control system, which can be used to optimize production processes, improve efficiency, and reduce downtime. Overall, AGVs can support the collaborative robots in a smart factory environment by providing them with the materials, tools, and equipment they need to perform their tasks, and by observing and monitoring the surrounding environment to ensure safe and efficient operations.

References

1. Sabattini L et al (2018) The PAN-robots project: advanced automated guided vehicle systems for industrial logistics. IEEE Robot Autom Mag 25(1):55–64
2. Piccarozzi M, Aquilani B, Gatti C (2018) Industry 4.0 in management studies: a systematic literature review. Sustainability 10(10):3821
3. Um I, Cheon H, Lee H (2009) The simulation design and analysis of a flexible manufacturing system with automated guided vehicle system. J Manuf Syst 28(4):115–122
4. Oztemel E, Gursev S (2020) Literature review of industry 4.0 and related technologies. J Intell Manuf 31(1):127–182
5. Chen G, Wang P, Feng B, Li Y, Liu D (2020) The framework design of smart factory in discrete manufacturing industry based on cyber-physical system. Int J Comput Integr Manuf 33(1):79–101
6. Arifin R, Egbelu P (2000) Determination of vehicle requirements in automated guided vehicle systems: a statistical approach. Prod Plan Control 11(3):258–270
7. Chavez Z, Hauge JB, Bellgran M (2022) Industry 4.0, transition or addition in SMEs? A systematic literature review on digitalization for deviation management. Int J Adv Manuf Technol 119(1–2):57–76

8. Elia V, Gnoni MG, Tornese F, Guglielmi A, Pellicci M, De Merich D, Campo G (2022) Applications of smart technologies for automatic near miss detection in the industrial safety. Procedia Comput Sci 200:1282–1287
9. Gnoni MG, Elia V, Bragatto P (2016) An IoT-based system to prevent injuries in assembly line production systems. In: 2016 IEEE international conference on industrial engineering and engineering management (IEEM), pp 1889–1892
10. Grau A, Indri M, Lo Bello L, Sauter T (2021) Robots in industry: the past, present, and future of a growing collaboration with humans. IEEE Ind Electron Mag 15(1):50–61
11. Hofmann E, Rüsch M (2017) Industry 4.0 and the current status as well as future prospects on logistics. Comput Ind 89:23–34
12. José DL (2022) Industry 4.0 and the small business. Serb J Manag 17(1):161–178
13. Kiel D, Müller JM, Arnold C, Voigt KI (2020) Sustainable industrial value creation: benefits and challenges of industry 4.0. Int J Innov Manag 21(8):1–34
14. Kesen SE, Baykoç OF (2007) Simulation of automated guided vehicle (AGV) systems based on just-in-time (JIT) philosophy in a job-shop environment. Simul Model Pract Theory 15(3):272–284
15. Ballai G, Sőrés MA, Vásárhelyi L, Szenti I, Kun R, Hartmann B, Sebők D, Farkas F, Zahoor A, Mao G, Sápi A, Kukovecz Á, Kónya Z (2023) Exploration of Li-ion batteries during a long-term heat endurance test using 3D temporal microcomputed tomography investigation. Energy Technol 11:2300207. https://doi.org/10.1002/ente.202300207
16. Sarcevic P, Kincses Z, Pletl S (2014) Wireless sensor network based movement classification using wrist-mounted 9DOF sensor boards. In: 2014 IEEE 15th international symposium on computational intelligence and informatics (CINTI), Budapest, Hungary, pp 85–90. https://doi.org/10.1109/CINTI.2014.7028654
17. Dominik C, Ákos O, Richárd P, József S, Peter S (2023) A novel orientation-based FSPL model parameter optimization method using PSO for indoor localization. In: IEEE 21st world symposium on applied machine intelligence and informatics SAMI (2023): proceedings, pp 201–205
18. Mészáros A, Sárosi J (2022) Soft robotics: state of art and outlook. Anal Tech Szeged 16(1):8–13. https://doi.org/10.14232/analecta.2022.1.8-13
19. Tamang D, Pozzebon A, Parri L, Fort A, Abrardo A (2022) Designing a reliable and low-latency LoRaWAN solution for environmental monitoring in factories at major accident risk. Sensors 22(6):2372

A Short Review of Polymer Friction Stir Welding

Róbert Gábor Stadler and Richárd Horváth

Abstract Polymers are today's most widely used materials. Accordingly, the joining technologies of these materials are a popular and extensively researched area. Of these technologies, welding procedures are the most important in industry and research. For polymers, one of the most promising welding technologies is friction stir welding (FSW). The technology has many advantages; it can be easily automated and it has a small impact on the environment. Also, the procedure can be used to weld thicker polymer sheets and thermoplastic polymer matrix composites. In this study, we present the technology of friction stir welding and its individual variants (procedure, tool modifications), based on recently published studies, and review current and future research directions.

Keywords Polymer welding · Friction stir welding (FSW) · Weld joint quality

1 Background

In recent decades, polymers have become today's most widely used materials. Today, we cannot find any industry where polymer or polymer composite materials are not used in some form. The popularity of polymers is due to their good manufacturability, low density, and good impact and chemical resistance, among other things. With the spread of polymers, the bonding technology of these materials became increasingly important, and it is still a popular and extensively researched field today.

R. G. Stadler (✉) · R. Horváth
Banki Donat Faculty of Mechanical and Safety Engineering, Obuda University, Budapest, Hungary
e-mail: stadler.robert@bgk.uni-obuda.hu

R. Horváth
e-mail: horvath.richard@bgk.uni-obuda.hu

R. G. Stadler
Doctoral School on Materials Sciences and Technologies, Obuda University, Budapest, Hungary

T. A. Kovács et al. (eds.), *The Impact of the Energy Dependency on Critical Infrastructure Protection*, Advanced Sciences and Technologies for Security Applications, https://doi.org/10.1007/978-3-031-78544-3_50

When two materials are joined, bond strength, automation, economy, and impact on the environment are important considerations. Taking these points into account among others, many technologies exist for joining polymer parts, but these technologies can be divided into the following three main groups: adhesive joining, mechanical joining, and welding [1]. Among the publications on the adhesive joining of polymers, the investigation of the treatment of the surfaces to be bonded is a popular research area (e.g. plasma surface treatment [2]), as well as the examination of the adhesive bonding of different materials to each other (polymer–metal [3] and polymer–polymer [4]). In the case of mechanical joints, metal–polymer joints with different mechanical joint technologies are also extensively studied [5].

In industry, welding technologies are the most often used when polymers or metals are joined. Among the welding processes for polymers, ultrasonic welding [6], laser welding [7], and hot gas welding [8] are most popular in industrial applications and research. In addition, friction stir welding (FSW) is considered a promising welding technology for polymers.

Friction stir welding was patented in 1991 by Thomson et al. [9]. The essence of the process is that a rotating tool is moved along the joint, and as a result, heat is generated from the friction between the parts and the tool. This heat, in turn, creates the welding seam. Figure 1 shows a schematic diagram of friction stir welding.

The FSW technology was specifically designed for welding materials that were difficult or impossible to weld with classical welding processes (e.g. aluminium and its alloys). Friction stir welding is currently used for welding aluminium on an industrial scale, and it is worth highlighting the number of related patents and publications [10]. At the beginning of the 2000s, the first studies on the friction stir welding of polymers were published [11, 12], then in 2004 Nelson et al. [13] filed a patent on the application of the technology to polymers. After that, weldability tests

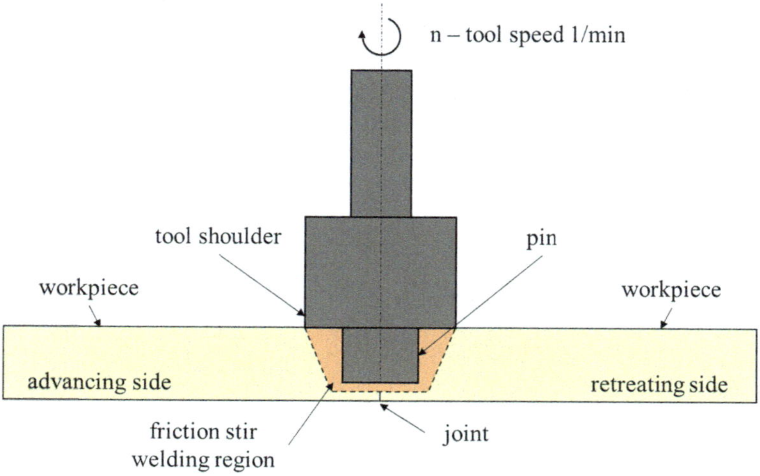

Fig. 1 A schematic diagram of stir welding

of various engineering polymers (PE [14], PP [15], ABS [16], PC [17], PA [18]) with stir friction welding were published. In addition to the many advantages of friction stir welding, the technology is also suitable for welding thicker polymer sheets and fiber-reinforced thermoplastics [19].

2 Process Parameters

Many process parameters can be distinguished in the friction stir welding (FSW) of polymers. The Ishikawa diagram (Fig. 2) describes the parameters that influence the mechanical properties and appearance of the weld.

The individual process parameters can be classified into three groups based on Fig. 2, such as material, tool and machine. Therefore, the effectiveness of friction stir welding (FSW) is greatly influenced by the type of polymers to be welded. Among the material properties, thermal properties stand out, such as melting point and the thermal conductivity coefficient. As has already been mentioned, the weldability of the most important engineering polymers with FSW has been investigated. Then, more and more publications were published in which the weldability of different types of polymers was investigated. Among other things, the tests examined the weldability of the following polymer material pairs:

- acrylonitrile butadiene styrene (ABS) and polymethyl methacrylate (PMMA) [20]
- polyethylene (PE) and polypropylene (PP) [21, 22]
- acrylonitrile butadiene styrene (ABS) and high-density polyethylene (HDPE) [23, 24]
- acrylonitrile butadiene styrene (ABS) and polypropylene (PP) [25, 26].

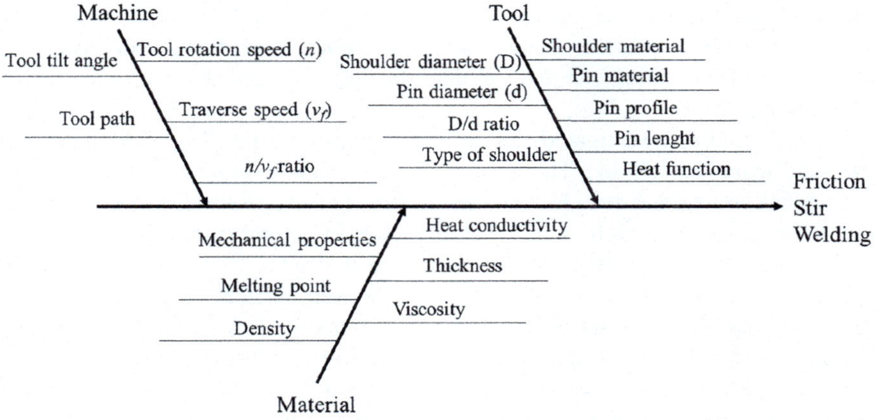

Fig. 2 The effect of process parameters on the friction stir welding of polymers

Welding resulted in successful bonding in all cases, and several publications concluded that an additional advantage of friction stir welding is that in many cases it can bond incompatible polymer material pairs.

In addition to the welding of different polymer materials, similarly to other welding technologies [27, 28], the welding of polymer-metal pairs has become a popular research area as well. In the literature, classical engineering polymers are also widely used in these tests, while the metal is usually an aluminium alloy; the friction stir welding of the following material pairs have been investigated:

- AA5059 and high-density polyethylene (HDPE) [29]
- AA7075 and polycarbonate (PC) [30]
- AA5052 and polypropylene (PP) [31]
- AA6061 and polycarbonate (PC) [32]
- AA5754 and poly(methyl methacrylate) (PMMA) [33]
- AA7475 and polyphenylene sulfide (PPS) [34].

Although successful welding was successful in every case, and friction stir welding is considered a potential method for joining metals and polymers, joint efficiency is still low.

The goal of several publications was to find the optimal welding parameters for the polymer material(s) investigated [35–37]. Among welding parameters, tool rotational speed (n) and tool feed (v_f) are most often examined, but some publications have examined tool tilt angle [38, 39], as well as the n/v_f ratio [40]. In recent years, there have been many review articles where the authors reviewed the results of individual tests and summarized the optimal welding parameter ranges of individual polymer materials [41, 42].

Although optimal welding parameters have been determined for the specific conditions used in the article, it is difficult to compare the results, because in the individual tests, tool construction, pin geometry, welding machine and material thickness are different, and these are all variable factors that greatly influence the success and quality of welding. Of these factors, the most important are the tool and its associated parameters (heating, pin and shoulder diameter, pin and shoulder material and shoulder type).

Several studies investigated the influence of the pin profile on the quality of the weld seam [22, 43, 44]. The most frequently examined pin profiles are straight cylindrical, conical and threaded cylindrical. Among these, the threaded profile stands out, which resulted in high-quality welding joints in several studies [22, 44].

In terms of tool construction, it is important to highlight the shoulder part; its primary task is to keep the molten material in the welding zone. Among the studies, we have found find one where the shoulder (D) and tool diameter (d) ratio (D/d) was investigated for the welding seam [45]. There are two distinct types of shoulders concerning rotation; the rotating shoulder design, where the shoulder rotates together with the pin, and the stationary shoulder design, where the shoulder does not rotate. The former is a simpler tool design, because in this case the tool is one unit, while the latter has a more complicated structure, because the pin, shoulder, and other elements (e.g. bearing) are separate parts. Finally, the heating function of the shoulders is important, too. During welding, friction causes heat, but the tool is also equipped with a heater, as a result of which heat can be introduced into the welding zone from an external source. Naturally, in this case, (as in the case of the non-rotating shoulder design) the tool will be more complex as a result. However, in the tests where a heated shoulder design was used, tool temperature had a great effect in all cases—in several studies, better quality welds were produced than by welding without heating [21, 23].

3 Quality Testing of Friction Stir–Welded Joints

During all bonding technology tests, the most frequently examined output parameter, which characterize the bond, is tensile strength. Correspondingly, the tensile strength of the welded joint is also the most often tested parameter in friction stir welding tests. From tensile strength, joint efficiency (JE) can be calculated with the following relationship, with which friction stir welding can be characterized:

$$JE = \frac{\sigma_{base\ material}}{\sigma_{welding\ joint}} \cdot 100\% \tag{1}$$

In addition, the mechanical quality of welded joints is usually characterized by impact testing and hardness measurement [46, 47].

Various optical tests (with an optical microscope, or a scanning electron microscope—SEM) are also common during tests where welding defects characteristic of different technologies are examined. The following are the most common welding defects in the friction stir welding of polymers are [48]:

- Tunnel defect: there is a significant lack of material inside the weld seam.
- Thickness reduction: the thickness of the plate at the welding seam is smaller than the thickness of the plate elsewhere.
- Lack of surface material: there is a lack of material on the upper surface of the weld seam.
- Burr error: during welding, the molten material comes out of the welding seam in the form of a burr.
- Porous defect: several smaller parts lacking material are formed in the weld seam.

These defects all weaken the weld seam and result in significant deterioration in mechanical properties and impair appearance.

In the case of polymers, thermoanalytical tests such as differential scanning calorimetry (DSC) are very important when seams are tested. This test is mainly used with semicrystalline plastics. During the test, the crystalline portion of the seam and the original material can be determined. Several publications have reported that the crystalline ratio of the welding seam is lower than that of the base material [15, 49].

Appearance and mechanical properties are greatly influenced by the temperatures and forces occurring during welding. Accordingly, many publications report on force

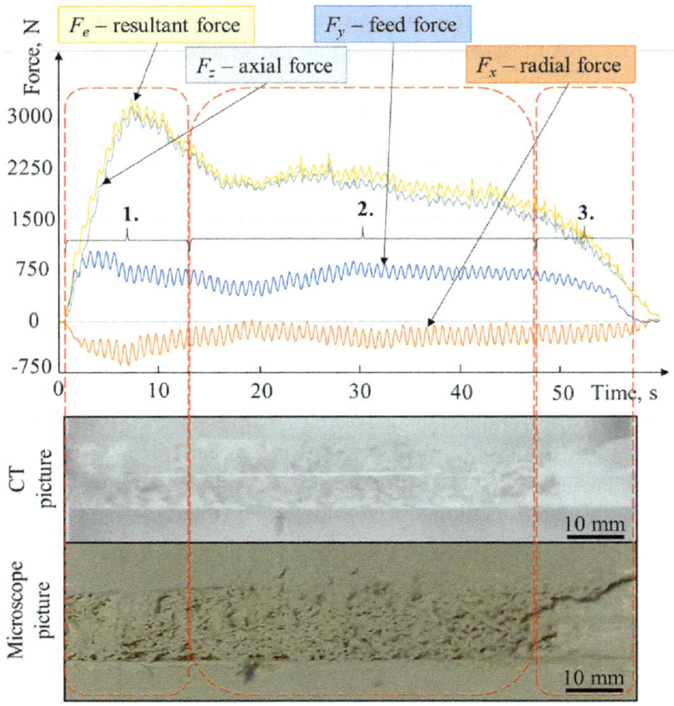

Fig. 3 Resulting welding force during welding and its effect on the welding seam (polypropylene) [40]

measurement [40, 50, 51] and temperature measurement (with thermal sensors and thermal cameras) [50, 52] during welding.

Based on the force measurements, the quality of the welding seam can be deduced from the evolution of the registered forces [40]. Force measurement during welding may be suitable for the adaptive control of the technology in the future, which can result in the production of higher quality and more reliable joints. This is further confirmed by the fact that in the case of aluminium, friction stir welding with adaptive control is now being investigated more widely [53, 54]. Figure 3 shows how the resulting welding force develops during welding and its effect on the welding seam.

4 Conclusion

Nowadays, polymers are widely used, therefore the binding of these materials has become very important. Over the decades, a number of technologies have been invented with which polymers can be bonded quickly, economically and with good quality. Among these technologies, the article reviews the research trends and latest research results of friction stir welding. In addition to the detailed presentation of the technology, the study discusses the input parameters (e.g. tool speed (n), feed speed (v_f)) whose effect on the welded joint are still being investigated. The article also reviews the main output parameters of welding which characterize the seams (e.g. tensile strength, bonding efficiency, hardness), as well as the parameters characterizing the welding process (e.g. temperature and forces). The individual chapters also touch upon the research directions which are still important areas of research today.

Acknowledgements Supported by the ÚNKP-23-3 New National Excellence Program of The Ministry for Culture and Innovation from the Source of The National Research, Development and Innovation Fund.

References

1. Silva LRR, Marques EAS, da Silva LF (2021) Polymer joining techniques state of the art review. Weld World 65(10):2023–2045
2. Sundriyal P, Pandey M, Bhattacharya S (2020) Plasma-assisted surface alteration of industrial polymers for improved adhesive bonding. Int J Adhes Adhes 101:102626
3. Moldovan E, Tierean MH, Stanciu EM (2017) Overview of joining dissimilar materials: metals and polymers. Bull Transylvania Univ Brasov Ser I Eng Sci 39–46
4. Tiwary VK, Padmakumar A, Malik V (2022) Adhesive bonding of similar/dissimilar three-dimensional printed parts (ABS/PLA) considering joint design, surface treatments, and adhesive types. Proc Inst Mech Eng C J Mech Eng Sci 236(16):8991–9002
5. Lambiase F, Scipioni SI, Lee CJ, Ko DC, Liu F (2021) A state-of-the-art review on advanced joining processes for metal-composite and metal-polymer hybrid structures. Materials 14(8):1890
6. Unnikrishnan TG, Kavan P (2022) A review study in ultrasonic-welding of similar and dissimilar thermoplastic polymers and its composites. Mater Today Proc 56:3294–3300
7. Kumar N, Kumar N, Bandyopadhyay A (2021) A state-of-the-art review of laser welding of polymers—part I: welding parameters. Weld J 100(7):221–228
8. Ahamad H, Alam M, Haque MS, Pandey SN, Amanuddin M (2016) Welding of plastics through hot gas technique: a review. Glob Res Dev J Eng 1(6)
9. Thomas WM, Nicholas ED, Needham JC, Murch MG, Templesmith P, Dawes CJ (1995) Friction stir butt welding. G.B. patent application No. 9125978.8, Dec 1991; U.S. patent No. 5460317, Oct 1995
10. Magalhães VM, Leitão C, Rodrigues DM (2018) Friction stir welding industrialisation and research status. Sci Technol Weld Join 23(5):400–409
11. Strand S (2003) Joining plastics—can friction stir welding compete? In: Proceedings: electrical insulation conference and electrical manufacturing and coil winding technology conference (Cat. No. 03CH37480), Sept 2003. IEEE, pp 321–326
12. Strand SR (2004) Effects of friction stir welding on polymer microstructure. Brigham Young University
13. Nelson TW, Sorenson CD, Johns CJ (2004) Friction stir welding of polymeric materials. Google patents
14. Arici A, Sinmazçelýk T (2005) Effects of double passes of the tool on friction stir welding of polyethylene. J Mater Sci 40(12):3313–3316
15. Kiss Z, Czigány T (2007) Applicability of friction stir welding in polymeric materials. Period Polytech Mech Eng 51(1):15–18

16. Bagheri A, Azdast T, Doniavi A (2013) An experimental study on mechanical properties of friction stir welded ABS sheets. Mater Des 43:402–409
17. Vidakis N, Petousis M, David C, Sagris D, Mountakis N, Moutsopoulou A (2024) The impact of process parameters and pin-to-shoulder ratio in FSW of polycarbonate: welding forces and critical quality indicators. Int J Adv Manuf Technol 130(11):5457–5477
18. Husain IM, Salim RK, Azdast T, Hasanifard S, Shishavan SM, Eungkee Lee R (2015) Mechanical properties of friction-stir-welded polyamide sheets. Int J Mech Mater Eng 10:1–8
19. Czigány T, Kiss Z (2011) Friction stir welding of fiber reinforced polymer composites. In: Proceedings of the 18th international conference on composite materials (ICCM), Jeju, South Korea, pp 21–26
20. Dashatan SH, Azdast T, Ahmadi SR, Bagheri A (2013) Friction stir spot welding of dissimilar polymethyl methacrylate and acrylonitrile butadiene styrene sheets. Mater Des 45:135–141
21. Eslami S, Ramos T, Tavares PJ, Moreira PMGP (2015) Effect of friction stir welding parameters with newly developed tool for lap joint of dissimilar polymers. Procedia Eng 114:199–207
22. Hajideh MR, Farahani M, Alavi SAD, Ramezani NM (2017) Investigation on the effects of tool geometry on the microstructure and the mechanical properties of dissimilar friction stir welded polyethylene and polypropylene sheets. J Manuf Process 26:269–279
23. Gao J, Li C, Shilpakar U, Shen Y (2015) Improvements of mechanical properties in dissimilar joints of HDPE and ABS via carbon nanotubes during friction stir welding process. Mater Des 86:289–296
24. Gao J, Li C, Shilpakar U, Shen Y (2016) Microstructure and tensile properties of dissimilar submerged friction stir welds between HDPE and ABS sheets. Int J Adv Manuf Technol 87:919–927
25. Hajideh MR, Farahani M, Ramezani NM (2018) Reinforced dissimilar friction stir weld of polypropylene to acrylonitrile butadiene styrene with copper nanopowder. J Manuf Process 32:445–454
26. Miranda MS, Almaraz GD, López JV, Vilchez JR (2022) Dissimilar joining of ABS and PP using friction stir welding (FSW) and mechanical properties evaluation. Procedia Struct Integr 39:161–172
27. Temesi T, Czigany T (2023) The effect of surface treatment on the shear properties of aluminium-polypropylene joints manufactured by laser beam. Polym Test 117:107882
28. Al-Obaidi A, Majewski C (2019) Ultrasonic welding of polymer–metal hybrid joints. In: Transactions on intelligent welding manufacturing, vol II, no 2. Springer Singapore, pp 21–38
29. Khodabakhshi F, Haghshenas M, Sahraeinejad S, Chen J, Shalchi B, Li J, Gerlich AP (2014) Microstructure-property characterization of a friction-stir welded joint between AA5059 aluminum alloy and high density polyethylene. Mater Charact 98:73–82
30. Moshwan R, Rahmat SM, Yusof F, Hassan MA, Hamdi M, Fadzil M (2015) Dissimilar friction stir welding between polycarbonate and AA 7075 aluminum alloy. Int J Mater Res 106(3):258–266
31. Pabandi HK, Movahedi M, Kokabi AH (2017) A new refill friction spot welding process for aluminum/polymer composite hybrid structures. Compos Struct 174:59–69
32. Patel AR, Kotadiya DJ, Kapopara JM, Dalwadi CG, Patel NP, Rana HG (2018) Investigation of mechanical properties for hybrid joint of aluminium to polymer using friction stir welding (FSW). Mater Today Proc 5(2):4242–4249
33. Derazkola HA, Simchi A (2019) An investigation on the dissimilar friction stir welding of T-joints between AA5754 aluminum alloy and poly(methyl methacrylate). Thin-Walled Struct 135:376–384
34. Sandeep R, Natarajan A (2022) Application of machine learning approaches to predict joint strength of friction stir welded aluminium alloy 7475 and PPS polymer hybrid joint. Proc Inst Mech Eng C J Mech Eng Sci 236(16):9003–9011
35. Bilici MK (2012) Application of Taguchi approach to optimize friction stir spot welding parameters of polypropylene. Mater Des 35:113–119
36. Jaiganesh V, Maruthu B, Gopinath E (2014) Optimization of process parameters on friction stir welding of high density polypropylene plate. Procedia Eng 97:1957–1965

37. Lenin K, Shabeer HA, Panneerselvam K (2014) Process parameters optimization for friction stir welding of polypropylene material using Taguchi's approach
38. Saeedy S, Besharati Givi MK (2010) Experimental application of friction stir welding (FSW) on thermo plastic medium density polyethylene blanks. In: Engineering systems design and analysis, vol 49187, pp 841–844
39. Saeedy S, Givi MB (2011) Investigation of the effects of critical process parameters of friction stir welding of polyethylene. Proc Inst Mech Eng Part B J Eng Manuf 225(8):1305–1310
40. Stadler GR, Szebényi G, Horváth R (2023) Investigation of weld forces and strength of friction stir welded polypropylene. Period Polytech Mech Eng 67(3):183–189
41. Pereira MA, Amaro AM, Reis PN, Loureiro A (2021) Effect of friction stir welding techniques and parameters on polymers joint efficiency—a critical review. Polymers 13(13):2056
42. Iftikhar SH, Mourad AHI, Sheikh-Ahmad J, Almaskari F, Vincent S (2021) A comprehensive review on optimal welding conditions for friction stir welding of thermoplastic polymers and their composites. Polymers 13(8):1208
43. Derazkola HA, Simchi A (2018) Experimental and thermomechanical analysis of the effect of tool pin profile on the friction stir welding of poly(methyl methacrylate) sheets. J Manuf Process 34:412–423
44. Bilici MK, Yükler AI (2012) Influence of tool geometry and process parameters on macrostructure and static strength in friction stir spot welded polyethylene sheets. Mater Des 33:145–152
45. Sadeghian N, Givi MKB (2015) Experimental optimization of the mechanical properties of friction stir welded acrylonitrile butadiene styrene sheets. Mater Des 67:145–153
46. Miloud MH, El Bahri OC, Abdallah L (2019) Mechanical behavior analysis of a friction stir welding (FSW) for welded joint applied to polymer materials. Frattura Integr Strutt 13(47):459–467
47. Derazkola HA, Simchi A, Lambiase F (2019) Friction stir welding of polycarbonate lap joints: relationship between processing parameters and mechanical properties. Polym Test 79:105999
48. Huang Y, Meng X, Xie Y, Wan L, Lv Z, Cao J, Feng J (2018) Friction stir welding/processing of polymers and polymer matrix composites. Compos A Appl Sci Manuf 105:235–257
49. Moreno-Moreno M, Macea Romero Y, Rodríguez Zambrano H, Restrepo-Zapata NC, Afonso CRM, Unfried-Silgado J (2018) Mechanical and thermal properties of friction-stir welded joints of high density polyethylene using a non-rotational shoulder tool. Int J Adv Manuf Technol 97:2489–2499
50. Paoletti A, Lambiase F, Di Ilio A (2016) Analysis of forces and temperatures in friction spot stir welding of thermoplastic polymers. Int J Adv Manuf Technol 83:1395–1407
51. Vidakis N, Petousis M, David C, Sagris D, Mountakis N, Moutsopoulou A (2024) Polycarbonate hybrid friction stir welding joints: welding forces and quality indicators on 3D printed and solid sheets single and two-axis seams. Int J Adv Manuf Technol 134(1):431–444
52. Eslami S, Francisco Miranda J, Mourão L, Tavares PJ, Moreira PMGP (2018) Polyethylene friction stir welding parameter optimization and temperature characterization. Int J Adv Manuf Technol 99:127–136
53. Davis TA, Shin YC, Yao B (2010) Observer-based adaptive robust control of friction stir welding axial force. IEEE/ASME Trans Mechatron 16(6):1032–1039
54. Meyer SP, Bernauer CJ, Grabmann S, Zaeh MF (2020) Design, evaluation, and implementation of a model-predictive control approach for a force control in friction stir welding processes: validation of different force-control approaches for friction press joining of aluminum with thermoplastics. Prod Eng 14(4):473–489

Approach of By-Design-Risk-Analysis on the Cybersecurity of Advanced Engineered Technologies

Kristóf Stölczer, Erzsébet Cserta, and Tamás Szádeczky

Abstract Critical infrastructures of the energy sector play a central role in the appropriate operation of our society and economy. In order to ensure the essential energy security and the economic development, there is a strong demand for the use of advanced energy technologies and their defense against physical and especially cyber-attacks. These forefront technologies have their own typical vulnerabilities as well that have to be unveiled and managed as quick as possible. The preferred scenario is when the resilience of a system can be proofed still during the test operation. In this article, we focus on revealing the cyber vulnerability and resilience associated with engineered energy industrial processes through identifying potential gaps, hazards and risks by design already to protect both the critical infrastructure system and the integrating technologies before the system integration process.

Keywords Secure-by-design · Critical infrastructure · Cyber-risk assessment · Resilience · Industrial cybersecurity

1 Abstract. Introduction of the Nuclear Hydrogen Production with Small Modular Reactor Technology

Since nuclear fission reactor electricity produces no greenhouse gas emissions, it lays with an increasing tendency in the focus of the worldwide intention to stop climate change [1]. The smaller nuclear reactors have the advantage that they are

K. Stölczer (✉)
Doctoral School on Safety and Security Sciences, Obuda University, Népszínház str. 8, Budapest 1081, Hungary
e-mail: stolczerk@icloud.com

E. Cserta
Hungarian Mobility Development Agency Zrt., Dávid Ferenc Street 4-6, Budapest 1113, Hungary

T. Szádeczky
Department of Cybersecurity, Faculty of Public Governance and International Studies, Ludovika University of Public Service, Budapest 1441, Hungary

T. A. Kovács et al. (eds.), *The Impact of the Energy Dependency on Critical Infrastructure Protection*, Advanced Sciences and Technologies for Security Applications, https://doi.org/10.1007/978-3-031-78544-3_51

less expensive to develop, as well. Small modular reactors (SMRs) are determined by the World Nuclear Association as nuclear reactors generally 300 MWe equivalent or less, designed with modular technology using module factory fabrication, pursuing economies of series production and short construction times [2]. The International Atomic Energy Agency (IAEA) defines 'small' as under 300 MWe, and up to about 700 MWe as 'medium'—including many operational units from the twentieth century [3]. A subcategory of very small reactors—vSMRs—is proposed for units under about 15 MWe, especially for remote communities [1].

Moreover, 'SMR' is used more commonly as an acronym for 'small modular reactor', designed for serial construction and collectively to comprise a large nuclear power plant. Generally, modern small reactors for power generation, and especially SMRs, are expected to have greater simplicity of design, economy of series production largely in factories, short construction times, and reduced siting costs [1, 2, 4, 5]. All SMRs features a high level of passive or inherent safety components [3].

Several types of SMRs models are based on different technical principles and focused on different functions [6]. There is a distinction between different reactor concepts. The classification can be made according to the technique as well as the possible application. Here, the most frequent categorization of SMRs found in the literature is given with relevant examples of them.

Pressurized Water Reactors (PWRs)

PWRs are similar to those used in large-scale nuclear plants, but they are smaller and modular [7]. They use water as a coolant and a neutron moderator to slow down neutrons and enhance fission [8]. The BWRX-300 model developed by GE Hitachi is a 300 MWe boiling water reactor (BWR) that uses various passive/naturally driven systems. Primary pumps are not required, and safety systems are passive [9, 10]. Another example is the Rolls-Royce SMR [11]. It is a compact three-loop pressurized water reactor (PWR). The cooling water is circulated by three reactor coolant pumps to three vertical U-tubular steam generators.

Micro-range Nuclear Reactors (MNR)

Ideal micro-nuclear reactors are so small they can fit on the back of a truck or a standard 40-foot shipping container. Aside from being compact, micro-reactors must be designed to operate safely for many years—perhaps several decades. They can also be self-controllable, operating without the need for a constant human presence [5].

Leading SMR company Westinghouse Electric propelled to the forefront of the nuclear technology industry with its transportable eVinci™ micro reactor. More of a nuclear battery than a traditional nuclear reactor, the high-temperature heat pipe reactor can generate 5 MW of electricity and up to 13 MW of heat from its 15 MW solid thermal core. It is based on a closed-core microreactor and uses solid TRistructural ISOtropic (TRISO) fuel [3, 12].

The question arises whether a nuclear battery can be handled as an extra-micro (already "nano") version of NMRs, since, it works by using the energy released by the decay of nuclear isotopes due to absorbing and converting it into electrical energy. The Beijing Betavolt New Energy Technology Co., Ltd. has successfully developed a miniature atomic energy battery, which uses nickel 63 nuclear isotope decay technology [13]. Among the main characteristics/expectations of an SMR, the miniaturization, modularization and low cost of construction is successfully realized in the development of these atomic energy batteries.

High-Temperature Gas Reactors (HTGRs)

HTGRs use helium gas as a coolant and graphite as a moderator. They are capable of operating at very high temperatures, which makes them suitable for a variety of applications, including **hydrogen** production and industrial processes. The world's first high-temperature gas-cooled reactor is the HTR-PM constructed in China [14]. The modular pebble bed reactor was developed by the Institute of Nuclear and New Energy Technology, Tsinghua University [15]. It is a demonstration unit for electricity production. The HTR-PM consists of two NSSS modules coupled to one 210 MWe steam turbine, each with independent protection systems and shared non-safety assistance systems. Each NSSS module contains a reactor with a reactor pressure vessel, a steam generator and a hot gas channel (helium).

Molten Salt Reactors (MSRs)

MSRs use liquid fuel, typically a mixture of fluoride salts containing fissionable materials, as both the fuel and the coolant [4]. They operate at high temperatures and can be used for a range of applications, including electricity generation and hydrogen production. In these technologies the major concern is corrosion because the radiation inside a nuclear reactor can speed up the rate at which the salt corrodes and weakens those metal components [10]. MIT researchers have demonstrated that some metal alloys will corrode not more but less when they're exposed to radiation under some conditions. The team devised and built an experimental setup that emulates conditions inside a molten salt-based nuclear reactor. Tests with various alloys show that certain combinations of elements become more resistant to corrosion when they are also subjected to radiation. The researchers provide simple guidelines for designers and operators to follow when selecting commercial alloys for their molten salt-based nuclear reactors.

One of the most innovative solutions is the Kairos Power FHR (KP-FHR) that is an advanced modular reactor (AMR) with a net electrical capacity of 140 MW, based on a mix of molten salt reactor and the reactors using TRISO fuel, which is made from tiny uranium-containing kernels that can be embedded in graphite casings [16, 17]. This type of AMR is being developed by the American company Kairos Power. Inventive design features include among other things, the use of spherical TRISO fuel elements, with which the reactor is continuously refilled, with a molten salt as coolant in contrast to a gas in other designs around TRISO fuel elements. A molten salt as a coolant means that normal operation can be done at low pressure. Other features related to nuclear safety are a relatively small surplus of reactivity and the elimination of the risk of a meltdown, due to the use of TRISO fuel elements and their continuous replenishment.

Liquid Metal-Cooled Reactors (LMRs)

LMRs use liquid metals, such as sodium or lead, as a coolant. They are capable of operating at very high temperatures and can be used for a variety of applications, including electricity generation and nuclear propulsion for ships.

The Westinghouse Lead-cooled Fast Reactor (LFR) program is at the conceptual design stage for up to the capacity of 450 MWe as a modular pool-type unit, simple, scalable and with passive safety [1]. Its fast reactor operating at high temperatures—eventually 650 °C. The concept has inherent safety features due to, among other things, the physical properties of lead (high boiling point) that allow operation under practically atmospheric conditions. The design features an integrated energy storage system, where the reactor power remains unchanged, for load following.

In this chapter, the technical background of the SMR technology was briefly discussed. Since nuclear can be by far the most reliable, abundant, and clean energy source available for a cleaner future, the gap between the variable output of renewables can be closed with it and speed up the transition to net-zero [4]. As the future part of the critical infrastructures, it is essential to ensure the nuclear energy sector's cyber safety with a special attention given to the SMR technology. Parallel investment and innovation in the IT sector connecting to the SMRs are the key factors of being able to protect this particular infrastructure system in the cyber space. The challenge of this parallel development is analyzed in the next chapters.

2 The Challenges of Nuclear Hydrogen Production's Cyber Risk Analysis

The production and use of nuclear (yellow) hydrogen through the use of electricity generated by SMR's is interesting from several points of view, as it can be an innovative, economical and green way to add a component to a national energy mix. However, this is a complex technology from a cyber security perspective, because it is a combination of several different and also hazardous technologies. Therefore, it requires serious assessment of cyber security risks from the point of view of electricity production by nuclear technology. Furthermore, the production of hydrogen, where the explosive nature of hydrogen already requires an industrial safety risk analysis—combined with cyber security have to be analyzed considering the cyber security risks that potentially arise from its transport by pipeline. Both the nuclear technology and the hydrogen production and its transportation by pipeline have those security concerns, which the experts must to evaluate individually and together as well. The following chapters are referring to these approaches from the most crucial perspective. Firstly, we introduce the industrial safety concerns connecting to cybersecurity aspects and risks to define basic methodologies. Then the nuclear technologies cyber risk management is briefly discussed. To finish this chapter, the cybersecurity concept of the hydrogen transportation through pipeline is also detailed.

Overview of the Industrial Safety Related Cybersecurity Consideration

The integration of advanced technologies has fundamentally reshaped industrial processes in the modern, hyperconnected and digitalized world, bringing incredible improvements in effectiveness and competitiveness. However, this transition has also opened up critical infrastructures to new and ever-evolving risks, making cybersecurity a critical concern. Simultaneously, businesses are operating highly sophisticated processes, such as chemical, petrochemical and manufacturing industries, depending on methodologies such as Hazard and Operability Assessments (HAZOP) and Process Hazard Analysis (PHA) to identify and mitigate their potential safety-related vulnerability. This chapter examines the interconnectedness between cybersecurity and HAZOP/PHA, emphasizing the necessity of a comprehensive approach to provide security and safety for industrial systems.

These systems, built of interconnected devices and control systems, are vulnerable to cyber threats that can undermine the integrity, availability, and confidentiality of critical operations data. Similarly, HAZOP and PHA methodologies are focused on identifying process-related threats, providing safety for personnel, assets and the environment. The convergence of these processes and technologies requires a common approach to managing both cybersecurity and process safety concerns. Both cyber security and HAZOP/PHA aim to identify and manage risks, although in different areas. Cybersecurity targets the securing of digital assets and preventing unauthorized access, while HAZOP/PHA focuses on assessing physical processes to identify potential threats and their consequences [18]. This combination of the efforts empowers companies to develop comprehensive risk management strategies that cover aspects of both digital and physical processes, assuring a more scalable and resilient industrial infrastructure. In sectors where data integrity is critical to guaranteeing process safety, the relationship between cybersecurity and HAZOP/PHA becomes even more clear. Cyber-attacks are targeting industrial control systems and they can potentially manipulate data, leading to incorrect process decisions and potentially resulting in dangerous situations. Integrating cybersecurity controls into HAZOP/PHA helps preserve data integrity, assuring that decision-making processes based on these analyses will remain accurate and predictable. Critical infrastructure, such as power plants, chemical facilities and manufacturing establishments, are frequently targeted by cyber-attackers due to their potential for extensive disruption. The crossroads of cybersecurity and HAZOP/PHA becomes the key to defend these critical assets. By incorporating cybersecurity controls into the design and operation of industrial processes, companies can prevent cyber threats from escalating into physical harm, preserving the security and functionality of critical infrastructure. The governance framework for both cybersecurity and process safety is constantly developing, and regulators have identified the necessity for a more comprehensive approach. The integration of cybersecurity controls into HAZOP/PHA analyses is in accordance with regulatory requirements and demonstrates a strong commitment to securing digital and physical assets. This integrated compliance approach encourages a culture of safety and security in industrial organizations [19, 20].

Overview of the Nuclear Cybersecurity Considerations

The development and operation of nuclear power plants are vital components of a national energy infrastructure. As the world increasingly becomes more digitalized, it is necessary to adopt cyber risk-based design to guarantee the cyber resilience and safety of nuclear power plants. This chapter discusses the significance of integrating cyber risk considerations into the design phase of nuclear facilities, emphasizing the need for a blended proactive and adaptive perspective. Cyber risk-based design is the key cornerstone to provide cyber security and resiliency for nuclear power plants. By implementing cybersecurity principles early in the design phase, incorporating in-depth defense strategies, and adhering to international standards, nuclear power plants can reduce cyber risks more effectively. As the energy sector continuously advances, proactive management of cyber security during the design and implementation of nuclear power plants will be crucial to secure critical infrastructure and ensure the safety of nuclear power generation. In the followings, the most important cyber protection approaches are briefly discussed.

Cyber Risk-Informed Design in Nuclear

Designing for cyber risks includes the integration of cyber security principles at the earliest stages of nuclear power plant design. This approach takes into account the potential cyber threats and vulnerabilities from the very beginning, thus implementing security controls as an essential part of the plant design [21].

Threat Assessment and Risk Modelling: The use of threat assessment and risk analysis methodologies is a key aspect of risk mitigation based on cyber risk information. These processes identify potential cyber threats, vulnerabilities [22] and their potential impact on the operation of nuclear power plants. By performing a comprehensive risk analysis, engineers can customize security controls to mitigate specific threats, ensuring solid safeguards against cyber-attacks [23].

Defense-in-Depth Strategies: Implementing defense-in-depth strategies [24] is also fundamental to designing for cyber risks. This approach represents layering several security controls in the design of a nuclear power plant to create a robust and resilient security posture [25]. By integrating physical, technical and procedural safeguards, the plant is better prepared to resist and respond to cyber threats.

Adaptive Design for Emerging Threats: The dynamic nature of cyber threats [26] again requires an adaptive planning approach. Designers need to anticipate future cyber threats and develop architectures that are capable of addressing emerging challenges. This adaptability will ensure that nuclear power plants remain resilient throughout their operational lifecycle, even as cyber threats advance and become more sophisticated.

International Standards and Cooperation: International cooperation is essential to developing common standards and best practices for nuclear cyber security. Organizations such as the International Atomic Energy Agency [27] (IAEA) provide guidelines (NSS Computer Security Guidelines) and recommendations to Member

States, facilitating a global response to cyber risks in the nuclear sector. Sharing experiences and lessons between nations contributes to the development of more robust cyber security measures.

Legacy System Integration: Incorporating cyber risk-informed design can be challenging, particularly when dealing with legacy systems [23]. NPPs often have components with long lifespans, and integrating modern cybersecurity measures into existing infrastructure requires careful planning and implementation to ensure compatibility and effectiveness.

Human Factor and Training: The human factor remains a significant consideration in cybersecurity [23]. Designing NPPs with a focus on user awareness, training, and education is crucial. Ensuring that personnel are well-prepared to recognize and respond to cyber threats enhances the overall resilience of the facility.

Overview of the Pipeline Cybersecurity Considerations

Pipeline transportation serves as the backbone of modern energy distribution, efficiently transporting oil, gas, and other critical resources across vast distances. However, the integration of digital technologies into pipeline control systems has exposed the industry to a myriad of cybersecurity challenges. This chapter explores the unique cybersecurity landscape of pipeline transportation, examining the potential threats, vulnerabilities, and outlining strategies to ensure the resilience and security of these essential infrastructures. The integration of Supervisory Control and Data Acquisition (SCADA) systems, Industrial Control Systems (ICS), and Internet of Things (IoT) devices has ushered in a new era of efficiency and real-time monitoring in pipeline transportation. However, it has also created a complex cybersecurity landscape with unique challenges. Threat actors, ranging from state-sponsored entities to cybercriminals, are increasingly targeting these critical infrastructures, posing significant risks to the uninterrupted flow of energy resources [28].

Threats to Pipeline Transportation Cybersecurity: Pipeline transportation systems face a diverse range of threats [29], including but not limited to:

- Cyber Espionage: Nation-states may engage in cyber espionage to gather intelligence on energy infrastructure, aiming to gain a strategic advantage or disrupt the energy supply of rival nations [30].
- Ransomware Attacks: Cybercriminals may deploy ransomware to encrypt critical systems, demanding payment in exchange for restoring normal operations. Such attacks can lead to operational downtime and financial losses [31, 32].
- Insider Threats: Malicious or unintentional actions by employees or contractors with access to critical systems can pose a significant risk. This includes intentional sabotage or inadvertent exposure of sensitive information.
- Outdated Infrastructure: Legacy systems that lack modern security features are more vulnerable to exploitation. Upgrading and maintaining these systems pose logistical and financial challenges.

- Interconnected Networks: The interconnected nature of pipeline systems increases the attack surface. A compromise in one segment of the network can potentially affect the entire pipeline infrastructure.
- Supply Chain Vulnerabilities: Dependence on third-party suppliers for equipment and software introduces additional risks. Compromised components can be used as entry points for cyber attacks.

Cybersecurity Best Practices and Strategies [33]: Addressing the cybersecurity challenges in pipeline transportation requires a multifaceted approach:

- Risk Assessment: Conducting comprehensive risk assessments helps identify vulnerabilities and prioritize security measures. Understanding the specific threats facing the pipeline system is crucial for developing effective mitigation strategies.
- Encryption and Authentication: Implementing robust encryption protocols and multi-factor authentication enhances the security of data transmitted across pipeline networks, preventing unauthorized access.
- Regular Training and Awareness: Training personnel on cybersecurity best practices and raising awareness about potential threats reduce the likelihood of insider incidents and improve overall cybersecurity hygiene.
- Continuous Monitoring: Implementing continuous monitoring systems allows for real-time detection of anomalies and potential security breaches. Timely response to incidents can mitigate the impact of cyber-attacks.
- Regulatory Frameworks and Compliance: Governments and regulatory bodies play a crucial role in ensuring pipeline cybersecurity. The development and enforcement of cybersecurity standards, such as those outlined by organizations like the American Petroleum Institute (API) or the International Electrotechnical Commission (IEC), create a baseline for industry-wide cybersecurity practices.

Securing pipeline transportation is a multifaceted challenge that requires collaboration between industry stakeholders, governments, and cybersecurity experts. By understanding the evolving threat landscape, identifying vulnerabilities, and implementing proactive cybersecurity measures, the pipeline transportation sector can strengthen its defenses and ensure the continued safe and reliable transportation of critical energy resources. In an era where digital connectivity is essential, cybersecurity becomes an indispensable component in safeguarding the lifeline of pipeline transportation.

In comparison, this combination of technologies can functionate effectively, however the safety and security aspects are hard to align. The cybersecurity frameworks and the interest of the profitability, effectiveness of the whole production system is still too far from each other technically to be handled together.

3 Relevant General Risk Assessment Standards and Frameworks to Use

The aim of the following chapter is to describe the basic standards and frameworks that can assist in the risk evaluation of such a complex technology, the yellow hydrogen production. We will cover the policies, frameworks, that needed to ensure the cybersecurity, both in the IT environment and the industrial OT environment from different aspects of the technology stages.

ISA/IEC 62443 Standard Series

IEC 62443 is an international standard that provides a comprehensive framework for securing industrial automation and control systems (IACS). The standard emphasizes a risk-based approach to cybersecurity, guiding organizations through the process of identifying, analyzing, and mitigating potential risks within industrial environments. One of the key features of IEC 62443 is its zoning concept, which involves dividing industrial networks into security zones based on the criticality and sensitivity of assets. Defense-in-depth strategies, as recommended by IEC 62443, involve the implementation of multiple layers of security measures, including both technical safeguards and organizational policies. The standard advocates for a holistic, lifecycle approach to cybersecurity, emphasizing the importance of continuous monitoring and adaptation of security measures throughout the industrial system's life. IEC 62443 compliance certifications serve as a valuable demonstration of an organization's commitment to adhering to best practices in industrial cybersecurity. Incident response planning, a critical aspect outlined in IEC 62443, focuses on developing effective strategies to mitigate the impact of cybersecurity incidents on industrial processes. The standard underscores the significance of security training and awareness programs to ensure that personnel are well-equipped to address and prevent cybersecurity threats. IEC 62443 provides clear guidelines on integrating its recommendations with existing cybersecurity standards, allowing organizations to leverage their current security measures [34].

ISO/IEC 27000 Standard Series

ISO 27000 is a family of international standards that addresses information security management systems (ISMS) [35]. The ISO 27000 series provides a systematic and comprehensive framework for organizations to manage and secure their information assets. ISO 27001 is the central standard in the series and specifies the requirements for establishing, implementing, maintaining, and continually improving an ISMS. It is designed to be adaptable to various types and sizes of organizations, ensuring its relevance across different industries and sectors. ISO 27002 complements ISO 27001 by providing a set of guidelines and best practices for implementing specific security controls within the ISMS. The standards within the ISO 27000 series cover a wide range of aspects, including risk management, incident response, business continuity, and compliance. Certification to ISO 27001 is often sought by organizations to demonstrate their commitment to information security and build trust with stakeholders. ISO 27000 standards are regularly updated to reflect the evolving landscape of information security threats and technologies. The framework encourages a risk-based approach, requiring organizations to assess and prioritize potential security risks to their information assets [36].

NIST 800-82 Framework

The primary focus of NIST 800-82 is to provide comprehensive guidance on securing ICS, which are systems used to control and monitor industrial processes in sectors such as energy, manufacturing, and utilities. These systems are critical for the functioning of various infrastructures, making them attractive targets for cyber adversaries. The guide emphasizes the importance of implementing robust security measures to protect against potential cyber-attacks that could compromise the integrity, availability, and confidentiality of ICS components. One of the key aspects of NIST 800-82 is its risk management framework tailored for ICS environments. The document advocates for a proactive approach to identify and assess risks, develop mitigation strategies, and continuously monitor and update security measures. It also emphasizes the need for collaboration among stakeholders, including system operators, engineers, and IT professionals, to create a holistic security posture [37].

API RP 1164—Pipeline Control Systems Cybersecurity (American Petroleum Institute)

The primary objective of API RP 1164 is to establish a systematic approach to integrity management, ensuring the safe and reliable operation of pipeline systems throughout their lifecycle. The document addresses various aspects of pipeline integrity, encompassing risk assessment, inspection, maintenance, and incident response. By adopting the principles outlined in API RP 1164, operators can proactively identify, assess, and mitigate potential threats to pipeline integrity, thereby reducing the risk of incidents and ensuring the safety of personnel and the environment. One key component of API RP 1164 is the risk assessment process, which involves evaluating and prioritizing potential threats to pipeline integrity. This includes factors such as corrosion, external interference, material defects, and other environmental considerations. The document provides methodologies for assessing the likelihood and consequences of these threats, enabling operators to focus resources on the highest-risk areas. Through a risk-informed approach, API RP 1164 aids in optimizing the allocation of resources to address the most critical integrity issues, enhancing the overall effectiveness of integrity management programs. Another significant aspect covered by API RP 1164 is the establishment of a comprehensive inspection and monitoring program. The document outlines the importance of utilizing various inspection techniques, including both in-line and external methods, to assess the condition of pipelines. Regular inspections help identify potential issues before they escalate, allowing operators to implement timely maintenance and repair activities. Additionally, API RP 1164 emphasizes the need for advanced technologies and data management systems to enhance the accuracy and efficiency of inspections. Maintenance practices play a pivotal role in ensuring the long-term integrity of pipeline systems. API RP 1164 provides guidelines for developing and implementing effective maintenance programs, taking into account the specific characteristics of different pipeline segments. The document encourages a proactive approach to maintenance, emphasizing the importance of addressing potential integrity threats before they result in failures or incidents [38].

4 Summary

As discussed in this paper, analyzing cybersecurity risks and taking steps as soon as possible, even at the planning stage of energy system construction, is extremely important. The reason is that some components of complex technologies may carry various consequences in the event of a cybersecurity incident. In comparison, this combination of technologies can work effectively, but the security and protection aspects are difficult to harmonize. The technological gap between the cybersecurity framework and the interests of the profitability and efficiency of the whole production system is still too far to be managed together. The above-named nuclear hydrogen technology has great potential both in green power generation and from a national energy security perspective. In the form of an isolated power plant, if properly connected to the national power transmission system, it can temporarily compensate for the power loss/shortfall caused by a cyber-attack during a national incident. Thus, any research effort to identify and prepare for the cybersecurity of this technology before similar installations are implemented is definitely beneficial and forward-looking.

References

1. US firm unveils game-changing small nuclear reactor that can power 300,000 homes (interestingengineering.com). https://interestingengineering.com/innovation/us-firm-unveils-game-changing-small-nuclear-reactor
2. Small nuclear power reactors. World Nuclear Association (world-nuclear.org). https://www.world-nuclear.org/information-library/nuclear-fuel-cycle/nuclear-power-reactors/small-nuclear-power-reactors.aspx
3. What are small modular reactors (SMRs)? | IAEA. https://www.iaea.org/newscenter/news/what-are-small-modular-reactors-smrs
4. Small modular reactors could be key to achieving a secure energy future (c3newsmag.com). https://c3newsmag.com/small-modular-reactors-could-be-key-to-achieving-a-secure-energy-future/
5. Micro-nuclear reactors: up to 20 MW, portable, safer. Energy Post. https://energypost.eu/micro-nuclear-reactors-up-to-20mw-portable-safer/
6. Types of small modular reactors | List of 10 SMR designs (small-modular-reactors.org). https://small-modular-reactors.org/types-of-small-modular-reactors/
7. Rolls-Royce SMR power plant introduction, rolls-royce-smr.com; gda.rolls-royce-smr.com
8. NuScale Power | Small modular reactor (SMR) nuclear technology. https://www.nuscalepower.com/en

9. BWRX-300 small modular reactor | GE Hitachi Nuclear (gevernova.com). https://www.gevernova.com/nuclear/carbon-free-power/bwrx-300-small-modular-reactor
10. BWXT awarded engineering contract from GE Hitachi for BWRX-300 small modular reactor. https://www.bwxt.com/news/2023/03/21/BWXT-Awarded-Engineering-Contract-from-GE-Hitachi-for-BWRX-300-Small-Modular-Reactor
11. Documents | Rolls-Royce SMR. Generic Design Assessment (rolls-royce-smr.com). https://gda.rolls-royce-smr.com/documents
12. Five of the world's leading small modular reactor companies (c3newsmag.com). https://c3newsmag.com/five-of-the-worlds-leading-small-modular-reactor-companies/
13. Beta volt has successfully developed a civilian atomic energy battery. https://www.betavolt.tech/359485-359485_645066.html
14. China starts up world's first high-temperature gas-cooled reactor. Global Construction Review. https://www.globalconstructionreview.com/china-starts-up-worlds-first-high-temperature-gas-cooled-reactor/
15. China's pebble bed reactor finally starts commercial operation. NextBigFuture.com. https://www.nextbigfuture.com/2023/12/chinas-pebble-bed-reactor-finally-starts-commercial-operation.html
16. How hot salt could transform nuclear power. MIT Technology Review. https://www.technologyreview.com/2024/01/17/1086736/how-hot-salt-could-transform-nuclear-power/
17. Homepage. Kairos Power. https://kairospower.com/
18. Iaiani M (2023) Process hazard and operability analysis of BPCS and SIS malicious manipulations by POROS 2.0. Process Saf Environ Prot 176:226–237
19. Iaiani M (2023) Identification of cyber-risks for the control and safety instrumented systems: a synergic framework for the process industry. Process Saf Environ Prot 172:69–82
20. Sliwinski M (2018) Integrated functional safety and cyber security analysis. IFAC PapersOnLine 51(24):1263–1270
21. Eggers SL, Anderson RS (2022) Cyber-informed engineering for nuclear reactor digital instrumentation and control. Idaho National Laboratory, Idaho Falls, Idaho
22. Eggers S (2021) A novel approach for analyzing the nuclear supply chain cyber-attack surface. Nucl Eng Technol 53(3):879–887
23. Ayodeji A (2023) Cyber security in the nuclear industry: a closer look at digital control systems, networks and human factors. Prog Nucl Energy 161
24. Bansal S (2021) Investigating the implementation of the safety-diagnosability principle to support defence-in-depth in the nuclear industry: a Fukushima Daiichi accident case study. Eng Fail Anal 123
25. Rehman Z (2024) Proactive defense mechanism: enhancing IoT security through diversity-based moving target defense and cyber deception. Comput Secur 139
26. Zimba A (2018) Multi-stage crypto ransomware attacks: a new emerging cyber threat to critical infrastructure and industrial control systems. ICT Express 4(1):14–18
27. IAEA homepage. https://iaea.org
28. Synchrony Systems. Inc. 5 ways your legacy systems may add to cybersecurity risks. https://sync-sys.com/5-ways-your-legacy-systems-may-add-to-cybersecurity-risks/
29. CISA homepage. https://cisa.gov. Pipeline Cybersecurity Initiative. PCI cyber risk infographic. https://www.cisa.gov/resources-tools/resources/pci-cyber-risk-infographic
30. Benmalek M (2024) Ransomware on cyber-physical systems: taxonomies, case studies, security gaps, and open challenges. Internet Things Cyber-Phys Syst 4:186–202
31. Transportation Security Administration (2018) Pipeline security guidelines. Transportation Security Administration, USA
32. How one company refused to let its cyberattackers win | TIME. https://time.com/6080293/norsk-hydro-ransomware-attack/
33. The Norsk hydro attack. Industrial Cyber. https://industrialcyber.co/features/the-norsk-hydro-attack/
34. Tripwire homepage. https://tripwire.com, https://www.tripwire.com/state-of-security/isa-iec-62443-framework

35. EWON homepage. https://ewon.biz
36. IT Governance homepage. https://itgovernance.co.uk
37. National Institute of Standards and Technology. Special publication, 800-82 rev. 3, guide to operational technology (OT) security
38. American Petroleum Institute. API STD 1164: pipeline control systems cybersecurity

Preventing Attacks on Key Installations or Recapturing Them by Forced Entry

László Szalkai

Abstract Protecting critical facilities is key to keeping society safe and functioning. This paper summarizes the tools and options for the active and passive protection of civil and military critical infrastructure. We highlight the protection systems that can provide solutions for the complex security of high-profile facilities. We examine the risk factors and examine the options for protection against blast attacks, and summaries the security changes that can be made to each structure to greatly improve the resilience of high-profile facilities and the survival of their occupants. We present the devices and explosive techniques that can be used to forcibly open high-security doors and windows, which can help to recapture the protected object and to realize the perpetrators.

Keywords Critical infrastructures · Protection systems · Explosives · Forced entry · Realization

1 Introduction

Critical infrastructure protection legislation was conceived in 2004. As part of this, the European Union has developed a proposed Critical Infrastructure Protection Programme (CIP), which is designed to protect critical infrastructure, both to protect Member States' critical systems and to address the problems they face at global level [1].

Recent decades have presented significant challenges to law enforcement and military organizations. Increased security challenges have created new challenges for professionals. Both civil and military critical infrastructures need to be protected by an appropriate level of protection systems, supplemented by specialized solutions where necessary, to ensure the complex security of installations [2].

L. Szalkai (✉)
Doctoral School of Military Engineering, Ludovika University of Public Service, Budapest Hungária Str. 9-11, 1011, Hungary
e-mail: szalkai.laszlo81@gmail.com

T. A. Kovács et al. (eds.), *The Impact of the Energy Dependency on Critical Infrastructure Protection*, Advanced Sciences and Technologies for Security Applications, https://doi.org/10.1007/978-3-031-78544-3_52

A distinction is made between civilian and military installations according to their purpose. The purpose of the facility has a major influence on the physical appearance of the area, the physical means of protection and the security measures to be adopted to suit the tolerance level of the people working and visiting the facility [3]. There may also be differences in the design of the protection and in the regime rules applied.

2 Options for the Physical Protection of Objects

Basic requirements for a physical protection system.

Mechanical protection must be able to deter intruders until the intervention units, which withdraw in response to an electronic protection signal, are able to stand between the intruders and the object to be protected. In other words, the physical protection system must ensure that there is a delay of sufficient duration to allow the retreating units to intercept and neutralize the offenders before they reach the protected target area.

The design of the physical protection system shall be implemented in such a way that:

- minimize the possibility of sabotage by an insider offender,
- prevent unlawful theft by an insider offender,
- operate effectively in all weather conditions, at all times of day, and at all stages of application, processing, storage and transport,
- only environmentally certified equipment is used in the design and construction of the system.

The physical protection system shall ensure: effective interaction of the physical protection functions of deterrence, detection, delay and response.

To achieve deterrence means to induce the offender to prepare or abandon the act. This deterrence can take the form of: warning signs, signs, artificial barriers, sound and light signals, hazard awareness information, record-keeping checks.

In terms of detection, the perpetrator's activity must be detected, its authenticity evaluated, its exact location identified and the forces deployed to prevent it must be alerted. The combination of systems that can be used to achieve detection—video surveillance and evaluation, intrusion and attack detection, guard center and personal detection, access control—can be used to prepare effectively for the prevention of possible intrusions. In addition, it must be ensured that the intrusion is detected in good time to allow sufficient time to verify its authenticity.

These access control systems must be designed and implemented in such a way as to restrict access by authorized persons only to the extent necessary, but to delay the entry of unauthorized persons and to indicate the exit of unauthorized devices.

The essence of the delay is to prevent the perpetrator from carrying out the given act for the time necessary for the arrival of the intervention of the forces of law enforcement.

Active movable and passive mechanical barriers, structures, safety storage devices, plate cabinets may be used as a system element of delay.

The units used to carry out the evacuation activity are site guards, armed security guards, police and extraction services. Their task is to intercept intruders, neutralize them, prevent theft and sabotage.

Developing an appropriate level of guarding culture is one of the cornerstones of a complex security system. Both external and internal staff should give appropriate priority to activities related to physical security. In order to prevent vulnerability of the security system to internal perpetrators, the credentials of persons with clearance or special authority should be subject to national security risk controls. Access to facilities through security systems may provide additional protection to the complex. Such equipment is known as baggage screening, explosives detection and metal detection equipment.

Traditional cluster X-rays can be used to search for individual features and components, while newer X-ray images that can be viewed in three dimensions can be used to achieve even higher detection success rates.

Explosive trace detection devices are currently available for the detection of explosives, although these devices can already detect real explosive material, the length of the evaluation process makes these devices unsuitable for large numbers of entries [4].

There are two types of metal detectors: the manual metal detector and the full-height metal detector gate. These devices are not really suitable for the detection of improvised explosive devices (IED), as the material of IEDs does not necessarily require a significant metal composition. Other materials can be substituted for the destructive, shrapnel-enhancing effect [5, 6].

"Improvised Explosive Devices are devices improvisedly assembled and/or placed in a manner designed to destroy, incapacitate or confuse, using military explosive devices, military or industrial explosives, or homemade materials capable of detonation, or a combination of these, which may contain lethal pyrotechnic or flammable chemical substances or CBRN components which are harmful to health" [7].

The use of body scanners can greatly assist in the detection of explosives and explosive devices. Of course, there have been innovations in this area as well, and although they are still in a few places, they have already been used in test operations. These machines eliminate the need to stop at the test cabin and carry out the protocol previously required. In addition to the continuous passage, they monitor the human body to detect inappropriate discharges and indicate this by means of light and sound signals and on-screen displays [8].

In order to increase the level of security, additional protection against the harmful effects of blast can be provided by defining appropriate protection distances for the priority object. The variation of pressure decreases with distance in inverse and hyperbolic proportion. Although the priority objects have some level of pre-established perimeter protection, the legislator may still provide for other restrictions and protection measures [9] to ensure their road access during qualifying periods. These may include the installation of temporary barriers and cordons to restrict pedestrian traffic or the rapid installation of concrete elements to control vehicular traffic.

In the case of explosive acts inside a building, explosive acts can be expected to occur in particular in the premises designated for access. Blasting in an enclosed environment may result in different values compared to blasting in the open air. The effects resulting from the propagation and reflection of shock waves in some areas of the premises may be several times the primary maximum pressure instead of a multiple of the primary maximum pressure [10, 11].

This is a factor that we should definitely take into account in the event of an intrusion by explosives. The people in the room can influence the way and means of carrying out the intrusion.

The blast resistance of entry points can also be made more resistant by retrofitting. These modifications may include strengthening wall surfaces, door and window frames, and window glass surfaces.

In the case of new buildings or parts of buildings, the mechanical reinforcement can be carried out to a greater extent by professionals. The design of new buildings now takes into account the proportion of volume components and the multi-directional articulation of the space to dissipate the pressure changes generated during detonation. The protection system of the installation must be divided into protection zones. Their protection level should be increased as the area to be protected is approached, both in terms of physical resistance and by reinforcing other additional optical and electrotechnical devices. A risk-based approach is needed to prepare for the expected attack scenarios and to develop the protection and detection capability accordingly. The defence system should be designed with a risk analysis prior to its development and should be designed to be 100% resilient to the expected attacks [12].

The philosophy of security, to which the technical and security equipment of the access areas must be adapted, must be examined.

The location and number of mechanical and technical devices must be planned in such a way that, in the event of an intrusion, they are able to warn and intercept the intruder in good time until the initial intervention forces arrive between the attackers and the area to be protected, in response to the electronic protection signal.

Diverse protection solutions should be used to secure the priority installation and its surroundings, adapting to the environmental conditions and providing redundancy. Such technical solutions may include: infra-passive infra-sensors, camera thermal imaging, pressure sensor-rupture detection equipment and other technical solutions. All of these are designed to ensure the prediction of intrusion in any weather conditions. The tasks, technical capabilities and training of intervention forces vary from one protection sector to another.

3 Attack Against Priority Installations

A complex security system should be based on prevention, with the main objective of preventing the introduction of explosives and structures into a building and the entry of unauthorized persons. According to the principle of defence in depth, as we move towards the "center to be protected", we encounter increasingly high technical and mechanical barriers. Such physical protection barriers, by keeping intruders out, also challenge the response forces. Opening them is sometimes only possible by using forcible entry techniques.

The repertoire of tools used in such cases can be very diverse:

Manual opening tools: ram, hammer, Halligan, etc.
Mechanical tools: gasoline-powered rapid chipper; hydraulic tensioning tools.

Opening of doors and windows by blasting.

The means that can be used may differ from one law enforcement unit to another. In the present study, we will only deal with the forced opening of high-security doors, since this type of doors is the main type used in high-security facilities. As a principle of proportionality, it can be stated that the means of opening must be proportionate to the objective to be achieved, so that "conventional" metal entrance doors can be attacked by hydraulic door opening devices (Fig. 1).

The opening of the hatch shall be in accordance with the tactical concept of the operational unit. The means used to open the hatch determines the outcome of the realization in each individual case. An incorrectly chosen device, which may be due to the existing equipment, the quality of training, lack of experience or disregard for the opinion of the technical specialist, can lead to fatal events. Another method used to forcibly open high security hatches may involve the use of explosives.

Although the methodology for this type of door opening may differ between the military and law enforcement, the principle is the same. In brief, the process is as

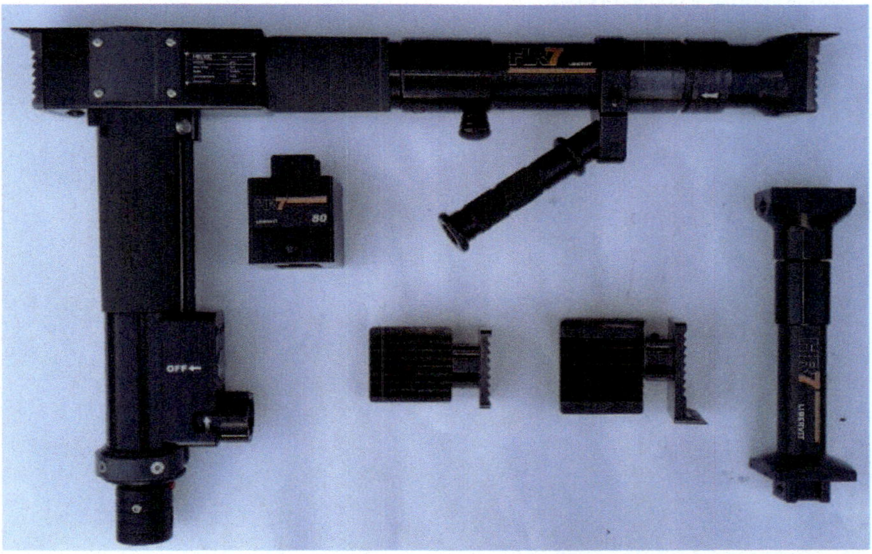

Fig. 1 Hydraulic door opener equipment. Libervit HR7 type hydraulic door opener working cylinder. By the author

follows: the required amount of explosive is applied to the surface of the door leaf, according to the physical parameters of the opening, using the method previously mastered, and is activated by retracting into cover after initialization.

In special cases, the opening of a door by this method may be accomplished by the assembly of a cutting charge or by the use of a pressure charge.

Explosive fuses (hereinafter referred to as RGYZS), plastic explosives (SEMTEX 1A) are mainly used to carry out blasting operations. There may be differences for international law enforcement, military units.

4 Application of the Cutting Charge

The purpose of a cutting charge is to use it to blast through steel or concrete (reinforced concrete) structures. The application of the cutting charge on homogeneous steel plate has a different penetration capacity. It can also be used for demolition work, and is suitable for underwater demolition work. Today, the most common application is in the field of 3D printed cumulative cutting inserts [13].

The main charge, when detonated, forms a cumulative cutting beam from the cone of the charge, which, depending on the size, net explosive content and type of the selected charge, will have a cutting power at a specified depth on the target (Fig. 2).

The linear cumulative beam is formed by a well-dimensioned spacer, which provides the required amount of clearance from the target surface and also performs a target surface locking function. In order to successfully initiate the main charge, a so-called booster is used (Fig. 3).

Fig. 2 Cutter-loaded openings. By the author

Fig. 3 Civilian-type use of a holding charge. By the author

This main charge initiator can be prepared in a prefabricated form or can be made on site from plastic explosive material mounded around the primer.

The cutting charge is fixed to the target surface with self-adhesive tape. Its application may vary from one type of door to another, depending on the structure, material quality and thickness of the door. Its positioning is determined on the basis of the information obtained during the technical reconnaissance prior to blasting, taking into account the characteristics of the quality of the high security door or window. On wet, heavily contaminated surfaces, other fixing methods and materials may be appropriate.

The cutting charge may be initiated by fire, electric ignition or nonel initiation system. The explosive material of the cutter charge is a plastic explosive (e.g. Semtex 1A), therefore the handling is governed by the specifications for Semtex type explosives.

5 Possible Applications of Pressure Fillers

This type of modern and effective linear explosive system is often used in law enforcement because it uses a small number of explosives. Reducing the use of explosives has many advantages: the blasting operation can be carried out without damaging irrelevant property, it is less polluting, and there is less potential for injury to objects and innocent people.

This opening method can be used to open a multi-point metal security door that opens inwards and outwards by blasting. Its effectiveness is based on the principle of the incompressibility of water. The required amount of detonating cord is placed between the liquid-filled bodies and the bodies are then secured by turning them inside each other (Figs. 4 and 5).

Fig. 4 Installing a pressure filling on a metal door

Fig. 5 Post-operational status

Practice opening a metal security door by blasting.

The direction in which the door opens depends on the amount of detonating cord placed in the unit, since the door opening outwards will also move inwards as a result of the explosion, but this requires additional energy. The difference between the amount of detonating cord required to open a metal security door inwards and outwards is 40%. The assemblies are assembled to form a rigid structure for the blasting task. The assembled device is then fixed to the surface of the door and connected to the non-electrically triggered blasting network.

It has the particular advantage that the safety distance can be significantly reduced during its use, making it suitable for use by operational units. In principle, the solutions for opening the illustrated security door meet the requirements of the times.

6 Door Structures of Protected Installations

In principle, the increased reinforcement of perimeter-protected objects also fits the intended use of the protected facility in terms of the capacity of the opening devices. In my scientific research [14], I have not yet come across any results that have examined the opening systems of high-priority installations from the point of view of the use of forced entry techniques. This would provide a theoretical illustration of interventions against such hatches (Fig. 6).

VA door, frontal pressure protection: 0.5 MPa.

The material, wall thickness and locking mechanism of such hatches, in the case of a pressurized charge on the center of the shaft, would require the use of explosives in such quantities to achieve the desired objective that the primary, secondary and tertiary effects of the explosion would preclude their use. Another potential attack surface would be possible by attacking the anchorage points of the door frame in the concrete wall [16]. Access to the anchorage points could be achieved by filling and detonating explosives in the holes made at the anchorage points, but this method is

Fig. 6 Protected installation internal door structure door VA [15]

Fig. 7 VGA door, in a secure facility [15]

also doomed to failure due to possible frame deformation and the length of the task execution [15] (Fig. 7).

VGA door protection class: 0.5 MPa.

7 Conclusion

Protecting critical facilities is key to keeping society safe and functioning. The security and protection systems used in such facilities are responsible for anticipating and preventing attacks. A well-functioning complex security system guarantees the integrity of the facility, but it is also necessary to be prepared for the eventuality that the protected facility is penetrated.

Blasting open doors is just one of the many violent intrusion techniques. It is not a technique that can be used in all cases, as there are circumstances that prevent its use. Blast openings in a closed environment have a much greater detrimental effect on the human body. The gases, shrapnel, debris and smoke generated by the explosive can sabotage the safe penetration process. When used, the net explosive used in an explosive detonation must be proportional to the volume of air available on site. In some cases, the blast may deform the hatch in such a way as to create additional barriers to opening it. In addition, there may also be circumstances—proximity to utilities, flammable materials, etc.—which inherently preclude opening by blasting. I am convinced that the problem of forcibly opening the doors of high-priority facilities must be addressed, because it is a conceivable target for attack from the point of view of terrorists and it is a social expectation that they should be brought to justice.

References

1. 65/2013. (III. 8.) Korm. rendelet a létfontosságú rendszerek és létesítmények azonosításáról, kijelöléséről és védelméről szóló 2012. évi CLXVI. törvény végrehajtásáról. Forrás: net.jogtar.hu/jogszabaly?docid=A1300065.KOR
2. Daruka N (2022) Critical infrastructure protection in the production and use of explosives industry products. In: Security-related advanced technologies in critical infrastructure protection. NATO science for peace and security series C: environmental security. Springer, Dordrecht, pp 297–313. https://doi.org/10.1007/978-94-024-2174-3_26
3. Zsuzsanna B (2013) Objektumok robbantásos cselekmények elleni védelmének lehetőségei. Doktori (Ph.D.) értekezés, Budapest. Forrás: http://archiv.uni-nke.hu/downloads/konyvtar/digitgy/phd/2013/balogh_zsuzsanna.pdf
4. László L, Zsuzsanna B (2018) Turisztikai létesítmények, események robbantásos fenyegetettsége (Blasting menaces of touristic places, events). Műszaki Katonai Közlöny XXVIII. 2018/2, pp 238–251
5. Kovacs Z (2016) Explosion of improvised explosive device on structures. In: Hadmérnök (2016) 1788–1919, vol 11, no 1, pp 56–63. http://www.hadmernok.hu/161_06_kovacsz.pdf
6. Norbert D (2012) Bűnös célú robbanószerkezetek alkalmazásának és hatástalanításának sajátosságai, "Fúrás-robbantástechnika 2012". In: XI. Nemzetközi Konferencia Balatonkenese 2012, HU, Szept 19–21. ISSN 1788-5671, pp 109–118
7. Norbert D, Gyula KJ (2020) A kínai termékek hatása a nemzetközi hadszíntéren felbukkanó házilagosan készített robbanószerkezetek kialakítására. In: Trhacia Technika 2020. Slovenska spolocnost pre trhacie a vrtacie prace, Slovak Republik, pp 79–89. ISBN 978-80-89914-08-1
8. Ákos B, József D, Tamás F (2022) Személybeléptetés során alkalmazható robbanóanyag és robbanószerkezet felderítésére alkalmas biztonsági ellenőrzési eljárások összehasonlítása. In: I. Fúrás-Robbantástechnika Nemzetközi Szimpózium 2022, Velence, pp 217–231. ISBN 9786150160092
9. Fundamental law of Hungary. Dordrecht. https://net.jogtar.hu/jogszabaly?docid=a1100425.atv
10. András D (2019) Kiemelten védett objektumok robbantás elleni védelmének kiegészítése a biztonsági szint növelése érdekében. Műsz Ktn Közlöny 29(1):155
11. Zoltán S (1992) Zárt robbantóterek nyomásviszonyainak számítógépes meghatározása. Előadás az Építéstudományi Egyesület (ÉTE), Robbantástechnikai szakbizottság, 5. Nemzetközi Robbantástechnikai Kollokviumán, Budapest, 08–11 Sept 1992. Megjelent a konferencia kiadványában
12. Lilla R (2018) Létfontosságú rendszerek és létesítmények védelmével kapcsolatos kockázatelemzési módszertan szakirodalmának bemutatása, Védelem tudomány, III: évfolyam 3. szám 1–21 old
13. István E (2022) Modern kumulatív töltetek hatékonyságának vizsgálata. Haditechnika 56(6):15–20
14. Richárd P (2017) Objektumok védelmének eszközei és lehetőségei a bűnös célú/terror jellegű robbantásokkal szemben. Doktori (PhD) értekezés, Budapest. Forrás: http://lib.uni-obuda.hu/sites/lib.uni-obuda.hu/files/Peto_Richard_ertekezes.pdf
15. Ferenc K (2023) Védett létesítmények előadás (PPT) Nemzeti Közszolgálati Egyetem Katonai Műszaki Doktori Iskola, 20 Dec 2023
16. Ákos B (2023) Nukleáris létesítmények bűnös szándékú robbantásos cselekmények elleni védelmének nemzeti szabályozása. In: II. Fúrás-Robbantástechnika Nemzetközi Szimpózium 2023, Budapest, pp 118–129. ISBN 9786150187471

Protection Management Against Drones in Industrial Facilities

István Szávay and Arnold Őszi

Abstract Examining the complex security challenges posed by the growing use of drones in industrial installations, risk management and effective drone defence. Identifying the various threats with implications for industrial facilities. The study analyses passive and active drone detection and countermeasure technologies, highlighting their advantages and limitations. Due to the rapid proliferation of drones, the need for effective detection and protection mechanisms is growing. Emphasis will be placed on the importance of detection technologies, risk assessment and response plans to counter threats posed by drones. A detailed risk analysis framework is presented, including a risk matrix for assessing the likelihood and impact of drone-related incidents. The research emphasises the importance of proactive measures, including incident preparedness and response planning, to minimise the risks associated with unlicensed drone activities. The study concludes with recommendations for continuous updating of security plans and prioritisation of security resources. This comprehensive approach aims to enhance the resilience of industrial installations to the threats posed by drone technology.

Keywords Threats · Drone detection · Defense strategies · Risk assessment · Non-conventional defense

I. Szávay (✉)
Doctoral School on Safety and Security Sciences, Obuda University, Budapest, Hungary
e-mail: szavay.istvan@phd.uni-obuda.hu

A. Őszi
Banki Donat Faculty of Mechanical and Safety Engineering, Obuda University, Budapest, Hungary
e-mail: oszi.arnold@bgk.uni-obuda.hu

T. A. Kovács et al. (eds.), *The Impact of the Energy Dependency on Critical Infrastructure Protection*, Advanced Sciences and Technologies for Security Applications, https://doi.org/10.1007/978-3-031-78544-3_53

1 Introduction

The explosion of drone use began in the last decade and continues today. The protection of industrial installations from the harmful effects of drones is particularly important, as these drones have the potential to change traditional security approaches and toolkits, and raise new issues of both protection and prevention. It is of paramount importance for industrial installations to understand and manage these risks.

Drones have the potential to breach traditional security frameworks and pose a threat of such magnitude that the risks are many and varied. The techniques for detecting and managing drones are diverse. The use of passive and active defence systems against them has become a necessity. When planning the protection of industrial installations against drones, priority is given to risk analysis and the appropriate development and ongoing management of protection plans.

2 The Threats Posed by Drones

2.1 *Observation, Industrial Espionage*

Nowadays, drones are available on the internet in a wide range of types and configurations, allowing the average user to discreetly record camera footage, listen in on sensitive conversations or even carry out industrial espionage. The importance of camera and audio recordings made by drones is particularly significant for companies in a competitive business environment. With the help of these tools, a company may be able to gain technological insights and knowledge about its rivals' remote areas in order to gain an advantage in the market competition. Entering the market early or disrupting competitors through such information can provide significant economic benefits. This trend is expected to continue given the continued and intense competition between technology firms.

2.2 *Unauthorised Use of Airspace, Negligence, Armed Attack*

Today, with enough determination and research on the Internet, it is possible to get the idea of building a drone, describing its components, capable of carrying and using a live fire weapon. Advanced control systems perform counter-steering when firing a shot, thus reducing instability caused by recoil. Although such a device may be inaccurate in motion, it can hit a stationary figure at short range [1].

In addition, a wooden drone can be built from ordinary parts, even in the home, using garden tools, with a load of cargo under its wing to carry out an attack [2].

Careless incidents are usually unpredictable. Drone flights into prohibited areas can not only cause material damage, but in more serious cases can also lead to personal injury or, in extreme cases, death. Preventing the accidental or deliberate entry of drones into airspace where there is a risk of injury to the general public is of paramount importance [3].

These events underline the need for increased attention to aviation safety, particularly with regard to drones as a potential safety risk.

3 Passive Drone Detection

Acoustic detection is used to detect the presence of drones by taking advantage of the distinctive sound emitted by the engine, which can be detected even in noisy environments. Detection consists of three main components: long-range microphones, a processor that processes sound waves, and a sound pattern database. The microphones can be either short-range or long-distance microphones. Sound samples are identified by computer processing and, after successful identification, the system can generate alarms or take action [4–6].

Radar detection is done by surveillance radars with different antennas that use radio waves to detect flying objects. The types of radar include fixed and rotating radars. Frequency modulated continuous wave radar and Micro-Doppler radar are used to detect small cross-section objects. Monostatic pulse radar has strong signal transmission and reception capabilities, but also includes blind spots due to switching between transmission and observation [6–8].

Radio frequency detection is an effective way to detect drones. An RF-based drone detection system can monitor the communication frequency spectrum to detect drones by measuring the frequency of radio waves and the length of communication packets. However, these methods are not always accurate because other devices emit similar signals, which can cause false detections. Another WiFi-based method identifies drones based on their WiFi characteristics. It is important to note that some drones use a special frequency-shifting technique, which also needs to be taken into account during detection. The disadvantage is the limited ability to detect many drones at the same time, as well as interference caused by obstructions to the line of sight [6, 9].

Electro-optical (EO) sensors operate in the visual and infrared spectrum. EO sensors are not primary detection sources, but rather provide visual confirmation. Thermal sensors use the non-visible electromagnetic spectrum and are particularly useful for detection independent of ambient illumination. However, thermal cameras can be more expensive and have lower resolution compared to optical cameras [10].

To achieve effective detection of drones, a combination of different sensor technologies is recommended, especially in complex environments. The advantages and disadvantages of each technology are summarised in Table 1, emphasising that the use of a single sensor technology alone is not sufficient for full detection.

Table 1 Comparison of passive counter measures

Method	Benefits	Disadvantages
Radar	Detect and track multiple objects simultaneously (360°)	The most expensive drone sensors
	Operation in adverse weather conditions	Unable to locate the drone operator's position
	Long range and accuracy	The rotating antenna requires maintenance
	Separating drones and birds	Can't see behind the obstacles
RF detection	Their operation is not affected by the time of day or the weather	RF signal required, cannot detect autonomous flying drones
	Ability to locate the UAV's remote pilot	Interference and vision loss impair perception
	High accuracy detection	Detection restrictions on the drone swarm
	Triangulation is possible with multiple RF sensors	Less effective in strong radio frequency environments
Acoustic	Low cost	Limited range
	Recognises commercially available drones	Future drones may be quiet
	Does not require direct visibility	Unsuitable for detection of fixed-wing drones
	Installation can be fixed or mobile	Sensitive to rain and wind, high background noise
Electro-optical	Long range	Performance weakens in bad weather conditions
	Classification and detection in software	Many pixels are needed to cover a wide field of view
	Night vision is possible with IR	High maintenance requirements due to lenses

4 Active Countermeasures

Today, there are many innovative and effective solutions available in the field of drone deterrence technologies that can prevent unwanted drone intrusions in various ways. In both civil and military applications, security and legal compliance are important considerations.

The primary and perhaps most preventive method is to inform and raise awareness of the no-fly zones and restricted flight zones.

Jamming the communication signal is another effective means of interrupting the data exchange between the drone and its operator. In this case, the purpose of jamming is to trigger the drone to activate manual or automatic return routines [11,

12]. However, the use of these technologies may be subject to legal restrictions, especially in civilian life.

Coordinate spoofing, which uses fake signals aimed at the GPS system to interfere with drone navigation, is also an effective countermeasure. However, long-range jamming can cause environmental interference that can affect other devices.

By taking control of the drone, or using a hijacking method, control of the drone can be gained, thus preventing its original purpose [11–13]. This technique is particularly relevant in the field of cyber security.

Mechanical defeat methods include the use of net-launching weapons, which allow an intruding drone to be intercepted without crashing. Kinetic countermeasures, such as the use of conventional weapons and laser guns, although effective, raise legal and ethical issues, especially in civilian applications [14].

This suggests that there are a number of ways of deterring drones, the choice between which depends on the purpose, the legal framework and environmental factors. Technological progress promises further innovations in this field (Table 2).

5 Protection Against Drones in Industrial Facilities

The risk factors for drone incidents can be grouped as follows:

- Incidents due to negligence: in these cases, the perpetrator either does not know or does not respect the rules and regulations governing the operation of drones. As a result, they often fly their drones in prohibited areas without intentionally disturbing the order or integrity of industrial facilities.
- Deliberate incidents: include cases where offenders deliberately ignore the rules for personal or professional gain. This category also includes incidents where the aim is to disrupt the operation of the establishment without endangering human life.
- Crime-related incidents: this category includes incidents where the perpetrators, whether or not they are aware of the rules, actively seek to neutralise protection measures and cause harm. These cases often involve explicit threats to human life and property, such as terrorist acts.

A primary consideration when developing protection strategies is to identify and distinguish between different types of incidents. The effectiveness of protection measures is determined not only by technological sophistication, but also by a properly designed and managed response plan. This includes the preparedness of participants and the correct execution of tasks.

Table 2 Comparison of active detection technologies

Method	Benefits	Disadvantages
Signal jamming and management monitoring	The least destructive operation	Does not work on custom drones
	Widely available technology and knowledge	Correct flight heading data is required for guidance to a safe place
	Landing control enables a safe stop	Drone protection against hijackings could be strengthened in the future
Drone acquisition	Low miss rate when getting close to the drone	The net launcher must be relatively close to the drone
	Possibility to intercept multiple drones with one net	Ineffective against fast and agile drones
	Mobility	During an operation, only one mesh size
	No range limitation	The drone is likely to be damaged during the operation
Conventional weapons	Mesh ammunition that can be fired from a conventional handgun	Crashing drones are a safety hazard
	High mobility	Poor performance against moving drones
	Easy to learn and use	Difficult to hit fast moving targets
	Immediate feedback on success	May not be used for destruction in civil applications
	Availability of existing technology and methodology	Very short range
Laser weapons	Long range and accuracy	High energy consumption
	Speed of light, fast reuse	High costs
	Can be mounted on a stand or on a vehicle	High accuracy even from a moving vehicle

5.1 *Preventive Measures and Incident Preparedness*

To prevent this, a comprehensive risk assessment is required, taking into account, among other things, the location, size and operational aspects of the installation. The risk assessment is the basis for the development of the protection strategy, which requires regular review. It is important to develop cooperation with local authorities and aviation organisations and to introduce restrictions against unauthorised drone operations, for example by designating "No Drone" zones.

An effective physical security system should also take into account internal security measures, such as reducing the possibility of access through windows. To protect external areas, it is important to use appropriate shielding and covering techniques.

The development of drone surveillance mechanisms is key, including the involvement of civilians. This includes procedures and protocols for detecting, identifying and responding to possible drone presence. It is important to map the possible locations of pilots in advance so that they can be identified more quickly should an incident occur.

5.2 *Incident Handling*

When a drone incident occurs, a prompt and effective response is essential. Rapid processing of the information gathered during the sighting helps to accurately determine the situation. Continuous evaluation of the position and movements of suspected unlicensed drones is necessary to determine the risk and threat level. The use of responses appropriate to the threat level and coordinated action between security forces and defence management will ensure an effective response. Once the incident has been dealt with, it is important to restore normal operations as soon as possible.

5.3 *Aftermath of the Event: Documentation and Analysis*

After the incident has been closed, evaluation and analysis of the available information is essential to better prepare for future incidents. On the basis of this information, the security concept should be revised and updated if necessary.

Notifying the authorities and handing over documentation is an important step in initiating official procedures, which can also act as a deterrent to prevent future incidents. Public information campaigns can help to prevent unintentional drone operations and to promote public assistance and cooperation in detection and reporting.

6 Risk Analysis and Protection Design

Activity-based risk analysis is key to maintaining the safety of industrial installations. These facilities can take a variety of forms, such as being open or adjacent to populated areas, which can involve different risk factors. Taking these environmental characteristics into account in the risk assessment is essential. An integrated approach that takes into account the company as a whole and not just parts of it is essential to effectively manage the different risks.

Identifying risks involves assessing sources, scopes and consequences. To determine the level of risk, the risk matrix provides an appropriate result [15], which can take into account the specific characteristics of different industries. In this model, risk can be interpreted as the product of probability and consequence, where the assessment of both factors may be subjective but can be refined by data collection.

For an industrial facility, the probability elements are defined in Table 3 with 5 levels to categorise the probability of risk based on previous observations of events. This categorization is done in five levels, from the "Sure" level, indicating frequent events at semi-annual intervals, to the "Rare" level, where the occurrence of events is not typical for the industry.

The impact of the events on the industrial establishment and its consequences are assessed in Table 4, which gives the weighting of the consequence in terms of impact areas, i.e. the consequences of the risks in three key areas: business, finance and brand/reputation. Four levels of consequence—"Severe", "High", "Moderate", and "Low"—detail the impact that different levels of risk can have on an organisation's operations, financial stability, and market perception.

Of course, the scope can be modified and extended according to the industry and the activity carried out. The probability of occurrence and its expected consequences

Table 3 Probability scale for risk analysis

Level	Probability	Comment
5	Sure	In this industry or similar, 5–6 detections in 0.5 years
4	Probably	2–3 sightings in 1 year in this industry or similar
3	Possible	4–5 sightings in 5 years in this industry or similar
2	Unlikely	2–3 detections in 10 years in this industry or similar
1	Rare	Not found in this industry or similar

Table 4 Consequence scale for risk analysis

Level	Result	Boundary areas		
		Business activity	Finance	Brand/fame
4	Serious	Complete stop of operation; stop of essential operations	Severe financial loss; significant loss of revenue	Loss of customer loyalty; failure to win new customers
3	High	Shutdown of key operations; service delays	Financial loss; significant loss of revenue	Loss of trust from existing and new customers
2	Moderate	Reduced, lower power operation	Financial loss, but the impact can be reduced by reallocating resources	Slight loss of customer loyalty; damage to corporate image
1	Low	The existence of the organisation is not at risk	Minor financial loss; unlikely to affect budget or business activities	Little or no impact on customer loyalty

are summarised in the risk matrix presented in Table 4. The matrix defines the level of risk as the probability of the event occurring multiplied by the consequence of the event, i.e.:

$$\text{Risk} = \text{Probability} \times \text{Consequence}$$

The risk matrix combines probability and severity of consequences. The matrix provides quantified values to characterise the overall impact of the risk, allowing prioritisation of different levels of risk and planning of appropriate responses. The severity of the levels is indicated by the colouring, with orange indicating severe and light blue indicating low (Table 5).

Based on the levels identified in the risk assessment, the organisation must decide on strategies to manage acceptable and unacceptable risks. Unacceptable risks may require immediate action, while others can be addressed later.

Table 6 provides a risk management system that defines the classification of the severity of risks and how they are managed based on the value of the risk matrix levels and their colour code. The system distinguishes between four different risk classifications based on the severity of the risk, which categorise risks as ‘Severe’, ‘High’, ‘Moderate’ and ‘Low’. Each category represents a specific risk score range derived from the risk assessment matrix.

Risk management is a continuous process in the life of a facility, which includes monitoring potential threats and preparing for prevention and response. As technology evolves, security managers must constantly be prepared for new threats and take appropriate action.

Protecting civilian facilities from drone attacks poses technical and legal challenges. Threat analysis and the development of an appropriate response plan are essential, including the design of defense in depth and the elimination of potential detection blind spots.

Table 5 Risk matrix

Result		Probability				
		1 Rare	2 Unlikely	3 Possible	4 Probably	5 Sure
4	Serious	4	8	12	16	20
3	High	3	6	9	12	15
2	Moderate	2	4	6	8	10
1	Low	1	2	3	4	5

Table 6 Risk classification for assessment

Risk classification	Degree	Risk management
12–20	Serious	Immediate corrective action required
8–10	High	Corrective action is needed; risk needs to be monitored and reassessed
3–6	Moderate	Corrective action required within 1 month; risk to be monitored and reassessed at a later date
1–2	Low	No remedial action required at present; monitoring of risk

Planning for the prioritised protection of installations is also an important step, involving the identification of sensitive areas and the appropriate allocation of protection resources. Technological, legal and environmental factors must be taken into account in the selection and planning of protection means.

Managing the risk of drone threats in different industrial facilities requires a unique approach that takes into account the specific circumstances and challenges of the facility. The design of protection measures requires an analysis of potential launch points, such as open or high altitude areas, and the optimal placement of detection equipment. This will help to identify and intercept aircraft before they encounter sensitive areas of the installation.

Updating the facility security plan, including drone response measures, is also important. This ensures that all personnel know and understand security procedures. The deployment of an integrated protection system and the use of an associated plan can facilitate in-depth detection and reduce response time in the event of an incident.

As illustrated in Fig. 1, the principle of gradualism helps to prioritise the protection of individual installations, so that the potential damage caused by a drone incident can be minimised by focusing protection resources on the most critical areas.

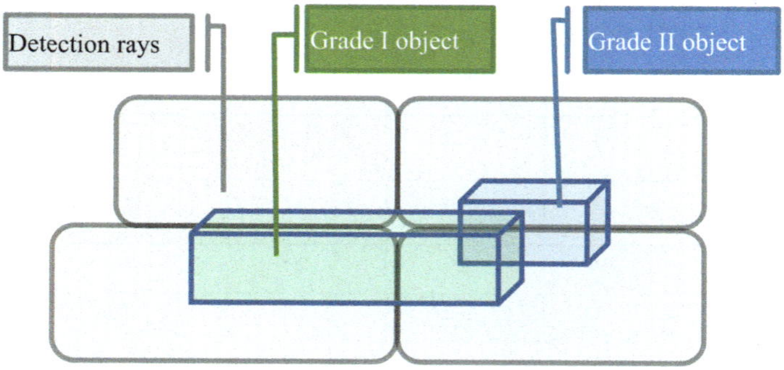

Fig. 1 Illustration of building classification and area protection

In determining the protection risk level of each object and, through this, in planning the placement of physical protection equipment, each level is determined according to the need for protection:

- I. The object or area within the facility where the likelihood of a drone incident occurring is greatest or which is most vulnerable to the threat,
- II. The security-sensitive part of the facility,
- III. The remaining parts of the facility.

Grading can indicate where to focus detection and protection resources to minimise the potential damage caused by a drone incident.

When selecting and designing protection devices, it is important to take into account environmental and technological factors, such as potential radio interference between radars and other sensors, as well as legal regulations and the influence of landmarks. A coordinated application of all these will ensure effective protection of facilities against drone threats, while maintaining operational safety and efficiency.

7 Summary

Drones are extremely versatile and can perform a wide range of tasks. Low cost, easy availability and reduced risk to human life have led to the mass adoption of commercial drones. However, these features also make them attractive for malicious use.

The frequency, complexity and severity of incidents involving drones in industrial installations are expected to increase in proportion to their proliferation and the development of their capabilities. However, effective risk management and flexible measures can reduce the negative impact of these incidents.

Protecting industrial installations against unwanted drone activity is a broad and deep problem. The variety of protection solutions available means that facility security experts are expected to build a cost-effective technological and decision-making response that is in line with legislation and proportionate to the scale and scope required based on risk assessment.

Two groups of countermeasures have been identified based on the characteristics of the measures they can implement: passive and active solutions.

For the implementation of effective countermeasures, the time factor between detection and response is decisive, i.e. the earlier the detection, the more effective the countermeasure response and thus the more effective the drone deterrence system.

All detection technologies have their drawbacks and therefore, when used alone, they are not a complete solution for drone detection. Only combined sensors have the potential to complement each other to form a coherent system of passive devices with active countermeasure systems for drone defence.

However, the introduction of state-of-the-art technology will not achieve the expected results unless a drone incident response plan is designed and implemented based on the results of an industrial facility risk analysis.

References

1. Trackimo homepage. Turning normal drones into weaponized drones. Accessed 10 Sept 2023. https://trackimo.com/weaponized-drones/
2. Trevithick J. Thedrive homepage. Russia offers new details about Syrian mass drone attack, now implies Ukrainian connection. Accessed 23 Oct 2023. https://www.thedrive.com/the-war-zone/17595/russia-offers-new-details-about-syrian-mass-drone-attack-now-implies-ukrainian-connection
3. Goudlijian A. Maxifoot homepage. CdF: quand un drone interrompt Vannes-PSG. Accessed 23 Nov 2023. https://www.maxifoot.fr/football-rss360837/cdf-quand.php
4. Park S, Kim HT, Lee S, Joo H, Kim H (2021) Survey on anti-drone systems: components, designs, and challenges. IEEE Access 9:42635–42659. https://doi.org/10.1109/ACCESS.2021.3065926
5. Khan MA, Menouar H, Eldeeb A, Abu-Dayya A, Salim FD (2022) On the detection of unauthorized drones—techniques and future perspectives: a review. IEEE Sens J 22(12):11439–11455. https://doi.org/10.1109/JSEN.2022.3171293
6. Wang J, Liu Y, Song H (2021) Counter-unmanned aircraft system(s) (C-UAS): state of the art, challenges, and future trends. IEEE Aerosp Electron Syst Mag 36(3):4–29. https://doi.org/10.1109/MAES.2020.3015537
7. Riabukha VP (2020) Radar surveillance of unmanned aerial vehicles (review). Radioelectron Commun Syst 63:561–573. https://doi.org/10.3103/S0735272720110011
8. Zhang X, Kusrini K (2021) Autonomous long-range drone detection system for critical infrastructure safety. Multimed Tools Appl 80:23723–23743. https://doi.org/10.1007/s11042-020-10231-x
9. Basak S, Rajendran S, Pollin S, Scheers B (2022) Combined RF-based drone detection and classification. IEEE Trans Cogn Commun Netw 8(1):111–120. https://doi.org/10.1109/TCCN.2021.3099114
10. Elsayed M, Reda M, Mashaly AS, Amein AS (2021) Review on real-time drone detection based on visual band electro-optical (EO) sensor. In: 2021 tenth international conference on intelligent computing and information systems (ICICIS), Cairo, Egypt, pp 57–65. https://doi.org/10.1109/ICICIS52592.2021.9694151
11. Shashok N (2017) Analysis of vulnerabilities in modern unmanned aircraft systems. Tuft University, pp 1–10
12. Pojsomphong N, Visoottiviseth V, Sawangphol W, Khurat A, Kashihara S, Fall D (2020) Investigation of drone vulnerability and its countermeasure. In: 2020 IEEE 10th symposium on computer applications & industrial electronics (ISCAIE), Malaysia, pp 251–255. https://doi.org/10.1109/ISCAIE47305.2020.9108835
13. Dey V, Pudi V, Chattopadhyay A, Elovici Y (2018) Security vulnerabilities of unmanned aerial vehicles and countermeasures: an experimental study. In: 2018 31st international conference on VLSI design and 2018 17th international conference on embedded systems (VLSID), Pune, India, pp 398–403. https://doi.org/10.1109/VLSID.2018.97
14. Robbe C, Papy A, Nsiampa N (2018) Using kinetic energy non-lethal weapons to neutralize low small slow unmanned aerial vehicles. Hum Factors Mech Eng Def Saf 2(1):1. https://doi.org/10.1007/s41314-018-0008-y
15. Duijm NJ (2015) Recommendations on the use and design of risk matrices. Saf Sci 76:21–31. https://doi.org/10.1016/j.ssci.2015.02.014

Quantum Computers and IT Security: Unlocking the Future, Exploring the Questions

Sára Kornélia Szatmáry

Abstract The evolution of secure communication has transitioned from classical cryptographic systems to the innovative realm of quantum cryptography, leveraging principles of quantum mechanics. Quantum computers pose a significant threat to traditional cryptographic systems, leading to a transformative phase known as post-quantum cryptography. Quantum key distribution (QKD) protocols, central to quantum cryptography, offer a fundamentally different approach to secure communication, providing unparalleled security for the EU's sensitive communications, financial transactions, and critical infrastructure. Quantum cryptography's resilience against quantum computing threats positions the EU as a cybersecurity leader. Investing in research and development can bolster technological leadership and global competitiveness, revolutionizing digital network security. Practical challenges in quantum cryptography implementation, such as specialized hardware requirements and security considerations, must be addressed. The development of post-quantum cryptographic algorithms and global collaboration efforts are essential for navigating the quantum technology landscape and ensuring a secure quantum era. In conclusion, while quantum cryptography holds immense promise, realizing its full potential requires concerted efforts to address challenges, standardize resilient algorithms, and foster global collaboration. Through strategic collaboration and innovation, the EU can position itself at the forefront of quantum technology, safeguarding communication networks and driving advancements in cybersecurity for years to come.

Keywords Quantum cryptography · Quantum · Cryptographic algorithms

S. K. Szatmáry (✉)
Obuda University, Budapest Bécsi út 96/B, 1034, Hungary
e-mail: szatmary.sara@stud.uni-obuda.hu

T. A. Kovács et al. (eds.), *The Impact of the Energy Dependency on Critical Infrastructure Protection*, Advanced Sciences and Technologies for Security Applications, https://doi.org/10.1007/978-3-031-78544-3_54

1 Introduction

The evolution of secure communication has transitioned from classical cryptographic systems, reliant on mathematical principles, to the innovative realm of quantum cryptography, rooted in the principles of quantum mechanics. While classical systems have long been trusted for their resilience, the advent of quantum computers poses a significant threat due to their potential to break widely-used public-key cryptographic systems. In response, the field of cryptography is undergoing a transformative phase known as post-quantum cryptography, aiming to develop algorithms capable of withstanding the computational power of quantum machines.

Quantum cryptography, leveraging quantum properties like photon polarization, offers a fundamentally different approach to secure communication. Quantum key distribution protocols, a cornerstone of quantum cryptography, create shared secret keys between parties, with security underpinned by quantum principles, making eavesdropping detection feasible [1]. The EU stands to benefit greatly from quantum cryptography, as it provides unparalleled security for protecting sensitive communications, financial transactions, and critical infrastructure.

One of the key advantages of quantum cryptography lies in its resilience against quantum computing threats, unlike traditional cryptographic systems vulnerable to such attacks. By embracing quantum cryptography early, the EU can solidify its position as a cybersecurity leader, safeguarding digital infrastructure against emerging threats and ensuring communication network resilience. Moreover, investing in quantum cryptography research and development can bolster the EU's technological leadership, fostering innovation and collaboration across academia, industry, and research institutions.

On a global scale, quantum cryptography is pivotal for maintaining the EU's competitiveness, as nations worldwide race to deploy quantum-based solutions. Integration of quantum cryptography into communication infrastructure can revolutionize digital network security, offering highly secure methods for exchanging cryptographic keys through QKD protocols.

This paper will tell, how, by-embracing quantum technologies and investing in research and development can position the EU as a pioneer in the field, reaping economic, strategic, and societal benefits. Through collaboration and innovation, the EU can navigate the digital landscape with confidence, ensuring security, prosperity, and leadership in the digital era.

2 Classical Cryptography

Cryptography is the practice and study of techniques for secure communication in the presence of third parties, often referred to as adversaries. The goal of cryptography is to ensure the confidentiality, integrity, and authenticity of information. There are two main types of cryptography [2, 3].

2.1 *Symmetric-Key Cryptography*

Symmetric-key cryptography, also referred to as secret-key or private-key cryptography, operates on the principle of using a single secret key for both encrypting and decrypting messages. The secrecy of this key is paramount, as possessing it grants the ability to decipher encrypted messages [4].

This method finds practical application in secure communication within closed systems or networks, where the management of key exchange is more straightforward. However, its effectiveness is somewhat limited in open network scenarios due to challenges in securely distributing the secret key. In such cases, public-key cryptography tends to be the preferred choice.

2.2 *Public-Key Cryptography*

Public-key cryptography, also recognized as asymmetric cryptography, introduces a unique approach to secure communication by employing a pair of keys: the public key and the private key. These keys, intricately linked through mathematical relationships, serve distinct roles, collectively enhancing security beyond the capabilities of symmetric-key cryptography.

This sophisticated cryptographic technique finds extensive application in diverse scenarios. It plays a pivotal role in securing online communication, facilitating digital signatures, powering key exchange protocols, and ensuring secure access to online services. The hallmark of public-key cryptography lies in its proficiency to establish secure communication over open networks, positioning it as an indispensable component within contemporary cryptographic systems [5].

3 Classical Versus Quantum Cryptography

Classical cryptography and quantum cryptography represent two distinct paradigms in securing communication, each with its own set of principles, technologies, and implications. Classical cryptography relies on unproven assumptions of computational complexity, where security is based on the difficulty of solving mathematical problems such as factoring large numbers or solving discrete logarithms. However, this reliance on computational complexity leaves classical cryptographic systems vulnerable to potential advancements in algorithms and computing technology. As computing power increases, classical cryptographic systems may become susceptible to attacks that exploit weaknesses in their underlying mathematical foundations [6].

In contrast, quantum cryptography operates on the principles of quantum mechanics, providing a fundamentally different approach to cryptographic tasks. Quantum cryptography leverages quantum phenomena such as superposition and entanglement to establish secure communication channels that are theoretically immune to undetectable eavesdropping. Unlike classical cryptography, which relies on mathematical complexity assumptions, the security of quantum cryptography is rooted in the laws of physics, offering a higher degree of protection against emerging threats, including those posed by quantum computers.

One key distinction between classical and quantum cryptography lies in their respective approaches to key generation. In classical cryptography, key generation must account for potential advancements in algorithms and computing technology throughout the entire duration of secrecy requirements, necessitating a forward-looking approach to system design. In contrast, quantum key generation only needs to consider the technology available at the precise moment of key generation, as the security of quantum keys is guaranteed by the laws of physics.

The potential threat posed by quantum computers to classical cryptography is significant. Quantum computers have the potential to efficiently solve certain mathematical problems that underpin widely used cryptographic algorithms, such as factoring large numbers or solving discrete logarithms. Notably, algorithms like Shor's algorithm and Grover's algorithm could compromise the security infrastructure of classical cryptographic systems, rendering many existing encryption methods vulnerable to attacks.

In summary, classical and quantum cryptography offer distinct approaches to securing communication, with classical cryptography relying on mathematical complexity assumptions and quantum cryptography leveraging the principles of quantum mechanics. While classical cryptography is vulnerable to potential advancements in algorithms and computing technology, quantum cryptography provides a more secure alternative, especially in the face of emerging threats posed by quantum computers. As the field of quantum computing continues to advance, the importance of quantum cryptography in ensuring secure communication channels becomes increasingly apparent.

4 Quantum Cryptography

The emergence of quantum computers presents both opportunities and challenges for the field of quantum cryptography, a field at the intersection of quantum mechanics and cryptography, which is underpinned by the foundational principles of quantum mechanics. Understanding these principles is crucial for comprehending the novel approach that quantum cryptography offers to secure communication.

At its core, quantum mechanics describes the behavior of particles at the quantum level, where classical physics breaks down. One of the fundamental principles of quantum mechanics is superposition, which states that particles can exist in multiple states simultaneously until measured, collapsing into a definite state upon observation. Another principle is entanglement, where particles become correlated in such a way that the state of one particle is dependent on the state of another, regardless of the distance between them.

4.1 Quantum Key Distribution (QKD)

Quantum Key Distribution represents an innovative approach to secure communication, harnessing the fundamental principles of quantum mechanics to facilitate the creation of a shared, randomly generated secret key between two parties. The bedrock of its security lies in the core tenets of quantum mechanics, rendering it theoretically resilient against specific types of attacks, even those that could be mounted by quantum computers [7, 8].

In the realm of quantum cryptography, or specifically QKD, a distinct paradigm unfolds. This methodology utilizes the quantum properties of particles, such as photons, to orchestrate the establishment of a shared secret key between communicating entities. The security assurance of QKD comes from the inherent principles of quantum mechanics, where any attempt to intercept the key introduces disturbances to the quantum states, promptly alerting the involved parties to potential eavesdropping [9, 10].

This quantum-inspired strategy not only offers a cutting-edge alternative to classical cryptographic methods but also introduces a level of security deeply rooted in the enigmatic behavior of quantum particles, showcasing the potential for highly secure communication channels in the quantum realm.

In a typical QKD protocol, such as the BB84 protocol developed by Charles Bennett and Gilles Brassard in 1984 [11], Alice wishes to securely communicate with Bob. To begin, Alice prepares a stream of qubits (quantum bits) encoding random bits of information, using various quantum properties like photon polarization or particle spin. She then sends these qubits to Bob over a quantum channel.

However, due to the principles of quantum mechanics, any attempt by an eavesdropper, Eve, to intercept and measure these qubits would inevitably disturb their quantum state. This disturbance introduces errors in the qubits received by Bob, allowing Alice and Bob to detect Eve's presence. Therefore, if Alice and Bob observe a high level of error in their received qubits, they can infer that Eve attempted to intercept the transmission.

Following the transmission of qubits, Alice and Bob engage in a public discussion to compare a subset of their transmitted and received qubits. By openly revealing which qubits were sent and received, without disclosing their actual values, they can identify and discard any qubits that may have been intercepted by Eve.

Through this process, Alice and Bob can distill a subset of error-free qubits that form the basis of their shared secret key. This key can then be used to encrypt and decrypt messages securely, without fear of interception by Eve or any other potential eavesdropper.

Key Generation Process in Quantum Key Distribution

1. Quantum Superposition: QKD relies on the principles of superposition and uncertainty in quantum mechanics. The sender (often called Alice) prepares a stream of quantum bits, or qubits, that can exist in multiple states simultaneously.
2. Quantum Entanglement: QKD may also use the concept of quantum entanglement. Pairs of qubits can be entangled, meaning the state of one qubit is directly related to the state of the other, regardless of the physical distance between them.
3. Transmission: Alice sends the qubits to the receiver (often called Bob) over a quantum communication channel. Due to the principles of quantum mechanics, any attempt to intercept or measure the qubits by an eavesdropper (often called Eve) would disturb their states, and this disturbance can be detected.
4. Quantum Measurement: Bob receives the qubits and performs measurements based on the quantum states he receives. Since he knows the entanglement or the quantum states prepared by Alice, he can make measurements that allow him to obtain a random set of bits.
5. Quantum Key Verification: Alice and Bob then communicate over a classical channel (non-quantum communication channel) to compare a subset of their bits. If these bits match, it indicates that the key exchange was successful and secure.
6. Error Detection: QKD systems also include error detection and correction mechanisms to account for potential errors introduced during transmission.

In summary, the theoretical foundations of quantum cryptography are rooted in the principles of quantum mechanics, particularly superposition and entanglement. Quantum key distribution protocols leverage these principles to establish secure communication channels by detecting any attempts at interception. By harnessing the unique properties of quantum mechanics, quantum cryptography offers a promising avenue for achieving unprecedented levels of security in communication.

4.2 Current Trends in Quantum Cryptography

Current trends in quantum cryptography reveal a landscape characterized by significant advancements and promising developments in key areas. Quantum Key Distribution is experiencing a notable surge in deployments, indicating a growing emphasis on leveraging quantum technologies for secure communication. Research and commercial initiatives are actively engaged in practical implementations of QKD across various contexts. Moreover, efforts are underway to integrate QKD systems with existing classical communication networks, fostering the emergence of practical quantum-safe communication infrastructures [12, 13].

In the realm of post-quantum cryptography, there is a concerted push towards standardization led by organizations such as the National Institute of Standards and Technology (NIST). The focus is on evaluating and selecting cryptographic algorithms resilient to attacks from both classical and quantum computers. Concurrently, ongoing research and development endeavors aim to create and test new cryptographic primitives capable of withstanding quantum threats. Cryptographers are exploring diverse approaches, including lattice-based cryptography, code-based cryptography, and hash-based cryptography [14].

Quantum communication networks are witnessing notable advancements, with a particular spotlight on quantum repeaters. Research in this domain is gaining momentum as quantum repeaters play a pivotal role in extending the range of quantum communication, laying the foundation for efficient long-distance quantum networks. Simultaneously, various research initiatives are dedicated to shaping the future of quantum internet, with the ultimate goal of enabling secure quantum communication on a global scale [15].

Experimental realizations in quantum cryptography are marked by achievements in achieving longer quantum communication distances. Researchers are making substantial progress in mitigating challenges such as quantum noise and losses in communication channels, contributing to the feasibility of long-distance quantum communication. Moreover, there is a discernible trend towards increasingly sophisticated experimental setups, resulting in higher entanglement fidelities and overall improved performance in quantum communication protocols. These developments collectively underscore the dynamic and promising trajectory of quantum cryptography, pointing towards a future where quantum technologies play a pivotal role in shaping secure communication paradigms [16].

4.3 Advantages of Quantum Cryptography for the EU

The advantages of quantum cryptography for the European Union (EU) are manifold, offering heightened security, protection against quantum computing threats, technological leadership, and enhanced global competitiveness. Quantum cryptography, rooted in the principles of quantum mechanics, provides theoretical unhackability due to its reliance on quantum properties like superposition and entanglement. This heightened security is crucial for safeguarding sensitive government communications, financial transactions, and critical infrastructure within the EU. Moreover, quantum cryptography offers resilience against quantum computing threats, unlike traditional cryptographic systems vulnerable to such attacks [14].

Investing in research and development, fostering collaboration among researchers, academia, and industry, and promoting the commercialization of quantum technologies can position the EU as a hub for quantum cryptography expertise. This technological leadership drives economic growth, job creation, and enhances the EU's standing as a global leader in cutting-edge technologies. Furthermore, embracing quantum cryptography bolsters the EU's global competitiveness, ensuring that its industries and institutions stay competitive on the global stage.

Integrating quantum cryptography into EU infrastructure offers substantial economic and strategic benefits. Economically, it fosters innovation, drives technological advancement, and creates job opportunities within the EU. Strategically, it underscores the importance of strategic collaborations and partnerships in advancing quantum technologies and cybersecurity. Moreover, it revolutionizes communication networks, enhances cybersecurity, and safeguards critical infrastructure against cyber threats, enhancing national security and resilience [17].

In conclusion, integrating quantum cryptography into EU infrastructure is essential for driving economic growth, enhancing cybersecurity, and safeguarding critical infrastructure. By embracing quantum technologies and investing in research and development, the EU can solidify its position as a pioneer in the field, reaping economic, strategic, and societal benefits in the process. Through collaboration and innovation, the EU can navigate the digital landscape with confidence, ensuring security, prosperity, and leadership in the digital era.

4.4 Limitations of Quantum Cryptography

The integration of quantum cryptography into the European Union (EU) faces a multitude of challenges and limitations that must be overcome for widespread adoption. One significant obstacle is the requirement for complex infrastructure, involving the deployment of specialized hardware capable of handling quantum states, which is both intricate and financially demanding. Real-world environments exacerbate these challenges, with issues such as quantum noise, decoherence, and the need for robust quantum communication channels imposing constraints on system performance and reliability.

Within Quantum Key Distribution (QKD), challenges persist, particularly concerning channel loss and noise, potentially compromising the quality of quantum states. Security assumptions in quantum cryptography also raise concerns, particularly regarding the security of quantum devices involved in communication. Additionally, uncertainties surround the theoretical nature of quantum-safe algorithms in post-quantum cryptography development, requiring validation and confirmation of practicality [18].

Cost and scalability pose formidable barriers, with initial infrastructure costs being high and scalability issues arising from accommodating a large user and device base while maintaining security. Global standardization efforts face limitations due to the absence of global standards, potentially impeding interoperability and widespread adoption.

Addressing these challenges will require collaborative efforts across academia, industry, and government, along with continued investment in research, development, and standardization initiatives. By confronting these obstacles, the EU can pave the way for the widespread adoption and realization of the transformative potential of quantum cryptography.

In conclusion, the potential of quantum cryptography is undeniable, but its widespread adoption hinges on addressing practical challenges and navigating the evolving quantum landscape strategically. The emergence of quantum computers poses a significant threat, necessitating exploration of post-quantum cryptography and standardization of resilient algorithms [19].

The development of post-quantum cryptographic algorithms is underway to safeguard against evolving quantum threats, with institutions like the National Institute of Standards and Technology (NIST) actively evaluating and selecting resilient algorithms. However, practical challenges such as infrastructure updates and balancing security and performance considerations remain.

Global collaboration is essential in navigating the quantum technology landscape, with stakeholders needing to establish standards and regulations for secure integration of quantum technologies into information infrastructure. Through collaborative efforts and innovation, the EU can position itself as a leader in quantum technology, safeguarding communication networks and driving advancements in cybersecurity.

5 Conclusion

In conclusion, the potential of quantum cryptography to revolutionize secure communication technologies is undeniable, yet its widespread adoption hinges on addressing practical challenges and navigating the evolving quantum landscape strategically. The emergence of quantum computers poses a formidable threat to traditional cryptographic systems, necessitating the exploration of post-quantum cryptography and the standardization of resilient algorithms, led by institutions like the National Institute of Standards and Technology.

The practical implementation of quantum cryptography presents hurdles such as specialized hardware requirements, quantum noise mitigation, and ensuring physical security for quantum key distribution systems. As quantum computers advance, the urgency of transitioning to quantum-safe algorithms becomes apparent, emphasizing the need for algorithms resilient against both classical and quantum attacks.

In parallel, the development of post-quantum cryptographic algorithms is underway to safeguard sensitive information against evolving quantum threats, with NIST actively evaluating and selecting resilient algorithms to meet diverse security needs. However, the migration to quantum-safe algorithms entails practical challenges, including infrastructure updates and the delicate balance between security and performance considerations.

Quantum technologies extend beyond cryptography, finding applications in secure communication, quantum teleportation, and quantum networking, which leverage entanglement for building secure communication infrastructures. The societal implications of quantum computing underscore the importance of proactive measures, such as quantum-safe blockchain solutions and ensuring the security and privacy of machine learning models and datasets.

In navigating the quantum technology landscape, global collaboration is essential. Governments, industries, and academia must collaborate to establish standards and regulations guiding the secure integration of quantum technologies into our information infrastructure. The journey toward a secure quantum era demands adaptive and collaborative efforts from all stakeholders, ensuring that the promise of quantum cryptography translates into tangible benefits for society.

In conclusion, while quantum cryptography holds immense promise, realizing its full potential requires concerted efforts to address challenges, standardize resilient algorithms, and foster global collaboration. Through strategic collaboration and innovation, the EU can position itself at the forefront of quantum technology, safeguarding communication networks and driving advancements in cybersecurity for years to come.

Acknowledgements Supported by the ÚNKP-23-1 new national excellence program of the Ministry for Culture and Innovation from the source of the National Research, Development and Innovation fund.

Új Nemzeti Kiválóság Program

MINISTRY OF CULTURE AND INNOVATION

NATIONAL RESEARCH, DEVELOPMENT AND INNOVATION OFFICE HUNGARY

References

1. Rivest RL (1990) Cryptography. In: Algorithms and complexity. Elsevier, Amsterdam, pp 717–755
2. Zbinden H, Bechmann-Pasquinucci H, Gisin N, Ribordy G (1998) Quantum cryptography. Appl Phys B Lasers Opt 67(6)
3. Chandra S, Paira S, Alam SS, Sanyal G (2014) A comparative survey of symmetric and asymmetric key cryptography. In: 2014 international conference on electronics, communication and computational engineering (ICECCE). IEEE, pp 83–93
4. Bellare M, Desai D, Jokipii E, Rogaway P (1997) A concrete security treatment of symmetric encryption. In: Proceedings of the 38th annual symposium on foundations of computer science, FOCS'97. IEEE, pp 394–403
5. Pointcheval D (2002) Asymmetric cryptography and practical security. J Telecommun Inf Technol 4:41–56
6. Buchmann J, Braun J, Demirel D, Geihs M (2017) Quantum cryptography: a view from classical cryptography. Quant Sci Technol 2(2):020502
7. Bennett CH, Brassard G (1984) Quantum cryptography, public key distribution and coin tossing. In: Proceedings of IEEE international conference on computers, systems, and signal processing, Bangalore, 175 pp
8. Xu F, Ma X, Zhang Q, Lo HK, Pan JW (2020) Secure quantum key distribution with realistic devices. Rev Mod Phys 92(2):025002
9. Biham E, Boyer M, Brassard G, van de Graaf J, Mor T (2002) Security of quantum key distribution against all collective attacks. Algorithmica 34(4):372–388. arXiv:quant-ph/9801022

10. Biham E, Mor T (1997) Security of quantum cryptography against collective attacks. Phys Rev Lett 78:2256–2259. arXiv:quant-ph/9605007
11. Bennett CH, Brassard G (1984) An update on quantum cryptography. In: Workshop on the theory and application of cryptographic techniques. Springer Berlin Heidelberg, Berlin, Heidelberg, pp 475–480
12. Renner R, Wolf R (2023) Quantum advantage in cryptography. AIAA J 61(5):1895–1910
13. Meher KM (2023) Quantum-safe cryptography. In: Intelligent approaches to cyber security. Chapman and Hall/CRC, pp 181–191
14. Das K, Sadhu A (2022) Challenges and trends on post-quantum cryptography. In: Internet of things: security and privacy in cyberspace. Springer, Singapore, pp 271–293
15. Pirandola S, Andersen UL, Banchi L, Berta M, Bunandar D, Colbeck R, Englund D, Gehring T, Lupo C, Ottaviani C, Pereira JL, Razavi M, Shaari JS, Tomamichel M, Usenko VC, Vallone G, Villoresi P, Wallden P (2020) Advances in quantum cryptography. Adv Opt Photon 12(4):1012–1236
16. Sharma N, Ketti Ramachandran R (2021) The emerging trends of quantum computing towards data security and key management. Arch Comput Methods Eng 1–14
17. Manzalini A (2020) Quantum communications in future networks and services. Quant Rep 2(1):221–232
18. Bennett CH, Brassard G, Ekert AK (1992) Quantum cryptography. Sci Am 267(4):50–57
19. Kumar N, Agrawal A, Chaurasia BK, Khan RA (2020) Limitations and future applications of quantum cryptography. IGI Global

Trust and Security: How Trust in Autonomous Systems Shapes Decision-Making

Rozalia Szatmary

Abstract The article explores the intricate interplay between trust in autonomous systems and its profound influence on security-conscious behavior. In an era marked by the pervasive integration of automation and artificial intelligence, comprehending how our levels of trust, or lack thereof, in these systems affect our approach to security is of paramount importance. The study delves into the data, shedding light on the diverse degrees of trust in autonomous systems, their perceived capability to make accurate decisions, and individual preferences regarding human involvement in decision-making processes. Notably, the data underscores substantial disparities in trust dynamics, with particular reservations evident among specific demographic groups, including females and the elderly. Whether the central theme addressed is a healthy dose of skepticism, or distrust, can actually contribute to a more vigilant and safety-conscious society. By examining the intricate relationship between trust in autonomous systems and security-conscious behaviors, this presentation emphasizes the need for a nuanced and well-informed perspective on automation's role in our lives, ensuring a harmonious coexistence of human judgment and machine capabilities.

Keywords Autonomous systems · Trust · Human–robot-interaction

1 Introduction

The evolution of robotics has transcended beyond mere automation, venturing into the intricate realm of Human–Robot Interaction (HRI), which necessitates a nuanced understanding of both technological capabilities and human factors [1, 2]. This article delves into the essence of HRI, emphasizing the human element as its cornerstone. By exploring various forms of cooperation, situational and individual factors influencing HRI, alongside considerations of technology, ethics, law, and infrastructure, this

R. Szatmary (✉)
Obuda University, John von Neumann Informatics Faculty, Budapest 1034, Hungary
e-mail: szatmary.rozalia@stud.uni-obuda.hu

T. A. Kovács et al. (eds.), *The Impact of the Energy Dependency on Critical Infrastructure Protection*, Advanced Sciences and Technologies for Security Applications, https://doi.org/10.1007/978-3-031-78544-3_55

analysis aims to unravel the dynamics of control versus trust, perception of benefits, risk perception, and the impact of background knowledge and experiences with HRI.

2 Literature Review

Human–Robot Interaction (HRI) stands as a crucial area of inquiry within the field of information technology, aiming to seamlessly integrate robots into human environments for cooperative, supportive, or substitutive purposes. The crux of successful HRI lies in fostering engaging interactions between humans and robots, where both parties actively participate [1, 3]. This engagement hinges not only on the robot's capacity to comprehend and respond to human actions and intentions—necessitating a sophisticated blend of technological advancement, ethical considerations, legal compliance, and infrastructural support [4]—but also on the individual's initial perceptions regarding the ease of use of autonomous systems, associated risks, potential benefits, and prior experiences with and knowledge about technological systems [5]. Current paper places emphasis on the human element and the form of interaction, recognizing the significant impact of situational and individual factors on the nature and effectiveness of HRI.

2.1 Forms of Interactions in HRI

The integration of robots into human environments marks a significant frontier in both technological advancement and societal development within the realm of Human–Robot Interaction (HRI). This interaction between humans and robots transcends mere functionality, offering a platform for multifaceted cooperation across various dimensions. As HRI evolves, understanding the distinct forms of interaction—additional, supportive, and substitutive—becomes paramount, each presenting unique implications, advantages, and challenges that merit a thorough analysis [6].

Additional cooperation in HRI refers to scenarios where robots are integrated into human tasks to augment capabilities without replacing human roles [7]. This form of cooperation is particularly prevalent in environments where human expertise is irreplaceable yet can be significantly enhanced through robotic assistance. For example, in surgical procedures, robots can offer precision and stability that augment the surgeon's skill, leading to outcomes that might not be achievable through human capability alone [8]. The primary advantage of this form lies in its ability to combine the best of both worlds: the nuanced judgment and emotional intelligence of humans with the precision, consistency, and endurance of robots [9]. However, the challenge in additional cooperation lies in maintaining a seamless interface between human and robot, requiring sophisticated communication systems that can interpret and predict human intentions accurately [10]. Moreover, there's the risk of over-reliance on robotic assistance, potentially atrophying human skills that are not regularly exercised [11].

Transitioning to supportive cooperation, this form underscores the role of robots in providing necessary support to humans, particularly in tasks that are repetitive, hazardous, or beyond human capabilities [12]. Robots deployed in disaster response scenarios exemplify supportive cooperation, where they can navigate through debris or hazardous environments, providing real-time information and performing rescue operations without putting human lives at risk [13]. The advantages of supportive cooperation are evident in enhanced safety and efficiency, relieving humans from the burden of dangerous tasks, and potentially saving lives in scenarios where human presence is perilous. Nevertheless, the limitations of this form of cooperation often revolve around the current technological constraints in autonomy and decision-making under unpredictable circumstances [14], necessitating ongoing human oversight and intervention. Moreover, there's the challenge of ensuring that these robots can operate in diverse and unstructured environments, requiring advances in artificial intelligence and robotics [15].

Substitutive cooperation, where robots replace human effort in tasks deemed too dangerous, mundane, or requiring levels of consistency and endurance beyond human capacity, presents a form of HRI with profound implications. Manufacturing environments have long benefited from this form of cooperation, with robots taking on repetitive tasks, thus freeing humans to focus on more complex, creative, or strategic activities [16]. The benefits of substitutive cooperation include increased productivity, consistency in quality, and the reduction of human exposure to potentially harmful conditions. However, the implications of substitutive cooperation extend beyond the workplace, touching on broader societal and ethical concerns. The potential for job displacement and the erosion of certain skill sets in the workforce pose significant challenges, necessitating thoughtful approaches to workforce retraining and education [17, 18]. Furthermore, the psychological impact of substitutive robots, where human roles are entirely replaced, raises questions about identity, purpose, and social integration [19].

As robots become increasingly integrated into daily life, the complexity of the interface between human and machine grows. This evolution underscores the need for a multidisciplinary approach to HRI, encompassing engineering, computer science, psychology, sociology and philosophy [20]. By fostering collaboration across disciplines, balancing technological innovation with human-centric considerations, and addressing issues such as user-friendly design, ethical programming, and legal frameworks [21], HRI can pave the way for responsible and beneficial interactions between humans and robots in society. Furthermore, ongoing research and development efforts must prioritize the enhancement of user experiences, ensuring that robots are seamlessly integrated into various domains while respecting individual autonomy, privacy, and dignity [22]. Ultimately, HRI has the potential to enhance human well-being, foster innovation, and shape a future where humans and robots coexist harmoniously [23].

2.2 The Human Factor in HRI

The advancement of technology and the development of intelligent systems are inherently aimed at assisting individuals and improving their quality of life, underlining the fundamental requirement for both technology users and surrounding individuals to feel safe in the proximity of robots [24]. However, for robots and autonomous systems to find their place in industrial, commercial, and personal spheres, mere technological infrastructure is insufficient. While the majority of people have come to accept the necessity of machines, or at least acknowledge their potential to aid in daily life, there remains a reluctance to fully trust intelligent systems to make decisions without human intervention or to relinquish control to these devices [25]. The willingness to engage with intelligent systems is influenced by various factors related to the perception of technology, shaped by individuals' personal experiences with technology and the experiences of others they trust [26]. To make the decision to utilize robots, individuals must possess certain knowledge regarding the use, functionality, usefulness, and reliability of intelligent systems [5, 27].

As the landscape of human–robot interaction (HRI) continues to evolve, the significance of the human factor emerges as a central focus of debates and analyses. As robots become increasingly integrated into everyday life and workspaces, the complexity of their interaction with humans extends beyond technological innovation into the realms of ethics, law, and individual and societal acceptance [27, 28]. This chapter argues for the indispensable role of the human factor in the successful and ethical development of HRI, highlighting the importance of situational factors such as technology, ethics, law, and infrastructure, as well as individual perspectives, including the perception of positive impacts, risk perception, and the influence of background information and past experiences, in the examination of HRI.

Situative Factors

The technological prowess of a robot significantly determines its capacity for meaningful engagement with humans. Advancements in artificial intelligence (AI) and machine learning are crucial for developing robots capable of understanding and responding to human behaviors [29]. However, the argument often presented against this technology-centric view is that it might lead to an overemphasis on technical capabilities at the expense of understanding human-oriented needs [30]. It is argued that focusing solely on technological advancement neglects the importance of designing robots that are empathetic and responsive to the human emotional spectrum, potentially leading to interactions that feel mechanical and impersonal. This critique underscores the need for a balanced approach that considers both technological innovation and the nuances of human emotion and social interaction [31].

The requirement for humans and surrounding individuals to feel safe in the presence of robots forms the basis of interaction with robots [32]. However, this approach presents challenges. Firstly, safety can be compromised due to countless unforeseen circumstances, potentially resulting in harm to the current user. The finite number of programmed codes cannot account for every possible risk, rendering the robot incapable of consistently ensuring individual protection. Secondly, decision-making situations may arise where all options jeopardize human lives. In such cases, the robot is unable to make "good" decisions, leaving users and bystanders vulnerable [33]. Despite these uncertainties, adherence to Isaac Asimov's three (or four) laws of robotics, formulated in 1941 during the inception of human–robot interaction (HRI), prioritizes user safety, especially concerning potentially hazardous robotic tools, while aiming to minimize harm to others [34].

Ethical considerations in HRI, encompassing respect for autonomy, privacy concerns, and the avoidance of deception, further illustrate the centrality of the human factor [35]. Ethical robotics necessitates that interactions are grounded in moral principles, ensuring that robots act in ways that are beneficial and not harmful to humans. The critique here revolves around the potential for ethical guidelines to be vague and subject to interpretation, leading to inconsistencies in application [4]. Moreover, the rapid pace of technological advancement may outstrip the development of comprehensive ethical frameworks, raising concerns about the adequacy of current ethical standards in addressing future challenges. This critique highlights the ongoing need for dynamic and responsive ethical guidelines that evolve in tandem with technological advancements [36].

Legal frameworks play a critical role in guiding the development and deployment of robots, addressing issues such as liability, data protection, and safety standards [36]. The critique in this domain focuses on the lag between technological innovation and legal regulation, with existing laws often ill-equipped to address the novel challenges posed by advanced robotics. Furthermore, there is a concern that overly stringent regulations might stifle innovation, while too lax an approach could leave individuals unprotected against potential harms [37]. This tension underscores the importance of developing legal frameworks that are both forward-looking and flexible, capable of adapting to the rapid pace of technological change while safeguarding human interests.

Infrastructure, including sophisticated communication networks and energy sources, is vital for the integration of robots into human environments [38], however, international literature argues that the focus on infrastructure development might lead to disparities in access to advanced robotics, with well-resourced areas benefitting disproportionately. This raises concerns about digital divides and inequalities in the benefits derived from HRI [39], highlighting the need for equitable infrastructure development that ensures broad and inclusive access to robotic technologies.

Individual Factors

The importance of establishing trust is widely recognized as a fundamental requirement for the widespread adoption of autonomous systems within academic discourse [40]. However, as intelligent systems become increasingly complex, users' sense of vulnerability grows rapidly. Consequently, building trust in such intricate and less transparent systems presents a significant challenge.

While trust involves consciously accepting the vulnerability of agents (ourselves) in exchange for expected benefits and can function as a substitute for lack of control in certain circumstances [41], there is no straightforward approach for effectively nurturing or enhancing trust in artificial intelligence or autonomous systems [42, 43]. Furthermore, individuals' initial inclination towards trusting unfamiliar entities plays a significant role in their willingness to depend on them as well [44, 45]. Those inclined towards dispositional trust, as referred to in the literature, are more inclined to engage with robots [46], enabling both machine and human participants to gain insights into each other and acquire additional information to enhance future interactions. Therefore, experiential knowledge contributes to the development of human trust in intelligent systems. As interaction experiences accumulate, trust in robots grows, reducing individuals' perceived need to control the actions of the robot and facilitating smoother task execution [47].

External environmental factors exert influence on users or consumers through two mediators. Specifically, perceived positive outcomes such as increased efficiency and convenience, alongside perceived ease of use, have the potential to bolster trust in robots [48]. However, this optimistic perspective is often offset by apprehensions regarding potential malfunctions, breaches of privacy, and loss of workplaces, which cultivate unease and impact the willingness to interact with robots [49].

The quality of a robot's performance during human use and interaction greatly influences individual satisfaction and, consequently, trust in the smart system [50]. Experiential knowledge is not only bolstering the human trust but based on initial experiences with robots and the information acquired through using intelligent machines, individuals also decide for themselves whether incorporating robots into their daily lives is worthwhile. Since robots are not the first technological contraptions humans have used, individuals may have basic technology related feelings—, such as technostress (stress related to technology), and a general evaluation of technological devices, gadgets, the notion of autonomous systems—that are also influencing factors of the acceptance of and trust in autonomous systems [47].

Past experiences play a vital role in shaping expectations and attitudes towards robots. Individuals with positive experiences with technology are more likely to trust and embrace robots, while those with negative experiences may be more cautious or skeptical. Basic technological affinity also impacts the acceptance of intelligent systems for everyday use, since those capable of keeping pace with technological advancements, possessing fundamental knowledge of, background knowledge about technology, are more likely to harbor positive sentiments towards robots and may even have reduced fears associated with robot use [51].

Every computer communicating with another computer or accessed by another computer is vulnerable to cyberattacks. In recent years, numerous incidents of hackers breaching various databases and infecting computers with viruses have been reported. Autonomous systems and robots are no exceptions to this; intelligent vehicles have been used as tools for terrorist acts [52]. Moreover, autonomous vehicles, as a type of autonomous system, may be particularly vulnerable to such misuse, as hacking into the system could grant unauthorized access to all data and control over the vehicle's functions. However, providing more information does not necessarily improve the situation. In fact, ordinary individuals may trust autonomous vehicles less when they possess detailed knowledge about these systems [53]. Users aware of risks, such as the unpredictability of robots when used in “hazardous” conditions, are much less likely to trust them and may even refuse to cooperate with intelligent systems, especially when human lives are at stake [54, 55]. Furthermore, since individuals may not fully comprehend the information received about the system, their perceived level of control in emergency situations may also diminish. Therefore, while transparency and communication about autonomous systems are essential, achieving an appropriate level of available information depends heavily on rigorous market research [56]. Improving public awareness of autonomous systems and robots in general is necessary to avoid this information trap, reduce stress and anxiety induced by these systems, and increase perceived knowledge, thereby fostering readiness to integrate autonomous systems to not only the spheres of human lives, where individuals are forced to, but into their private lives as well [57].

The success of HRI initiatives greatly depends on individuals’ biased perception of robots [58], which highlights the need for educational and outreach efforts to provide balanced and accurate information about the capabilities and limitations of robots, fostering informed and nuanced perspectives. Nevertheless, a great challenge lies in striking a balance between fostering positive perceptions and addressing legitimate concerns and risks associated with robotics. Establishing trust and garnering acceptance among the general populace are vital for ensuring enthusiastic rather than resistant integration of robots into society [59].

3 Research Questions and Methodology

In order to examine the impact of trust on decision-making processes concerning autonomous systems, a quantitative study was initiated. In the context of my questionnaire seeking opinions on autonomous systems, it was necessary to specify a particular type of autonomous system due to the inherent difficulty individuals may face in reflecting on the broad concept of autonomy. Given the current widespread interest and discourse surrounding self-driving cars [60], they presented a compelling focus for my survey, as individuals were more likely to have formed opinions about them. Furthermore, self-driving cars, as they currently exist, predominantly serve as additive elements in human–robot interaction, with human intervention still necessary in certain situations. However, they also hold the potential to function as substitutive entities, where human intervention is not required, based on the vehicle’s experiential and sensor technologies [61]. The inclusion of self-driving cars in my survey not only allowed for the exploration of specific issues and concerns related to this particular application but also reflected the evolving nature of human–robot interaction. By specifying self-driving cars as a focal point, my questionnaire could better captured respondents’ attitudes and perceptions towards emerging technologies, thereby enhancing the depth and relevance of the research findings.

The survey, distributed via an online Google Forms platform, utilized convenience and snowball sampling techniques. It is noteworthy that while convenience sampling was employed, efforts were made to encompass diverse age groups and educational backgrounds. Additionally, although not entirely controllable, attempts were made to ensure balanced gender representation and inclusion of individuals from varying sizes of residential areas, aiming for a representative sample. These demographic variables, as highlighted in the international literature, play a significant role in shaping trust perceptions towards autonomous systems.

Since I did not want to measure different kind of levels of understanding or trust, that influence different kind of decisions, I decided to apply an interpretative research approach in order to heighten my understanding of the very topic as well as to discover deeper connections between my variables. Therefore, the primary research question was exploratory in nature, devoid of explicit hypotheses regarding anticipated outcomes.

Q1: How does trust affect decisions related to autonomous systems?

Acknowledging the recurring emphasis in international literature on the importance of demographic variables in shaping trust perceptions, another research question investigated whether demographic disparities exist between individuals exhibiting trust and those who do not.

Q2: Are those who trust different from those who do not?

4 Introduction of the Sample

Among the demographic variables, respondents were queried regarding their gender, age, place of residence, and highest level of education attained. Additionally, respondents were asked about their employment status, general attitude towards technology, level of awareness, and possession of a driving license. In relation to the topic, opinions on autonomous vehicles were elicited using various statements, which were evaluated on a 5-point Likert scale.

A total of 485 respondents completed the questionnaire. Among them, 290 were male and 188 were female, 7 respondents opting not to disclose their gender. While the sample is not balanced in terms of gender distribution, this deviation will not be considered as a basis for relationship among variables.

Despite the aim of achieving a broad spectrum of respondents, the selected snowball method and convenience sampling did not fully realize this goal, as evidenced by the predominantly young age distribution of respondents. As depicted in Fig. 1, approximately 80% of respondents were under the age of 25, 12% fell between the ages of 26 and 40, while the remaining constituted those aged 41 and above. The average age of respondents was 27.366 years, though the standard deviation of 10.978 suggested a relatively diverse age distribution, as the questionnaire reached individuals spanning from 12 to 70 years old,

The distribution of respondents by their highest level of education is illustrated in Fig. 2. The majority of individuals had a secondary school level of education, 44 respondents reported having a bachelor's degree as their highest qualification, while 33 respondents indicated having a master's degree. The graph fundamentally reflects

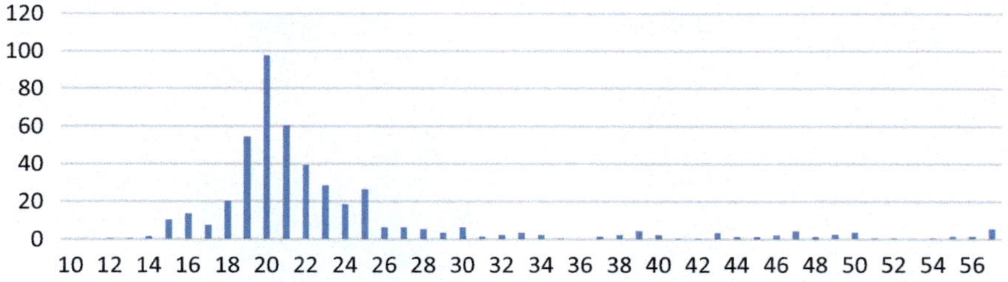

Fig. 1 Distribution of the respondents by age

the age distribution of respondents as well, as younger individuals typically possess lower levels of education. Regrettably, the highest level of education has become a demographic variable in my sample, whose relationship with other variables needs to be evaluated cautiously, considering age-related correlations. Clearly, the highest level of education is also linked to the respondents' current educational status. Those who have completed primary school are currently in high school, and those who have completed high school are pursuing a bachelor's degree. Interestingly, only 8% of respondents indicated that they are no longer studying, while 40% are employed. This suggests that a significant proportion of individuals either work while studying or pursue further education while employed.

A noteworthy aspect of the research is that despite the convenience sampling initially attracting respondents primarily from urban areas, particularly the capital city, the snowball method ultimately led to success. This is evident in the diverse range of responses obtained from individuals residing in areas of varying sizes depicted in Fig. 3, including not only large cities but also towns and villages. The place of residency reflects a situate factor mentioned in the literature review, namely, the aspect of infrastructural development. While Budapest is typically an urban centre with advanced infrastructure, other cities as well as towns in the country are considerably smaller and less developed in terms of infrastructure. This factor could influence residents' perceptions of the feasibility of autonomous systems in such environments.

Fig. 2 Distribution of the respondents by highest education

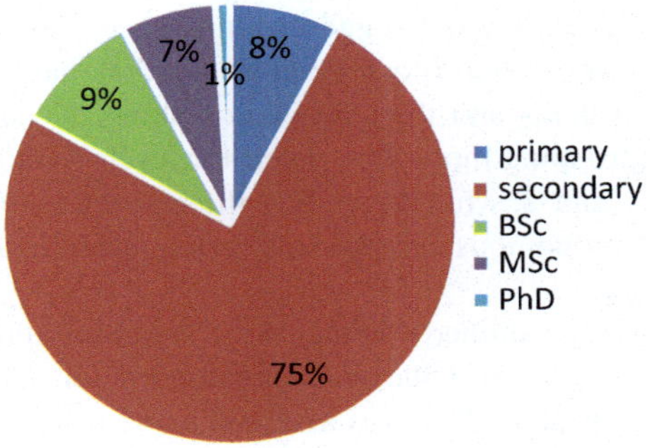

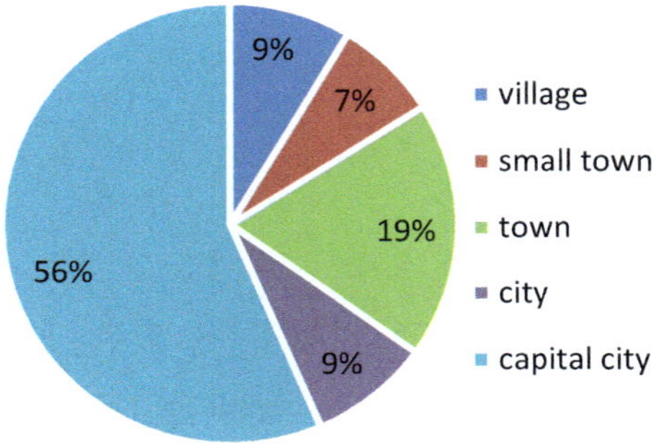

Fig. 3 Distribution of the respondents by place of residency

Regarding driving licenses, 64% of respondents possess one, 30% do not have one but intend to acquire it in the future, and the remaining 6% neither have nor wish to obtain a license. This aspect is particularly interesting in the context of autonomous vehicles, as driving is considered an enjoyable activity by many, and therefore, some may be less inclined towards autonomous vehicles simply due to the pleasure derived from driving. Thus, among those with driving licenses, it is possible that it is not fear or reluctance towards technology that influences perceptions or purchasing decisions regarding autonomous vehicles, but rather this enjoyment factor.

Respondents generally exhibited varied attitudes towards autonomous vehicles. Assessing their trust in autonomous vehicles using the six autonomy levels defined by SAE International (2016), ranging from level 1 autonomy, where human agents control everything, to level 6 autonomy, where robotic agents control everything, the following distribution can be observed. The intermediate levels, namely 2, 3, 4, and 5, each add or remove a layer of control, altering the sense of being in charge. These levels and their meanings were explained in the survey to ensure comprehension among respondents unfamiliar with autonomous systems.

As depicted in Fig. 4, majority of respondents were only interested in vehicles with lower levels of autonomy, indicating a lack of trust in autonomous systems. Only 3.5% expressed interest in using fully autonomous vehicles (voting for level 6), where the system has complete control, while 12.6% indicated a desire to use autonomous vehicles (voting for level 5) where they could regain control if necessary. Levels 1 and 2 suggest minimal acceptance of autonomous decision-making, while levels 5 and 6 can be considered as having a certain level of trust in autonomous systems and their decision-making. Henceforth, I will utilize this categorization, grouping those who voted for levels 5 and 6, as well as those who voted for levels 1 and 2, while excluding the intermediate layers, to identify individuals who trust autonomous systems to a certain extent and those who do not trust these systems at all.

I examined whether there is a discernible pattern that could identify individuals who support the widespread adoption of autonomous vehicles versus those who oppose it. This was assessed using responses to the statement "I would support the introduction of autonomous vehicles." The distribution of respondents regarding

Fig. 4 Distribution of respondents by level of autonomy acceptance

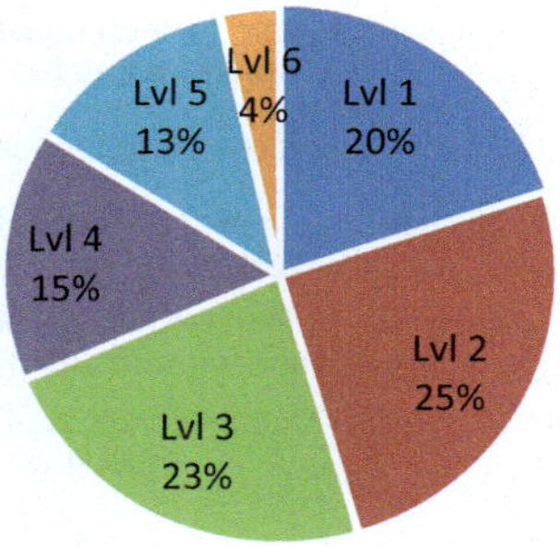

self-driving cars as autonomous systems and their perceptions of such autonomy systems is highly variable as it can be seen on Fig. 5. Approximately 44% of respondents (perhaps not directly as users/owners) expressed support for the introduction of autonomous cars, while only a quarter of respondents were against it. Interestingly, a significant proportion of respondents, the majority, opted for an intermediate response ("Maybe") indicating uncertainty regarding their stance on the introduction of autonomous vehicles.

Approximately three-quarters of respondents expressed interest in trying out autonomous vehicles, however when it came to purchasing decisions, only 11% is sure to be willing to purchase an autonomous system and an additional 16% would be in favour of, but still contemplating. As observed on the graph, people tend to have relatively less apprehension towards autonomous vehicles in their surroundings. Following this pattern, respondents generally believe that the positive consequences associated with the introduction of autonomous vehicles are more likely to materialize than not, although many were unable to clearly articulate their stance on these statements.

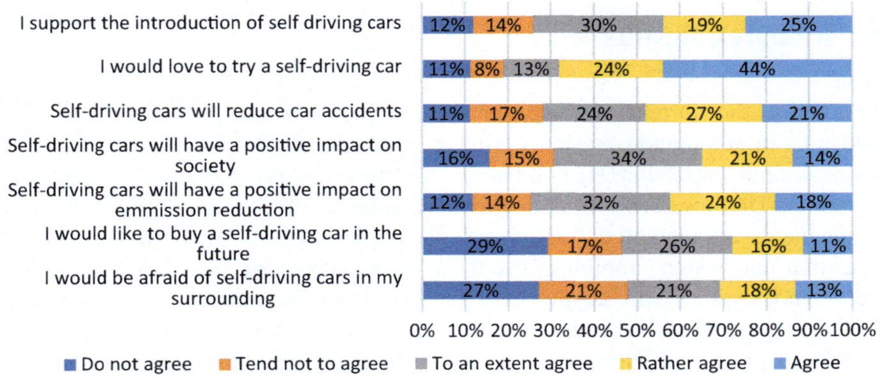

Fig. 5 Distribution of respondents regarding self-driving cars as autonomous systems and their perceptions of such autonomous systems

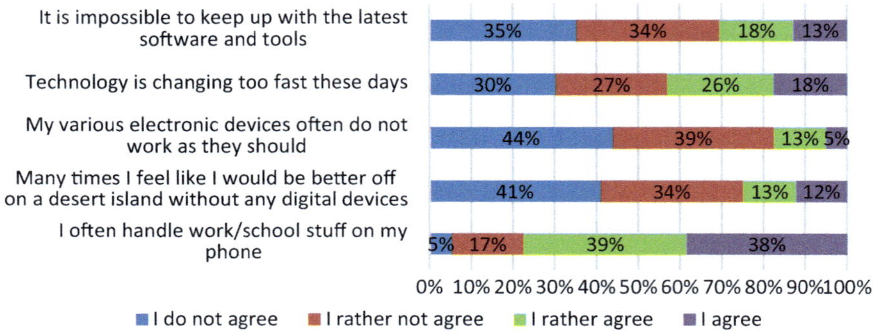

Fig. 6 Distribution of respondents by basic technological affinity

Respondents' general attitudes towards technology was also examined. According to Fig. 6, the vast majority of respondents do not perceive malfunctions in their electronic devices, and while they do acknowledge the rapid pace of technological change, they can relatively easily keep up with its development. Only a quarter of respondents indicated that life without smart devices and all kinds of technology would be more ideal for them on a deserted island, which may be an indicator of a high level of technological invasion and their negative relationship with technology. However, most respondents often handle work/school-related matters and likely also engage with electronic devices for personal use and leisure activities. Based on these data, it can be concluded that the individuals in the sample are not bothered by technology in their daily lives; in fact, they willingly use electronic devices. Therefore, they likely possess a certain level of knowledge and familiarity with technology.

5 Research Results

Examining the correlations between variables concerning personal perceptions and trust in autonomous systems provides valuable insights into the dynamics of human–robot interaction (HRI) and the decision-making processes surrounding their adoption. The observed strong positive correlation between the desire to own an autonomous vehicle and the willingness to try one suggests an alignment between intention and action, implying that those who express an interest in purchasing autonomous vehicles are significantly more likely to also hold positive perceptions of their capabilities, thus being more inclined to engage with them firsthand.

Moreover, several factors demonstrate significant correlations with both intentions to purchase and willingness to try autonomous vehicles. Notably, individuals who perceive the potential positive impact of autonomous vehicles on the environment, on society as well as road safety are significantly more inclined to try out as well as purchase a self-driving car. The positive significant correlation between the number of smart devices used on a daily basis and the intention to try out, or to purchase an autonomous vehicle in the future, suggests that technological familiarity is indeed an influencing factor of the perceptions of autonomous systems.

The negative correlations observed between fear of hacking into autonomous vehicle systems, general fear of technological advancements, and intentions to purchase or try autonomous vehicles highlight the critical role of trust and security concerns in decision-making. Individuals who harbour fears of potential security breaches or possess general apprehensions towards technological advancements may lack trust in the safety and reliability of autonomous systems, thereby hindering their willingness to try or purchase them.

Overall, these correlations highlight the complex relationship between personal perceptions, trust, and decision-making regarding autonomous systems. Addressing both positive perceptions and concerns, particularly regarding security and trust, is essential to promote positive attitudes and facilitate the integration of autonomous vehicles into society (Table 1).

I attempted to identify differences between individuals who trust and do not trust different autonomous systems by creating a proxy variable in a binary manner—those who completely trust and those who don't trust at all—while excluding intermediate variables. I conducted t-tests for these independent samples, and the results indicate significant differences between those who trust and those who don't trust autonomous systems in terms of gender, age, place of residency, and student status. It is unsurprising that there are significant differences between males and females, as international literature suggests a general disparity between the two genders regarding the willingness to integrate autonomous systems into their personal lives and their overall attitude towards new technologies. Additionally, younger individuals are significantly more trusting than older individuals, likely due to differences in technological exposure during upbringing. Moreover, residents of the capital city are more inclined to accept and trust autonomous systems compared to those living in smaller settlements, a difference attributed to infrastructural disparities.

However, surprisingly, there were no significant differences observed based on the highest level of education, work experience, or possession of a driver's license. One might expect individuals without a driver's license to be more interested in utilizing autonomous systems instead of public transportation, but our dataset did not provide evidence of such significant difference. Furthermore, the highest level of education could serve as an indicator for technological knowledge, which technological knowledge is in fact an influencing factor of the trust and attitude towards autonomous systems [47]; however, as the educational level of respondents was influenced by the age distribution of the sample, definitive conclusions cannot be drawn (Table 2).

In Fig. 7, I have highlighted that while the two groups—those who trust and those who don't trust autonomous systems—can significantly differ in regards to some demographic variables, even those who trust autonomous systems perceive certain levels of fear, and conversely, those who completely do not trust autonomous systems

Table 1 Correlation between purchasing intentions, willingness to try out self-driving cars and other attitudinal variables

	I would like to buy a self-driving car in the future	I would love to try a self-driving car
Many times I feel like I would be better off on a deserted island without any digital devices	− 0.153**	− 0.119**
It is impossible to keep up with the latest software and tools	− 0.145**	− 0.126**
How many smart devices do you use on a daily basis?	0.169**	0.128**
I would like to buy a self-driving car in the future	1	0.558**
Self-driving cars will have a positive effect on emissions	0.446**	0.405**
Self-driving cars will have a positive impact on society	0.549**	0.436**
Self-driving cars will reduce car accidents	0.527**	0.390**
I would love to try a self-driving car	0.558**	1
Fear of hackers getting into the car's system	− 0.145**	− 0.117*
Fear of the self-driving car breaking down	− 0.266**	− 0.241**
Fear of new technology	− 0.354**	− 0.373**
Fear of people losing their jobs because of it (e.g. taxi drivers)	− 0.304**	− 0.232**
Fear that control cannot be taken back	− 0.293**	− 0.225**
Fear of losing the joy of driving	− 0.287**	− 0.194**
Fear of lack of security of personal data	− 0.192**	− 0.188**

*The correlation is significant at the 0.05 level (2-tailed)
**The correlation is significant at the 0.01 level (2-tailed)

can perceive some positive side effects of autonomous systems, here in this sample, regarding self-driving cars.

For instance, concerning the factor "I am afraid that my personal data are not secure when using self-driving cars," the rating of 2.49 on a Likert scale of 1–5 indicates a moderate level of concern, suggesting that even supporters of autonomous systems are aware of potential data breaches. Similarly, regarding fears of potential malfunctions and hackers accessing the car's systems, individuals who trust autonomous systems perceived relatively high levels of fear, with ratings of 3.23 and 3.32.

Table 2 Significant differences between those who trust and do not trust

	Levene's test for equality of variances			t-test for equality of means			
		F	Sig.	t	df	Sig. (2-tailed)	Mean diff.
What is your gender?	*	90.69	0	3.759	291	0	0.241
How old are you?	*	10.68	0.001	− 2.493	296	0.013	− 2.893
What is your place of residence?	*	26.198	0	− 3.58	296	0	− 0.635
Do You still study?	*	14.07	0	3.407	296	0.001	0.471

*Equal variances assumed

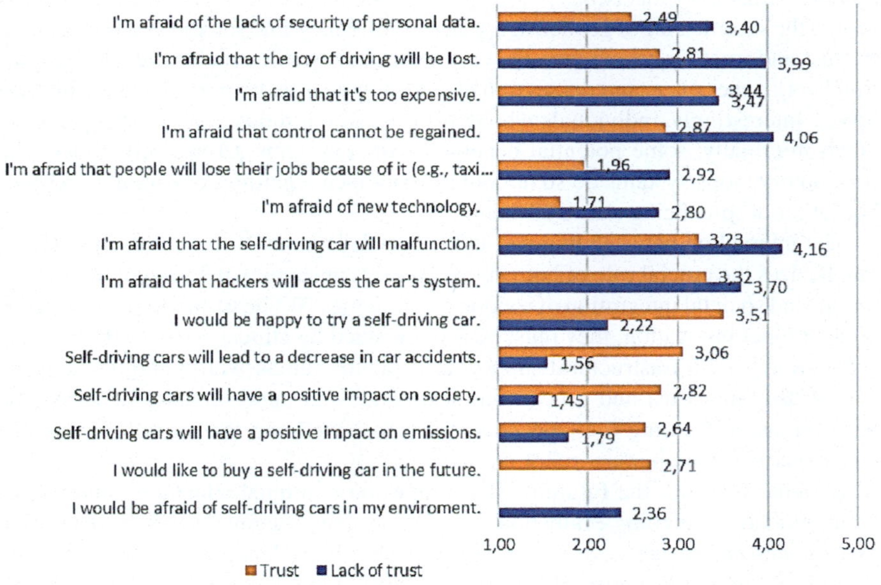

Fig. 7 Perception of positive and negative outcomes of autonomous systems of the trusting and not trusting group

On another note, individuals who acknowledge that self-driving cars might decrease the number of accidents can still lack trust in the systems, as indicated by the rating of 1.56, suggesting that they recognize the potential positive effects of autonomous systems to some extent. Similar perceptions of positive effects can be observed concerning the impact on the environment and society. Though not at the same level as those who trust autonomous systems, individuals who do not trust autonomous systems would also be willing to try self-driving cars, indicating a certain level of interest despite their reservations.

Figure 7 also reveals that the most significant concerns of those who do not trust autonomous systems revolve around the possible malfunction of self-driving cars and the fear of not being able to take back control if desired. Interestingly, the loss of joy while driving was not only one of the biggest concerns among those who don't trust autonomous systems but was also a relevant fear (with a value of 2.81) among the group trusting autonomous systems.

6 Conclusion

In summary, despite the prevalence of young respondents in the sample, more than half expressed distrust towards autonomous systems, with only 78 indicating complete trust in the decisions made by such systems. The results of the t-test underscored the significance of gender, age, place of residency, and student status in relation to trust and trusting behaviours. However, certain variables such as level of education, work experience, and possession of a driving license did not correlate with trust levels. Interestingly, individuals who don't trust autonomous systems still perceive, albeit minimally, some potential benefits of such systems. Conversely, those who trust autonomous systems do so not out of ignorance regarding potential drawbacks but rather despite being aware of them.

The interaction between humans and robots in Human–Robot Interaction (HRI) emphasizes the crucial role of human involvement in shaping the design, implementation, and societal integration of robotic technologies. While technological advancements propel innovation, they must be accompanied by ethical considerations, legal regulations, and infrastructural support prioritizing human well-being and societal values. Building trust and acceptance necessitates an understanding of individual perceptions, addressing risk concerns, and acknowledging the impact of humans' past experiences on human–robot interactions.

Looking forward, the future of HRI hinges on continual adaptation and refinement, guided by ongoing evaluation of human–robot dynamics. Drawing from past experiences and insights, we can navigate the evolving landscape of HRI, striving towards collaborative partnerships that enhance human well-being and propel societal progress. Ultimately, the synergy between human values and technological innovation will determine the trajectory of HRI, underlining the intrinsic connection between human factors and the advancement of robotics.

References

1. Prati E, Peruzzini M, Pellicciari M, Raffaeli R (2021) How to include user experience in the design of human–robot interaction. Robot Comput Integr Manufact 68:102072

2. Lyons JB (2013) Being transparent about transparency: a model for human–robot interaction. In: 2013 AAAI spring symposium series
3. Weidemann A, Rußwinkel N (2021) The role of frustration in human–robot interaction—what is needed for a successful collaboration? Front Psychol 12:707
4. Wullenkord R, Eyssel F (2020) Societal and ethical issues in HRI. Curr Robot Rep 1:85–96
5. Lazanyi K (2018) Readiness for artificial intelligence. In: 2018 IEEE 16th international symposium on intelligent systems and informatics (SISY). IEEE, pp 000235–000238
6. Breazeal C (2004) Social interactions in HRI: the robot view. IEEE Trans Syst Man Cybernet Part C (Appl Rev) 34(2):181–186
7. Su H, Qi W, Chen J, Yang C, Sandoval J, Laribi MA (2023) Recent advancements in multimodal human–robot interaction. Front Neurorobot 17:1084000
8. Kim S (2022) Working with robots: human resource development considerations in human–robot interaction. Hum Resour Dev Rev 21(1):48–74
9. Wang T, Zheng P, Li S, Wang L (2023) Multimodal human–robot interaction for human-centric smart manufacturing: a survey. Adv Intell Syst 2300359
10. Galin RR, Meshcheryakov RV (2020) Human–robot interaction efficiency and human–robot collaboration. In: Robotics: industry 4.0 issues & new intelligent control paradigms. Springer International Publishing, Cham, pp 55–63
11. George AH, George AS (2023) From pulse to prescription: exploring the rise of AI in medicine and its implications. Partn Univ Int Innov J 1(6):38–54
12. Zhang M, Xu R, Wu H, Pan J, Luo X (2023) Human–robot collaboration for on-site construction. Autom Constr 150:104812
13. Humphrey CM, Adams JA (2009) Robotic tasks for chemical, biological, radiological, nuclear and explosive incident response. Adv Robot 23(9):1217–1232
14. Abbass HA, Scholz J, Reid DJ (2018) Foundations of trusted autonomy. Springer Nature
15. Soori M, Arezoo B, Dastres R (2023) Artificial intelligence, machine learning and deep learning in advanced robotics, a review. Cognit Robot
16. Faccio M, Granata I, Menini A, Milanese M, Rossato C, Bottin M, Rosati G et al (2023) Human factors in cobot era: a review of modern production systems features. J Intell Manuf 34(1):85–106
17. Rawashdeh A (2023) The consequences of artificial intelligence: an investigation into the impact of AI on job displacement in accounting. J Sci Technol Policy Manag
18. Kumar A, Krishnamoorthy B, Bhattacharyya SS (2023) Machine learning and artificial intelligence-induced technostress in organizations: a study on automation-augmentation paradox with socio-technical systems as coping mechanisms. Int J Organ Anal
19. McLeay F, Osburg VS, Yoganathan V, Patterson A (2021) Replaced by a robot: service implications in the age of the machine. J Serv Res 24(1):104–121
20. Bartneck C, Belpaeme T, Eyssel F, Kanda T, Keijsers M, Sabanovic S (2020) Mensch-roboter-interaktion. Eine Einführung
21. Cain W (2023) AI emergence in education: exploring formative tensions across scholarly and popular discourse. J Interact Learn Res 34(2):239–273
22. Alakwe K (2023) Human dignity in the era of artificial intelligence and robotics: issues and prospects. J Humanit Soc Sci Stud 5(6):87–97
23. Blessing E, Klaus H (2024) The impact of robotics on society and civilization
24. Vasic M, Billard A (2013) Safety issues in human–robot interactions. In: 2013 IEEE international conference on robotics and automation. IEEE, pp 197–204
25. Szikora P, Szatmáry R (2023) Autonomous cars-what lies behind the lack of readiness. Interdisc Descrip Complex Syst INDECS 21(4):399–411
26. Paluch S, Tuzovic S, Holz HF, Kies A, Jörling M (2022) "My colleague is a robot"—exploring frontline employees' willingness to work with collaborative service robots. J Serv Manag 33(2):363–388
27. Potočan V, Lazányi K, Nedelko Z (2024) Organizations and Industry 4.0: management solutions beyond technology. Taylor and Francis Group

28. Chromjakova F, Trentesaux D, Kwarteng MA (2021) Human and cobot cooperation ethics: the process management concept of the production workplace. J Compet 13(3):21–38
29. Dautenhahn K (2007) Socially intelligent robots: dimensions of human–robot interaction. Philos Trans Roy Soc B Biol Sci 362(1480):679–704
30. Moniz AB, Krings BJ (2016) Robots working with humans or humans working with robots? Searching for social dimensions in new human–robot interaction in industry. Societies 6(3):23
31. Sugiyama S, Vincent J (2013) Social robots and emotion: transcending the boundary between humans and ICTs. Intervalla 1(1):1–6
32. Rubagotti M, Tusseyeva I, Baltabayeva S, Summers D, Sandygulova A (2022) Perceived safety in physical human–robot interaction—a survey. Robot Auton Syst 151:104047
33. Nagar S (2016) Human robot interaction—a psychological perspective. https://doi.org/10.13140/RG.2.1.2035.1763
34. Asimov I (1940) I, Robot. Narkaling Productions
35. Koumpouros Y (2020) Major metrics, concerns, and assessment strategy for mobility assistive devices. In: Quality assurance in the era of individualized medicine. IGI Global, pp 171–203
36. Babikian J (2023) Justice in flux: evolving legal paradigms in response to technological advancements. J Soc Sci Stud 1(1):1–16
37. Kolt N (2023) Algorithmic black swans. Wash Univ Law Rev 101
38. Oğul S (2024) In tune with ethics: responsible artificial intelligence and music industry. Reflektif Sosyal Bilimler Dergisi 5(1):139–149
39. Vasilescu MD, Serban AC, Dimian GC, Aceleanu MI, Picatoste X (2020) Digital divide, skills and perceptions on digitalisation in the European union—towards a smart labour market. PLoS ONE 15(4):e0232032
40. Dirsehan T, Can C (2020) Examination of trust and sustainability concerns in autonomous vehicle adoption. Technol Soc 63:101361
41. Pinney J, Carroll F, Newbury P (2022) Human–robot interaction: the impact of robotic aesthetics on anticipated human trust. PeerJ Comput Sci 8:e837
42. Wortham RH, Theodorou A (2017) Robot transparency, trust and utility. Connect Sci 29(3):242–248
43. Nesset B, Robb DA, Lopes J, Hastie H (2021) Transparency in HRI: trust and decision making in the face of robot errors. In: Companion of the 2021 ACM/IEEE international conference on human–robot interaction, pp 313–317
44. Kraus JM (2020) Psychological processes in the formation and calibration of trust in automation. Doctoral dissertation. Universität Ulm
45. Rossi A, Dautenhahn K, Koay KL, Saunders J (2017) Investigating human perceptions of trust in robots for safe HRI in home environments. In: Proceedings of the companion of the 2017 ACM/IEEE international conference on human–robot interaction, pp 375–376
46. Lazányi K, Hajdu B (2017) Trust in human–robot interactions. In: 2017 IEEE 14th international scientific conference on informatics. IEEE, pp 216–220
47. Hancock PA, Billings DR, Schaefer KE, Chen JY, De Visser EJ, Parasuraman R (2011) A meta-analysis of factors affecting trust in human–robot interaction. Hum Factors 53(5):517–527
48. Koul S, Eydgahi A (2018) Utilizing technology acceptance model (TAM) for driverless car technology adoption. J Technol Manag Innov 13(4):37–46
49. Salem M, Lakatos G, Amirabdollahian F, Dautenhahn K (2015) Would you trust a (faulty) robot? Effects of error, task type and personality on human–robot cooperation and trust. In: Proceedings of the tenth annual ACM/IEEE international conference on human–robot interaction, pp 141–148
50. Abd M, Gonzalez I, Nojoumian M, Engeberg E (2017) Impacts of robot assistant performance on human trust, satisfaction, and frustration. In: RSS: morality and social trust in autonomous robots
51. Uggirala A, Gramopadhye AK, Melloy BJ, Toler JE (2004) Measurement of trust in complex and dynamic systems using a quantitative approach. Int J Ind Ergon 34(3):175–186
52. Khan SK, Shiwakoti N, Stasinopoulos P, Chen Y (2020) Cyber-attacks in the next-generation cars, mitigation techniques, anticipated readiness and future directions. Accid Anal Prev 148:105837

53. Jiang X, Yu W, Li W, Guo J, Chen X, Guo H, Chen T et al (2021) Factors affecting the acceptance and willingness-to-pay of end-users: a survey analysis on automated vehicles. Sustainability 13(23):13272
54. Helldin T, Falkman G, Riveiro M, Davidsson S (2013) Presenting system uncertainty in automotive UIs for supporting trust calibration in autonomous driving. In: Proceedings of the 5th international conference on automotive user interfaces and interactive vehicular applications, pp 210–217
55. Stormont DP (2008) Analyzing human trust of autonomous systems in hazardous environments. In: Proceedings of the human implications of human–robot interaction workshop at AAAI, pp 27–32
56. Koo J, Kwac J, Ju W, Steinert M, Leifer L, Nass C (2015) Why did my car just do that? Explaining semi-autonomous driving actions to improve driver understanding, trust, and performance. Int J Interact Des Manuf (IJIDeM) 9(4):269–275
57. Gromet DM, Kunreuther H, Larrick RP (2013) Political ideology affects energy-efficiency attitudes and choices. Proc Natl Acad Sci 110(23):9314–9319
58. Baker AL, Phillips EK, Ullman D, Keebler JR (2018) Toward an understanding of trust repair in human–robot interaction: Current research and future directions. ACM Trans Interact Intell Syst (TiiS) 8(4):1–30
59. Akalin N, Kristoffersson A, Loutfi A (2022) Do you feel safe with your robot? Factors influencing perceived safety in human–robot interaction based on subjective and objective measures. Int J Hum Comput Stud 158:102744
60. Fisher DL, Lohrenz M, Moore D, Nadler ED, Pollard JK (2016) Humans and intelligent vehicles: the hope, the help, and the harm. IEEE Trans Intell Veh 1(1):56–67
61. Lugano G (2017) Virtual assistants and self-driving cars. In: 2017 15th international conference on ITS telecommunications (ITST). IEEE, pp 1–5

Technological Challenges and Self-Driving Cars: Acceptance and Stress

Péter Szikora

Abstract The proliferation of self-driving cars is creating a number of contradictory feelings among people, which are further complicated by the phenomenon of technostress. The extent to which technostress causes mental distress is unclear from current research and further studies are needed to describe the benefits and drawbacks in more detail. Examining acceptance and analysing attitudes is key to understanding and accepting technological change. In addition, clarification of ethical and legal issues is important to better integrate new technologies into our daily lives. The research methodology combines primary and secondary data, and the use of a questionnaire survey with convenience sampling is a reasonable approach in the light of available resources. The interpretative insights and identification of patterns can help shape future action plans and shape the direction of technological development, thus facilitating the adoption and diffusion of self-driving cars.

Keywords Autonomous cars · Self-driving cars · Technostress

1 Introduction

The uptake and adoption of self-driving vehicles is a dynamic and multifaceted process that is fundamentally reshaping both the automotive industry and people's daily lives. As these technological developments become more and more part of society, the challenges we face become more pronounced. The emergence of self-driving vehicles is not only changing the way we travel, but is also transforming people's habits, safety and lifestyles.

This article provides an overview of the challenges and opportunities of self-driving vehicles and the factors that influence their adoption in society. We examine the steps being taken at government level to integrate autonomous transport and how people are responding to new technological developments. Furthermore, we will look

P. Szikora (✉)
Keleti Károly Faculty of Business and Management, Obuda University, Budapest, Hungary
e-mail: szikora.peter@kgk.uni-obuda.hu

T. A. Kovács et al. (eds.), *The Impact of the Energy Dependency on Critical Infrastructure Protection*, Advanced Sciences and Technologies for Security Applications, https://doi.org/10.1007/978-3-031-78544-3_56

in more depth at the ethical, safety and technological challenges that self-driving cars face, as well as the responses and possible solutions to these. In addition, we will present the social and economic impacts that the uptake of self-driving vehicles could have and what steps can be taken to promote their adoption. By looking at the relationship between the revolutionary potential of self-driving vehicles and their social acceptance, our aim would be to get a clear picture of how these innovations are changing our lives and how we should respond to them in the future.

2 Self-Driving Cars

Technological advances and innovations are part of our everyday lives, but not everyone embraces them with equal enthusiasm. In particular, the spread of self-driving cars has provoked mixed feelings among people [1]. However, few studies are currently available and further research is needed to describe the benefits and drawbacks in more detail. Examining the acceptance of technology and analysing attitudes is key to understanding and adopting technological developments. In addition, it is important to clarify ethical and legal issues in order to better integrate new technologies into our daily lives.

Although self-driving vehicle technology is already available, many questions remain about its reliability and the need for widespread deployment. Self-driving vehicles offer modes of transport where human supervision is not required to perform specific functions. The term "self-driving" best describes this situation and is sometimes used as a synonym for "autonomous" cars. SAE International (2016) defines six different levels of autonomy, with level 0 being full autonomy and level 5 being the highest, where a human driver is no longer required [2] (Fig. 1).

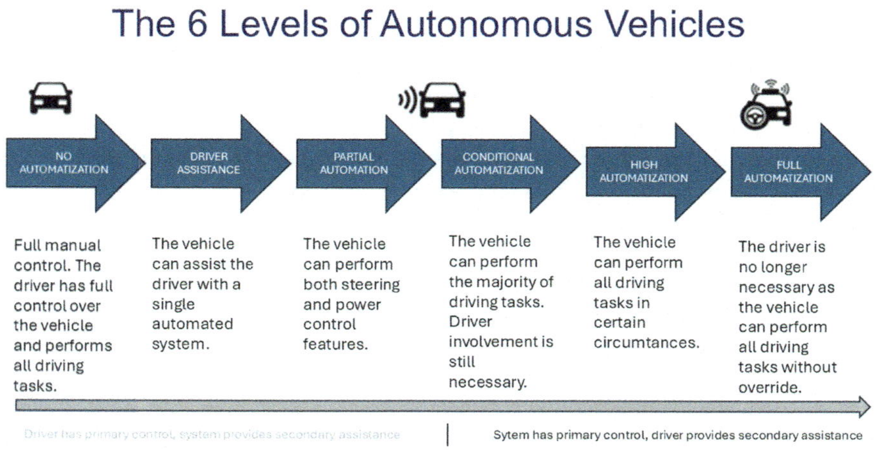

Fig. 1 Automation levels based on [2]

These autonomous levels require varying degrees of human intervention to operate the vehicle. Although autonomous technology is available, legal and moral issues remain and few people are prepared to trust an autonomous system completely. The legal framework governing the introduction of autonomous cars in the EU has not yet moved forward significantly, but there are signs that this situation may change in the near future. In the US, the uptake of self-driving cars is rapidly approaching. The US National Highway Traffic Safety Administration (NHTSA) has issued new guidelines for self-driving cars, emphasising the importance of safety standards for both passengers and road users. Overall, the proliferation of self-driving cars raises complex issues that require further research and legislative action to ensure the safe and effective introduction of the technology [3].

The challenges to the deployment of autonomous systems are based on a number of factors. Not only high prices, but also a number of personal fears are hampering the wider adoption of these technologies [4]. Legal and ethical uncertainties are further barriers to the uptake of self-driving cars. According to the literature, individuals' perceptions of technology are influenced by a number of demographic characteristics, such as age, gender, education and attitudes towards technology [5]. However, it is important to stress that software companies have an interest in the widest and fastest possible adoption of self-driving vehicles. The problem of the acceptance of self-driving cars cannot be addressed in a uniform way, and therefore the use of an interpretative perspective can help to understand the complexity of the phenomenon. This requires an overview of the regulatory background, ethical issues and a deeper understanding of human–robot interaction.

The technological development of self-driving cars is long underway, but mass adoption is still a long way off. This is due, among other things, to the concerns and reliability issues surrounding the technology. Although autonomous vehicles are designed to be able to perform tasks without human intervention, full autonomy is not yet ensured in vehicles or other systems. Defining the different levels of autonomy is necessary to understand the role of humans in vehicle control. These levels describe in detail the extent to which human intervention is required in the operation of the vehicle. This helps to clearly distinguish between autonomous and self-driving vehicles. Legal and ethical issues continue to hinder the widespread adoption and use of autonomous systems. Social concerns must also be taken into account, as many people are not comfortable with handing over full control to a machine [6].

In the European Union, the biggest barrier to the uptake of autonomous vehicles is the old road traffic conventions, such as the Vienna Convention, which requires a human driver to be in the vehicle. However, some countries have already amended these conventions to adapt to new technologies and allow the development and use of autonomous systems under certain conditions. In the United States, the era of self-driving cars is already near. The National Highway Traffic Safety Administration (NHTSA) has issued new regulations for self-driving vehicles that provide precise guidelines for the development and production of such vehicles. Safety is a key concern, and automakers must ensure that self-driving cars are as safe as conventional vehicles, even if they don't have steering wheels or pedals. When autonomous vehicles are introduced, there is a lot of fear and concern among people, which is a natural reaction to new technologies. Car developers and manufacturers must pay particular attention to safety and reliability to make these vehicles acceptable for road use. People are often afraid to hand over full control to a system whose operation is not yet fully understood and regulated. Aviation history has shown similar challenges with the proliferation of aeroplanes, and people have similar concerns about the safety and reliability of self-driving cars. Car manufacturers are spending significant sums on developing self-driving cars, but it is important to recognise that people are more concerned than enthusiastic about the emergence of new technology. Studies show that most drivers do not want to use fully autonomous cars, but would like to see some automated features in their vehicles. Therefore, the car industry needs to take people's concerns and preferences into account when developing self-driving technology [7, 8].

According to 2014 surveys, the majority of people in the US, the UK and Australia expressed a number of concerns about autonomous vehicles, in particular their cost and the reliability of the technology. However, when it came to comfort and driving experience, many people would welcome a higher level of autonomy in their cars without paying extra for it. In contrast, research conducted by Kyriakidis and colleagues in 2015 showed that the majority of respondents were optimistic about the future of self-driving cars, believing that by 2050 a significant proportion of vehicles on the road would be autonomous [9]. However, research conducted by Kettles and Van Belle in 2019 suggested that the majority of people would not be interested in using or buying self-driving cars within the first six months of their local launch. However, people responded positively to the high performance of autonomous vehicles and the enjoyable experiences they provide [10]. People's attitudes towards self-driving cars are much more positive when it comes to public transport. Most people don't care whether the public transport they use is autonomous or not, as long as it is clean and comfortable. All in all, although the technology is available, people are still not sufficiently prepared to use autonomous vehicles. It is important to take people's opinions and attitudes into account in order to understand the barriers to the uptake of the technology and to accelerate its take-up. As with any new technology, people's reactions are variable and may change over time. It is important to consider the human side in the development and integration of autonomous technology [11–13].

When it comes to technological developments, it is not only the technology itself that is important, but also the readiness of users and external observers. If the level of readiness is low, this can lead to lower adoption, demand and penetration of the technology. For this reason, trust is one of the most important aspects to be taken into account when presenting and developing a technology. This is particularly true for self-driving cars. Trust is a key driver for the uptake of self-driving cars. People perceive their vulnerability in complex systems, which increases the lack of trust. Building trust in such complex systems is not easy and there is no clear way to effectively encourage or improve it in autonomous vehicles. The widespread emergence of self-driving cars raises ethical and societal dilemmas that need to be taken into account. Autonomous vehicle software and hardware providers are likely to stick to approaches that best protect their own and their companies' reputations, but it is important that people feel comfortable around self-driving cars on the roads. Car passengers may also be concerned that they will lose control of the vehicle and that the car will make decisions that may be contrary to their well-being and physical safety. These concerns pose additional challenges to the adoption and uptake of self-driving cars [14, 15].

Self-driving cars and the computer programs that control them have to make many difficult moral choices, especially in extreme situations. One such example is when a child suddenly runs out in front of the car and a collision is unavoidable. The car is faced with the dilemma of whether to jerk the steering wheel, endangering the occupants, or to hit the child on the road. Such ethical issues can have a major impact on the acceptance of autonomous vehicles in society. Research into the ethical decisions of self-driving cars aims to determine how cars should be programmed to make decisions in the event of unavoidable accidents. For example, whether to minimise the loss of life, even if passengers have to be sacrificed or the occupants must be protected at all costs. The answers to such ethical decisions are important for the social acceptance of self-driving cars. Research shows that people generally accept the idea that cars should reduce the number of casualties in an accident [16]. However, this opinion may change when people imagine themselves in the car in a given situation. Regular software updates and improvements are essential to ensure that self-driving cars work. The software must be prepared for any situation and be able to make life safety decisions. It is also important that it can recognise and distinguish roadside objects, such as traffic signs, and make decisions about the vehicle's behaviour based on these. Overall, the development and adoption of autonomous vehicles poses many challenges, but the discussion of ethical issues and the continuous development of software are key steps in shaping the future of autonomous transport [16–19].

When it comes to security risks associated with autonomous vehicles, it is important to understand that any computer that communicates with or provides access to another is potentially vulnerable to cyber hacking. Autonomous cars are no exception to this, and there have been cases of hackers manipulating these vehicles to carry out terrorist attacks [20]. One such example was when two hackers took over the UConnect system of a Jeep Cherokee for testing in 2015. The journalist in the car sat helplessly while the hackers controlled the car, including the pedals and steering wheel. These cases have highlighted that the deployment of safety systems in autonomous vehicles is a huge challenge for both car manufacturers and consumers [21]. In addition, autonomous vehicles face not only ethical dilemmas at the societal level, but also technological uncertainties. This includes issues such as data security and the possibility of attacks by hackers. Autonomous vehicles must always prioritise security and be prepared for potential attacks. The acceptance of autonomous vehicles is influenced by people's perception of technological security and privacy. According to the technology acceptance model, people are willing to accept a new technology if they perceive its usefulness and ease of use. However, people may be willing to trust autonomous vehicles less if they are given too much information about how the system works and do not fully understand it. Therefore, communication and transparency about autonomous vehicles should be provided to reduce people's concerns and increase their confidence. Overall, addressing safety risks and increasing technological safety are key steps to promote the uptake and adoption of autonomous vehicles. It is important that the manufacturers and developers of autonomous vehicles take appropriate measures to ensure safety and that people are properly informed about how the technology works and how safe it is. This will enable them to trust autonomous vehicles and be open to using them in the future.

3 The Research

I chose an anonymised quantitative questionnaire as the data collection tool for the research. Anonymity allows participants to answer the questions honestly without worrying about negative consequences. This is particularly important when the regulatory environment and societal pressures influence consumer choices, and a clear 'right' answer is expected, which may differ from actual behaviour. Anonymised questionnaires allow a large sample size, increasing the statistical reliability and generalisability of the research. They allow for rapid data collection and analysis while being cost-effective. Online questionnaires also allow participants to take part in the research at their own convenience.

It is important to note that the anonymous questionnaire method has limitations and disadvantages. The low response rate may distort the sample and may not always allow causality to be explored. Bias may also occur and it is difficult to verify or confirm the reliability of responses. Furthermore, anonymous questionnaires do not provide participants with immediate feedback or clarification of questions, which can cause problems if questions are unclear or ambiguous.

Although anonymised questionnaire surveys can be effective, it is important to consider their limitations and disadvantages. In the convenience sampling procedure I used, I did not randomly select participants, and research results may be biased as a result. Convenience sampling may result in additional representativeness gaps as it is difficult to ensure randomness and objectivity in sampling.

Presentation of the Sample

For demographic variables, I asked respondents about gender, age, place of residence and highest level of education. In addition, I was interested in the respondent's current level of education, employment status, attitudes towards technology in general, level of awareness, and whether they have a driver's license. Related to the topic, I sought to elicit respondents' opinions on self-driving cars by making various statements, which I rated on a 5-point Likert scale.

The gender composition of the sample (men: 1266, women: 1206) indicates that the sample includes a significant number of both sexes and can therefore be considered balanced in this respect. On the basis of these data, it does not seem relevant for the analysis whether there is a gender difference between the sample members, and I therefore no longer consider the gender difference to be a determining factor in the analysis of the relationship between the variables. Individuals who have a non-binary gender identity or different sexual orientation may face their own unique challenges in emotional expression and experience. Social attitudes and discrimination may affect their emotional well-being and emotional expression. The sample included a total of 18 individuals who did not represent either a female or male gender identity. While this fact alone should not be taken as a rigid conclusion, this low figure of 0.72% suggests that the largest proportion of respondents identified with their biological gender identity. The questionnaire was completed by 2490 respondents. Of these, 1266 were male and 1206 female, with 18 respondents choosing not to identify their gender. Although the sample is not balanced in terms of gender, the number of respondents of both genders allows us to examine the difference in their opinions on self-driving cars based on this demographic variable.

The average age of the respondents was 25.25, which is considered to be the young adult age group. The average age of the respondents is shown in the graph in Fig. 2, where the highest peak at 20 years indicates that it was the most frequently (n = 363) responded year. Although the average age of respondents was 25.15 years, indicating that the majority of respondents were young, the standard deviation was 9.2534, indicating that the questionnaire reached a relatively wide range of ages—we managed to reach people aged between 12 and 70. The relatively young average age of the respondents is probably a consequence of the snowball method chosen and the convenience sampling, as this is the age range I reached (Fig. 3).

As the graph shows, around 80% of respondents were under 25, 12% were aged between 26 and 40, and the remainder were over 41. Although the sample cannot be said to be representative in terms of age, I nevertheless thought it would be worthwhile to investigate age-related differences in opinions about self-driving cars and age-related correlations.

The distribution of respondents by highest level of education is shown in Fig. 4. The majority of people have the highest education at secondary school level, with 479 people responding that BSc is their highest education and 140 people with MSc as their highest education. The graph basically reflects the age distribution of the respondents, as those who are young can only have a lower level of education. Obviously, the highest level of education is also related to the current level of education of the respondents; those who have completed primary school are now in secondary school and those who have completed secondary school are now studying for a Bsc.

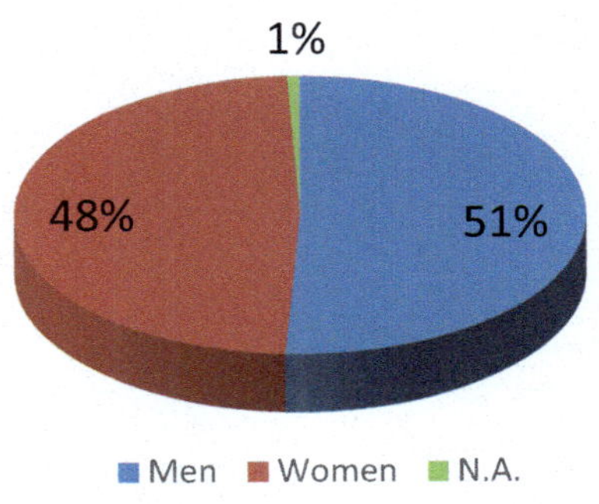

Fig. 2 Distribution of respondents by gender

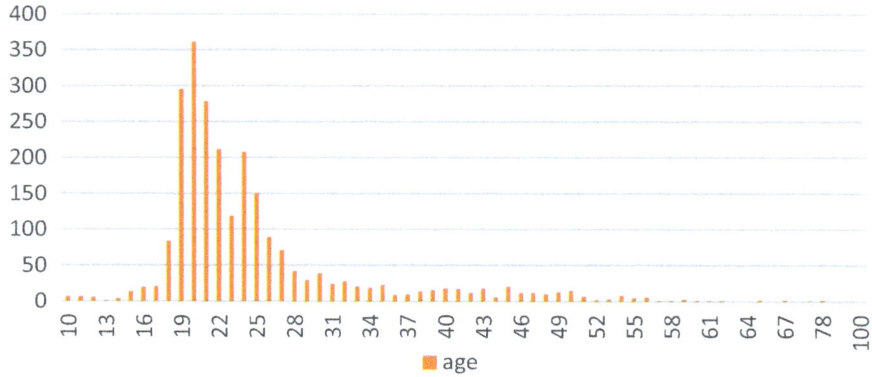

Fig. 3 Age distribution of respondents

What is interesting to note is that less than 1% of respondents indicated that they are not studying, but 56% of respondents are working, which means that a relatively large number of people are studying while working or working while continuing their studies (Fig. 5).

What is interesting about the survey is that although the first respondents were mainly from the capital due to convenience sampling, the snowball method was basically successful, as I managed to get responses from people living in different sizes of dwellings, reflecting not only from Cities and Towns, but even from Villages (Fig. 6).

Interestingly, 69% of respondents have a driving licence, while 31% do not. This means that those who have experience of driving are in the majority. While the option of driving is open to everyone, and thus all opinions are valuable on the subject of self-driving cars, it is conceivable that those who have a driving licence have more experience in this area. However, my research showed no significant difference in attitudes between those who have a driving licence and those who do not. This 31%

Fig. 4 Distribution of respondents by educational level

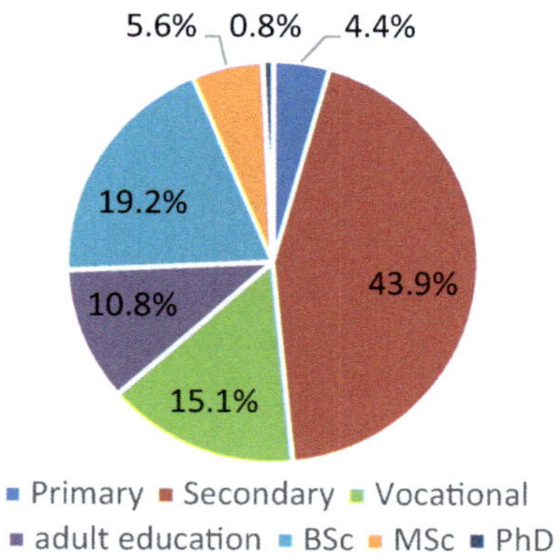

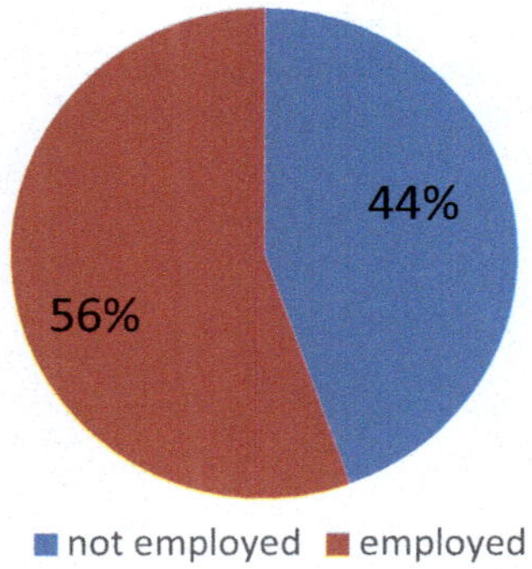

Fig. 5 Distribution of respondents by whether they are working or not

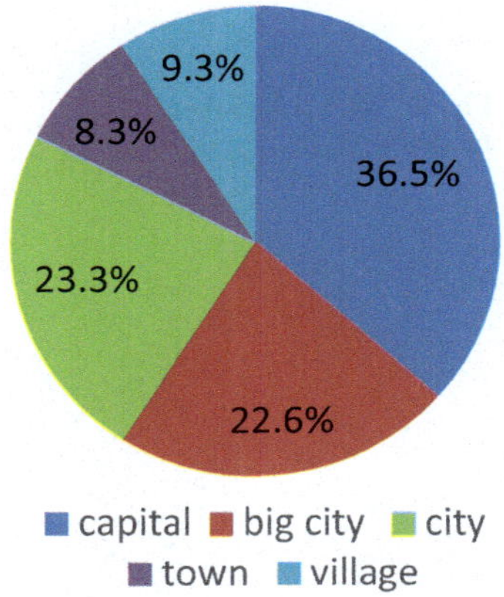

Fig. 6 Distribution of respondents by place of residence

figure is probably due to the fact that many respondents under 18 who have not yet obtained their driving licence and some who do not wish to drive were also included in the survey.

In general, respondents had different attitudes towards self-driving cars. When assessing their confidence in autonomous vehicles using the 6 levels of autonomy given in the literature, the following distribution emerges As can be seen in Fig. 7, the majority of respondents were only interested in the lower level of autonomy of vehicles. Only 6.1% would be interested in using autonomous vehicles where the system is fully in control, and 22.3% would like to use a self-driving car where they can regain control of the vehicle if necessary. These data are in line with the international findings mentioned earlier [15]; the majority of respondents in our sample are not yet ready for self-driving cars, their introduction or their potential use (Fig. 8).

I also looked at whether there is a pattern that clearly identifies those who support self-driving vehicles and those who oppose their use on the roads. The responses to the statement "I would support the introduction of autonomous vehicles" can be used to measure this. The distribution of respondents is shown in Table 1. Approximately

Fig. 7 Distribution of respondents whether they have a driving licence or not

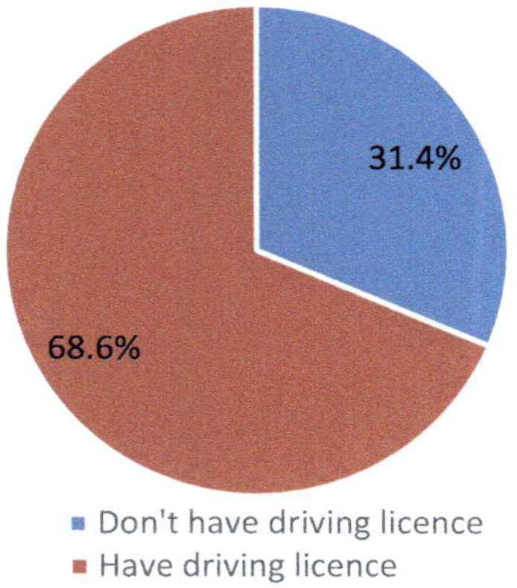

Fig. 8 What self-driving mode would be supported

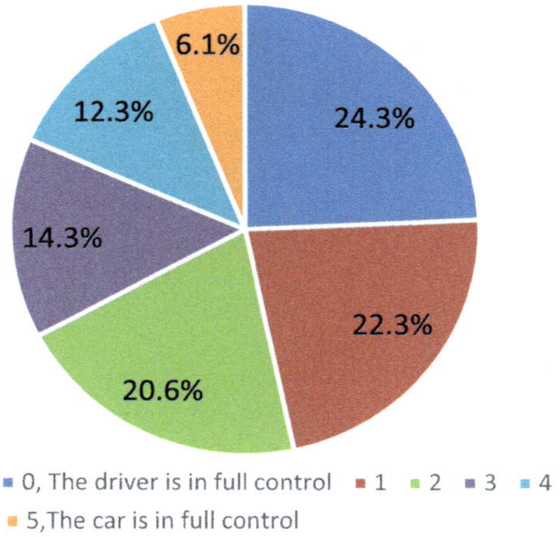

43% of respondents (perhaps not as direct users/owners, but) would support the introduction of self-driving cars, compared to only a quarter of respondents who were against, meaning that the responses to the middle option, "Both yes and no", were the most frequent, meaning that a very high percentage of respondents could not decide whether they would support or not the introduction of self-driving cars. The responses of those who chose the middle option, neither for nor against the introduction of autonomous vehicles, are excluded from the following analysis and only a sample of 1655 respondents is considered. Interestingly, as shown in the cross-tabulation below, female respondents were more opposed to the introduction of autonomous vehicles, while the number of men in favour of such vehicles was almost twice as high as the number of men against.

We find that men and women have different opinions about self-driving vehicles, which is in line with the results of previous international research [6]. Women are

Table 1 support for self-managers

I support the introduction of self-driving cars			
	Male (%)	Female (%)	Total (%)
I do not support	27	37	26
I support	43	28	43

generally more opposed to the introduction of self-driving vehicles than men. This result also confirms the findings in the literature that women are more opposed than in favour, i.e. women tend to be more opposed to the introduction of self-driving cars than in favour. This phenomenon may be influenced by a number of factors, such as women's sense of security and distrust of new technologies, as well as general doubts and fears about autonomous vehicles. On the basis of these observations, it would be important to investigate in detail gender differences and attitudes in order to develop more effective communication strategies and policies to promote the adoption of self-driving vehicles (Fig. 9).

According to the survey results, nearly half of the respondents are open to trying autonomous vehicles, but only a quarter plan to buy one in the near future. The graph also suggests that people are relatively less concerned about the potential environmental impact of self-driving cars. Therefore, the majority of respondents

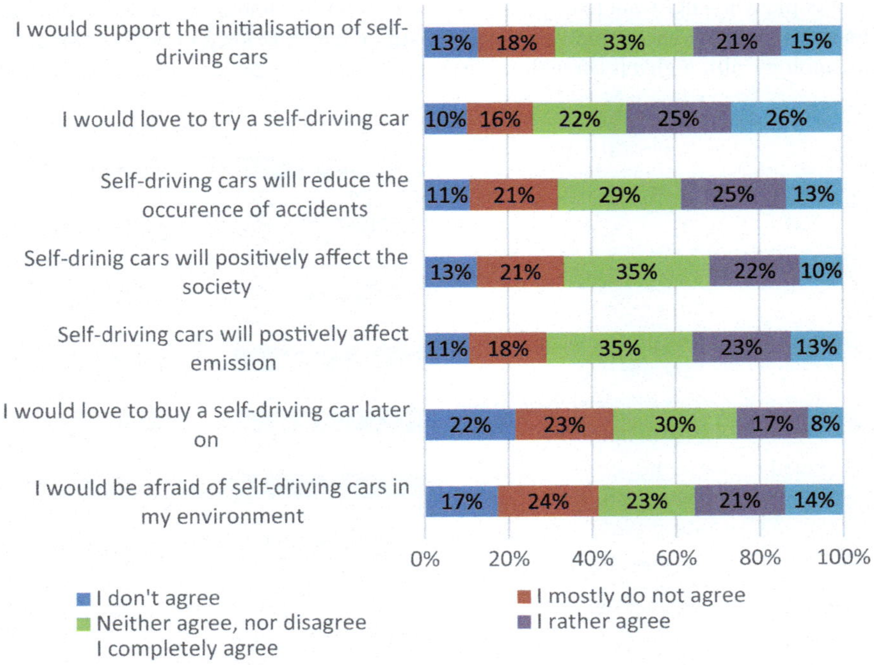

Fig. 9 Opinions on self-driving cars

are optimistic about the introduction of autonomous vehicles and believe that they will have a positive impact on both transport and the environment. However, it is worth noting that many of them could not form a clear opinion about them, probably due to their novelty and unfamiliarity. It is therefore important to promote social dialogue and education about autonomous vehicles so that people have a clearer picture of the benefits and risks of the technology. It is also worth examining if we exclude neutral responses (those who could not make up their minds) and only look at whether or not the question is specific to the respondents. In this case we are likely to obtain more interesting results. For example, it can be observed that, although "I would like to buy a self-driving car" is not considered to be very typical, trying self-driving cars is an attractive option for the vast majority, which may suggest that people are open to trying new technologies even if they do not plan to buy them. For the other questions, there are generally similar proportions of typical and non-typical responses, but it is interesting to note that while there is less fear of self-driving cars, many people see their potential in reducing accidents and environmental impacts. These latter findings suggest that further research on the impacts and benefits of self-driving technologies would be particularly important to ensure a wider societal understanding and acceptance of these innovations and their effective integration into everyday life. Better awareness and communication would help to dispel doubts and misgivings about self-driving vehicles, thus facilitating their acceptance and uptake by society (Fig. 10).

People who later want to buy a self-driving car tend to find life without technology on a desert island less attractive. They also find it easier to keep up with advances in technology, although it is often the case that the electronic devices of such individuals

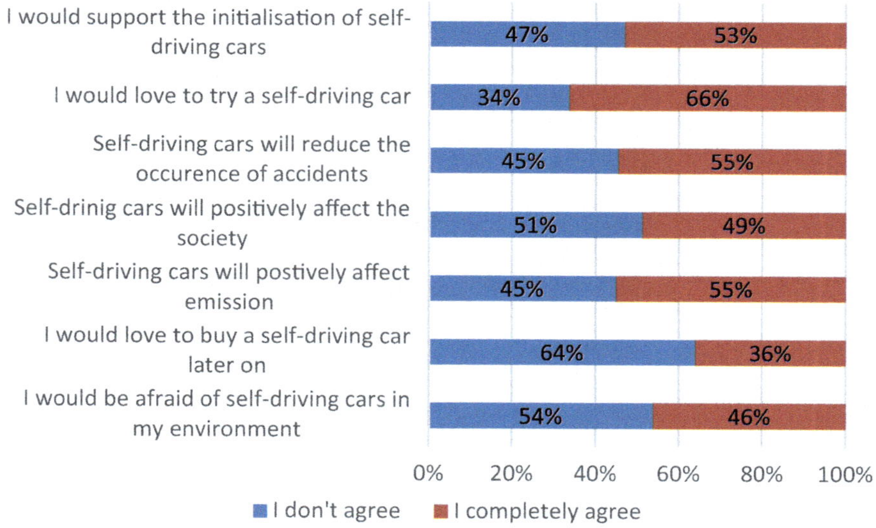

Fig. 10 Opinions on self-driving cars, only extremes

do not always work properly, although the correlation is low. This may be because such people tend to have more electronic devices than those who are not considering buying a self-driving car. A self-driving car is an expensive piece of equipment, so those who can afford it are likely to be inclined to buy other smart, technologically advanced devices. The more electronic devices someone has, the more likely it is that one of them will malfunction or fail. This may be explained by the fact that people considering self-driving cars tend to have a higher number of electronic devices, but of course there may be other explanations for these correlations and statistical error should be taken into account as the sample is not very large.

The statement "I would like to buy a self-driving car later" is strongly positively correlated with the introduction of self-driving cars and their expected positive effects. Accordingly, those who would like to have such a car believe that it will not only reduce emissions and accidents, but also have a positive impact on society as a whole. Those who would buy such a car are likely to be keen to test drive it, which suggests a strong correlation in the Table 2. Future potential buyers of self-driving cars are not overly concerned about car breakdowns or technological risks. They are also not concerned that they may not be able to take back control of the car or that drivers may lose their jobs due to the introduction of self-driving cars. So, in general, those who would buy such a car have a positive attitude towards the technology, believe in the potential positive effects of self-driving cars and are less concerned about the potential risks.

The statement "I would like to try a self-driving car" is similarly correlated with other statements such as "I would like to buy a self-driving car later". Those who are enthusiastic about trying self-driving cars tend to have a higher level of technological knowledge and do not feel that they would be better off without technology. However, they also experience, albeit less frequently, that their electronic devices malfunction. They also believe in the positive impact of the introduction of self-driving cars. Those who would like to try self-driving cars are less worried about car breakdowns or technological risks. The more curious someone is about trying self-driving cars, the less afraid they are of the potential dangers.

Positive beliefs about the emissions of self-driving cars are moderately strongly correlated with the intention to buy self-driving cars. The positive impact of self-driving cars on society and on the number of car accidents is also strongly correlated with purchase intention. This may mean that people who believe in the positive effects of self-driving cars are more likely to buy such cars. Fear of car breakdowns and concern about loss of control are negatively correlated with enthusiasm for self-driving cars, although the strength of the fear is greater than the fear of car breakdowns.

Table 2 Correlation table

		I would love to buy a self-driving car later on	I would love to try a self-driving car
I would be afraid of self-driving cars in my environment	*	– 0.141	– 0.161
	**	0.000	0.000
I would love to buy a self-driving car later on	*	1.000	0.478
	**		0.000
Self-driving cars will postively affect emission	*	0.422	0.388
	**	0.000	0.000
Self-drinig cars will positively affect the society	*	0.530	0,436
	**	0.000	0.000
Self-driving cars will reduce the occurence of accidents	*	0.448	0.464
	**	0.000	0.000
I would love to try a self-driving car	*	0.478	1.000
	**	0.000	
I would support the initialisation of self-driving cars	*	0.601	0.601
	**	0.000	0.000
The driver is in full control—perceived safety	*	– 0.082	– 0.039
	**	0.000	0.053
The driver is in control, the car is either parking or heeding the speed limit on its own—perceived safety	*	0.092	0.244
	**	0.000	0.000
The car can take control of the pedals and the steering wheel at the same time, but the driver keeps an eye on it and intervenes if necessary—perceived safety	*	0.307	0.405
	**	0.000	0.000
The car can change lanes and react to the environment. The task of the driver is to pay attention; take back control—perceived safety	*	0.376	0.397
	**	0.000	0.000
The car is in control, but can ask for human help—perceived safety	*	0.375	0.408
	**	0.000	0.000
The car is in full control—perceived safety	*	0.432	0.276
	**	0.000	0.000
Preferred level of autonomy	*	0.287	0.287
	**	0.000	0.000
The owner should be responsible	*	– 0.007	– 0.077
	**	0.727	0,000

(continued)

Table 2 (continued)

		I would love to buy a self-driving car later on	I would love to try a self-driving car
The one sitting in the car should be responsible	*	0.035	0,069
	**	0.084	0,001
The producer of the car should be responsible	*	– 0.079	0.021
	**	0.000	0.305
The producer of the software should be responsible	*	– 0.026	0.037
	**	0.202	0.067
The government should be responsible	*	0.010	– 0.059
	**	0.615	0.003
The insurance company should be responsible	*	0.048	0.009
	**	0.016	0.653
Who should the self-driving car save/protect in an accident situation, where stopping is not an option anymore?	*	– 0.063	0.005
	**	0.002	0.808
On the basis of what principle shall the self-driving car decide?	*	– 0.082	– 0.027
	**	0.000	0.180
If 1 are the driver, who should the car save?	*	– 0.060	– 0.037
	**	0.003	0.062
Hackers tamper with the car	*	– 0.046	0.019
	**	0.023	0.342
The self-driving system breaks down	*	– 0.180	– 0.024
	**	0.000	0.229
The car decides different from how I would like it to	*	– 0.159	– 0.072
	**	0.000	0.000
Fear from the new technology	*	– 0.049	– 0.172
	**	0.015	0.000
People (in various professions, such as chauffeurs) will lose their jobs	*	– 0.102	– 0.128
	**	0.000	0.000
Control cannot be regained	*	– 0.158	– 0.056
	**	0.000	0.005
Too expensive	*	0.007	0.096
	**	0.735	0.000
I loose the joy of driving	*	– 0.229	– 0.049
	**	0.000	0.015

(continued)

Table 2 (continued)

		I would love to buy a self-driving car later on	I would love to try a self-driving car
The security of personal data cannot be provided	*	– 0.084	– 0.069
	**	0.000	0.001
What is your gender?	*	– 0.078	– 0.128
	**	0.000	0.000
How old are You?	*	0.107	0.001
	**	0.000	0.947
What is your nationality	*	0.099	– 0.106
	**	0.000	0.000

*Correlation Coefficient, **Sig. (2-tailed)

4 Summary and Conclusions

In our article, we showed that although self-driving vehicles are increasingly present in public transport, their private use is still in its infancy. Despite the availability of adequate hardware and software and the growing number of semi-autonomous vehicles, public acceptance remains low. Fear is a key factor in the use of self-driving vehicles, especially among women. Research data presented in this article shows that those who are not open to self-driving vehicles tend to feel more fear of possible negative consequences, such as hacker attacks, system failures or lack of control. On the other hand, those who support automated vehicles believe they can have positive impacts, for example by reducing accidents and promoting environmental protection. Overall, the article helps to understand the factors associated with trust in self-driving systems and provides lessons for manufacturers and policymakers to address concerns and integrate self-driving vehicles into the private transport system of future smart cities.

References

1. Dragano N, Lunau T (2020) Technostress at work and mental health: concepts and research results. Curr Opin Psychiatry 33(4):407–413
2. SAE International: 2016 U.S. DoT chooses SAE J3016 for vehicle autonomy policy guidance. https://articles.sae.org/15021. Accessed 20 Jan 2023
3. Atiyeh C (2021) Self-driving cars' look, feel is clearer through final U.S. safety rules. https://www.caranddriver.com/news/a35247978/us-autonomous-car-safety-rules-finalized. Accessed 20 Jan 2023
4. Gál L, Sipos T (2018) The impact of autonomous vehicle penetration on specific economic loss values. Transp Sci Rev 68(4):74–82

5. Jaradat M, Jibreel M, Skaik H (2020) Individuals' perceptions of technology and its relationship with ambition, unemployment, loneliness and insomnia in the Gulf. Technol Soc 60:101199
6. Schoettle B, Sivak M (2015) Motorists preferences for different levels of vehicle automation. [Online]. Available:http://deepblue.lib.umich.edu/bitstream/handle/2027.42/114386/103217.pdf?sequence=1&isAllowed=y
7. Consumers in US and UK Frustrated with Smart Devices That Frequently Crash or Freeze New Accenture Survey Finds. http://newsroom.accenture.com/article_display.cfm?article_id=5146. Accessed 27 May (2023)
8. Yvkoff LM (2024) Car buyers show interest in autonomous car tech. http://reviews.cnet.com/8301-13746_7-57422698-48/many-car-buyers-show-interest-inautonomouscar-tech. Accessed 27 May 2023
9. Kyriakidis M et al. (2015) Public opinion on automated driving: results of an international questionnaire among 5000 respondents. Transp Res Part F: Traffic Psychol Behav 32:127–140. https://doi.org/10.1016/j.trf.2015.04.014,
10. Kettles N, Van Belle JP (2019) Investigation into the antecedents of autonomous car acceptance using an enhanced UTAUT model. In: International conference on advances in big data, computing and data communication systems, 2019. IEEE, Winterton. https://doi.org/10.1109/ICABCD.2019.8851011
11. Szikora P, Szatmáry R (2023) Autonomous cars-what lies behind the lack of readiness. Interdiscip Descr Complex Syst: INDECS 21(4):399–411
12. Lazányi K (2018) Are we ready for self-driving cars-a case of principal-agent theory. In: 2018 IEEE 12th International symposium on applied computational intelligence and informatics (SACI). IEEE, pp 000251–000254
13. Lazanyi K (2018) Readiness for artificial intelligence. In: 2018 IEEE 16th international symposium on intelligent systems and informatics (SISY). IEEE, pp 000235000238
14. Zhang Y, Chen X, Ma J, Yu L (2024) Environmental impact of autonomous cars considering platooning with buses in urban scenarios. Sustain Cities Soc 101:105106
15. Cugurullo F, Acheampong RA (2023) Fear of AI: an inquiry into the adoption of autonomous cars in spite of fear, and a theoretical framework for the study of artificial intelligence technology adoption. AI SOCIETY, 1–16
16. Servin C, Kreinovich V, Shahbazova SN (2023) Ethical dilemma of self-driving cars: conservative solution, in recent developments and the new directions of research, foundations, and applications: selected papers of the 8th world conference on soft computing, Baku, Azerbaijan, vol II. Cham, Springer Nature Switzerland, pp 93–98
17. Woollard F (2023) The new trolley problem: driverless cars and deontological distinctions. J Appl Philos 40(1):49–64
18. Paulo N (2023) The trolley problem in the ethics of autonomous vehicles. Philos Q, pqad051
19. Königs P (2023) Of trolleys and self-driving cars: what machine ethicists can and cannot learn from trolleyology. Utilitas 35(1):70–87
20. Padmaja B, Moorthy CV, Venkateswarulu N, Bala MM (2023) Exploration of issues, challenges and latest developments in autonomous cars. J Big Data 10(1):61
21. Miller C, Valasek C (2015) Remote exploitation of an unaltered passenger vehicle. Black Hat USA, p 91

Military 5G as a Critical Information Infrastructure

Andras Toth

Abstract Fifth-generation wireless technology, 5G, promises a seismic shift in military operations by vastly improving data transfer rates. This enables seamless communication and intelligence sharing. This document meticulously examines how 5G capabilities could fortify military prowess combined with the Internet of Battlefield Things (IoBT). Potential boons and obstacles are highlighted. Advanced 5G enables unprecedented data transmission speeds, revolutionizing battlefield communications and intelligence networks. This paper thoroughly assesses leveraging 5G's potential through integration with the Internet of Battlefield Things (IoBT), including operational benefits and challenges. This research investigates the most recent advancements in 5G network technology. It examines the growth of technology, addressing security concerns, and the expenses involved in applying 5G for defense uses. The study further delves into the relationship between 5G and the Internet of Battlefield Things. It underscores how 5G and IoBT can help raise awareness, enhance command, and quicken operations. The research demonstrates that the role of 5G military networks is vital for effective communication and data exchange among military members. Thus, defense organizations must allocate funds to build the necessary infrastructure for incorporating 5G features into their operations. Military forces gain a major edge by using 5G networks on the battlefield. These high-speed, responsive networks enable more effective communication and data sharing. This streamlines decision-making and boosts overall mission success. With 5G's cutting-edge capabilities, armed forces can outmaneuver rivals in modern combat situations. The potential to transform how troops coordinate actions and access intel is groundbreaking. Critical infrastructure powered by next-generation networks revolutionizes warfare dynamics. Troops leverage rapid data transfers to make split-second choices.

Keywords Military 5G · Critical information infrastructure · Enhanced operational capabilities · Internet of battlefield things · Situational awareness · Critical information infrastructure

---

A. Toth (✉)
Ludovika University of Public Service, Budapest, Hungary
e-mail: toth.hir.andras@uni-nke.hu

T. A. Kovács et al. (eds.), *The Impact of the Energy Dependency on Critical Infrastructure Protection*, Advanced Sciences and Technologies for Security Applications, https://doi.org/10.1007/978-3-031-78544-3_57

1 Introduction

5G technology for the military is a revolutionary advance in critical information infrastructure that will significantly improve operational efficiency. The technology offers advanced features such as high positioning and non-destructive analytics that help transform military forces. It provides several benefits, including high data latency and low latency. It also enhances connectivity, provides instant status visibility, and increases mobility and flexibility, allowing security personnel to operate quickly in complex and remote locations. The Army's 5G technology also provides superior cybersecurity, ensuring the integrity and security of military data and information is properly maintained. Therefore, 5G technology in the military is delivering cutting-edge advancements that will provide unparalleled benefits to defense forces and transform their operations and the defense of their country.

The fundamental role of 5G technology in military infocommunications and operations is paramount, providing unprecedented speed, lower latency, and improved capacity. It enables real-time fact-changing, advanced surveillance and reconnaissance, and the seamless operation of unmanned and self-sustaining devices that are vital to prevail in the high-stakes environment of naval operations. The low latency of the 5G era will ensure that unmanned and autonomous systems can operate seamlessly, improving military surveillance and reconnaissance. With the 5G era, unmanned systems can send and receive information in real-time, taking into account more accurate and reliable data collection, which is critical in military operations. Furthermore, the overcapacity of 5G technology can support the simultaneous operation of multiple devices, enabling seamless information exchange between specialized systems through increased operational efficiency and coordination. This method will allow differentiating systems to talk to each other seamlessly, accounting for faster decision-making and more efficient data communication.

Today's military operations mechanisms require advanced technologies to ensure successful missions and the protection of troops. With its remarkable advances, the 5G generation is proving to be a key tool for infocommunications, surveillance, and reconnaissance. Its high data rates, low latency, and ability to support massive machine-type communications make it a perfect candidate for military applications. Devices exploiting the 5G era can drastically improve situational awareness and operational effectiveness in complex military operations. For example, 5G-equipped unmanned aerial vehicles (UAVs) can transmit high-resolution video feeds in real-time to ground control stations, enabling the unique targeting, monitoring, and assessment of enemy movements without compromising troops. In addition, 5G technology supports network-centric operations, integrating and coordinating distinctive military assets that include ships, aircraft, and ground vehicles. This capability creates an extra-connected and effective military force that can adapt quickly to dynamic conditions. Therefore, adopting 5G generation in the defense sector is a huge step towards enhancing the capabilities of defense forces worldwide. The technology's ability to provide fast, reliable, and secure communications and assistance to various military assets makes it a vital tool for today's military.

2 5G and Internet of Battlefield Things

The Internet of Battlefield Things (IoBT) is a revolutionary framework that integrates various connected devices, sensors, and systems to enhance military capabilities on the battlefield. IoBT includes networked battlefield devices with two-way communication, capable of transmitting operational battlefield data, information, and operational situational awareness to other devices and sharing it in near-real time using a specific technology (databases, file sharing, cloud-based systems) to support decision-making at the tactical level [1]. This framework includes wearables, sensors on military vehicles, autonomous weapons systems, and a communications environment that creates a seamless and responsive military environment.

The IoBT will be enriched with state-of-the-art algorithms for statistics evaluation, artificial intelligence for autonomous selection, and advanced surveillance structures for intelligence gathering. These technologies are crucial for processing the large amounts of data generated by IoBT devices and transforming them into actionable intelligence information. IoBT sensors can detect enemy movements, and device algorithms analyze this information to provide informed responses. As technological advances continue to evolve, the future of IoBT could also pave the way for more autonomous and sensitive battlefield systems. This could lead to a paradigm shift similar to warfare. IoBT implementation may also include the deployment of UAVs, robots, and other superior technologies capable of operating autonomously in the field. Such applications offer promising opportunities in future military operations, including superior situational awareness, enhanced logistical advances, and the ability to execute missions with greater precision and efficiency [2].

Ultimately, the future of IoBT depends on balancing its capabilities and mitigating its threats, while also ensuring the security of IoBT structures through advanced encryption, encrypted communication protocols and continuous monitoring of potential vulnerabilities. In this way, IoBT can be leveraged to create decisive advantages on the battlefield, increasing the competencies of armies and enhancing the safety of military personnel.

2.1 Characteristics of the Interconnection of 5G and the Internet of Battlefield Things

Integrating 5G technology and the IoBT represents a major change in modern warfare. This integration can provide superior operational capabilities and technological advantages for military forces. 5G networks provide the infrastructure for a vast network of connected devices, facilitating real-time data exchange and conversation between battlefield devices [3]. This unique combination of technology provides situational awareness, accuracy, and operational tempo, enabling commanders to acquire, process, and act on significant amounts of data in real-time.

The 5G-powered IoBT (5G-IoBT) ecosystem will enable a network-centric process of warfare in which all devices are interconnected, contributing to a unified and dynamic military force. The excessive bandwidth of 5G networks enables the transmission of large information-gathering data streams of high-resolution video from surveillance drones. At the same time, low latency is essential for real-time control of remote weapon systems. In addition, 5G's ability to support multiple devices simultaneously ensures that IoBT can scale to the demands of complex military operations without compromising performance.

However, the networked battlefield's accelerated attack surface and complexity increase critical security issues. Cyber security becomes even more important, requiring superior cryptographic solutions, robust secured exchange protocols, and strong countermeasures against digital warfare techniques. Ensuring the security and resilience of these structures is of the utmost importance to preserve operational integrity and protect against the possibility of hostile interference.

The confluence of 5G and IoBT has far-reaching implications for the future of warfare. It offers a revolution in processes and methods in which autonomous systems play a significant role, and human decision-making is deeply intertwined with artificial intelligence. Addressing ethical concerns is imperative to ensure that the use of advanced systems is consistent with international law and the concepts of armed conflict [4].

2.2 *The Opportunities of Cloud Computing*

Integrating cloud computing and 5G-IoBT has significantly increased the military's technological capabilities and operational efficiency. A centralized cloud computing platform allows the military to collect data from various IoBT sources, facilitating fast and efficient decision-making. The superior analytics, artificial intelligence processing, and machine learning algorithms provided through cloud computing will deliver previously unavailable analytics and capabilities. With these capabilities, the military can discover patterns, predict consequences, and comprehensively understand the battlefield and enemy activity. This data can help operational leaders make informed decisions, improve situational awareness, and save lives. In addition, the scalability and flexibility of cloud-based responses in the 5G-IoBT framework enables efficient resource allocation. The military can remotely introduce new applications and services without modifying its physical infrastructure. This capability saves time, resources, and money, allowing the armies to respond quickly to changing circumstances [5].

The evolution of technology has led to cloud computing becoming an essential element of military strategic processes. However, combining cloud computing must be approached cautiously, focusing on security and flexibility to reduce risk and ensure mission success. Proper planning and meticulous implementation are important to maximize benefits while minimizing risks. These solutions can provide armies various benefits, including enhanced situational awareness, real-time collaboration between units, advanced communications and coordination, and a robust and flexible information storage platform. Integrating cloud computing responses with 5G-IoBT provides the military a significant opportunity to boost their technological capabilities and operational performance. Cloud computing solutions with 5G-IoBT can also provide valuable insight and intelligence to militaries, helping them make informed decisions and complete missions. Cloud-based analytics can examine large amounts of data and provide real-time insights into various components of Army operations, including logistics, supply chain management, and troop activities. Cloud computing can also facilitate real-time collaboration between individual assets, regardless of location, when time is of the essence, including in military operations.

In addition, cloud-based communication and coordination tools can ensure that each device has access to the same data, reducing the chances of miscommunication, misunderstandings, or confusion. Finally, it offers a stable, secure, and flexible platform for storing data that can be accessed anywhere. This can be particularly beneficial for military personnel stationed in remote locations.

The topology shown in Fig. 1 illustrates a complex network system that highlights the interconnection between 5G technology and IoBT. The network's core is 5G technology, surrounded by different elements with various characteristics. These include the core control plane, the user plane, individual network services, and shared databases. Fig. 1 shows the IoBT concept with the general public network on the left and the military private network on the right. This concept demonstrates the interconnection of devices and structures that can be used in modern warfare.

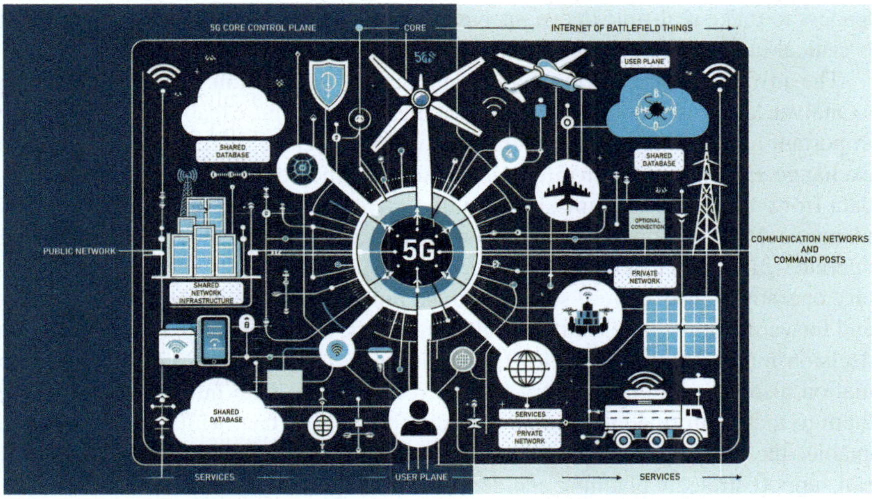

Fig. 1 5G private network infrastructure

3 5G as a Critical Information Infrastructure in Military Operations

5G technology has become a critical information infrastructure that delivers high-speed connectivity, low latency, massive network capacity, increased reliability, and technological innovation. With data rates of up to 10 Gbps, it enables faster downloads and uploads, more effective online operations, and extremely low latency for real-time applications, autonomous driving, and the Internet of Things (IoT). In addition, 5G networks can support more connected devices per unit area, making integrating smart devices and industrial IoT devices less challenging [6].

In modern military operations, 5G technology has become an essential information infrastructure, in particular, because of its high-speed information transmission, reduced latency, and enhanced command and control systems. This technology demonstrates several excellent capabilities that enable real-time intelligence, surveillance, and reconnaissance (ISR) to manage the accumulation of accurate and relevant data, which is essential in military operations. The low latency of 5G technology ensures that critical information can be transmitted in real-time, enabling military leaders to respond quickly to changing environments on the battlefield. This feature is critical to protecting military personnel and civilians in conflict zones. In addition, 5G technology's advanced command and control systems provide military leaders with a better understanding of the evolving situation, enabling them to make well-informed decisions. Its capabilities include enhanced situational awareness, better communication networks, and superior data analytics. These capabilities enable military leaders to make real-time decisions based on fully accurate, up-to-date data, which is critical in a rapidly changing battlefield environment.

The advantage of 5G technology in military applications, in general, is its ability to analyze large amounts of information at high speeds. This capability is particularly important for modern military operations, which increasingly rely on real-time data exchange. Examples of such information exchange include video from drones, sensor data from reconnaissance missions, and immediate command and control communications. With the low latency of 5G networks, command and control systems can operate with minimal delay, a vital issue in the rapidly changing environment of military operations where every second counts. Furthermore, 5G networks can process and forward records in real-time to relevant events, an important feature for efficient decision-making within the field. The ability to quickly examine and process information also ensures that military leaders receive up-to-date information, enabling them to make informed decisions. In addition, the high bandwidth of 5G networks enables the exchange of large data sets, such as aerial surveillance imagery, which can support strategic planning and decision-making.

Using 5G solutions in military operations is a major development that could redefine defense mechanisms' strategic and tactical dynamics worldwide. Next-generation wireless technologies provide enhanced connectivity through the above-mentioned capabilities, which will provide revolutionary advances in military communications and operational effectiveness. However, the integration of 5G presents complex challenges regarding key security aspects, including confidentiality, integrity, availability, and common security challenges. The 5G technology in military operations will transmit and receive significant amounts of data in real time, facilitating faster and more robust responses to changing circumstances. However, the scale of information transmitted over 5G networks poses significant security risks. Malicious actors can also exploit system vulnerabilities to access sensitive data. Therefore, it is paramount that 5G networks deployed in military infrastructures are robust, reliable, and resilient enough to withstand cyber-attacks and other security threats [7]. The deployment of 5G as a critical information infrastructure in military operations raises significant security issues that require careful consideration. Three pillars of security must be addressed to ensure the protection and success of military operations. The first pillar is confidentiality, paramount in preventing unauthorized access and espionage. Encryption and privacy measures must be robust to protect sensitive information transmitted over 5G networks. However, the dynamic nature of 5G networks, relying on software-defined networking and virtualization of network functions, introduces new vulnerabilities that attackers can exploit to intercept sensitive information. As such, it is paramount to enforce superior security features to ensure confidentiality. The second pillar of security is integrity, which requires 5G networks to ensure that the information transmitted through their channels is correct and unaltered. Any manipulation or falsification of data can have extremely dangerous consequences, as inaccurate decisions based on falsified information can be made. Comprehensive measures are therefore needed to ensure the integrity of information to prevent such situations. The third cornerstone of security is availability, which ensures that operations run properly, as the necessary information is always available in the right place, form, and time. Any disruption or downtime within the 5G network could critically impact the military's ongoing operations, potentially jeopardizing mission success. Therefore, it is crucial to put measures in place to ensure the availability of 5G networks avoid any interruption.

Based on the above, 5G technology in military operations can open up new opportunities for military personnel by enabling them to perform their tasks more efficiently during operations. However, this can only be achieved if security challenges are properly addressed. Military leaders and technologists need to work together to expand and implement advanced security measures to ensure the safe and stable deployment of 5G technology in military operations, improving military skills and saving lives. The integration of 5G technology into operations is a huge milestone that can eliminate communication and interoperability problems at the international level, speed up operations, support joint task execution, and greatly contribute to avoiding friendly fire by sharing information in real time with allies.

4 Conclusion

Integrating 5G solutions into military operations has become critical in modern warfare. It offers unprecedented data rates, reduced latency, and an improved command and control system, revolutionizing military operations. High-speed data transmission enables fast and efficient information exchange between military personnel, resulting in faster and more robust decision-making. It also enables real-time intelligence, surveillance, and reconnaissance, which is crucial in a combat environment. Ultimately, military personnel can obtain accurate and up-to-date data from the battlefield, enabling them to make informed decisions quickly.

In addition, the introduction of 5G has enabled the expansion of superior autonomous systems, including unmanned vehicles, drones, and robots that can be controlled remotely. These systems can carry out dangerous or difficult tasks for people, such as reconnaissance missions or transporting materials to troops on the front walls. This ensures the military staff can focus on performing their duties without endangering their lives.

5G technology has presented complex challenges in military infrastructure that require careful network security considerations. Protecting sensitive data is crucial in military operations, and robust encryption and privacy measures are important to protect against unauthorized access, espionage, and cyber-attacks. The dynamic nature of 5G networks introduces new vulnerabilities that attackers can exploit to seize sensitive information. Therefore, it is very important to ensure that the transmission of information remains secure and private while ensuring the integrity of the information. Regular audits and vulnerability tests are needed to mitigate security risks and identify weaknesses in network architecture, protocols, and software programs. Advanced encryption algorithms are important to protect sending and resting data and strictly install controls to prevent unauthorized access to sensitive information. In addition, monitoring techniques should be implemented to meet and respond directly to any cyber attack involving any effort.

Furthermore, ensuring the accuracy and authenticity of the data transmitted over 5G networks is important. Military decisions and actions rely heavily on the accuracy of the data acquired, and the integrity of the information assets must be maintained to ensure successful operational effects. Data authentication and verification strategies can be used with digital signatures and other authentication mechanisms to ensure data authenticity and avoid tampering.

To address these challenges, military decision-makers and technologists must expand and implement superior security measures tailored to the specific needs of 5G networks. This includes deploying end-to-end encryption, continuously monitoring for signs of intrusion or tampering, and designing network architectures that can continue to operate under adverse conditions. Also important are the ongoing research and development efforts to improve the security and defense capabilities of 5G technology, thereby strengthening its reliability as the backbone of future operations for militaries.

As the complexity of military operations continues to evolve, the role of 5G technology as a critical information infrastructure will become increasingly important. The deployment of 5G technology guarantees improved situational awareness, better decision-making, and extra operational flexibility. Figuring out this potential, however, depends on a concerted effort to conquer the full defense and security demands of using 5G technology in a military environment. In order to fully exploit the capabilities of 5G technology, military organizations must prioritize the growth and deployment of advanced security features to ensure the safety and effectiveness of their operations. The benefits of the 5G generation include faster data transmission and evaluation, real-time monitoring and response capabilities, and reduced threats to military personnel. However, deploying 5G technology in a military environment presents unique defense and security challenges that must be addressed. For example, protecting the confidentiality, integrity, and availability of sensitive information, ensuring the reliability and resilience of networks, and preventing cyber-attacks are all essential issues. In order to mitigate these challenges, military organizations need to employ a layered security approach that includes strong encryption, authentication, and intrusion detection and prevention mechanisms. By implementing these measures, militaries can significantly reduce the chances of security breaches and ensure the successful deployment of 5G technology.

References

1. Toth A (2022) A Katonai Egységes Felhőalapú Eszközrendszer fogalmi rendszerének meghatározása. In: Hadtudomány: A Magyar Hadtudományi Társaság Folyóirata (1215-4121 1588-0605) 32(4):112–125 (2022). https://doi.org/10.17047/Hadtud.2022.32.4.112.
2. Stocchero JM et al (2023) Secure command and control for internet of battle things using novel network paradigms. In: IEEE communications magazine, vol 61, no 5, pp 166–172. https://doi.org/10.1109/MCOM.001.2101072
3. Toth A (2022) The use of 5G in military cloud of things solutions. Acad Appl Res Mil Public Manage Sci 21(3):5–20. https://doi.org/10.32565/aarms.2022.3.1
4. Feng Y, Li M, Zeng C, Liu H (2020) Robustness of internet of battlefield things (IoBT): a directed network perspective. Entropy 22(10):1166. https://doi.org/10.3390/e22101166
5. Cole S, Next-gen 'Internet of Battlefield Things' on the way. https://militaryembedded.com/comms/communications/next-gen-internet-battlefield-things-the-way. Last accessed 21 Nov 2023
6. Bastos L, Capela G, Koprulu A, Elzinga G (2021) Potential of 5G technologies for military application. In: 2021 International conference on military communication and information systems (ICMCIS). The Hague, Netherlands, pp 1–8. https://doi.org/10.1109/ICMCIS52405.2021.9486402
7. Mohan JP, Sugunaraj N, Ranganathan P (2022) Cyber security threats for 5G networks. In: IEEE international conference on electro information technology (eIT). Mankato, MN, USA, pp 446–454. https://doi.org/10.1109/eIT53891.2022.9813965.

Effect of Heat Treatment on the Microstructure and Mechanical Properties of 3D Printed Ti-6Al-4V Alloy

László Tóth, Enikő Réka Fábián, Endre Ákos Ködöböcz, and Tünde Anna Kovács

Abstract In terms of biocompatible materials, titanium and its alloys stand out from other materials used in medicine. Titanium alloys 3D-printed for implants offer patients a long life span, customizability and excellent material properties to ensure freedom from irritation [1]. It was investigated the heat treatment-affected properties of Ti-6Al-4V $\alpha + \beta$ titanium alloy made by SLM (Selective Laser Melting) technology. The experiments aimed to reduce the elastic modulus since the elastic modulus of the dental implant installation environment elastic modulus (mandible E = 12.6–21 GPa) and the elastic modulus of the base material (Ti-6Al-4V) of the implant (110–130 GPa) show a significant difference. The elastic modulus is very important in the implantation because the implant several times replace a human bone [2]. In addition, keeping the corrosion resistance against environmental influences was also an important consideration. The specimens were heated in similar steps to equilibrate the core-to-surface temperature to temperatures below Tβ (650 °C for 10 min, 850 °C for 10 min, and 950 °C for 25 min). It was then cooled to room temperature at high speed using high-pressure N_2 gas for 18 min. The three specimens were age-hardened at different temperatures (450 °C or 600 °C, for 130 min). The results of the microscopic examination clearly show that the SLM fabricated specimens developed an α' needle martensitic microstructure, which transformed into an $\alpha + \beta$ Widmannstätten microstructure upon heat treatment. The Vickers Hardness test result confirmed the microstructure transformation. A hardness reduction of almost 20% was observed in the heat-treated specimens. The elastic modulus determined in the tensile test did not decrease significantly (103–132 GPa). Increased corrosion resistance to HCl was observed in the corrosion tests. The heat treatments applied improved the properties of titanium alloys made by SLM technology in several aspects.

Keywords Biomaterials · Titanium-alloys · Heat-treatment · Microscopy · Microhardness · SEM · Corrosion test

L. Tóth · E. R. Fábián · E. Á. Ködöböcz · T. A. Kovács (✉)
Banki Donat Faculty of Mechanical and Safety Engineering, Obuda University, Budapest, Hungary
e-mail: kovacs.tunde@bgk.uni-obuda.hu

T. A. Kovács et al. (eds.), *The Impact of the Energy Dependency on Critical Infrastructure Protection*, Advanced Sciences and Technologies for Security Applications, https://doi.org/10.1007/978-3-031-78544-3_58

1 Introduction

Titanium is a very special metal in terms of materials technology. From a metal physics point of view, it is on the borderline between heavy and light metals, with a density of only 4.506 g/cm^3, which is almost 60% of that of iron [3]. Another important property of titanium is that it is passivated, like chromium or aluminium, so the oxide layer on its surface is continuous and adheres well to the metal [4, 8]. This makes it biocompatible and, with the exception of hot concentrated acids, highly corrosion resistant. The oxide layer is mainly composed of the compound TiO_2, the most common modifications of which are rutile and anatase. Both modifications have a tetragonal structure, are extremely stable and have a high hardness ($\Delta G \sim 890$[kJ/mol]; ~ 2600–3300 HV), with a melting point exceeding that of titanium [9]. There are two allotropic modifications of titanium at atmospheric pressure: the hexagonal close-packed (HCP) lattice α phase and the body-centered cubic (BCC) β phase. Pure titanium is a β-phase from the melting point up to 882 °C, below which it appears as α-phase titanium after a transformation involving a change in species volume. The different titanium alloys can be classified into three groups: α, α + β and β alloys [3, 10].

Ti-6Al-4V is a typical α + β titanium alloy. When α + β titanium is heat-treated, the β transformation temperature, $T_\beta \sim 980$ °C for Ti-6Al-4, plays a key role in the change of the microstructure. This is the temperature that separates the pure β region from the two-phase α + β region. T_β can be calculated from an empirically determined Eq. (1) as follows:

$$T_\beta = 882 + 21,1[\mathrm{Al}] - 9,5[\mathrm{Mo}] + 4,2[\mathrm{Sn}] - 6,9[\mathrm{Zr}] - 11,8[\mathrm{V}] - 12,1[\mathrm{Cr}] - 15,4[\mathrm{Fe}] + 23,3[\mathrm{Si}] + 123,0[\mathrm{O}] \rightarrow T_\beta =\sim 980\ ^\circ\mathrm{C} \quad (1)$$

When titanium alloys are cooled below the temperature T_β, single-phase β titanium can transform into different equilibrium and non-equilibrium phases based on the cooling rate and chemical composition. When cooled in oil or water, the β phase undergoes a martensitic transformation, which, like steel, is a transformation without diffusion at a rate equal to the speed of sound. However, unlike steel, the result for titanium can be either an α' (hexagonal) phase or an α'' (orthogonal) phase. The ratio of α' to α'' is determined by the amount of β stabilizers present in the alloy, such that as the amount of β stabilizers increases, an increase in the ratio of the α'' phase is observed [11, 12].

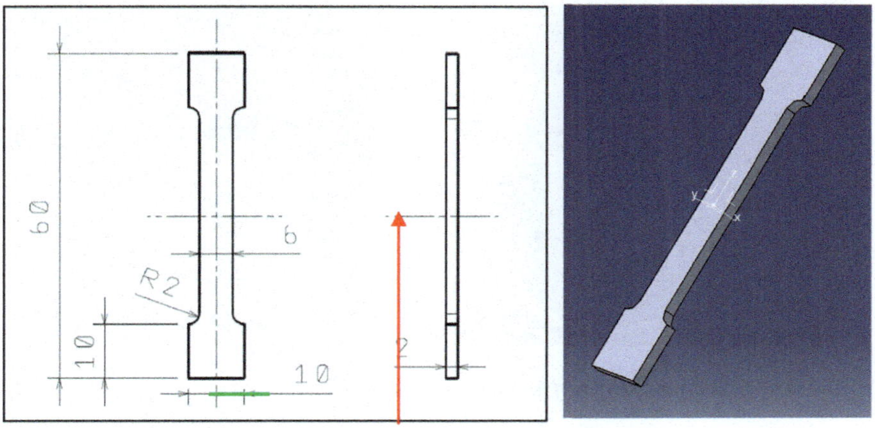

Fig. 1 Technical drawing (left) and CAD model (right) of the tensile test specimen

Table 1 Chemical composition of the test specimens

Alloy	Average chemical composition (wt.%)						
	Ti	Al	V	N	C	O	Fe
Ti-6Al-4V	Remainder	6 ± 0.5	4 ± 0.5	0.05	0.08	0.2	0.8

2 Materials and Methods

2.1 Material Preparation

For the experiments, Ti-6Al4V alloy was used and the tensile test specimens were fabricated by SLM technology with the dimensions and design shown in Fig. 1. The test specimens were made by the Sisma MySint 100 SLM. To melt the titanium powder, a 55 μm diameter laser beam with a power of 200W was used to build up the specimen in 20 μm layers. Fabrication was performed in an inert shielding gas (Argon) [13]. The specimen was printed in the stationary position as shown in Fig. 1, where the green line indicates each layer, and the red arrow indicates the direction of printing of the specimen. The chemical composition of the used alloy is shown in the Table 1.

2.2 Thermal Treatment

For the heat treatments, we used an IU72/1F 2RV 60 × 60 × 40 10 bar CP Schmetz-type vacuum furnace at the Lőrinci plant of Titan 94 Ltd. The furnace can be used

Fig. 2 Furnace (left) and specimen basket with specimen (right)

for heat treatments in the temperature range between + 1350 °C and − 150 °C. A vacuum of 10^{-3} bar was used for heating and a nitrogen protective gas for cooling. Therefore, the consequences of O_2 sequestration at high temperatures typical for Ti alloys are not to be expected. The hardening process by the injection of high-pressure gas leads to a cooling rate of nearly 1 K/s. The temperature in the immediate vicinity of the test specimens was measured with a Pt-PtRh thermocouple [1, 14]. The setup of the furnace shown in Fig. 2.

2.3 *Microhardness*

The measurement was carried out on a QNESS 60 M Vickers hardness tester manufactured by QATM Qness GmbH. The machine can measure in the range of $HV_{0.00025}$–HV_{30}. To determine the average hardness of the specimen, $HV_{0.5}$ measurements were taken at several points almost equidistant from each other, from the edge to the centre of the specimen.

The microstructure was analysed under a 10 × magnification eyepiece and the corresponding 20 × and 50 × magnification objective lenses. The images at 200 × magnification are for the $HV_{0.5}$ measurements, while in the case of the $HV_{0.02}$, it was used 500 × magnification.

2.4 *Microstructural Characterization Methods*

The specimens preparation was made, an eight-step grinding (with P80, P180, P280, P360, P600, P1000, P2000, P4000) and a 2-step polishing process followed using the Struers Tegramin-25 polishing machine (with diamond particle 3 μm és 1 μm). The microstructure of polished specimens can be seen in polarised light without etching,

Table 2 Chemical composition of the etching fluid

Component materials			
Destillated H_2O (ml)	HCl (37%) (ml)	HNO3 (65%) (ml)	$C_{19}H_{19}N_7O_6$ (40%)
18	50	16	16 1

but a clear contrast image can only be obtained by chemical etching. Several etching agents are used in industry to etch titanium alloys, the two most common being Kroll reagent and Weck reagent. The important difference between the two etching agents is that the β-phase appears dark brown on specimens etched with Weck's reagent, whereas the α-phase reacted with oxygen becomes visible (white) only under the influence of bifluoride, and this oxide layer is formed even at room temperature [15].

For etching, we used a special solution with 30 s submersion time as recommended by the materials testing laboratory (Table 2).

2.5 *Scanning Electron Microscopy*

Scanning electron microscopy (SEM) was performed on the JEOL JSM 5310 at the university according to MSZ EN ISO 9220:2022.

2.6 *Corrosion Test*

The corrosion test was carried out in two different corrosive media. One corrosive medium was placed in HCl solution (pH 1). To prepare the hydrochloric acid solution, I used 37% hydrochloric acid and distilled water. The other typical corrosive medium is sodium fluoride. The solution was prepared as a mixture of 200 ml distilled water and 100 ml toothpaste containing NaF (1450 ppm F). The specimens were left in the two corrosive media for 487 h. The experiment was carried out in a Weiss WKS3 180/70/5 climate chamber at 36 °C for 487 h.

Before and after the test, the mass of the specimens was measured on a Mettler Toledo AB204 analytical balance. Prior to the mass measurement after corrosion loading, the specimens were subjected to ultrasonic cleaning.

The corrosion rate (v_{corr}) was determined as the ratio of the surface area (A) of the specimens under test to the mass loss (Δm) of the specimens (2) [16].

$$v_{corr} = \frac{\Delta m}{A}\ (\mathrm{g/cm^3}) \tag{2}$$

3 Results and Discussions

3.1 Heat Treatment Processes

During the heat treatment, the specimens were heated to 960 °C in steps, with 10–10 min dwell time at 650 and 850 °C for each step. At 960 °C, a 25 min dwell time was applied. Test specimen HT1 was 130 min ageing heat treatment at 450 °C in nitrogen at 2 bar. Test specimen HT2 was cured at 600 °C with 130 min of annealing. As well as for HT3, only intensive cooling after solution annealing was performed. The heating and cooling phases lasted 50–50 min at 1.5 bar pressure (Fig. 3) [14].

Table 3 summarises the heat treatment process and the heat-treating parameters found in Table 4.

Nitrogren gas was used for cooling during the heat treatments.

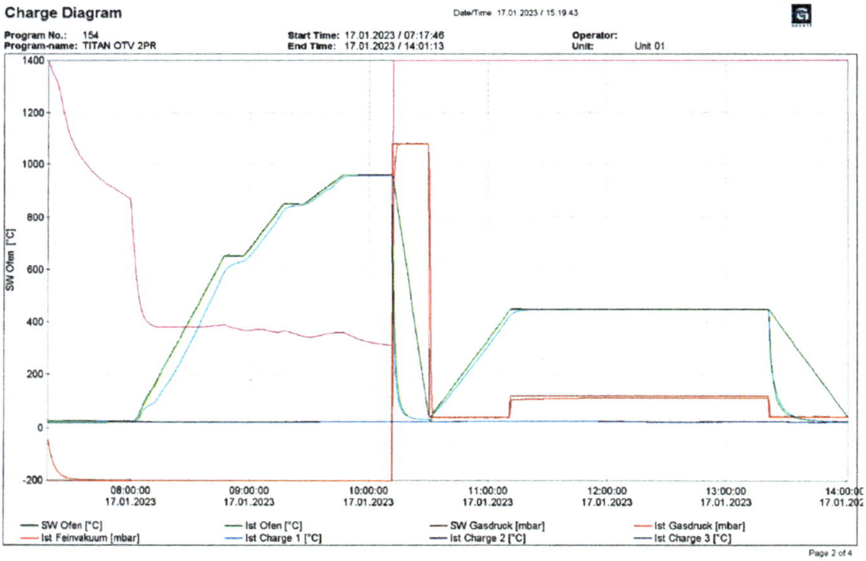

Fig. 3 Heating diagram of the heat-treatment process (green—programmed and measured temperature, purple—vacuum, red—nitrogen pressure)

Table 3 The heat treatment of the test specimens

Heat treatment group index	Solution treatment	Quenching to RT	Ageing
HT0	–	–	–
HT1	960 °C	With N_2 gas	450 °C
HT2	960 °C	With N_2 gas	600 °C
HT3	960 °C	With N_2 gas	–

Table 4 Heat treatment parameters

Charge diagram					
Step no.	Name of Step	Time (min)	Step no.	Name of step	Time (min)
1	Vacuum	45	7	Dwell on time at 960 °C	25
2	Heating to 650 °C	50	8	Cooling to room temperature	18
3	Dwell on time at 650 °C	10	9/a	Heating to 450 °C	50
4	Heating to 850 °C	20	9/b	Heating to 600 °C	
5	Dwell on time at 850 °C	10	10	Dwell on time at 960 °C	130
6	Heating to 960 °C	20	11	Cooling to room temperature	50

3.2 *Tensile and Hardness Test Results*

There are no significant differences between the tensile test results of the four specimens in terms of either tensile strength or elastic modulus. The tensile strength and elongation of HT0 is different than the results of the heat-treated specimens (Fig. 4).

To determine the average hardness of the specimens, $HV_{0.5}$ measurements were taken at a number of equally spaced points from the edge to the centre of the specimen.

The average hardness of the HT0 specimen was 392 HV. The average of the values measured here was 465 HV, a change in hardness of almost 19%. The hardness measured on the HT1 specimen is 338 HV averaged over the points in the centre of the specimen. The values of the martensitic crust at the edge of the specimen are

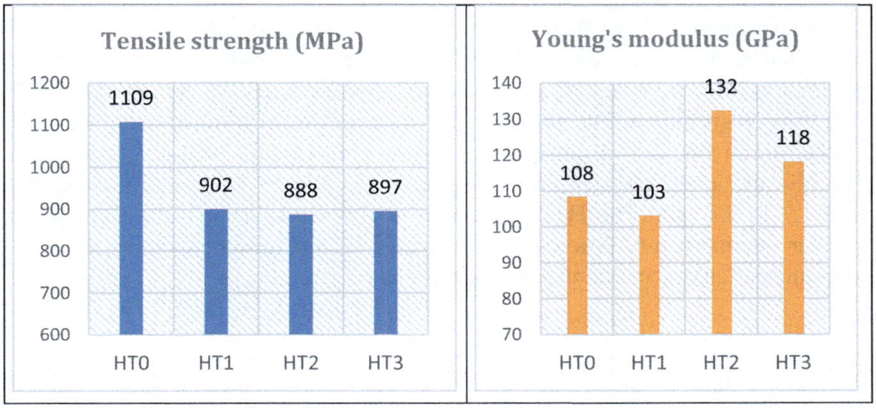

Fig. 4 Results of the tensile test

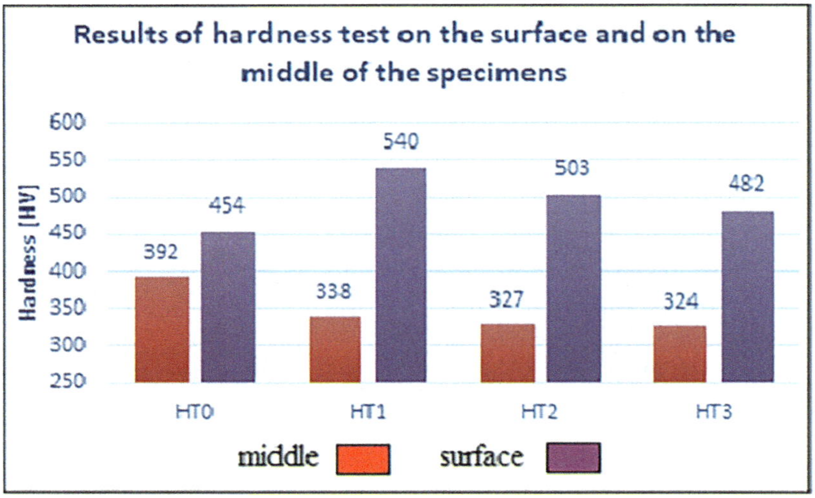

Fig. 5 Hardness test results

well above the values measured for HT0 measured an average hardness of 540 HV. The average of the hardness values measured in the HT2 test was 327 HV. For this specimen, the average of the values measured in the extreme 50 μm band is 503 HV with a standard deviation of near 58 HV (Fig. 5). Vickers hardness results shows higher hardness on the surface than in the middle of the investigated specimens. The heat treating caused a hardening of the specimen's surface.

3.3 Microstructure

The microstructure of the HT0 specimen is martensitic with α' needles both in the centre and at the edges (Fig. 6a). The microstuctre analysis of each heat-treated specimens (HT1, HT2 and HT3) revealed of approximately 30–50 μm thick at the edge of each specimen. The surface layer is composed of α' martensite, but in the middle, it can find $\alpha + \beta$ lamellar microstructure. In case of the specimen HT1 (Fig. 6b), the length of each flake varies from 33.3 to 59.9 μm and the thickness from 1.7 to 2.7 μm. The HT2 (Fig. 6c). microstructure shown the same $\alpha + \beta$ lamellar microstructure. Microscopic image of HT3 (Fig. 6d). showing the $\alpha + \beta$ Widmannstätten mycostructure with α colonies.

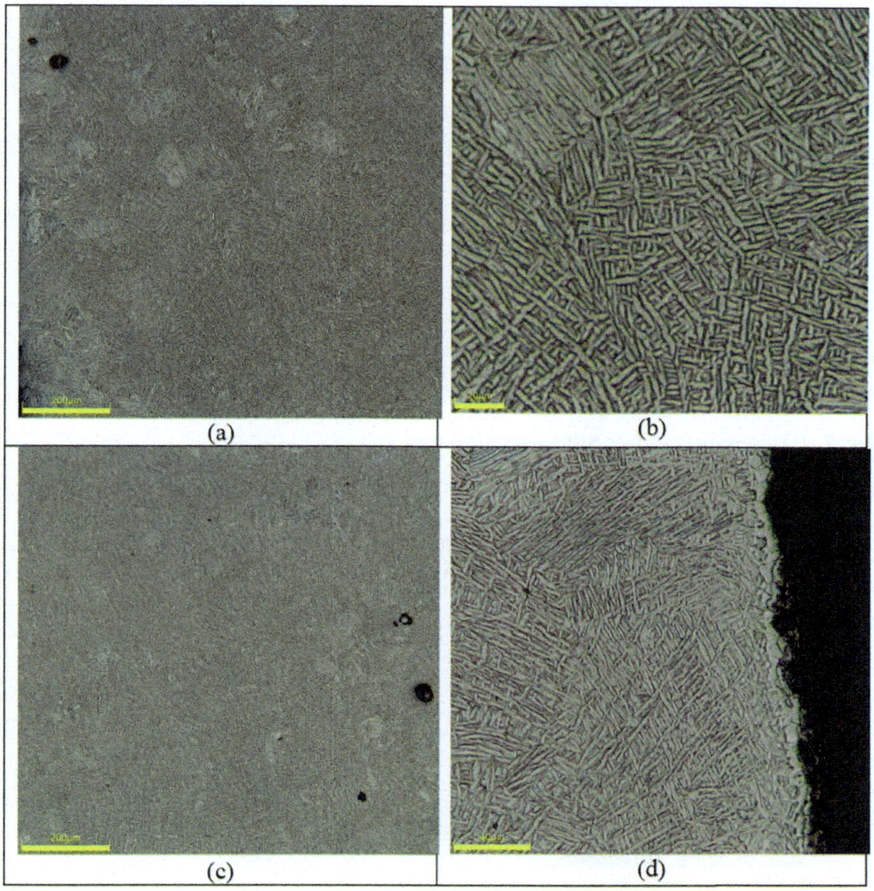

Fig. 6 Microstructure analysis of Ti-6Al-4 V alloys with different heat treatment **a** HT0, **b** HT1, **c** HT2, **d** HT3

3.4 SEM Results

Based on the EDS image, the study area contains 93.2% titanium, 3.9% aluminium and 2.9% vanadium by weight (Fig. 7b). It can find some porosities in the Fig. 7a.

The HT1 (Fig. 8a) it can see an intermetallic phase precipitation and the ESD test results (Fig. 8b) can see the chemical composition of the precipitation which one is $TiAl_3$.

The HT2 SEM and EDS analysis resuts shown on Fig. 9 on the results of the EDS it can conclude that the precipitation is $TiAl_3$.

In the case if the HT3 test specimen it was detected by EDS and SEM test the $TiAl_3$ precipitations (Fig. 10 and Table 5).

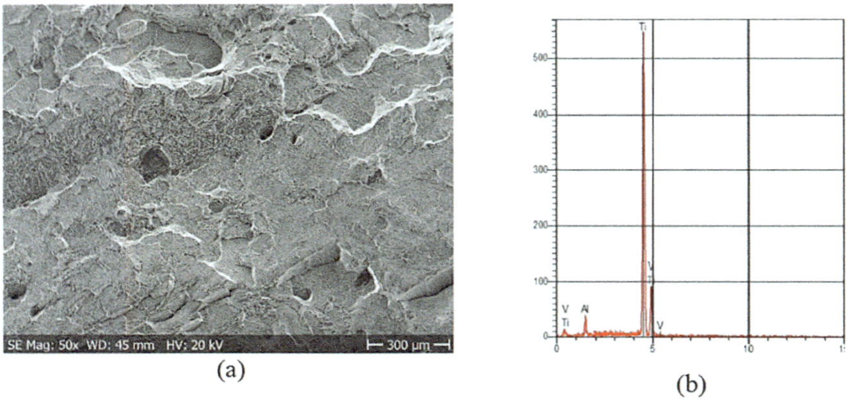

Fig. 7 HT0 specimen surface SEM and ESD results

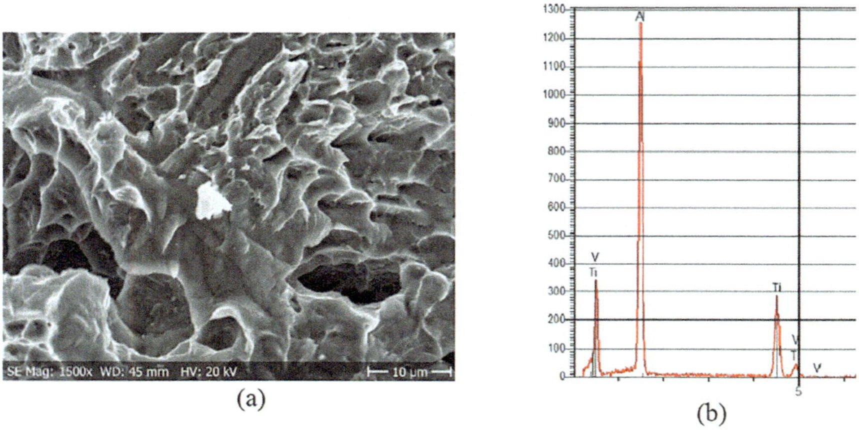

Fig. 8 4–5 μm diameter intermetallic phase precipitation on HT1 specimen (**a**) and EDS analysis (**b**)

Fig. 9 HT2 specimen broken surface with detected precipitation

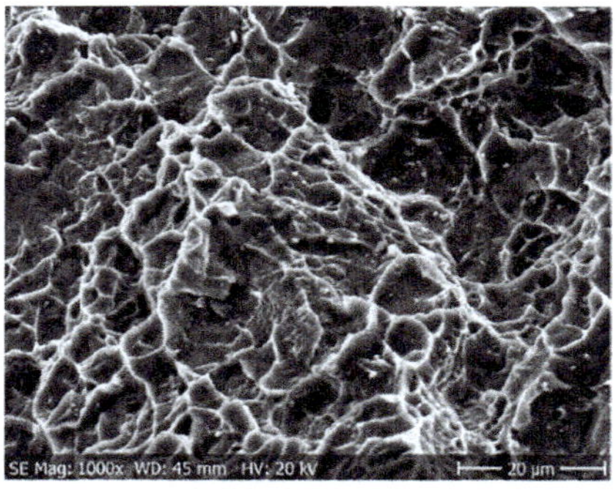

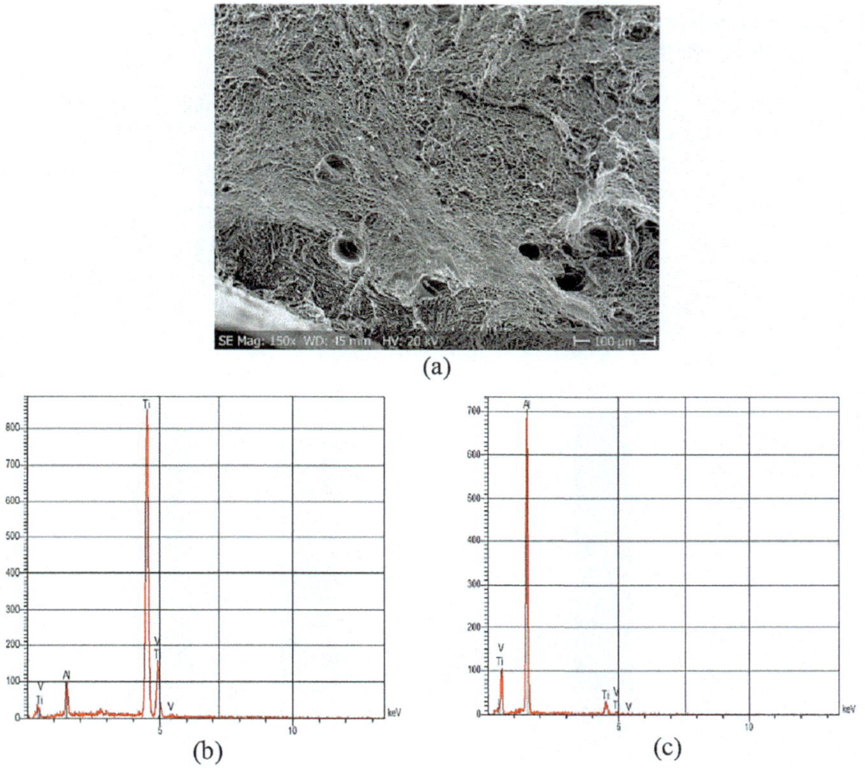

Fig. 10 SEM image of the HT3 specimen (**a**), the EDS analysis of the base metal (**b**) and EDS analysis of the $TiAl_3$ precipitate (**c**)

Table 5 Quantitative results of the SEM test of the TH3 specimen

Chemical component	Tömeg %
Al	92.1
Ti	7.88
V	~ 0

3.5 Corrosion Test

The corrosion rate was determined as a function of weight loss during the corrosion tests. Specimen HT0 lost 22.8 g of its mass in hydrochloric acid, corresponding to a loss of 4.11% of the original mass. Specimens HT3 and HT2, like HT0, lost about 4% of their mass (HT2 = 3.8%; HT3 = 4.3%), compared to only 2.34% for HT1 (Table 6. and Fig. 11).

Table 6 Initial values of the specimens in the corrosion test

HT0	HT1	HT2	HT3
Original surface			
198.7 mm^2	188.6 mm^2	194.3 mm^2	201.3 mm^2
Original weight of the specimens			
0.5549 g	0.5336 g	0.5458 g	0.5469 g
Weight of the specimens after the corrosion test			
0.5321 g	0.5211 g	0.5250 g	0.5234 g

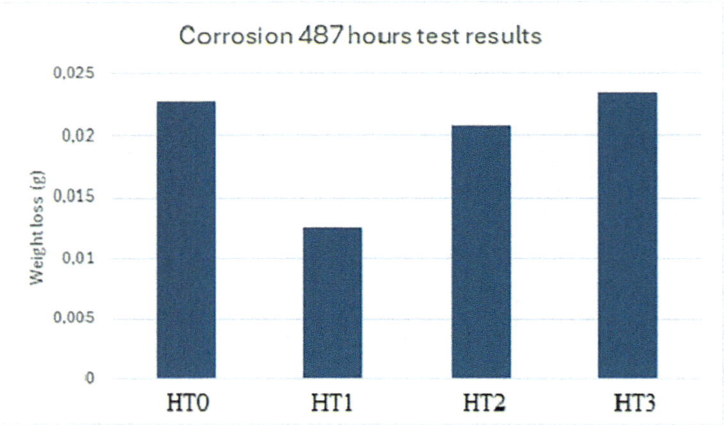

Fig. 11 Results of the corrosion tests

4 Conclusions

It was investigated Ti-6Al-4V specimens made by additive manufacturing technology (SLM) to find out which heat treatment can be used to achieve the best conditions for dental implants. The first of the objectives of the experiments was to get the modulus of elasticity as close as possible to the modulus of the installation environment. The second is to increase the corrosion resistance to relevant environmental influences.

From a material structure point of view, it was detected a homogeneous α' martensitic structure before heat treatment, a tougher $\alpha + \beta$ lamellar Widmanstätten structure for all heat-treated specimens, and a high hardness martensitic crust with an average thickness of 50 μm.

The tensile test results show that the tensile strength decreased as a result of the heat treatment, but Young's modulus decreased from 108 to 103 GPa only for the specimen heat-treated with the HT1.

Microscopic analysis and SEM analysis showed that the microstructure of all the heat-treated specimens contains αTi colonies and Al-containing precipitates.

Vickers hardness results hows higher hardness on the surface than in the middle of the investigated specimens. The heat treating caused a hardening of the specimens surface. For the corrosion test, I performed an elevated temperature accelerated life test in hydrochloric acid solution.

The resistance to the acid environment was significantly increased by heat treatment with the HT1.

On the base of the investigation results it can be concluded that the HT1 was the best way to optimize the mechanical and corrosion properties of Ti-6Al-4V specimens made by SLM technology according to the requirements of dental implants.

Acknowledgements The authors would like to thank to János Kónya and the colleagues of Dent-Art-Technik LTD for providing the SLM samples for testing.

References

1. Bey V, Lore T, Jean-Pierre K (2012) Heat treatment of Ti6Al4V produced by selective laser melting: microstructure and mechanical properties. J Alloy Compd 177–185
2. Tamatsu Y, Kaimoto K, Arai M, Ide Y (1996) Properties of the elastic modulus from buccal compact bone of human mandible. Bull Tokyo Dent Coll. 1996 May;37(2):93–101. PMID: 9151581.
3. Pinke P, Zitnansky M, Caplovic L, Réger M (2004) Biocompatible titanium alloys. In: First Hungarian conference on biomechanics, Budapest ISBN 963-420-799-5, pp 374–382
4. Zenou M, Grainger L (2018) Additive manufacturing of metallic materials. Addit Manuf 53–103
5. Haize G, Lados DA (2006) Effects of the microstructure and porosity on properties of Ti-6Al-4V ELI alloy fabricated by electron beam melting (EBM). Addit Manuf 47–57
6. Donachie Jr MJ (2001) Heat treating titanium and its alloys. Heat Treat Process 47–57
7. Lee DB, Kim MJ, Chen L, Bak SH, Yaskiv O, Pohrelyuk I, Fedirko V (2011) Oxidation of nitride layers formed on Ti-6Al-4V alloys by gas nitriding. Metals Mater Int 471–477
8. Zhang F, Chen SL, Chang YA, Kattner UR (1997) A thermodynamic description of the Ti-Al system. Intermetallics 471–482
9. Németh R (2016) Az előállítási körülmények hatása nanokristályos titán-dioxid szerkezetére. ELTE Fizikai Intézet, Budapest
10. Mehl MJ (2002) Tight-binding study of high-pressure phase transitions in titanium: alpha to omega and beyond. Europhys Lett 60:248–254
11. Jan S, Waldemar Z (2013) Microstructure and mechanical properties of high strength two-phase titanium alloys. In: Titanium alloys—advances in properties control, pp 69–80
12. Vydehi AJ (2006) Titanium alloys an atlas of structures and fracture features. Taylor & Francis Group, Boca Raton
13. Kónya J (2022) Lézerhegesztés alkamazása additív technológiával gyártott orvostechnikai eszközök korrekciós és utómunkálatainál. Hegesztés Technika 41–47
14. Madalina SB, Petrica V, Tóth L (2022) Effect of heat treatment on some titanium alloys used as biomaterials. Appl Sci 1–15
15. Taylor B (2020) Metallographic preparation of titanium. Struers ensuring certainty/application notes, Glasgow. 23 Jan 2020

16. Mogoda AS, Ahmad YH, Badawy WA (2004) Corrosion behaviour of Ti–6Al–4V alloy in concentrated hydrochloric and sulphuric acids. J Appl Electrochem 873–878
17. Adamiak M., et al. (2024) Surface Modification of Metallic Materials Using Laser and Plasma Technologies In: ICCECIP, Budapest, 2022, ISSN 1613-5113. Springer, Heidelberg, pp 81–92 https://doi.org/10.1007/978-3-031-47990-8

Brittle Structures and Hydrogen Damage of Steels

Mehmet Alican Polat and Tünde Anna Kovács

Abstract The advantages of Hydrogen are well known; what about the disadvantages? What are the storage challenges? What are the hydrogen embrittlement mechanisms and other damage mechanisms? 70–90% of the existing hydrogen pipelines are made of steel. It is crucial to understand the hydrogen damage mechanisms, their differences, and possible failure scenarios. This article identified and compared different damage mechanisms, evaluated real-world examples and failures, and analyzed the failures behind them. HEDE, HELP, AIDE, HIC, SSC, SCC, and HTHA all have different kinds of relations with steel. Good practices and, most importantly, correct material selection are lifesaving. After identifying the damage mechanisms, the article focused on the industry standards for best practices.

Keywords Hydrogen storage · Nickel · Steel · Martensitic · Polyester · Chromium · Molybdenum

1 Introduction

It is well known that the world's energy needs are increasing every day, leading to a rise in carbon footprint. This has raised efforts to find new energy solutions, such as using hydrogen as energy carriage. Hydrogen offers advantages such as high energy density, the ability of easy combination with other elements or combustion energy.

Hydrogen production can be achieved through various methods categorized as grey, blue, or green, with each colour designation representing its environmental impact. However, the most used method for hydrogen production remains steam

M. A. Polat
Doctoral School on Safety and Security Sciences, Obuda University, Budapest, Hungary
e-mail: mehmet.alicanpolat@uni-obuda.hu

T. A. Kovács (✉)
Banki Donat Faculty of Mechanical and Safety Engineering, Obuda University, Budapest, Hungary
e-mail: kovacs.tunde@bgk.uni-obuda.hu

T. A. Kovács et al. (eds.), *The Impact of the Energy Dependency on Critical Infrastructure Protection*, Advanced Sciences and Technologies for Security Applications, https://doi.org/10.1007/978-3-031-78544-3_59

reforming, conducted at temperatures ranging from 700 to 900 °C and pressures between 3 and 35 bars. This process involves the conversion of methane into carbon monoxide and hydrogen [1].

Another method, which is more expensive but still commonly used, is coal gasification, which involves the partial oxidation of coal at high temperatures and pressures, leading to the production of H_2 and CO_2.

The most researched and open-to-improvement area of hydrogen production these days is green hydrogen production. This method can be achieved through the electrochemical separation of hydrogen and oxygen using electric energy applied to water. Unlike other methods, green hydrogen production produces carbon-free hydrogen.

Besides the advantages of hydrogen, such as green hydrogen with carbon-free production, there are drawbacks as well; one of the drawbacks is that it is mostly stored either as compressed gas at pressures up to 800 bar or cryogenically at − 253 °C. Material selection becomes crucial under these extreme conditions, as hydrogen can cause material damage due to high hydrostatic stress (Internal Hydrogen Embrittlement) and loads of hydrogen sulfide gases (Hydrogen Environment Embrittlement) [2]. Also, these two terms are called HAC hydrogen-assisted cracking; two mechanisms of HAC are bond weakening by hydrogen and the other one is hydrogen-induced strain localisation [3]. Studies are being done to produce lighter but stronger metals to reduce fuel consumption. One such technique is press hardening, which involves the use of 22MnB5 boron steel. Hydrogen is used as a reducing agent to protect the steel from oxidation during this process. However, this can lead to embrittlement, known as hydrogen-assisted cold cracking (HACC) [4]. Besides embrittlement, blistering and hydrogen attacks are part of hydrogen damage. These mechanisms are going to be reviewed further.

2 Hydrogen Storage

Hydrogen storage can be done in various ways, considering the future use of hydrogen; this can include gaseous hydrogen, liquid hydrogen and material-based storage by using adsorption or absorption. Adsorption techniques include activated carbons, and absorption techniques include metal hybrids. Several methods are available for hydrogen storage [5]. Different kinds of techniques can be seen in Fig. 1.

2.1 Compressed Gaseous Hydrogen

Compressing hydrogen is one of the most conventional methods of storing or transferring it. This method is the physical storage of hydrogen.

To be able to succeed in the storage of hydrogen in gas form, a storage compartment and a compressor are needed, as pressure should go up to 800 bars. The volumetric

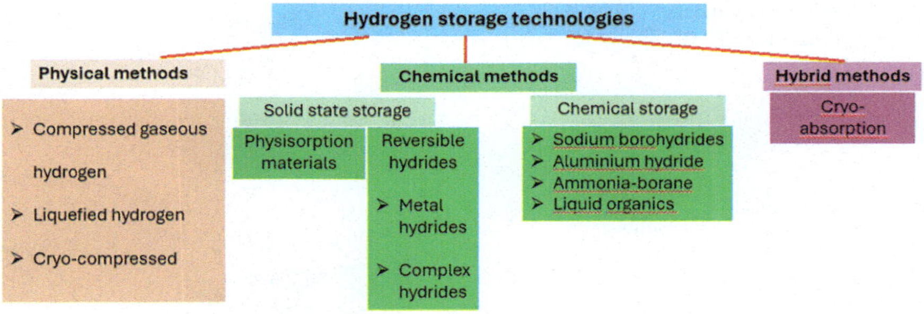

Fig. 1 Hydrogen storage technologies (by author)

density of gaseous hydrogen can go up to 40 (kg H_2)/m^3; this is low compared to other techniques. Hydrogen can be stored in four types of vessels. Metallic pressure vessels are known as type 1, type 2 vessels have fiber resin inside, type 3 vessels have carbon fibers embedded, and type 4 vessels mainly consist of a polymer liner or an extremely thin metal liner [6]. Pressure vessels are classified into cylindrical and spherical vessels. Cylindrical vessels can be in horizontal and vertical form and have advantages such as being easier to find in the market and cheaper. However, stress can affect the joints due to their geometric shape. Meanwhile, spherical vessels have a larger surface area, allowing for better distribution of pressure. However, it is also known that spherical vessels are still expensive [7]. Hydrogen cylinders can seen in Fig. 2.

2.2 *Liquefied Hydrogen*

Hydrogen shows a reverse Joule–Thomson effect, implying that any gas with an inversion temperature higher than its critical temperature displays a negative Joule–Thomson effect. However, this doesn't preclude the application of the Thomson effect to hydrogen. Hydrogen can be cooled below its inversion temperature by using nitrogen, after which the Joule–Thomson effect can be applied. The volumetric density of hydrogen can reach up to 70 (kg H_2)/m^3, which is more advantageous compared to compressed hydrogen.

The most important feature of liquid hydrogen storage tanks is having multilayer insulation to be able to keep the hydrogen in its liquefied temperature which is around − 253 °C.

Another important thing with liquid hydrogen is shape of tank, cylinder tanks are good for volume ratio but takes too much space, therefore comfortable tanks are being using but their thermal performance is not such good and manufacturing is hard [8].

Fig. 2 Hydrogen cylinders in cement plant for CEMC

So called boil-off of LH_2 describes the cryogenic liquid in tank starts being evaporate by time and this loose cannot be prevented totally even though vacuum super insulations [8].

2.3 Material Based Hydrogen Storage

Besides physical storage of hydrogen at extreme low temperature or extreme high pressure, there are some other methods using absorption and adsorption techniques.

While the adsorption technique relies on van der Walls bonds, absorption of hydrogen relies more on covalent and ionic bonds. One of the techniques of adsorption-based material-based storage is activated carbons, which extend the surface area and allow for storing more hydrogen. Absorption techniques can be used to produce ammonia as chemical storage and metal hydrides can absorb and release the hydrogen. Metal hydrides can be advantageous, but the materials themselves are hefty.

As can be seen, each technique has its advantages and disadvantages, and their usage can vary depending on the area of application, logistics, and distance.

Meanwhile, gaseous and liquid hydrogen are the most common storage techniques and steel and aluminium are the most common materials.

3 Hydrogen Damage Types

It was previously mentioned that materials are subject to hydrogen embrittlement even under atmospheric conditions. This embrittlement can be by internal hydrogen, which is part of the process, and external hydrogen, which is introduced under stress. SCC terminology, which means stress corrosion cracking is the terminology to describe the absorption of hydrogen produced by corrosion; this absorption usually occurs by fasteners, bolts, or screwing materials [9]. The protection method can be coating fasteners with appropriate material so this material will corrode, but not fasteners, or methods such as acid cleaning or baking might be part of hydrogen removal.

Understanding Hydrogen Embrittlement mechanism is difficult, but it is known that it involves three key steps: first, hydrogen atoms enter the material and spread through diffusion; second, they interact with material structure creating defects such as voids, vacancies, cracks, and dislocations; finally, this leads to the initiation, nucleation, evolution, and progression of damage caused by hydrogen [10].

Besides embrittlement, there are many other hydrogen damages such as HAC (Hydrogen Assisted Cracking), Hydrogen Induced Blistering, Sulfide Stress Cracking, Hydrogen Induced Cracking, Stress Corrosion Cracking, Hydrogen Attack at high temperatures, and Metal Hydride formation. It has many different mechanisms, such as Lattice decohesion, internal hydrogen pressure, hydrogen-induced plasticity, and hydride formation. The hydrogen-metal reaction has many forms, it can be dissolved hydrogen in a metal lattice, causing a brittle form of the material, which is known as embrittlement, it can cause cracking once formed with methane; it can form hydrides with manganese and this will cause brittle material, the hydrogen reacting with manganese or carbon is also capably reacting with Sulphur, Sulphur is part of petrochemistry, H_2S is not unavoidable in the petrochemistry, the H_2S with water causing reduced PH meaning corrosion will be accelerated causing more hydrogen production and this can happen vice versa, hydrogen can react with sulfur and creates H_2S causing cracking in the metal, also the cathodic protection is the source of hydrogen, cathodic protection is used for corrosion has 2 sides, anode and cathode, the electrons fed flows from anode to cathode and in cathode side the excess electrons combines with hydrogen atoms and creates hydrogen gas. While these happens, some interactions can be helping hydrogen trap, such as nickel and chromium [11].

3.1 Hydrogen Embrittlement

Hydrogen embrittlement occurs in atomic form; absorbed hydrogen causes a reduction of ductility of the metals, increases the cracking, and decreases the fracture toughness. The loss of ductility decreases in toughness and causes failure under stress [12]. HE can occur at room temperature, and even low hydrogen pressure is enough for Hydrogen Embrittlement to happen; most steels are subject to HE such as steels at gas pipelines, but even Nickel based metals can be susceptible to HE; it should be known that low strength steels are less susceptible to HE then high strength steels [13].

The most important 2 mechanisms suggested about Hydrogen Embrittlement are HEDE (Hydrogen Enhanced Decohesion) and HELP (Hydrogen-enhanced localized plasticity).

Hydrogen-Enhanced Decohesion Mechanism (HEDE).

This model, proposed by Troiano, suggests that hydrogen reduces the cohesive strength of lattice planes or interface boundaries. The mechanism involves hydrogen electrons entering the unfilled 3D shells of iron atoms, which increases the inter-atomic repulsive forces and thus decreases the cohesive strength [14, 15]. The HEDE mechanism is based on the weakening of interatomic bonds at crack tips due to hydrogen absorption, leading to the formation and spread of these dislocations.

Hydrogen-Enhanced Localized Plasticity (HELP).

HELP mechanism was first mentioned by Beachem in 1971, and proposed based on fracture surface analyses as a function of intensity factor in tempered steels and the flow stress comparison was done with or without hydrogen [16]. The main idea of this mechanism is that the presence of hydrogen lowers the movement barrier of dislocations in solid solutions, increasing their mobility and facilitating deformation. The fracture process here is not embrittlement but high-locality plastic fracture, which occurs due to a decrease in material strength as the region localized by hydrogen softens [11].

Adsorption Induced Dislocation Emission (AIDE).

Hydrogen atoms are easily absorbed into free surfaces, the AIDE mechanism affects the surface by absorbed H. It is basically a model describing the interaction between the surface and H atoms. Unlike HELP where the failure is on crack tips, AIDE model is on the surface [10]. In AIDE mechanism, crack growth are mainly due to emission of dislocations.

As per the descriptions above, it is important to understand the difference between decohesion and dislocation to be able to understand the difference between HEDE, HELP and AIDE mechanisms. Decohesion is refers to the separation or weakening of atom bonds within a material, leading separation of atom bonds and atomic planes such as HEDE mechanism so Hydrogen weakens the bonds at crack tips causing the material break apart easily due to loss of cohesion between atoms leading the crack initiation and propagation. Once we talk about dislocation, it is a type of defect in crystal structure and there is misalignment of atoms along the line, so it leads the plastic deformation, the movement of dislocations goes through the lattice, so lattice is permanently deformed but not necessarily broken and causing crack growth through ductile fracture mechanism. Therefore, HELP and AIDE mechanisms are part of dislocation.

3.2 Hydrogen Induced Cracking (HIC)

It is well known that Hydrogen that causes damage can be part of the environment or might be present in the process already. Hydrogen source in HIC is the corrosion reactions or H_2S.

The mechanism is the hydrogen atoms are diffused in the metal and create hydrogen molecules, these molecules causing cracks in the plane and surface of steel component. HIC is usually does not need applied tensile stress and associated with MnS, non-metallic inclusions such as MnS in the metal makes hydrogen flow through them and localize there and increase the concentration. Once atomic hydrogen is diffusing into material, then in the microscopic void there are blisters formed, and these blisters generate high pressure leading high stress which also leads to cracking. Therefore, the path is usually, first the formation of internal blister cracks on inclusions, then extension of cracks, then plastic deformation. Avoiding HIC might be achieved by alloying steels with Cu or eliminating the inclusions such as MnS [17, 18].

As summary, HIC might happens at around 200–250 °C in high corrosive atmosphere with no applied stress during cool down of the welds and the atomic hydrogen that becomes hydrogen molecule gets bigger in size and causes pressure, and nonmetallic impurities significantly effects the mechanism.

3.3 Sulphide Stress Cracking (SSC)

H_2S rich environment, usually in oil and gas industry causing the Sulfide Stress Cracking, the difference of sulfide stress cracking from hydrogen induced cracking is that SSC is in atomic form and lower corrosivity with tensile stress (Fig. 3). As we know, hydrogen is usually produced during the cathodic reaction, therefore this hydrogen reduces the steels' ability.

As can be seen in Fig. 4, the oil and gas industry operate in a sour environment, meaning an H_2S environment. Corrosion occurs as a result, producing hydrogen atoms. These hydrogen atoms enter the steel, form molecules, and cause damage. In this case, two types of cracking might happen: HIC and SSC. HIC is when hydrogen enters the steel and causes voids without stress, while SSC usually happens at welds because it requires high yield stress [19, 20]. SSC can be avoided by eliminating these stress factors and reducing H_2S environment.

Considering all the factors, we can claim that the difference of SSC from HIC is that stress effects SSC, location can be anywhere, and SSC can happen even with low corrosion environment, with HIC the molecular hydrogen creates the pressure and causes crack but SSC is causing the hydrogen generation by corrosion and then diffuses into metal and diffused hydrogen accumulates on the high metal harness areas (welding points) of the metal and embrittles the steel, therefore hydrogen embrittlement is involved with SSC.

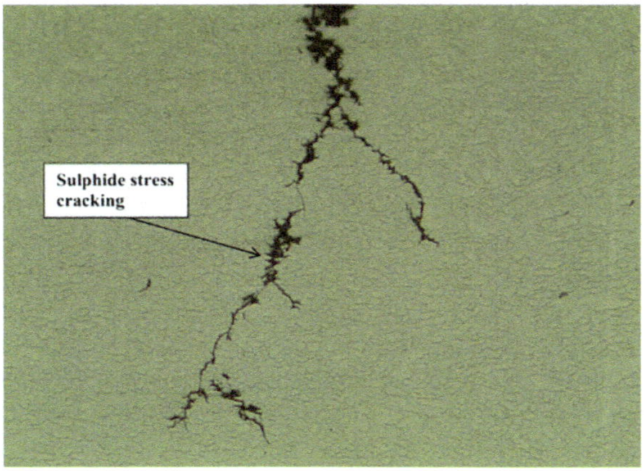

Fig. 3 Sulfide stress cracking

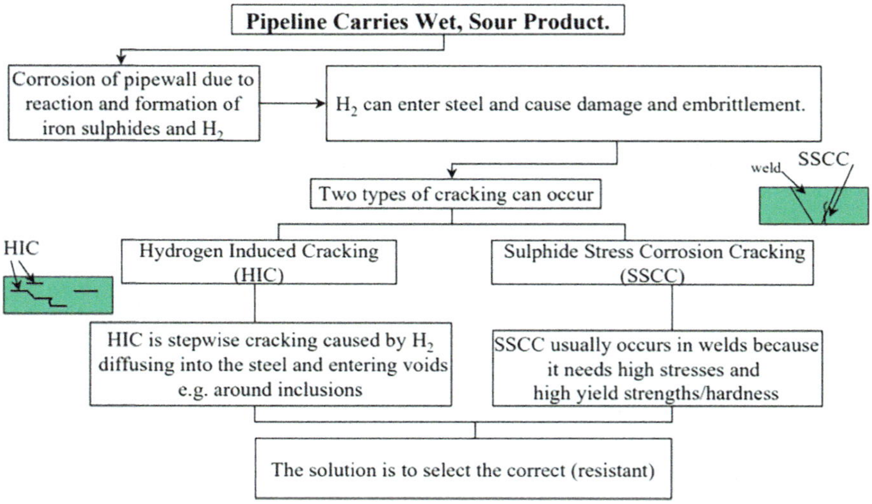

Fig. 4 Environmental cracking in line pipe carrying sour product (by author)

3.4 Stress Corrosion Cracking

Stress Corrosion cracking is like SSC, the only difference at mechanism is that SCC needs high corrosion environment. Once we talk about stress corrosion cracking, we can talk about the chemistry is moving from metal rich to oxygen rich, this is the point where problem starts. It is the combination of corrosion and stress, and different PH values might affect the metal in different forms. It is also claimed that the ductile effect happens outside of the corrosive media, but brittle effects happen around the corrosive media [20, 21].

Stress corrosion cracking occurs under tensile stress and factors causing corrosion such as water, sulfur or chlorides.

SCC is a combination of corrosion, tensile stress and susceptible material. It will be increasing by temperature, so the question is how hydrogen is involved, hydrogen is not directly involved in this process, but it can be effect as secondary factor, if hydrogen exists in the environment, it will be absorbed into metal and cause ductility and hydrogen embrittlement will happen as secondary effect.

As we discussed before, increasing the nickel content of alloy will help minimize the SCC. The other options can be selecting ferritic steel, double stainless steel, adjusting temperature to lower level, annealing. Cathodic protection must be carefully utilized as it might oxidize with S in H_2S, and hydrogen will release causing the SSC.

3.5 High Temperature Hydrogen Attack

Many oil and gas industries, refinery and fuel conversion processes require high temperature and high pressure. One of the reasons for high temperature and high pressure in these industries is removing impurities such as nitrogen, metals, oxygen, sulfur compounds from liquid petroleum fractions. One of the methods that can be used in refineries is hydrodesulphurization. On this purification process hydrogen has to be purged to be able to bind sulfur with hydrogen and create hydrogen sulfide. For example, it is well known that 14 ppm sulfur in naphtha can be found, and this must be removed.

As it can be seen high temperature is needed for this process, therefore heat exchanger is needed. Another important point to mention is that hydrogen molecules purged in the system will break their atoms at higher temperatures such as 200 °C will increase the kinetic energy and this kinetic energy will cause the breaking down of the covalent bonds of H_2, around 200–250 °C is enough to breakdown the covalent bond of H_2.

Coming back to heat exchangers, they are widely used in the industry such as in power generation, chemical processing, automotive, food beverage, oil and gas. High temperature and pressure during these processes might cause unintentional decarburization meaning carbonized iron molecules will release their carbon and hydrogen in the system will be combined resulting in methane molecules. This can be easily defined as tiny hydrogen atoms accumulating into steel and then reacting with methane creating bigger methane molecules (Eq. 1) and they are not able to diffuse out of steel and this big molecule causes stress in the material. If this is a heat exchanger, it will not withstand the stress anymore and cracks will grow resulting in catastrophic failures such as in the Tesoro Facility in 2010 causing 7 deaths and serious injuries to more people.

$$Fe_3C + 4H_2 \rightarrow 3Fe + CH_4 \tag{1}$$

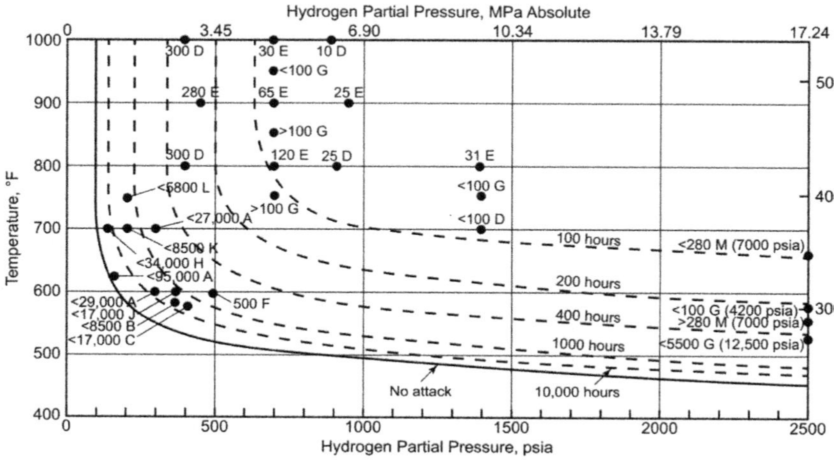

Fig. 5 Incubation time for high-temperature hydrogen attack damage of carbon steel (Non-welded or welded with post weld heat treatment) in high-temperature hydrogen service [34]

Decarburization can be defined as the removal of carbons and therefore steel loses strength. We mentioned it happens in hot environments and under pressure during the process. So, decarburization becomes dangerous when associated with HTHA. It is important to know that decarburization can be decreased by alloy steels with chromium and molybdenum and API RP 941 should be followed for steels operating under high-temperature hydrogen environments.

The curves seen in Fig. 5 from API 941 show the combination of pressure and temperature of hydrogen-safe operating limits. The system parameters under these curves are safe and above are claimed as targets to HTHA.

Table 1 gives a short summary of the different types of damage mechanisms and their well-known properties (Table 1).

4 Lessons Learned

In this section, I will provide some real-world examples of equipment failures related to hydrogen damage. The primary source of these examples is H_2Tools.

4.1 Real-Life Examples and Lessons Learned

Table 1 Summary of damage mechanisms

Mechanism	Form of hydrogen	Stress	Temperature	Environment	Strength level
Embrittlement	Atomic form	Yes	Room temperature	Low corrosion	High strength level materials
Hydrogen induced cracking	Molecular	No applied stress needed	Lower than 100 °C − 20 °C to room temperature [23]	Non-metallic inclusions especially MnS and high corrosive atmosphere [17]	Low strength [17]
Sulphide stress cracking	Atomic [24]	Occurs on welds, therefore needs high stress and yield stress [19]	Room temperature and slightly less Decreases with increased temperature [25]	Corrosive environment, H_2S etc. [25]	High yield strength steels (ultimate tensile above 700 Mpa) are more susceptible [24]
Stress corrosion cracking	Atomic [20]	Residual stress and applied stress	Increases with temperature [19]	Corrosion	Medium and high-strength carbon and alloy steels [26]
High temperature hydrogen attack	Atomic to molecular then to methane [27]	High applied or residual stresses [28]	Higher temperature, more than 205 °C and more than 150 bars. [34]	Corrosion, decarburization. [28]	Low and medium strength materials are more effected. [27]

(continued)

(continued)

Area	Equipment	Reason	Explanation	Lessons learned
Area	Equipment	Reason	Explanation	Lessons learned
Power plant	Valve	Equipment, Failure, Embrittlement	Two globe solenoid valve was found malfunction springs inside the valve went into catastrophic failure, spring with 25 coil ended up with only 1–2 coils due to hydrogen embrittlement. Luckly this was recognized during inspection therefore no ignition and damage or injuries occurred [30]	This incident shows potential incompetency of end user/ manufacturer problem. End user has to be completely competent about their process specifications and they have to inform the manufacturer with these specifications and manufacturer has to be competent to be able to provide correct material susceptible to embrittlement under these specific application and service conditions
Government facility	Valve	Equipment Failure, Embrittlement	The shaft of check valve is blown out causing million dollars lost and damage and also some injuries. The reason was hydrogen embrittlement, system had high hydrogen content and hydrocarbons under 20 bars. More the system parameters more this kind of damages and failures can happen [31]	Facilities should review their process and determine the most correct material for their process, and they should do risk analyses and hazard analyses to be able to see the possible scenarios. Facilities should have enough competency and training to be able to achieve good communication between end user and manufacturer. Also periodic inspections are important including necessary maintenance and repairs

(continued)

(continued)

Area	Equipment	Reason	Explanation	Lessons learned
Chemical plant	Vessel, Burst disk, Venting system, Vent line	BLEVE (Boiling liquid expansion vapor explosion)	The rupture disk blew out in a 20-ton liquid hydrogen tank, and the only way that gaseous hydrogen could go out was through the exhaust vent. Some liquid hydrogen was removed by emergency responders, and they sprayed water into the tank. However, this caused very cold hydrogen gas, around -253 degrees Celsius, to go through the exhaust vent, which was the only remaining outlet for the hydrogen. This gas froze in the vent and blocked it. As a result, the remaining hydrogen in the tank was unable to escape and became pressurized due to the rising temperature, causing an explosion [32]	BLEVE was mentioned before in [33] It is very serious terminology about hydrogen therefore, each liquid hydrogen tank should have sign warning that cold water should not put on vent stack. Also, there can be one back up vent stack. It is also important that employees are competent and fire fighter, emergency responders are well trained
Tesoro refinery	Heat exchanger	High temperature hydrogen attack	Tesoro refinery used carbon steel equipment for their heat exchangers, over time hydrogen diffused into steel, leading to the formation of methane and microcracks. As previously mentioned in Sect. 3.5 methane molecule increased the stress on the material and caused failure on heat exchanger leading catastrophic failure	Lack of competence. Importance of trainings and using appropriate materials, such as chrome-molybdenum steel. Importance of inspections, maintenance to be able to keep equipment in their own shape, repair if needed to be able to bring back to their shape

(continued)

(continued)

Area	Equipment	Reason	Explanation	Lessons learned
Refinery	Piping, Heat exchanger	Stress corrosion cracking	Fire occurred on desulphurization unit, fire was result of hydrogen leak related to SSC in the drain valve connecting pipe to hydrogen heating furnace, SSC happened due to high chlorine in industrial water. When plant restarted the remained chlorine caused temperature increase leading cracking on pipe [33]	SSC occurred due to high chlorine concentration in the water used during maintenance, this crack caused hydrogen leak and hydrogen leak caused the fire. Therefore, underlaying problem was SSC due to high chlorine concentration in the industrial water, Changing industrial water to pure water will be correct action. The response in this accident to the initial fire was key point, protected further possible catastrophic results

Most of the hydrogen incidents, for now, happening in the laboratory, but also plants, fuel stations, logistics, refineries, and power stations are affecting big role. Looking at the reasons, we can clearly see that, equipment failure plays a very big role and the second biggest reason behind hydrogen failures is the human factor. Human factors are also causing problems with bad design, not following standards, and not following maintenance schedules. Luckily for now the incidents caused property damage and minor injuries, but bigger failures might cause catastrophic failures and loss of human life.

5 Good Practice in the Industry

For all the reasons mentioned above, good practice is important in the industry for hydrogen to be able to reduce malfunctions, material failures, human errors, design and planning mistakes, and personal incompetence.

Hydrogen-involving systems, risk should be addressed. Then, good practice in operation should consider minimizing quantity, reducing system parameters such as temperature or pressure (if possible), zoning the areas therefore people will not be in contact, proper and efficient ventilation, using detectors and alarms, forcing personnel with PPE, keeping maintenance schedules and housekeeping.

5.1 Material Selection

Material selection is one critical part as well. As previously mentioned, embrittlement and high-temperature hydrogen attack risks can be reduced by correct material selection. As an example, for HTHA, decarburization can be controlled by alloy steels with chromium and molybdenum.

5.2 Purging System

Purging System is also a prevention technique, but it is important to know that nitrogen cannot be used with liquid hydrogen, better to utilize helium as helium gas has a lighter density than liquid hydrogen.

5.3 Ignition Sources

An ignition source is to be eliminated, as known, an explosion is a chemical reaction when a gas or dust mixture self-propagates its flames and creates big pressure waves, and oxygen is needed with an ignition source for an explosion to happen. Therefore, eliminating an ignition source is a very important step in the prevention of hydrogen hazards. This can be achieved by eliminating the ignition sources such as lights, switches, electrical control panels etc. or replacing them with Ex-proof versions.

5.4 Hydrogen Gas Detections

Using hydrogen detectors is an important part of the process, but the location and efficiency are important too; they should be located at the exhaust of the system, considering the properties of hydrogen such as density, LFL, and HFL.

Other good practices are increasing the competency of people, using barriers where hydrogen exists, purging systems, good venting, using appropriate alarms, proper PPE, and emergency procedures should be designed at the early steps of design.

5.5 Good Practice for High-Temperature Hydrogen Attack

The mechanism of High Temperature Hydrogen Attack was previously mentioned, using alloy steels such as chromium and molybdenum will increase carbide stability.

API RP 941 curves should be used to identify the maximum pressure and temperature of the system, but it should be noted that these curves might change over the year.

Periodic inspections are also part of the good practice; conventional inspections such as NDE are usually not helpful if a crack has not reached the surface; therefore, so far, the most effective ones are AUBT, TOFD and PAUT inspections [26].

5.6 Good Practice for Hydrogen Embrittlement

To be susceptible to HE, hard-strength steels have to be harder than Rockwell hardness number HRC 22. Hydrogen concentration must be maintained, and alloy compositions must be used. Dry and low-hydrogen electrons should be used during welding. In corrosive services, a protective lining is required. Avoid cadmium for high-strength bolts. Inspections are important, and PT, MT, and WFMT inspection and monitoring can be used [26].

5.7 Good Practice for Hydrogen Blistering, HIC, SOHIC, SSC

Carbon steels, low-allow steels, and martensitic stainless steels are usually affected materials; critical factors are level of H_2S (more the worse), PH (minimal at seven and more at lower and higher PH), temperature (blistering HIC, SOHIC at ambient temperature but SSC most at 20 °C), metal properties (MnS inclusions increases), contaminants (such as salt will change the pH value or HCN). Hard steels such as Cr–Mo steels are susceptible to SSC; therefore, good practice might be limiting the strength [26].

Barrier coatings for HIC or SONIC, keeping pH at 7, post-weld treatment for SOHIC, limiting hardness of welds for SSC, and specialized corrosion inhibitors [26].

Inspections are needed to be able to identify the wet environment, such as H_2S, and follow up on the corrosion. Visual testing can catch blistering, and SOHIC inspections focus on nozzles. Cracks caused by HIC can be identified visually, but for earlier steps, magnetic particle inspection, eddy current testing, and electromagnetic inspection can be done. High-resolution NDE methods, angle beam UT, and AET also can be options [26].

6 Conclusion

In this paper, the disadvantages of hydrogen from storage to damage types, embrittlement and its mechanisms, the accidents around the world and the reasons for these accidents considering the failure mechanism, wrong material selection or human errors assessed, after mentioning the common reasons for these human errors and material problems, the good practices are assessed. One important consideration in selecting materials as per the tests is limiting the strength level of steel. It is possible high high-strength steels can become more brittle compared to low-strength steels. Also, the gas pressure of hydrogen, temperature and heat treatment (quenching, tempering) are important considerations [32, 33].

Fe–Cr-Mo alloys seem to be the best but their strength level, temperature, and carbon content are also important.

As per ISO TR-15916 Aluminium, copper and Nickel alloys are less susceptible to embrittlement but steels such as Maraging steel, 17-7P are extremely embrittled meanwhile A286 is defined as the least affected [36].

Tensile stress is the force divided by the material cross-section, but yield stress is when the material starts deforming and the material will no longer turn its original form, and elongation is the fact that is calculated considering the tensile stress, yield stress, material length and materials final length after force. The bigger the elongation the material is more ductile meaning that can withstand hydrogen embrittlement more. The same alloy might have 110 ksi tensile stress and 40 ksi yield stress ending up having 60% elongation, but the same material with a different process can end up having 252 ksi tensile stress, 245 ksi yield stress meaning having just 3% elongation even though it is very strong, but it will not be able to withstand the embrittlement anymore as its ductility is very low. Usually, FCC crystal structures are more ductile such as nickel, aluminium, copper, gold, etc.

It is important to experiment with each steel and its alloy to be able to see its sustainability. Hydrogen solubility, crystal structure, oxide layers, hydrogen infusibility etc., are important but also it is important to apply good practices mentioned in Sect. 5; inspections, material selection, personal competency, emergency responses and so on.

References

1. Hermesmann M, Müller TE (2022) Green, Turquoise, Blue, or Grey? Environmentally friendly hydrogen production in transforming energy systems. Prog Energy Combust Sci 90:100996. https://doi.org/10.1016/J.PECS.2022.100996
2. Lynch SP (2011) Hydrogen embrittlement (HE) phenomena and mechanisms. Stress Corros Crack: Theory Pract 90–130. https://doi.org/10.1533/9780857093769.1.90
3. McEvily AJ, Le May I (1991) Hydrogen-assisted cracking. Mater Charact 26:253–268. https://doi.org/10.1016/1044-5803(91)90016-W
4. Wiberg S, Laumen C (2009) European patent application (43)

5. Demirocak DE (2017) Hydrogen Storage Technologies. In: Chen Y-P, Bashir S, Liu JL (eds) Nanostructured materials for next-generation energy storage and conversion: hydrogen production, storage, and utilization. Springer, Berlin Heidelberg, Berlin, Heidelberg, pp 117–142
6. Barthelemy H, Weber M, Barbier F (2017) Hydrogen storage: recent improvements and industrial perspectives. Int J Hydrog Energy 42:7254–7262. https://doi.org/10.1016/J.IJHYDENE.2016.03.178
7. Cheng Q, Zhang R, Shi Z, Lin J (2024) Review of common hydrogen storage tanks and current manufacturing methods for aluminium alloy tank liners. International J Light Mater Manuf 7:269–284. https://doi.org/10.1016/J.IJLMM.2023.08.002
8. Eberle U, Arnold G, Von Helmolt R (2006) Hydrogen storage in metal–hydrogen systems and their derivatives. J Power Sources 154:456–460. https://doi.org/10.1016/J.JPOWSOUR.2005.10.050
9. www.iso.org. Published in Switzerland (2019)
10. Sun B, Zhao H, Dong X et al (2024) Current challenges in the utilization of hydrogen energy-a focused review on the issue of hydrogen-induced damage and embrittlement. Adv Appl Energy 14:100168. https://doi.org/10.1016/J.ADAPEN.2024.100168
11. Sergeev NN, Sergeev AN, Kutepov SN et al (2019) Mechanism of the hydrogen cracking of metals and alloys, part II (Review). Inorg Mater Appl Res 10:32–41. https://doi.org/10.1134/S2075113319010271
12. Qayyum F, Umar M, Dölling J et al (2024) Mechanics of new-generation metals and alloys. Compr Mech Mater 31–57. https://doi.org/10.1016/B978-0-323-90646-3.00020-4
13. Shin HS, Kang S, Pascua R et al (2023) Effect of hydrogen pressure and punch velocity on the hydrogen embrittlement susceptibility of pipeline steels using small punch tests under gaseous hydrogen environments at room temperature. Metals (Basel) 13. https://doi.org/10.3390/met13121939
14. Oriani RA (1972) Mechanistic theory of hydrogen embrittlement of steels. Ber Bunsenges Phys Chem 76(8)
15. Troiano AR (2016) The role of hydrogen and other interstitials in the mechanical behavior of metals. Metallogr Microstruct Anal 5:557–569. https://doi.org/10.1007/s13632-016-0319-4
16. Beachem CD (1972) A new model for hydrogen-assisted cracking (hydrogen "embrittlement")
17. Elboujdaini M, Revie RW (2009) Metallurgical factors in stress corrosion cracking (SCC) and hydrogen-induced cracking (HIC). J Solid State Electrochem 13:1091–1099. https://doi.org/10.1007/s10008-009-0799-0
18. From Wikipedia the free encyclopedia hydrogen embrittlement
19. Hopkins P (2003) The structural integrity of oil and gas transmission pipelines
20. Contreras A, Albiter A, Salazar M, Pérez R (2005) Slow strain rate corrosion and fracture characteristics of X-52 and X-70 pipeline steels. Mater Sci Eng A 407:45–52. https://doi.org/10.1016/j.msea.2005.07.028
21. Jones DA (1996) Principles and prevention of Corrosion, 2nd edn.
22. API recommended practice 941 eighth edition F 2016 (2016) Steels for hydrogen service at elevated temperatures and pressures in petroleum refineries and petrochemical plants
23. Ormellese M (2018) Corrosion in oil and chemical industry. In: Encyclopedia of interfacial chemistry: surface science and electrochemistry, pp 145–154. https://doi.org/10.1016/B978-0-12-409547-2.13434-1
24. Albiter A (2020) Sulfide stress cracking assessment of carbon steel welding with high content of H_2S and CO_2 at high temperature: a case study. Engineering 12:863–885. https://doi.org/10.4236/eng.2020.1212061
25. Nogara J, Zarrouk SJ (2018) Corrosion in geothermal environment: part 1: fluids and their impact. Renew Sustain Energy Rev 82:1333–1346. https://doi.org/10.1016/J.RSER.2017.06.098
26. ANSI/API Recommended Practice 571 Third Edition 2020 (2020) Damage Mechanisms Affecting Fixed Equipment in the Refining Industry

27. Ali M, Ul-Hamid A, Alhems LM, Saeed A (2020) Review of common failures in heat exchangers—part I: mechanical and elevated temperature failures. Eng Fail Anal 109:104396. https://doi.org/10.1016/J.ENGFAILANAL.2020.104396
28. H2Tools (2024) Failure of stainless steel valves due to hydrogen embrittlement. https://h2tools.org/lessons/failure-stainless-steel-valves-due-hydrogen-embrittlement
29. H2Tools (2024) Check valve shaft blow-out due to hydrogen embrittlement. https://h2tools.org/lessons/check-valve-shaft-blow-out-due-hydrogen-embrittlement
30. H2Tools (2024) Liquid hydrogen tank boiling liquid expanding vapor explosion (BLEVE) due to water-plugged vent stack. https://h2tools.org/lessons/liquid-hydrogen-tank-boiling-liquid-expanding-vapor-explosion-bleve-due-water-plugged-vent
31. H2Tools (2024) Stress corrosion cracking leads to a hydrogen leak and fire at a refinery. https://h2tools.org/lessons/stress-corrosion-cracking-leads-hydrogen-leak-and-fire-refinery
32. ISO (2015) Basic considerations for the safety of hydrogen systems Copyright Protected Document
33. Somerday BP, What is hydrogen embrittlement?—causes, effects and prevention—TWI. https://www.twi-global.com/technical-knowledge/faqs/what-is-hydrogen-embrittlement. Accessed 11 Dec 2023
34. Moody NR, Baskes MI, Robinson SL, Perra MW (2001) Temperature effects on hydrogen-induced crack growth susceptibility of iron-based superalloys. Eng Fract Mech 68:731–750. https://doi.org/10.1016/S0013-7944(00)00122-3

The manufacturer's authorised representative in the EU is Springer Nature Customer Service Centre GmbH, Europaplatz 3, 69115 Heidelberg, Germany. If you have any concerns regarding our products, please contact ProductSafety@springernature.com

Printed and bound by CPI Group (UK) Ltd, Croydon, CR0 4YY
30/09/2026
02225593-0018